Vanished and **Vanishing Parrots**

Profiling Extinct and Endangered Species

F. KNIGHT
17

Vanished and **Vanishing Parrots**

Profiling Extinct and Endangered Species

Joseph M. Forshaw
Illustrated by Frank Knight

CSIRO PUBLISHING

Comstock Publishing Associates
a division of
Cornell University Press
Ithaca and London

This book is dedicated to researchers, fieldworkers and volunteers working tirelessly in programs in many countries to protect endangered parrots.

First published in Australia in 2017 by CSIRO Publishing

National Library of Australia Cataloguing-in-Publication entry

Forshaw, Joseph M. (Joseph Michael) author.

Vanished and vanishing parrots : profiling extinct and endangered species / Joseph M. Forshaw and Frank Knight.

9780643096325 (hardback)
9780643106499 (ePDF)
9780643106505 (epub)

Includes bibliographical references and index.

Parrots – Identification.
Parrots – Conservation.
Rare birds – Geographical distribution.
Extinct birds – Identification.

Knight, Frank, 1941– illustrator.

CSIRO Publishing
Locked Bag 10
Clayton South VIC 3169
Australia

Telephone: +61 3 9545 8400
Email: publishing.sales@csiro.au
Website: www.publish.csiro.au

First published in the United States of America in 2017 by Cornell University Press

Library of Congress Cataloging-in-Publication Data
Names: Forshaw, Joseph Michael, author. | Knight, Frank, 1941- illustrator.
Title: Vanished and vanishing parrots : profiling extinct and endangered species / Joseph M. Forshaw ; illustrated by Frank Knight.
Description: Ithaca, New York : Comstock Publishing Associates, a division of Cornell University Press, 2017. | Includes bibliographical references and index.
Identifiers: LCCN 2017008328 | ISBN 9781501704697 (cloth : alk. paper)
Subjects: LCSH: Parrots--Identification. | Extinct birds--Identification. | Rare birds--Identification.
Classification: LCC QL696.P7 F67 2017 | DDC 598.7/1--dc23
LC record available at https://lccn.loc.gov/2017008328

Frontispiece:
Restoration drawing of three extinct parrots known only from subfossil remains recovered in the Mascarene Islands, and from accounts by early travellers to those islands.
TOP Leguat's Parrot *Necropsittacus rodericanus* ♂;
MIDDLE Mascarene Grey Parakeet *Psittacula bensoni* ♂;
BOTTOM Broad-billed Parrot *Lophopsittacus mauritianus* ♂ (front), ♀ (rear).

Front and back cover:
Carolina Parakeets (Frank Knight)

Inside back flap:
Photograph of Joseph Forshaw courtesy of Julie Clark; photograph of Frank Knight courtesy of Melanie Knight

Set in Frutiger light
Cover design by Derrick I. Stone Design
Typeset by Derrick I. Stone Design
Printed in China by 1010 Printing International Ltd

FOREWORD

Among the many orders of living birds, the nearly worldwide parrot order (Psittaciformes) has been one of the most vulnerable to species endangerments and extinctions. According to the current Red List of the International Union for Conservation of Nature, only 55 per cent of the 414 recent members of this group can be considered at low risk of extinction, while an astonishing 45 per cent are either threatened (27%), near threatened (14%), or already extinct (4%). These alarming statistics have been worsening rapidly in modern times and reflect the extreme seriousness of threats faced by parrots.

In part, the discouraging prospects of parrots have resulted from gunfire to reduce depredations on crops and provide food for rural families. In part they are a result of widespread unregulated capture of species popular as captives. But to highlight these particular stresses is not to imply any lack of vulnerability to a great variety of other detrimental influences, many of them other direct and indirect impacts of our own species. Here, for example, it is perhaps especially important to recognize the role of humanity in producing habitat degradation and loss, and in moving predators and wildlife diseases to new locations where native species have had no effective defences against them. Most endangered parrots have faced stresses of multiple sorts. Conserving these species in the short term depends on valid identification and reduction of the most pressing proximate threats, especially the threats most susceptible to correction.

However, determining the major immediate causes of endangerment of wildlife species is often difficult because so many factors can affect the viability of natural populations, either directly or indirectly, and many of these factors can interact with one another. Guesses about causes made from the armchair often prove wide of the mark, and to the extent they become dogmas, they can sometimes hinder the collection of the crucial data needed to allow reliable discrimination among alternatives. Sometimes a major cause of endangerment (e.g. poisoning from ingestion of lead ammunition fragments in the case of the California Condor *Gymnogyps californianus*) can lie hidden from view for decades, yet can demand solutions very different from the habitat conservation and prevention of human predation commonly pursued in species conservation.

Identification of major causes of decline is also of considerable interest for extinct species and may aid in determining the causes of endangerment of other species still extant that share similar ecological roles. And if it ever becomes possible to 're-create' extinct species through manipulations involving genetic material from preserved specimens, it may still be impossible to reintroduce such species to wild environments without first developing effective cures for the threats causing extinction in the first place. This presumes correct knowledge of what these threats are. Unfortunately, the task of cause identification is far more difficult with species already extinct than with species still in existence. Without living individuals, it is often impossible to test alternative explanations in any rigorous way.

Such difficulties are well illustrated by the example of the very colorful Carolina Parakeet *Conuropsis carolinensis*, the only parrot species ever indigenous to the eastern United States. To launch a general discussion of the complexities in identifying threats to endangered parrots I want to start with this extinct species, as back in the 1970s and 1980s I had an opportunity to become closely familiar with its plight in Florida through interviews of multiple people with personal memories of it as a living species. Following this discussion, I'll consider some instructive contrasts provided by conservation efforts for two other parrot species I've been involved with over the years – studies concerning the still-extant Puerto Rican Parrot *Amazona vittata*, now found only on that island in the West Indies, and the still-extant Thick-billed Parrot *Rhynchopsitta pachyrhyncha*, presently limited to northwestern Mexico but historically also found in the southwestern United States.

THE CAROLINA PARAKEET: A PUZZLING DISAPPEARANCE

Never given any rigorous first-hand biological study while it still existed, the Carolina Parakeet often has been regarded as extinct by the early 20th century, although judging from quite credible eyewitness accounts, it may have endured as late as the 1930s or even 1940s. Decline of the species has often been attributed to habitat destruction and gunfire, but in fact its disappearance may well have been due largely to other factors – including one very plausible factor rarely even considered in popular accounts of the species – exposure to exotic diseases.

Historical writings on the Carolina Parakeet were examined in fascinating detail in a long series of perceptive papers in the late 20th century by Daniel McKinley. McKinley managed to locate early accounts of the parakeet in a bewildering array of often unrelated historical writings, but perhaps wisely he never offered a definitive explanation for the extinction of the species. Nevertheless, he did express considerable doubt that shooting to protect crops was really a major factor in the decline, despite the oft-quoted early descriptions by Alexander Wilson and John James Audubon of such activities. Sadly, by the time of McKinley's studies, the parakeet evidently was gone, and he never had the good fortune to locate and interview anyone who remembered the species from life. Moreover, the information available in early writings leaves many gaps in our understanding of the species. There are, for example, almost no detailed accounts of Carolina Parakeet nestings, and the ones that exist have often been dismissed as fraudulent.

Consequently, when I learned in the late 1970s from my good friend Rod Chandler of the Audubon Society that he was aware of a number of elderly people who once knew the Carolina Parakeet personally, the opportunity arose to significantly expand knowledge of this species. The senior citizens involved had all spent their youth in the Kissimmee Prairie region of central Florida in the early 20th century, and together with Rod in the late 1970s and early 1980s, I was able to hear their personal parakeet memories, including their thoughts on the causes of the species' disappearance. Their accounts are more fully discussed in my book – *The Carolina Parakeet: Glimpses of a Vanished Bird* (2004), and they revealed many unexpected facets to the biology of the parakeet, although no one interviewed had any idea why the species disappeared.

Perhaps most surprisingly, none of the dozen people interviewed could recall any direct persecution of the parakeet by humans and all doubted that such persecution had been occurring in the region, even though some, as former plume hunters, had been personally engaged in wildlife exploitation in their own careers. The species evidently was not a significant crop pest in central Florida, and in fact was generally valued for its frequent feeding on noxious plants such as cockleburs and sandspurs. Further, none of the people interviewed had any knowledge of substantial pressures on the species from the pet trade or feather trade. In addition, it was not considered large enough to be worth shooting for food when much more attractive game such as deer, wild turkeys, feral pigs, and waterfowl were abundant in the region. And since ammunition was relatively expensive for residents in those days, it was not generally wasted on inappropriate targets. Moreover, habitat destruction was never more than localized in the region, and seemingly always left abundant potential nest sites and foods for the species. The foods of the parakeet were clearly diverse and included fruits and seeds from numerous common plants, including ubiquitous weeds.

The Carolina Parakeet simply does not appear to have been a habitat specialist tied to pristine forests, and although Daniel McKinley has suggested the possibility of significant detrimental competition for nest sites from the introduced European honeybee *Apis mellifera*, a species long established in the wild in Florida, nest site preferences of the honeybee are known to be strong for cavities that are much larger than would presumably have sufficed for nesting parakeets. With an abundance of sizeable woodpeckers continuously present in the region and creating cavities too small for bees but probably large enough for the parakeets, a general scarcity of good nest sites for the parakeets seems doubtful, although there could have been some honeybee impacts on communal parakeet roosts in relatively large natural cavities. No other secondary-cavity-nesting birds are known to have gone extinct or suffered endangerment from nest-site scarcity in the region.

Nevertheless, the parakeet did disappear, so it presumably must have been under significant stress from some factor or factors. Moreover, the interviews suggested the potential importance of one factor closely tied to early human settlement of the region. Specifically, the interviews revealed that the parakeets of the Kissimmee Prairie region often practised overnight roosting in barns, something that had never been reported anywhere historically, but nevertheless could have been going on elsewhere as well. This habit immediately suggests possible fortuitous close exposure of the parakeets to diseases of domestic livestock, perhaps especially poultry.

To this day, domestic poultry have proved vulnerable to a great variety of avian diseases, including diseases of foreign origin to which the native birds of North America have no historic immunities. The Carolina Parakeet was presumably capable of dealing with the indigenous diseases of the continent, but may well have had serious problems with exotic diseases. This susceptibility may have underlain the scattering of reports in early times of captive Carolina Parakeets dying of a malady referred to as 'apoplexy', apparently with birds experiencing fits and sudden death. But no one was carefully studying parrot diseases in those days, and what specific diseases the Carolina Parakeet may have suffered from is highly speculative. One example of serious exotic disease of poultry and parrots that could potentially have been involved is Newcastle Disease, which evidently became established in North America early in the 20th century.

What is known about the pattern of disappearance of the parakeet in central Florida and elsewhere seems consistent with the potential importance of disease. Here there evidently were several small populations that persisted for a number of years into fairly recent times, then abruptly disappeared completely. Such a pattern does not suggest stress of a continuous nature, but stress more episodic in time and space, although disease is not the only factor that could produce such a pattern. In particular for small populations roosting in barns, exposure to serious disease might have occurred only with introductions of new stocks of domestic poultry, but then could plausibly have spread quickly and fatally throughout the highly social groups of the parakeets that were in proximity to these birds. At the same time other unexposed social groups in other locations may have remained completely unaffected, at least in the near term.

Another feature seemingly consistent with potential stress from disease that emerged from interviews is the fact that several parties remembered seeing juveniles (green-headed individuals) among the adults in the last flocks, and certain of the very last reports involved actively breeding pairs. These facts, to the extent they may have been typical, suggest that mortality factors, rather than reproductive factors, could have been predominant in the decline.

Could there be biochemical traces of pathogens in the many (more than 800) Carolina Parakeet specimens that exist in museums? Since the great majority of these specimens were presumably healthy birds at the time of collection, one would not expect a high incidence of pathogen-affected specimens, even if disease had been an important factor overall. Most parakeets suffering from serious disease in the wild would likely have quickly become victims of natural predators or scavengers, rather than specimen collectors. However, it would be of exceptional interest if the few Carolina Parakeet specimens that were known to have died of apparent disease (all captives) could be given comprehensive retrospective study of a sort that might reveal traces of specific diseases.

A focus on exotic disease as a potential stressor for the parakeets in central Florida says little about the potential importance of disease in other parts of the species' range. Under an argument that barn-roosting may have been a crucial facilitating trait leading to disease exposure, the absence of reports of the species roosting in barns elsewhere could be considered evidence for relative freedom from this stress in many other parts of the species' range. However, it remains possible that barn-roosting also may have been occurring elsewhere, but may simply have escaped documentation through the hit-or-miss nature of historical accounts. Until our interviews, it escaped such documentation in Florida, the state with the most extensive overall literature on the species.

In any event, since species can sometimes face different limiting factors in different geographic regions, it is essential not to generalize too far from geographically limited data. This caveat also applies to the lack of evidence for persecution of the parakeets for crop depredations in central Florida, a force that may well have been important in at least some other locations, judging from historical accounts.

What we are left with in the case of the parakeet is simply a lack of enough information to allow definitive specification of the causes of extinction in any region, and while the sorts of information available for central Florida suggest a potential importance of disease problems and a general absence of other commonly suggested stresses, these characteristics may or may not have relevance for other regions. Whether new evidence may ever emerge to improve this situation remains to be seen. With no living members of this species left to study, and potentially no one still alive today who remembers the species from life in any region, our opportunities for gaining better understanding are now quite limited.

THE PUERTO RICAN PARROT: A REVIVING SPECIES

My first exposure to the world of parrot conservation came through involvement in efforts to unravel another very puzzling case of endangerment – that of the Puerto Rican Parrot. To provide some additional feeling for the complexities that can be involved in safeguarding parrots, it is worthwhile reviewing certain of the surprising developments that have characterized the conservation efforts for this species.

Intensive studies of the biology and conservation of the Puerto Rican Parrot were begun in the 1950s under José Rodríguez-Vidal, and from the late 1960s through to the present, many additional researchers have participated in research and conservation efforts for this species. The main organizations involved have been the United States Fish and Wildlife Service, the United States Forest Service, and the Commonwealth of Puerto Rico. Only through the combined efforts of numerous individuals and a number of organizations have many of the mysteries surrounding the plight of this species begun to disappear.

At the time Columbus discovered Puerto Rico, the island was virtually completely forested, and there may well have been more than a half million Puerto Rican Parrots in existence. But by 1972, when conservation efforts became truly intensive, the wild population had dropped to just 16 individuals, down precipitously from some 200 individuals known in the 1950s and ranking the species as one of the most highly endangered wildlife species in the world. The surviving individuals were all located at the eastern end of the island in well-protected high-elevation virgin and near-virgin forests of the Luquillo Mountains, a substantial region known to residents of the island as El Yunque. These were virtually the only old-growth forests left on Puerto Rico after centuries of deforestation.

Decline of the wild population continued to a low point of just 13 individuals in early 1975, during the period I was personally participating in efforts on behalf of this species, and at this point it appeared likely that only a very few years might be left before the species would be gone (for details, see Snyder *et al*. 1987). Yet causes of the continuing decline remained obscure, and the best ways to reverse the decline were highly speculative. A decision had been made in 1972 to begin last-ditch captive breeding of the species, but at the same time efforts were continued to preserve a potential for recovery of the existing wild population by leaving some birds in the wild for greatly expanded research on the nature of the threats they faced. It was not at all clear at the time these decisions were made which approach might produce the most benefit, but it seemed plausible, and has proved true, that these two general approaches might each be valuable, while together they might increase the options that might develop for recovery.

To allow both approaches, the formation of a captive flock was limited largely to eggs taken from wild pairs, while pairs were still allowed to fledge at least some of their youngsters in the wild. Basic biological studies were focused initially on efforts to locate all the remaining breeding pairs in the population and efforts to observe their activities in detail. Discovering and understanding the main stresses to the species was not a simple or predictable process, and in part has involved detailed studies of the natural enemies of the parrot and observation of many unexpected events occurring over many years.

The forces threatening the Puerto Rican Parrot have been numerous. Among the most important historic factors have been threats such as habitat destruction, capture for the pet trade, and shooting for food and to reduce crop depredations. But continuing into the present, the species has also been stressed by the exceedingly high rainfall of the Luquillo Mountains, depressing both nest availability and egg and chick viability, and by exceedingly high populations of natural enemies, most importantly predacious Red-tailed Hawks *Buteo jamaicensis*, cavity-nesting and predacious Pearly-eyed Thrashers *Margarops fuscatus*, and parasitic *Philornis* warble flies. All of these factors have been important to the parrot and have merited effective management countermeasures where feasible. Nevertheless, their comparative importances have turned out to vary greatly as a function of location and time period, and it has become increasingly clear that these are not the only important factors still stressing the species.

Much attention was given in early conservation efforts to increasing the availability of favorable cavity nest sites, an availability that was directly determined by extensive cavity surveys to be extremely low, especially because of the high rainfall characteristic of the last range of the species (sometimes more than 200 inches per year). Such extremely wet conditions simply do not favor the formation of dry tree cavities of adequate size. Available data suggested that the wild population was suffering significant reproductive stress tracing to this scarcity, despite the old-growth character of the habitat – stress that was apparently depressing both nesting success and nesting effort.

Nevertheless, despite provision of many dry artificial cavities and drainage of many otherwise suitable natural cavities, reproductive effort of the wild parrot population remained low through the 1970s and 1980s, with usually only four pairs attempting breeding in any year. However, at the same time the wild population was showing greatly increased nesting success in the pairs that were breeding, thanks especially to diverse and intensive nest-guarding and nest-management efforts of Jim and Beth Wiley. The total wild population was also gradually increasing, in part from fostering of captive-produced young into wild nests, and reached nearly 50 individuals by the late 1980s. In itself, this increase was highly encouraging and surely did not suggest that it might be time to give up on efforts to conserve the wild Luquillo population.

But why was the number of breeding pairs also not increasing in parallel with the population increase? Many good nest sites were now available to the population, but virtually none were being adopted by new parrot pairs. Even worse, they were frequently being occupied by a cavity-nesting natural enemy and competitor of the parrots – the Pearly-eyed Thrasher. Not only was this presumably increasing the population and direct threats of this species to the parrot, but the high population of the thrashers appeared, by its own susceptibility to parasitic warble flies, to be likely serving as a major sponsor of substantial rates of infestation of parrot nestlings with fly maggots. In sum, while various nest management efforts on behalf of the parrot had clearly stopped the population decline and were producing obvious improvement in nesting success of these birds, they were not producing any perceptible increase in parrot nesting effort and may actually have been increasing certain stresses to the parrot population, intensifying the need for such actions as effective anti-thrasher and anti-warble fly measures.

It was only the occurrence of Hurricane Hugo in 1989 that at last seemed to offer some explanation for chronic low breeding effort in the population. This massive storm scored a near direct hit on the last habitats of the wild population and caused the immediate loss of approximately half of the wild parrot population. The forest was badly damaged, losing much of its foliage, and indeed looked as though a nuclear bomb might have detonated nearby. Yet despite the devastation and despite the parrot population crash, the number of wild breeding pairs of parrots did not crash in parallel, and in fact actually increased by 50 per cent by the second breeding season after the storm! Where there had usually been only four wild egg-laying pairs for the past two decades, now there were six, and the excellent reproduction of these birds largely made up for the population losses produced by the storm within just a few years! Yet within a few more years, as the forest recovered, the breeding population dropped back down to just four pairs. Moreover, the wild population of the Luquillo Mountains has remained in the low dozens ever since, despite the continued application of a great variety of intensive management efforts, including releases of many captives into the population.

The new pairs breeding immediately following Hurricane Hugo were quite surely comprised of individuals that had existed before the storm, but had not been breeding. Why had they not bred earlier? They apparently had not been facing any scarcity of good nest sites at that time, and the most reasonable conclusion in retrospect seems to be that the storm caused some beneficial environmental change or changes unrelated to nest availability that allowed these birds to improve their reproductive status. Further, the poor breeding effort before the storm was not likely attributable to genetic problems because the same parrot individuals were involved before and after the storm.

However, though a major reproductive increase was seen also in Pearly-eyed Thrashers immediately after the storm, potentially due to similar factors, exactly what environmental factor or factors might have led to these increases was uncertain. One possibility is that many rainforest trees of the West Indies are indeed adapted to greatly enhance fruit and seed production after hurricanes, but whether the increased food that followed Hugo may have underlain the breeding enhancement of the parrots and thrashers was not established definitively. Other possibly important changes produced by the storm could have been temporary reductions in densities of certain predators or parasites or changes in vulnerability to these and other natural enemies because of changes in forest structure.

In any event, after nearly a half century of intensive conservation efforts, it was finally becoming quite clear that the Luquillo Mountains population of Puerto Rican Parrots, while a natural population in impressive old-growth forest, is probably best regarded as a 'sink' population that just does not seem to have the potential to be intrinsically self-sustaining in most years, not just because of extremely high populations of natural enemies and extremely high rainfall, but evidently in part because of some unidentified environmental factors apart from nest availability that somehow depress reproductive effort severely. Low reproduction in the population was evidently alleviated temporarily by some positive effects of Hurricane Hugo, but it may well be that populations of the species in these mountains may always be marginal at best, even given major management assistance from our own species. As argued by Tom White, a current leader of conservation efforts, El Yunque may actually be one of the worst places in Puerto Rico for the parrot. 'The parrots didn't end up in El Yunque because they chose to. They ended up there because that was the last refuge left for them when Puerto Rico was deforested.' Evidently, the last habitats occupied by endangered species may sometimes be less than adequate for these species in their overall characteristics even when they are well-protected virgin habitats.

Other regions of Puerto Rico that once were inhabited by the parrots seemingly offer relief from many of the stresses characteristic of the Luquillo Mountains. Red-tailed Hawk populations, which reach their highest known density in the world in El Yunque, are far less vigorous in some of the second-growth commonwealth forests regenerating farther west in Puerto Rico, and rainfall is also far less in these other regions. Moreover, research of the 1970s revealed that Pearly-eyed Thrasher populations and warble-fly threats were also much less substantial in these locations.

One of these locations – Rio Abajo Commonwealth Forest – was known from interviews of senior citizens in the 1970s to have once hosted a large and vigorous parrot population before widespread deforestation early in the 20th century. Accordingly, efforts were begun early in the 21st century to reestablish a wild parrot population in Rio Abajo with release of birds from the captive population that was first established in the Luquillo Mountains in the 1970s and was later expanded to an aviary in Rio Abajo.

The results in Rio Abajo have so far proved sufficiently successful that it appears likely that at last the Puerto Rican Parrot may be on the road to true recovery as a wild creature. By 2015 there were on the order of a hundred or more parrots flying free in the region and exhibiting good levels of reproduction. And with rapid success in Rio Abajo, release efforts have now also been started in Maricao Commonwealth Forest. All evidence suggests that the major factor allowing the success in Rio Abajo appears to be the much more favorable habitat existing in this location than is offered by the species' last natural range in the Luquillo Mountains. Further, the strength of reproduction and population expansion in the Rio Abajo birds appears to provide persuasive evidence against the fears of some that the Luquillo population was likely suffering mainly from terminal genetic deterioration, as the source of the Rio Abajo birds has been the Luquillo population itself.

The Puerto Rican Parrot has not been an easy species to understand and would surely have been lost without the many conservation efforts that have been made over the decades. Fortunately, these efforts always have been diversified so that success of the overall enterprise has not rested entirely on the success of any particular approach. It is also fortunate that the species has received continuing research and conservation attention from so many skilled and devoted personnel over the years – personnel too numerous to mention individually here, but all contributing ultimately to some long-overdue and sustained momentum toward what appears to be an incipient real recovery.

Finally, it is worth noting that the Puerto Rican Parrot is one of the parrot species for which captive breeding has had a crucially positive role in leading to recovery of wild populations, but this would not likely have been the case without the simultaneous intensive research that has been conducted with the wild populations. Attempting to reestablish wild populations from captivity poses many challenges, and parrots vary widely in their suitability for such efforts, as we will see in consideration of the next species.

THE THICK-BILLED PARROT: ANOTHER SURVIVOR

Prospects for recovery have been encouraging also for a third and last species of parrot I want to consider here – the Thick-billed Parrot of northwestern Mexico and formerly of extreme southwestern United States. This moderately large species, a specialist on pine seeds for food, was once abundant throughout the high-elevation pine forests of the Sierra Madre Occidental of Mexico as far south as the state of Michoacán and as far north as the Chiricahua Mountains of southeastern Arizona. But coincident with the lumbering of nearly all pine forests in western Mexico, populations of this species have contracted largely to a few remnant pockets of relatively good habitat in the Sierra Madre Occidental in Chihuahua and Durango. Under intensive study for the past two decades by Ernesto Enkerlin-Hoeflich and his students and associates, the most important limiting factors for these populations now seem relatively well understood, and future prospects for this species appear relatively good, at least in Mexico.

In the 1980s and 1990s, I was involved in an attempt to reestablish this species in the Chiricahua Mountains of Arizona for the Arizona Game and Fish Department, and that attempt yielded some initial success in creating small breeding populations from releases of confiscated birds that had been wild-caught in Mexico (see Snyder *et al.* 1994). At the same time, parallel efforts with captive-bred birds were thoroughly unsuccessful, largely because behavioral deficits in these birds proved very difficult to correct. Nevertheless, even the initially successful efforts with wild-caught birds faltered after a few years, apparently due largely to local food scarcity problems caused by the onset of long-term drought conditions, but due also to cryptic disease contamination of many of the confiscated birds used in releases. The drought conditions have led also to massive fires and widespread destruction of pine-forest habitat in southeastern Arizona in the past two decades. Success in reestablishment may depend on a return to more mesic rainfall conditions, coupled with recovery of the main populations of the species in Mexico and restriction of resumed releases to wild-caught birds kept free of the cryptic disease threats so prevalent in captive stocks.

Within Mexico, the recent effective limitation of the species as a breeding bird to pockets of well developed pine, fir, and aspen forest has suggested either nest-site limitations or food limitations or both as primary limiting factors. Pine cone abundance is clearly related to the maturity of pine stands, but so is natural cavity abundance and quality. Lumbering could be exerting stress on the species in either or both ways. However, the specific studies that Enkerlin-Hoeflich and his associates have conducted on a vigorous population of the species in the region of Madera in Chihuahua have offered some preliminary resolution of the uncertainties over this question.

Thick-billed Parrots in the Madera region nest in tree cavities, just as do other populations, but unique to this region is a very strong tendency to nest in mature aspen trees *Populus tremuloides*, which are largely limited to this region within Mexico. Other thick-bill populations nest most commonly in cavities in Douglas fir *Pseudotsuga menzesii* or various pines. Perhaps in part because of their very limited numbers and distributions, neither aspens nor Douglas firs are normally lumbered in the Sierra Madres, and both species are generally allowed to grow to an age of senescence, when cavities become most frequent.

Pine forests surrounding the aspen groves near Madera are under intensive forestry management and include no old-growth stands and very few natural cavities. Evidently the parrots in this population have a favorable nest-site availability situation, owing to the mature aspen stands, and are able to find enough food in surrounding areas of pines to sustain vigorous nesting activity, despite intensive lumbering and the lack of fully mature pines. These results are most consistent with a view that the main limiting factor for these parrots may be nest-site availability, not food availability, a view that is supported by a high degree of recent success achieved in getting these parrots to adopt artificial nestboxes and good nest success in pairs using these boxes. So where the artificial nestbox approach did not prove successful for Puerto Rican Parrots in early efforts in the Luquillo Mountains, it is proving highly successful in very recent efforts for the Thick-billed Parrots of the Sierra Madre Occidental.

The kind of forestry management practised in the Sierra Madre Occidental does not include clear-cutting, but neither does it provide for preservation of widespread old-growth stands. Instead it involves selective cutting at intervals of essentially all pine-forested regions – a regime that results in continuous forest cover of pines of young to moderate ages throughout the pine-forest ecosystems. While this forestry management does not provide abundant nest sites for the parrots, and this may be the main reason for the historic decline of parrot populations, it evidently does provide enough food for viable populations if nest sites can be provided by some other means, either uncut aspen or Douglas fir stands or artificial nestboxes. Thus it appears likely that recovery of wild populations can become compatible with existing lumbering practices so long as a favorable nest-site situation is ensured simultaneously. Wild populations of the species have been quite stable in recent decades and may well be poised for significant increase.

Except for a brief period of heavy illegal capture of these parrots for avicultural markets in the United States in the 1980s, Thick-billed Parrots have not been popular as captives, perhaps especially because of their ear-splitting vocalizations. Nor have they been suffering from excessive shooting in Mexico, though they are highly vulnerable to shooting because of their loud vocalizations and approachability. Such pressure was evidently much more important in destruction of wild populations of this species in Arizona, where interviews, early publications, and early photographs have indicated massive shooting of this species early in the 20th century. Much of this shooting was apparently curiosity shooting, rather than shooting for food, and the species has never been known to be a crop pest. Shooting of parrots was no longer a problem during experimental releases of the species in this region in the 1980s and 1990s.

SOME GENERAL CONCLUSIONS

The three parrot species touched upon in this foreword illustrate some of the differing proximate stresses faced by endangered species, but they also illustrate some realities in common. For one thing, they all provide evidence that limiting factors for an endangered species may vary widely across the range of the species. They also provide evidence that limiting factors can change dramatically as a function of changing environmental conditions and human pressures. As a result, success in conservation can often necessitate different measures in different portions of a species' range at different times. There are no panaceas to achieve successful reduction of immediate conservation threats, and successful programs must be tailored to the specific needs of individual species. Programs that advocate only familiar approaches, such as habitat protection or captive breeding, and fail to address the unique ecological difficulties faced by individual populations, often will fall short of success in maintaining wild populations.

Choices need to be made in conservation actions, and reliable selection of the most promising choices cannot normally be made in the absence of good data on specific causes of a species' endangerment. Quality research is a necessary part of efforts. Unfortunately, many endangered species, especially little known species with modest public appeal, never receive adequate study before they become extinct or reach such low numbers that gaining a comprehensive understanding of their problems is very difficult. Endangered species often have quite complicated ecologies, and the most successful programs on their behalf often are ones that recognize that mistakes are often made in identifying crucial limiting factors, but that the penalties of mistakes can sometimes be minimized if a variety of approaches are pursued simultaneously and if programs can somehow remain flexible in responding to new information, quickly adopting new strategies when appropriate and quickly abandoning strategies that are not working.

How many of the numerous currently threatened parrot species will be saved from extinction? This surely is a question that no one can answer, and success may well differ in the short and long terms. In the short term, the currently good results being achieved in efforts for species such as the Puerto Rican Parrot, the Thick-billed Parrot, Lear's Macaw *Anodorhynchus leari*, the Kakapo *Strigops habroptila*, the Mauritius Parakeet *Psittacula eques*, and the Golden-shouldered Parrot *Psephotellus chrysopterygius* have shown that many very seriously threatened species can be conserved if enough effort and persistence can be provided to counter immediate threats. In the long term, things are much more uncertain and will likely depend importantly on what our species achieves in terms of controlling our own population size and reversing worrisome planetary trends such as global warming, ocean acidification, and rising sea levels. These are quite basic forces that underlie most of the immediate threats faced by endangered species and also must be addressed successfully if we are ever to achieve and maintain a truly sustainable biosphere.

Noel F. R. Snyder
Portal, Arizona, United States
31 December 2016

CONTENTS

PREFACE

In May 1971, near Vacaria, northeastern Rio Grande do Sul, southern Brazil, I and William Belton in the company of some local persons watched very large flocks of Red-spectacled Amazons *Amazona pretrei* coming to a communal nighttime roost in a grove of *Araucaria* trees surrounding a marsh in a valley and, to the accompaniment of an incredible volume of screeching, many thousands of birds circled overhead before settling for the night. This was one of the most memorable of my field encounters with parrots, so I was dismayed to learn some 10 years later that most of this and adjoining tracts of *Araucaria* forest had been destroyed in landclearing for pastures, and very few Red-spectacled Amazons remained in the district. Since that time, I have become increasingly aware of the impact of habitat loss or degradation on parrot populations in my own country and elsewhere. The concept of addressing the plight of parrots as one of the most endangered groups of birds has been with me for some time, but acquired a new impetus in recent years when there has been a marked deterioration in the status of two species with which I have been involved – the Orange-bellied Parrot *Neophema chrysogaster* in southeastern Australia and the Norfolk Island subspecies of the Red-fronted Parakeet *Cyanoramphus novaezelandiae cookii*, sometimes treated as a separate species. This book is a fulfilment of that concept, and its purpose is to focus on seriously threatened species, with the possibility that in this century some could be placed alongside the Carolina Parakeet *Conuropsis carolinensis*, the Paradise Parrot *Psephotellus pulcherrimus* and other species that we lost in the 1900s.

In selecting species for inclusion in this book, I have, for the most part, adopted listings in the *HBW and Birdlife International Illustrated Checklist of the Birds of the World*, Volume 1: Non-passerines (del Hoyo and Collar 2014), which in turn are based on the 2014 Red List of the International Union for Conservation of Nature (IUCN). In addition to species listed as endangered, I have included some vulnerable species which, in my opinion, are particularly threatened, and most occur on islands. For this reason two vulnerable *Vini* lorikeets – the Blue Lorikeet *V. peruviana* and the Henderson Island Lorikeet *V. stepheni*, from islands in the South Pacific Ocean, and the Horned Parakeet *Eunymphicus cornutus* from New Caledonia are included, and it is noteworthy that closely related species in both genera are endangered. The polytypic Red-fronted Parakeet *Cyanoramphus novaezelandiae* from New Zealand and nearby islands, and Philippine Hanging Parrot *Loriculus philippensis* are included because again some island forms are extinct and others are threatened. The 2016 Conference of Parties to the Convention on International Trade in Endangered Species of Wild Fauna and Flora (CITES) accepted a proposal from Gabon to list on Appendix I the two *Psittacus* parrots from Central Africa – the Grey Parrot *P. erithacus* and the Timneh Parrot *P. timneh*, and presumably this will result in both being upgraded to endangered status in the IUCN Red List, although numbers of both in the wild remain reasonably high; at the time of writing neither species is listed as endangered so is not included in this book.

PLAN OF THE TEXT

This book is planned as a reference work, and the text is set out so that information can be located easily and quickly. Information pertaining to a species is included in relevant sections of the text for that species, even though the behaviour and habits of closely allied species often are quite similar. I am of the opinion that in a reference work some repetition is more desirable than generalised comparisons that force readers to search back and forth to find the required information. Use of headings and subheadings in the text is intended also to facilitate a prompt finding of information.

References are incorporated in the text in much the same format as is adopted customarily in scientific literature because that procedure is in keeping with my objective of bringing together important published and unpublished data. Authors of published works must be credited appropriately and are responsible for the accuracy of their published observations. I prefer to use the Harvard system, which gives the date with the author's name, because that incorporates a time perspective; in other words, when a reference dated 1999 is cited, the reader knows immediately that the data were recorded almost 20 years ago. The date is given with a reference when that reference is first mentioned in a text section, but not if the same reference is quoted again in the same text section; for example if 'Chisholm (1922)' is mentioned in the STATUS section, then each time 'Chisholm' without a date is repeated in that same section the '1922' also applies, but if both 'Chisholm (1922)' and 'Chisholm (1924)' are used anywhere in the text for a species, then each time either is repeated the date also is repeated. Full details of each reference are given in the REFERENCES CITED (p. 297).

ENGLISH NAMES With few exceptions, I have used English names adopted in the *HBW and Birdlife International Illustrated Checklist of Birds of the World*, Volume 1: Non-passerines (del Hoyo and Collar 2014). I have included OTHER NAMES where those names are widely used in ornithological or avicultural literature. Both 'Amazon' and 'Parrot' are used widely for *Amazona* species, so both are listed, but use of the term 'Conure' for *Aratinga*, *Psittacara* and *Pyrrhura* species is no longer used widely, even in avicultural literature, so I have not included it in OTHER NAMES for these species.

DESCRIPTION All descriptions are taken from museum specimens. For polytypic species, the initial description is of a nominated endangered or vulnerable subspecies, preferably the nominate subspecies where applicable. Length is from a preserved specimen and is merely an indication of size, not a precise measurement. Weights have been extracted from published records, museum labels, and unpublished field reports. Included in the description are standard measurements from a stated number of museum specimens; these measurements are:

wing length from the carpal joint to the tip of the longest primary of the folded right wing flattened against a rule,
tail length from the base of the central feathers to the tip of the longest feather,
exposed culmen (exp. cul.) length in a straight line from the tip of the upper mandible back to the anterior edge of the feathered or unfeathered cere,
tarsus (tars) length of the tarsometatarsus *in situ*; it is virtually impossible to obtain accurate tarsus measurements from museum specimens because the legs are shrivelled, and often are set permanently in an awkward position.

The measurements are listed as a range from minimum to maximum, with the mean in parenthesis.

Specimens or digital images of specimens have been obtained from the following institutions:

AM Australian Museum, Sydney, Australia
AMNH American Museum of Natural History, New York, USA
ANSP Academy of Natural Sciences, Philadelphia, USA
ANWC Australian National Wildlife Collection, Canberra, Australia
CUMZ Cambridge University, Museum of Zoology, Cambridge, UK
MNHN Museum National d'Histoire Naturelle, Paris, France
NHM Naturhistorisches Museum, Vienna, Austria
NMS Royal Museums of Scotland, Edinburgh, UK
SAM South Australian Museum, Adelaide, Australia
UKNHM Natural History Museum, Tring, UK
YPM Peabody Museum, Yale University, New Haven, USA.

DISTRIBUTION The overall range of the species is given, and if there are no subspecies the range is set out in more detail. If there are endangered or vulnerable subspecies, the overall range is delineated only in broad, general terms and detailed information is included in the text for each of those subspecies.

Distribution maps, based on published data, are included for all species. Although there are valid reasons for criticising distribution maps, they are helpful in showing clearly and at a glance the approximate range of a species. While acknowledging that within its range a species can be expected to occur only in areas of suitable habitat, and at times a species may be recorded extralimitally or even in districts well outside its normal range, my objective has been to show the normal range, so each map is a conservative indicator of distribution. Former ranges of extinct species are shown in red.

SUBSPECIES For polytypic species the distinguishing features, measurements and distribution of endangered or vulnerable subspecies are given. All comparative phrases in descriptions refer back to the initially described subspecies; for example, the phrase 'breast darker brown; less crimson on hindneck' means that a darker brown breast and less crimson on the hindneck are two differences from the initially described subspecies. The numbering of subspecies is no indication of a priority listing or any degree of differentiation, but is merely to show at a glance the number of subspecies under discussion.

STATUS In a book dealing with endangered species, information on status obviously is of key importance, and I have endeavoured to ensure that coverage of all aspects is as comprehensive as possible. Known and perceived threats are identified, and declines in both regional and local populations are documented. Attention is focused on actions being taken to address threats and, where they have been initiated, recovery programs are discussed. Population estimates come almost exclusively from the online database of Birdlife International.

HABITATS Following generalised statements on habitat preferences, there are more detailed accounts from various parts of the range with particular mention made of any regional differences in habitat preferences. Many endangered parrots are threatened because they are habitat specialists, often being dependent on certain types of forest or woodland, sometimes at well delineated altitudinal levels or often dominated by specific trees preferred for feeding or nesting, and loss or degradation of these habitats is a primary threat.

For the most part, I have adopted self-explanatory, simplified descriptive terms for habitats, such as lowland forest, open woodland, savanna and cultivation, but explanations are needed for the following, sometimes peculiarly regional, descriptive terms:

caatinga (northeastern Brazil): semiarid to arid scrubland with sparse groundcover of few grasses, and dominated by cacti and deciduous, often spiny trees and bushes remaining leafless for many months and with characteristically pale grey branches,
cerrado (inland Brazil): semiarid scrubland with sparsely scattered low trees and bushes having characteristically gnarled or twisted branches bearing thick, grooved bark and leathery leaves,
elfin forest (closely spaced trees) or elfin woodland (widely spaced trees): stunted forest or woodland at high elevations, often at or near the treeline, and typically on exposed ridges or slopes,
gallery forest or woodland: narrow strips of forest or woodland bordering watercourses, and usually in open country,
mallee woodland (closely spaced trees) or mallee scrubland (widely spaced trees): in inland southern Australia a low woodland community dominated by scrubby *Eucalyptus* trees varying in height from 2 m to 8 m and with multiple stems arising from large lignotubers,
terra firme forest (Amazonia, South America): humid or wet lowland forest on elevated dry ground and not subject to inundation,
varzéa forest (Amazonia, South America): humid or wet lowland forest seasonally flooded for several months once or twice each year, and situated mostly on floodplains of major rivers.

In South America, altitudinal distribution is particularly notable in the Andes and associated mountain ranges, where species often are associated with one or more of three forested zones – tropical (up to 1000 m), subtropical (1000 m to 2500 m), or temperate (2500 m up to the treeline at about 3500 m).

MOVEMENTS In Australia, seasonal migration between Tasmania and the southeastern mainland is undertaken by three parrots, and two of these – the Orange-bellied Parrot *Neophema chrysogaster* and Swift Parrot *Lathamus discolor* – are endangered. Elsewhere, parrots may undertake regular or irregular seasonal movements, usually in response to seasonal changes in food availability, or altitudinal movements in response to weather conditions, and local wandering or changes in abundance are not uncommon. Reference is made to records of movements, and consideration is given to factors that do, or could influence these movements.

HABITS A general account of habits is sourced from published and unpublished accounts together with personal observations. Special mention is made of habits that may in some way increase the vulnerability of a species. An example of this would be the oft quoted claim that Carolina Parakeets *Conuropsis carolinensis* were easily killed in large numbers because of their gregarious habits.

CALLS I am well aware of the inadequacy of describing calls as onomatopoeic representations of the sounds by using terms such as screech, whistle, cackle etc, when calls can be heard and interpreted differently by different observers. I see no practical alternatives. For most readers, sonograms on the printed page fail to convey the actual sounds, though they well may reflect structures of calls. Recorded calls of many parrots, including endangered species, now can be accessed online, and this certainly is recommended to readers.

FEEDING A composite description of the diet is given, and mention is made of favoured foods, especially where a species may be a specialist feeder. Among such species are the endangered *Rhynchopsitta* parrots, which feed primarily on *Pinus* seeds in the highland coniferous forests of Mexico, and clearing or degradation of those forests consequently poses a serious risk to both species. Observations and reports on feeding are quoted, and records of stomach contents are reported.

BREEDING I have included in this section all available information on breeding, and that takes in gonadal development as recorded on specimen labels or reported in published accounts. The term 'breeding season', as used in this book, refers to the entire period during which the birds are engaged in nesting activities, normally commencing with courtship displays or selection of nesting sites and ending with the fledging of chicks from renestings. Decreased availability of nesting sites, usually resulting from deforestation or selective logging of mature trees, has been identified as a major threat to many endangered parrots, and providing artificial sites is a key aspect of many recovery programs.

Where available, recorded measurements of eggs are included, and registration numbers are given to identify measured eggs in museum collections.

ACKNOWLEDGEMENTS

Many persons are involved in one way or another in production of a book of this kind, and we have been most fortunate to have received generous assistance from many sources. First and foremost, we extend a special expression of gratitude to Noel Snyder and Walter Boles for their most valued participation in this project. Tribute is paid to authors of published works which form the basis of the text, and also we have been pleased to be able to include unpublished data provided by Caroline Blanvillain (Société d'Ornithologie de Polynésie), Alan Lieberman and Gerald McCormack from field studies of *Vini* lorikeets, Stephen Murphy from field studies of Night Parrots *Pezoporus occidentalis*, Carl Jones and Hazel Jackson (Durrell Wildlife Conservation Trust) from field studies of Mauritius Parakeets *Psittacula eques*, Julian Hume (UKNHM) from studies of extinct parrots of the Mascarene Islands, and Fabio Olmos and Robert Ridgely from field studies of *Pyrrhura* parakeets. Assistance with references, including access to published papers, was provided by Robert Heinsohn (Australian National University), Julian Hume, Hazel Jackson, Carl Jones, Leo Joseph (ANWC), Mary LeCroy (AMNH), and David Waugh (Loro Parque Fundación). Special thanks go to staff at the Australian National Library, Canberra, and the State Library of New South Wales (Mitchell Library), Sydney, for providing access to historical works and for supplying digital images of drawings from those works. Comments on, and amendments to draft texts were provided by Julian Hume, Carl Jones, Leo Joseph and Noel Snyder.

For making available specimens or digital images of specimens in their collections, we thank Sandy Ingleby (AM), Paul Sweet and Thomas Trombone (AMNH), Nate Rice (ANSP), Leo Joseph and Robert Palmer (ANWC), and Hans-Martin Berg (NHM). Information about the specimen of *Psittacula eques* in the National Museums of Scotland, Edinburgh, was provided by Robert McGowan, and additional data from specimens in the AMNH Collection was generously provided by Mary LeCroy. During a visit by Frank Knight to the Naturhistorisches Museum, Vienna, to prepare working drawings from specimens of extinct parrots, generous assistance was given by museum staff.

Preparing digital images of the paintings for publication was undertaken most capably by Robert and Sonja Millis at Arthead Digital Imaging, Mossvale, and by Aaron and Jason Cook at Cie-Elle Digital Imaging, Merfield.

Research pertaining to this project was undertaken by Joseph Forshaw as a Research Associate in the Department of Ornithology, Australian Museum, Sydney. Trustees, Director and staff at the museum generously facilitated that research, and this is very much appreciated.

Again, it has been particularly rewarding to work with Derrick Stone in preparing this book for publication. Special thanks go to John Manger and colleagues at CSIRO Publishing for bringing the project to fruition.

JOSEPH M FORSHAW
FRANK KNIGHT
Canberra, ACT, Australia
31 December 2016

INTRODUCTION

I do admit to experiencing a sense of melancholy, or even sadness, when examining museum specimens of extinct parrots. While holding a specimen of the Carolina Parakeet *Conuropsis carolinensis*, it is very easy to visualise a flock of these striking birds in flight through a forest in the Appalachians, and to ask the question – Why did we lose them? There rarely, if ever, is a straightforward answer to this question, because the pressures leading to endangerment or extinction usually are complex and interrelated.

The first of two recent studies looking at status and population trends in parrots as one of the most threatened bird families is that undertaken by Marsden and Royle (2015) who reviewed literature to identify gaps in data on population densities and to assess how habitat and landuse changes affect their abundance. Density estimates were found for only 25 per cent of species, and threatened species were no more likely to have a density estimate than non-threatened species, and were less likely to have estimates of abundance change. Of the 90 species for which at least one density estimate was available, most had density estimates from only a single study, but 22 species had estimates from two studies, 12 from three studies, and the Eclectus Parrot *Eclectus roratus* had estimates from five studies. Paired abundance levels across habitat or across sites were available for 179 habitat/site pairs, the most common comparisons being between primary forest and selectively logged forest and across different primary forest types/sites. Non-threatened species were more likely to have a comparison of abundances across habitats than were threatened species. Overall, the lack of data was found to be serious for all parrot groups, and acute for some, leading the authors to question whether it ever will be possible to gather enough data on parrots to give an understanding of the effects of environmental change, and there seems to be insufficient capacity to track the abundance of parrot species in response to environmental changes.

To address a lack of in-depth quantitative analyses of the nature and trends in threats to parrots, the intrinsic, extrinsic, global and regional factors associated with endangerment and extinction were subjected to comparative analyses, which included biological, life history and ecological attributes, together with external anthropogenic threats and socio-economic variables associated with countries where the parrot species occur (Olah *et al.* 2016). Most importantly, these analyses have confirmed that parrots are more threatened than other comparable taxonomic groups, with consistently negative trends in extinction risk during the past 25 years, and conservation successes have been outweighed by the number of species upgraded to higher levels of threat. The highest proportion of threatened species occurs in the Neotropics, where 37 per cent of species are threatened, followed by Australasia-Oceania with 22 per cent, the Afrotropics with 19 per cent, and Indomalaya with 14 per cent. These analyses revealed also that there is no significant effect of phylogeny on threatened status, with extinction risk usually being distributed non-randomly with respect to phylogeny. Findings from these analyses are extremely important because, for the first time, we have a comprehensive overview, both globally and regionally, of threats to parrot populations, and that will facilitate a priority listing of conservation measures addressing all pressures impacting adversely on populations at risk or in decline. Pressures associated with extinction risk in parrots were summarised in two categories – biological attributes and anthropogenic and socio-economic factors.

BIOLOGICAL ATTRIBUTES

It is pointed out that, as for other taxonomic groups, large-bodied parrots are more prone to extinction risk as large body size correlates with many known traits leading to extinction, including low population densities, lower recruitment levels, and a higher vulnerability to hunting (in Olah *et al.* 2016). Risks associated with these traits obviously would be higher in a species occurring in a small range, especially on an island, or with highly specialised habitat requirements. Among recently extinct species, the Norfolk Island Kaka *Nestor productus* was a large-bodied parrot, presumably occurring in relatively low numbers on Norfolk Island and two quite small outlier islands, and with a low recruitment rate preventing recovery from intense hunting for food by colonists and convicts facing an acute food shortage. I suspect that the same attributes contributed primarily to the early extinction of larger parrots on the Mascarene Islands, especially the very large Broad-billed Parrot *Lophopsittacus mauritianus*, which succumbed quickly to wholesale killing by the first voyagers seeking to replenish food stocks on their ships. *Amazona* parrots now confined to single islands in the Lesser Antilles exhibit the same biological attributes, and I would identify the Imperial Amazon *A. imperialis* from the mountain forests of Dominica as being particularly vulnerable. Large-bodied parrots with specialised habitat requirements include the cliff-nesting Indigo Macaw *Anodorhynchus leari* from inland Brazil and Red-fronted Macaw *Ara rubrogenys* from river valleys in Bolivia, both of which are dependent on the near proximity of foraging habitat to nesting sites, and the Blue-throated Macaw *Ara glaucogularis*, which occurs only in seasonally-flooded savanna with natural or artificial elevated islands of forest dominated by motacú palms *Attalea phalerata*, their most important food source; all three macaws are endangered.

Ecological specialisation is another attribute that can increase the risk to parrot populations. Thick-billed Parrots *Rhynchopsitta pachyrhyncha* and Maroon-fronted Parrots *R. terresi* are large-bodied parrots with strong dietary preferences for *Pinus* seeds and are confined to highland conifer forests in Mexico, where they are endangered by logging and landclearance for agriculture. Extensive logging of pine forests in Sonora and Chihuahua, northern Mexico, and a resulting barrier preventing northward movements of Thick-billed Parrots have been identified as causes of their disappearance from southwestern United States, but persistent reports of widespread shooting suggest that excessive persecution may have contributed to their extirpation north of the Mexican border (Snyder *et al.* 1999). Few parrots are more specialised than the Glossy Black Cockatoo *Calyptorhynchus lathami* from eastern Australia, where it feeds almost exclusively on *Allocasuarina* and *Casuarina* seeds, the massive bill being modified to extract these seeds from the cones, and although not endangered in much of their range, landclearance has resulted in the extirpation of some populations.

ANTHROPOGENIC AND SOCIO-ECONOMIC FACTORS

The main anthropogenic threats to parrots are habitat loss, degradation or fragmentation driven by agriculture, logging and commercial or residential development, and hunting and trapping (Olah *et al.* 2016). Most parrots are forest dependent because of their nesting and feeding requirements, and these forest dependent species are more likely to be threatened. At least 70 per cent of species nest in pre-existing tree hollows, the availability of which is lost or decreased by the destruction or degradation of forests, so impacting severely on recruitment levels. A loss of key food resources also can result from deforestation, especially if clearance is on a large scale, because many parrots rely on extensive tracts of suitable habitat to provide year-round sources of seasonally variable seeds and fruits. Total loss of forests or woodlands can occur on smaller islands when all native vegetation is cleared, often to make way for plantations or village gardens, but landclearance normally results in habitat fragmentation with remnant stands confined to reserves or less accessible sites, such as steep hillsides, and sometimes surviving only as scattered trees left standing in pastures or farmlands. Sangihe Island, in the Indonesian Archipelago has been almost totally converted to coconut and nutmeg plantations resulting in extirpation of the endemic population of the endangered Red and Blue Lory *Eos histrio*. Similarly in the central Philippine Islands, two subspecies of the Philippine Hanging Parrot *Loriculus philippensis* have disappeared from islands where deforestation has been almost complete. There have been no confirmed records of the highly distinctive Cebu Hanging Parrot *L. p. chrysonotus* on Cebu since about 1930, and in the early 2000s only seven small patches of degraded native forest remained on that island. Searches undertaken in the 1950s failed to find any trace of the endemic *L. p. siquijorensis* on Siquijor, where, in the early 1990s, four remaining patches of forest covered a total of only 781 ha.

Although usually resulting from human activities, fragmentation and degradation of habitats can be brought about by natural forces. In April 1980, during a brief visit to Dominica, in the Lesser Antilles, I observed the devastating damage to mountain forests caused in the previous year by Hurricane David, and came to the conclusion that there must have been a very serious post-hurricane depletion in numbers of endangered Imperial Amazons *Amazona imperialis*. Many mature gommier trees *Dacryodes excelsa*, which are favoured for nesting and feeding, were uprooted and those left standing in sheltered valleys had been stripped of foliage and fruits, so there must have been an acute food shortage. It was suspected that at the time as few as 50 birds survived. The total population of the critically endangered western subspecies of the Ground Parrot *Pezoporus wallicus* along the coast of southwestern Australia recently was estimated at less than 140 birds, with the majority confined to Cape Arid National Park, where a devastating wildfire in November 2015 destroyed vegetation in almost 90 per cent of the park, leaving only two unburned pockets of habitat, where an unknown number of parrots survived. A combination of human induced and natural forces was identified as posing a threat to populations of the vulnerable Palm Cockatoo *Probosciger aterrimus* on Cape York Peninsula, northern Australia (Murphy and Legge 2007). Modification of the nesting habitat in woodlands adjoining rainforest, mostly by mismanaged fire, could result in long-term vegetation changes, or loss of nesting trees, or both. At a study area, between 1999 and 2005, the fate of 61 nest-trees was followed, with an average of 48 trees being monitored each year, and an average 5.2 per cent of nest-trees was lost each year. Losses were due mainly to fire, followed by wind and then decay. Offsetting these losses was the episodic creation of potential new nesting hollows when high winds associated with severe cyclones break branches from trees, so exposing the inside of the tree to rot and termite activity. Murphy and Legge concluded that because nest-trees are a limiting resource, their loss to fire is significant, so should be monitored, and fire management should focus more on maintaining an adequate recruitment of large trees, which could be impacted adversely by excessively frequent burning.

Fragmentation of habitat can have less obvious though equally damaging impacts on parrot populations. One less obvious impact is the disruption of dispersal pathways leading to a loss of genetic connectivity between remnant tracts of habitat and to genetic depletion in small isolated populations (in Klauke *et al.* 2016). In eastern Australia, terrestrial Ground Parrots have suffered greatly from destruction of their specialised habitats – coastal and subcoastal heathlands, the result of extensive urban and agricultural development along the seaboard, and formerly they were plentiful in areas now occupied by major cities. They continue to be threatened because of low numbers, and more importantly because the loss of habitat means that their range is discontinuous with marked fragmentation of the genetic pool available to most populations. In New South Wales and Victoria, some protected pockets of ideal habitat, with a past history of frequent burning or grazing, have not been recolonised, probably because alienation of surrounding lands precludes any immigration of birds. Randomly amplified DNA fingerprinting was used to determine levels of genetic diversity in populations at three breeding sites of Ground Parrots in southeastern Queensland (Chan *et al.* 2008). One site was approximately 50 km distant from another two sites, with non-heath bushland and major urban infrastructure in the intervening area, and the two nearest sites were connected by a 7 km corridor of non-heath vegetation. Blood samples were taken from 20 parrots captured at the sites and, despite the small sample size, there was sufficient evidence to suggest that the low levels of genetic variation in the sampled populations at least equal those found in the lower end of diversity for endangered species. These results emphasised the need for caution to be exercised to ensure that genetic diversity is not reduced further through loss of habitat.

Between 2009 and 2012, similar investigations were undertaken in western Ecuador to assess the genetic divergence between populations of the endangered El Oro Parakeet *Pyrrhura orcesi* at three sites separated by small geographic distances, but characterised by a heterogeneous habitat structure (Klauke *et al.* 2016). Results revealed the presence of genetic differentiation between all three sites even though separating distances were quite small, varying from 3 km to 17 km, and individual genotype assignment revealed similar genetic divergence across a 3 km wide valley compared to about 13 km of continuous mountain range. Observational data compiled during the study, together with estimations based on genetic data, indicated that dispersal rates were low. It was suspected that the complex social system of El Oro Parakeets contributes to the recorded genetic structure, and low dispersal rates coupled with slow breeder turnover entails that immigrants face long reproductive cues within flocks, so reducing the rate of effective dispersal events. It was noted also that unlike other sympatric parrots, such as the Red-masked Parakeet *Psittacara erythrogenys* and Bronze-winged Parrot *Pionus chalcopterus*, El Oro Parakeets usually fly through rather than above the forest canopy, so probably are more susceptible to habitat loss and fragmentation.

Species which require a near proximity of nesting and foraging habitats are particularly vulnerable to landclearance. During the years 1970-1975, the effects of woodland clearing on recruitment in breeding populations of endangered Carnaby's Black Cockatoos *Calyptorhynchus latirostris* were assessed at two localities in the wheatbelt of Western Australia (Saunders 1977). At one locality large tracts of foraging habitat surrounded the stand of eucalypt woodland in which nesting occurred, and clearing had been carried out only during the previous 20 years. In this study area there were 75 breeding pairs, and each pair fledged two chicks every three years, or 0.6 chicks per year. The mean fledging weight of chicks was approximately 97 per cent of the mean adult weight. Nesting birds foraged quite near to their nesting sites and their behaviour was as would be expected, with both brooding females and newly-hatched chicks depending on the males for food. It was apparent that the local food supply was reliable and quite adequate during the breeding season. A very different situation prevailed farther south at the second locality, where pairs nested in a water-catchment reserve and foraged in surrounding cleared farmland with numerous small patches of uncleared land around homesteads, along some roads and bordering sections of the railroad. At this locality, during the course of the study, each pair fledged one chick every three years, or 0.3 chicks per year, and the mean fledging weight of chicks was 83 per cent of the mean adult weight. The behaviour of nesting pairs was strongly indicative of a food shortage, and contrasted markedly with the behaviour of pairs at the other locality. Females were observed feeding by themselves or with their mates while they still had eggs in the nest, thus exposing the eggs to predators or to usurpation of the nesting hollow by competitors. Quite often the female left the nestling all day during its first 10 days, a further exposure to predators. Visits by parents to the nest to feed the chicks were infrequent, and at times in hot weather they did not visit the nest at all between leaving at dawn and returning at dusk. Shortage of food seemed to be the cause of this deterioration of parental care, the parents being obliged to spend more time foraging.

In southeastern Australia, two vulnerable *Polytelis* parrots are dependent on the near availability of foraging habitat to nesting areas in riverine forests of river red gums *Eucalyptus camaldulensis*. In some districts within the range of the Superb Parrot *P. swainsonii*, suitable breeding habitat is not occupied because all foraging habitat in open box-eucalypt woodland within 10 km of the rivers has been converted to pasture or farmland. Similarly, the eastern subspecies of the Regent Parrot *P. anthopeplus* has abandoned stands of breeding habitat in river red gums along the Murray River and its tributaries because all adjoining mallee woodland, its preferred foraging habitat, has been replaced by irrigated orchards, vineyards and gardens.

SECONDARY PRESSURES

Direct human interference with parrot populations are what I term secondary pressures, and these include persecution as pests, hunting for food and trapping or nest-robbing for the live-bird market. In addition to landclearance, these secondary pressures can be poverty driven in some developing countries, so the severity of their impact can be correlated with the socio-economic status of a country and, not unexpectedly, human population density has been found to be closely related to the proportion of threatened species in a country (in Olah *et al.* 2016). The significance of these secondary pressures increases when impacting on parrot populations with restricted ranges or already threatened by primary pressures, especially habitat loss. Persecution of Indigo Macaws *Anodorhynchus leari* in Brazil because of damage caused to maize crops or of Red-fronted Macaws *Ara rubrogenys* in Bolivia to protect peanut crops obviously is far more damaging than is any shooting of widespread, abundant Galahs *Eolophus roseicapilla* by farmers in Australia.

Predation and disease impact naturally on populations, as occurred recently when eggs were predated in the first confirmed nest of the critically endangered Night Parrot *Pezoporus occidentalis*, and DNA testing of shell fragments revealed that the predator was a native King Brown Snake *Pseudechis australis*. It is difficult to assess levels of natural predation and disease, but the damaging impact of predation can be exacerbated very significantly, even resulting in the extirpation of local populations, when new predators are introduced, and probably nowhere has the impact of introduced predators been more devastating than in New Zealand, where the introduction of mammalian predators, especially mustelids, has resulted in the near extinction of parrot populations. Subfossil bones show that prior to Maori settlement in New Zealand, approximately 1300 years ago, the flightless, nocturnal Kakapo *Strigops habroptila* was widespread, but the range contracted substantially after Maori settlement. Dogs were used by Maori hunters to capture and kill the parrots for meat, skins and feathers. In addition to dogs, the Pacific Rat or kiore *Rattus exulans* was introduced with Maori settlement, and so began a long history of predation by introduced mammals. After 1880 there was a dramatic decline following the introduction and establishment of three species of mustelids, Black Rats *Rattus rattus*, several species of deer, and Australian Brush-tailed Possums *Trichosurus vulpecula*, and predation by mustelids undoubtedly was the main cause of this decline. Extinction of the Kakapo appeared imminent when, in 1977, a population was found on Stewart Island, where mustelids are not present, and birds from this population were translocated to three predator-free offshore islands, where an intense management program is being implemented to secure the future survival of these unique parrots. Predation by these same introduced mammals, particularly the mustelids, has been implicated also in dramatic declines in populations of *Nestor* parrots and *Cyanoramphus* parakeets. It is estimated that in the past 100 years there has been a decline of approximately 60 per cent in numbers of New Zealand Kakas *Nestor meridionalis*, and the present population is estimated at 1000–5000 mature individuals. Formerly widespread in New Zealand and present on a number of subantarctic islands and islands in the South Pacific Ocean, *Cyanoramphus* parakeets have been extirpated in the Society Islands, on some subantarctic islands, and on Lord Howe Island, and have disappeared from much of their former range in New Zealand, largely as a result of predation by introduced predators, especially Black Rats *Rattus rattus*. Formerly widespread and common throughout the main North and South Islands in New Zealand, Red-fronted Parakeets *Cyanoramphus novaezelandiae* are particularly susceptible to mammalian predation because they often feed on the ground, so now are common only on predator-free islands, and the natural range of the endangered Orange-fronted Parakeet *C. malherbi* has been reduced to just three river valleys in South Island. On Norfolk Island, a critically endangered subspecies of the Red-fronted Parakeet *C. n. cookii* is virtually confined to the small remnant tract of native forest in the National Park, and is a classic example of an island population declining under a combination of pressures, notably loss of habitat, competition from introduced species, predation by rats and feral cats, and disease. Other subspecies have been extirpated by the same pressures on Lord Howe Island and subantarctic Macquarie Island.

After spending four months visiting 28 islands in French Polynesia to collect data on endemic landbirds, habitats and introduced animals, focusing particularly on factors involved in population declines and extinction, Seitre and Seitre (1992) concluded that, in addition to habitat destruction and hunting, introduced predators play a major role, with the introduced Black Rat *Rattus rattus* being the most dangerous. They point out that some 2000 years ago Polynesians brought domestic chickens, pigs, dogs and Pacific Rats *Rattus exulans* to the islands, and Europeans subsequently introduced goats, sheep, cattle, horses, rabbits and cats, as well as Black Rats, Ship Rats *Rattus norvegicus* and House Mice *Mus musculus*. Although not documented, these introductions of European pests could have occurred earlier than is generally believed and could partly explain why at least five native species, including the Raiatea Parakeet *Cyanoramphus ulietanus*, were never seen again after the expeditions of Captain James Cook in the late 18th century. As pointed out by Walter Boles in his contribution on the fossil history of parrots, small lorikeets, both living and extinct, are prominent in bird remains from the Polynesian islands. Two extinct *Vini* species – *V. vidivici* and *V. sinotoi* – are known from remains found in the Marquesas, Society and Cook Islands and extant species were more widespread than at the present time. Apart from the more widespread Blue-crowned Lorikeet *V. australis*, which is secure in some parts of its range, all existing *Vini* species are at risk.

Perhaps nowhere have island parrot populations been more severely affected by hunting and the introduction of predators than on the Mascarene Islands, in the Indian Ocean. Cheke and Hume (2008) present a fascinating description of undisturbed ecosystems on these islands, where there occurred a remarkable array of parrots, but arrival of the first European travellers in the late 1500s and early 1600s brought about widespread deforestation, excessive hunting and the introduction of competitors and predators, eventually resulting in appalling levels of extinction. From a few museum specimens, fossil remains and contemporary accounts by early travellers, six endemic parrots have been described, but only the Mauritius Parakeet *Psittacula eques* now survives. Hume (2007) notes that because of the paucity of fossil parrot remains, the number of species may well have exceeded the number currently recognised.

Parrots in the West Indies fared a little better after the arrival of European colonists, with a number of species surviving to the present time, though possibly on fewer islands than in the past, but again there have been significant losses. Two species – the Cuban Macaw *Ara tricolor* and Puerto Rican Parakeet *Psittacara maugei* – are known from museum specimens and several other species are known from subfossil remains and accounts from early travellers or colonists, but determining the number of species that may have occurred originally is not possible, for many of the records are highly doubtful, and there is little or no evidence to support the existence of some species that have been described in the literature. Parrots were hunted for food and were kept as pets by Amerindians and subsequently by European settlers, so it is highly likely that live birds were traded between islands, and possibly were brought from the mainland.

As secondary pressures, hunting and persecution as pests are reported to have been implicated in the extirpation of at least two mainland species – the Carolina Parakeet *Conuropsis carolinensis* in North America and the Glaucous Macaw *Anodorhynchus glaucus* in South America. Snyder (2004) notes that although data are insufficient to confirm the accuracy of early accounts of very large flocks of Carolina Parakeets, it is apparent that there was little interaction with Native American culture, so it is reasonable to assume that the overall population was relatively stable prior to European settlement. Consequently, demise of the species can be attributed both directly and indirectly to the activities of European settlers. Persistent persecution has been identified as one of the causes of the extirpation of these parakeets, but McKinley (1966) pointed out that it is too easy to claim that they were such pests of fruit and grain crops that they were relentlessly exterminated. Perhaps the significance of persecution in their extirpation has been unduly emphasised, but I would caution against its being underrated because there are repeated references to their gregarious habits contributing to the ease with which very large numbers could be shot. It has been suggested that, when first reported by naturalists, the relict population of Glaucous Macaws already was extremely local and declining, and the restricted natural range of *Butia* palms, the presumed staple food source, would have supported only relatively small numbers. However there is evidence of excessive exploitation for food and for the pet market contributing to their demise. During the 18th and 19th centuries, Jesuit missionaries organised the taking of chicks from nests for use as pets, and there is evidence of these macaws being shipped to Europe for private zoos. Nests in most riverbank cliff-faces would have been easily accessible, and there are accounts of riverboat travellers shooting macaws for food. Such depredations on both adults and chicks could have extirpated entire colonies.

Conflict between parrots and agriculture have been reported from a number of countries, but to date there has been little objective evaluation of the problem. I am most familiar with the situation in Australia, where any assessment of the issue is compromised by campaigns promoting short-term or 'quick-fix' responses, which certainly offer no effective solutions. Damage to crops in Australia is not of national economic significance, but losses experienced by farmers locally can be severe and there is a need to protect crops. In other countries where subsistence food crops may be at risk of damage there is a stronger need to minimise attacks by parrots, and farmers often rely on shooting or poisoning, which can pose a significant threat if rare or endangered species are involved. Already, I have mentioned the persecution of Indigo Macaws *Anodorhynchus leari* to protect maize crops and of Red-fronted Macaws to protect peanut crops, actions which pose additional threats to these endangered species. Persecution of the Philippine Cockatoo *Cacatua haematuropygia* in agricultural areas has contributed to its disappearance from many parts of its range. In Australia, I often hear of sunflowers or other crops highly attractive to parrots being planted near tree-lined watercourses, a preferred habitat for many parrots, and high damage levels are the inevitable consequence. In such situations, where severe damage will occur, the substitution of less attractive crops, such as canola or soybeans, becomes a viable option. The likelihood of damage by wildlife, including parrots, must be considered in the initial assessment when planning any crop-growing enterprise and remedial measures formulated as an integral component of management practices. In his assessment of Neotropical parrots as agricultural pests, Bucher (1992) also gives priority to crop substitution and bird repellency over lethal techniques, pointing out that promotion of unrestricted trapping is neither an efficient or justifiable method for reducing agricultural damage, and may lead to the extinction of some species.

POACHING FOR THE LIVE-BIRD MARKET

Referring to the popularity of parrots as cagebirds, I have heard it said that 'We have loved them to death!', and there is some truth in this statement because very large numbers are involved in the live-bird trade, both nationally and internationally, with high levels of mortality featuring in the processes of capture and shipment. There is ample evidence that in many parts of the worldwide range, parrots were held as pets and traded by native peoples. In the South Pacific Ocean, *Vini* lorikeets and *Prosopeia* parrots were transported between islands by Polynesian voyagers, and Amerindians in the West Indies transported live parrots between the islands and probably also from the mainland to the islands. However, trafficking in live parrots took on a new impetus with the early voyages of discovery, and this is highlighted by the retrieval of two small bones referred to as an *Aratinga-Psittacara* or *Pionus* parrot from a Spanish shipwreck of 1622 off the Florida Keys (Cooper and Armitage 2013).

Concern has been expressed at the level of trade in parrots, both domestic and international, especially when it involves the taking of chicks from nests as this frequently results in tree hollows being chopped open, so rendering the sites unsuitable for future use. A study of the international trade in parrots listed by the Convention on International Trade in Endangered Species of Wild Fauna and Flora (CITES) revealed that 1.2 million parrots were exported between 1991 and 1996, with the majority of birds coming from the Neotropics, and these figures are presumed to be a gross underestimate of the actual numbers taken from the wild because they exclude pre-export mortality, which has been estimated to reach 60 per cent of all birds harvested (in Wright *et al.* 2001). Levels of domestic trade and illegal international trade are not taken into account in these figures, and both are thought to be substantial, though not well quantified. In an analysis of ecological and socio-economic factors affecting extinction in parrots, it was found that threatened species used for pets tend to belong to the lower categories of endangerment and this is supported by other recent studies showing that the vast majority of species in domestic and international trade are non-threatened species (Olah *et al.* 2016). I have misgivings about the appropriateness of this conclusion, for it tends to mask the often serious local or regional impacts of trapping and the poaching of nests.

An analysis was made of 23 field studies examining nesting success in 21 parrot species in 14 Neotropical countries, with a total of 4204 nesting attempts being recorded between 1979 and 1999 (Wright *et al.* 2001). The mean nest poaching rate across all field studies was 30 per cent, and some level of nest poaching was present in most studies. Approximately 20 per cent of nests poached was reported in 13 studies, and more than 70 per cent of poaching was reported in studies of four species – the Yellow-naped Amazon *A. auropalliata* in Guatemala, the Yellow-crowned Amazon *A. ochrocephala*, the White-chinned Amazon *Amazona kawalli*, and the endangered Vinaceous-breasted Amazon *Amazona vinacea*. An absence of nest poaching was recorded for only six species. Also of concern was the finding that a 57 per cent mortality of eggs and nestlings was higher for species that experienced poaching than the 32 per cent for species that experienced no poaching. Higher levels of nest poaching were found in mainland studies, contributing to a greater overall nesting failure there than in studies from the West Indies. Of the 21 species included in the analysis, 13 are classified at some level of risk, with four – the Vinaceous-breasted Amazon, the Lilac-crowned Amazon *Amazona finschi*, the Green-cheeked Amazon *A. viridigenalis* and the Yellow-headed Amazon *A. oratrix* – now listed as endangered. The mean poaching level of 30 per cent for species at risk did not differ significantly from the 49 per cent for non-threatened species. In summary, the analysis found that nest poaching is a widespread and significant cause of nesting failure in Neotropical parrots, and for species experiencing nest poaching, failure due to poaching was significantly greater than failure due to natural causes. It is pointed out that low rates of reproduction experienced by many parrot species strongly suggest that poaching levels in excess of 70 per cent, as found in four of the species studied, will lead to severe declines in natural populations, but some degree of uncertainty exists as to whether poaching levels of between 10 and 30 per cent, as reported in almost half of the studies, would adversely affect the maintenance of viable populations.

Desenne and Strahl (1991) report that an assessment of trade in parrots throughout Venezuela in 1988–1989 revealed alarmingly high internal and international levels, with main outlets for national trade being in major cities, where it was compounded by the use of feathers for Indian artifacts sold to tourists. International trade involved illegal export chiefly from the Río Orinoco delta, with the majority of birds, estimated at between 65 000 and 75 000, being destined for Guyana. The large macaws suffered badly from both levels of trade but, because of its restricted range, the near endemic Yellow-shouldered Amazon *Amazona barbadensis* was perhaps the most critically threatened species.

Another insight into regional domestic trade in the Neotropics comes from Herrera and Hennessey (2007) who monitored the illegal pet trade in Los Pozos pet market in Santa Cruz de la Sierra, Bolivia, from August 2004 to July 2005. During this period, 7279 individuals of 31 parrot species were recorded in trade, and these included two endangered Indigo Macaws *Anodorhynchus leari* and 26 endangered Red-fronted Macaws *Ara rubrogenys*. If the other four markets in Santa Cruz trade similar numbers of parrots then it is estimated that the number of parrots recorded in the Los Pozos market represents only 20 per cent of trade in Santa Cruz city, and the city of Cochabamba is expected to have similar trade numbers, so giving an estimate of 22 000 parrots being illegally traded in 2004–2005.

Scarce local or regional populations of generally non-threatened species also can be put at risk by trapping or nest poaching. Solano-Ugalde (2011) notes that Rose-faced Parrots *Pyrilia pulchra* have declined in northwestern Pichincha, northern Ecuador, where, in 2009–2010, during informal surveys of illegal pet markets, two cages with juveniles of these parrots were seen. Capture of this species for the pet trade is a previously unknown threat and raises the possible need to reassess its conservation status in Ecuador.

Trapping and nest poaching represent a major threat to bird species in much of the Indonesian Archipelgo (Eaton *et al.* 2015). That catastrophic declines in many populations of the critically endangered Yellow-crested Cockatoo *Cacatua sulphurea* are due almost entirely to trade is indicated by an increase in numbers of the endemic subspecies on Sumba following the cessation of legal trade in 1993. Elsewhere these cockatoos have disappeared entirely from some islands, are near to extinction on others, and in Sulawesi, where formerly they were widespread and common, they no longer occur in some districts even though forested habitat remains. Concerns have been expressed that a lack of information precludes any assessment of the impact of trade on other species thought to be at risk, and included in these species are at least two subspecies of the Red-breasted Parakeet *Psittacula alexandri* – *P. a. alexandri* on Java and Bali and *P. a. dammermani* on Karimunjawa (in Eaton *et al.* 2015). In July–August 2012, interviews and field surveys were undertaken to assess the parrot trade in seven villages on Obi Island, Maluku Utara province, and estimates were made of the annual harvest for the three most commonly trapped species – Chattering Lory *Lorius garrulus*, Violet-necked Lory *Eos squamata* and Eclectus Parrot *Eclectus roratus* (Cottee-Jones *et al.* 2014). The estimate for the minimum annual harvest of the vulnerable Chattering Lory on Obi was substantially higher than previous estimates, so suggesting that this species, and in particular the distinctive *flavopalliatus* subspecies, is more threatened than previously assumed.

It has been suggested that the endemic Philippine Cockatoo *Cacatua haematuropygia* may be the most threatened species in the Philippine Islands (Kennedy *et al.* 2000). Excessive exploitation for the live-bird market and persecution as pests, coupled with widespread deforestation, has brought this species to the brink of extinction, and the surviving population is estimated at less than 1500 birds. On some islands, the same nests have been poached year after year, so preventing any recruitment into the local population and eventually leading to total extirpation of that population.

High levels of trade in African parrots have long been a feature of international trafficking, with *Agapornis* lovebirds and the vulnerable Grey Parrot *Psittacus erithacus* featuring prominently, and the Grey Parrot is a striking example of a species at risk from trade. Despite an absence of meaningful data on the status of wild populations, Grey Parrots continued to be among the most heavily traded birds, with up to 50 000 individuals being exported each year, and undoubtedly there was a high rate of pre-shipment mortality. Data suggested that approximately 21 per cent of the wild population is harvested annually, and the total number of birds taken from the wild between 1982 and 2014 may have been about 1.3 million, with perhaps some 100 000 birds being captured each year in Cameroon during the late 1990s and early 2000s (in Birdlife International 2016). Concerns about these high levels of international trade gave rise to a campaign by conservation organisations, and especially by the World Parrot Trust, for restrictions to be implemented, and this culminated in acceptance by the 2016 Conference of Parties to the Convention on International Trade in Endangered Species of Wild Fauna and Flora (CITES) to a proposal by Gabon to list the Grey Parrot on Appendix I.

In the light of this listing of the Grey Parrot on CITES Appendix I, it is timely to pose the question – How effective are CITES controls in restricting international trade in this, or in any other, parrot species? As I have stated previously, trade in parrots and other live birds is utilisation of wildlife and, as such, must be based on sound principles of sustained yield harvesting. Before trade in any species is undertaken it is necessary to ascertain whether wild populations of that species can sustain trade and, if so, at what level. When those data are at hand, regulated trade could be permitted, with periodic checks put in place to monitor the effects of trade. Such monitoring procedures should incorporate assessments of recruitment rates to ensure that harvesting is directed at appropriate age cohorts of the population, usually the younger non-breeding birds, some of which may never have the opportunity to breed. Wasteful and inhumane methods of capture, such as felling nesting trees or chopping into hollows to secure nestlings or shooting adults in the hope that one or two will recover from their injuries, must be eliminated. Harvesting programs need to be sufficiently flexible to enable action to be taken promptly to restrict, or even halt altogether, the trade if monitoring studies detect declines in the population. Studies along the lines that I am advocating have been undertaken on the Green-rumped Parrotlet *Forpus passerinus* in Venezuela showing that, because of its small body size, low age at first reproduction, high clutch size combined with hatching asynchrony, and possible limitation by nest sites, this species is a successful candidate for sustainable harvest (see Beissinger and Bucher 1992). These studies can and should be undertaken prior to the determination of trade quotas. Cameroon currently has a quota of 3000 Grey Parrots per annum, and until January 2016 the Democratic Republic of Congo had an annual quota of 5000 parrots for trade, but the question needs to be asked as to whether these quotas were determined on the basis of sound data? I strongly believe that the effectiveness of CITES in controlling trade levels would be enhanced by the adoption of reverse listing, with species being listed only when there are data supporting the capability of wild populations to sustain harvesting and at levels determined in accordance with those data.

Another consequence of trade in live parrots is the potential of feral populations becoming established in importing countries where there can arise competition with native species or conflicts with agriculture. Populations of invasive Rose-ringed Parakeets *Psittacula krameri* and Monk Parakeets *Myiopsitta monachus* have become established in a number of countries, and both have the potential to impact adversely on ecosystems in those countries. The threat of competition often is more significant when an introduced species becomes established on an island. Introduction of the Rose-ringed Parakeet to the island of Mauritius certainly poses a threat to the critically endangered Mauritius Parakeet *Psittacula eques*, and the introduction of a number of parrot species, including other *Amazona* parrots, to Puerto Rico could hinder or even thwart efforts to ensure long-term survival of the critically endangered Puerto Rican Amazon *A. vittata*. On Norfolk Island, a critically endangered subspecies of the Red-fronted Parakeet *Cyanoramphus novaezelandiae cookii* is threatened by competition from the more aggressive Crimson Rosella *Platycercus elegans*, which was brought to the island from Australia. Conversely, the establishment of feral populations at times can have beneficial consequences, as is the case with critically endangered Yellow-crested Cockatoos *Cacatua sulphurea* in Hong Kong, where a breeding population in parklands and gardens may be less threatened than any population in the Indonesian Archipelago. Similarly, strong feral populations of endangered Green-cheeked Amazons *Amazona viridigenalis* are well established in the suburbs of Los Angeles and elsewhere in southern California, where it has been estimated that up to 25 per cent of the world population now occurs.

CLIMATE CHANGE

In his Foreword, Noel Snyder has listed climate change as one of the basic forces that underlie most of the immediate threats faced by endangered species. There are virtually no data on the likely effects of climate change on parrot populations, but there are indications that some species could be affected quite seriously. I shall mention some early signs of possible adverse effects detected in Australia, for that is where I am most familiar with events. It is predicted that in Australia there will be increased summer temperatures and more prolonged heatwave conditions giving rise to more frequent and more intense wildfires, all of which will impact significantly on the landscape. Vegetation changes could deprive specialist species of required food sources, and species associated with temperate forests or woodlands could lose much habitat. In a comparison of findings from a second atlassing of Australian birds, carried out between 1998 and 2001, with published results from the initial Atlas, undertaken during 1977 to 1981, the Gang Gang Cockatoo *Callocephalon fimbriatum* is among species that have declined since the 1980s, and it has been speculated that exceptionally warm weather through the 1990s may have affected survival or breeding success in this species (in Barrett *et al*. 2007). The restricted distribution of this cockatoo coincides with cool temperate vegetation, the extent and nature of which may be altered by climate change, and this vulnerability to the impact of climate change is one of the factors cited by the New South Wales Scientific Committee in its formal listing of the species as vulnerable in New South Wales. In January 2010, in southwestern Australia, deaths of flocks of 65 and 145 endangered Carnaby's Black Cockatoos *Calyptorhynchus latirostris* were recorded during a severe heatwave, and the deaths seemed to be caused solely by the extremely hot conditions. Also, in southwestern Australia, in November 2015, a devastating wildfire destroyed heathland vegetation in almost 90 per cent of Cape Arid National Park, one of the last remaining strongholds of the critically endangered western form of the Ground Parrot *Pezoporus wallicus*. Coastal and subcoastal heathland is particularly vulnerable to wildfire, and any frequency in burning certainly would pose a serious risk to the specialised Ground Parrot. I strongly suspect that the near-extinct Orange-bellied Parrot *Neophema chrysogaster* is another specialised species highly vulnerable to the effects of climate change.

Much attention has been focused on the effects of rising sea levels on low-lying islands in the South Pacific Ocean, where the inundation of beachfront areas is occurring, and many of these islands are frequented by endangered *Vini* and *Charmosyna* lorikeets.

RECOVERY PROGRAMS

In a number of countries recovery programs for endangered parrots are being undertaken, with varying levels of success being achieved. I do commend the dedication of all persons involved in these programs, especially of volunteers who often participate without reimbursement. Perhaps the program most demanding of effort from all participants is management and monitoring of the last surviving populations of the critically endangered Kakapo *Strigops habroptila* in New Zealand. Being the only nocturnal flightless parrot and the only species that has a lek mating system, the Kakapo is highly deserving of the effort being put into its survival. This unique parrot now is extinct in its natural range, and all surviving birds are in intensively managed populations on three predator-free, offshore islands. In 1995, the population reached its lowest level, with just 30 males and 21 females surviving, but in that year the introduction of new management procedures, including the eradication of Pacific Rats *Rattus exulans* from Codfish Island, the provision of food supplements to nesting females, and intensive monitoring with intervention when necessary, brought about a significant increase in nesting success (Elliott *et al*. 2006). Furthermore, the survival of adults on these islands has been remarkably high, averaging only about 1.3 per cent mortality per annum. In November 2005, the population comprised 45 males and 41 females, with 40 or 45 per cent being reared on the islands. At the time of writing the population of 125 birds comprises many younger birds and an equal ratio of males and females. All females are from Stewart Island, and the resulting lack of genetic diversity has given rise to high levels of infertility, which is a major problem inhibiting the recovery effort. On average only 60 per cent of eggs are fertile, and in the 2014 breeding season the fertility level dropped to 40 per cent. Artificial insemination is being used in an effort to improve the fertility rate and, if the behaviour of sitting females causes concern, their eggs are taken for artificial incubation. Wherever possible chicks are reared by their mother or by a foster mother, but they will be handreared if necessary.

Since 1973 an intensive recovery program has been undertaken in Black River Gorges National Park, on Mauritius in the Indian Ocean, to save the critically endangered Mauritius Parakeet *Psittacula eques*, which is the only surviving member of a fascinating assemblage of psittaculine parrots that formerly occurred in the Mascarene Islands. Early attention was given to improving the low level of recruitment, attributed largely to a scarcity of suitable nesting cavities in large native trees, and there was strong competition for available sites, with other cavity-nesting species taking over hollows in which the parakeets had commenced nesting. Also, some nests were susceptible to predation, particularly from Black Rats *Rattus rattus* and Crab-eating Macaques *Macaca fascicularis*, which took eggs and chicks or at times killed sitting females. The provision of nestboxes initially was unsuccessful, but subsequently birds released from the captive breeding program used them more readily, eventually being followed by wild pairs, and during the 2009–2010 breeding season 78 per cent of nesting attempts occurred in artificial nestboxes. The fitting of predator-deterrent collars to the trunks of nesting trees has been highly effective in eliminating predation by rats, and natural hollows have been deepened and fitted with baffles to prevent the macaques from reaching eggs or chicks. Spectacular success was achieved in boosting the level of recruitment, and release into the wild of captive-bred birds has augmented the population to an estimated 500 birds in 2010 (Richards 2010). The increased number has highlighted the shortage of food in the small remnant of native forest, much of which is degraded by exotic plants, so supplementary feeding has become a key element of the conservation program. Before the introduction of supplementary feeding, food shortages during the breeding season were a major factor limiting population growth, but breeding pairs now are able to meet their nutritional requirements for successful nesting. The population continues to increase, and at the time of writing it is approaching 700 birds. In response to numbers nearing the maximum carrying capacity within Black River Gorges National Park, birds have been released in the mountains of eastern Mauritius to establish a second population. The prospect of reintroducing birds to Réunion has been raised, and introduction to the Seychelles, where another *Psittacula* parrot formerly occurred, also has been mentioned as another possibility.

In his Foreword, Noel Snyder recalls his involvement with the recovery program for the critically endangered Puerto Rican Amazon *Amazona vittata*, the surviving population of which had declined to only 13 birds in the mid 1970s, and all were in the eastern part of Puerto Rico in virtually the only remaining old-growth forest in the Luquillo Mountains. At this time an intensive conservation program was initiated, the objectives being to undertake intensive field studies of the wild population with a view to implementing a recovery strategy and to establish a captive-breeding program so that captive-bred birds could be released into the wild population. From this program captive-produced chicks were fostered into wild nests, and this fostering technique, together with intensive guarding and management of wild nests brought about an increase in the population, which numbered almost 50 birds by the late 1980s. Captive-bred birds subsequently were released to augment the wild population in the Luquillo forests. Despite these actions recovery of the population was slow, and there were strong indications that the Luquillo forests may be suboptimal habitat for the parrots because of the high rainfall and high numbers of Red-tailed Hawks *Buteo jamaicensis*, a predator of adults and juveniles, and Pearly-eyed Thrashers *Margarops fuscatus*, an aggressive competitor for nesting hollows and a predator of eggs and chicks. It was becoming increasingly evident that the last population of Puerto Rican Amazons was confined to the Luquillo forests not because these forests were ideal habitat, but because they were the only remaining stands of old-growth forest. Preparations were made to establish wild populations at other sites known to have been frequented by the parrots before deforestation in the early 1900s. Captive-bred birds were released in protected forests at two sites in eastern Puerto Rico, both resulting in the successful establishment of wild populations, and birds will be released at a site in the west of the island. At the time of writing more than 100 Puerto Rican Amazons are being monitored in the wild, and nearly 400 birds are held in captivity.

I see some similarities between the recovery program for the Puerto Rican Amazon and the program being undertaken to save the near extinct Orange-bellied Parrot *Neophema chrysogaster* in southeastern Australia. The release of captive-bred birds to augment the wild population initially resulted in little increase in numbers of amazons in Puerto Rico and similarly has not arrested the alarming decline in the population of Orange-bellied Parrots in southeastern Australia. Success was achieved in Puerto Rico when releases of captive-bred birds were made at other sites, which appear to be more typical of habitat originally preferred by the parrots. Likewise with the Orange-bellied Parrot, perhaps optimal nesting habitat may not be present in southwestern Tasmania, where the last surviving population is breeding. There is evidence that Orange-bellied Parrots formerly were more widespread in southeastern Australia, and there may have been populations that nested elsewhere in Tasmania. Overwintering on the mainland also appears to have been along southern sections of the east coast, where occasional occurrences of overwintering birds are still recorded. Also, there is an early unconfirmed nesting record from the vicinity of Sydney, so indicating that a breeding, possibly resident population formerly may have occurred in that district however unlikely it may now seem (in North 1911). There seems to be little purpose in continuing to release captive-bred birds in southwestern Tasmania, and some consideration could be given to establishing populations elsewhere in the former range.

In New Zealand, a longstanding practice adopted very successfully by wildlife authorities has been to establish new populations of endangered species on predator-free offshore islands, and endangered parrots have featured in these projects. The translocation of all surviving Kakapo *Strigops habroptila* to three islands has been mentioned, and new populations of endangered *Nestor* and *Cyanoramphus* parrots, originating either from translocated birds or releases of captive-bred birds, have been established on some islands. The practice has been adopted also on some islands in the South Pacific Ocean, notably on Atiu in the southern Cook Islands, where the translocation of Rimatara Lorikeets *Vini kuhlii* from Rimatara, in the Austral Islands, appears to have been successful, but in the Marquesas Islands initial success with an attempt to establish a population of endangered Ultramarine Lorikeets *Vini ultramarina* on Fatu Hiva was reversed when Black Rats *Rattus rattus* reached the island and the new population was extirpated. I believe that there is scope for this approach to be adopted more widely, and I know of two projects that should be initiated in Australia to return *Cyanoramphus* parakeets to parts of their former range. If the program to remove rats and introduced Tasmanian Masked Owls *Tyto castanops* from Lord Howe Island succeeds, there would be an opportunity to translocate Red-fronted Parakeets *C. novaezelandiae* from Norfolk Island to establish a second population of this endangered subspecies. The parakeets now could be reintroduced to subantarctic Macquarie Island following the successful eradication of all predators and competitors from that island. Terrestrial *Cyanoramphus* parakeets on the Antipodes Islands differ very little, if at all from those that formerly occurred on Macquarie Island, so translocation of birds from the treeless Antipodes Islands would be the appropriate means of re-establishing these parakeets on similarly treeless Macquarie Island. This would require the cooperation of New Zealand, Australian and Tasmanian authorities and, because of their long experience with translocating endangered birds to islands, officers from the New Zealand Wildlife Service would be well equipped to handle the transfer of birds to Macquarie Island.

Rehabilitation of preferred habitat, mostly by planting favoured food and nesting trees to replace exotic vegetation or to reclaim cleared lands, and providing nestboxes to address a shortage of natural nesting sites are key elements of many recovery programs, and in some instances these actions certainly have assisted in the recovery of endangered populations. In western Colombia and northwestern Ecuador, endangered Yellow-eared Parakeets *Ognorhynchus icterotis* are dependent on *Ceroxylon* wax palms for food and nesting sites, and in Colombia the large-scale destruction of stands of these palms is a threat that has been addressed in a successful conservation program, with habitat restoration and reforestation exceeding the objectives. Also in Colombia, following rediscovery of the critically endangered Indigo-winged Parrot *Hapalopsittaca fuertesi*, a concerted conservation effort was initiated promptly to protect the small surviving population, and this included protecting some

1500 ha of montane forest in a newly-established nature reserve and setting up more than 200 nestboxes to overcome a critical shortage of mature trees with cavities. The shortage of suitable nesting hollows in mature trees became most evident when five of the nestboxes were occupied almost immediately. Additional reserves have been established, so that approximately 70 per cent of the known population now is protected. The provision of nestboxes has featured prominently also in recovery programs for endangered *Pyrrhura* parakeets in South America, and again the quick occupation of these by nesting pairs attests to the shortage of natural tree hollows brought about by landclearance.

Finally, there is a need for attention to be directed at the role of captive breeding in the conservation of endangered parrots. Too often it is promoted as the ideal recovery process without due consideration given to the many pitfalls. Derrickson and Snyder (1992) point out that captive breeding is only one of many possible conservation strategies, and its potential contribution to the preservation of endangered parrots is limited by a variety of significant difficulties:

(i) obtaining consistent reproduction in some species,
(ii) controlling diseases in confinement,
(iii) avoiding detrimental genetic and behavioural changes,
(iv) meeting long-term financial and logistical requirements, and
(v) securing administrative continuity and commitment.

Ideally, the direct role of captive breeding in species preservation should be restricted to short-term situations where alternative management options have proved ineffective, or can be enhanced significantly. Where practical alternatives exist, they generally are far preferable to captive breeding.

When released, captive-reared birds are disadvantaged in many ways, and experimental releases of captive-reared Hispaniolan Amazons *Amazona ventralis* in the Dominican Republic have shown that subsidy generally is crucial to their survival in the period after release (in Wiley *et al.* 1992).This seems to be especially true if they are released where there is no wild population. Captive-reared birds are totally ignorant of many aspects of survival outside their cages, and often are in less than hardened condition, so their release needs to be undertaken by way of a comprehensively managed, gradual transition so that problems, such as failure to find wild sources of food and water can be compensated by subsidy at the release sites. Handreared birds are most unsuitable for release because of their almost total lack of fear of humans, and their reluctance, or perhaps inability, to socialise with wild-caught or parent-reared birds of the same species. Serious deficiencies in flocking tendencies, abilities to find food in the wild environment, vigilance against raptors, and interest in joining wild flocks were observed in releases of handreared Thick-billed Parrots *Rhynchopsitta pachyrhyncha* in Arizona, despite intensive pre-release training to counter these problems (Snyder *et al.* 1989).

Establishing and maintaining a captive breeding program for release of birds to the wild is both costly and labour intensive, and the desirable setting up of aviaries in the range of the wild population, sometimes at a remote location, can increase costs significantly. Breeding the numbers of birds needed for successful release may take some years, and that again adds to costs. Derrickson and Snyder (1992) note also that, over a period of generations, captive-bred birds can be expected to become less suitable for successful release because of behavioural and genetic selection for the captive environment, so captive-bred birds to be released should be as near as possible to wild stocks in time and generations. There is a real risk of diseases being introduced to wild populations with the release of captive-bred birds, especially if the captive birds have not been housed in isolation from other avian species or have not been sourced from known disease-free facilities.

Despite the difficulties associated with captive breeding for the release of birds into the wild, it can be a valuable component of a recovery program if undertaken as an integral component of that program, and with good disease control in place. It is being used successfully in a number of recovery programs, notably in Puerto Rico for recovery of the Puerto Rican Amazon *Amazona vittata* and on Mauritius for recovery of the Mauritius Parakeet *Psittacula eques*. Conversely, it does not appear to have been successful in the recovery program for the near extinct Orange-bellied Parrot *Neophema chrysogaster* in southeastern Australia.

FUTURE PRIORITIES

In the 1900s, we lost four parrots, or five if continued searches for the Red-throated Lorikeet *Charmosyna amabilis* in the Fijian Islands are unsuccessful. Spix's Macaw *Cyanopsiita spixii* presumably survives only in captivity, and that possibly could soon be the status of the Orange-bellied Parrot *Neophema chrysogaster*. This prompts consideration of the number of species that could be lost in this century. I am optimistic that we can lower significantly the level of extinction because we are better informed of threats to endangered populations, and are better equipped to determine and implement actions to counteract those threats. Natural forces, such as hurricanes or cyclones devastating island ecosystems, have the potential to extirpate populations, and possible unforeseen consequences of global warming certainly are cause for concern. Because loss or degradation of habitat remains the main threat coming from human activities, conservation priorities will need to be concentrated in the Indonesian Archipelago and the Neotropics, where landclearance is more prevalent. On a number of occasions, I have expressed my pessimistic outlook for parrots in the Indonesian Archipelago, and I see little likelihood of current population declines being arrested or reversed. Deforestation continues unabated, even in reserves and national parks, and there is little or no enforcement of protection measures to curb trapping and poaching of nests for domestic and international live-bird markets. Similarly in the Neotropics, deforestation continues to be a major threat, particularly in the Amazon River basin and in the Andean uplands, but gains have been made in habitat protection and rehabilitation, largely with international funding support to non-government environmental agencies, and hopefully this positive development will continue. Also in a number of countries in Central and South America, gains have been made in generating an awareness of native wildlife and a sense of national pride in protecting that wildlife.

In southeastern Australia, outstanding success has been achieved in improving the status of the Superb Parrot *Polytelis swainsonii* and the eastern subspecies of the Regent Parrot *P. anthopeplus* by implementing management programs at an early stage when the first signs of declining numbers became apparent. By intervening at this early stage, changes in landuse management and forestry practices could be introduced in a way that minimised impacts on landholders and related interests. The management programs have been in place for more than 40 years, and populations of these parrots have increased quite dramatically, with the breeding range of the Superb Parrot being expanded. Furthermore, the costs in both expenditure and labour input incurred by wildlife authorities have been very significantly lower than are required for intensive recovery programs. I believe that success achieved with these two species demonstrates the strong advantages of implementing management programs at the early signs of any decline in numbers instead of waiting until a species becomes endangered, and adopting this approach should be afforded high priority in the conservation of parrots throughout their worldwide range. This is in accord with the recommendation made by Derrickson and Snyder (1992) that the identification of limiting factors and practical management alternatives should be a priority for research, and should be accomplished before populations decline below minimum viable size.

FOSSIL HISTORY OF PARROTS

Modern parrots are among the most distinctive birds in the world – there is little else with which they can be confused. One of the most obvious characters is the short, mobile decurved bill. This varies in its proportions and length among species, but obviously is parrot-like in all. Even without the keratin sheath covering, the underlying bone retains the characteristic shape. Another trait is the zygodactyl toe arrangement. Most birds have the anisodactyl condition in which three toes point forwards and the fourth, where present, backwards. Among living birds, only a few groups exhibit the zygodactyl arrangement: parrots, cuckoos (Cuculiformes) and most of the Piciformes (woodpeckers, puffbirds, jacamars, toucans, etc.). In these groups, the fourth (lateral) toe is fully and permanently directed posteriorly. In several other bird groups, the feet are semizygodactyl (the fourth toe points laterally, rather than backwards) or facultatively zygodactyl (the fourth toe can be moved into and out of a backwards position).

This arrangement is reflected in the tarsometatarsus (the last long bone of the hindlimb, actually the foot bone). The distal end of the tarsometatarsus has protrusions, or trochlea, for the attachment of the toes. The trochlea for the fourth toe in zygodactyl birds has a laterally directed, wing-like extension and a posteriorly placed accessory trochlea or *Sehnenhalter*. The latter serves as an articular surface for the reversed fourth toe. Its morphology differs among each group of zygodactyl birds. In modern parrots, there is a deep furrow between the accessory trochlea from the fourth trochlea. The presence of an accessory trochlea in a fossil is a good indication of that bird's toe arrangement and is crucial for working out its relationships.

While the bill shape and toe arrangement are major features of parrots, they certainly are not the only ones that are useful for recognising parrot relationships, but other features are much less obvious. These include such disparate characters as the development of projections and presence of incisions in the sternum, configuration of the articulations on the pectoral girdle, presence of openings for nerves in the coracoid, shape of the crest on the humerus for attachment of the pectoral flight muscle, arrangement of canals and ridges on the backside (hypotarsus) of the tarsometatarsus, relative lengths of the pedal phalanges (toe bones), among others.

Modern parrots are so distinctive that it has proved difficult to determine their closest living relatives among other birds and a number of different groups have been nominated, particularly pigeons. Recent molecular analyses have produced a novel and quite unexpected result. Studies by a range of different workers have been in agreement that the nearest living relative of parrots is the passerines, or songbirds (Passeriformes). Songbirds, which comprise about 60 per cent of all living bird species, are quite different in appearance. Other than for a few species with superficially curved bills, there is no resemblance to the characteristic parrot bill. Further, all songbirds are anisodactyl, with none having a toe condition approaching that of parrots.

A clue to clarifying this seemingly anomalous connection between two very different looking groups came with identification of a fossil group of zygodactyl birds, the Zygodactylidae. First described in 1969, these small birds were known only from isolated and fragmentary leg bones from the early and middle Miocene (34–17 million years ago [mya]) of Europe. They now are known from North America and Europe, with a temporal range of early Eocene to middle Miocene (48–17 mya). In some deposits, these are among the most abundant small birds.

Although relationships of this group were uncertain, most authors placed them with the lineage containing woodpeckers (Piciformes). The discovery of new specimens, including nearly

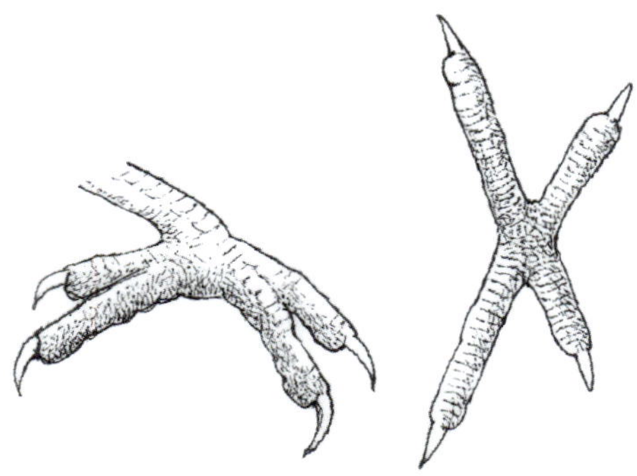

Fig. 1 Zygodactyl toe arrangement in modern parrots.

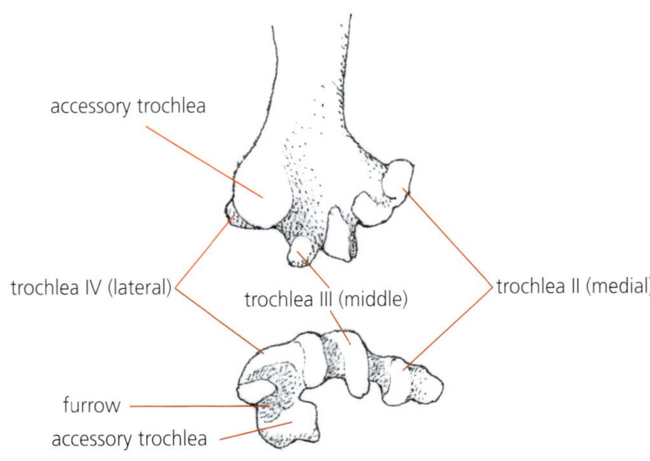

Fig. 2 Distal end of the tarsometatarsus of a modern parrot showing the accessory trochlea and furrow (after Mayr 2002b).

complete, articulated skeletons from the early Oligocene of France, permitted a reassessment of this idea. The skeleton of zygodactylids was very similar to that of songbirds, with some elements being almost indistinguishable. The long, slender and quite unparrot-like tarsometatarsus, however, exhibited a fully zygodactyl toe arrangement. More and better preserved specimens have led to the conclusion that this family is the sister-group to songbirds. These also start to bridge the morphological gap between the parrots and songbirds, which diverged many millions of years ago. The name Psittacopasseres has been created for this lineage of parrots, zygodactylids and songbirds.

Modern parrots belong to crown-group Psittaciformes, that is, all living parrots, their common ancestor and all its descendants, including extinct forms. In the early section of the lineage leading up to the crown-group were a number of basal and now extinct taxa. These are known as stem-group parrots (or simply stem-parrots). They exhibit earlier stages in the development of characters that make crown-group parrots so distinctive. The assignment of some of these fossils as stem-parrots is still controversial because some differences from living parrots are quite marked.

Some authors restrict the use of the name Psittaciformes to the crown group. For the crown and stem groups together, they employ the term Pan-Psittaciformes. The recognition of stem-parrots has been subject to considerable ongoing revisions as a result of several factors. Partial fossils, sometimes comprising only a few bones or even one fragmentary bone, can be difficult to place in the avian classification. There are characters in parrots that are shared also by other groups in which these have been acquired independently. The history of stem-parrot classification has seen taxa originally described as parrots subsequently being shown to belong to other groups. Conversely, other taxa first regarded as belonging to another bird group have since been regarded as psittaciforms. Some of the fossils discussed here remain controversial and their placement as parrots is still tentative.

Parrots are one of the most characteristic components of modern avifaunas in southern continents. It has been speculated that the group had its origins in the southern supercontinent of Gondwana, but this may not necessarily be so. Many recent bird families occurring only in the Southern Hemisphere now are known to have fossil records from the Northern Hemisphere. Although prominent in the Southern Hemisphere at the present time, parrots are not well represented in the fossil record of these continents, particularly in the older time periods. The record is sparse for much of the Tertiary (65–1.6 mya), and offers limited clues about the development and radiation of modern parrot groups. Most fossils are known only from the Quaternary (1.5 million years to the present) and comprise modern genera and even species.

The past few decades have been notably productive in the discovery of fossil parrots. Lambrecht (1933) in *Palaeornithologie* listed only five taxa under Psittaciformes, two of which were driven to extinction by Europeans during the Pleistocene. Brodkorb (1971) in his *Catalogue of Fossil Birds* included only two pre-Quaternary species, five extinct Quaternary species and a number of records of modern species recorded as fossils. Both authors included a bird now regarded as a stem-parrot but then classified elsewhere. There now are more than 12 stem-parrot species recognised, as well as more than six extinct crown-group genera and a growing number of fossil species in modern genera.

The oldest putative record comes from the late Cretaceous (Maastrichtian: 74–65 mya) Lance Formation in Niobara County, Wyoming, United States. Stidham (1998) based this identification on a tip of the lower bill, approximately 15 mm in length; this was a toothless, completely fused symphysis of the left and right mandibles, with small holes or foramina along the middle of the lingual surface, an absence of grooves or ridges on the walls, and a specific pattern of the neurovascular canals. The shape of the tip was judged by Stidham to be most similar to that of lorikeets among living parrot groups. In addition to being the earliest known occurrence of parrots, this specimen would be the earliest record of a modern 'terrestrial' bird group. Dyke and Mayr (1999) queried the assignment of this fossil as a parrot, pointing out that characters used in its identification are found in other groups of avian and non-avian vertebrates. Additionally, the morphology of this fossil differed from that of early Tertiary parrots from Europe, so it was suggested that at present the recognition of parrots from the Cretaceous should be considered tentative. In response, Stidham (1999) noted that, although some of the characters are found in other groups, the particular states observed in the fossil occur only in parrots and, indeed, the combination of characters is unique to this group of birds. This intriguing fossil has not been restudied, nor have any more specimens been recovered that have a bearing on its identification.

In several instances, birds that originally were described as parrots subsequently have been shown not to belong to this group. Harrison (1982) described 11 incomplete bones, putatively from a single bird, recovered from the early Eocene (57–52 mya) London Clay at Walton-on-the-Naze, in Essex, Britain, and to the same species he referred an incomplete distal tarsometatarsus from the middle Eocene (52–40 mya) of Hampshire. Notably, the referred tarsometatarsal fragment was missing the fourth trochlea. These fossils were regarded as representing a parrot similar in size to the living Senegal Parrot *Poicephalus senegalus*, and the

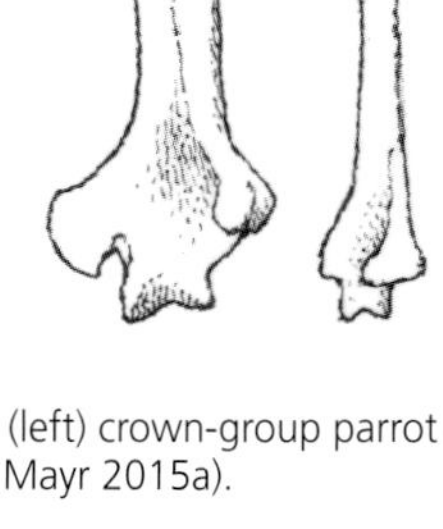

Fig. 3 Tarsometatarsi of (left) crown-group parrot and (right) zygodactylid bird (not to scale) (after Mayr 2015a).

Fig. 4 Putative Cretaceous parrot bill (after Stidham 1998).

species was named *Palaeopsittacus georgei*. From a comparison with skeletons of modern parrots, Harrison interpreted the morphology of the fossil bird to be more generalised than in recent forms, and suggested that the wings were proportionately longer than in most living parrots, with the flight being generally less vigorous. Features of the leg, including position of attachment for the tendons used to pull the body towards the foot while climbing, prompted Harrison to propose 'that climbing ability and a tendency to clamber among the branches was poorly developed or absent' in *Palaeopsittacus*. Other features suggested that the foot was capable of a more restricted range of movements, thus lacking the dexterity prominent in modern parrots. Olson (1985) noted that these fossils lacked most of the important features that characterise modern parrots, so referral of the specimens to the Psittacidae required confirmation.

Mayr and Daniels (1998) reported on new material of *Palaeopsittacus* from the Walton-on-the-Naze site, including a complete tarsometatarsus, that demonstrated that this bird was actually anisodactyl or, at best, facultatively zygodactyl and so was not a member of the Psittaciformes. This bone further showed that the distal tarsometatarsal fragment referred by Harrison did not belong to this taxon. Based on an analysis of relative sizes of bones originally described by Harrison, Dyke and Cooper (2000) concluded that a proximal fragment of ulna was not associated with the other specimens. They commented that this taxon should be 'considered to be of uncertain status at present because of problems with the questionable association of the original specimens and their absence of any characters regarded as diagnostic at the ordinal level'. A nearly complete, articulated post-cranial skeleton of *Palaeopsittacus* from Messel was described by Mayr (2002a). This provided further evidence that this genus was not related to parrots, but it did show some similarities to the frogmouths (Podargidae).

Another case of misallocation to parrots occurred when Waterhouse *et al*. (2008) announced two new birds from the Lower Eocene Fur Formation of Denmark. One was an indeterminate genus and species of the stem-parrot family Pseudasturidae (see below). For the other bird, the authors created a new genus *Mopsitta*. (The name is derived from the 'Mo Clay' from which this bird was recovered.) The species *Mopsitta tanta* was described on the basis of a humerus, which Waterhouse *et al*. considered to differ from other stem-parrots by its larger size – about that of a medium-sized cockatoo – and features such as the greater curvature of the shaft. While they noted some similarities of the humerus to that of the early ibis *Rhynchaeites*, these were dismissed as not significant. The authors considered *Mopsitta* to be more similar to crown-parrots than to stem-parrots and, further, that it likely was a member of Psittacidae s.l. (e.g. crown-group parrots). Mayr and Bertelli (2011) questioned the identification of *Mopsitta* as a parrot. They were unable to confirm a number of the characters identified by Waterhouse *et al*. in support of this putative relationship, but they themselves noted features that they considered precluded the fossil being psittaciform. In contrast to Waterhouse *et al*., these authors noted a close resemblance between the bones of *Mopsitta* and those of *Rhychaeites* from Messel. Mayr and Bertelli considered *Mopsitta* to be a stem-ibis, not a parrot.

STEM-GROUP PARROTS

Most stem-parrots are known from the Eocene (56–34 mya), mainly from the Northern Hemisphere. There are only a handful of Eocene localities producing bird fossils, of which only nine have yielded fossil parrots, and three of these are particularly important, together having produced most of the fossils. London Clay, England (early Eocene 56-49 mya) has yielded bird fossils from two main localities in the southeast of the country: Walton-on-the-Naze, Essex, and the Isle of Sheppey, Kent. Fossils from these sites are found as associated but disarticulated bones in clay nodules. The Green River Formation, Wyoming, United States

Eocene sites from which stem-group parrot fossils have been recovered: map by W. Boles

(53.5–48.5 mya) and Gruße Messel, western Germany (48 mya) both preserve animal and plant remains on slabs. These often are complete, although compressed, and may retain feather impressions. Stem-parrot representatives are known also from sites in France, Denmark, India, Namibia, eastern United States, and elsewhere in Germany.

Birds of comparable age to *Palaeopsittacus* were recorded by Mourer-Chauviré (1992) from the late Eocene (40–36 mya) Phosphorites du Quercy, at Le Bouffie, France, citing them at first only as Psittacidae. She subsequently described them in detail from the coracoid, carpometacarpus, tibiotarsus and tarsometatarsus (Mourer-Chauviré 1992). These birds had a combination of characters of modern parrots, such as the zygodactyl foot, and other more primitive features. She recognised two new species, *Quercypsitta sudrei* and *Q. ivani*, for which she created the new, extinct family Quercypsittidae, to which also was referred *Palaeopsittacus georgei*, then still considered to be a parrot. Two interpretations of the Quercypsittidae were considered. The first was that these birds were the direct ancestors of modern parrots. The second was that they were vicariant relatives of modern parrots, sharing a common ancestor, but evolving in the Northern Hemisphere, while other relatives evolved in the Southern Hemisphere. Eventually the quercypsittids became extinct, while the southern birds survived and became successful. Mourer-Chauviré expressed a preference for the second of these two possibilities.

Mayr and Daniels (1998) reported on parrot-like birds from the London Clay and the middle Eocene deposits at Messel, Hessen, Germany. From the Messel specimens, they named *Psittacopes lepidus*, a bird that was about the same size as a *Loriculus* hanging parrot. One of the most striking features was its bill shape; compared to that of modern parrots, the bill was short and rather shallow, particularly the maxilla, with large nostrils, and an overall superficial resemblance to the bill of the African mousebirds (Coliidae) was noted. This led Mayr (2015a) to propose that this bird fed on soft plant material, rather than seeds. The family Psittacopedidae was created for the genus (Mayr 2015a). Three well-preserved specimens comprising mainly associated bones from the London Clay seem to be closely related to this species, but as all are held in a private collection, Mayr and Daniels chose not to formally name them. One, informally designated London Clay A, the largest of the three taxa, is known from most major bones and a skull. London Clay B and C are less well represented. They are similar in size and morphology, but the latter has a broader tarsometatarsus. There also are some isolated bones, including a skull, that cannot be assigned to either B or C with certainty. Like *Psittacopes*, these unnamed species resembled crown-group parrots in their tarsometatarsus and other bones, but also had differences in the sternum, shoulder girdle, carpometacarpus and relative lengths of the wing bones, prompting Mayr and Daniels to suggest that their flight characteristics differed from those of modern parrots. The two London Clay skulls also shared a rather coly-like bill.

Hindlimb elements similar to those of *Quercypsitta sudrei* also were found in the London Clay and differed from other specimens from this site in several features of the tarsometatarsus. Mayr and Daniels considered *Psittacopes* and the three unnamed taxa to comprise the sister-group to modern parrots, while the Quercypsittidae was sister-group to these. They concluded that parrots appeared to be a very ancient group, possibly diverging from other birds in the Palaeocene (66–57 mya) or earlier.

Several important fossils initially were not regarded as having parrot affinities. It has only been as additional specimens of these taxa have been discovered and studied that a better appreciation of the skeleton has been possible. Further, these make possible more useful comparisons between taxa not previously considered related.

Mayr (1998) described several flattened and articulated specimens from the Messel site, several largely complete. The beak was short and robust, the supraorbital processes (the projection of bone above the orbits in modern raptors that contributes to their 'scowling' stare) were large and the tarsometatarsus had a *Sehnenhalter* present. It was unsure if this bird was completely or facultatively zygodactyl, but the tarsometatarsus differed from all living birds, showing similarities to *Quercypsitta*. At least two, possibly three, species were present, for which Mayr created the generic name *Pseudastur*. It was realised later that the name *Pseudastur* was preoccupied and so it was replaced with the name *Pseudasturides* (Mayr 2003). One species was formally named at the time as *Pseudastur macrocephalus* (now *Pseudasturides macrocephalus*), another remained unnamed and a third provisionally was assigned to this genus. It was uncertain to what other group *Pseudastur* was related, but similarities to the skull of *Messelastur*, another fossil bird from Messel (see below), particularly the robust supraorbital processes, were noted. An incomplete tarsometatarsus was described by Mayr (2002c) from the Geiseltal, Sachsen-Anhalt, Germany. This fossil, from a layer about the same age as Messel (48 mya), was very similar to, and probably conspecific with, *P. macrocephala*. Additional pseudasturid fossils from a slighter younger layer at this site (46 mya) came from a second and unnamed species in this family.

The family Primobucconidae was erected by Feduccia and Martin (1976) to accommodate several North American Eocene taxa that purportedly were zygodactyl and regarded as components of the Piciformes (these are now known also from Europe). Following further preparation of the specimens, however, it was realised that the family as constituted was not a natural assemblage (Houde and Olson 1988) and eventually the genera were reallocated to three different orders. Only *Primobucco* remained in the Primobucconidae and is recognised as an early lineage of the rollers (Coraciiformes) (Ksepka and Clarke 2012).

One species initially placed in *Primobucco*, *P. olsoni*, was found to be unrelated to the other members. Whereas the type species of *Primobucco* was anisodactyl, *P. olsoni* was zygodactyl. When Mayr (1998) erected the Pseudasturidae, he noted that '*Primobucco*' *olsoni* was closely related to *Pseudasturides* and belonged in this family. The specimen on which the name is based was preserved in two slabs containing an almost complete skeletal impression plus some remnants of feathering. The head appears

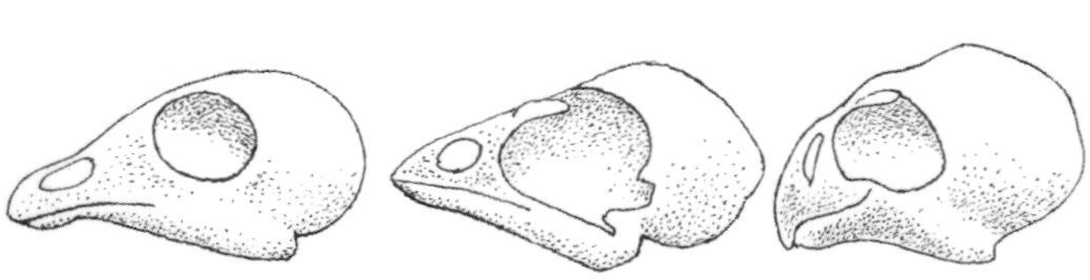

Fig. 5 Skulls of *Psittacopes*, *Pseudasturides* and *Messelastur*.
Left: *Psittacopes* – tracing merged from two images using updated bill shape (after Mayr and Daniels 1998; Mayr 2015a).
Centre: *Pseudastur* [*Pseudasturides*] (after Mayr 1998).
Right: *Messelastur* – tracing from photograph (after Mayr 2011).

to have raised feathers in what might have been a crest (Ksepka *et al.* 2011).

Dyke and Cooper (2000) described another fossil from the London Clay – *Pulchrapollia gracilis* (*Pulchrapollia* loosely translated means 'pretty polly'). From their analyses, Dyke and Cooper considered that this taxon was the sister-group to modern psittacines, being even more closely related to living parrots than *Psittacopes*. The partial associated skeleton of *Pulchrapollia* exhibited, among other similarities with modern parrots, a fully zygodactyl foot. Dyke and Cooper also placed *Quercypsitta* between *Pulchrapollia* and *Psittacopes* in their phylogeny. After consideration of the fossil record, they believed that it could be 'taken as strong evidence for a Tertiary radiation of Psittaciformes' (Dyke and Cooper 2000). Mayr (2001) showed that *Pulchrapollia* had several characters that were not found in any extinct or living parrots but agreed with the pseudasturids. He concluded that *Pulchrapollia* represented a taxon within this extinct family, whose relationships to other birds had not yet been established.

A tarsometatarsus, the most complete known for any pseudasturid bird, plus some tentatively referred wing bones, were found in the early Eocene (53.6–52.8 mya) Nanjemoy Formation in northeastern Virginia, United States (Mayr 2016a). Although not given a name, this bird closely resembles *Pulchrapollia*. Mayr *et al.* (2004) placed '*Primobucco*' *olsoni* in *Pulchrapollia*, but Martin (2010) later coined the generic name *Cyrilavis* for this taxon. In an analysis conducted by Ksepka and Clarke (2012), *olsoni* did not link closely with *Pulchrapollia*, supporting its transfer to a new genus. Mayr (2000a) named another member of this family from new Messel specimens. *Serundaptus pohli* had a fully zygodactyl foot. It closely resembled *Pseudasturides*, but was larger than species of this genus or *Cyrilavis*, and had a shorter, stouter tarsometatarsus, sporting long raptor-like claws. Other features resembled those of a climbing bird, so *Serandaptus* might have been scansorial.

Further London Clay specimens were reported by Dyke (2001), but were not identified beyond family level (Pseudasturidae). Mayr (2002b) was able to redescribe the skeleton of *Pseudasturides* based on newly acquired specimens, and the additional material added further support to psittaciform relationships of this group. While examining comparative fossils in the British Museum, Mayr (2002b) recognised another pseudasturid among specimens referred to the enigmatic London Clay genus *Precursor*. Three species of *Precursor* were described

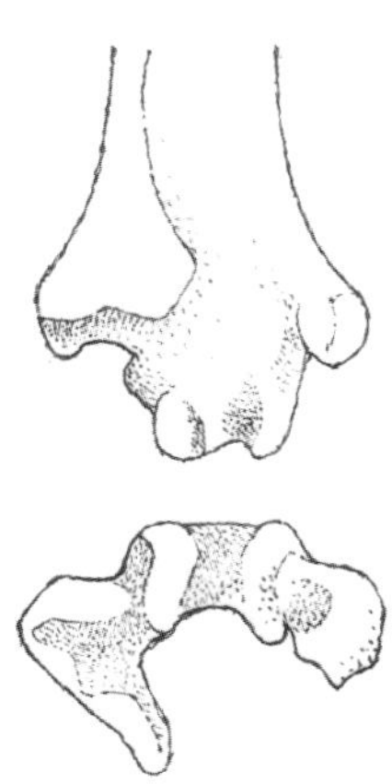

Fig. 6 Distal tarsometatarsus of *Pseudasturides* showing the incompletely developed accessory trochlea (after Mayr 2002b).

by Harrison and Walker (1977) as pratincole-like birds. It now is considered that this genus comprises skeletal elements of several unrelated groups of birds. Among the remains, Mayr (2002b) recognised several that were, in fact, those of a form of the Pseudasturidae. These remain unnamed, and the relationships of the remaining specimens of *Precursor* are unresolved.

Koenig (1825) named *Larus toliapicus* on the basis of a partial skull, missing a beak and comprising mostly a damaged braincase. This was illustrated with a rather rudimentary sketch, but it was sufficient to make it the earliest named fossil bird. Koenig regarded it as a gull, but subsequently Owen (1846) considered it to be close to the kingfishers (Halcyonidae) and gave it the new generic name *Halcyornis*. There was no further study of this specimen until it received additional preparation, after which it was re-examined by Harrison and Walker (1972). They interpreted its greatest similarity to lie with the Coraciiformes, particularly Coracii (rollers, etc.) and placed this genus in its own family, the Halcyornithidae. Nothing more was done with it until Mayr (2007), while studying new, well-preserved specimens of *Pseudasturides*, found similarities with *Halcyornis* that he regarded as demonstrating a relationship between the two. These birds were close enough to be placed in the same family. A consequence of this was that the older family name Halcyornithidae took priority over Pseudasturidae. Micro-CT scans of the original skull of *Halcyornis* conducted by Walsh and Milner (2011) showed that this bird possessed an essentially modern type brain. It probably had well-developed senses: high visual acuity, hearing perception similar to that of modern birds and some reliance on smell. Walsh and Milner (2011) felt that the structure of brain contradicted psittaciform relationships for *Halcyornis*, but made no further comment on its relationships.

Based on two specimens, Ksepka *et al.* (2011) described a second species of *Cyrilavus* from the Green River Formation. It differed from *C. olsoni* by its larger size (20% longer) and small morphological characters. Like other specimens from this site, the new fossils were preserved flattened in slabs. These showed that *Cyrilavis* contrasted with other genera by having longer beaks (slightly less than 50 per cent of the skull length) and a claw on the alula. There also were prominent supraorbital processes. At present, the Halcyornithidae comprises five named genera – *Pseudasturides*, *Halcyornis*, *Pulchrapollia*, *Serundaptus* and *Cyrilavus* – and several unnamed forms, such as that originally included in *Precursor*.

Peters (1994) gave the name *Messelastur gratulator* to two skulls and a few associated vertebrae found at Messel. This bird had a decurved bill and obvious supraorbital processes. *Messelastur* tentatively was associated with the diurnal raptors (Accipitridae) in the original paper. A subsequent, partial post-cranial skeleton led Mayr (2005a) to propose that *Messelastur* was a possible link between owls and falcons. He also noted a probable relationship with a bird from the Green River Formation, *Tynskya eocena*, although the only specimen of the latter was not well preserved and lacked some skeletal elements (Mayr 2000b). Fossils similar to *Tynskya* also were recovered from the London Clay (unfortunately, these are part of a private collection and little else has been published on them). The affinities of *Tynskya* were not apparent, although it resembled owls and raptors in some aspects of its morphology.

A new, well-preserved and nearly complete specimen of *Messelastur* was discovered at Messel (Mayr 2009), which, among other things, preserved the feet of this bird, which up to that time had not been known. This demonstrated, among other aspects, that the tarsometatarsus shared a number of characters with *Tynskya*. This and other comparisons supported the relationship between these two taxa, while also showing that *Messelastur* had larger feet, more like a raptor. In addition to supporting the placement of *Messelastur* and *Tynskya* together in the same family (Messelasturidae), the new evidence supported a relationship between this group and the Halcyornithidae, since corroborated by subsequent studies.

The oldest fossil birds recovered from the Indian subcontinental plate were reported by Mayr *et al.* (2007). These are significant because this area was a section that broke off from the giant southern landmass Gondwana, in contrast to other stem-parrots known at the time, all of which came from the Northern Hemisphere. The specimens come from the Vastan Lignite Mine, Gujarat, India. They comprise incomplete coracoids and scapulae, quite different from any other birds of this age then known. Mayr *et al.* (2007) gave these the name *Vastanavis eocaena*. At the time it was not obvious to what other group the fossils were most closely related. Although a resemblance between the coracoid and that of bustards was noted, it was not considered definitive. Further bird remains found at the site, including a tarsometatarsus, also were referred to *Vastanavis* by Mayr *et al.* (2010). The new material helped reveal the parrot affinities of this genus. Some differences in morphology and size led to recognition of a second species, *V. cambayensis*. A number of additional bones obviously belonged in this genus, but could not be determined to species. The structure of the tarsometatarsus showed that species of *Vastanavis* were semi-zygodactyl, and in overall morphology they were similar to *Quercypsitta*. Mayr *et al.* (2010) created a new family, Vastanavidae, for these taxa. Some phylogenetic analyses indicated that the family was sister to the remaining Pan-Psittaciformes. Mayr *et al.* (2013) subsequently described additional remains, which now represented most of the post-cranial skeleton, but the skull and beak remain unknown. A small coracoid in the new material may represent another species of *Vastanavis*, but this was not named owing to the fragmentary nature of the specimen. London Clay specimens that originally were assigned to *Quercypsitta* by Mayr and Daniels (1998) were considered closer to *Vastanavis* (Mayr *et al.* 2010).

Another stem-parrot, *Avolatavis tenens*, came from the Green River Formation (Ksepka and Clarke 2012). The new specimen comprised a pelvis, hindlimb and caudal vertebrae. It differed from similar forms by its wider pelvis. The zygodactyl or semi-zygodactyl toe arrangement point to an arboreal lifestyle. Other morphological features suggest differences from other stem-parrots, including a large pygostyle that could mean that the tail feathers had been larger and more developed than in other stem-parrots. It has not been allocated to any of the currently recognised families. Stem-parrots were diverse in the Green River Formation – this was the fourth species known from this site – but not particularly common.

The only stem-parrot thus far known from Africa is *Namapsitta praeruptorum*, based on a distal end of a tarsometatarsus and referred complete ulna from a middle Eocene site from Eocliff, Namibia (Mourer-Chauviré *et al.* 2015). The accessory trochlea has the configuration of modern parrots, but other features of these specimens differ from extant groups. *Namapsitta* exhibits some similarities with *Psittacopes*, but differs from this and other stem-parrots in a number of characters.

The unusual *Eurofluvioviridavis robustipes* from Messel, Germany, combined a flycatcher-like beak, long legs, short toes and robust claws (Mayr 2005b). Some similarities between the tarsometatarsus of this species and that of *Quercypsitta* were identified by the author, who noted that the *Eurofluvioviridavis* was probably only facultatively or semi-zygodactyl. A later analysis that incorporated *Eurofluvioviridavis*, as well as stem-parrots not included in the previous studies, recovered a tentative grouping of this taxon with *Avolatavis* and *Vastanavis* (Mayr 2015b), suggesting that it too may be a stem-parrot, albeit one with a bill morphology unlike any other known member of Psittaciformes past or present.

Only a handful of these stem-parrot fossils include the skull and beak. What is clear is that the characteristic beak of modern parrots had not yet evolved. For those stem-parrot fossils that include a beak there are some clues as to their ecology. The large supraorbital processes and somewhat hooked beaks of the pseudasturids, and especially the messelasturids, has led some authors to propose relationships with raptorial birds, a connection that still has some support. It is possible that these birds were carnivorous early relatives of parrots, a suggestion strengthened by the raptorial foot in the Messelasturidae.

There is little consensus on relationships of any of these stem-parrot groups to modern crown-group parrots, nor to each other. *Psittacopes* or *Quercypsitta* (sometimes together with *Avolatavis*) most often have been regarded as the sister taxon to crown-group parrots. *Vastanavis* usually is recovered as one of the earliest diverging lineages. One of the few aspects that is becoming generally accepted is an alliance between Halcyornithidae and Messelasturidae, although the placement of these as an early psittaciform lineage remains tentative, and some analyses maintain a placement with birds of prey.

As new taxa are discovered, existing ones re-examined, and additional characters described, incorporation of new information into phylogenetic analyses alters the conclusions, hopefully resulting in an increasingly refined outcome. An example of how the inclusion of new taxa in an analysis can change the outcome is the systematic position of *Psittacopes*. Initially, it was considered to be the closest of these fossil families to crown-group parrots. When representatives of the early passerine relatives Zygodactylidae, as well as *Psittacopes*, were incorporated into an analysis, the latter shifted from a position in the Pan-Psittaciformes to one as a stem-taxon in the passerine-zygodactylid lineage, together with some other odd zygodactyl fossil genera. Most of the characters that Mayr and Daniels (1998) initially had identified as linking *Psittacopes* to parrots occur also in the passerine-zygodactylid lineage. In addition, it also shared with that group some other features not found in parrots. Mourer-Chauviré *et al.* (2015) commented on similarities between *Namapsitta* and *Psittacopes*, and Mayr (2016b) recommended that the former taxon be further studied to see if it might be better considered as another passerine-zygodactylid stem taxon.

CROWN-GROUP PARROTS

There is an intriguing gap through the Oligocene in which almost no fossil records of stem- or crown-parrots are known. Before this time, all records are those of stem-group parrots. After it, all are modern-type parrots, probably but not definitively crown-group parrots.

MIOCENE PARROTS

From Miocene until about 10 000 years ago, the record of parrots is patchy, and while there are only a small number of fossils from scattered sites, they indicate that these birds were widespread and diversified by the early to middle Miocene (23–11 mya). Some fossils were placed in, or at least associated with, modern genera by their early describers (although most of these allocations have been questioned), while other fossils have had new genera erected for them.

The earliest known record of modern-type parrots is from the late Oligocene/early Miocene (15–13.5 mya) of the Mainz Basin, Germany. A distal tarsometatarsus represents a parrot about the size of the modern Superb Parrot *Polytelis swainsonii* and was given the name *Mogontiacopsitta miocaena* (Mayr 2010). Two other parrots have been recovered from this site. Both are considered indeterminate, although one of these also could belong to *M. miocaena* but, because the specimen is from another part of the skeleton, no definite assignment can be made. A fragmentary tarsometatarsus from Baikal Lake in Siberia was similar to that of *Mogontiacopsitta*, but too damaged to make such allocation with certainty (Zelenkov 2016). In size, this bird was intermediate between a lovebird and a smaller conure. The structure of this bone indicated that the species was more derived than cockatoos or kakas. It also has the distinction of being the first record of a crown-parrot from Asia, as well as the northernmost record of this group.

The first fossil parrot to be identified as such came from the early Miocene Saint-Gerand-le-Puy, France (22–20.5 mya), based on specimens of the humerus, tibiotarsus and tarsometatarsus (Milne-Edwards 1870). This originally was named *Psittacus verreauxi*. Lydekker (1891) remarked that, despite being considerably smaller, the tarsometatarsus came nearest to that of the African Grey Parrot *Psittacus erithacus*, while showing 'some resemblance to *Palaeornis* [=*Psittacula*]'. Lambrecht (1933) created for it the new genus *Archaeopsittacus*. Ballman concurred with Milne-Edwards on similarities of these fossils to *Psittacus* (pers. comm. to Olson 1985), and Mlikovsky (1998) assigned this species to the Psittaculini. Pavia (2014) recognised a humerus from Sansan, France, as belonging to *Archaeopsittacus*, although it was smaller than that of *A. verreauxi*. He assigned it to this genus, but did not make an allocation to species. This record extends the temporal distribution of this genus from the early Miocene to the middle Miocene.

The slightly older *Xenopsitta feiferi* came from the early Miocene (19–18 mya) of Merkur, Czech Republic (Mlikovsky 1998). This species was named on a damaged tarsometatarsus. The bone is short and stout and shows similarities with the African taxa *Psittacus* and *Poicephalus*. Mlikovsky regarded *Xenopsitta* to belong to this group of parrots (Psittacinae), comprising its earliest record and the first outside Africa.

Ballman (1983, pers. comm. to Olson 1985) stated that fossil parrots had been recovered from middle Miocene (16–11 mya) deposits at Nördlinger Ries, Germany, and Sansan, France. Heizmann and Hesse (1995) also indicated the presence of indeterminate parrots at Nördlinger Ries and the similar-aged Steinheimer Beckens, Germany. Mayr and Göhlich (2004) named the parrot from Nördlinger Ries (15–13.5 mya) as *Bavaripsitta ballmanni*, the holotype of which is an almost complete tarsometatarsus, and also tentatively includes a referred humerus. In size, the tarsometatarsus agrees with the Budgerigar *Melopsittacus undulatus*, with proportions and shape approaching that of some smaller platycercines.

From a collection of bones from Sansan, Milne-Edwards (1869–71) described a rail *Rallus dispar*, for which Lambrecht (1933) subsequently created the genus *Pararallus*. As part of making an emended diagnosis of this species, Cracraft (1973) designated a distal humerus as the lectotype. Cheneval (2000) later discovered that specimens of *Pararallus dispar*, except for this lectotype, belong to an already known fossil rail, *Palaeoaramides beaumonti*. The lectotype humerus, he found, belonged to a small parrot. This meant that the name *Pararallus dispar* now applied to a member

Miocene sites from which crown-group parrot fossils have been recovered: map by W. Boles

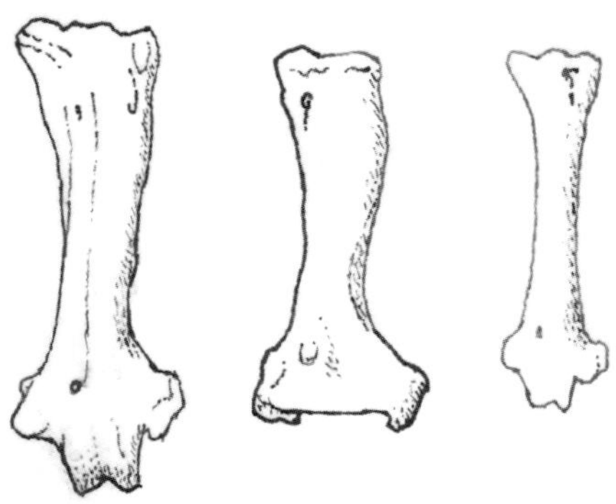

Fig. 7 Diversity of tarsometatarsal morphologies in Miocene fossil parrots. Left to right: *Archaeopsittacus*, *Xenopsitta*, *Bavaripsitta* (after Mayr and Göhlich 2004).

of the Psittacidae. The suggestion that there are nomenclatural problems with Cheneval's interpretation have been dismissed (see discussion in Mlikovsky 1998 and Mayr and Göhlich 2004). *Pararallus dispar* now is known from the coracoid, ulna, radius, tibiotarsus and tarsometatarsus.

It has not been possible to confidently assign any of these European parrots to a modern group, although the original authors frequently attempted to do this. What was demonstrated was the diversity of parrots that occurred at the time, with at least five different taxa now known from the early and middle Miocene of Europe. Fossil parrots from other parts of the world have had taxonomic assignments made with greater assurance.

Based on a humerus from the middle Miocene-aged Snake Creek quarries of Sioux County, Nebraska, United States, Wetmore (1926) named *Conuropsis fratercula*, remarking on its resemblance to the quite recently extinct and somewhat larger Carolina Parakeet *C. carolinensis*: 'in form [it] reproduces so closely the humerus of the modern bird . . . that the two would be inseparable were it not for their disparity in size'; the fossil species was about three-quarters the size of *C. carolinensis*. Despite this assertion, Olson (1985) did not regard the generic identification as positively established.

The first record of modern parrots from Australia is of comparable age to those from the Northern Hemisphere. This is based on the maxilla of a parrot from the early Miocene RSO Site at Riversleigh, in northwestern Queensland. A comparison of this fossil with specimens of more than 40 genera of modern psittaciforms looked at characters of the outline and lateral angle of the naso-frontal hinge, configuration of the cere region, size, position and orientation of the nostrils, width of the inter-narial septum, curvature in lateral profile, shape of the tomium and ratio of length to width (Boles 1993). Different groups of parrots could be separated on their maxillary characters. The fossil could be recognised easily as a cockatoo and, in turn, as a white cockatoo. This made it the oldest record of cockatoos for the world. It is so similar to the maxilla of the modern Little Corella *Cacatua sanguinea* and Galah *Eolophus roseicapilla* that it could not be differentiated for purposes of naming, so it remains as *Cacatua* indeterminate. Because of the resemblance of this fossil to the small white cockatoos, it was speculated that, in addition to the small bill, the bird probably shared with them a short, uncoloured rounded crest and somewhat pointed wings.

A diversity of parrots has been recovered from the early Miocene (19–16 mya) St Bathans Fauna of Otago, New Zealand (Worthy *et al.* 2011). Four species are present, all of which are members of the Nestorinae (known today from two living species of *Nestor*) and showing no close relationship to *Strigops* (Kakapo), the other living New Zealand taxon. The most commonly represented fossil species is *Nelepsittacus donmertoni*, known from 60 specimens of most of the major bones and mandible. This bird was about the size of a Crimson Rosella *Platycercus elegans*. A second species, *N. minimus*, approximately the size of *Cyanoramphus* species, is represented by 17 specimens and the third species, *N. daphneleeae*, as large as the living Kea *Nestor notabilis*, known from six specimens. A fourth unnamed taxon, larger than any of these species, is also present but too poorly represented for formal description.

By the early Miocene, the record of parrots shows that these birds were much more widespread than at present. Europe, currently home to no parrots, had a substantial diversity in taxa, with a range of tarsometatarsal morphologies. It is not certain to which modern group of parrots each of these genera was most closely related, but in some cases, external similarities between fossil and extant groups may indeed be valid indicators. Elsewhere, in North America, Australia and New Zealand, Miocene-aged fossils confidently can be assigned to living groups of parrots. Together, all these early crown-group parrots confirm that many (and probably most) of the divergences of the major parrot lineages had already taken place by this time. Zelenkov (2016) interpreted the European Miocene record as indicative of dispersal across a connection between North America and Asia via the Bering Strait (Beringia) around 18–16 mya, most likely from west to east. He also considered that ancestors of modern genera in Africa spread there from Europe.

PLIOCENE PARROTS

The number of fossil parrots known from the Pliocene (3.5–2.6 mya) is not great, and these are close to living taxa.

Wetmore (1944) noted the occurrence of an indeterminate genus of parrot from the Pliocene-aged (3.3 mya) Rexroad Formation, Meade County, Kansas. This bird was between a Thick-billed Parrot *Rhynchopsitta pachyrhyncha* and White-fronted Amazon *Amazona albifrons* in size.

A new species of *Nandayus* (this genus now merged into *Aratinga*), *N. vorohuensis*, was named by Tonni and Noriega (1996) from a largely complete skull and mandible found in late Pliocene deposits south of Buenos Aires, Argentina. This bird was very similar to the extant Black-hooded Parakeet *Aratinga nenday*, but was somewhat more robust. The skull was sufficiently well preserved to allow reconstruction of its jaw muscles and bite force (Carril *et al.* 2014).

At Ahl al Oughlam, a late Pliocene (2.5 mya) site in Morocco, Mourer-Chauviré and Geraads (2010) named the new species, *Agapornis atlanticus*, from a collection of bones. This bird was larger than any recent or fossil species of lovebird. No extant *Agapornis* species occurs this far north in Africa and evidence from some of the bones indicates that this bird was breeding in the near vicinity. Rich (1980) noted the presence of two types of parrots in the early Pliocene (5.2 mya) fossil deposits of Langebaanweg, South Africa, while Stidham (2006) reported three species – a large bird about the size of the African Grey Parrot *Psittacus erithacus*, a medium-sized about the size of a *Psittacula*, and a small lovebird, smaller than any other known member of the genus. The last was described by Manegold (2013) as *Agapornis attenboroughi*, and he considered that it was likely to be a stem lineage of this group.

In several instances, fossils of *Agapornis* have been among the avian remains recovered from sites producing early human ancestors. At the early Pliocene (4.4 mya) site of *Ardipithecus ramidus* in the Afar Desert, Ethiopia, parrots comprise about 36 per cent of all avian fossils, representing a minimum of 22 individuals. These appear to have been accumulated through owl predation. Specimens of *Agapornis* have been found in late Pliocene (1.84 mya) deposits at Olduvai Gorge, Tanzania (Brodkorb 1985; Prassack 2010). Two species of *Agapornis*, one large, one small, were reported from Kromdraai, South Africa (late Pliocene–early Pleistocene, 2.0–1.6 mya) (Pocock 1969). Stidham (2009) considered that their presence gave clues to the palaeoenvironment in which *Australopithecus robustus* lived, but Perrin (2011) rebutted this, pointing out that the different species of modern lovebirds occupy such a broad range of habitats that no such conclusion could be made.

Manegold (2013) named the *Psittacula*-sized parrot from Langebaanweg as *Khewenena leopolinae*. This is the only instance of a parrot of this age or younger that did not belong to a living genus. The only other genus of parrot reported from an African Pliocene site is an indeterminate *Poicephalus* noted by Louchart *et al.* (2009) from the early Pliocene Afar Desert site.

QUATERNARY PARROTS

All fossil parrots from the Quaternary (Pleistocene 1.5 mya–10 000 years ago (kya) and Holocene 10 kya to present) are modern species or their close but now extinct relatives. Many parrot bones of this age came from archaeological sites. These frequently are subfossils, that is, there has not been complete mineralisation of the bone and organic material.

North America and Mexico

Most parrots in North America occur in the southern half of Mexico. Only a few extended into northern Mexico and the United States – the Thick-billed Parrot *Rhynchopsitta pachyrhyncha*, which was restricted to the southwest, and the much more widespread, but now extinct Carolina Parakeet *Conuropsis carolinensis*. Most remains of these and other parrots have been recovered from archaeological sites.

The Carolina Parakeet *Conuropsis carolinensis*, which went extinct in the early 20th century, had the northernmost distribution of any New World species. Bones and beaks of this species have been found in several Native American middens and refuse sites in the United States (for example, see Parmalee 1958, 1967 for such occurrences in Illinois).

From the late Pleistocene San Josecito Cave, Nuevo Léon, northeastern Mexico, Steadman *et al.* (1994) recovered remains of the Thick-billed Parrot *R. pachyrhyncha* and others that were tentatively referred to the Maroon-fronted Parrot *R. terrisi*, both extant species. A larger extinct relative, *R. phillipsi*, also was recovered from this cave, and it differed from its congeners in size and shape of the bill (Rea 1997). These records were made through natural processes. Human activity resulted in the presence of *R. pachyrhyncha* in a number of archaeological sites in northern Mexico, Arizona and New Mexico. Often associated with *Rhynchopsitta* remains were those of Scarlet Macaws *Ara macao*, and there is a single record of the Military Macaw *A. militaris*. Unlike *Rhynchopsitta*, these did not occur naturally in this area, so were acquired by trade from southeastern Mexico by transport for hundreds of kilometres on foot. Alternatively, large captive breeding areas could have been established at some sites to provide birds. These macaws were important in ceremonies, and frequently were sacrificed as part of rituals, an important practice by at least 850 AD. The skeletal remains of both species have been unearthed at several Amerindian sites, sometimes in large numbers. Macaws disappeared from this region around 1130 AD (Hargrave 1939, 1970; Olsen and Olsen 1974; Minnis *et al.* 1993; Creel and McKusick 1994; Somerville *et al.* 2010; Watson *et al.* 2015).

West Indies

The assessment of fossil parrots from the West Indies is complicated because the birds were traded by the native population and European colonisers. Species were transported between islands or even from the South American mainland. The early reports of explorers and colonists mention a number of putative parrot species, especially macaws, for which there is no specimen evidence. Some parrot material has been found in archaeological sites, but there often is no way to determine with confidence that they represent species indigenous to the area rather than birds transported to the location. Prehistoric fossils are of particular value in determining the presence of parrots. The absence of suitable comparative skeletal material of identified species prevents confident assignment of these remains to known taxa.

The now extinct Cuban Macaw *Ara tricolor*, known from 19 skins, also is represented by three palaeontological records, including a worn cranium (Wetmore 1938; Arredondo 1984; Olson and Suárez 2008). These come from deposits in springs, caves or sinkholes in central and western Cuba and are of Quaternary age, probably Pleistocene.

The extinct Puerto Rican Parakeet *Psittacara maugei* is known from a few skins and mounts, but has been found also as an archaeological specimen in south-central Puerto Rico (Olson and Maíz 2008) somewhat younger than 300 AD, and a few fossils from caves just north of the centre of this island. Based on the state of preservation, Olson (2015) considered most of the latter to be early Holocene to late Pleistocene age (~5–15 kya) in age, with those from one site being considerably older: 'almost certainly one of the glacial periods of the late or even middle Pleistocene (~30 kya–300 kya)'. They demonstrated that this parrot, known with certainty only from skins from offshore Mona Island, occurred also on Puerto Rico.

The macaw *Ara autochthones* originally was described by Wetmore (1937) from a single tibiotarsus of a young bird recovered from an archaeological site on St Croix in the Virgin Islands. Olson and Maíz López (2008) referred to this species several skeletal elements found in association in another archaeological site in Puerto Rico. In size, this macaw was intermediate between large-bodied and smaller-bodied macaws. This verified the presence of another endemic macaw from the West Indies but, owing to the regular trade in these birds between islands, the authors could not determine which island this species originally inhabited, though they considered that it was unlikely to have been indigenous to St Croix, where it was first discovered.

There is no evidence from skins or mounts for the bird named *Ara guadeloupensis* (Guadeloupe Island), with the name based on multiple early accounts and an illustration that may be attributable to it. An ulna, slightly smaller than that of *A. macao*, found at an archaeological site on Marie Galante was tentatively referred to this species (Williams and Steadman 2001). Olson and Suárez (2008) thought this bone was best considered to belong to the Imperial Amazon *Amazona imperialis*. Gala and Lenoble

(2015) referred a terminal phalanx from a Pleistocene-aged site (c. 10700 years before present [ybp]) on Marie-Galante to a large *Ara* macaw. Because this deposit predated the arrival of humans on the island, it provided evidence of a now-extinct macaw on Guadeloupe.

Fossil remains of amazon parrots include several records of the Cuban Amazon *Amazona leucocephala* from five islands in the Bahamas (Todd and Worthington 1911; Wetmore 1938; Brodkorb 1959; Olson and Hilgartner 1982; Carlson 1999 cited in Williams and Steadman 2001; Steadman *et al.* 2007) and a small number of bones of the Puerto Rican Amazon *A. vittata* from Antigua and Barbuda (Steadman *et al.* 1984; Pregill *et al.* 1988; Pregill *et al.* 1994; Williams and Steadman 2001).

Archaeological sites on the West Indies have yielded parrot bones that can be assigned to a genus, but owing to their isolation or condition cannot be identified to species level. It is uncertain if they come from living species or represent now extinct ones or whether they were indigenous to the island in which they were found. An undescribed *Amazona* larger than *A. leucocephala* has been recovered from an archaeological site on Grand Turk, Bahamas, where it occurred together with that species (Carlson 1999 cited in Williams and Steadman 2001). A tibiotarsus, from a large *Amazona* parrot, possibly *A. violacea*, was found at an archaeological site on Marie Galante, and a bone from an unidentified *Amazona* parrot was reported from La Désirade (Williams and Steadman 2001; Grouard 2002). From Montserrat come several bones of a small amazon (Reis and Steadman 1999; Williams and Steadman 2001), as well as a nearly complete coracoid that fits within the range of measurements of several species of macaw such that its identity beyond being a species of *Ara* cannot be determined (Williams and Steadman 2001). An undescribed *Aratinga*, known from a quadrate and sternum from a cave on Barbuda, was larger than any living species (Pregill *et al.* 1994; Williams and Steadman 2001).

South America

The fossil record of parrots in South America is restricted to the Pliocene and Quaternary, and all extinct species have been assigned to living genera. Spillman (1942) named *Protoconurus roosevelti* on the basis of a number of bones from the late Pleistocene of Santa Elena Peninsula, Ecuador. Brodkorb (1971) placed this in the genus *Aratinga*, as did Campbell (1976), who commented that he could not detect sufficient differences between the fossils and living species of *Aratinga* to recognize a different genus. The latter author had a large collection of parrot bones from La Carolina, not far from Spillman's locality, that he referred to this species. Owing to the lack of adequate comparative material, Campbell (1976) could not definitely conclude that *A. roosevelti* was indeed a valid extinct species and not an extant one. A smaller bone among his collection of fossils might be that of a different species.

The first reported fossil parrot from South America was identified by Ameghino (1891) as *Conurus* sp. (most species are now placed in *Aratinga* or *Psittacara*) from deposits near Buenos Aires, Argentina, of Pleistocene age. Holyoak (1971) re-examined the specimen, an upper mandible and part of a skull, and concluded that 'there is insufficient evidence for assigning it to a particular genus'. He considered that the specimen could belong to any of a number of New World parrots.

The genus *Cyanoliseus* is represented in the fossil record by three species, two extinct and the other still living. Cattoi (1957) described a new species from a near complete humerus from the early Pleistocene of Buenos Aires Province, Argentina, naming it *Pionus ensenadensis*. The specimen subsequently was restudied by Tonni (1972), who concluded that it was misclassified and transferred the species to *Cyanoliseus*. This was the smallest member of the genus. Additional fossil material of this species was discovered in Uruguay from somewhat younger deposits (Acosta Hospitaleche and Tambussi 2006). Another fossil species of *Cyanoliseus* was the largest member of the genus. *Cyanoliseus patagonopsis* exceeded in size the extant Burrowing Parrot *C. patagonus* by about a third and was more than twice the size of *C. ensenadensis* (Acosta Hospitaleche and Tambussi 2006). It was recovered from somewhat younger deposits in the Buenos Aires Province. Remains of some indeterminate species of *Cyanoliseus* have been found also in late Pleistocene sites in this Province. Fossils of the extant *Cyanoliseus patagonus* are known also from this Province, as well as the Pleistocene of Uruguay (Acosta Hospitaleche and Tambussi 2006; Tambussi *et al.* 2009).

Other living species of parrots known from fossil and subfossil remains were found in Peru, from where Reichenow (1880) cited the Blue and Yellow Macaw *Ara ararauna* and Mealy Amazon *Amazona farinosa* from Ancón, and Campbell (1979) reported the Pacific Parrotlet *Forpus coelestis* from the Talara Tar Seeps. The most productive site is the cave system Lapa da Escrivania in Goiás, Brazil. From this late Pleistocene deposit, Winge (1888) identified Green-winged Macaw *Ara chloropterus*, Blue-winged Macaw *Primolius maracana*, undetermined macaws, one larger and one smaller than the preceding species, White-eyed Parakeet *Psittacara leucophthalmus*, Peach-fronted Parakeet *Eupsittula aurea*, Maroon-bellied Parakeet *Pyrrhura frontalis*, Yellow-chevroned Parrot *Brotogeris chiriri* and undetermined amazons, one the size of the Orange-winged Amazon *Amazona amazonica* and one smaller.

Africa

A lovebird came from Plovers Lake, South Africa (Pleistocene, 1 mya), a non-hominid-bearing fossil site. Like many of the Pliocene *Agapornis* remains, it could not be assigned to a living species, nor confidently be diagnosed as a new fossil one. Subfossil remains of the Vasa Parrot *Coracopsis vasa* from Sirabé, central Madagascar (Lambrecht 1933) comprise the only other Quaternary record of a parrot from Africa.

Indian Ocean

On the Mascarene Islands in the Indian Ocean lived several now extinct parrots mentioned in early travellers' narratives, for which no modern specimens exist but were formally described from subfossil remains. The Broad-billed Parrot *Lophopsittacus mauritianus* survived on Mauritius until around 1680, and early accounts of its occurrence are quoted elsewhere in this book. A subfossil mandible found in the Mare aux Songes swamp was described by Owen (1866). Additional material was discovered from Le Pouce (Newton and Gadow 1893), and most major bones now are known, representing at least 28 individuals (Hume 2007). The Rodrigues Parrot or Leguat's Parakeet *Necropsittacus rodericanus* from nearby Rodrigues Island became extinct in the late 18th century, and also was named officially from a partial subfossil upper bill (Milne-Edwards 1867). Additional specimens, from at least four individuals, have since been discovered in caves on Plaine Corai and in Caverne Tortue (Hume 2007).

The Mascarene Parrot *Mascarinus mascarin* became extinct when the last captive bird died in about 1800 or earlier. Known from two specimens and a number of subfossil remains, it

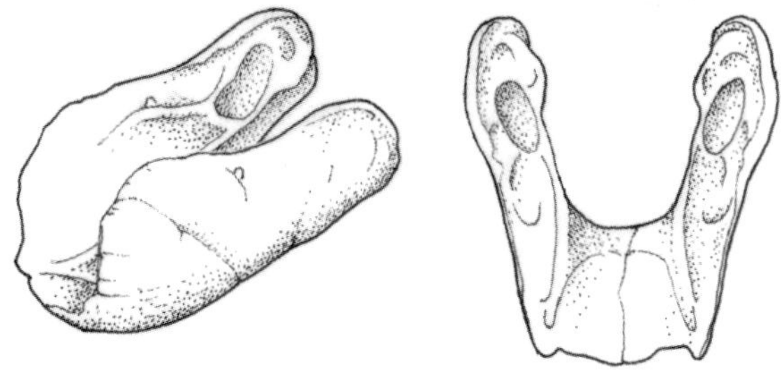

Fig. 8 Lower mandible of the Broad-billed Parrot *Lophopsittacus mauritianus* held in the collection of the Natural History Museum, Tring, UK.

occurred on Réunion, where bones were found in several small caves (Mourer-Chauviré *et al.* 1999; Hume 2013). Subfossil remains of three species of *Psittacula* parrots have been found – extinct Mascarene Grey Parakeet or Thirioux's Grey Parrot *P. bensoni* on Mauritius, extinct Newton's Parakeet or Rodrigues Parakeet *P. exsul* on Rodrigues, and the extant Echo Parakeet *P. eques* on Mauritius). There are two specimens of the Rodrigues Parakeet from Rodrigues and one specimen of the Echo Parakeet from Réunion, where it no longer occurs.

Based on bones from at least three individuals, Meijer *et al.* (2013) identified the presence of *Geoffroyus*, likely the Red-cheeked Parrot *G. geoffroyi*, from Liang Bua, Flores, Indonesia, the site from which remains of *Homo floresiensis* (the Hobbit) have been recovered.

Australasia and southwest Pacific

The early Pleistocene Rackham's Roost Site (originally thought to be early Pliocene in age) from Riversleigh, Queensland, has yielded bones indistinguishable from those of the living Budgerigar *Melopsittacus undulatus* (Boles 1998). This fossil accumulation consists mainly of the prey remains of predatory *Macroderma* ghost bats, which today are represented in Australia by a single species. Numerous small vertebrates, including birds, are part of their diet, and prey is brought to the roost cave to be consumed, with discarded body parts being dropped to the cave floor. Bones of *Melopsittacus* found to date include limb bones, three carpometacarpi and a tarsometatarsus. The presence of this species suggests that the palaeoenvironment at Riversleigh during this time was open woodland, similar to the habitat still frequented by these parrots throughout the region.

All other records of fossil parrots in Australia come from the Quaternary (1.6 mya–present). Our knowledge of these birds through this period is based on the work of Robert Baird, who has studied accumulations in cave deposits along the southern edge of the Australian mainland. He has documented the presence in these deposits of the Glossy Black Cockatoo *Calyptorhynchus lathami*, Red-tailed Black Cockatoo *C. banksii*, Yellow-tailed Black Cockatoo *C. funereus*, Gang Gang Cockatoo *Callocephalon fimbriatum*, Galah *Eolophus roseicapilla*, Slender-billed Corella *Cacatua tenuirostris*, Rainbow Lorikeet *Trichoglossus haematodus*, Musk Lorikeet *Glossopsitta concinna*, Little Lorikeet *Parvipsitta pusilla*, Purple-crowned Lorikeet *P. porphyreocephala*, Australian King Parrot *Alisterus scapularis*, Regent Parrot *Polytelis anthopeplus*, Crimson Rosella *Platycercus elegans*, Eastern Rosella *P. eximius*, Western Rosella *P. icterotis*, Red-capped Parrot *Purpureicephalus spurius*, Port Lincoln Parrot *Barnardius zonarius*, Bluebonnet *Northiella haematogaster*, Mulga Parrot *Clarkona varia*, Elegant Parrot *Neophema elegans*, Blue-winged Parrot *N. chrysostoma*, Scarlet-chested Parrot *N. splendida*, Swift Parrot *Lathamus discolor*, Budgerigar *Melopsittacus undulatus*, Ground Parrot *Pezoporus wallicus* and Night Parrot *P. occidentalis* (see Baird 1985, 1991a, 1991b, 1992, 1993). At some deposits, parrots make up a significant proportion of avian remains. At Devil's Lair, Western Australia, 42 per cent of the minimum number of individuals represented by bird bones belonged to the Purple-crowned Lorikeet.

Although all these parrots belong to known species, the records provide information on past distributions and allow inferences to be made about the environment around the caves at the time of deposition. The Glossy Black Cockatoo, currently restricted to the east coast of the mainland and Kangaroo Island, has been found at Green Waterhole Cave (Pleistocene), in southeastern South Australia (Baird 1985, 1986). At several sites the avian species assemblage is characteristic of areas more arid than occur at present. Remains of Budgerigars are known from Clogg's Cave, near Buchan in eastern Victoria, and Mulga Parrots have been found at Devil's Lair, in far southwestern Western Australia. The presence of these species implies that at the time of deposition during the last glacial period the environment was drier and more open than at present. The large sample of specimens of the Slender-billed Corella from Green Waterhole Cave average 4 per cent larger than living birds, and this parallels similar Pleistocene gigantism exhibited by other species from this deposit.

The fossil history of parrots on Australia's external territories is limited to records from Norfolk Island, where Meredith (1991) noted the occurrence of the Red-fronted Parakeet *Cyanoramphus novaezelandiae* and the extinct Norfolk Island Kaka *Nestor productus* in coastal sand dunes estimated to be only a few thousand years old. These finds confirm that the Kaka became restricted to Philip Island only in more recent times, having formerly occurred on, but subsequently lost from, Norfolk Island. To date, there are no fossil records of parrots from Lord Howe Island.

For many years the only records from New Zealand were restricted to existing species in late Holocene deposits. The Kea *Nestor notabilis*, New Zealand Kaka *N. meridionalis*, Kakapo *Strigops habroptila* and *Cyanoramphus* parrots are known as subfossil remains from a number of sites on both the North and South Islands and Chatham Islands (Holdaway and Worthy 1993; Millener 1991; OSNZ 2010). Several populations of Red-fronted Parakeets are represented, as is the Yellow-fronted Parakeet *Cyanoramphus auriceps*. 'Identifications for fossil bones of *Cyanoramphus* are tentative because most bones (particularly isolated ones) cannot be reliably assigned to species due to overlap in size' (OSNZ 2010).

Evidence of an extinct species of nestorid parrot, closely related to *N. meridionalis*, was discovered on the Chatham Islands (Wood *et al.* 2014). The fossil material of this bird, given the name *Nestor chathamensis*, came from late Quaternary midden deposits and demonstrates that this parrot survived after the arrival of Polynesians (13–16 centuries AD) before eventually becoming extinct as a result of human activities.

Intriguing past distribution patterns of parrots across the southwestern Pacific Ocean have been discovered by David Steadman and others. Working with fossil and archaeological material, they have found several comparatively recently extinct species, as well as populations of some extant species now restricted to one or a few small islands, but which formerly were more widely distributed. These extinctions seem to have followed immediately the arrival of humans to the islands.

Fossil remains of *Cacatua* white cockatoos have been found at several sites. On New Ireland, in the Bismarck Archipelago, Papua New Guinea, there has been found a larger cockatoo that may be the same as, or similar to, those surviving on other islands, probably 'an undescribed species or subspecies in the *C. galerita* species group' (Steadman *et al.* 1999). A similar and possibly identical *Cacatua* cockatoo has been found in the Mussau Islands, also in the Bismarck Archipelago (Steadman and Kirch 1998). Steadman (2006a) noted the presence at Grande Terre, New Caledonia, of a bird the size of the Blue-eyed Cockatoo *C. ophthalmica,* although it was uncertain if the remains represented this species.

Lorikeets are by far the most commonly recovered group of parrots in the Pacific. Larger species are known from an indeterminate species of *Lorius* from New Ireland, from where a medium-sized Loriinae species also is known, and these specimens represent locally extirpated species (Steadman *et al.* 1999). Prehistoric evidence of the Cardinal Lory *Chalcopsitta cardinalis* has been found at Buka, in the Solomon Islands (Steadman 2006a). The presence of the Rainbow Lorikeet *Trichoglossus haematodus* in deposits at Me´ Aure´ Cave, New Caledonia, appear to come from the pellets of Barn Owls *Tyto alba*, which would have preyed on these species (Boyer *et al.* 2010). The records are quite young, from around 200 years before modern times. This species has been found also in archaeological deposits on Efate, Erromonga and Malekula (Steadman 2006a). The Pohnpei Lorikeet *T. rubiginosus* is one of the few prehistoric records of parrots from Micronesia (Pohnpei) (Steadman 2006a).

Small *Charmosyna* and *Vini* lorikeets are more widely distributed through the islands. Remains of the Palm Lorikeet *C. palmarum* are known from the island of Santo, Vanuatu (Steadman 2006a). From the Teouma archaeological site (3000–2500 ybp), Efate, Vanuatu, was recovered a small lorikeet, the size of this species, which occurs there today (Worthy *et al.* 2015). Anderson *et al.* (2010) recorded the presumed extinct New Caledonian Lorikeet *C. diadema* from Pindari Caves, New Caledonia. A locally extirpated *Charmosyna* species has been found in fossil deposits in New Ireland (Steadman *et al.* 1999).

Small lorikeets, both living and extinct, are prominent in bird remains from the Polynesian islands. Steadman and Zarriello (1987) described two extinct species of small lorikeets, *Vini vidivici* and *V. sinotoi* from Ua Huka and Tahuata in the Marquesas Islands. Both were larger than any species alive today, *V. vidivici* only somewhat, but *V. sinotoi* considerably so. Subsequently, both species were documented from Huahine, in the Society Islands (Steadman and Pahlavan 1992) and Mangaia, in the Cook Islands (Steadman and Kirch 1990; Steadman 2006a).

Evidence of several extant species has been found. Remains of the Collared Lory *Phigys solitarius,* which now occurs only on Fijian islands and islands in the Lau Archipelago, have been discovered in deposits on 'Eua and Lifuka and 'Uiha in the Ha'apai group (Steadman 1993, 1995, 2006a). In the Marquesas Islands, remains of the living Ultramarine Lory *V. ultramarina* have been found on islands where the species was not known to occur (Steadman and Zarriello 1987). At 'Eua, Tonga, there was evidence of the extirpation of the Blue-crowned Lory *V. australis*, still found on some neighbouring islands (Steadman 1993, 1995). Small lorikeets from Mangaia, in the Cook Islands, are referrable to Kuhl's Lory *V. kuhlii*, now found naturally only in the Tubuai or Austral Islands (Steadman and Kirch 1990). Remains of this species have been found also in deposits on Aitutaki and Atiu, in the southern Cook Islands (Steadman 1991). The endemic Henderson Lorikeet *V. stepheni* is known from subfossil remains on that island (Schubel and Steadman 1989; Wragg and Weisler 1994).

The fossil parrot *Eclectus infectus* is known from material found in late Holocene archaeological sites and late Pleistocene palaeontological sites in Tonga, on the islands of 'Eua, Lifuka and 'Uiha (Steadman 2006b). It differed from the living Eclectus Parrot *E. roratus* by having somewhat larger cranial and hindlimb bones and proportionally shorter wings. A few bones from the Malua Bay archaeological site (late Holocene), Malakula, Vanuatu, were referred to this species (Steadman 2006b). An unidentified large parrot, possibly this species, is represented by a single tibiotarsal fragment recovered from Payapai Cave (c. 1000 years ago), Rota, in the Mariana Islands (Steadman 1992, 1999, 2006a).

The Red-fronted Parakeet *Cyanoramphus novaezelandiae*, and Horned Parakeet *Eunymphicus cornutus* were recorded in material from Pindari Caves, New Caledonia (Anderson *et al.* 2010). The presence of *C. novaezelandiae* in deposits at Me´ Aure´ Cave, New Caledonia, appear to come from the pellets of Barn Owls *Tyto alba*, which would have preyed on these species (Boyer *et al.* 2010). The records are quite young, from around 200 years before modern times.

On the basis of a single bone each, Steadman (1995) listed two parrots from Easter Island as 'cf. Psittacidae sp. nov. 1' and 'cf. Psittacidae sp. nov. 2'. Although no more precise identification could be made, the morphology of the fossils pointed to Pacific, rather than South American, relationships.

Fossil occurrences of modern genera demonstrate that distributions were once greater than at present. For example, *Agapornis* lovebirds previously inhabited regions of Africa much farther north than seen today. Similarly, through the Pacific islands prehistoric and archaeological remains indicate that a number of parrots lived in areas where they are absent today. While some of these became extinct before the arrival of humans on the islands, in far more instances human-related activities are implicated in the loss of these birds.

Despite its many gaps, the known fossil record of parrots and possible parrot relatives has increased significantly in the last few decades of the 20th century and start of the 21st century, even if it remains far less revealing than desirable. From the growing record have been gained important new insights into the early history and diversification of this group, past distributions at various stages of the record and the dramatic impact on island populations of first encounters with humans. At the present time, much information remains tantalisingly out of reach, so all new discoveries certainly will assist our understanding of the long history of parrots.

Walter E. Boles
Australian Museum
31 December 2016

AUSTRALASIAN DISTRIBUTION

The Australasian Distribution, as determined for this book, extends from the Philippine and Indonesian Archipelagos south to Australia and New Zealand, including offshore and subantarctic islands, and east to Henderson Island in eastern Polynesia. It is the only part of the worldwide distribution of parrots where all three superfamilies are represented, Strigopoidea and Cacatuoidea being endemic. Australia is the dominant landmass, and is where the highest number of parrot species occurs.

AUSTRALIA

In the 16th and 17th centuries, European traders coming to the 'Spice Islands' learned of a great land to the south where parrots were abundant, and early maps depicted *Terra Psittacorum* – the Land of Parrots, as an unknown landmass at the bottom of the world. The first European to record the abundance of parrots in the southern land was the English adventurer William Dampier, when on 22 August 1699 he visited one of the small islands in what is now the Dampier Archipelago, off the coast of northwestern Australia. In his *Voyage to New Holland*, published in 1703, Dampier described the birds seen on the island as '… some Cormorants, Gulls, Crab-catchers etc., a few small Land Birds, and a sort of White Parrot, which flew a great many together'. These 'White Parrots' were Little Corellas *Cacatua sanguinea*, which still occur on the islands.

'Land of Parrots' remains a term often applied to Australia, and seems appropriate because one-sixth of the world's species occur there, with no other country having such a richness and diversity of forms. This strong diversity of forms, which so strongly differentiates Australia from other regions in the worldwide distribution of parrots, is a legacy of the Gondwanan ancestry of parrots, with focus on Australasia as the most likely centre of origin (Cracraft 2001). Molecular analyses support a Gondwanan origin of parrots, most probably during the Cretaceous, some 82 million years ago when New Zealand split from Gondwana and after the separation of Africa and the India-Madagascar block (Wright *et al.* 2008).

Since the time of European settlement, biologists and observers have been fascinated by Australian parrots. In his *Handbook to the Birds of Australia*, published in 1865, John Gould wrote:

> *No group of birds gives to Australia so tropical and foreign an air as the numerous species of this great family by which it is tenanted, each and all of which are individually very abundant.*

The comment made by Gould about the abundance of parrots in Australia remains valid today, although there have been changes in the status of some species. Parrots are a very prominent component of the avifauna of the continent, and in every region they can be observed in good numbers.

Fortunately, the appalling extinction rate for Australian mammals is not mirrored in the loss of Australian birds, but since European settlement four parrots have been lost – the Norfolk Island Kaka *Nestor productus*, the Paradise Parrot *Psephotellus pulcherrimus* and two forms of the Red-fronted Parakeet *Cyanoramphus novaezelandiae* on Lord Howe and Macquarie Islands. Conversely, a few generalist species, notably the Galah *Eolophus roseicapilla*, the Sulphur-crested Cockatoo *Cacatua galerita* and the Little Corella *C. sanguinea*, have prospered since European settlement, but a number of more specialised species and populations with restricted ranges have declined, sometimes dramatically. Under the Commonwealth *Environment Protection and Biodiversity Conservation Act 1999* the Orange-bellied Parrot *Neophema chrysogaster* and the Western Ground Parrot *Pezoporus wallicus flaviventris* are listed as critically endangered, and listed as endangered are Carnaby's Black Cockatoo *Calyptorhynchus latirostris*, the Golden-shouldered Parrot *Psephotellus chrysopterygius,* the Swift Parrot *Lathamus discolor*, the Night Parrot *Pezoporus occidentalis*, the southeastern population of the Red-tailed Black Cockatoo *Calyptorhynchus banksii graptogyne*, the Kangaroo Island population of the Glossy Black Cockatoo *C. lathami halmaturinus*, the Norfolk Island population of the Red-fronted Parakeet *Cyanoramphus novaezelandiae cookii* and Coxen's Fig Parrot *Cyclopsitta coxeni*. Baudin's Black Cockatoo *Calyptorhynchus baudinii* meets criteria of the International Union for the Conservation of Nature (IUCN) for an endangered listing so is treated as such in this book (see Garnett *et al.* 2011).

A recent analysis of the ecological and socio-economic factors affecting extinction risk in parrots revealed that the most important threats to parrots are agriculture, hunting and trapping, logging, climate change and severe weather, invasive alien species, and residential and commercial development (Olah *et al.* 2016). All of these threats have affected or are affecting parrot populations in Australia, but particularly prominent are agriculture and logging. For more than two centuries, Australia has been a primary-producing nation supplying the traditional markets of Europe. Though often disclaimed, the phrase 'Australia lives off the sheep's back' is a telltale statement with some validity. This prominence of primary production has resulted in widespread, sometimes dramatic changes to the natural environment, especially in humid coastal areas where agricultural and urban development has been concentrated. Recher (1985) points out that within 100 years of settlement in 1788, most of the Australian continent had been colonised and converted to farming or grazing lands. Landclearance has been excessive and wasteful, with widespread degradation, especially through salination, attesting to the futility of vegetation removal from marginal lands incapable of supporting profitable agricultural enterprises. Regrettably, this excessive landclearance continues, particularly in western Queensland. In recent times, mining has become increasingly prevalent in transformation of the Australian landscape, with large-scale open-cut extraction of minerals and coal resulting in significant local losses of natural vegetation. Logging has contributed to declines in populations of the Superb Parrot *Polytelis swainsonii* and the Regent Parrot *P. anthopeplus*, and Heinsohn identifies commercial logging, even on a small scale, on Cape York Peninsula as a potential threat to Eclectus Parrots *Eclectus roratus* (Garnett and Crowley 2000; in Curtis *et al.* 2012).

Since the time of European settlement there has been disturbance of all habitats. None has been destroyed completely, but some, including tropical or subtropical rainforest and mallee shrublands, have been gravely depleted. Consequently, parrot species have been affected in one way or another by human-induced changes. It appears that a few species, such as the Galah *Eolophus roseicapilla*, Little Corella *Cacatua sanguinea* and Port Lincoln Parrot *Barnardius zonarius*, have benefited, and others seem to have been little affected, but certainly there are signs that a number of species, especially those that are specialised in any way, now are declining. Summarising the abundance and population trends of Australian parrots, Smith (1978) concluded that 72 per cent of species appear to be holding their own, but only 60 per cent have stable or increasing populations. Noting a great increase in detailed ecological research on Australian parrots during the 1970s and 1980s, Joseph (1988) emphasises the need not only to continue that research and extend the number of species studied, but to conserve existing natural habitats in the face of often increasing pressures of agricultural and industrial development.

During field studies of the breeding biology of Carnaby's Black Cockatoo *Calyptorhynchus latirostris* at two sites in Western Australia, it was found that, although each pair attempted to breed every year, the production of chicks per breeding unit at one site was only half that of the other site. Observations indicated that poor breeding success at the second site was due to a shortage of food, the parents being obliged to spend more time foraging in the scattered feeding areas. Some areas of useful food were not visited by the birds from one year to the next, so it seemed that the total amount of food in the area was sufficient, but because of its patchy distribution the birds did not have sufficient time to gather it. This important study demonstrates that breeding populations may be endangered by fragmentation of nearby feeding areas, despite a continuing sufficiency in total food resources.

At the same study sites in Western Australia, annual rates of loss of nesting hollows were 4.8 per cent and 2.2 per cent, and these exceeded the rates of formation of hollows. Losses were caused by trees being blown over, bulldozed or burnt in clearing operations, and by sections breaking away or collapsing into the hollows. At a study area within and adjacent to Iron Range National Park, on Cape York Peninsula, north Queensland, between 1999 and 2005, an average 5.2 per cent of nest-trees of Palm Cockatoos *Probosciger aterrimus* was lost each year, and these losses were due mainly to fire, followed by wind and then decay. The loss of mature, hollow-bearing trees is a problem throughout much of the continent, largely because of landclearance for agriculture and an increasing adoption of clear-felling practices in forests. Woodland on private property is generally over-mature, trees having been left to provide shelter for stock, and grazing prevents regeneration. Similarly, clear-felling in forests, particularly during wood-chipping operations, removes old trees, and harvesting is often repeated too frequently to allow replacement. I am concerned also about losses due to 'dieback', a term loosely applied to the effects of the root-attacking fungus *Phytophthora cinnamomi* and to severe outbreaks of defoliating insects. Such losses can be quite significant where woodlands already are fragmented because of large-scale landclearance or in specialised habitats such as woodlands of river red gums *Eucalyptus camaldulensis* on riverine floodplains.

There is ample evidence of wildlife populations being affected adversely by losses of hollow-bearing trees. In recent years, there have been increasing instances of multiple sequential, or even simultaneous occupation of hollows, and eviction of occupants by more aggressive species. In eastern and southern Australia, smaller parrots must contend with intense competition for nesting sites from introduced House Sparrows *Passer domesticus,* Common Starlings *Sturnus vulgaris* and Common Mynahs *Acridotheres tristis*. Rapid utilisation of hollow logs provided for Red-tailed Black Cockatoos *Calyptorhynchus banksii* at a breeding site in southwestern Victoria suggests that a shortage of suitable nesting hollows is an important factor contributing to the decline of that endangered population. Artificial nesting hollows are being utilised also by the critically endangered Orange-bellied Parrot *Neophema chrysogaster* in southwestern Tasmania and by Glossy Black Cockatoos *Calyptorhynchus lathami* on Kangaroo Island, South Australia.

Ground-feeding parrots are dependent on an abundance of seeding grasses and herbs, particularly during the breeding season, so changes in groundcover are potentially damaging to local populations. It has been suggested that in the early 1900s, severe drought, coupled with a rapid expansion of the cattle industry, brought about fundamental changes to groundcover vegetation in parts of central Queensland, and this probably contributed to extinction of the Paradise Parrot *Psephotellus pulcherrimus* and disappearance of the Turquoise Parrot *Neophema pulchella* from the northern sector of its range. I am of the opinion that similar changes, brought about by grazing and persistent dry-season burning, are responsible for the disappearance of Golden-shouldered Parrots *Psephotellus chrysopterygius* and Hooded Parrots *P. dissimilis* from parts of their ranges, and predation of adults and fledglings by Pied Butcherbirds *Cracticus nigrogularis* is a secondary pressure now assuming greater significance because of lower population levels.

The effects of secondary pressures, particularly disease, predation and hunting or trapping, are significantly increased when an affected species is already threatened by habitat loss. I see a pattern of disease and habitat loss combining to bring about near-extinction of the Orange-bellied Parrot *Neophema chrysogaster* in southeastern Australia. Again in southeastern Australia, recent studies have revealed that numbers of Swift Parrots *Lathamus discolor* are declining at an alarming rate, and the species is severely threatened by loss of habitat coupled with high predation at nests (Heinsohn *et al.* 2015). It was revealed also that in Tasmania survival of nests of these parrots is a function of mature forest cover in the surrounding landscape, with the likelihood of predation by introduced Sugar Gliders *Petaurus breviceps* increasing with decreasing forest cover brought about by logging operations. Trapping often is cited as a factor contributing to the decline of parrot populations. While not wishing to undermine the importance of ensuring adequate legal protection for parrots, I stress that nothing will be gained by prohibiting trapping if little is done to counteract the effects of habitat destruction – the two are complementary! I have no evidence of any Australian species currently being at risk from trapping, but chopping into nesting hollows to illegally remove eggs or chicks continues to threaten local populations of some *Calyptorhynchus* black cockatoos and Major Mitchell's Cockatoo *Lophochroa leadbeateri*. I have no doubt that the taking of nestlings of Golden-shouldered Parrots *Psephotellus chrysopterygius* contributed significantly to the dramatic decline of that species during the 1960s and 1970s.

Of concern are the consequences of domestic trade in large numbers of live parrots or cockatoos. Since the 1970s, feral populations of some species, notably Little Corellas *Cacatua sanguinea* and Slender-billed Corellas *C. tenuirostris,* have become established in areas well outside the normal ranges, and many of these populations are expanding dramatically, especially in and around major urban centres. Also, there is a thriving population of the eastern subspecies of the Galah *Eolophus roseicapilla* in the Perth area, and genetic swamping of local populations of the distinctive western subspecies already is underway. Again in the Perth area, expansion of feral populations of Sulphur-crested Cockatoos *Cacatua galerita* and Rainbow Lorikeets *Trichoglossus haematodus* can be expected to impact on populations of endemic species and increase the likelihood of crop damage.

In Australia, damage to crops is not of economic significance at the national level, but losses experienced by farmers locally can be severe and there is a need to protect crops. Any assessment of the issue is compromised by campaigns promoting short-term or 'quick-fix' responses, which certainly offer no effective solutions. Repeatedly, it has been demonstrated that shooting, trapping or poisoning are ineffective in reducing crop damage, yet these actions continue to be promoted. Trapping and export of so-called 'pest species' frequently is suggested as a solution, and I am amazed that any credence at all is given to this suggestion for export certainly would not alleviate conflicts with agriculture in Australia and could pose very serious environmental and economic risks in importing countries. The deliberate release of Rainbow Lorikeets *Trichoglossus haematodus* in the suburbs of Auckland, New Zealand, in 1992, posed a threat to native nectar-feeding species, and prompted wildlife authorities to initiate an eradication program, which succeeded in eliminating the population by 2002 (Heather and Robertson 2015). It is unlikely that crop damage can be eliminated, but damage levels can be reduced by modification of farming practices or by adopting crop protection measures based on sound ecological principles. Beeton (1977) noted that shooting has been found to be ineffective in controlling damage to sorghum crops by Little Corellas *Cacatua sanguinea* in the Kimberley division of Western Australia, and he recommended that ripening of the crops be timed to coincide with lowest density levels in the annual population cycle of the cockatoos, with initial plantings being in plots overflown first by flocks on their regular flight paths; these plantings would ripen first, and the birds could be encouraged to feed there instead of moving to other parts of the crop. Ford (1990) points out that decoy crops are a more effective and economical means of protecting oilseed crops than shooting or scaring, and Garnett (1999) records outstanding success with a decoy crop to protect peanut crops from damage by Red-tailed Black Cockatoos *Calyptorhynchus banksii* at Lakeland, on Cape York Peninsula, north Queensland. Too often, I hear of sunflowers or other crops highly attractive to parrots being planted near tree-lined watercourses, a preferred habitat for many parrots and cockatoos, and high damage levels are the inevitable consequence. In such situations, where severe damage will occur, the substitution of unattractive crops, such as canola or soybeans, becomes a viable option. Investigations have shown that erecting covering nets to protect fruit crops becomes cost-effective in new, intensively-grown orchards, but is unlikely to be cost-effective for old orchards (Sinclair 1990). The likelihood of damage by birds must be considered in the initial assessment when planning any crop-growing enterprise and remedial measures formulated as an integral component of management practices.

When determining conservation needs, it must be recognised that parrots are long-lived birds, especially the large cockatoos, so without adequate recruitment a population may remain apparently stable in overall numbers, though there is a uniform ageing of individuals, and eventually this brings about a dramatic and sudden decline. Saunders (1979b) warned that unless landuse practices in Western Australia follow sensible guidelines for the management of woodlands, in order to maintain a continuing supply of mature trees, Carnaby's Black Cockatoo *Calyptorhynchus latirostris* and some other hollow-nesting species are likely to be threatened with extinction. Studies of the Superb Parrot *Polytelis swainsonii* and the Regent Parrot *P. anthopeplus* in southeastern Australia have highlighted these same needs for nesting habitat. Indeed, the requirements are applicable in most parts of Australia, and there is urgent need to safeguard habitats through sound landuse practices. Additionally, there always is the need to adopt special protective measures for species that are threatened, either throughout or in parts of their ranges.

INDONESIAN AND PHILIPPINE ARCHIPELAGOS

In the Indonesian Archipelago, the adverse impact on forest birds, particularly parrots, of a combination of deforestation and trapping or nest-robbing for the live-bird market is most evident, and perhaps more than elsewhere these two pressures acting in concert are responsible for catastrophic declines in many populations, especially on smaller islands. With its long tradition of bird-keeping and its high levels of biological diversity under strong pressure from habitat loss, Indonesia holds the greatest number of globally threatened bird species for which trade is registered as a significant threat (Eaton *et al.* 2015). All populations of the critically endangered Yellow-crested Cockatoo *Cacatua sulphurea* have suffered dramatic declines because of widespread landclearance coupled with unsustainable trapping for the live-bird market, and the cockatoos now have disappeared from some parts of the former range and are near extinction on some smaller islands. An apparent recovery of the population on Sumba, in Nusa Tenggara, following an imposition of international trade controls in 1994, has been cited as evidence of the impact of the live-bird trade, but I would point out that trapping still continues, the recovery in numbers has not been strong, and it remains questionable whether the cockatoos can persist long-term in the fragmented forest patches (see Cahill *et al.* 2006).

Although the Yellow-crested Cockatoo may be the 'flag' species in any discussion on the decline of parrot populations in Indonesia, there are other species being pushed to near extinction and, as favoured species disappear, others are targeted as next-best substitutes. The threatened status of the White-crested Cockatoo *Cacatua alba* has been upgraded to endangered, and the Salmon-crested Cockatoo *C. moluccensis*, which probably now is extinct in parts of its former range, remains vulnerable.

I expect that pressures on both of these cockatoos will increase as numbers of Yellow-crested Cockatoos continue to decline. Similar patterns are emerging in pressures on *Eos* and *Lorius* species, which also are very popular as pet birds. Having disappeared from much of their former range, Red and Blue Lories *Eos histrio* now are confined to the Talaud Islands, where they survive almost exclusively on Karakelang, and in 1999 research suggested that 1000–2000 birds were being taken from this island each year (Birdlife International 2016). Widespread and intense trapping for the live-bird market, together with ongoing deforestation brought about by logging, oil drilling and hydroelectric power schemes, are suspected to be causing a rapid and continuing decline in numbers of Purple-naped Lories *Lorius domicella*, which seem always to have occurred at low densities, and I suspect that, as these lories become less readily available to trappers, trafficking pressures will increase on the vulnerable Chattering Lory *Lorius garrulus*. Replacement species are appearing in the markets, and in recent years some forms of the widespread, ubiquitous Rainbow Lorikeet *Trichoglossus haematodus* have been exported in large numbers, resulting in the decline or probable loss of populations on small islands.

The Blue-fronted Lorikeet *Charmosyna toxopei* and Wallace's Hanging Parrot *Loriculus flosculus* are endangered species about which very little is known, and I suspect that, as more information becomes available, other little-known species could be categorised as endangered. Other *Loriculus* species are considered to be near threatened, but on Sangihe Island, where extensive deforestation has contributed to the apparent extinction of Red and Blue Lories *Eos histrio*, long-term survival of the Sangihe Hanging Parrot *L. catamene* could be at risk. Confined to Buru, southern Maluku, the poorly known Black-lored Parrot *Tanygnathus gramineus* is said to be locally common in mountainous areas in the northwest of the island, but that assessment is well outdated, and the parrots are seldom recorded, so their present status is uncertain.

I cannot be optimistic about future prospects for parrot populations in the Indonesian Archipelago because human population growth is impacting so severely on the natural environment. Deforestation continues unabated, and excessive exploitation for the live-bird market continues because there is little or no enforcement of protection measures. There seems to be little understanding of the principles of sustained harvesting, with no awareness of the need for recruitment levels to be higher than rates of harvesting, and the same nests of Yellow-crested Cockatoos or Purple-crowned Lories are robbed annually. The viability of national parks and reserves is compromised by illegal logging and the encroachment of settlement, often aided by greed and corruption at all administrative levels.

In the Philippine Archipelago, parrots feature prominently among avian frugivores that are seriously threatened by massive loss of habitat and direct exploitation through hunting and the pet trade (Española *et al.* 2013). As in the Indonesian Archipelago, the viability of national parks and reserves is compromised by illegal logging and the encroachment of settlement, so their effectiveness in curtailing deforestation is limited. The impact of habitat loss is well highlighted by extinction of distinctive forms of the widespread Philippine Hanging Parrot or Colasisi *Loriculus philippensis* on Cebu and Siquijor, two islands where native forest has been almost totally cleared. Surveys undertaken at 14 sites across the island of Luzon, northern Philippines, between December 2009 and September 2010, revealed that on this large island parrots are indeed rare, and no regional density estimate for any of the *Tanygnathus* or *Prioniturus* species exceeded 3.5 birds per km², which was significantly lower than similar density estimates for large parrots in Wallacea, Indonesia, and Papua New Guinea (Española *et al.* 2013). Elsewhere, the Blue-naped Parrot *Tanygnathus lucionensis* formerly was more widespread and common than at present, though still found in good numbers on Palawan, and the Blue-backed Parrot *T. sumatranus* formerly was fairly widespread but now is very rare throughout most of its range, except in the Sulu Archipelago (Kennedy *et al.* 2000).

The adverse impact of habitat loss is compounded by excessive exploitation for the live-bird market, and the endemic Philippine Cockatoo *Cacatua haematuropygia* is most at risk. These cockatoos have been targeted by trappers and nest-robbers, even with 'ownership' of nests being claimed so that chicks can be taken year after year, resulting in the widespread extinction of local populations. They formerly occurred on 52 islands in the archipelago, but in the period 1989–1994 were found on just eight islands (Birdlife International 2001). Some success has been achieved in arresting the decline on Palawan and offshore islets, but the species remains critically endangered.

Recommendations stemming from the surveys undertaken in Luzon would seem to be applicable throughout the Philippine Archipelago (Española *et al.* 2013). In response to habitat loss it is recommended that research be undertaken to determine whether land adjacent to existing reserves could be rehabilitated as part of the national forest estate. To assess the impact of hunting and trapping, comparative studies at sites with different levels of hunting and trapping could be carried out, complemented by interviews with local people.

NEW ZEALAND AND OFFSHORE ISLANDS

The adverse impact on landbirds of habitat loss coupled with predation by introduced mammals, especially rats and mustelids, has been particularly severe in New Zealand, and parrots have featured very prominently in catastrophic population declines. This is particularly distressing, for the larger species, namely the Kakapo *Strigops habroptila* and the two *Nestor* species, are so distinctive as to warrant differentiation in a separate superfamily, and molecular analyses have revealed that they comprise the basal clade with a sister relationship to all other parrots, having diverged in the Lower Eocene (approximately 50 million years ago) or possibly the Cretaceous (82 million years ago) when New Zealand broke away from Gondwana (Wright *et al.* 2008; White *et al.* 2011).

Decline of the Kakapo commenced with the arrival of Maori colonists some 1300 years ago. Tipa (2006) notes that dogs were used by Maori hunters to capture and kill the parrots for meat, skins and feathers. The meat was regarded as a great delicacy, feathers were used for head decorations, and skins with the feathers intact were softened to make dress capes and cloaks for wives and daughters of leading chiefs. Dogs became feral and, together with Pacific Rats or kiore *Rattus exulans*, probably initiated a long history of predation by introduced mammals. At the time of European colonisation during the late 18th century, Kakapo still occurred in parts of central North Island, while on South Island they were extinct in eastern regions, but remained common in some higher rainfall areas in northern, western and southern regions. After 1880, the situation changed dramatically when the introduction and establishment of three species of mustelids, Black Rats *Rattus rattus*, several species of deer, and Australian Brush-tailed

Possums *Trichosurus vulpecula* brought about a rapid decline. Predation by mustelids undoubtedly was the main cause of this decline, with the spread of Black Rats, which would have taken eggs and chicks as well as competed with adults for food, being a contributing factor. By the early 1900s, Kakapo had become extinct in North Island and had disappeared from northern parts of South Island, with the last remaining birds confined to remote subalpine valleys in Fiordland in southwestern South Island, where a few aged males were scattered in less accessible parts of a few remote valleys (Merton 1985; Butler 1989). In 1977, discovery of a population on Stewart Island provided a new opportunity to save the parrots from extinction and, following evidence of predation by feral cats, the remaining birds, 38 males and 24 females, were translocated to predator-free offshore islands. Kakapo now are extinct in their natural range, and all surviving birds are in intensively managed populations on three offshore islands where, at the time of writing, the population comprised 125 birds.

New Zealand Kakas *Nestor meridionalis* were widely hunted by Maori for food and for feathers used to make cloaks, but they remained widespread and common at the time of European settlement. However, it again was widespread landclearance following European settlement and the introduction of mammalian predators that were responsible for dramatic declines in numbers. Clearfelling of native forests, competition from introduced Brush-tailed Possums and rats for fruits, competition from introduced *Vespula* wasps for honeydew, and predation by introduced mustelids have been identified as major causes of strong declines in local populations. Nesting females are killed more easily by mustelids and possums, so producing the very skewed sex ratio in favour of males recorded in many populations on the main islands (Heather and Robertson 2015). On North Island, Kaka now are either absent or rare in most regions, with remnant populations restricted to larger tracts of podocarp-hardwood forest in central districts and some predator-free offshore islands, and on South Island, mostly west of the Southern Alps, they are widespread, though in declining numbers, through larger tracts of *Nothofagus* and podocarp-hardwood forests (in Powlesland *et al.* 2009). While intensive and sustained pest control has dramatically improved the density and sex ratio of populations in a few districts where mammalian pest control is carried out, Kaka are declining throughout the remainder of the range. Subfossil remains of *Nestor*, possibly of this species, were recorded on the Chatham Islands, where the birds are thought to have become extinct before 1871.

Kea *Nestor notabilis* are confined to the highlands of South Island, though subfossil remains indicate that they formerly occurred also in North Island, and in past years they were heavily persecuted. For almost a century landholders claimed that the parrots attack and kill sheep, a claim that repeatedly was unduly emphasised in the media. On the basis of this claim a government bounty was paid on them, and an estimated 150 000 birds were killed between 1880 and 1971 (Heather and Robertson 2015). After a review of the evidence was undertaken in the 1960s, it was concluded that although Kea do attack sheep trapped in snow, sick or injured sheep, and sheep they mistake as dead, they are not an important predator and the payment of bounties could not be justified. They were afforded partial protection in 1971, and were fully protected in 1986.

Closely related to *Nestor* are the four extinct *Nelepsittacus* parrots described from fossil remains recovered from early Miocene Saint Bathans Fauna from the Lower Bannockburn Formation in Otago, South Island (Worthy *et al.* 2011). Flora from the Saint Bathans Fauna fossil beds indicate that these parrots occurred in subtropical rainforest.

Closely related to platycercine parrots in Australia and with an apparent centre of origin in New Zealand, *Cyanoramphus* parakeets are believed to have dispersed across ocean barriers to many islands, and their historical range extends from islands in the South Pacific Ocean south to subantarctic islands. This range has contracted following the extinction of populations on some islands, and habitat degradation, exacerbated by predation by introduced mammals, particularly rats and mustelids, is impacting severely on many remaining populations, including those in mainland New Zealand. The Red-fronted Parakeet *C. novaezelandiae* is the most widespread species, and it remains common on Stewart Island and some other islands, including those in the subantarctic, but in New Zealand it is common only on offshore islands, having been almost totally extirpated on North and South Islands by introduced rats and mustelids. Yellow-crowned Parakeets *C. auriceps* have fared a little better on North and South Islands, presumably because of their more arboreal foraging and nesting habits, but nowhere are they particularly abundant, and they also are more numerous on offshore islands. The closely allied, possibly conspecific Chatham Islands Parakeet *C. forbesi* is the most critically endangered of New Zealand parakeets, with phenotypical or 'pure' birds surviving only as a very small population in remnant forest on Little Mangere, in the Chatham Islands, and as a low proportion, estimated at 50 to 120 birds in 1996, of 'yellow-crowned' birds on adjacent Mangere Island, where hybridisation with Red-fronted Parakeets continues to pose a very serious threat. Wild populations of Orange-fronted Parakeets *C. malherbi* survive in only three river valleys in South Island, where predator-control measures are being implemented, but the birds remain vulnerable to predation. Captive-bred birds have been used to establish populations of Orange-fronted Parakeets on predator-free offshore islands, and this program seems to have been successful.

New Zealand wildlife authorities have undertaken translocation programs to establish populations of a number of endemic bird species, including parrots, on predator-free offshore islands, and for the most part these programs have been very successful. Efforts are being made also to reintroduce species in reserves or sanctuaries on the main North and South Islands, where predator-control measures are being undertaken, and as part of this project Kaka and Red-fronted Parakeets have been introduced successfully to Zealandia Wildlife Sanctuary near Wellington.

TROPICAL PACIFIC OCEAN ISLANDS

Landbirds occurring on oceanic islands are particularly vulnerable, and this is most evident in the plight of landbirds, including parrots, found on islands in the tropical Pacific Ocean. Steadman (1989) points out that in eastern Polynesia more species of landbirds have become extinct since human arrival some 2000 years ago than now survive in the region, and nearly all species that have survived no longer occur on most of the islands that previously made up their natural or 'pre-human' range. Fossils from archaeological sites have extended the ranges of many extant species by hundred to thousands of kilometres, and species traditionally regarded as endemic to one or several islands were widespread until human influences caused their disappearance from island after island. Steadman identifies the extinction-causing processes as:

(i) habitat alteration, especially the clearing of forests, browsing and grazing by introduced herbivores, and the introduction of non-native plants;
(ii) predation by humans;
(iii) predation by human-introduced mammals;
(iv) diseases introduced by non-native birds.

Habitat alteration, particularly large-scale burning of forests by prehistoric peoples, occurred nearly throughout the Pacific, and European colonists, with their metal tools, were able to cut down forests. Pigs were the only large herbivores kept by prehistoric peoples, but other large grazing or browsing herbivores, including goats, sheep, cattle and horses were brought by European colonists, and all were particularly destructive in native forests. The first Polynesians and Melanesians introduced non-native plants, including coconuts, bananas, yams and breadfruit, but a vast array of non-native plants have been introduced in the past 200 years, and now it is difficult to find tracts of native forest not infiltrated by non-native plants.

Throughout Polynesia and Melanesia, parrots were among landbirds favoured as food items by prehistoric peoples, and they were hunted also for their feathers to be used in making ceremonial cloaks and head adornments. *Vini* lorikeets and *Prosopeia* parrots were targeted for their red feathers and, in the course of trade, live birds were transported to islands well outside the natural range, often across vast expanses of ocean. Populations of Rimatara Lorikeets *Vini kuhlii* in the Line Islands, Kiribati, and of Maroon Shining Parrots *Prosopeia tabuensis* on islands in the Tongatapu Group, Tonga, originate from such prehistoric introductions. The presence on Aitutaki, in the southern Cook Islands, of probably the largest population of Blue Lorikeets *Vini peruviana* is thought to have originated with pet birds brought from the Society Islands.

Pacific Rats *Rattus exulans*, dogs and pigs arrived with the first peoples nearly throughout the Pacific Ocean islands and, although there are few specific studies, these mammals are likely to have been particularly destructive to birds that nested on or near the ground. The most profound impact of mammalian predation came with the arrival of Black Rats *Rattus rattus*, Norway Rats *R. norvegicus* and cats that reached the islands with Europeans, and Black Rats have been implicated in the disappearance of *Vini* lorikeets from many islands as well as the probable extinction of the Red-throated Lorikeet *Charmosyna amabilis* on some Fijian islands. It is possible that predation by rats and cats contributed also to extirpation of *Cyanoramphus* parakeets on Tahiti and Raiatea, in the Society Islands, though the Raiatea Parakeet *C. ulietanus* was extremely scarce at the time of Captain Cook's second voyage in 1773 or 1774, when the only two specimens were collected, so Polynesian hunting may have been the primary cause of its disappearance. Holyoak and Thibault (1984) suggest that disappearance of Blue Lorikeets from much of the Society Islands correlates with spread of the introduced Swamp Harrier *Circus approximans* but, as pointed out by Seitre and Seitre (1992), the lorikeets have become extinct or are declining on islands where the harriers have not occurred, so predation by rats is the main danger.

Steadman acknowledges that the role of introduced diseases in the decline of island birds may be an important one, but much research remains to be done. Chickens were the only continental birds brought to the Pacific Ocean islands prehistorically, but in recent times there have been other introductions, most notably of Common Mynas *Acridotheres tristis* which are widespread and well established in many island groups. Populations of these mynas and of other introduced species should be investigated as possible sources of avian diseases. Holyoak (1973a) raised the possibility of avian malaria being introduced with alien mosquitoes transported in aircraft.

Conservation concern has focused on the plight of *Vini* lorikeets. Fossil remains reveal that these spectacularly plumaged parrots formerly were widely dispersed on island groups throughout the tropical Pacific Ocean, but only Blue-crowned Lorikeets *V. australis* remain widespread and fairly common, though in some parts of their range numbers are declining. *V. sinotoi* and *V. vidivici* have been described from bones collected in archaeological sites in the Marquesas Islands, where they probably occurred sympatrically with the extant Ultramarine Lorikeet *V. ultramarina* until after the first Polynesian arrival some 2000 years ago (Steadman and Zarriello 1987). Both of these extinct species are known also from fossil remains obtained on Huahine, in the Society Islands (Steadman 1989). Steadman (1985) reports that among fossil remains of birds recovered on Mangaia in the southern Cook Islands is an almost complete sternum inseparable from that of the Rimatara Lorikeet *V. kuhlii*, which occurs naturally some 480 km to the east, on Rimatara in the Austral Islands, and also it cannot be distinguished with certainty from the sternum of the Blue-crowned Lorikeet, the nearest occurrence of which is on Niue Island, approximately 1300 km to the west-northwest. Of particular concern are two species, each of which now is confined to one island and so is highly vulnerable to the accidental introduction of Black Rats. Henderson Island is without human habitation, and its remoteness provides some safeguards for the Henderson Island Lorikeet *V. stepheni*, but there always is the possibility of rats reaching the island with unauthorised landings of yachts. Far more threatened is the Ultramarine Lorikeet, which formerly occurred on five of the Marquesas Islands, but now survives only on Ua Huka. Establishment of a second population by translocating birds to another island was undertaken in the Marquesas Islands in the early 1990s, when 29 lorikeets were taken to Fatu Hiva, and initially this was successful, but after Black Rats reached Fatu Hiva in 2000 the translocated population declined rapidly to disappear entirely by 2007. A successful translocation of Rimatara Lorikeets from Rimatara, in the Austral Islands, to Atiu, in the southern Cook Islands, was undertaken in 2007, with successful breeding soon being recorded on Atiu, and within 15 months of the translocation birds were seen on the neighbouring island of Mitiaro, some 50 km from Atiu (Lieberman and McCormack 2008).

The Ouvéa Parakeet *Eunymphicus uvaeensis* also occurs only on a single, fairly small island, being restricted to Ouvéa, in the Loyalty Islands, off New Caledonia, and attempts were made to establish another population on Lifou, also in the Loyalty Islands. These attempts failed, and subsequent confirmation of the presence of Black Rats on Lifou has ruled out any further translocations. Despite the mixed results achieved so far with translocation projects, I suggest that establishment of a translocated population of Rimatara Lorikeets on Atiu, in the southern Cook Islands, demonstrates that this strategy can be successful if the introduction of rats can be avoided and it should remain a management option.

SUPERFAMILY
STRIGOPOIDEA Bonaparte

There is longstanding recognition of the distinctiveness of three genera of large parrots confined to New Zealand and outlying islands, including Norfolk Island, and this high distinctiveness is confirmed by molecular analyses revealing that they comprise the basal clade with a sister relationship to all other parrots (Wright *et al.* 2008; White *et al.* 2011). It is hypothesised that their divergence dates to the Lower Eocene (approximately 50 million years ago), or possibly to the Cretaceous (82 million years ago) when New Zealand broke away from Gondwana.

Of the described species, only the Kea *Nestor notabilis* survives in reasonable numbers on the South Island of New Zealand. Of the other two *Nestor* species, one is extinct and the other is endangered. The sole *Strigops* species is critically endangered, and *Nelepsittacus* species are known only from subfossil remains.

FAMILY **STRIGOPIDAE** Bonaparte

Morphologically and behaviourally, the sole species in this family is well differentiated from other parrots. It is by weight the largest parrot, is nocturnal and flightless, and has a lek mating system (Merton *et al.* 1984). The most distinctive anatomical characteristic is the presence of only a rudimentary keel to the sternum, rendering the bird flightless, but the broad, rounded wings are used for balance while climbing, running or leaping back to the ground from a low tree or shrub.

GENUS ***Strigops*** G. R. Gray

Strigops G. R. Gray, *Gen. Bds*, **2**, 1845, p. [426], pl. CV. Type, by monotypy, *Strigops habroptilus* G. R. Gray.

Diagnostic features of this genus are the same as for the family. The distinctively 'mottled' green, yellow and brownish plumage colouration features a prominent facial disc with bristle-like feathers at the base of the bill giving an 'owl-like' appearance. Features of the stout, blunt bill are prominent transverse ridges on the underside of the upper mandible and a strongly ridged lower mandible. The conspicuously swollen cere is unfeathered. The sexes can be differentiated by weight, males being significantly heavier than females, and juveniles are duller than adults.

Strigops is endemic to New Zealand.

ENDANGERED

Kakapo

Strigops habroptila G. R. Gray

Strigops habroptilus G. R. Gray, *Gen. Bds*, **2**, 1845, p. [427], pl. CV. 'One of the islands in the South Pacific' = New Zealand, (restricted to Dusky Sound, South Island, by Mathews and Iredale, *Ibis*, 1913, p. 426–427.)

OTHER NAMES Owl Parrot, Night Parrot (inappropriate as it is the name of an unrelated Australian species).

DESCRIPTION Length 64 cm. Weight males 1.5–3.0 kg, females 950 g–1.6 kg.
ADULTS Upperparts bright green irregularly barred and streaked dark brown and yellow; underparts greenish-yellow irregularly barred paler lemon-yellow and light brown; variable yellow superciliary band; forehead and facial disc yellowish-brown, feathers with paler centres and darker tips; wings and tail dull green barred brown and dull yellow; bill pale grey becoming horn-coloured on lower mandible; iris dark brown; legs bluish-grey in males, pale pinkish-grey in females.
10 males: wing 265–284 (272.6) mm, tail 203–250 (222.6) mm, exp. cul. 34–43 (38.0) mm, tarsus 45–57 (51.8) mm.
10 females: wing 252–285 (269.1) mm, tail 205–238 (220.3) mm, exp. cul. 35–37 (36.4) mm, tarsus 46–56 (49.6) mm.
JUVENILES Like adults, but general plumage colouration duller; forehead and less prominent facial disc more brownish, less yellowish; pointed, not rounded tips to outermost primaries; bill pale yellowish-white.

DISTRIBUTION New Zealand, where formerly widespread on North, South and Stewart Islands, but now present on only three offshore islands – Anchor and Codfish Islands in the south, and Little Barrier Island in the north.

GENERAL NOTES Oliver (1955) noted that the Kakapo was mentioned by Ernst Dieffenbach in his *Travels in New Zealand*, published in 1843, but he had not seen a specimen, and the first

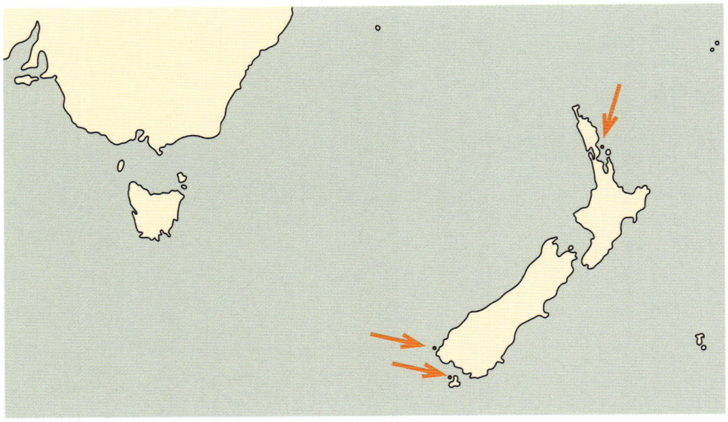

specimen to reach Europe was collected at Dusky Sound, South Island, and forwarded to England in 1845. Another specimen collected at Milford Sound, South Island, in 1851, was presented to the British Museum and, in 1861–1862, Julius von Haast and James Hector found the parrots to be quite numerous along the west coast of South Island. In the second edition of his *A History of the Birds of New Zealand*, published in 1888, Walter Buller referred to the Kakapo as being remarkable, and made reference to the comment made by Philip Sclater, then Secretary of the Zoological Society of London:

> *This is one of the very remarkable forms peculiar to New Zealand, and has been appropriately termed an Owl Parrot. Dr Sclater refers to it as 'one of the most wonderful, perhaps, of all living birds'. As its name* Stringops *indicates, its face bears a superficial likeness to that of an Owl. In all the essential characteristics of structure it is a true Parrot; but in the possession of a facial disk (in which it differs from all other known parrots), in the soft texture of its plumage, and especially in its decidedly nocturnal habits, it betrays a striking resemblance to the Owl tribe.*

The more we learn about this amazing parrot, the more appropriate becomes the statement by Sclater. That the Kakapo is unlike any other parrot in appearance and behaviour was brought home to me in September 1976, on Maud Island, where I observed males that had been brought to the island from Fiordland. Their large size, strong legs and feet, and cryptic plumage colouration are features indicative of a terrestrial lifestyle and, while watching a bird moving through the undergrowth in a peculiarly shuffling manner, I was left with the impression that here indeed was a parrot that had survived virtually unchanged from ancient times in an isolated environment without predators.

STATUS Subfossil bones show that prior to Maori settlement in New Zealand, approximately 1300 years ago, Kakapo were widespread on North and South Islands, being most common in high-rainfall areas west of the main divide where rimu *Dacrydium cupressinum* and *Nothofagus* beech trees dominated the forest canopies (Worthy and Holdaway 2002). Subfossil remains have not been found on Stewart Island, so it has been suggested that the population there may have originated from liberations by Maori in pre-European times or from Fiordland birds liberated there in the 1880s (Powlesland *et al.* 2006). The range contracted substantially after Maori settlement, especially in North Island. Tipa (2006) notes that dogs were used by Maori hunters to capture and kill the parrots for meat, skins and feathers. The meat was regarded as a delicacy, feathers were used for head adornments, and skins with the feathers intact were softened to make dress capes and cloaks. Dogs became feral and, together with Pacific Rats or kiore *Rattus exulans*, probably initiated a long history of predation by introduced mammals. Widespread burning of forest, scrublands and tussock grasslands by Maori may have caused local extinctions, especially in drier eastern and central regions of South Island (Powlesland *et al.* 2006).

At the time of European colonisation during the late 18th century, Kakapo still occurred in parts of central North Island, and were reported from the Hunua Range, south of Auckland, as late as 1912 (McKenzie 1979). They were extinct in eastern regions of South Island, but remained common in some higher-rainfall areas in northern, western and southern regions (Lloyd and Powlesland 1994). Landclearance by European settlers after the early 1800s probably did not impact significantly on the population because by then the parrots already were confined to scattered remote areas, and predation by feral cats, together with habitat destruction by feral herbivores, probably were responsible for only local declines. After 1880, the situation changed dramatically when the introduction and establishment of three species of mustelids, Black Rats *Rattus rattus*, several species of deer, and Australian Brush-tailed Possums *Trichosurus vulpecula* brought about a rapid decline. Predation by mustelids undoubtedly was the main cause of this decline, with the spread of Black Rats, which would have taken eggs and chicks as well as competed with adults for food, being a contributing factor. Competition for food and habitat degradation by deer and possums would have inhibited successful breeding. By the early 1900s, Kakapo had become extinct in North Island and had disappeared from northern parts of South Island, with the last remaining stronghold confined to remote subalpine valleys in Fiordland in southwestern South Island. Decline in the Fiordland population had been noted in the late 1800s, and it also succumbed to the same pressures of predation and habitat degradation, so that by the 1970s it was reduced to a few aged males scattered in less accessible parts of a few remote valleys (Merton 1985; Butler 1989). Of 18 males found in Fiordland during the 1970s, five were transferred to offshore islands, and none of the remaining 13 are known to have survived beyond 1987 (Clout and Craig 1995).

Earlier reports of sightings of Kakapo on Stewart Island were investigated in 1977, and a population estimated to comprise between 100 and 200 birds of both sexes was found in an area of some 8000 ha of modified scrubland and forest (Powlesland *et al.* 1995). There are no mustelids on Stewart Island, but feral cats are present, and evidence of predation by feral cats was recorded, with the remains of 15 Kakapo killed by cats being found during 1980–1982 (Best and Powlesland 1985). The predation rate on radio-tagged adult Kakapo reached 56 per cent during 1981–1982, so an intensive cat control program was commenced in 1982, and no further remains of cat-killed Kakapo were found. However, to ensure survival of remaining birds, 38 males and 24 females, the total number that could be found, were translocated to islands free of mustelids and cats, but not Pacific Rats.

The last accepted record for North Island was from the Huiarau Range in 1927, the last South Island record was of three males in

Plate 1
Kakapo *Strigops habroptila* (adult)

F. KNIGHT 08

Fiordland in 1987, and the last record for Stewart Island was of a female found and transferred to Codfish Island in 1997. Kakapo now are extinct in their natural range, and all surviving birds are in intensively managed populations on three offshore islands. In 1995, the population reached its lowest level, with just 30 males and 21 females surviving, but in that year the introduction of new management procedures, including the eradication of Pacific Rats from Codfish Island, the provision of food supplements to nesting females, and intensive monitoring with intervention when necessary, brought about a significant increase in nesting success (Elliott *et al.* 2006). Furthermore, the survival of adults on these islands has been remarkably high, averaging only about 1.3 per cent mortality per annum. In November 2005, the population comprised 45 males and 41 females. with 40 or 45 per cent being reared on the islands. At the time of writing the population of 125 birds comprises many younger birds and an equal ratio of males and females. All females are from Stewart Island, and the resulting lack of genetic diversity has given rise to high levels of infertility, which is a major problem inhibiting the recovery effort. On average only 60 per cent of eggs are fertile, and in the 2014 breeding season the fertility level dropped to 40 per cent. Artificial insemination is being used in an effort to improve the fertility rate and, if the behaviour of sitting females causes concern, their eggs are taken for artificial incubation. Wherever possible chicks are reared by their mother or by a foster mother, but they will be handreared if necessary.

The Kakapo is listed on Appendix I of the Convention on International Trade in Endangered Species of Wild Fauna and Flora (CITES).

HABITATS Williams (1956) noted that Kakapo occurred in higher-rainfall areas up to 1250 m, and primarily were birds of the mossy *Nothofagus* forest, especially where these forests adjoined open country along river flats or the subalpine scrub belt bordering meadows of *Danthonia* tussocks above the tree-line. Williams also cautioned that records of higher abundance at altitudes of 1000 m and above may be a reflection of signs of their presence, namely tracks, bowls and chewed vegetation, being more easily detected in less densely vegetated places. There were occasional sightings in predominantly podocarp forest, but such occurrences were few. Other authors make mention of a prevalence of Rimu *Dacrydium cupressinum*, a favoured food tree, in preferred habitats.

Atkinson and Merton (2006) report that in February–March 1974, ground traverses, line transects and aerial photographs were used to describe and map the home ranges, or more appropriately the 'core home ranges', of two Kakapo living in the Esperance River valley, Fiordland, South Island. The first home range of only 1.8 ha was at an altitude of approximately 700 m on a gently sloping river terrace on the right bank of the river, well above normal flood levels and on the more sunny side of the valley. Averaging a metre in height, scrub featuring snow totara *Podocarpus nivalis* and of varying composition covered most of this home range, with 90 per cent of the canopy comprising shrubs and the groundcover consisting of grasses, herbs, mosses and ferns. Some 51 per cent of the shrubs were species with succulent berries, including *Coprosma, Gaultheria* and *Coriaria* shrubs and weeping matipo *Myrsine divaricata*, with tall *Chionochloa* tussocks scattered throughout and herbaceous plants abundant in the groundcover. The more intensively used central part of the range was characterised by a significantly lower canopy, together with an increased prevalence of turpentine bush *Dracophyllum uniflorum* and a lower abundance of muttonbird scrub *Brachyglottis rotundifolia* and weeping matipo. The distribution of feeding signs, feathers and droppings, as well as nocturnal observations, indicated that this Kakapo was not penetrating the forest beyond a distance of 500 m, and large tracts of the forest margin showed no evidence of the bird's presence. Some 4 ha in area, the second home range was at a distance of 0.5 km from the first home range, and at approximately 850 m altitude was sited on a very steep, northwest facing valley wall with shrubland dominated by kiokio fern *Blechnum novae-zelandiae* and low forest of silver beech *Nothofagus menziesii* being alternating parallel stands of vegetation up and down the slope. The fern shrubland averaged 0.7 m in height, with 40 per cent of the canopy comprising shrubs, 30 per cent consisting mostly of kiokio fern and mountain flax *Phormium cookianum*, and the remainder comprising herbs, grasses and sedges. The forest of silver beech varied in height from 21 m in the lower half of the range down to 6 m at the treeline and, of the many species in the understorey, *Pseudopanax simplex*, hūpiro *Coprosma foetidissima* and weeping matipo were particularly common. Where the canopy was more open, muttonbird scrub and juvenile silver beech were present. More than 50 per cent of the groundcover consisted of the mountain kiokio *Blechnum procerum*, and there were numerous fallen tree trunks, stumps and loose rocks. Foliose lichens were abundant on trunks of the beech trees, and mosses were prevalent on the northern side of these trunks. No part of this home range appeared to be used more intensively than other sections.

On Maud Island, in Pelorus Sound, South Island, seasonal habitat selection by translocated Kakapo was investigated between December 2000 and October 2001 (Walsh *et al.* 2006). Significant selection occurred in summer for lowland indigenous forest dominated by pepper tree *Macropiper excelsum*, kohekohe *Dysoxylum spectabile* and nikau *Rhapolostylus sapida*, and in autumn for treeland scrub dominated by five-finger *Pseudopanax arboreus*, rangiora *Brachyglottis repanda* and bracken *Pteridium esculentum*. A 15 ha plantation of Monterey pine *Pinus radiata* also was preferred in summer, but when foraging in this plantation during the night, the birds moved to daytime roosts in a narrow band of dense lowland scrub within the plantation. Other habitats, which mostly were more open and included managed and unmanaged pasture, were seasonally or totally avoided. Translocated Kakapo now are present on Codfish Island, some 3 km off the northwestern coast of Stewart Island, and here the dominant habitat is mixed podocarp-rata *Metrosideros umbellata* forest (Farrimond *et al.* 2006b). Moorhouse and Powlesland (1991) report that on Little Barrier Island, in Hauraki Gulf, North Island, translocated Kakapo prefer wetter, high-altitude areas, and again there was some seasonal change in habitat preferences, with birds occupying lowland kanuka *Kunzea ericoides*-manuka *Leptospermum scoparium* forest in winter and moving to higher-altitude northern rata *Metrosideros robusta*-tawa *Beilschmiedia tawa* forest during summer.

HABITS Kakapo are solitary, and throughout much of the year single birds remain in overlapping home ranges of varying sizes. Early reports of groups of up to six birds found roosting together are likely to have been of family groups or groups of juveniles (in Higgins 1999). On Maud Island, between December 2000 and October 2001, when translocated birds were radio-tracked at night in each of four seasons, home ranges were estimated for four adult males, three juvenile males and two juvenile females in each season, and for nine females in summer, and it was determined that home ranges varied from 1.8 ha to 145.0 ha, with seasonal variation being in response to patterning of vegetation, and smallest ranges were occupied in winter (Walsh *et al.* 2006). On Little Barrier Island, home ranges were estimated at 21 ha to 38 ha in area, and on Codfish Island, where the population was denser, mean home ranges of adult females and recently-fledged juveniles were estimated to be approximately 15 ha, with only one home range of an adult female exceeding 30 ha (in Powlesland *et al.* 2006). On Stewart Island, home ranges were estimated at 15 ha to 50 ha in area, or 50 ha for males and 100 ha for females (in Powlesland *et al.* 2006). It is pointed out that these estimates of home range areas were determined for remnant populations of Kakapo, probably at a lower density than would have occurred in similar habitats when the birds were more numerous prior to the introduction of mammalian predators.

In addition to being solitary, kakapo are highly unsociable, and birds of either sex maintain a distance of several hundred metres from their neighbours. Agonistic behaviour includes vigorous chasing, harsh vocalisation and fighting, the last being more prevalent among displaying males. Although vocal rather than visual signals seem to be utilised more commonly to repel approaching neighbours, there are a range of visual displays which have been observed to be given by a few displaying males (in Higgins 1999). When performing a threat display, a bird faces its opponent, raises and holds its wings over its back until almost touching, raises its head and neck feathers with the neck outstretched and the bill partly opened, while making a low, prolonged growl. In response, the defensive bird stands upright while facing its opponent, and has one foot raised in preparation for striking downwards.

Kakapo are nocturnal and, when not breeding, they normally retreat to daytime roosts about an hour before sunrise, becoming active about an hour after sunset. Routinely sited on steep, damp, shaded, densely vegetated slopes, often with a southerly aspect, favoured roost sites are dark, dry, sheltered from strong winds, sufficiently spacious to allow the bird to turn, and typically are on the ground under dense cover or in natural cavities, such as caves or hollow tree stumps and logs. Birds occasionally will roost on a horizontal branch in the forest understorey or canopy with dense overhead cover, and individual birds often show a preference for roosting either on or above the ground. Some sites may be used repeatedly or irregularly for lengthy periods, and in Fiordland desiccated droppings still remain on the floors of some former roost sites even though the birds died out there about 40 years ago (in Powlesland *et al.* 2006). There are occasional reports of birds seen feeding during daylight hours in the early morning and late afternoon, and it is probable that these birds were breeding females with month-old chicks in the nest or without access to good supplies of food. On Maud Island, well after sunrise, I watched a bird come to take supplementary food from a feeding tray.

Although flightless and predominantly terrestrial, Kakapo are excellent climbers, using their outspread wings for balance as they grasp branches with the bill and feet to advance up into the higher foliage of trees and shrubs in search of food, and they are quite capable of moving through the canopy from one tree to the next, again with much wing-flapping for balance. The outspread wings are used also to cushion their descent back to the ground. They walk along the ground with a rather rapid jogging gait, and when travelling downhill they can move with surprising speed. When walking, they adopt a near-horizontal posture with the lower rictal bristles in contact with the ground, and it is suspected that sensory perception may be important when traversing unfamiliar terrain in darkness and when feeding at night on certain foods, such as plants with spiny foliage. When travelling longer distances they follow regularly used, well-defined paths, and these paths are characteristic signs of the presence or former presence of the birds, often persisting in districts long after the local population has died out. Individuals of either sex occasionally walk up to several kilometres from their core home ranges to sites where they may remain for several days, weeks or even months, and such long-distance movements may be made to take advantage of a localised abundance of food or because snow cover forces a descent to a lower altitude.

CALLS Except during times of courtship, when displaying males are highly vocal, Kakapo seldom call, though they do have a varied repertoire of notes. In addition to a mechanical bill-clicking, they emit pig-like grunts and squeals, duck-like *warks* and a donkey-like braying, as well as a characteristic *skraark*, frequently heard from competing males during courtship, hissing and humming sounds, and harsh screeches (in Powlesland *et al.* 2006). Females are known to make a 'hoarse cough', and when handled females and juveniles produce a series of drawn-out vibrant, croak-like distress calls, whereas males usually are silent.

Most frequently heard is the loud, low-frequency, resonant 'booming' from displaying males during courtship. From a standing position, the male progressively lowers his head and inflates grossly the thoracic region, simultaneously producing three or four quiet measured grunts on a descending scale at intervals of approximately two seconds. At maximum thoracic inflation, the grunts give way to soft booms, which increase in volume so that maximum intensity is reached after six to eight booms. The loud booms are repeated 15 to 20 times, at intervals of about two seconds, after which there is a gradual decrease in volume during a further 10 to 20 booms. During booming the head is held low, and often is almost totally engulfed by the inflated air-sac. After each booming sequence, the male assumes an erect, alert and motionless stance for some 20 to 30 seconds before repeating the sequence. A local dialect has been detected in 'booming', with males in Fiordland consistently producing a disyllabic boom in contrast to the monosyllabic boom from males on Stewart Island (in Higgins 1999). Fiordland males also boomed more slowly and with greater resonance than males on Stewart Island.

DIET AND FEEDING Kakapo are exclusively herbivorous, and an early record of remains of a lizard being found in a crop sample has been discounted. Best (1984) points out that they are versatile feeders with highly variable feeding patterns, and many foods are taken opportunistically. This flexible feeding pattern also allows utilisation of a wide spectrum of seasonal foods, some of which may be available only for short periods or in intermittent years. Also, they are selective feeders, not only in their seasonal exploitation of food plants and parts of those plants, but also in selecting individual plants. One particular shrub or grass tussock often is selected, but the same adjoining shrubs or tussocks are ignored. The diet comprises leaves or leaf buds, twigs or stems, bark, nectar, fruits, seeds, fern pinnae, roots and rhizomes or tubers, and moss or fungi and, apart from seeds of some fruiting plants and grass tussocks, only finely ground material is ingested (in Powlesland *et al.* 2006). The blunt, strong bill, with transverse ridges on the underside of the upper mandible, and the short, thick tongue are adapted for cutting, crushing and grinding to extract juices from fibrous plant material (Kirk *et al.* 1993). A less musculature gizzard suggests that vigorous chewing is needed to break down coarse plant material into small fragments suitable for ingestion. Much fibrous material is ejected in the form of characteristic kidney-shaped, tightly compressed pellets known as 'chews', and these may stay attached to the plant or more usually litter the ground underneath, where they may remain intact for long periods, in due course becoming bleached by the sun to become telltale signs of the presence or former presence of the birds. Similarly, recent grubbings in the soil for roots and rhizomes can be identified by bill impressions and the remnant fragments of material that have been crushed, presumably to squeeze out the starchy contents. The bill is used also to strip away bark of trees and shrubs to gain access to the cambium layer and sap. The feet are used to hold food items but rarely to lift items to the bill. Blades of grass are eaten from the tip downwards, but tillers of snow tussock *Chionochloa rigida* and coarse leafy sedges are bitten from the plants and the bases eaten. Seeds of grasses are obtained by passing the stem through the partly opened bill. Berries, fruits and most seeds are pulped in the bill, but some hard seeds are swallowed intact (Powlesland *et al.* 2006).

Between June 1977 and April 1980, Best examined signs left by the parrots on Stewart Island and determined that favoured food plants were the club mosses *Lycopodiella ramulosa* and *Lycopodium fastigiatum*, narrow club fern *Schizaea fistulosa*, soft water fern *Blechnum minus* and mountain kiokio *B. procerum*, mingimingi *Cyathodes juniperiana*, inaka *Dracophyllum longifolium*, leatherwood *Olearia colensoi*, veined sun orchid *Thelymitra venosa*, the sedge *Oreobolus strictus*, saw sedge *Gahnia procera* and tall sedge *Carex appressa*. Although feeding occurred frequently throughout the year at all identified food plants, except *Oreobolus strictus* and *Gahnia procera*, variation in the utilisation of different parts of the plants occurred both seasonally and annually. Analyses of plant remains present in 648 faecal droppings collected on Stewart and Codfish Islands, between 1977 and 1998, revealed that females were more likely to have eaten podocarp fruits or leaves of trees and shrubs, while males favoured fern and *Lycopodium* rhizomes, monocots in breeding years, and fruits of manuka *Leptospermum scoparium* in non-breeding years (Wilson *et al.* 2006). Podocarp fruits were much more prevalent in the diet in breeding years than in non-breeding years, and when podocarp fruits were available in breeding years the birds were less likely to eat several other foods. Fronds of *Blechnum* ferns appeared more frequently in the droppings of females in breeding years than in non-breeding years. As amounts of podocarp fruits increased in the diets of both sexes during the summers of breeding years, the incidence of many other foods declined. Quantities of leaves of Hall's totara *Podocarpus hallii* in the diet of females increased during summer in non-breeding years, but decreased in breeding years. It is suspected that these findings probably reflect differences in foraging behaviour between the sexes, particularly in breeding years, when females gather food for their nestlings mainly in forest habitats, while males feed mainly in tussock and subalpine scrub in the vicinity of their track-and-bowl systems (in Powlesland *et al.* 2006).

There is strong evidence that irregular breeding by Kakapo is in synchrony with the heavy fruiting of favoured food plants, including rimu *Dacrydium cupressinum* and pink pine *Halocarpus biformis*. On Codfish Island, nesting has occurred only when more than 10 per cent of rimu branch tips bore green fruits in October (Harper *et al.* 2006). On the island, in the breeding season of 2002, chicks were fed almost exclusively on entire rimu fruits, each of which weighs approximately 0.1 grams, and in early April one female fed 90 grams of food to each of her two six-weeks-old nestlings, so she must have eaten about 16 fruits per minute during the 113 minutes spent feeding before coming to the nest (Cottam *et al.* 2006). On Stewart Island, in 1981, a female rearing two nestlings fed heavily on ripe rimu fruits (Powlesland *et al.* 1992). On Maud Island, in 1998, the diet of three chicks in a nest in a plantation of Monterey pine *Pinus radiata* comprised a high proportion of pine needles in addition to food supplements, and supplementary foods were present with *Blechnum* fern fronds and seeds of kauri *Agathis australis* in the diet of nestlings on Little Barrier Island in 1990–1991 (in Powlesland *et al.* 2006). In 1989, a program of supplementary feeding was introduced in the hope that it would increase the low rate of reproduction, and high-protein foods were chosen because it was found that podocarp fruits contain twice the level of protein present in other foods (Rubenheimer and Simpson 2006). Although supplementary feeding arguably has had a positive impact on reproduction, the desired increases have not been achieved, and it has been suggested that calcium may be a limiting factor and foods with low macronutrient-calcium ratios may be more effective.

BREEDING Courtship behaviour of the male is a lek display involving social displaying and 'booming' from excavated bowl-like depressions or 'courts' on an arena or traditional display ground (Merton *et al.* 1984). This highly specialised form of courtship behaviour has not been reported in any other parrot, in any other New Zealand bird, or in any other flightless bird. Each male typically has four to 12 bowls or 'courts', which invariably are against some form of reflecting surface, which must aid in projection of the booming call, enabling it to be heard up to 4 km away. Bowls are excavated and meticulously maintained by the male exclusively for courtship purposes. All bowls used by one male are linked by well formed tracks that are pruned of vegetation and may be spaced for 60 m or more along the crest of a ridge, or they may occupy an area only 10–15 m in diameter on a hilltop. A track-and-bowl system may be within the core home range of a male, but more often is located up to several kilometres away. At least one bowl in each system usually is at a commanding position, such as on a ridge crest overlooking a valley. Distances between systems of neighbouring males may vary from 15 m to several hundred metres, with groups of up to 50 systems extending over several km². Each male occupies and defends his system from one breeding season to the next. 'Booming' always is carried out in, or very near to a bowl, and is a long-distance communication to attract females. The intensity and duration of 'booming' is influenced by social stimulation and light intensity, with an apparently sole survivor at one locality heard calling for an average 5 hours 35 minutes on clear nights, whereas calling by a group of males at another locality averaged 9 hours on clear nights, up to 12 hours 40 minutes on overcast nights, and under extreme conditions continued for up to 17 successive hours.

'Booming' usually commences about an hour after nightfall and ceases about an hour before sunrise. Neighbouring males frequently duet, but it is not known whether this is by chance or is deliberate. 'Booming' undoubtedly serves to advertise the location of the calling male, and possibly stimulates and synchronises sexual activity in both sexes. In years when nesting occurs, males typical 'boom' nightly between late November and early April, but in years when there is no nesting 'booming' seldom lasts more than two or three months, mostly in January–February.

Nesting occurs in synchrony with abundant fruiting or seeding of preferred food plants, mostly podocarps, at intervals of two to seven years, and it was recorded in 12 of the 29 years between 1977, when female Kakapo were rediscovered on Stewart Island, and 2006 (Powlesland *et al.* 2006). Mating has occurred only in years when more than 90 per cent of males attended track-and-bowl systems, or males commenced 'booming' before 5 December (Eason *et al.* 2006). On Codfish Island, it is a profuse fruiting of rimu *Dacrydium cupressinum* that triggers nesting, and it is likely that females become sexually active when they commence feeding on green fruits in late October and early November, with mating and egg-laying taking place in January and February (Harper *et al.* 2006). Because it has not been practical to collect and store sufficient green rimu fruit, an attempt to induce nesting was made in a year of poor fruiting by feeding freeze-dried fruits of rimu and kahikatea *Dacrycarpus dacrydioides* to females, but nesting did not take place, and in another year nesting did take place when females were fed a supplementary diet of pine cones and green walnut, but this was not a conclusive result as there was also a moderate crop of rimu fruit at the same time. No nesting was recorded on Little Barrier Island in the seven years before supplementary feeding commenced in 1989, but subsequently was recorded in five of the next eight years and, although no information on the size of fruit crops on the island in those years is available, several potential food plants, including kauri *Agathis australis* and *Nothofagus* trees, are known to have produced above-average fruit or seed crops elsewhere in New Zealand during these same five years (in Harper *et al.* 2006).

Mating occurs mainly between early January and early March, and on Codfish Island the peak period of mating was between 15 January and 4 February. Lengthy periods of nocturnal observations at track-and-bowl systems on Stewart Island revealed that visits by females are brief, infrequent and asynchronous between individuals (Merton *et al.* 1984). In response to the presence of females, males at their track-and-bowl systems in Fiordland indulged in elaborate, ritualised displays and posturing, which included side-to-side rocking movements, and highlighting the exquisitely marked upperparts by walking backwards with the head slightly bowed and with vigorous flapping of the opened wings. Females may spend just a few hours in an arena and mate with only one male, or stay for up to five nights and mate with two or more males. There is no pair formation, and females leave to undertake all nesting activities without any participation by males.

All of the 67 nests found between 1980 and 2005 have been within or near to the home range of the nesting female, with 57 being in forest and 10 in open scrubland or tussock grasslands (Powlesland *et al.* 2006). On Stewart Island, where open scrubland and tussock grasslands are more common, three of the five nests were in this habitat, but on offshore islands this habitat is poorly represented, and all but seven of the 62 nests were in forest. Nests are in natural cavities at or below ground level and, of the 67 nests, 40 were in caves or burrows, 18 in standing tree-stumps, five under tussocks or other dense groundcover vegetation, and four in fallen tree-trunks. Nesting cavities averaged 500 mm in diameter, with a range of about 400 mm to more than 2 m, and heights averaged about 450 mm, with a range of 350 mm to more than 1 m. Each nest comprised a shallow bowl-shaped depression some 250–300 mm in diameter and 20-50 mm deep. Distances from the cavity entrance to the nest averaged about 500 mm, with a range of about 200 mm to more than 4 m. The nest is lined to a depth of about 50 mm with material gathered by the female on site, often from walls and ceiling of the cavity, and usually consists of dry rotted wood chips, bark, peat and soil. There are three instances of nests being reused by the same female.

Egg-laying takes place in January–February, or occasionally in March. Up to 2005, the median date of egg-laying on Little Barrier Island was 24 January, and on Codfish and Pearl Islands combined was 7 February (Eason *et al.* 2006). Clutches normally comprise two to four, or rarely only one or up to five eggs, with the first egg usually being laid about 10 days after mating, and then eggs are laid at intervals of about three days. Of 67 clutches found between 1981 and 2005, nine were single-egg clutches, 19 comprised two eggs, 36 comprised three eggs, and three comprised four eggs. Females that lay early in the breeding season are capable of laying a replacement clutch if the first clutch is lost or is removed for artificial incubation. Incubation averages 30 days in duration and commences with laying of the first egg, though initially the female sits intermittently. On Stewart Island, the behaviour of an incubating female was monitored by radio-tracking, and each night she left to feed, on 11 occasions leaving 1–2 hours after sunset and on three occasions at about midnight, with her time away from the nest averaging 108 minutes (Powlesland *et al.* 1992). On Codfish Island, the behaviour of incubating females was monitored by remote infra-red video systems, and prior to completion of the clutch females usually left the nest nightly, often for up to three hours at a time. Subsequently, they usually were away from the nest for less than 90 minutes on any night until hatching. During the first 10 to 14 nights they normally left the nest every second or third night to feed, but during the next two weeks feeding absences usually occurred nightly. When food was not readily available during the first two weeks, females left their nests every night, occasionally leaving before nightfall and not returning until after sunrise. In 1997, on Codfish Island, the behaviour of females with access to supplementary food was compared with the behaviour of females without access, and it was found that females with access to supplementary food were absent from their nests about one-third of the duration of females that relied on natural foods (in Powlesland *et al.* 2006).

Hatching occurs during late February to early April, and generally is at intervals corresponding to intervals of egg-laying, though occasionally eggs laid at intervals of three days will hatch at intervals of one or two days. At hatching, chicks are blind, helpless and weigh about 28 to 35 grams (Farrimond *et al.* 2006a). Growth and weight gain is rapid during the first three to eight weeks, after which weight continues to increase at a more modest rate until reaching a maximum just prior to fledging when their weights mostly declined (Eason *et al.* 2006). From the night they hatch, nestlings are left unattended for several hours each night while the female feeds. On Stewart Island, three to five weeks old nestlings in two nests received little brooding each night, and were fed nightly at about 2300 hours and again at 0200 hours, each feeding visit by the female lasting 10 to 40 minutes. Females returned to these nests at dawn, again fed the chicks, and then brooded throughout the day (Powlesland *et al.* 1992). At eight to 10 weeks after hatching, daytime brooding ceased, with the females roosting away from the nests and feeding visits of 10 to 20 minutes duration by the females took place once or twice during the night and again at dawn. Fledging occurs in late May to early June, and is at an average of 76 days after hatching, though it is a gradual process and from about nine weeks of age chicks may spend progressively longer periods a few metres away from the nest until finally not returning. On Codfish Island, the average weight of chicks at fledging was 1813 grams, with males averaging 1984 grams and females averaging 1715 grams (Farrimond *et al.* 2006a). When 11 or 12 weeks old, most fledglings roost on the ground more than 10 m away from the nest, and a month later they usually are roosting up to 100 m away from the nest. Young birds continue to receive some food from their mother for at least three months after leaving the nest, and they remain within or near to her home range until six to eight months of age, often roosting in close association with her.

Efforts to boost numbers of Kakapo are hindered by the low rate of nesting success. Of 166 eggs monitored during 1981 to 2005, 66 were infertile, 68 hatched, and 48 nestlings fledged, with all but one of these fledglings reaching independence (in Powlesland *et al.* 2006). It is suggested that the high level of infertility is likely to be a result of very low genetic diversity in the surviving population. Most of the 47 chicks that reached independence would not have survived without human intervention, with 17 being partially or entirely handraised, and at least 16 of the remainder receiving supplementary food from their mothers. Survival following independence is high, with the annual survival of juveniles in the managed population at 2005 estimated to be 0.91 (Elliott 2006). The age at which Kakapo reach sexual maturity is not known, but on Little Barrier Island five-years old males had established track-and-bowl systems, and their 'booming' was similar to that of older males. Three known-age females bred for the first time at 10 and 11 years but, as abundant fruit crops did not occur in preceding years, it is possible that they may have been sexually mature at an earlier age. It is suspected that males may reach sexual maturity at about five years of age, and females at about nine years of age. Kakapo are thought to be exceedingly long-lived. The oldest known bird, an adult male captured in Fiordland in 1975, died in December 2010, and Clout (2006) speculated that he may have been almost 100 years old.

EGGS Merton (1985) describes the eggs as being ovate and with a fine granular surface. Measurements of 122 eggs laid on offshore islands between 1990 and 2002 are 50.7 (46.3–55.8) × 38.3 (35.0–41.1) mm (Eason *et al.* 2006).

FAMILY **NESTORIDAE** Bonaparte

The genera *Nelepsittacus* and *Nestor* are included in this family. *Nelepsittacus* is known only from subfossil material found in Miocene deposits in New Zealand (Worthy *et al.* 2011). *Nestor* currently is confined to New Zealand, including some outlying islands, but formerly occurred also on Norfolk Island and nearby islands in the Tasman Sea.

GENUS ***Nestor*** Lesson

Nestor Lesson, *Traité d'Orn.*, livr. 3, 1830. Type, by monotypy and tautonymy, *Nestor novaezelandiae* Lesson = *Psittacus nestor* Latham = *Psittacus meridionalis* Gmelin.

Parrots belonging to this genus are large, robust birds with short squarish tails, and shafts of the tail-feathers project beyond the webs to form spine-like tips. The massive bill is longer than it is deep, and on the underside of the upper mandible there are longitudinal serrations. The cere is partly covered with hairy feathers, and the tongue is tipped with a hair-like fringe. Sexual dimorphism is slight, and juveniles resemble adults.

ENDANGERED

New Zealand Kaka

Nestor meridionalis (Gmelin)

Psittacus meridionalis Gmelin, *Syst. Nat.*, **1**, pt. 1, 1788, p. 333. (New Zealand = Dusky Sound, South Island, *ex* Latham.)

OTHER NAME Kaka.

DESCRIPTION Length 45 cm. Weight males 383–575 g, females 494–500 g.
ADULTS Forehead, crown and occiput pale greyish-white, almost white and feathers sometimes margined dull green; nape greyish-brown, feathers marked olive-brown; neck and abdomen brownish-red, noticeably more crimson on hindneck where feathers finely tipped yellow and dark brown; breast olive-brown; ear-coverts orange-yellow; back and wings greenish-brown, on mantle some feathers tipped red; tail-coverts crimson barred dark brown; underwing-coverts and undersides of flight feathers scarlet; tail brownish, tipped paler; brownish-grey bill longer and more curved in male; iris dark brown; legs dark grey.
12 males: wing 265–306 (289.3) mm, tail 151–185 (164.6) mm, exp. cul. 49–54 (51.7) mm, tarsus 35–44 (37.7) mm.
8 females: wing 267–304 (290.0) mm, tail 151–190 (165.4) mm, exp. cul. 42–54 (48.5) mm, tarsus 35–40 (37.3) mm.
JUVENILES Like adults, but overall plumage colouration duller; base of lower mandible yellow.

DISTRIBUTION New Zealand, where present on North, South and Stewart Islands, and some offshore islands; formerly on Chatham Islands.

SUBSPECIES

1. *N. m. meridionalis* (Gmelin)

The nominate subspecies, as described above, occurs on South Island, mostly west of the Southern Alps, on Stewart Island, and on some offshore islands, including Big South Cape and Codfish Islands and the Chetwode Islands.

2. *N. m. septentrionalis* Lorenz

Nestor septentrionalis Lorenz, *Verh. zool.-bot. Ges. Wien*, **46**, 1896, p. 198. (North Island, New Zealand.)
ADULTS overall plumage colouration duller; forehead, crown and

occiput dull grey; back and wings darker olive-brown, feathers edged darker brown; breast darker brown; less crimson on hindneck; slightly smaller size; weight males 320–555 grams, females 210–455 grams.
8 males: wing 267–272 (269.4) mm, tail 140–161 (152.1) mm, exp. cul. 41–50 (46.6) mm, tarsus 34–36 (35.0) mm.
7 females: wing 256–271 (263.0) mm, tail 140–154 (150.6) mm, exp. cul. 39–44 (41.3) mm, tarsus 33–37 (35.1) mm.
Occurs on North Island, and some offshore islands, including Hen and Chickens, Great Barrier, Little Barrier, Mayor and Kapiti Islands.

STATUS Subfossil material indicate that New Zealand Kaka were widespread and common throughout North Island, South Island and Stewart Island, and they were abundant at the time of European settlement, but by the early 1900s they had declined to localised flocks (Heather and Robertson 2015). Buller (1888) noted that they were 'met with, more or less in every part of the country', and it was claimed that thousands were killed when large flocks came to feed on favoured seasonal foods, such as nectar from flowering rata *Metrosideros robusta*. They were widely hunted by Maori for food and for feathers used to make cloaks, and principal methods of capture were with snared perches placed around a decoy bird or spearing and striking birds attracted by a decoy (in Oliver 1955). However, it is widespread landclearance following European settlement and the introduction of mammalian predators that were responsible for dramatic declines in numbers. Clearfelling of native forests, competition from introduced Brush-tailed Possums *Trichosurus vulpecula* and rats for fruits, competition from introduced *Vespula* wasps for honeydew, and predation by introduced mustelids have been identified as major causes of strong declines in local populations. Nesting females are killed more easily by mustelids and possums, so producing the very skewed sex ratio in favour of males recorded in many populations on the main islands (Heather and Robertson 2015).

Beggs and Wilson (1991) report that at Big Bush State Forest, in the north of South Island, between November and early January in each year from 1985 to 1990, at a site in *Nothofagus* forest modified by introduced browsing mammals, field studies were undertaken to examine factors influencing the ability of Kaka to breed successfully where habitat quality has been reduced by introduced browsing mammals and *Vespula* social wasps, and where introduced predators are present. These investigations were undertaken because earlier studies of a remnant population on South Island had suggested that energy may be a limiting factor for the following reasons:

(i) Kaka may expend more energy than they gain while obtaining one of their major protein foods, larvae of Kanuka Longhorn Beetles *Ochrocydus huttoni*, which are extracted from the trunks of living mountain beech *Nothofagus solandri*,
(ii) *Nothofagus* forests have a small range of foods with a high net energy return, and this range is reduced further by browsing mammals such as possums,
(iii) the availability of the major remaining source of energy, honeydew produced by the scale insect *Ultracoelostoma assimile* is greatly reduced by introduced *Vespula* wasps at certain times of the year.

Of the 31 birds fitted with radio transmitters during the six years of the study, only two pairs attempted to breed, and only one of these attempts was successful with two chicks fledging. The only successful nest was protected by wrapping aluminium sheeting around the trunk of the tree to exclude mustelids, but the female from this nest used a different site the next year and was killed, presumably by a Stoat *Mustela erminea*. Not only were Kaka absent when wasps were numerous, no birds fitted with radio transmitters were seen feeding on honeydew, so for about four months of the year the presence of wasps so reduced the energy value of honeydew, and probably the potential feeding rate, that honeydew was no longer a worthwhile source of energy for the parrots. It is suspected that this loss of honeydew as an energy source was the reason for only one successful nesting in the six years of the study. Beggs and Wilson point out that, although the life expectancy of Kaka is not known, successful breeding by only one pair of 31 birds in five years is a low reproductive rate for even a long-lived species, and is unlikely to compensate for mortality by predation.

On North Island, Kaka now are either absent or rare in most regions, with remnant populations restricted to larger tracts of podocarp-hardwood forest in central districts and some predator-free offshore islands, and on South Island, mostly west of the Southern Alps, they are widespread, though in declining numbers, through larger tracts of *Nothofagus* and podocarp-hardwood forests (in Powlesland *et al*. 2009). While intensive and sustained pest control has dramatically improved the density and sex ratio of populations in a few districts where mammalian pest control is carried out, Kaka are declining throughout the remainder of the range. It is estimated that in the past 100 years there has been a decline of approximately 60 per cent in the total population and in habitat area, and the present population is estimated at 1000–5000 mature individuals.

Subfossil remains, possibly of this species, were recorded on the Chatham Islands, where the birds are thought to have become extinct before 1871 (in Higgins 1999).

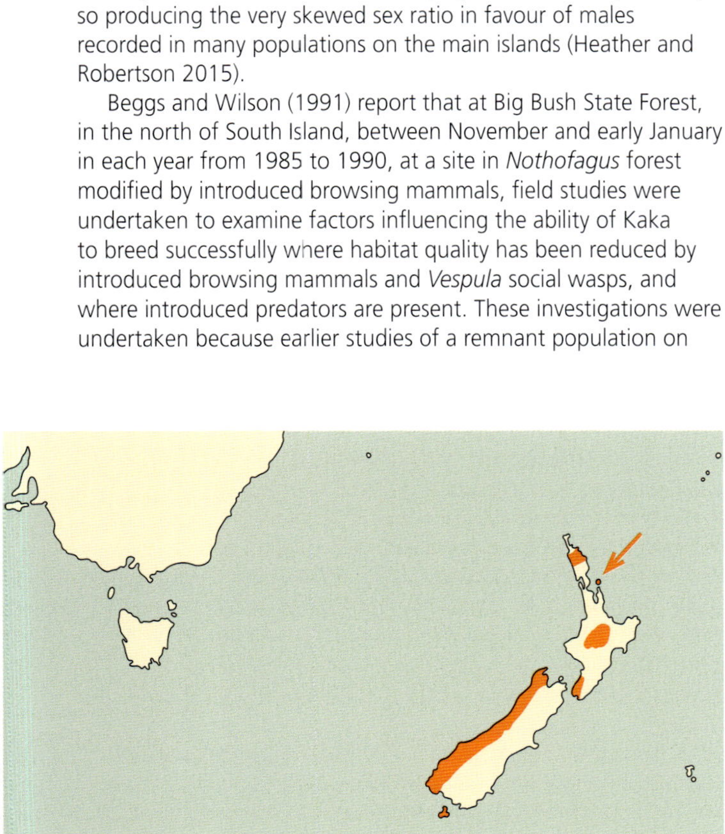

HABITATS Kaka are forest birds, frequenting mostly unmodified indigenous, temperate rainforest, but occurring also in low broadleaf forests comprising trees such as tawa *Beilshmiedia tawa*, hinau *Elaeocarpus dentatus*, kohekhe *Dysoxylum spectabile*, rata *Metrosideros excelsa* and kamahi *Weinmannia racemosa*, usually with much epiphytic growth and a diverse undergrowth featuring a prevalence of ferns, and tall podocarp forest dominated by totara *Podocarpus totara*, rimu *Dacrydium cupressinum*, rata, miro *Prumnopitys ferruginea* and matai *P. taxifolia* (in Higgins 1999). They have been recorded also in *Nothofagus* forests with a sparse understorey or mixed podocarp-broadleaf-beech forests, but rarely in *Leptospermum* scrubland or logged areas, including stands of dense regrowth. They usually do not adapt to altered landscapes, but small numbers may remain in remnant patches of forest, and occasionally venture into farmlands, orchards, and urban gardens or parklands, the last being utilised more commonly on Stewart Island.

Between 1996 and 2002, studies of nesting behaviour were undertaken at two sites on North Island and two sites on South Island (Powlesland *et al.* 2009). The first study site on North Island was in a stand of dense podocarp forest at Pureora Forest Park, where the tall forest cover consisted of emergent podocarps, particularly rimu, kahikatea *Dacrycarpus dacrydioides* and matai, over a canopy of mainly tawa. At the second study site in Whirinaki Forest Park, the forest comprised mainly tall stands of podocarp and hardwood trees, such as emergent rimu, with scattered kahikatea, matai and miro, again over a canopy of mainly tawa, while on some exposed ridges and slopes the podocarp-hardwood forest was replaced by a mixture of rimu, miro, red beech *Nothofagus fusca* and Hall's totara *Podocarpus hallii*. At the first study site on South Island, comprising 825 ha of beech forest on the western side of the St Arnaud Range and bordered partly by farmland and Lake Rotoiti, the lower slopes were dominated by red beech and silver beech *Nothofagus menziesii*, with mountain beech *N. solandri* and kanuka *Kunzea ericoides* at sites with poor drainage, and upper slopes by silver beech and mountain beech grading to pure mountain beech at the tree-line. In Fiordland National Park, the second study site on South Island was in the Eglinton River valley, a steeply-sided valley with a 0.5–1.0 km wide flat floor, where open, grassy areas were near the river, and elsewhere, up to 1000 m elevation, the covering forest comprised pure stands of silver beech along the margins giving way to stands of red beech farther into the forest, with mountain beech occurring occasionally in the canopy at low altitudes and becoming more common with increased altitude. At all four study sites, an important component of the forest canopy was formed by masting tree species which produce a super-abundance of fruits or seeds at irregular intervals of one to seven years, but little or no fruits or seeds in intervening years.

Moorhouse (1997) points out that on Kapiti Island, where foraging studies were undertaken, a diverse mosaic of indigenous forest and shrubland covers 81 per cent of the island, and dominant tree species include five-finger *Pseudopanax arboreus*, kanuka *Kunzea ericoides*, kohekohe *Dysoxylum spectabile* and tawa *Beilschmiedia tawa*. Also common are hinau *Elaeocarpus dentatus*, karaka *Corynocarpus laevigatus* and mahoe *Melicytis ramiflorus*, with northern rata *Metrosideros robusta* and pukatea *Laurelia novae-zelandiae* occurring as scattered emergents.

MOVEMENTS Kaka wander about, though this was more regular in earlier times when they were more common and widespread, and then they could appear suddenly in an area from which they had been absent for some time. Roberts (1953) reported that the lightkeeper on Burgess Island, in the Mokohinau Group, claimed that they passed through the island at a certain time of the year and perched near the lighthouse for only one day before moving south the following night. They commute freely between the northern islands, and wander to coastal forests and towns on North Island (Heather and Robertson 2015). Also on North Island, translocations to Zealandia Wildlife Sanctuary and Pukaha Mount Bruce National Wildlife Centre have been successful, with the local population currently estimated at 200 to 250 birds, and now Kaka commonly are encountered in suburbs of nearby Wellington. One bird reared in Zealandia Wildlife Sanctuary dispersed to Pukaha Mount Bruce National Wildlife Centre, 100 km to the northeast, before returning to the Sanctuary (Miskelly *et al.* 2005). On South Island, birds occasionally come east of the Southern Alps into coastal Canterbury and Otago.

HABITS In pairs or small parties of up to 10 birds, Kaka are highly conspicuous when in flight above the forest canopy, but when sitting quietly in the treetops or feeding alone amongst the foliage they can be difficult to detect, their presence often being betrayed only by occasional calling or by discarded food scraps falling to the ground underneath. Beggs and Wilson (1991) report that at Big Bush State Forest, in the north of South Island, the daily activities of 31 birds fitted with radio transmitters were monitored, and it was found that they were active from at least 30 minutes before sunrise to 30 minutes after sunset. For males, time spent sleeping was almost 10 hours in summer and slightly more than 12 hours in winter, time spent perching was six hours in summer and almost three hours in winter, time spent walking was one hour in summer and 30 minutes in winter, time spent flying was only 30 minutes in summer and 10 minutes in winter, time spent extracting beetle larvae from tree-trunks was three hours in summer and four hours in winter, and time spent foraging for other foods was five hours in summer and very slightly longer in winter. For females, time spent sleeping was nine hours in summer and a little more than 12 hours in winter, time spent perching was five hours in summer and about 90 minutes in winter, time spent walking was about 90 minutes in summer and about 40 minutes in winter, time spent flying was approximately 40 minutes in summer and about 20 minutes in winter, and time spent foraging for other foods was about eight hours in summer and approximately nine hours in winter, but no females were observed extracting beetle larvae from tree-trunks.

Heather and Robertson (2015) note that Kaka delight in acrobatics and aerobatics, jumping through the trees and using the bill as an extra appendage when climbing, and tumbling through the air for enjoyment. There are reports of highly conspicuous pre-roosting flights becoming common towards nightfall, and Sibson (1947) noted that at each evening on Little Barrier Island, while some birds stayed behind and called continuously from the forest, parties of eight to 12 birds would fly from the flat high over the ridges and sometimes out to sea and back. On Kapiti Island, these parrots are remarkably tame, and will accept food from the hands of visitors to the sanctuary. Males can be intolerant of a near approach of other males, especially in the vicinity of the nest. Threat displays usually are subtle, such as turning to face the oncoming bird, but at times this is accompanied by harsh calling and lifting the wings to display the red underwing-coverts (in Higgins 1999).

CALLS Kaka are more vocal in the early morning and at dusk, and outside the breeding season little calling is heard at other times. Their calls comprise a wide variety of melodious whistling notes and harsh grating sounds. The most frequently heard call is a loud, harsh *ka-aa* usually given in flight, and contact calls comprise high-pitched, usually monosyllabic whistles and warbles ranging in volume from loud to muted. A harsh, loud *kraak* is

---

Plate 2
New Zealand Kaka *Nestor meridionalis* (adults)
UPPER *N. m. meridionalis*
LOWER *N. m. septentrionalis*

F. KNIGHT '15

given when alarmed, and a harsh, but muted *ngaak-ngnaak* in disputes between feeding birds. Feeding may be accompanied by low musical *karrunk* notes. When indicating a potential nesting site to the female, a male gives a high-pitched squeaking *tsee-tsee-tsee*, and when soliciting food from the male, a female utters a loud, harsh whining *kree-kree*, interspersed with a guttural *aa-aa* (in Higgins 1999).

DIET AND FEEDING The varied diet comprises fruits, seeds, nectar and honeydew, sap from tree-trunks, and insects and their larvae. Kaka use their strong bills to crush seeds and to tear away loose bark or dig into decaying tree-trunks to extract larvae of wood-boring insects or to feed on exposed sap, and use their brush-tipped tongue to take honeydew excreted by scale insects or nectar from flowers (Heather and Robertson 2015).

Moorhouse (1997) reports that on Kapiti Island, between March 1991 and January 1992, the food of Kaka was recorded while observing the foraging activities of nine radio-tagged birds. The proportion of time these birds spent foraging relative to other activities varied considerably from month to month, increasing from March to June and from September to November, and there was marked seasonal variation in the diet. Although most birds foraged primarily for wood-boring insect larvae, all birds located in March were feeding primarily, and in most cases exclusively, on seeds of hinau *Elaeocarpus dentatis*. Two males continued to feed primarily on hinau seeds until June, but all females stopped eating them after March. Most birds spent about 30 per cent of their observed foraging activity feeding on nectar or pollen from five-finger *Pseudopanax arboreus* in August and nectar and pollen from a variety of sources between November and January. Foraging for seeds of tawa *Beilschmiedia tawa* also increased in this same November to January period. Little of the observed foraging activity was devoted to taking fruits and, although most birds fed on fruits of hinau and tawa later in the year, they usually spent less than 12 per cent of their observed foraging activity on these foods. Only one bird was seen to take fruits of five-finger, but nestling faeces frequently contained these seeds, indicating that it was fed to nestlings. Invertebrates eaten on Kapiti Island included coleopteran and lepidopteran larvae, probably of kanuka longhorn beetle *Ochrocydus huttonii* and puriri moth *Charagia virescens*, extracted from dead branches and live wood, larvae of an undescribed gall midge extracted from galls, six-penny scale insects *Ctenochiton viridis* gleaned from leaves, and unidentified invertebrates, possibly including nymphs of *Hemideina* tree-weta extracted from dead twigs.

Beggs and Wilson (1987) report that during field studies undertaken at Big Bush State Forest, in the north of South Island, Kaka were found to spend 35 per cent of their feeding time digging into the trunks of mountain beech *Nothofagus solandri* to extract the larvae of kanuka longhorn beetles and an unknown amount of time searching for these larvae. When extracted, a majority of these large larvae were at their most vulnerable stage, packed into pupal chambers and unable to escape into lower portions of their tunnels. They were also more easily detected at this stage by the plugged exit hole on the surface of the tree-trunk. To extract a single larva, a parrot spent up to two hours of vigorous activity while clinging to the vertical tree-trunk, so it was hypothesised that more energy was expended obtaining one larva than was gained by eating it, and that the parrots would need to balance their energy budget by gaining energy from other more energetically economic food such as seeds, fruits or honeydew. At Big Bush State Forest, the main alternative food was honeydew excreted by the scale insect *Ultracoelostoma assimile*, and in spring, when the average feeding time for Kaka was nine hours, the three hours of this time spent taking honeydew was sufficient to give them most of their daily energy requirements (Beggs and Wilson 1991).

Sap-feeding by Kaka is concentrated in late winter and spring, when very few nectar sources are available, temperatures are lower and energy demands are high. O'Donnell and Dilks (1989) point out that the parrots use two distinct techniques to feed on sap: They strip bark from a branch or trunk to expose the surface cambium, and then lick the sap exudate from the surface. With the second, more specialised technique, a parrot starts by peeling and discarding loose bark from a tree-trunk. It then uses the lower mandible to prise a 'trapdoor' through the remaining bark and to gouge a series of tiny holes into the superficial layer of yellow cambium some 6–10 mm below the surface. At times, when hanging upside down, a bird will lever the 'trapdoor' downwards. The resulting horizontal marks are very distinctive, occurring on trunks and large branches from ground level up to high in the canopy, and they persist for a very long time, prompting the suggestion that old scars on large trees could date from pre-European times. O'Donnell and Dilks recall that in early August 1984, in South Westland, South Island, sap-feeding by two parrots was observed for almost an hour, and spells of prising away bark averaged 6.6 minutes, with intervening times spent revisiting older scars to extract sap which had leaked from the wounds. An average 1.8 minutes was spent licking up sap, and one new scar was visited at least four times in the period of observation. Sap-feeding has been recorded at both native and introduced trees, and Charles (2012) notes that trees with very few scars have been found near to heavily-scarred trees, suggesting that the parrots may test a number of trees before selecting a preferred tree for feeding. In native forests, sap-feeding has been recorded mostly at Myrtaceae trees, southern rata *Metrosideros umbellata* and *Metrosideros* hybrids, kanuka *Kunzea ericoides*, rimu *Dacrydium cupressinum*, matai *Prumnopitys taxifolia*, tawa *Beilschmiedia tawa*, totara *Podocarpus totara* and conifers, but it is not known why these trees are preferred, with a higher sugar concentration or easier access to sap being suggested as possible reasons.

In the Wellington district, North Island, tree damage from sap-feeding by Kaka in the population originating from successful translocation has been recorded predominantly in introduced conifers and eucalypts, and mainly in large mature trees. Charles and Linklater (2014) report that in order to determine the characteristics that make trees prone to sap-feeding by Kaka, a total of 282 trees were sampled in 45 groups at 15 sites across Wellington city, and these sampled trees were of 31 different species, 12 of which were not native to New Zealand. Damage was recorded on 85 trees of 10 species, and exotic trees, particularly conifers, were significantly more likely to be damaged than native trees. In this study, half of the species observed with damage were conifers, and more than 80 per cent of individual trees were conifers, including macrocarpa *Cupressus macrocarpa*, Lawson's cypress *Chamaecyparis lawsoniana* and Japanese cedar *Cryptomeria japonica*. The extent of damage ranged from an estimated 2.5 to 60 per cent of bark-cover, but predominantly was low, with damage to 64 per cent of damaged trees between 2.5 and 10 per cent. Most damaged trees had bark removed from the trunk only and damage occurred mostly on the upper third of the tree. Diameter at breast height was the most informative characteristic for predicting damage and was highly correlated with tree-height. Damage from sap-feeding was more likely to occur on taller trees with a wider girth. Topographic exposure was the second most influential characteristic for predicting damage, with trees on more exposed sites, such as ridges or steep hillsides, more likely to be damaged than trees in valleys. Charles (2012) points out that in this urban situation tree damage is of concern because there is a higher risk of branches falling from damaged trees, especially during high winds.

BREEDING What is known of the breeding biology of Kaka comes mostly from field studies undertaken between 1996 and 2002 in central North Island, at Waipapa Ecological Area of Pureora Forest Park and in Whirinaki Forest Park, and in the Rotoiti Nature Recovery Area, in the north of South Island, and in the Eglinton River valley in Fiordland National Park, in the far south of South Island (Powlesland *et al.* 2009). The proportion of radio-tagged females that nested at a site in a given year varied from none to all, with most nesting occurring in years of mast-fruiting or seeding by preferred food trees, and most nests were in natural hollows in the trunks of live canopy or emergent trees. At the two sites in North Island, 75 of the 97 nests were in matai *Prumnopitys taxifolia* or rimu *Dacrydium cupressinum* trees with a mean height of 34.1 m and a mean trunk diameter at breast height of 128.8 cm. Nests were at a mean height of 12.6 m, and mean dimensions of hollow entrances were 551 mm in height and 120.6 mm in width. Nesting chambers were at a mean depth of 780.3 mm in the hollows, and the mean diameter of these chambers was 912.2 mm at one site, but only 473.2 mm at the other site. At the two sites in South Island, all but two of the 71 nests were in the dominant *Nothofagus* trees with a mean height of 24.6 m and a mean trunk diameter at breast height of 111.5 cm. Nests were at a mean height 10.4 m, and mean dimensions of hollow entrances were 452 mm in height and 88.3 mm in width. Nesting chambers were at a mean depth of 666.6 mm in the hollows, and the diameter of these hollows was measured at only one site, where it was 573.1 mm. Each female energetically grubbed to a depth of 50–100 mm into dry rotten wood on the floor of the chamber, and supplemented this material with live or dead wood chewed from walls of the cavity, biting and chewing large pieces to form a well-aerated dry tilth in which she scraped a nest-bowl about 300 mm in diameter and 20 mm deep.

At the Rotoiti site, northern South Island, prior to egg-laying, on seven occasions over two days, a male was seen to enter the nesting hollow while the female was inside, each time remaining inside the hollow for one to two minutes, and this was the only record at any of the study sites of a male entering an occupied hollow, including during periods of incubation and care of the nestlings. Copulatory behaviour was observed on several occasions at each of the study sites. Prior to mounting, the male walked up to the female several times, gently nudging her with his bill and forehead or with a foot. In response, the female assumed a soliciting posture, with the back flattened and head lowered while the wings were slightly drooped. The male then mounted and, with flapping wings to maintain balance, assumed the copulatory position, cloaca to cloaca. For up to 15 minutes, the male maintained a slow, rhythmic wing-flapping together with unsteady movements of his body. During copulation, the female raised slightly her head and tail, and the male knocked his bill against hers during each unsteady body movement. Although the male remained mounted throughout copulation, cloacal contact probably was only brief during each unsteady body movement. The female often uttered a soft, high-pitched squeaking throughout copulation. After dismounting the male perched motionless beside the female for several minutes before flying away.

The commencement of egg-laying at each study site varied by up to a month. At the two sites in North Island and at the Rotoiti site in South Island there were very few records of egg-laying commencing in October, and peak times were in November and December, but at the Eglinton site in Fiordland, far southern South Island, the first records were in December and peak times were in February. Of 83 clutches, there was one clutch of a single egg, five clutches of two eggs, 17 clutches of three eggs, 24 clutches of four eggs, 29 clutches of five eggs, three clutches of six eggs, one clutch of seven eggs and three clutches of eight eggs. Video-taping of laying of a clutch of four eggs revealed that the interval between laying of the first and second eggs was 3.8 days, between the second and third eggs was 3.2 days, and between the third and fourth eggs was 3.7 days. The moment of laying was preceded by pelvic movements, hunching of the back, and fanning and depression of the tail. Soon after each egg was laid, the female circled around and, with her bill, gently moved it under her body.

Incubation by females lasted 20 to 23 days, and video monitoring of four incubating birds revealed that females left the nest a mean of 13 to 17 times, with eight to 12 of these absences being in daytime. On average, females incubated for 73 to 109 minutes between absences, and were absent for a mean of four to six minutes per absence, with a range of one to 62 minutes. Males occasionally alighted at the nest entrance and looked inside, but did not enter the hollow. The male of a nesting pair normally approached to within about 20 m of the nesting tree and called, causing the sitting female to emerge from the nest, and then both birds flew to a perch some 50 m from the nest, where the female was fed regurgitated food by the male. Females invariably returned to their nests after only a brief absence. The time of hatching for each egg was determined for a clutch of four eggs and for a clutch of five eggs and, assuming that eggs hatched in the order that they were laid, intervals between hatching of the first and second eggs were 0.3 and 0.4 days, between the second and third eggs were 1.2 and 0.3 days, between the third and fourth eggs were 1.3 and 3.3 days, and between the fourth and fifth eggs was 1.0 day, to give a mean hatching interval of 1.1 days.

Newly-hatched chicks are covered with white down and, when not being brooded by the female, they huddled closely together, raising their heads only when stimulated by the female. While the nestlings were less than 10 days old, females were absent from the nest two or three times during the night and 12 or 13 times during the day for periods of seven to 25 minutes, but the chicks were not fed at night even though females often were absent from the nest for extended periods. Between 11 and 20 days, nestlings developed a dense covering of grey down and the eyes were partially open. At this time, the mean number of absences of females at night ranged between two and six of 23 to 124 minutes duration, and during the day ranged between nine and 16 of 35 to 93 minutes duration. After 20 days, there was little brooding of the chicks by females, and between 21 and 30 days eyes of the chicks opened fully and flight feathers began to extend beyond the down, with little change in absences from the nest by females from those of the previous 10 days. At 31 to 40 days, the chicks were partially feathered, and females were absent from the nests for 49 to 82 per cent of the time. At 41 to 50 days, the chicks were fully feathered and, assuming that chicks fledged in the order of hatching, the fledging age for a brood of four chicks was calculated at 64.8, 70.3, 72.1 and 69.5 days, and for a brood of two chicks at 67.9 and 68.2 days. Fledglings initially were poor fliers, sometimes ending up on the ground and remaining there for three to four days, usually attempting to hide under low vegetation or fallen logs, before climbing into the canopy or flying away.

Nestling survival rates at the two sites on North Island and the Rotoiti site on South Island were similar, ranging from 48.4 to 57.9 per cent, but there was considerable variation between breeding seasons at each site. A much higher success rate of 80.9 per cent was recorded at the Eglinton site in Fiordland, and there was very little variation between the two seasons for which data were available. Although an occasional brood died of starvation or exposure when a cavity was flooded during heavy rain, the main cause of losses was mammalian predation of eggs and chicks, and video footage was obtained of chicks being killed by Stoats *Mustela erminea* and Brush-tailed Possums *Trichosurus vulpecula*. It was suspected that the much higher success rate at the Eglinton site in Fiordland could be attributed to the low density of possums, stoats and rats during the two seasons of the study.

The benefits of effective predator controls on nesting success have been demonstrated at other localities. At Zealandia Wildlife Sanctuary, near Wellington, North Island, where pairs nested within a site fenced to exclude mammalian predators and where supplementary food was supplied, 57 per cent of 84 eggs produced fledglings, and at nearby Pukaha Mount Bruce Wildlife Centre, where nesting also took place at predator-free sites and with the provision of supplementary food, 40.7 per cent of 167 eggs produced fledglings (in Powlesland *et al.* 2009).

EGGS Eggs are slightly oval in shape and with a slightly rough surface. Measurements of 12 abandoned or infertile eggs of *N. m. septentrionalis* taken from seven nests in Whirinaki Forest Park, North Island are 41.5 (40.2–44.4) × 31.5 (30.1–33.0) mm (Powlesland *et al.* 2009).

EXTINCT

Norfolk Island Kaka

Nestor productus (Gould)

Plyctolophus productus Gould, *Proc. Zool. Soc., London*, 1836, p. 19 (no locality = Philip Island, off Norfolk Island).

OTHER NAME Norfolk Island Parrot.

DESCRIPTION Length 38 cm.

ADULTS Much variation in plumage; forehead, crown and nape brown-grey to brown; lores, cheeks and ear-coverts yellow to orange, sometimes lightly tinged red; throat and upper breast yellow-orange to grey-brown; lower breast yellow or yellow-orange; abdomen, thighs and under tail-coverts dark orange or dull red, feathers marked with greyish-brown; sometimes a greenish-yellow band across grey-brown hindneck; mantle and back dark greyish-brown; wings brown, washed dull green on upper coverts; rump and upper tail-coverts dark orange or dull red, feathers edged greyish-brown; underwing-coverts dull yellow; tail brown with yellow to orange margins on inner webs of lateral feathers; bill brownish-grey; iris dark brown (Gould); legs olive-brown.

1 male: wing 259 mm, tail 130 mm, exp. cul. 48 mm, tars. 35 mm.
1 female: wing 261 mm, tail 128 mm, exp. cul. 42 mm, tars. 33 mm.
5 unsexed: wing 232–259 (248.4) mm, tail 121–145 (137.6) mm, exp. cul. 40–49 (45.0) mm, tars. 32–38 (35.2) mm.

JUVENILES (Gould) Like adults, but breast olive-brown.

DISTRIBUTION Formerly occurred on Norfolk Island and nearby Philip and Nepean Islands, in the Tasman Sea; now extinct.

STATUS Hume and Walters (2012) note that first mention of the Norfolk Island Kaka was made by J.R. Forster on the second voyage to the South Pacific by Captain James Cook, after discovery of Norfolk Island in 1774. First European settlement on the island came in 1788 with establishment of a penal colony that persisted until 1814, and I suggest that references to both *parrots* and *paroquets* in the writings of visitors to the island during this time can be interpreted as differentiation between the Norfolk Island Kaka and the Red-fronted Parakeet *Cyanoramphus novaezelandiae* by these writers. In a letter written from Norfolk Island on 28 October 1788, and printed in *The Kentish Gazette* at Deal, England, on 6 June 1789, it is stated:

> *The wild fowl are pigeons, quails, doves, plovers, corlieus, parrots, paroquets and a black bird with a web foot that burrows in the ground, with a great variety of smaller birds …*

In a letter written in August 1790 by a non-commissioned officer on Norfolk Island, there is clear differentiation between the *parrot* and the *paroquet*:

> *There is some little variety of birds. A Parrot, Paroquet, Pidgeon, Dove, Hawk, and Owl, with many small birds. The Parrot is very curious; the Paroquet common, with no great share of beauty; the Pidgeon beautiful; the Dove delicate; the Owl and the Hawk quite common.*

In an account from 1793, it is noted that Lieutenant King found the parrots to be '….numerous, and the ugliest bird of the kind that I ever heard of; this added to the harshness of their voice, makes them a very disagreeable bird' (in Tennyson and Martinson 2006).

After the loss of HMS *Sirius* at Norfolk Island on 19 March 1790, Captain John Hunter was marooned on the island until

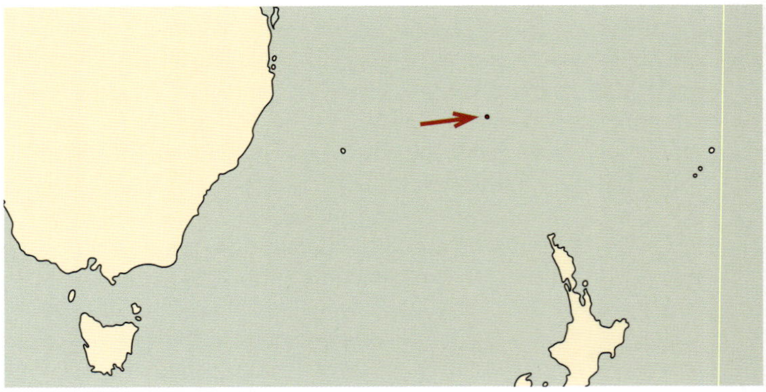

Plate 3
Norfolk Island Kaka *Nestor productus* (adults)

F. KNIGHT '07

February 1791, and during that time he made drawings of 'birds, flowers and fishes'. Included among these drawings are figures of the extinct Norfolk Island Ground-Dove *Alopecoenas norfolkensis* and Norfolk Island Kaka and, having been made by an artist who observed living birds on the island, they are the only known visual records of these birds (see Groom 2012).

A second penal settlement operated on Norfolk Island between 1825 and 1854, and it seems that the Norfolk Island Kaka disappeared before or soon after establishment of this settlement. Notes on 11 of the more conspicuous birds on Norfolk Island were made by Ensign Abel D. W. Best, who was stationed on the island between August 1838 and some time soon after 11 April 1839, the date of the last entry from Norfolk Island in his diary. A diary entry by Best noting that he shot 'five or six Lowries' in the bush at Ball's Bay on 21 September 1838 can be interpreted as referring to Red-fronted Parakeets or introduced Crimson Rosellas *Platycercus elegans*, but no mention is made of the larger parrots. Moore (1985) points out that we may draw some conclusions from a lack of any reference by Best to species which might reasonably have been expected to attract his attention. It seems likely that he would have observed and commented on the Norfolk Island Kaka had it been present, and his lack of reference to it probably suggests that it was already extinct, at least on Norfolk Island. Gould (1865) refers to its reported survival on nearby Philip Island after having been extirpated on Norfolk Island:

> *I regret to state that the native haunts of this fine bird have been so intruded upon, and such a war of extermination been carried on against it, that it is now entirely extirpated, and a few stuffed specimens are the only mementos of its having existed. Until lately it still lived on Philip Island (an islet lying off Norfolk Island), but in this small domain it is no longer to be found.*

Although there is no supporting evidence of its survival on either Philip or Nepean Island after its disappearance on Norfolk Island, I have no doubt that the parrots would have moved freely between all three islands, which are only a few kilometres apart.

Norfolk, Philip and Nepean Islands are not large islands, so I suspect that numbers of large, presumably long-lived Norfolk Island Kakas never were very high, being determined by available food resources and nesting sites. Moore points out that considerable adverse impact on native birds occurred during extensive modification of the island while it was a penal colony, with much of the area being cleared for building and for agriculture, and the remainder indirectly affected by the deliberate liberation of domestic animals. Feral cats were well established by the time of Best's arrival, and he reports hunting them with dogs. In common with other large birds on the island, notably the Providence Petrel *Pterodroma solandri* and 'Wood Quest' or Norfolk Island Pigeon *Hemiphaga novaeseelandiae spadicea*, these parrots would have been killed for food by convicts and early settlers, especially after the wreckage of HMS *Sirius* when there was an acute shortage of food. Tennyson and Martinson highlight the ease with which the parrots were killed:

> *A hunter 'would invariably find them on the ground' but, although they were 'very curious', they would retreat by running to Norfolk Island pine trees* (Araucaria heterophylla) *and 'haul themselves up by the bill'. There they would 'remain till they were shot'.*

Such intense hunting, coupled with extensive clearing of the native forest and predation by feral cats would have brought about their early extinction.

At least one of the museum specimens has a grossly overgrown upper mandible, prompting Flannery and Schouten (2001) to speculate that this may have resulted from inbreeding in a small population. I suggest that a more likely explanation is that it occurred in captive birds held in less than satisfactory conditions not allowing natural abrasion of the mandibles by chewing into wood or rubbing against rocks. Some birds were kept as pets and, according to Greenway (1967) the last living bird died in captivity in London sometime after 1851.

HABITATS This species is said to have been a forest bird like the very closely allied New Zealand Kaka *Nestor meridionalis* (Greenway 1967). Gould was informed by Mrs Anderson that on Philip Island it 'was found among the rocks and upon the loftiest trees'.

HABITS Virtually nothing is known of the habits of Norfolk Island Kakas, though presumably they closely resembled habits of the closely-allied New Zealand Kaka *Nestor meridionalis*. Early reports refer to their being tame or 'very curious', and to rarely using their wings 'except when close pressed' (in Tennyson and Martinson 2006).While in Sydney in 1839 or again in 1840, Gould observed the behaviour of a captive bird, and on another occasion saw a captive bird in England, recording both encounters as follows:

> *During my stay at Sydney I had an opportunity of seeing a living example in the possession of Major Anderson, and was much interested with many of its actions, which were so different from those of every other member of its family, that I felt convinced they were equally different and curious in a state of nature. This bird was not confined to a cage, but permitted to range over the house, along the floors of which it passed, not with the awkward waddling gait of a Parrot but in a succession of leaps, precisely after the manner of the* Corvidae. *Mrs. Anderson, to whom I am indebted for the little I could learn respecting it, informed me that it was found among the rocks and upon the loftiest trees of Philip Island, that it was so tame as to be readily taken alive with a noose, and that it fed upon the blossoms of the white-wood tree, or white* Hibiscus, *sucking the honey of the flowers: the mention of this latter circumstance induced me to examine the tongue of the bird, which presented a very peculiar structure, not, like that of the true honey-feeding Parakeets (the* Trichoglossi), *furnished with a brush-like termination, but with a narrow horny scoop on the under side, which, together with the extremity of the tongue, resembled the end of a finger with the nail beneath instead of above: this peculiarity in the structure of the organ is doubtless indicative of a corresponding peculiarity in the nature of the food upon which the bird subsists. Mrs. Anderson told me that it lays four eggs in the hollow part of a tree, but beyond this I was unable to ascertain anything respecting its nidification. I may mention that I once saw a living example of the bird in England. It was in the possession of Sir. J. P. Millbank, Bart., who informed me that it evinced a strong partiality to the leaves of the common lettuce and other soft vegetables. and that it was also very fond of the juice of fruits, of cream and butter.*

CALLS The call was described as a 'hoarse, quacking, inharmonious noise, sometimes resembling the barking of a dog' (in Tennyson and Martinson 2006).

DIET AND FEEDING The 'blossoms of the whitewood tree or white *Hibiscus*' identified as a food by Mrs Anderson in her comments to Gould presumably are the flowers of *Lagunaria patersonia*. I agree with the remark made by Tennyson and Martinson that the size and strength of its bill suggests that the Norfolk Island Kaka must have eaten other foods, and I am sure that, in common with the New Zealand Kaka, the bill was used to dig into dead or decaying timber to extract insects and their larvae.

SPECIMENS AVAILABLE Greenway (1967) notes that specimens of *Nestor productus* are held in museums at Amsterdam, Besançon-France, Berlin, Dresden, Florence, Leiden, Liverpool, London, Melbourne, New York, Philadelphia, Prague, Vienna and Washington. There is also a desiccated carcass at the Australian Museum, Sydney.

Tennyson and Martinson (2006) report that fossil bones were recorded from Norfolk and Nepean Islands in the 1980s, and excavations at a prehistoric archaeological site at Emily Bay, on Norfolk Island, in 1996 and 1997 recovered 21 bones representing at least three individuals. In total there are fossil remains of at least nine individuals.

Although not particularly lifelike, this drawing by John Hunter is the first drawing of a Norfolk Island Kaka, and is the only drawing by an observer of wild birds on the Island.

SUPERFAMILY
CACATUOIDEA G. R. Gray

There is a consensus, emanating principally from biochemical and chromosomal studies, that the cockatoos are quite distinct from other parrots, and molecular analyses have identified them as constituting the second major lineage or clade in the evolutionary history of the parrots, with divergence estimated to have occurred during the Eocene, approximately 40 million years ago (White *et al.* 2011).

FAMILY **CACATUIDAE** G. R. Gray

The most conspicuous external characteristic of cockatoos is a movable head-crest, which the birds usually raise immediately after alighting and when alarmed or excited. Anatomical features described by Smith (1975) include a complicated formula of the carotid arteries, an ossified orbital ring in the skull, together with a fused bridge across the temporal fossa, and presence of a gall-bladder. Powder-down patches are particularly prominent, but the feathers have no 'Dyck-texture', a structural composition of the feather barbs, which through back-scattering of light produces the plumage colours common to most species of parrots (see Dyck 1971). An examination of the morphology of spermatozoa from three species of parrots has revealed a structural difference between that of cockatoos and the spermatozoa of other parrots (Jamieson *et al.* 1995).

In marked contrast to the consensus surrounding distinctiveness of the cockatoos are differing taxonomic arrangements proposed to reflect affinities within the family. Most arrangements recognised three groupings loosely categorised as the 'black cockatoos', the 'white cockatoos' and an assemblage of genera with uncertain affinities (see Adams *et al.*1984; Brown and Toft 1999). Joshua and Parker pointed out that the karyotype of the Palm Cockatoo *Probosciger aterrimus* differs greatly from that of both the 'black cockatoos' and the 'white cockatoos', so raising the possibility of a fourth principal lineage (in Low 1993b). Findings that I consider to be convincing now have come from recent molecular analyses indicating that diversification within the Cacatuidae commenced during the Oligocene, approximately 28 million years ago, and it is suspected that a driving force behind the speciation of cockatoos was a transformation from mainly mesic to xeric habitats featuring fire-adapted woodlands and grasslands, which occurred during the middle Miocene, approximately 15 to 20 million years ago (White *et al.* 2011). It is suspected also that this diversification was driven, in part, by bill adaptations and specialisations that enabled the birds to colonise previously unoccupied niches, and plumage colour, body size, wing shape and bill morphology have evolved in parallel or convergently across lineages. The three cockatoo lineages that were identified are:

(i) monotypic *Nymphicus*,
(ii) a calyptorhynchine lineage of *Calyptorhynchus*,
(iii) a speciose cacatuine-type lineage of *Probosciger*, *Callocephalon*, *Eolophus*, *Cacatua and Lophochroa*.

There was no clear separation of 'black' and 'white' cockatoos, and the large 'cacatuine' lineage comprised a mixture of white, grey, pink and black cockatoos with at least five sub-lineages. The three principal lineages are treated as subfamilies in accordance with the arrangement advocated by Joseph *et al.* (2012).

Cockatoos bathe by fluttering among wet foliage of a tree or during rain showers by flying about or by hanging upside-down from their perches. When cornered or aroused, they emit a peculiar hissing noise. Courtney (1996) points out that cockatoos appear to be unique among parrots in possessing a food-swallowing vocalisation, which is a series of short, rapidly repeated notes uttered by young birds when actually engaged in swallowing food that is being transferred from the bill of a parent. Also the straight, wheezing food-begging calls, that are mostly lacking in structure, set cockatoos apart from other parrots.

Sexual dimorphism is pronounced in a few species, slight in some species and absent in others. In most species, the sexes share incubation. Newly-hatched chicks of most species are covered with long yellow down.

Cockatoos are confined to the Papuo-Australasian region, being distributed from the Solomon Islands west to Nusa Tenggara, Indonesia, and the Philippines, and south to southern Australia. Brown and Toft (1999) point out that a biogeographic analysis of results from molecular studies supports the hypothesis that cockatoos originated in Australia, with a combination of dispersal and secondary speciation contributing to diversification of *Cacatua* in two separate radiations to the northern island regions. With 14 species, Australia has by far the strongest representation, and elsewhere in the range one, two or rarely three species occur. In Australia, the two white-tailed black cockatoos from the southwest are endangered because of extensive landclearance, but other species remain common and some are increasing, most notably the Galah *Eolophus roseicapilla,* Sulphur-crested Cockatoo *Cacatua galerita* and Little Corella *C. sanguinea*. Elsewhere, landclearance and capture for the live-bird trade have brought about declining numbers in most species, with the Yellow-crested Cockatoo *Cacatua sulphurea*, White-crested Cockatoo *C. alba* and Philippine Cockatoo *C. haematuropygia* now listed as endangered.

SUBFAMILY **CALYPTORHYNCHINAE** Bonaparte

The 'black cockatoos' belonging to this subfamily are readily recognised by their coloured tail-bands and short to moderately long, recumbent crests. There is no bare facial skin, though some species have a prominent bare eyering, and the thighs are feathered. The cere is unfeathered. The slightly rounded tail is rather long. The bill is large and robust, but there are structural modifications enabling the species to exploit different foods. In the yellow-tailed and white-tailed black cockatoos the narrow, protruding bill with sharp pointed mandibles is well suited to digging into timber to extract wood-boring insect larvae or to extracting seeds from hard, woody nuts. In the Red-tailed Black Cockatoo *Calyptorhynchus banksii* the broad, blunt bill is ideal for crushing seeds or hard nuts, with further modification evident in the Glossy Black Cockatoo *C. lathami*, which has a protruding bulbous bill with an exceptionally broad lower mandible, an adaptation for extracting seeds from *Allocasuarina* and *Casuarina* cones.

These structural modifications in the bills, together with marked differences in sexual dimorphism, separate the species into two clearly discernible groupings which here are differentiated subgenerically.

A subspecies of the Glossy Black Cockatoo is listed as endangered, but subspecific differentiation is highly doubtful. The two similar-plumaged white-tailed black cockatoos from southwestern Australia are endangered. In eastern Australia, the Yellow-tailed Black Cockatoo *Calyptorhynchus funereus* is widespread and common.

GENUS ***Calyptorhynchus*** Desmarest

Calyptorhynchus 'Horsfield', Desmarest, *Dict. Sci. Nat.*, ed. Levrault, **39**, 1826, pp. 20, 117. Type. By subsequent designation, *Psittacus banksii* Latham (G. R. Gray, *List Gen. Bds*, 1840 p. 53).

Differentiating features of this genus are the same as for the subfamily. The genus is restricted to Australia, including Tasmania.

SUBGENUS ***Zanda*** Mathews

Zanda Mathews, *Austr. Av. Rec.* **1**, 1913, p, 196. Type, by original designation, *Calyptorhynchus baudinii tenuirostris* Mathews = *Calyptorhynchus baudinii* Lear.

Narrow, protruding bills with sharp, pointed mandibles distinguish species in this subgenus. Sexual dimorphism is slight, with yellow or white tail-bands and pronounced pale yellow or white margins to the body feathers being present in adults and juveniles of both sexes. The two endangered white-tailed black cockatoos from southwestern Australia belong to this subgenus, and both have been studied extensively.

ENDANGERED

Carnaby's Black Cockatoo

Calyptorhynchus latirostris Carnaby

Calyptorhynchus baudinii latirostris Carnaby, *West. Aust. Nat.*, **1**, 1948, p. 137 (Hopetoun, Western Australia).

OTHER NAMES Short-billed White-tailed Black Cockatoo, Short-billed Black Cockatoo.

DESCRIPTION Length 55 cm. Weight males 540–760 g, females 560–790 g.
ADULT MALE General plumage grey-black, duller and slightly brownish on lower underparts; feathers of neck and underparts broadly margined dusky white to give scalloped appearance, remaining body feathers narrowly edged dusky white; ear-coverts dusky white; central tail-feathers grey-black, lateral feathers grey-black with broad subterminal band of white; bill grey-black; iris dark brown; naked eyering flesh-pink; legs and feet greyish-brown, paler on undersides of feet.
21 specimens: wing 357–382 (369.6) mm, tail 242–273 (256.3) mm, exp. cul. 41–46 (44.3) mm, tars. 28–35 (31.1) mm.
ADULT FEMALE Like male, but ear-coverts more extensively clear white; feathers of underparts, including underwing-coverts, more broadly margined with clear white; bill horn-colored with grey at tip of upper mandible; naked eyering grey; legs paler brown tinged pink.
23 specimens: wing 351–382 (371.1) mm, tail 247–285 (265.5) mm, exp. cul. 42–46 (44.7) mm, tars. 28–32 (31.7) mm.
JUVENILES Similar to adult female, but with narrower, slightly shorter flight and tail feathers, dull white ear-coverts sometimes tinged yellow; narrower white tail-band sometimes with irregular dark blotches, more prevalent on inner webs of tail-feathers; bill horn-coloured variably suffused grey.

DISTRIBUTION Carnaby's Black Cockatoo is one of two white-tailed black cockatoos confined to southwestern Australia, where their allopatric breeding ranges are for the most part separated by the 750 mm annual isohyet (Blakers *et al.* 1984). This species is distributed mainly in the drier region between the 750 mm and 300 mm annual isohyets, mostly south of lat. 29°S and west of

long. 120°E. Breeding occurs mostly in the eastern sector, east to longitude 121°E and between latitudes 27°30′S and 29°30′S. Since the 1950s, as a consequence of large-scale landclearance, the centre of distribution of Carnaby's Black Cockatoo has shifted considerably westwards and southwards, with the non-breeding range extending into higher-rainfall areas of the extreme southwest, which are within the breeding range of Baudin's Black Cockatoo *Calyptorhynchus baudinii* (Storr 1991).

STATUS Because of large-scale landclearance throughout the wheatbelt of southwestern Western Australia, there has been a dramatic contraction of the breeding range, and Carnaby's Black Cockatoo has suffered a decline in numbers, estimated at approximately 50 per cent of the population in the past 60 years. Though landclearance has all but ceased, the consequences are continuing, with declining numbers and contraction of the breeding range expected to persist for some time. Saunders (1991) points out that between the 1970s and 1990s, the species disappeared from more than one-third of its breeding range, and a continuing contraction could result in survival in as little as half of the former range. Population estimates have ranged from 10 000 to 60 000 birds, but is likely to be about 40 000 birds, but with a warning that it could be as low as 10 000 birds (Mawson and Johnstone 1997; in Garnett *et al.* 2011).

The population is ageing, and it is low recruitment levels that are putting this species at risk. During the years 1970–1975, the effects of woodland clearing on recruitment in breeding populations were assessed at two localities in the northern wheatbelt (Saunders 1977). At the first locality, where large tracts of uncleared sandplain country surrounded the eucalypt woodland in which nesting occurred there were 75 breeding pairs, and each pair fledged two chicks every three years, or 0.6 chicks per pair per year. Nesting birds foraged quite near to their nest sites and it was apparent that the local food supply was reliable and quite adequate during the breeding season. A very different situation prevailed at the second locality, where the cockatoos nested in a 458 ha water-catchment reserve, and each pair fledged only one chick every three years, or 0.3 chicks per pair per year. At this locality the behaviour of nesting birds was strongly indicative of a shortage of food and contrasted markedly with the behaviour of birds at the first locality. Females were seen out feeding by themselves or with their mates while they still had eggs or young chicks in the nest, thus exposing the eggs or chicks to predators or the nesting hollow to usurpation by competitors. Shortage of food seemed to be the cause of this deterioration of parental care, the parents being obliged to spend more time foraging. There were areas of useful food not visited by the cockatoos from one year to the next, so it would seem that the total amount of food in the area was sufficient, but its patchy distribution prevented the birds from utilising it. The breeding population at this second locality was extinct by 1977 (Saunders 1986).

Changes brought about by agriculture, that have affected adversely Carnaby's Black Cockatoo, have benefited the Galah *Eolophus roseicapilla* and the Little Corella *Cacatua sanguinea,* both of which have expanded their ranges into the wheatbelt, where they and resident Western Corellas *C. pastinator* compete with black cockatoos for nest sites (Saunders *et al.* 1985). Another threat exposed during field studies is interference at nests by persons seeking eggs or chicks for the live-bird market, and in the past nesting trees were felled or hollows chopped open to gain access to the contents, so further exacerbating the continuing loss of nesting trees through landclearance. Hopefully, this threat has been lessened significantly by increased surveillance of nesting sites by wildlife authorities and a legal supply to licensed aviculturists of microchipped and DNA profiled young birds from a strictly controlled harvesting of second eggs from two-egg clutches. The objective of this program, which also involves Perth Zoo, is to maintain a captive breeding program, and the genetic diversity of the captive population is enhanced by the inclusion of rehabilitated birds injured in the wild.

In recent years, deaths of significant numbers of birds from disease and what could be considered natural causes have been documented. In October 2009, an unknown disease killed at least 23 females in a single breeding colony, a heatwave in January 2010 killed flocks of 65 and 145 birds, and in March 2010 a hailstorm was responsible for the deaths of 57 birds and injuries to 24 birds (in Garnett *et al.* 2011).

Initiated in 1999, the Carnaby's Black Cockatoo Recovery Project is a collaborative scheme involving government and non-government agencies together with private landholders. It is focused primarily on the protection of critical nesting habitats, rehabilitation of degraded foraging areas, and monitoring of known nesting hollows (Pitman *et al.* 2007). Each year counts are made by volunteers at roosting sites in the wintering range, and numbers counted provide some indication of trends in the population. Counts indicated that a very significant fall in numbers occurred after 2010, when 6700 birds were recorded, and then in 2013 a count of 5800 birds represented a 44 per cent increase on the count from 2012.

Carnaby's Black Cockatoo is listed as endangered under the Australian *Environment Protection and Biodiversity Conservation Act 1999.*

HABITATS Breeding takes place almost exclusively in wetter parts of the semiarid sandplains between the 750 mm and 300 mm annual isohyets, where proteaceous scrubs and heathlands are utilised by these cockatoos for foraging, while adjoining mixed eucalypt woodlands provide nesting sites. At times, presumed non-breeding birds wander eastwards into drier habitats, including mallee and arid scrublands, while to the west there is a regular post-breeding movement into *Pinus* plantations throughout the humid, forested zone of the extreme southwest, which is the breeding range of Baudin's Black Cockatoo *Calyptorhynchus baudinii.*

Saunders (1979b) described vegetation communities at four localities where studies of nesting behaviour were undertaken. At Coomallo Creek, in the northern wheatbelt, isolated patches of woodland dominated by wandoo *Eucalyptus wandoo* are scattered through sandplain heath with low dense scrub. Approximately 100 km to the southeast, at Manmanning, lands have been extensively cleared for agriculture and the remnant vegetation is woodland dominated by salmon gums *Eucalyptus salmonophloia* with some wandoo and gimlet *Eucalyptus salubris,* but also there are patches of sandplain heath with mallee in isolated thickets. Well to the south, near the western boundary of the breeding range at Tarwonga, woodlands of marri *Corymbia calophylla* dominate, but nesting took place almost exclusively in wandoo growing in the valleys or on alluvial white sand. Extensive landclearance has taken place also at Moornaming, where the remnant woodland is dominated by wandoo with swamp yate *Eucalyptus occidentalis* occurring on the wetter alluvial flats.

Outside the breeding season, these cockatoos can be seen foraging in a wider variety of habitats. In the far southwest, they commonly are encountered in *Pinus* plantations, but additionally they frequent eucalypt forests, remnant trees in farmlands, orchards, and parks or gardens in urban centres, at times being seen in Kings Park near Perth city centre.

MOVEMENTS In higher-rainfall districts with substantial tracts of native vegetation Carnaby's Black Cockatoo is resident, but to the east in much of the wheatbelt and in drier sandplain country it is a breeding migrant, undertaking regular seasonal movements between inland breeding areas and the coastal non-breeding range (Saunders and Ingram 1995). Results of long-term ecological studies carried out at localities in the wheatbelt, banding-return data, and answers to questionnaires circulated to schools have given us a good understanding of seasonal movements undertaken by these cockatoos. Post-breeding departure from inland nesting areas usually commences in December, and during July–August pairs return to immediately commence cleaning out their nesting hollows in preparation for egg-laying. I suspect that, because of large-scale landclearance and a consequent decline in food availability, the annual exodus of birds from the wheatbelt now is more widespread and comprehensive than in the early 1900s. It is during the non-breeding season that wandering flocks of Carnaby's Black Cockatoos come into contact with Baudin's Black Cockatoo *Calyptorhynchus baudinii.*

HABITS Because of difficulties in differentiating in the field the two species of white-tailed black cockatoos, observers rarely identify the species under observation. Birds seen in the wheatbelt and inland sandplain country are presumed to be Carnaby's Black Cockatoos, but usually there is doubt concerning the identity of birds encountered in the southwestern forests. When observing birds in the far southwest, I made a special effort to identify them, and on occasions have been able to approach closely enough to detect the different bill shapes, but birds seen at a distance often remain unidentified.

Like most *Calyptorhynchus* species, Carnaby's Black Cockatoos are noisy and conspicuous, especially when encountered in large flocks, and during the non-breeding season flocks coming to feed in *Pinus* plantations sometimes comprise hundreds of birds, though now rarely building up to the very large numbers seen in past years. A strong pair-bond is evidenced by mated birds remaining together throughout the year, the pair normally being apart only when the female is brooding, and pairs and their dependent offspring usually are quite discernible within flocks. In the non-breeding season, departure of flocks from the roosting trees takes place soon after first light, and daily activity peaks are in the early morning and late afternoon. During the middle of the day, especially in hot weather, the birds cease feeding and retreat to nearby trees, where they rest amidst leafy branches and also spend much time preening. During cool, overcast weather, I have observed birds feeding in the middle of the day.

At times, when left undisturbed while feeding in close proximity to human habitation, these cockatoos may become quite confiding, but normally they are rather wary, flying off to the accompaniment of loud calling if approached. While the flock is feeding, one or two birds often remain perched in nearby trees, and at the approach of danger they give alarm screeches, to which the entire flock responds by flying off. Feeding is mainly arboreal, though often quite low down in stunted *Banksia, Hakea* or *Grevillea* bushes in heathland, and my impression is that this species comes to feed on the ground more readily than does Baudin's Black Cockatoo *Calyptorhynchus baudinii.* Just before nightfall, the cockatoos return to the roosting trees, and once settled they remain quiet during the night.

The strong, buoyant flight is characterised by slow, flapping wingbeats. When moving between stands of trees, the birds normally fly above treetop level, typically spiralling down on motionless, down-curved wings to alight, but at times I have seen flocks fly quite long distances at low height just above the tops of stunted heathland shrubs. When alighting, a bird raises its crest and spreads the tail.

CALLS Saunders (1979a) points out that all but one of the calls given by Carnaby's Black Cockatoo fall into the category of call-notes defined as being mostly monosyllabic or disyllabic and practically never comprising more than four or five notes. The majority of calls apparently are associated with maintenance activities. Five categories of calls are recognised, the most commonly heard being the familiar, loud wailing *wy-lah* contact cry given by birds in flight, and often it is heard before the birds come into view. It is given by both sexes in a wide variety of situations, and a shortened version commonly is emitted by a bird just prior to taking flight. Also given by both sexes, and often in conjunction with the contact call, is the 'interrogative call', which is a more prolonged *wy-lah* with an inflection at the end (Saunders 1983).

The disyllabic 'whistle call', sounding like *whee-whee,* and a shortened version – the 'short whistle call' – have been heard only from adult females, the first being uttered at various levels ranging from soft to very loud, generally when the bird is perched, but the second usually is of low intensity and is audible only quite near to the calling bird. Males utter the low-intensity 'half call' when in the same situation as females giving the whistle calls. Given by all birds when alarmed or disturbed is the monosyllabic 'alert call', which is a harsh scream lasting approximately a half second. A rapid repetition of shrill squeaks and a series of *ah-ah-ah-ah-ah* notes are 'sexual calls' given by males during courtship, and always are directed at females. Among 'agonistic calls' is a very harsh, squeaky chattering, which usually is quite loud and normally is emitted during aggressive encounters between males. A very harsh squawk of low or very high intensity often is given when birds are handled or are in conflict.

'Begging calls' given by the adult female prior to being fed by the male usually comprise harsh grating sounds, and vary from low, brief notes to loud, continuous calls that go on for minutes. A similar call emitted by nestlings is described by Courtney (1996) as a slow repetitive rasp or squeak. Juveniles and dependent immatures also utter a prolonged buzzing or grating sound when near to their parents. Accompanying the transfer of regurgitated food from male to female or adult to young is a similar *ark-ark-ark-ark-ark* call (Saunders 1979c). Included among miscellaneous calls is a low-intensity note resembling *chuck,* which can be heard only quite near to the calling bird, and it is uttered by both sexes, usually when the birds are alone. Another low-intensity call is the 'low whistle' repeated several times, and it has been heard only from solitary adult females.

DIET AND FEEDING Seeds of native plants, especially proteads, are the main food of Carnaby's Black Cockatoo, but seeds of *Pinus* and other introduced species also are taken, together with fruits, blossoms, buds and insect larvae, and it is largely with these supplementary foods that dietary overlap occurs with Baudin's Black Cockatoo *Calyptorhynchus baudinii. Banksia, Dryandra,*

Grevillea and *Hakea* seeds feature prominently in the diet, and a majority of foraging observations are of birds feeding in these shrubs. Seeds, flowers and buds of eucalypts also are important food items, especially in the non-breeding range. During mid July to mid August in 2004 and 2006 foraging observations were undertaken at eight sites along a straight-line distance of 625 km and an annual mean precipitation gradient from 1050 mm in the west to 230 mm in the east and, of 18 observations of these cockatoos, 56 per cent were of feeding in trees and 44 per cent of feeding on the ground (Bell *et al.* 2010).

In the 1970s, during breeding seasons in the northern wheatbelt, almost 90 per cent of feeding observations were made on native vegetation, with the single most important food being seeds of honeybush *Hakea lissocarpha* (Saunders 1980). The only non-native item was seeds of wild geranium *Erodium* sp., which occurred in patches on cleared farmland at the southern end of the study area, and insect larvae from flowers or stems of some plants were found in the crops of nestlings in such quantities to suggest that they had been collected deliberately by the parents. Since the late 1990s, canola *Brassica napa* and *B. juncea* has been grown as a commercial crop throughout much of the higher-rainfall areas of the wheatbelt, and the cockatoos commenced feeding on mature canola seed immediately it was available (Saunders *et al.* 2014b). Because it is available during most of the nestling period, canola seed is a valuable replacement food offsetting the loss of remnant native vegetation, and now is being fed to most nestlings on a daily basis. In their non-breeding range, the cockatoos again fed mainly on seeds of native plants, with the most important food source being marri *Corymbia calophylla* that occurs in narrow strips along major watercourses, but seeds of jarrah *Eucalyptus marginata,* and slender banksia *Banksia attenuata* also featured prominently in the diet. Seeds of lupins and doublegee *Emex australis* were important non-native foods, and the birds fed extensively in *Pinus* plantations on the coastal plain.

Saunders (1974) presented data from analyses of crop contents of both *Calyptorhynchus latirostris* and *C. baudinii* collected at a locality some 33 km east of Perth, between April 1971 and October 1972. The same foods were available to both species at the same time, though it should be stressed that *baudinii* is a wandering visitor to the district. The analyses show differences in food preferences of the two species, with *latirostris* feeding mainly on seeds of *Dryandra, Hakea* and introduced *Pinus,* whereas *baudinii* took wood-boring insect larvae and seeds of marri. Both species occurred together in mixed flocks, and on three occasions *latirostris* was seen feeding in a *Pinus* plantation, while *baudinii* was feeding in marri trees at the edge of the plantation, but none of the *baudinii* collected had *Pinus* seeds in the crop. Though both eat marri seeds, the two species differ in the manner of attacking the fruits; *latirostris* gets at the seeds by breaking open the rim with its shorter upper mandible, but *baudinii* efficiently prises out the seeds with its long upper mandible, doing very little damage to the rim of the capsule.

Scott and Black (1981) report that at a locality some 25 km south of Perth, Carnaby's Black Cockatoos were observed selectively attacking immature fruits of *Banksia attenuata* that were infested with larvae of the seed-eating weevil *Alphitopis nivea.* These larvae occurred mostly in large fruits, and it was suspected that the birds deliberately attacked larger fruits because of this higher incidence of infestation. One or two larvae were taken from each fruit. There are records also of wood-boring larvae being extracted from acacias.

Mawson (1995) describes methods employed by these cockatoos when feeding on nectar from flowers of bottlebrush *Callistemon viminalis* in a suburban garden. If a flower spike was easily accessible the cockatoo merely opened its bill and poked its tongue into each flower, without grasping the spike in its bill. Flower spikes not easily accessible were bitten off and transferred to the foot to be held while the flowers were probed with the tongue in the same manner. After discarding a flower spike, the cockatoo ran its tongue around the inside of the bill, presumably to collect any nectar residue inside the bill. Only fully opened flower spikes were selected, one at a time, and there was little wastage.

BREEDING As expected for a species with a strong, permanent pair-bond, the courtship display of the male is simple. He raises the short crest, fans wide his tail to show the white band and, while giving the loud *ah-ah-ah-ah-ah* call, he struts along the branch towards the female, then bows while swishing his head quickly from side to side in a 'figure 8' movement. Saunders (1982) notes that from observations of 22 males and 20 females known to have changed mates there was only one case where the former mate was seen again, so it is likely that all other changes were made in response to death of the mate.

Saunders (1982) recalls that in the northern wheatbelt, birds arrived in the breeding area between July and September, and immediately pairs commenced selection and preparation of nesting hollows. Selection was undertaken around mid morning, and initially involved the pair flying to a patch of woodland, where the female moved from one tree to the next examining different hollows. She peered into hollows, walked around entrances and chewed briefly at these entrances, but did not enter any hollow. The male sat nearby and followed the female wherever she flew. Females were first seen near their chosen nesting hollow from five to 40 days before egg-laying, and the pattern of visits varied greatly between females. Nests were in hollow limbs and holes in the trunks of eucalypts, and appeared to be well distributed without any indication of 'clumping'. Distances between neighbouring nesting trees averaged 174 m, but in smaller patches of woodland, nests sometimes were quite near to each other, and up to three nests were found in the same tree.

Pairs defend the nesting hollow, and it is only at this time that intraspecific aggression is conspicuous. Saunders (1982) points out that not only are agonistic encounters between females more numerous than at other times, but intruding females are chased farther away. The intruder usually retreats away from an occupied hollow after the resident female gives a mildly aggressive display involving a turning of the head towards the intruder and the emission of loud squawks. If there is no retreat by the intruder, the defending female repeats the display while partly raising the wings, and then attempts to bite the other bird. Withdrawal of the intruder at this time usually results in an aerial chase, with the chasing defender squawking loudly. If there is no retreat and the intruder attempts to defend herself a fight may develop, but this lasts only very briefly and ends with the intruder flying off. Males confront intruding males, but take no part in actions against females.

Plate 4
LEFT Carnaby's Black Cockatoo *Calyptorhynchus latirostris* (adult ♂)
RIGHT Baudin's Black Cockatoo *Calyptorhynchus baudinii* (adult ♂)

F. KNIGHT '14

At a study site in the northern wheatbelt, between 1969 and 2013, hollows used for nesting were in two species of trees, with 228 of 252, or 98.7 per cent, being in wandoo *Eucalyptus wandoo*, 21 being in unidentifiable dead stags, and only three being in powderbark wandoo *Eucalyptus accedens*. Nesting trees had a mean trunk diameter of 580 mm at breast height, and hollows were at a mean height of 4.7 m above the ground (Saunders *et al.* 2014a). Nesting hollows averaged 1.2 m in depth, with an average floor diameter of 407 mm, and average dimensions of hollow entrances were 268 mm in width and 271 mm in height. By the 2013 breeding season, 40.1 per cent of nesting trees had been lost or damaged in such a way that the hollows no longer were suitable for nesting, and of extant trees only 22.2 per cent had hollows suitable for occupation by the cockatoos. At this same study site, between 1969 and 1984, dates for laying of the first egg of the season varied from 9 July to 26 August, with 50 per cent of eggs being laid by some time in August or September, depending on the season.

Although these cockatoos normally lay clutches of two eggs, there usually is a prolonged interval between laying of the first and second eggs, so the second nestling rarely survives more than a few days after hatching. At the study site in the northern wheatbelt, the interval between laying of the first and second eggs varied from one to 16 days, with the mean being eight days. The second chick survived more than three days in only 11 out of 222 clutches where both eggs hatched, and both chicks fledged from eight of those 11 clutches (Saunders 1982).

Incubation lasts 29 days, and only the female broods. Again at the study site in the northern wheatbelt, the sitting female was fed by the male at mid morning and in the evening. At dawn the female occasionally spent some time out of the nest with the male, and also went to drink if surface water was available within 700 m of the nest. The longest recorded period of absence in mid morning was 11 minutes and at dusk was 42 minutes. Nestlings were closely brooded for the first two weeks, and then were brooded only at night for up to six or seven weeks, after which time all brooding ceased (Saunders 1982). While being closely brooded, the chick was fed only by the female, but thereafter the male commenced to feed the nestling at mid morning and at dusk. Growth of nestlings, as determined by Saunders (1982) using increases in wing length, tail length, and weight, was quite consistent, though with a slightly faster increase early and a late slowing down, the latter particularly evident in weights, which showed a decline after about 55 days. Wing length increased from 25 mm at 10 days to 95 mm at 30 days, 225 mm at 50 days, 325 mm at 70 days, and 340 mm at 80 days. Tail length increased from 15 mm at 20 days to 40 mm at 30 days, 125 mm at 50 days, 230 mm at 70 days, and 260 mm at 80 days. Weight increased from 250 g at 15 days to 400 g at 30 days, 575 g at 40 days, and 620 g at 50 days, but then declined to 600 g at 70 days and 575 g at 77 days.

Approximately 80 days after hatching, the young bird leaves the nest, but remains with the parents until the next breeding season, and sometimes longer. It is fed by both parents until independent at about 7 months of age. Stark differences in nesting success at two study areas were attributed by Saunders (1977) to availability of foraging habitat. At the study site in the northern wheatbelt, the success rate was 0.6 young per pair per year, whereas at another site, where fragmented foraging habitat restricted food availability, it was only 0.3 young per pair per year. It was noted that females would relay in a different hollow after the loss of eggs or small chicks, and the interval between failure and relaying was 19 to 22 days (Saunders 1982). One female relaid twice, the first two attempts failing when the young chicks died, but the third attempt was successful.

Saunders (1986) points out that during the years 1970 to 1976, the known nesting success at the site in the northern wheatbelt was 65 per cent, but at the other site with fragmented foraging habitat it was only 35 per cent and subsequently breeding ceased altogether. Also at the northern wheatbelt site, by the end of 1984, the observed annual survival of tagged breeding females was 61.9 per cent. There was a trend for the probability of survival to increase with time after the initial tagging, but numbers were low and only four survived to the 10th year after tagging; two of these birds were estimated to be at least 14 and 15 years old. Data were not available for males because they spent little time around the nesting hollow, and so were difficult to survey.

Breeding studies undertaken between 1969 and 2012 have confirmed a strong, though not significant negative relationship between nesting success and percentage loss of native vegetation within 6 km and 12 km of nest-sites, but there was a significant negative relationship between the health of nestlings and percentage loss of native vegetation around nesting hollows (Saunders *et al.* 2014b). While both eggs hatched in 77 per cent of two-egg clutches, successful fledging of both chicks usually occurred only when older, more experienced females nested in areas where more native vegetation had been retained. It is possible that availability of canola seed during the nestling period at Coomallo Creek may improve the successful fledging of both chicks from two-egg clutches (Saunders *et al.* 2014b).

EGGS The ovate to elliptical-ovate eggs are without gloss. Measurements of four eggs from two clutches are 47.6 (46.1–49.0) x 35.8 (35.1–36.5) mm (in Forshaw and Cooper 2016).

Baudin's Black Cockatoo

Calyptorhynchus baudinii Lear

Calyptorhynchus baudinii Lear, *Illust. Psittac.*, pt 12, 1832 [= pl.6 of bound volume], (no locality = Geographe Bay, southwestern Australia, by designation of Saunders, 1979).

OTHER NAMES Long-billed White-tailed Black Cockatoo, Long-billed Black Cockatoo.

DESCRIPTION Length 56 cm. Weight males 610–720 g, females 620–760 g.

ADULT MALE Differs from *C. latirostris* by having a narrower bill with a markedly elongated tip to the upper mandible; general plumage grey-black, duller and slightly brownish on lower underparts; feathers of neck and underparts broadly margined dusky white to give scalloped appearance; remaining body feathers narrowly edged dusky white; ear-coverts dusky white; central tail-feathers grey-black, lateral feathers grey-black with broad subterminal white band; bill grey-black; iris dark brown; naked eyering flesh-pink; legs and feet greyish-brown, paler on undersides of feet.

16 specimens: wing 361–390 (381.3) mm, tail 255–291 (273.3) mm, exp. cul. 51–55 (52.7) mm, tars. 29–36 (32.3) mm.

ADULT FEMALE Similar to male, but ear-coverts more extensively clear white; feathers of underparts, including underwing-coverts, more broadly margined clear white; bill horn-coloured with grey at elongated tip of upper mandible; naked eyering grey; legs paler brown tinged pink.
17 specimens: wing 363–401 (382.6) mm, tail 248–301 (267.5) mm, exp. cul. 50–58 (53.6) mm, tars. 30–38 (33.4) mm.
JUVENILES Like adult female, but with narrower, slightly shorter flight and tail feathers; dull white ear-coverts sometimes tinged yellow; narrower white tail-band sometimes with irregular dark blotches, more prevalent on inner webs of lateral tail-feathers; bill horn-coloured variably suffused grey and with grey tips to both mandibles; shorter upper mandible.

DISTRIBUTION Baudin's Black Cockatoo is one of two white-tailed black cockatoos confined to southwestern Australia, where their allopatric breeding ranges are for the most part separated by the 750 mm annual isohyet (Blakers *et al.* 1984). This species is distributed mainly west of the 750 mm isohyet in higher-rainfall regions of the far southwest in the Darling Range and on the coastal plain, from Gidgegannup, Hoddy Well and Clackline south to Augusta and Cape Leeuwin National Park, and east to Mount Helena, Wandering, Williams, Kojonup, the Stirling and Porongurup Ranges, Albany and Waychinicup National Park. Breeding occurs in the extreme southwest, north to the Whicher Range and Lowden, with an isolated record at Serpentine, and east to Kojonup and near Albany (Johnstone and Kirkby 2008).

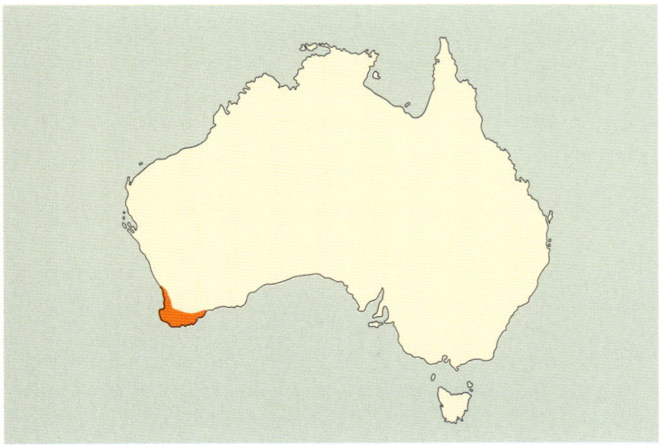

STATUS Though still distributed throughout much of their presumed pre-European range, Baudin's Black Cockatoo has suffered a dramatic decline in numbers, estimated at more than 50 per cent over three generations spanning approximately 60 years (Garnett *et al.* 2011). The long-term survival of this species is of increasing concern, and Mawson notes that surveys undertaken between 1995 and 2004 suggest that the current population probably comprises 10 000–15 000 birds, with only about 10 per cent breeding in any year (in Garnett *et al.* 2011). Declines are continuing at an unquantified rate, with principal threats being an acute shortage of suitable nesting hollows, loss of forest habitat, and illegal shooting by orchardists. Mawson and Johnstone (1997) point out that these cockatoos no longer occur in up to 25 per cent of their former habitat, which has been cleared for agriculture, and landclearance for agriculture and mining is expected to continue. Suitable nesting hollows form only in trees that are 130 to 220 years old, and these have been preferentially felled in the course of forestry operations (Abbott and Whitford 2002). A resulting acute shortage of nesting hollows is further exacerbated by intense competition from other hollow-nesting species, including other cockatoos and Wood Ducks *Chenonetta jubatta* and from feral European honeybees *Apis mellifera*, all of which have been recorded successfully displacing Baudin's Black Cockatoos.

Storr (1991) points out that with an interval of seven years between generations and an annual reproductive rate of 0.6 chicks per pair, current recruitment cannot replace numbers illegally shot each year by orchardists. These cockatoos do cause damage to apple and pear crops, and Long (1985) estimates that this can reach 10 per cent of the net income of an orchard. Halse (1986) confirms that in past years hundreds of birds were shot by orchardists and, in my opinion, this was a tragedy which never should have been permitted to occur. Baudin's Black Cockatoos were not fully protected until 1996, an unacceptably late recognition of their threatened status, and illegal shooting by orchardists still occurs. Authorities concede that because flocks are mobile, shooting is not considered to be effective in preventing damage in orchards, so alternative control measures should be adopted. In the long-term, enclosing an orchard under hail-proof netting can be a cost-effective means of thwarting attacks by birds and, where damage can be attributed to a threatened or vulnerable species, growers should be eligible for low-interest loans to meet high initial costs of erecting netting enclosures.

Baudin's Black Cockatoo is listed as vulnerable under the Australian *Environment Protection and Biodiversity Conservation Act 1999*, but is listed as endangered by Garnett *et al.* (2011).

HABITATS Baudin's Black Cockatoos occur mainly in eucalypt forests, especially in associations of jarrah *Eucalyptus marginata* and marri *Corymbia calophylla* and forests of karri *Eucalyptus diversicolor*, less commonly in woodlands of wandoo *E. wandoo*, blackbutt *E. patens*, flooded gum *E. rudis* and yate *E. cornuta*, as well as in partly cleared farmlands and urban areas, including roadside trees and house gardens (Johnstone and Kirkby 2008). On occasions near Pemberton and in the vicinity of Denmark, in the far southwest, I have encountered birds in heavy forest dominated by jarrah, while to the north, near Harvey, in early September 1995, I observed a flock of approximately 30 birds in open farmland. At times, they form mixed flocks with Carnaby's Black Cockatoos, particularly at nighttime roosts during the non-breeding season, and may be found feeding in forest adjoining *Pinus* plantations, but they do not feed in the plantations, which are a favoured foraging habitat for *C. latirostris.* Johnstone and Kirkby note that local numbers of Baudin's Black Cockatoos are highest in the far southwest during the spring breeding season in September to December and in the northern Darling Range during April to August.

MOVEMENTS Depending on their region of origin, Baudin's Black Cockatoos are residents or postnuptial nomads and migrants (Johnstone and Kirkby 2008). Small numbers are resident in a few places, but in autumn the bulk of the population largely vacates the karri forests in the coldest parts of the range to move northwards and also has a strong tendency to wander from the interior towards the coast. There appears to be a definite shift westwards on to the southern Swan River coastal plain just prior to the flocks moving south to breed, and by mid October most birds are back in their nesting territories, or are heading there, and are in breeding condition.

In late September 2012, two rehabilitated females were fitted with tail-mounted satellite transmitters and released at a site near Perth (Yeap *et al.* 2015). One female travelled some 250 km south of the release site in the 123 days that its tracking device was active and, assuming that it was travelling with a flock, this sustained southern movement indicated possible migration to breeding habitat. Transmissions from the device fitted to the second bird continued to come from near the release site, and a ground search undertaken after three months located the transmitter attached to dropped tail-feathers, with wing-feathers located nearby so indicating that the bird had been killed by a predator.

HABITS In the field, it normally is extremely difficult to differentiate between the two species of white-tailed black cockatoos, so observers rarely identify the species under observation. This is especially applicable to forests in the far southwest, where either, or both species may be encountered at different times of the year, and the resulting uncertainty probably is one of the main reasons why we know so little of the habits of Baudin's Black Cockatoo. It is noisy and conspicuous when encountered in flocks in open country, but small groups in dense forest may remain undetected until they fly off to the accompaniment of loud alarm screeches. During the breeding season it usually is seen singly, in pairs, or in trios, the last comprising parents and a dependent youngster, but at times small flocks also are encountered, and these are thought to be immatures or non-breeding adults. In the non-breeding season, flocks of 30 or 40 birds commonly are observed, and these may build up to 200 or more at a favoured food source, where intermixing with Carnaby's Black Cockatoos *Calyptorhynchus latirostris* and Red-tailed Black Cockatoos *C. banksii* may take place. In early January 1975, along the Murray River, near Dwellingup, I saw flocks of 10 to 20 birds, including recently-fledged young, in the company of much greater numbers of Red-tailed Black Cockatoos feeding in marri trees *Corymbia calophylla*, and at any one time I could see numerous individuals of both species, but there was an obvious segregation of the species. In mid September 2001, near Nannup, I again encountered these cockatoos and Red-tailed Black Cockatoos in the same stand of marri woodland, and again there was a marked segregation of the two species; well in from the edge of the woodland, a pair of Baudin's Black Cockatoos was seen extracting larvae from the trunk of a tree, while some 10 to 20 Red-tailed Black Cockatoos were observed feeding at the woodland margin and in a lone tree standing in pasture at approximately 60m out from the margin. In early September 1995, near Harvey, I watched a flock of 20 to 30 Baudin's Black Cockatoos feeding on the ground in open farmland; as the birds walked across the pasture picking up seeds, individuals from the rear regularly fluttered up to alight at the front, and there was a steady interchange of birds from the ground with those perched on a nearby fence, possibly positioned to give warning of any approaching danger. In addition to watching birds on the ground, I have seen these cockatoos feeding quite low down in stunted *Banksia* bushes, but mostly they have been fairly high up in forest trees, and my impression has been that they are more arboreal than Carnaby's Black Cockatoo, though Johnstone and Kirkby (2008) note that they feed at all levels of the forest from the canopy to the ground.

Johnstone and Kirkby report that during field investigations undertaken between 2001 and 2007, daily activities were monitored at a number of communal nightime roosts occupied in the winter non-breeding season. Most roosts were in tall, emergent eucalypts, often near a watercourse and in sheltered gullies. Roosting trees need to be of a certain height and have a canopy with sufficiently dense foliage to shield the birds from the elements or possibly help them retain body heat. Such protection from adverse weather conditions is needed because these cockatoos roost individually, about 30 cm or more apart, in outermost thin branches of the canopy, where they would be exposed to strong winds and rain. This roosting preference is similar to that of Carnaby's Black Cockatoos *Calyptorhynchus latirostris*, but differs from that of Red-tailed Black Cockatoos *C. banksii*, which in the southwestern forests roost side-by-side in family groups and on thick, sheltered perches under the canopy of tall trees. Some roosts included in the field investigations had been monitored for more than 10 years, and larger roosts seem to be used annually, one having been occupied every year for at least 12 years and another every year for at least 21 years. Between August 1998 and March 2007, largest flocks recorded at roosts in the non-breeding season were 800 birds at a roost on 18 June 2006 and 680 birds at a roost on 27 August 1998. At times, up to 200 Carnaby's Black Cockatoos also were present at larger roosts.

Johnstone and Kirkby report that at nightime roosts activity commenced some 30 minutes before sunrise, when birds started calling and first calls were from begging juveniles and immatures. At first light, any small groups that roosted away from the main flock usually came into the main roost to join with the others in frequent calling. After departure from the roosts, birds usually congregated at nearby pre-roost or post-roost sites, which were in dead trees or on dead limbs of living trees that projected above the surrounding forest canopy. If suitable feeding areas, including stands of marri, are near to the roost, the flock may remain to feed in the area until mid morning, or it may remain near the roost for no more than an hour. While moving from the roost to feeding sites, the flock may make brief stopovers along the way to preen, socialise and feed, many birds staying to continue feeding at these stopover points so that the flock size at the most distant site is much reduced. Away from roosting sites, the largest groups usually are recorded within the first two hours of daylight, before dispersal of smaller family groups, and in the two hours before dusk, as flocks reform before returning to the roosts. Some birds can be observed feeding throughout most of the day, but there are obvious periods of resting, bill-cleaning, preening, allopreening and socialising. Parental feeding of juveniles occurs mostly between dawn and midday and again late in the afternoon.

Birds have been seen returning to the nightime roost between 90 and 25 minutes before sunset, again with feeding stops being made along the way so that it is almost nightfall before most birds reach the roost. There is no strict routine, with birds coming to the roost in small family groups or in loose aggregations of up to 200 birds. Counts were made as birds returned to one roost site, and a total of 3899 individuals comprised 100 single birds, 482 groups of two birds, 519 groups of three birds, 182 groups of four birds and 110 groups of five birds (Johnstone and Kirkby 2008). The only groups that could be defined with reasonable certainty were family groups of parents with a juvenile or a juvenile and an immature from a previous breeding, and groups of five also contained an immature or juvenile. Single birds probably were birds separated from a mate or family group and made up only 2.5 per cent of the total count. Groups of two birds could not be counted as breeding pairs because a proportion were presumed to be subadults and some probably were older birds that were no longer breeding.

Comprising slow, flapping wingbeats, the strong buoyant flight is similar to that of Carnaby's Black Cockatoo *C. latirostris.*

CALLS Although the calls are similar to those of Carnaby's Black Cockatoo *Calyptorhynchus latirostris*, Saunders (1979a) reported a notable difference between the familiar contact calls of the two species. The mean total duration of this call in five adult *C. baudinii* was 0.47 second and in 14 adult *C. latirostris* was 0.64 second, the difference being due mainly to a shortening of the third segment in the call of *C. baudinii*, and the resulting difference in overall length of the calls is audible, so assisting in identifying each species from its contact call. Johnstone and Kirkby (2008) describe the different contact calls as a much shorter *whicha-whicha* from Baudin's Black Cockatoo and a longer *wee-loo...wee-loo* from Carnaby's Black Cockatoo.

DIET AND FEEDING While there is overlap in the diets of this species and Carnaby's Black Cockatoo *Calyptorhynchus latirostris,* largely with supplementary foods, differences in their principal food preferences clearly reflect the morphological differences in bill shape. These differences in food preferences are evident from analyses of crop contents of both species collected at a locality some 33 km east of Perth, between April 1971 and October 1972 (Saunders 1974). The same foods were available to both species at the same time, though it must be noted that *baudinii* is a wandering visitor to the district. Baudin's Black Cockatoo took mainly seeds of marri *Corymbia calophylla,* which were present in 89 per cent of crops, and wood-boring insect larvae, which were present in 16 per cent of crops. Carnaby's Black Cockatoo fed predominantly on seeds of introduced *Pinus,* present in 81 per cent of crops, and seeds of *Dryandra* and *Hakea,* which respectively were present in 20 per cent and 19 per cent of crops. Both species occurred together in mixed flocks; on three occasions *latirostris* was seen feeding in a *Pinus* plantation, while *baudinii* was feeding in marri trees at the edge of the plantation, and none of the *baudinii* collected had *Pinus* seeds in the crop.

Although both species eat marri seeds, they employ different techniques when extracting seeds from the large bowl-like fruits. With its elongated, sharply pointed upper mandible, Baudin's Black Cockatoo expertly prises out the seeds, doing very little damage to the rim of the capsule, but Carnaby's Black Cockatoo gets at the seeds merely by breaking open the rim with its shorter, blunt upper mandible. Seeds of proteads, which are the staple food of *latirostris,* also are eaten by *baudinii,* though to a lesser extent, and both species readily come to the ground to feed on seeds of *Erodium* storksbill, an exotic weed widespread in pastures and alongside roads or railways. In early September 1995, near Harvey, south of Perth, I watched a flock of approximately 30 *baudinii* on the ground feeding on *Erodium* seeds.

Baudin's Black Cockatoo routinely uses its elongated upper mandible to dig into flower spikes or into trunks and branches of trees and shrubs to extract wood-boring insect larvae. In mid September 2001, near Nannup, I watched a pair extracting larvae from low down on the main trunk of a marri tree. The elongated upper mandible is used also to dig into apples to extract the seeds, and significant damage can be done in orchards, especially during times when production of marri seeds is low. Long (1985) reports that between 1973 and 1975, during a survey of damage to cultivated fruits by parrots, damage caused by these cockatoos was recorded opportunistically. It was noted that both green-skinned and red-skinned apples were attacked, though the latter in higher proportions, and overall damage amounted to 9.24 per cent, which exceeded considerably that caused by three species of smaller parrots.

Johnstone and Kirkby note that although marri undoubtedly is their primary food source, Baudin's Black Cockatoos have been observed feeding on a wide range of foods, including insects and insect larvae typically taken from under bark or extracted from the wood of living and dead trees, from galls and from flower spikes of *Xanthorrhoea* grass-trees.

Drinking has been recorded at various times throughout the day, but most records involve birds en route to or from their roost sites. Records are of a very wide variety of watering places, ranging from large catchment dams and open creeks in paddocks to roadside pools or pools on gravel tracks and garden fountains or ornamental ponds (Johnstone and Kirkby 2008).

BREEDING Few nests of Baudin's Black Cockatoo have been found, so little is known of the breeding behaviour, which presumably is similar to that of Carnaby's Black Cockatoo *Calyptorhynchus latirostris,* though nesting appears to take place a little later in the year. Johnstone and Kirkby (2008) note that these cockatoos are monogamous, display a strong pair-bond and probably mate for life. Other than when females are incubating or brooding, pairs stay together throughout the year. Both sexes participate in selection of the nesting hollow, but preparation or renovation of the hollow for nesting is undertaken only by the female, and involves chewing around the entrance and down one part of the interior wall. Males have been observed displaying to females in most months, but the behaviour is more frequent in August to October. Pairs have been seen inspecting prospective hollows in most months, and also outside the breeding range. A normal clutch comprises one or two eggs, and egg-laying has been recorded in August to December.

In mid September 2001, at Leeuwin Naturaliste National Park, in the far southwest of Western Australia, I watched a pair sitting on the topmost branch of a dead tree in marri woodland, and from a slightly higher position, the male was seen to feed the female by regurgitation, each interlocking of bills being preceded by a straightening of the body to a fully upright stance, while simultaneously the crest was raised and all facial feathers were conspicuously puffed out. Lower down in the same tree sat another pair with a begging juvenile, which presumably was the offspring from the previous year.

Studies of captive birds, especially a breeding pair at Perth Zoo, have provided some information on the nesting habits. By dragging the tip of the upper mandible through the sand, the male of this pair made variable 'line drawings' on the floor of the aviary, and at times these were so extensive that they covered the entire floor area measuring 6 m in length and 2.1 m in width (in Sindel and Lynn 1989). It has been suggested that these strange actions may be an unusual form of courtship display directed towards the female prior to, and during the breeding season. An aviculturist with an extensive collection of black cockatoos tells me that a male *baudinii* and a male *latirostris* in his aviaries both performed these actions though less extensively by *latirostris,* and he suspected that the birds were seeking mineral salts from the sand. This could be the purpose of these actions, but it is noteworthy that to date they are known to be performed only by males.

Bohner (1984) gives details of a successful breeding in captivity. There was an interval of two days between laying of the first and second eggs. Incubation by the female commenced after laying of the second egg and lasted exactly 28 days. Both eggs hatched on the same day, but one chick died two days later. The surviving chick fledged, but was unable to fly because of a damaged wing, and this probably explained why it had remained in the nest for 16 weeks after hatching. Hamilton (1996) reports that at Perth Zoo, a chick was fed by both parents and left the nest approximately 10 weeks after hatching, but immediately moved to a second hollow log in the aviary, where it remained until making its first flight at two and a half months of age.

EGGS The elliptical-ovate to ovate eggs are without noticeable gloss. In the Western Australian Museum, Perth, there is a single egg taken at Lowden, in the Preston River valley, and it measures 54.8 × 37.4 mm.

SUBFAMILY **CACATUINAE** G. R. Gray

This large subfamily is a mixture of white, grey, pink and black cockatoos with at least five sublineages (White *et al.* 2011). I concur with the suggestion that two tribes should be recognised – one for the distinctive Palm Cockatoo *Probosciger aterrimus* and one for other members of the cacatuine clade (see Joseph *et al.* 2012).

TRIBE **CACATUINI** G. R. Gray

Included in this large tribe are the predominantly white, pink and grey cockatoos. The broad, plain-coloured tails are proportionately shorter than in the *Calyptorhynchus* 'black cockatoos', but crests are quite variable. There is no bare facial skin, but most species have prominent bare eyerings, and the cere is feathered. Sexual dimorphism is slight or absent in most genera, and incubation is shared by the sexes.

GENUS ***Cacatua*** Vieillot

Cacatua Vieillot, *Nouv. Dict. Hist. Nat.*, **17**, 1817, p. 6. Type, by designation, *Cacatua cristata* Vieillot = *Psittacus albus* P. L. S. Müller.

The so-called 'white cockatoos' belong to this genus. Their plumage is predominantly white or pale salmon-pink, the latter being present in *C. moluccensis*. They are medium to large birds with short, squarish tails.

Within this genus there are two species groupings, each of which at times has been given subgeneric or even generic status. Differentiation at generic level was proposed by Mathews (1917) largely on the basis of variation in crest formation, but he failed to recognise that this variation is fully transitional. The crests of *C. galerita* and *C. sulphurea* comprise narrow, elongated, forward-curving feathers. In *C. ophthalmica* these feathers are broader, curve slightly backwards and are bordered anteriorly by elongated frontal feathers, a formation that is intermediate between the narrow, forward-curving and the broad, backward-curving types. The backward-curving crests of *C. alba* and *C. moluccensis* consist of very broad, elongated crown feathers. In the remaining species the backward-curving formation is the same, but both the broadening and elongation are much reduced, least in *C. haematuropygia* and most in *C. tenuirostris*.

Sexual dimorphism is absent or very slight and young birds resemble adults.

The genus is distributed from the Philippines, Sulawesi, the Moluccas and Nusa Tenggara (Sunda Islands), in the Indonesian Archipelago, to New Guinea and the adjacent islands, the Solomon Islands, and Australia, including Tasmania.

SUBGENUS ***Cacatua*** Vieillot

Species belonging to this subgenus are medium-sized to large cockatoos with broad, rounded wings, a robust black bill, and a prominent forward-curving or backward-curving crest of elongated feathers. Two species from the Indonesian Archipelago are endangered.

CRITICALLY ENDANGERED

Yellow-crested Cockatoo

Cacatua sulphurea (Gmelin)

Psittacus sulphureus Gmelin, *Syst. Nat.*, **1**, part 1, 1788, p. 330 (Moluccas, *i.e.* Celebes).

DESCRIPTION Length 33 cm. Weight unsexed 308–380 g.
ADULTS General plumage white; ear-coverts bright yellow; yellow bases to feathers of neck and underparts variably showing through; undersides of flight and tail-feathers strongly suffused yellow; narrow forward-curving crest yellow; prominent bare eyering white slightly tinged pale blue; bill grey-black; iris darkest brown in males, brownish-red in females; legs grey.
18 males: wing 215–245 (229.4), tail 98–115 (109.9) mm, exp. cul. 33–39 (37.0) mm, tars. 22–25 (23.5) mm.
20 females: wing 211–242 (222.6) mm, tail 99–113 (107.6) mm, exp. cul. 29–36 (33.2) mm, tars. 21–25 (22.9) mm.
JUVENILES Like adults.

DISTRIBUTION Timor-Leste and Indonesia, where occurs on Sulawesi and offshore islands, the Masalembu Islands in the Java Sea, and throughout Nusa Tenggara, from Nusa Penida east to Timor, though apparently now extinct on Lombok: birds seen occasionally on Bali and Java probably are escaped cagebirds

(MacKinnon 1988). Introduced to Singapore, Hong Kong and Taiwan.

SUBSPECIES The traditional practice has been to recognise four subspecies, as advocated by White and Bruce (1986), who note that some birds from Tanahjampea, Kalao and Tukangbesi Islands have smaller bills and have been separated as *djampeana*, but the difference is not constant, so *djampeana* must be synonymised with *sulphurea*. They note also that irregular size variation is common in island cockatoos and, although bills of birds on Timor and Semau are small, there is a westward clinal increase in bill size, so there is no reason to recognise the separation of birds from Lombok, Sumbawa and Flores as *occidentalis*. After comparing morphometric data from 136 sexed specimens from across the entire range of the species, Collar and Marsden (2014) reinstated both *djampeana* and *occidentalis* and described *paulandrewi* subsp. nov. for birds from the Tukangbesi Islands. I tend to prefer the traditional recognition of only four subspecies because some variation in bill size and tail length does appear to be clinal, and the differences are not striking. Irrespective of subspecific differentiation, all existing populations should be afforded effective conservation measures.

1 *C. s. sulphurea* (Gmelin)
The nominate subspecies, as described above, occurs on Sulawesi and offshore Muna and Buton Islands, and on the Tukangbesi and Tanahjampea Islands in the Flores Sea. Presumably it is the nominate subspecies that has been introduced to Singapore and Hong Kong, though Viney and Phillips (1988) note that in Hong Kong the intensity of yellow or orange in the crest is variable, indicating that at least two subspecies are present.

2 *C. s. abbotti* (Oberholser)
Kakatoe parvulus abbotti Oberholser, *Proc. U.S. Nat. Mus.*, **54**, 1917, p. 181 (Masalembu Besar Island).
ADULTS Differ from *sulphurea* by smaller pale brownish-yellow patch on ear-coverts and less yellow suffusion on bases to feathers of neck and underparts; larger size.
5 males: wing 263–273 (270.2) mm, tail 125–135 (129.8) mm, exp. cul. 34–38 (35.4) mm, tars. 25–27 (25.8) mm.
3 females: wing 258–268 (262.0) mm, tail 130–145 (137.3) mm, exp. cul. 33–35 (33.7) mm, tars. 22–26 (23.7) mm.
Survives only on Masakambing, in the Masalembu Islands in the Java Sea.

3 *C. s. parvula* (Bonaparte)
Plyctolophus parvulus Bonaparte, *Compt. Rend. Acad. Sci. Paris*, **30**, 1850, p. 139 (No locality = Semau Island).
ADULTS Differ from *sulphurea* by paler yellow ear-coverts and less yellow suffusion to feathers of neck and underparts.
15 males: wing 220–235 (227.6) mm, tail 103–119 (111.8) mm, exp. cul. 31–38 (35.8) mm, tars. 22–25 (22.8) mm.
14 females: wing 212–230 (222.6) mm, tail 104–120 (110.9) mm, exp. cul. 30–35 (32.9) mm, tars. 21–24 (22.4) mm.
Occurs on Nusa Penida and throughout Nusa Tenggara, from Lombok east to Pantar and Alor, and Timor, including nearby Roti and Semau Islands; on Sumba replaced by *citrinocristata*.

4 *C. s. citrinocristata* (Fraser)
Plyctolophus citrino-cristatus Fraser, *Proc. Zool. Soc. London*, 1844, p. 38 (Aviary specimen without locality = Sumba).
ADULTS Differ from *parvula*, by having crest and ear-coverts orange instead of yellow; larger size.
8 males: wing 244–257 (251.6) mm, tail 110–130 (120.9) mm, exp. cul. 35–39 (37.3) mm, tars. 24–27 (25.3) mm.
6 females: wing 231–254 (245.8) mm, tail 116–130 (122.3) mm, exp. cul. 31–33 (32.0) mm, tars. 23–25 (23.7) mm.
Known as the Citron-crested Cockatoo, this distinctive subspecies is confined to Sumba.

STATUS All populations of the Yellow-crested Cockatoo have suffered dramatic declines because of widespread landclearance coupled with unsustainable trapping for the live-bird market, and the cockatoos now have disappeared from some parts of the former range and are near extinction on some smaller islands. An uneven distribution in Sulawesi was reported as early as 1930–1932, when Heinrich found that in some regions they were plentiful, but in other regions were scarce and were absent altogether from many districts (in Stresemann 1940). Watling (1983) reported that during November 1978 to March 1981, in northern and central regions, they were widespread in small numbers and were favourite cagebirds in towns and villages. In 1981, when David Bishop first visited the Palu Valley and Lore Lindu Reserve, central Sulawesi, these cockatoos were relatively common and easily observed, and up until 1985 birds were recorded continuously along the road from Palu, on the coast to about 900 m at Dongi Dongi, wherever there were good stands of remnant forest (*in litt.* 1987). However, in August 1985 and again in 1986, intensive searches along the same road failed to locate any cockatoos, even though large areas of forest remained, and Bishop witnessed shipments of 200 or more birds being transported by air from Sulawesi to Bali and Jakarta. Nandika (2006) notes that although small populations may exist elsewhere in Sulawesi, these cockatoos currently are confined mostly to Rawa Aopa Watumohai National Park and Buton Island in the southeast and Pasoso Island in central Sulawesi, and it is likely that only a few hundred individuals survive in Sulawesi. A survey undertaken in Rawa Aopa Watumohai National Park in 2000 resulted in an estimated population of 100 birds or less, and during a repeated survey in 2005 only 37 cockatoos were observed at seven surveyed locations, yielding an extraordinarily low average density of 1.3 individuals per 1000 km². On Buton Island, the population was estimated to be about 150 birds in 1997, but no birds were located by a survey team in 2009 (Waugh 2013).

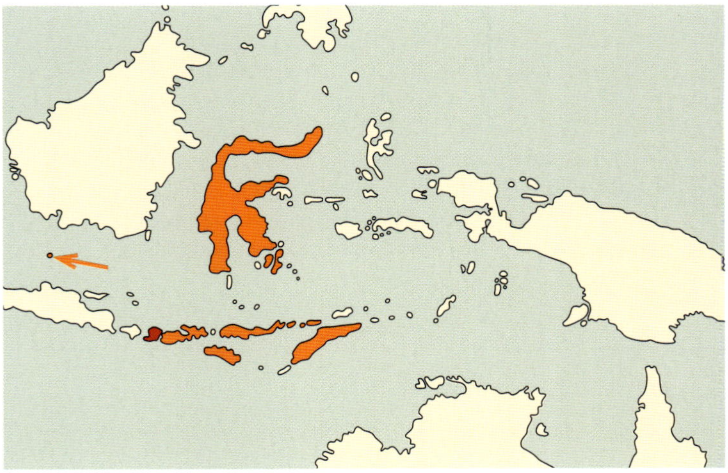

Sumba and Komodo Islands appear to support the largest surviving populations, although dramatic declines have occurred at both localities. At the end of the 19th century Yellow-crested Cockatoos were abundant on Sumba, but a serious decline in numbers during the 1970s and 1980s coincided with widespread landclearance and excessive trapping for the cagebird market. Only 10 per cent of the island is still forested, and this comprises 34 fragmented patches (Birdlife International 2016). Apart from Manupeu-Tanadaru National Park and Laiwangi-Wanggameti, which are important protected areas ensuring safer habitat in the long-term, forested areas are small, with only five of them exceeding 2500 ha in area (Cahill *et al.* 2006). In 1992, a population of 565 birds was estimated for four surveyed patches of closed-canopy forest to give an estimated island population of approximately 3200 birds or a density of about 2 birds per km^2. Before legal trade ceased in 1993 numbers of cockatoos leaving Sumba averaged approximately 1600 per year and, based on the 1992 estimated total population, this level certainly was not sustainable. To determine the effects of 10 years of international trade control, the same four forest patches were resurveyed in 2002, using the same methods, and a population of 1068 birds was estimated for the three patches where birds were observed, so extrapolating to an estimated island density of approximately 4 birds per km^2. Densities at two forest sites had increased considerably, at another the population was stable, but at one small patch no birds were observed. It was concluded that while the population has made a modest recovery, densities remain low compared to cockatoo populations elsewhere. There was evidence that trade continues, with signs of trapping being recorded and at one site many nests had ladders attached to them for the taking of chicks. In a more recent survey in 2012, extrapolations across remaining forested areas from counts made in six Important Bird Areas suggest a total population of only 563 birds (in Birdlife International 2016). It remains questionable whether the cockatoos can persist long-term in the fragmented forest patches, and their future survival is very much dependent on sound management of protected areas and effective protection against trade. On Komodo, where numbers declined by an estimated 60 per cent between 2000 and 2005, the current population is estimated at approximately 500 birds (in Birdlife International 2016). On Sumbawa, a count of 107 birds was made in a recent census, an estimated 40 to 70 birds are present on Flores, and another 100 birds on neighbouring Rinca Island (Birdlife International 2016). Strange (2013) points out that at the western extremity of the range, on Lombok, these cockatoos seem to have disappeared altogether, the last bird having been seen in 2001. At the eastern extremity of the range, the population on entire Timor Island is estimated at about 350 birds, a loose flock of about 18 birds was seen on Alor in 2009, one or two birds were captured on Pantar, and a tiny population of a few individuals survives on Roti Island. Indicative of the near extinct status of these cockatoos on small islands is the finding of a survey conducted in June–July 2008, on Masakambing, in the Masalembu Islands, when the last surviving population of *abbotti* was found to comprise only four males, four females and two juveniles, with repeat surveys finding eight birds in 2009, 11 birds in 2010, and 13 birds in 2011. On neighbouring Masalembu, the population is said to have numbered in the 'hundreds' during the early 1900s, but total extinction has been brought about by deforestation to make way for palm oil plantations, together with the shooting of birds for sport or export as souvenirs (Eaton *et al.* 2015).

The total population in the natural range is estimated at 5000 to 7000 birds, and a cooperative recovery program for the species has been adopted (in Birdlife International 2016). In addition to habitat loss and excessive trapping, psittacine circoviral disease (beak and feather disease) may be impacting on local populations, and birds showing classic symptoms of the disease have been observed. Populations occur in several national parks or reserves, and efforts are being made to enhance protective measures for these populations. At Rawa Aopa Watumohai National Park, in Sulawesi, local villagers have been employed as Forest Wardens to safeguard the birds against poachers and to monitor numbers. Also in this National Park, nests are being protected against predators by removing overhanging vegetation and fitting plastic collars around the trunks of nesting trees. Conservation awareness programs are being set up to generate support for protection measures. Mitigation of crop damage is being addressed with decoy crops, including sunflower, being planted to attract the birds away from subsistence crops. Long-term survival of Yellow-crested Cockatoos in their natural range will depend on the retention of expanses of habitat required to sustain local populations and improved enforcement of anti-trafficking measures, together with a community awareness of protection needs.

It is suspected that the feral population in Hong Kong originated from cagebirds released by British personnel at Flagstaff Hill in 1941 just prior to occupation by invading Japanese Forces, and now a self-sustaining population of 70 to 200 birds occurs on the north side of Hong Kong Island, where nesting has been recorded (Viney and Phillipps 1988). Introduction to Singapore is said to have taken place about 1927 and small feral populations were established. In May 1969, Ian Rowley observed small parties in a patch of tall rainforest in the Botanic Gardens (pers. comm. 1969). Subsequently, numbers declined rapidly and now very few birds are seen, with only four being recorded in the 2016 Annual Parrot Count (Jeyarajasingam and Pearson 2012; Strange *in litt.* 2016). Likewise in Taiwan, the feral population is very small, with records of single birds at three locations and a record of eight birds at another location listed among records for 1986 to 2000 in the Wildbird Federation Taiwan database (Lin and Lee 2006).

The Yellow-crested Cockatoo is listed on Appendix I of the Convention on International Trade in Endangered Species of Wild Fauna and Flora (CITES).

HABITATS Early reports from Sulawesi are of Yellow-crested Cockatoos ranging from the humid coastal lowlands and foothills up to about 500 m, where they frequented open woodland, cultivated fields and forest margins, though rarely penetrating into the forest interior, and around villages they were seen often in coconut palms (in Stresemann 1940). In the early 1980s, David Bishop recorded them up to about 900 m (*in litt.* 1987). In Nusa Tenggara, they have been recorded up to 800 m, occasionally ranging up to 1500 m. Nandika (2006) reports that during the survey conducted in 2000 in Rawa Aopa Watumohai National Park, southeastern Sulawesi, 28 of the 37 birds recorded were in forested areas and only nine were in cultivation, and in forested areas the birds showed a preference for the middle to

Plate 5
LEFT Yellow-crested Cockatoo *Cacatua sulphurea sulphurea* (adult ♂)
RIGHT Citron-crested Cockatoo *Cacatua sulphurea citrinocristata* (adult ♂)

F. KNIGHT '10

upper canopy. Mulyawati notes that on Sumba the availability of nesting trees is critical, but the cockatoos do not require large tracts of forest (in Birdlife International 2016). Such comments have prompted the suggestion that these cockatoos are less dependent on closed-canopy primary forest than previously claimed, but on Alor, in eastern Nusa, riverine and coastal swamp forests dominated by *Canarium* trees were identified as probably being the most important habitat because these large trees form hollows suitable for nesting and the large nuts are a preferred food item (Trainor *et al.* 2012). It seems to me that a strong dependence on forest habitat is real because tall forest trees are favoured for nesting and much foraging takes place at forest margins.

In Hong Kong, Yellow-crested Cockatoos frequent urban parks and gardens, and invariably they can be encountered in Hong Kong Park and at Victoria Barracks, while in Taiwan most sightings are in or near urban areas in western districts (Viney and Phillipps 1988; Lin and Lee 2006).

HABITS In early times, when they were much more numerous, Yellow-crested Cockatoos sometimes were encountered in quite large flocks, but most sightings now are of single birds, pairs or small parties. In Hong Kong, where the feral population is concentrated in urban parklands and gardens, larger flocks can be encountered (Viney and Phillipps 1988). They are noisy, conspicuous birds, especially when perching in the topmost branches of tall trees at the forest margin, where their white plumage contrasts strikingly against the green foliage. When daily activities were recorded in 2000 at Rawa Aopa Watumohai National Park, southeastern Sulawesi, morning activities consisted of perching in trees, accounting for 43 per cent of observed time, with 12 per cent of observed time for eating, 22 per cent of observed time for playful actions, and 22 per cent of observed time for other activities, including sunbathing in the treetops, moving about, preening and vocalising (Nandika 2006). In the afternoon, the observed time for eating increased to 30 per cent, with perching decreasing to 27 per cent of observed time, and observed time for other activities also declined, though socialising increased from 1 per cent to 9 per cent of that time. Preference was shown for the middle canopy, where birds were present for an average 61 per cent of observed time, followed by occupation of the top spatial canopy in 30 per cent of observed time for performing daily activities. All socialising occurred in the middle canopy, as did 51 per cent of feeding and 79 per cent of moving about. Perching on dry branches in the top canopy predominated during mornings, while the middle canopy seemed to provide protection from potential predators and shelter from the heat of the day. The lower canopy was used only for perching and limited movements. Also at Rawa Aopa Watumohai National Park, observed interactions with non-psittacine species included fighting with a Purple-winged Roller *Coracias teminckii* at a nesting hollow, pursuit of a cockatoo by an Asian Glossy Starling *Aplonis panayensis*, and pursuit of another cockatoo by a Sulawesi Serpent-eagle *Spilornis rufipectus*.

Bishop recalls that in southeastern Sumba, from a high position on a steep escarpment, small numbers of cockatoos observed sailing over a valley of undisturbed monsoon forest appeared to be searching for fruiting trees and, as soon as a few birds alighted to feed, other birds converged on the site. The strong flight comprises rapid, shallow wingbeats interspersed with gliding on slightly downcurved wings, and when travelling long distances the birds fly at a considerable height, gliding down to the trees in wide, sweeping circles. Also on Sumba, these cockatoos have been observed roosting with Eclectus Parrots *Eclectus roratus*.

CALLS In Sulawesi, recorded calls include a variety of loud, far-carrying single coarse, harsh screeches and a variety of less harsh whistles and squeaky notes (Coates and Bishop 1997). Calls recorded on Timor include a series of two to six hoarse, nasal, medium-pitched to high-pitched, slightly quavering screeches.

DIET AND FEEDING Yellow-crested Cockatoos feed mainly in the treetops, but will come to the ground, especially when attracted to crops or village gardens. At Rawa Aopa Watumohai National Park, southeastern Sulawesi, recorded foods include fruits of tompira *Vitex coffasus*, bitti *V. galabra*, tangkalase *Gmelina asiatica*, kayu besi *Instia bijuga*, kayu raja *Cassia* sp., soasoarate, and onangki, seeds of tampate *Lagerstoemia foetida*, flowers of bamboo *Bamboosa spinosa*, leaf buds of tahulo *Mallotus floribundus* and young leaves of kuiya *Alstonia scholaris* (Nandika 2006). Other likely foods include flowers and fruits of *Avicennia* mangroves, fruits of jambu bol *Eugenia malaccensis*, and flowers, fruits or seeds of introduced or cultivated plants, including bananas, mangoes, papaya, figs, guavas, custard apples, tamarind *Tamarindus indica*, and *Opuntia* prickly pear. There are reports of maize crops being raided.

BREEDING On Buton Island, specimens in breeding condition were collected in September and October, and on Komodo Island active nests were found in March and April (in van Bemmel and Voous 1951; Benstead 2005). During field studies undertaken at study sites on Sumba between September 2001 and May 2002, active nests were found in September through to March, with levels of nesting activity peaking in October, and this was linked significantly to warmer and drier weather conditions, there being a steady decline after December corresponding with decreasing temperatures and increasing rainfall (Walker *et al.* 2005). A strong preference for nesting in hollows in tetrameles *Tetrameles nudiflora* and *Chisocheton* trees was evident, with *T. nudiflora* being the most preferred nesting tree. Trees preferred by the cockatoos for nesting had an average trunk diameter of 1.16 m at breast height, and the smallest tree in which interest at a hollow by cockatoos was observed had a trunk diameter of approximately 0.68 m at breast height, so this was considered to be the minimum size acceptable to the birds. At one study site, trees with potentially active nesting hollows averaged 27.3 m in height, with an average diameter at breast height of 0.49 m, and on average there were another 2.04 hollows. The potentially active nesting hollows were at an average height of 30.2 m, at an angle of 8° to the vertical, and entrances averaged 0.25 m^2. Interspecific interactions were observed at all potentially active nesting hollows, with Eclectus Parrots *Eclectus roratus* and Great-billed Parrots *Tanygnathus megalorynchos* being the main competitors. The 2001–2002 breeding season, when the field study was undertaken, was the wettest for 10 years on Sumba, and that probably contributed to the very poor nesting success rate, with actual nesting attempts being made at only eight monitored hollows and just a single chick fledging.

Benstead reports that on Komodo Island, 30 potential nesting trees were found in March–April 2005, and there were 18 active nests. At one locality, the cockatoos tended to select banilad *Sterculia oblongata* as their nesting tree, and at another locality they favoured gebang palms *Corypha utan*. Nesting hollows were at a height of 7.2 m to 18.2 m above the ground. At

Rawa Aopa Watumohai National Park, southeastern Sulawesi, in 2000, six apparent nesting hollows were located, with at least two seeming to be active, and at one the mated pair with crests raised interacted energetically at the entrance; these nests were at heights of 10 m to 20 m in kuiya *Alstonia scholaris* or parinari *Maranthes corymbosa* trees (Nandika 2006).

What is known of nesting behaviour comes from breeding in captivity (in Forshaw and Cooper 1989). Up to three eggs were laid, and incubation lasted about 24 days. Both sexes incubated, the male sitting during the day, and the chicks fledged some 10 weeks after hatching.

EGGS The eggs are elliptical; measurements of seven eggs of *C. s. sulphurea* from Sulawesi are 41.2 (38.1–44.0) × 27.1 (25.7–28.4) mm, and of two eggs of *C. s. citrinocrista* from Sumba are 38.0 × 25.1 mm and 40.1 × 28.6 mm (in Schöwetter 1964).

ENDANGERED

White-crested Cockatoo

Cacatua alba (P. L. S. Müller)

Psittacus albus P. L. S. Müller, *Natursyst.*, suppl., 1776, p. 76 (Moluccan Islands).

OTHER NAME White Cockatoo (considered inappropriate as it is a commonly used collective term for many *Cacatua* species).

DESCRIPTION Length 46 cm. Weight unsexed 440 g.
ADULTS General plumage white, including broad, backward-curving crest; undersides of flight-feathers and tail-feathers basally suffused yellow; naked eyering cream-white; bill grey-black; iris darkest brown in male, reddish-brown in female; legs grey.
15 males: wing 272–312 (288.8) mm, tail 120–166 (148.8) mm, exp. cul. 34–45 (40.8) mm, tars. 27–32 (29.5) mm.
14 females: wing 252–301 (284.8) mm, tail 126–168 (146.1) mm, exp. cul. 36–45 (40.9) mm, tars. 26–33 (28.8) mm.
JUVENILES Like adults.

DISTRIBUTION Occurs on Halmahera, Bacan, Ternate, Tidore, Kasiruta and Mandiole Islands in North Maluku, Indonesia; records from Obi and Bisa Islands, North Maluku, probably of escaped cagebirds. Introduced breeding population in Taiwan.

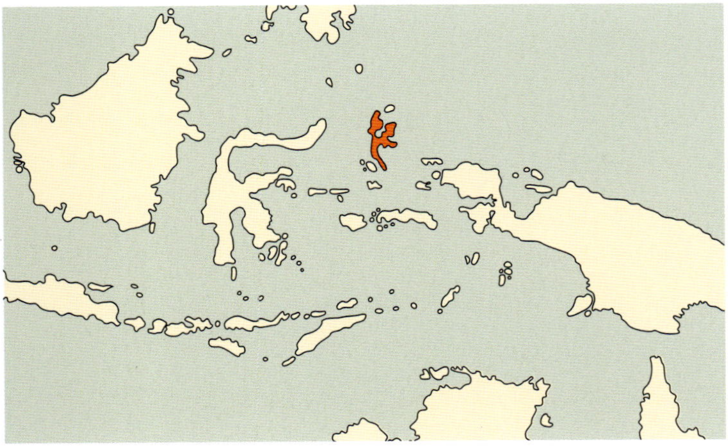

STATUS Although White-crested Cockatoos remain locally common, there has been a rapid population decline, due principally to unsustainable levels of trapping and nest-robbing for the live-bird trade, and declines are likely to increase with continued pressure from illegal trade adding to the adverse impact of forest clearance for plantations, cultivation and mining (in Birdlife International 2016). Lambert (1993a) reported that between October 1991 and February 1992, surveys were undertaken on Bacan, Kasiruta, Halmahera and Obi, and population estimates were based on approximate areas of broadly defined habitat types within the altitudinal range of the cockatoos, with the assumption that the birds were fairly evenly dispersed throughout their range and population densities recorded in different habitats were typical of densities in areas of suitable habitat that were not surveyed. Calculated total population estimates were a minimum of 7220 and maximum of 29 300 individuals for the Bacan Island Group and a minimum of 49 765 and maximum of 212 430 individuals for the Halmahera group. I would caution that these estimates were made almost 25 years ago, when forest cover on the islands was much more extensive than at present, so they may not reflect present population levels. Evidence supporting the suggestion that records from Obi Island are of escaped cagebirds was obtained during a survey undertaken on the island in July–August 2012, when White-crested Cockatoos were not recorded and local people were not familiar with the birds, though most parrot-trappers knew them as coming from Bacan, and there was a report from villagers at a northern locality of several white cockatoos living in the nearby hills (Mittermeier *et al.* 2013).

Lambert notes also that, by combining the number of birds exported with the proportion dying and those entering domestic trade, estimates of total numbers taken for the live-bird trade were 7110–7500 in 1990 and 3482–3674 in the first half of 1991. It was noted also that significant additional numbers enter illegal international trade, and Lambert concluded that trapping levels at that time were unlikely to be sustainable, there being inadequate biological data to confidently predict a realistic upper limit of sustainable exploitation. In recent years, annual harvests have declined in actual terms and as a proportion of the remaining population, but illegal trade continues, and is likely to have been underestimated (in Birdlife International 2016). Catch quotas have been exceeded by up to 18 times in some localities, indicating that in each year trappers were removing up to 17 per cent of the population.

Although forest remains relatively intact within parts of the range, logging has become quite intensive, and areas have been cleared for cultivation and mining. It is estimated that between 1990 and 2003, there was about a 20 per cent loss of forest within the range of White-crested Cockatoos, and a loss of approximately 65 per cent is projected to occur over the next three generations (in Birdlife International 2016). Significant changes in forest cover on Halmahera appear already to have brought about a decline in numbers of cockatoos. Habitat and, more importantly, nest-site availability are decreasing, and new logging roads greatly facilitate access for trappers.

The feral population in Taiwan is very small and probably not self-sustaining. During an assessment survey undertaken between September 1998 and December 2000, the only records were of three birds at one locality near Taipei and a single bird at the campus of National Sun Yat-Sen University, Kaohsiung (Lin and Lee 2006).

F. KNIGHT '15

HABITATS Lambert (1993a) reports that during field surveys undertaken between October 1991 and February 1992 on Obi, Bacan and Halmahera, White-crested Cockatoos were found to frequent primary and logged forest, and frequently they visited trees at the forest edge or tall trees within recently cleared lands. They appeared to favour flat or gently sloping areas in the lowlands, being scarce on steep slopes, and were not seen above about 600 m or rarely above 300 m on Bacan. Also on Bacan, they appeared to be absent from primary forest on the steep slopes of Mount Sibela, suggesting that the Mount Sibela Wildlife Sanctuary of 400 km² may protect insufficient cockatoo habitat to support a viable population. Contrastingly, they were relatively common between 300 m and 500 m on nearby Kasiruta Island, at a site with more gently sloping land than at surveyed sites at similar altitudes on Bacan.

These cockatoos occur also in mangroves, plantations, including coconut plantations, and agricultural lands, suggesting that they tolerate some habitat modification, but highest densities are in primary forest, and large trees are required for nesting and communal roosting (Birdlife International 2016).

HABITS In common with most cockatoos, White-crested Cockatoos are noisy and conspicuous, especially when perched in the tops of forest trees, where their white plumage contrasts strikingly against the green foliage and their discordant screeching attracts attention. Lambert (1993a) recalls that during surveys undertaken between October 1991 and February 1992 they were observed mostly in pairs or small groups, though loose flocks of up to 11 individuals also were encountered, and most frequently they were in the forest canopy or emergent strata, with few observations being at lower levels. Congregations of up to 50 birds can be encountered in the late afternoon, and they generally are fairly confiding, though more wary where trapping occurs (Coates and Bishop 1997). They often perch on large, shaded branches in the lower crown of a tall tree, but seldom on exposed dead branches. The fairly swift flight comprises rapid, shallow wingbeats interspersed with gliding on slightly downcurved wings.

CALLS The most frequently heard call is described as a short, loud, very nasal and high-pitched, penetrating screech repeated at intervals of 0.5 to 1.0 second, and birds in flight utter a rapid series of raucous, but slightly less high-pitched notes (Coates and Bishop 1997).

DIET AND FEEDING Lambert (1993a) points out that these cockatoos appear to be primarily arboreal feeders, taking seeds and fruits in the upper forest canopy. Bark-peeling actions observed on Bacan and frequent investigations of clumps of epiphytic vegetation suggest that larger invertebrates may be included in the diet.

BREEDING Little is known of nesting habits, but all available information indicates that hollows in the largest forest trees are selected, and Lambert (1993a) reports that, of two probable nest-sites observed on Bacan in October–November 1991, one was at an altitude of 90 m in a tree at the edge of recently cleared land bordering primary forest, and the other was in logged forest at an altitude of 70 m. Both hollows were at points where large branches had broken away from the main trunks. In Halmahera, villagers told Lambert that nests contain eggs in May, and trappers had a young bird that had been taken from a nest in a *Canarium* tree in August. A young bird found on Bacan probably was taken from a nest in March or April, and Smiet (1985) reported seeing several breeding pairs on Bacan in April. These records indicate that the breeding season commences in the early part of the year.

It is claimed that wild-caught birds do not breed until they are six years old, and local villagers told Lambert that nests contained one or two chicks (in Lambert 1993a). In captivity, clutches of two, or rarely three eggs are laid and incubation by both sexes lasts approximately 28 days, with fledging occurring at two to three months after hatching (in Forshaw and Cooper 1989).

EGGS Measurements of two eggs are 40.5 × 30.0 mm and 41.0 × 31.6 mm (in Schönwetter 1964).

SUBGENUS *Licmetis* Wagler

Licmetis Wagler, *Abh. K. Bayer. Akad. Wiss., Math. - Phys. Kl.*, **1**, 1832, p. 505. Type, by monotypy, *L. tenuirostris -Psittacus tenuirostris* Kuhl.

Members of this subgenus are small to fairly large cockatoos with short to very short, recumbent crests and proportionately small, horn-coloured bills, which in two Australian species are modified for subterranean foraging.

Distribution is centred on Australia, where three species occur, and all three are common, but at the western extremity, in the Philippine Islands, the sole species is critically endangered.

CRITICALLY ENDANGERED

Philippine Cockatoo

Cacatua haematuropygia (P. L. S. Müller)

Psittacus haematuropygius P. L. S. Müller, *Natursyst.*, Suppl., 1776, p. 77 (Philippine Islands).

OTHER NAMES Red-vented Cockatoo, Red-vented Corella.

DESCRIPTION Length 31 cm. Weight 343 g.
ADULTS General plumage white; ear-coverts tinged yellow-pink; bases to feathers of crest yellow and rose-pink; undersides of flight-feathers pale yellow; under tail-coverts orange-red; undersides of tail-feathers deep yellow; bare eyering white, sometimes tinged blue; bill greyish-white; iris dark brown in male, brownish-red in female; legs grey.
14 males: wing 209–231 (218.8) mm, tail 100–110 (105.1) mm, exp. cul. 25–29 (27.1) mm, tars. 22–24 (23.2) mm.

Plate 6
White-crested Cockatoo *Cacatua alba* (adults)

15 females: wing 201–230 (210.9) mm, tail 100–110 (103.1) mm, exp. cul. 23–28 (25.8) mm, tars. 22–24 (22.6) mm.
JUVENILES Like adults; iris brown.

DISTRIBUTION Philippine Islands, including the Palawan Group and the Sulu Archipelago, but now extirpated from much of the range.

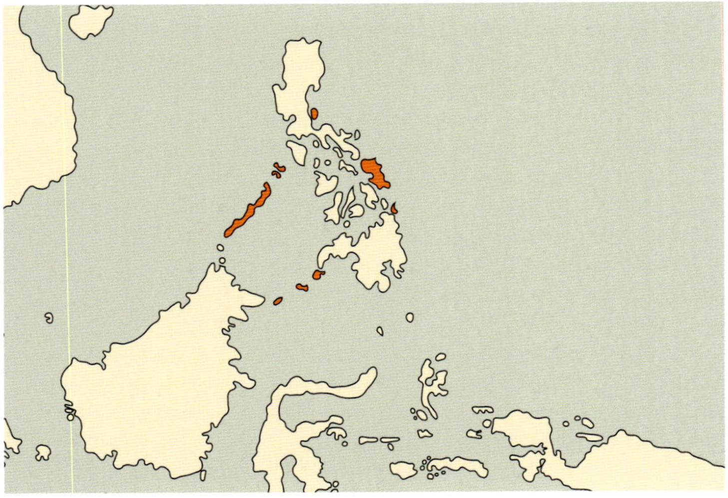

STATUS Widespread deforestation coupled with excessive exploitation for the live-bird trade have forced the Philippine Cockatoo to the brink of extinction, and the total population now is estimated at about 560 to 1500 individuals (Birdlife International 2016). A striking example of the rapid decline in numbers and effects of exploitation comes from Siquijor Island, southern Philippines, where only a single pair remained in 1991, and the nest had been robbed in each of the previous five years (Evans *et al.* 1993). Up until the 1950s, these cockatoos were common throughout much of the Philippine Islands, and in the early 1970s were recorded as being quite common in remnant forest on some islands, including Dinagat and Siargao which were visited by the Delaware Museum of Natural History and Mindanao State University Expedition (duPont and Rabor 1973a). From about the 1980s, extensive destruction of mangroves and lowland forest together with massive exploitation for the live-bird market, which included a near-total harvesting of nestlings, were responsible for a rapid decline estimated at approximately 80 per cent in 40 years, and leaving reasonable numbers only in the Palawan Islands and the Sulu Archipelago (in Birdlife International 2016). Also contributing to the decline was persecution of the cockatoos as pests in agricultural areas, and it is possible that typhoons may have accelerated the local extinction of already weakened populations, particularly in eastern regions where these weather conditions are more prevalent (Widmann and Widmann 2008). In the Palawan Islands, where the population is estimated at 440 to 700 birds, conservation efforts are focused strongly on Rasa and other satellite islands. An estimated 100 to 300 birds are thought to remain in the Sulu Archipelago, but a difficult security situation has prevented surveys to determine present numbers (Widmann and Widmann 2008). Elsewhere in the Philippines, recent surveys could confirm the existence of remnant populations only in the Polillo group of islands, east of Luzon, and in Samar, but these and any other subpopulations mostly are very small with little prospect of long-term recovery (Widmann and Widmann 2008; in Birdlife International 2016).

Boussekey (1995) reported that during a visit to Palawan Island in August 1995, more than 100 cockatoos were seen in four flocks at four localities, and these observations, together with information obtained from fieldworkers, indicated that some viable populations survived in the Palawan Faunal Region. In 1998, attention was focused on Rasa, a tiny coral island off the eastern shore of mainland Palawan, where the surviving population of 23 to 25 cockatoos was at risk of imminent extinction by rampant poaching and the conversion of coastal forest to coconut plantations (Widmann and Widmann 2008). The Philippine Cockatoo Conservation Program was set up to prevent the loss of this island population, and the single most important intervention in the program was securing the agreement of all active poachers to be trained and employed as wildlife wardens. In the next breeding season, and for the first time in many years, there was no harvesting of nestlings, resulting in a subsequent slight increase in numbers counted at the traditional roost site. In the meantime, Rasa Island was protected under local ordinances and nationally as a wildlife sanctuary under a presidential proclamation, so safeguarding more than 8 km² of core habitat in coastal and mangrove forests. After 2003, the conservation program was extended to additional sites with remnant cockatoo populations. On Dumaran Island, in the northeast of the Palawan Faunal Region, the last two forest patches with breeding cockatoos were only 1.5 km² and 0.6 km² in size, and the birds were threatened mainly by habitat destruction and persecution as agricultural pests. Exacerbating the loss of habitat was the frequent chopping down of nesting trees by poachers. Apart from nest protection, key activities on this island have been experimental habitat restoration and intensive conservation education to instill a sense of pride in having these cockatoos. In the south of Palawan Province, a nest protection scheme was started on Pandanan Island in 2008, and within four years the cockatoo population had increased from about 40 to 110 birds (in Birdlife International 2016). Other project sites have been selected in Palawan Province and on Patnanungan in the Polillo Islands (Widmann and Widmann 2008).

That increasing numbers on Rasa Island are putting an increasing pressure on food resources in the relatively small extent of habitat became apparent in 2005 when a scarcity of fruiting and flowering of food plants due to persistent drought resulted in several nestlings starving to death and the need to rescue 12 chicks for hand-rearing (Widmann and Widmann 2008). This setback recurred in 2010 when severe drought conditions again impacted seriously on nesting success, with 24 chicks being produced from 51 eggs in 15 of 25 occupied nests, the high percentage of infertile eggs possibly due to a scarcity of food available to adults prior to egg-laying, and 19 chicks died from starvation or mite infestation, leaving the remaining five to be rescued for supplemental feeding. Some of the rescued birds were used in an experimental translocation program undertaken at a resort island and, although the birds adapted well to conditions in the wild, their tameness resulted in unwelcome interactions with resort guests and they had to be recaptured. Sites again are being assessed for their suitability for translocation, and on Siargao Island, in southeastern Philippine Islands, which is being

Plate 7
Philippine Cockatoo *Cacatua haematuropygia* (adults)

F KNIGHT 10

prepared for a reintroduction attempt, a former poacher has been employed to patrol the potential release site and seedlings of cockatoo food plants were planted.

In July 2008, the number of cockatoos counted at the traditional roosting site on Rasa Island passed 200 for the first time, and foraging flocks were observed regularly in adjacent coastal areas on mainland Palawan. To mitigate against future food shortages hundreds of horseradish trees *Moringa oleifera* have been planted, and artificial nestboxes have been installed on the island and on mainland Palawan. If the success of the program on Rasa Island can be repeated at other project sites, chances for the long-term survival of Philippine Cockatoos will be enhanced significantly.

The Philippine Cockatoo is listed on Appendix I of the Convention on International Trade in Endangered Species of Wild Fauna and Flora (CITES).

HABITATS Philippine Cockatoos frequent mainly lowland, riverine and mangrove forests, but may occur also at forest margins, in open cultivation and at higher elevations (Kennedy *et al.* 2000). Boussekey (1995) recalls that during a visit to Palawan Island, southern Philippines, in August 1995, one of the four observed flocks was feeding in forest surrounding a small village, another was roosting near the seashore, another spent some 30 minutes feeding on the ground before leaving to roost in nearby trees, and the final flock was in a mangrove area. Widmann notes that these cockatoos are able to utilise regenerating forest for foraging and also for breeding if suitable trees are present (in Birdlife International 2016).

HABITS Philippine Cockatoos are noisy and conspicuous, especially while perched in the tops of forest trees, where their white plumage contrasts strikingly against the green foliage. During the breeding season they usually are seen singly, in pairs or small parties, but at other times normally in flocks of up to 30 birds, which wander widely in search of food. Lambert (1992) notes that on Palawan many pairs appear to roost near their nest sites, while others congregate in tall coconut palms on deserted beaches of offshore islands. On Rasa Island, between 1998 and 2000, activities were observed at a traditional nighttime roost in a forest of mangroves *Sonneratia alba*, near to the outer edge facing the ocean, and it was used throughout the year, though numbers varied, being highest in the non-breeding season and lowest when breeding pairs roosted at night in, or near to their nesting trees. (Widmann *et al.* 2001). The first birds arrived at this roost about 20 minutes before sunset, or later in the evening when strong winds prevailed, and they moved about between emergent trees before finally settling in a single tree for the night. The roosting site was occupied regularly also by Slender-billed Crows *Corvus enca*, and occasionally by *Ducula* imperial pigeons and White-breasted Woodswallows *Artamus leucorynchus*.

At places where they are trapped or persecuted these cockatoos are alert and wary, but elsewhere they can be quite confiding. Boussekey (1995) notes that on Palawan a flock seen feeding in forest surrounding a small village showed no fear of humans and was easily approached because the birds were protected by the local people despite some raiding of ricefields during harvest, and other birds at a seashore roost were not shy.

Potter (1953) noted that the flight features rapid wingbeats, and is swift and direct, though on three occasions he saw birds escape from pursuing hawks by darting and weaving. Flights of up to 8 km across open seas are said to be made between island roosting sites and feeding areas on the adjacent mainland.

CALLS The loud raucous calls have been described as varying from *eeeek* to *owwwwk* and *rouuuk*, and are particularly loud when several birds are calling together, especially in flight (Kennedy *et al.* 2000).

DIET AND FEEDING Seeds, nuts, fruits, berries and flowers make up the diet, and these are taken both on the ground and in the treetops. On Rasa Island, Palawan, between 1998 and 2000, most feeding observations were of two or more birds and, if undisturbed, they could remain in a food tree for long periods, with recorded bouts of feeding lasting from six minutes up to one hour or longer (Widman *et al.* 2001). On one occasion cockatoos were seen to carry the fruits of bagalunga *Melia dubia* out of the source tree and consume them in an adjacent tree about 30 m away. At times the cockatoos were seen feeding in the company of Blue-naped Parrots *Tanygnathus lucionensis*, *Ducula* imperial pigeons, Slender-billed Crows *Corvus enca* and Asian Glossy Starlings *Aplonis panayensis*. Food plants recorded on Rasa Island, together with the food items eaten, are listed in Table 1. Also on Rasa Island, horseradish trees *Moringa oleifera* have been selected as an important food source for propagation (Widmann and Widmann 2008). Flocks will raid ripening maize crops, sometimes causing significant damage.

Potter (1953) noted that white rectangular seeds 18 mm in length filled the crop of a specimen collected on Calicòan Island.

Food plants	Food items eaten
Burseraceae	
Garuga floribunda bogo	seeds
Caricaceae	
Carica papaya papaya	seeds, pulp
Fabaceae	
Erythrina variegata dapdap	flower, nectar ?
Pithecellobium dulce kamachile	seeds
Leucaena leucocephala ipil-ipil	seeds
Gliricidia sepium kakawate	seeds, bark
Pterocarpus indicus narra	seeds
Meliaceae	
Melia azedarach bagalunga	seeds ?
Moraceae	
Ficus spp. 'balete'	fruits
Moringaceae	
Moringa oleifera malunggai, horseradish tree	seeds
Sapindaceae	
Pometia pinnata malugai	fruits, seeds
Lythraceae	
Sonneratia alba pedada	fruits
Sterculiaceae	
Pterocymbium tinctorium taluto	flowers, seeds
Ulmaceae	
Trema orientalis anabiong	fruits, seeds
Arecaceae	
Cocos nucifera coconut	flowers
Poaceae	
Oryza spp. rice	unripe seeds
Zea mays corn	seeds

Table 1. Food plants recorded on Rasa Island in 1998–2000, together with the food items eaten (after Widmann *et al.* 2001).

BREEDING In 2000, nesting was monitored on Rasa Island, and it was observed that pair-bonding activities, including mutual preening at the roosting site, intensified in October (Widmann *et al.* 2001). Selection of nesting hollows commenced at the end of December, but the birds were not marked so it could not be determined if pairs were returning to nest-sites used in previous years. Attention to nesting hollows and the immediate surrounds included a widening of the cavities by chewing away wood from the entrance and internal walls, and biting off twigs or leaves interfering with access or visibility, the accumulated discarded twigs and leaves on the ground indicating occupancy of the hollow. Of 16 nests, 10 were in coastal forest and five were in mangroves, and all were in just five tree species – malugai *Pometia pinnata*, bogo *Garuga floribunda*, taluto *Pterocymbium tinctorium*, magtalisay *Terminalia foetidissima* and pagatpat *Sonnertia alba*. One nest in a dead taluto tree was destroyed because of decay. Almost all nesting trees were emergent above the forest canopy, and ranged from about 15 m to 48 m in height. Nesting hollows were at heights of 12 m to 35 m above the ground, and they varied between 10 cm and 25 cm in diameter, between 1.1 m and 2.0 m in depth, leading to an egg-chamber of 20 cm to 40 cm in diameter. Egg-laying was recorded between late February and early April, with a peak between the end of February and the beginning of March. Clutches comprised two or three eggs, but there was one clutch of four eggs, all of which survived. The duration of incubation was not determined, but chicks were first recorded in mid March, and fledging occurred about eight weeks after hatching. From eight nesting pairs, 18 chicks fledged successfully, with five nesting attempts being unsuccessful, and the outcome of two attempts in mangroves remained unknown. Fledglings and their parents remained in the near vicinity of the nest-site for about one week before dispersing.

In captivity, incubation by both parents lasts approximately 28 days, and fledging occurs about nine to 10 weeks after hatching (in Low 1993b).

EGGS A single egg laid in captivity and held in the Natural History Museum at Tring, UK, is elliptical and measures 37.7 × 26.7 mm (Harrison and Holyoak 1970).

SUPERFAMILY
PSITTACOIDEA Rafinesque-Schmaltz

Within this superfamily a major radiation among all other parrots is reflected in the systematic arrangement recognising three major groupings at family level rank (Joseph *et al.* 2012). In molecular analyses African *Psittacus* and *Poicephalus* have been recovered as sister to Neotropical parrots, and this Afro-Neotropical group making up the family Psittacidae is sister to all other parrots in the superfamily, and may be worthy of differentiation as another superfamily. *Coracopsis* from Madagascar and *Psittrichas* from New Guinea appear to be sister taxa, and separation of these two genera in the family Psittrichasidae reflects the biogeographically relictual nature of their relationship. Distribution within the family Psittaculidae is strongly centred in the Papuo-Australasian region, with only *Psittacula* and probably extinct *Mascarinus* ranging to Africa and the Mascarene islands.

FAMILY **PSITTACULIDAE** Vigors

Two major subgroupings have been identified within Psittaculidae. The so-called platycercines or 'broadtailed parrots' together with the lories and their allies make up a predominantly Australo-Papuan radiation, and the so-called 'coral-billed' parrots together with the *Micropsitta* pygmy parrots make up a more Afro-Asian radiation. The *Psittacella* tiger-parrots from New Guinea have been allied with both subgroups, but molecular analyses have revealed that they appear to be the sole extant representatives of an early divergence within the predominantly Australo-Papuan radiation. Subgroupings within Psittaculidae are differentiated as subfamilies.

SUBFAMILY **PLATYCERCINAE** Selby

The so-called 'broadtailed parrots' belonging to this subfamily include some of the best-known of Australian parrots. A broad, long graduated tail is present in all but a few specialised genera. The slight, usually blunt-tipped bill may be black, grey or horn-coloured, but is never red. The pronounced cere is naked or partly feathered. In most species sexual dimorphism is slight, but juveniles are duller than adults or may exhibit a distinctive plumage pattern. Females and juveniles of most species show an 'underwing-stripe', which comprises a series of variable white or yellow-white spots on the undersides of inner webs of all but the outermost flight feathers.

TRIBE **PLATYCERCINI**

A group of closely related 'rosella-like' genera belonging to this tribe have been identified as the 'core platycercines' (Christidis *et al.* 1991; Joseph *et al.* 2012).

The tribe is distributed from Fiji, Tonga, New Caledonia and New Zealand, including outlying islands, to Australia.

GENUS ***Psephotellus*** Mathews

Psephotellus Mathews, *Austr. Av. Rec.*, **2**, 1913, p. 57. Type, by original designation, *Platycercus pulcherrimus* Gould.

The three allopatric anthill-nesting parrots belonging to this genus are closely allied, and may be considered a single superspecies. Brown upperparts is a prominent feature of the distinctive plumage pattern of adult males, and there is a conspicuous yellow or red 'shoulder-patch'. Traditionally, they have been included in *Psephotus*, usually with subgeneric differentiation, but molecular analyses indicate that generic differentiation is more appropriate, and this is supported by morphological and behavioural characteristics. Lendon (1979) pointed out that, unlike the Red-rumped Parrot *Psephotus haematonotus*, they do not indulge in mutual preening and do not adopt the 'rosella-like' courtship display featuring prominent 'shoulder-squaring' and 'tail-wagging'.

Sexual dimorphism is not present in juveniles, which resemble the adult female. In fledglings, the bill is pale orange-yellow but soon changes to the colour of adults.

The genus is restricted to northern Australia. The three species are specialist feeders of small grass seeds, and all have suffered dramatically from changes to groundcover vegetation brought about by inappropriate fire regimes and the expansion of pastoral development. The Paradise Parrot *Psephotellus pulcherrimus* is extinct, and the Golden-shouldered Parrot *P. chrysopterygius* is endangered. Only the Hooded Parrot *P. dissimilis* survives in reasonable numbers, but there has been a significant contraction of its range.

ENDANGERED

Golden-shouldered Parrot

Psephotellus chrysopterygius (Gould)

Psephotus chrysopterygius Gould, *Proc. Zool. Soc. London*, 1857 (1858), p. 220 (Australia in lat. 18°S, long. 141° 30′E = *c.* 30 km south of Normanton and environs of Haydon *c.* 50 km southeast of Normanton, Queensland, determined by Schodde, 1997).

DESCRIPTION Length 26 cm. Weight male 56 g, female 54 g.
ADULT MALE Forehead, lores and below eye lemon-yellow, slightly tinged greenish-blue; crown, nape and above eyes black, merging into brownish-black on hindneck, where feathers faintly edged blue; chin and foreneck greyish; flanks, breast and sides of head turquoise-blue; abdomen, thighs and under tail-coverts orange-red with white bases and tips to feathers; back, mantle and lesser wing-coverts earth-brown; median wing-coverts yellow; under wing-coverts, bend of wing, outermost secondaries, primary-coverts and outer webs of primaries blue; inner secondaries greyish-brown, margined blue on outer webs; rump and upper tail-coverts turquoise-blue; central tail feathers above dull bronze-green, becoming bluish-black towards tips, lateral tail feathers greenish-blue, broadly tipped bluish-white; underside of tail bluish-white; underwing-stripe absent; bill greyish-horn; iris brown; legs greyish-brown.
4 specimens: wing 112–113 (112.3) mm, tail 138–157 (145.4) mm, exp. cul. 11–13 (12.3) mm, tars. 14–15 (14.5) mm.
ADULT FEMALE Forehead pale buff-yellow; crown and occiput dull bronze-green, faintly tinged brown; hindneck, mantle, back and wing-coverts dull yellowish-green; rump and upper tail-coverts pale turquoise-blue; chin and cheeks dull bronze-green, feathers margined dusky grey; sides of neck, throat and upper breast dull green, shading to palest turquoise-blue on lower breast and upper flanks; abdomen, lower flanks and under tail-coverts dull white, suffused pale greyish-blue; faint red margins to feathers at centre of abdomen; under wing-coverts, bend of wing, primary-coverts and outer webs of primaries and outermost secondaries pale blue; broad underwing-stripe of pale yellow present.
4 specimens: wing 105–110 (107.6) mm, tail 130–151 (137.0) mm, exp. cul. 11–13 (11.8) mm, tars. 14–15 (14.5) mm.
JUVENILES Resemble adult female; young males have brighter turquoise-blue cheeks, more pronounced brownish suffusion to crown and occiput, and more pronounced pink-red margins to feathers of abdomen; underwing-stripe present; bill orange-yellow in very young birds.

DISTRIBUTION Far northeastern Australia, where restricted to Cape York Peninsula, north Queensland, and formerly ranging from the Weipa district and the upper reaches of the Olive and Pascoe Rivers south to Clarina Creek, near Normanton, and the middle Lynd River, south of Bulimba (Weaver 1982; Storr 1984). Presently known to survive in two apparently isolated breeding populations in central Cape York Peninsula, the first occurring in the upper catchment of the Morehead River and adjacent headwaters of the Alice River, being centred on Artemis, Killarney and Dixie Stations, while the second occurs well to the south, primarily in the southeastern sector of Staaten River National Park and adjacent part of Bulimba Station to the northwest of Chillagoe. Only occasionally are sightings reported from outside these two locations, but there are no reports of sustained breeding at other localities (in Garnett *et al.* 2011).

STATUS An examination of published records reveals that since the early 1900s there has been a continual contraction of the range of Golden-shouldered Parrots, and this seems to have been particularly dramatic during the 1960s and 1970s, when large-scale trapping and nest-robbing are known to have occurred. Disappearance of the parrots from virtually all of the southern half of the range occurred prior to 1900 (Storr 1984). In the north, they remained common in the Stewart River drainage and north of Coen, but by 1970 they were confined mainly to an area approximately 120 km wide and 225 km long, centred on Musgrave Station. Contraction of the range continued, and is continuing steadily, leaving only the two, apparently isolated breeding populations. Based on surveys undertaken in 2009, the northern population is estimated at 1500 adults, and partial surveys carried out in 1999 and 2004 revealed an estimate of 1000 birds in the southern population (in Garnett *et al.* 2011).

In reviewing the findings of field investigations undertaken on Artemis Station, Garnett and Crowley (1997) were able to eliminate damage to nesting mounds, disease, predation by cats or cane toads, and moderate grazing by cattle as causes of substantial declines in numbers of Golden-shouldered Parrots. Capture of adults or the taking of nestlings for the live-bird trade may have contributed to local declines during the 1950s to 1970s, but no longer are considered to be major threats. Likewise, there probably has been a reduction in levels of predation by varanids as populations of these reptiles declined as a consequence of predation on venomous cane toads. Alteration of burning regimes and consequent effects on vegetation communities have been identified as the most likely cause of declines in the numbers of parrots (Garnett and Crowley 1997). Traditional burning practices employed by Aboriginal hunters maintained a mosaic of areas burnt at different times of the year, as well as unburnt areas protected from fires late in the dry season, and this favoured the parrots by providing open areas for feeding through the dry season. Also, spread of the broad-leaved tea-tree *Melaleuca viridiflora* into grassy woodlands was checked by the regular burning. Late fires in the early wet season were particularly beneficial because they exposed seeds on the ground and prompted

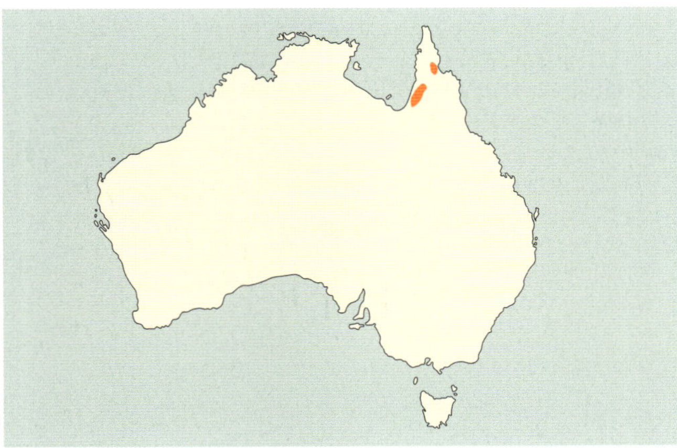

growth of some wet season food plants, while at the same time retarding regrowth of tea-trees. Altered burning regimes put into practice by pastoralists, together with grazing of grasses by cattle, have enabled tea-trees to spread into grassy woodlands, and under current fire regimes these woodlands are being lost to tea-trees at a rate of approximately 5 per cent per decade. In addition to impacting on food resources, colonisation of these woodlands by tea-trees seems to bring about significantly increased predation by butcherbirds of both adults and fledglings. Opportunities for butcherbirds to take parrots appear to be less in open grassy woodlands, especially where Black-faced Woodswallows *Artamus cinereus* provide effective warnings, but in denser thickets of tea-trees nesting parrots are more vulnerable. Garnett and Crowley believe that losses of adults is the major threat to remaining populations of Golden-shouldered Parrots, and Pied Butcherbirds *Cracticus nigrogularis* are the major predators. In the longer term, the loss of nesting sites may be a limiting factor, for monitoring has indicated that termitaria large enough for nesting are being lost faster than they are being replaced, largely from damage by cattle and feral pigs (in Garnett *et al*. 2011).

On Artemis Station, possibly the last major stronghold of this species, management practices are being implemented to assist the parrots. Paddocks will be spared periodically from cattle grazing and will be subjected to beneficial burning regimes. On an experimental basis, the surrounds of nests are being cleared to reduce predation by butcherbirds, and supplementary food is being provided in an effort to increase nesting success rates and to assist young birds in surviving through the wet season, when there is a shortage of natural food, with both measures seeming to be successful.

Under Queensland legislation the Golden-shouldered Parrot is categorised as an endangered species, and is afforded special protection. The species is listed as endangered under the Australian *Environment Protection and Biodiversity Conservation Act 1999* and also is listed on Appendix I to the Convention on International Trade in Endangered Species of Wild Fauna and Flora (CITES).

HABITATS Golden-shouldered Parrots frequent wet or dry, open eucalypt-*Melaleuca* woodlands with a groundcover of annual grasses, and seasonal preferences are determined by the presence of terrestrial termitaria required for nesting. There are early reports of sightings in other habitats, including at least one of birds being seen in coastal mangroves (Macgillivray 1918; Thomson 1935). Crowley notes that the range of the northern population in the Morehead River drainage comprises some 1380 km^2 and the range of the southern population centred on Staaten River National Park is at least 300 km^2 (in Curtis *et al*. 2012).

Garnett and Crowley point out that at Artemis Station, where extensive field studies have been undertaken, woodlands favoured during the dry season are dominated by eucalypts, including Darwin stringybark *Eucalyptus tetrodonta*, Cullen's ironbark *E. cullenii* and ghost gum *E. papuana*, together with ironwood *Erythrophleum chlorostachys* and bushman's clothes pegs *Grevillea glauca,* while fire grasses *Schizachyrium fragile* and *S. pachyarthron* feature prominently in the groundcover vegetation (*in litt*. 1998). In the wet season, birds move into low open woodlands on alluvial flats along drainage depressions of permanent or seasonal watercourses, where the dominant canopy trees are broad-leaved tea-tree *Melaleuca viridiflora,* golden grevillea *Grevillea pteridifolia* and smooth-leaved quinine *Petalostigma banksii*, while *Schizachyrium* fire grasses, glimmer grass *Planichloa nervilemma* and mintweed *Hyptis suaveolens* are prevalent in the groundcover, and termitaria preferred for nesting occur primarily at the margins of these grassy flats.

HABITS Much of what is known of the habits of Golden-shouldered Parrots comes from observations carried out on Artemis Station by Stephen Garnett and Gabriel Crowley (*in litt*. 1998). The parrots are monogamous, apparently forming lifelong pair-bonds, and during the breeding season a pair may be accompanied by an auxiliary young male. Outside the breeding season, young birds and second-year males associate in loose flocks of from 15 to 30 birds, and breeding pairs join these flocks immediately after nesting, but subsequently are seen with the flocks only infrequently. Flocks remain together while feeding, drinking, or resting during the middle of the day. When feeding, birds flutter to the ground in small groups, but then run about independently so that individuals or small groups often become separated. Likewise, approaches to waterholes are by small groups, with individuals independently walking to the water's edge. When disturbed, all birds fly up from the ground simultaneously to seek shelter amidst the foliage of a nearby tree, where they await the passing of danger before returning to the ground to resume feeding. During the middle of the day, they can remain undetected while at rest below the canopy in a shady tree. Chapman (1990) recalls finding four pairs resting in a *Melaleuca* thicket, but when he clapped his hands more than 30 parrots took flight, and so inconspicuous were these birds that they would not have been seen had not the noise been made. Nighttime roosts occupied by pairs or flocks are in the outer foliage of broad-leaved trees, usually cabbage gums *Corymbia confertiflora* or broad-leaved tea-trees *Melaleuca viridiflora*, and departure in the early morning usually is by small parties which follow each other through the treetops. After the first rains of the wet season there is a gradual dispersal of flocks, with pairs moving back to the proximity of their nesting mounds.

An association of Golden-shouldered Parrots and Black-faced Woodswallows *Artamus cinereus* is closest during the late dry season and early wet season. Each year, woodswallows nest in the same patches of woodland scattered at regular intervals approximately 3 km apart, and flocks of young Golden-shouldered Parrots, together with other seed-eating birds, can be found with most of these nesting groups. It is suggested that the parrots benefit from vigorous defence by the woodswallows of their nests against predators; any butcherbird or kookaburra that approaches the nests immediately is set upon by mobs of woodswallows and chased from the area. Whether woodswallows benefit from the association is not so obvious, though Garnett and Crowley suggest that they may take insects flushed from the ground by the foraging parrots.

The flight is rather swift and with only slight undulation. These parrots are strong fliers and when travelling longer

Plate 8
Golden-shouldered Parrot *Psephotellus chrysopterygius*
UPPER AND CENTRE adult ♂ LOWER adult ♀

F KNIGHT 09

distances they fly above the treetops, but shorter flights usually are through or below the canopies.

CALLS Len Robinson describes the normal contact call, which is given regularly in flight, as a disyllabic *fweep-fweep*, repeated two or more times in succession and sometimes prolonged to *few-weep . . . few-weep* (*in litt.* 1971). While perched, the parrots emit a sharp *weet*, often repeated two or three times, or a mellow *fee-oo . . . fee-oo*. Garnett notes that a descending trisyllabic whistle is given by the female when separated from her mate and, when disturbed, the birds utter a soft, almost inaudible *chirrup* just before taking flight (*in litt.* 1998). The alarm call is a loud, high-pitched whistle.

DIET AND FEEDING Garnett and Crowley (1994) point out that Golden-shouldered Parrots feed mainly on seeds of annual grasses. They extract unripe seeds from standing seedheads by pulling down stems and holding them under their feet or by reaching down to take the seeds while perching on a low overhanging branch. They switch to fallen seeds as soon as these become available, walking across the ground and pecking at a rate of 30 to 70 times per minute. Seeds of *Schyzachyrium* fire grasses are the staple food during the dry season, and the parrots show a preference for feeding in burned areas where fire has removed standing grasses to leave fallen seeds exposed. So abundant are these fallen seeds that the birds need to forage for only a couple of hours each day, spending the remaining time at rest in shady trees. With onset of the rains, seeds of glimmer grass *Planichloa nervilemma* become an important food, but are replaced late in the wet season by seeds of pendent milkdrop sedge *Scleria rugosa*, which are taken when unripe and while germinating. Seeds of cockatoo grass *Alloteropsis semialata* also are an important item in the diet during the wet season, which is a time of food shortage, and at this time these parrots have been observed chewing on flowers and new leaf-shoots of melaleucas. Occasionally, birds were seen chewing bark, and this was more prevalent during the breeding season. Leaf-shoots and flowers also are taken in the wet season (in Higgins 1999).

Weaver (1982) reports that pairs were seen foraging for seeds of Cuming's lovegrass *Eragrostis cumingii* and *Panicum mindanaese*, while seeds dropped by birds in and around an active nest included those of *Panicum* and *Seratia* grasses and unidentified monocotyledons. Roosting birds were observed chewing at flowers of golden grevillea *Grevillea pteridifolia*, and a female was seen nibbling at the bark of a broad-leaved tea-tree *Melaleuca viridiflora*, but on both occasions the ingestion of material could not be determined.

Garnett and Crowley note that at Artemis and Dixie Stations, seeds of *Desmodium* legumes were present in most of the crops of chicks that were examined (*in litt.* 1998).

BREEDING Onset of the breeding season is marked by an upsurge in activity in the vicinity of nesting mounds, often involving squabbling between adult males seeking to pair with widowed females or immature males chasing each other. During courtship, the displaying male makes short flights around the female and then, with the frontal feathers raised in a small crest and the breast feathers puffed out, he struts along the perch or the ground towards her. When extremely excited, females also may raise the frontal fathers. Copulation is initiated by the female sidling up to the perched male, raising her tail, drooping her wings and fluffing out the rump feathers and, after several approaches, the male, with frontal feathers raised, struts past the female and inspects her rump prior to mounting her for approximately 50 seconds.

Garnett and Crowley report that at Artemis and Dixie Stations, laying of the first egg was recorded between 1 March and 8 June, with the peak period being in early April, while the latest time for fledging of chicks was in mid August. The nest is in a chamber at the end of a tunnel excavated mainly by the female in a terrestrial termitarium. At the end of the wet season mounds remain damp for a considerable time, thus facilitating excavation, which is undertaken mostly in the mornings, the female digging with her bill and scraping out the loosened dirt with her feet. Excavation of the entire burrow lasts several days, and the first egg is laid within a week of completion. Weaver (1982) points out that contrary to earlier reports preference is shown for conical mounds constructed by *Amitermes scopulus,* and these are dominant in vast areas of low open woodland, though they occur also with the less favoured meridian mounds of *A. laurensis* along drainage depressions in the vicinity of Princess Charlotte Bay. Garnett and Crowley recall that of a total of 148 nests found on Artemis and Dixie Stations, 96.7 per cent were excavated in conical mounds of *Amitermes scopulus*, 2.8 per cent in meridian mounds of *A. laurensis*, and 0.5 per cent in turreted mounds of *Nasutitermes triodiae*, with most nests being in mounds on or beside drainage flats, some on gravelly slopes in hills, and a few on sand ridges (*in litt.* 1998). Weaver (1987) suspects that the preference for conical mounds may be associated with temperature regimes different from those of meridian mounds, and especially the longer retention of warmth in nesting chambers. Data compiled by Garnett and Crowley from an examination of nests on Artemis and Dixie Stations, and by Weaver (1987) from nests in the Musgrave area are summarised in Table 2.

At Artemis and Dixie Stations, 92.3 per cent of nests were in newly excavated burrows, with the remaining 7.7 per cent

Termite species in mound	Height of nest entrance (cm)	Depth of excavation (cm)	Length of tunnel (cm)	Height of tunnel (cm)	Width of tunnel (cm)	Directional aspect of tunnel entrance E	N	W (%)
A: Artemis and Dixie Stations								
Amitermes scopulus	68.0 (37.0–110.0)	32.0 (19.0–45.0)	12.5 (5.5–22.5)	4.5 (3.8–5.4)	4.4 (3.5–5.4)	16	29	24
B: Musgrave district								
A. scopulus	60.0 (46.0–91.5)	28.3 (20.5–37.5)						
A. laurensis	80.0 (49.0–125.0)	40.0 (20.0–61.5)						

Table 2. Data from an examination of nests on Artemis and Dixie Stations (A) by Garnett and Crowley (*in litt.* 1998) and in the Musgrave area (B) by Weaver (1987).

being in burrows excavated in previous years. Nests usually were colonised by *Trisyntopa scatophaga* moths, larvae of which eat excreta of the chicks, and some nests were occupied also by green tree-frogs *Litoria caerulea* and blowfly larvae. Host termites sometimes attack the chicks, but apparently without causing ill effects (Garnett and Crowley *in litt* 1998).

A normal clutch comprises four to seven, usually five or six eggs, which are laid at intervals of two days, or sometimes up to four days between the first two or last two eggs. Incubation by the female commences after laying of the second or third egg and lasts between 19 and 21 days. The male comes to the nest at approximately hourly intervals to feed the sitting female. Hatching usually takes place over about three days, or occasionally up to a week. Chicks are brooded by the female for up to a week after hatching, and are fed by both parents, with roles of the sexes in sharing feeding duties varying between pairs. Either parent can rear a brood of up to five chicks if the other parent dies. Young birds leave the nest approximately five weeks after hatching, and departure of a brood can be synchronous or at varying intervals over several days, with each fledgling flying directly from the nest for distances of more than 100 metres. Fledglings remain with their parents in family parties and are fed by the parents for at least five weeks after leaving the nest.

Thomson (1935) claimed that two broods may be reared in a season, but this has not been confirmed. Garnett and Crowley report that at Artemis and Dixie Stations, a clutch of eggs lost during incubation was replaced by the same female after five weeks, and on one occasion a second female added to the abandoned clutch of a female that had disappeared (*in litt.* 1998). Also on Artemis and Dixie Stations, 695 eggs were laid in 148 nests, with 475 or 68 per cent hatching, and 311 of 462, or 67 per cent of chicks fledging. Predation by reptiles was the major cause of losses of both eggs and chicks. Predation by butcherbirds was presumed to be the major cause of high post-fledging mortality.

For up to three months after fledging, young birds remain within a few kilometres of the nest, and then join with other young birds and second-year males in loose flocks that spend the dry season near to permanent waterholes. After the first rains, most young birds, particularly first-year males, come to traditional nesting sites of Black-faced Woodswallows *Artamus cinereus*, and breeding pairs return to their nesting sites. Post-breeding dispersal of the woodswallows brings about a dispersal of the attendant parrots with some young birds moving up to 30 km or more. Females breeding for the first time have been found at nests near to their natal sites, while others were at nests at least 12 km away.

EGGS The eggs are broad-elliptical and with slight gloss, but they quickly become nest-stained. A clutch of five eggs was taken at Coen, Cape York Peninsula, north Queensland, on 18 May 1923, and these eggs measure 21.2 (20.4–21.5) × 17.9 (17.3–18.1) mm (AM O.48652). On 11 July 1984, near Ebagoola, Cape York Peninsula, three eggs were taken from a deserted 36 cm long nesting tunnel some 40 cm from the ground in a mound of *Ametermes scopulus*, and these eggs measure 21.2 × 18.7 mm, 21.3 × 18.1 mm and 20.7 × 18.1 mm (ANWC E12485). Garnett and Crowley give 20.6 (18.8–24.0) × 17.8 (16.3–18.9) mm as measurements obtained from 80 eggs examined on Artemis and Dixie Stations, Cape York Peninsula (*in litt.* 1998).

Paradise Parrot

Psephotellus pulcherrimus (Gould)

Platycercus pulcherrimus Gould, *Ann. Mag. Nat. Hist.*, **15**, 1845, p. 115 (Darling Downs, Queensland).

DESCRIPTION Length 27 cm.

ADULT MALE Forehead brilliant red; crown and nape black, merging into dark brown on hindneck; mantle, back and wings earth-brown; lores and periophthalmic region yellowish with some white markings; rump, flanks and sides of abdomen turquoise-blue; cheeks, ear-coverts, breast and upper abdomen emerald-green, merging into turquoise-blue on sides of neck and lower abdomen; upper tail-coverts emerald-green suffused turquoise-blue; under tail-coverts, thighs and centre of abdomen bright red; median and outer lesser wing-coverts bright red; underwing-coverts, bend of wing, primary-coverts and outer webs of primaries and outermost secondaries deep blue; central tail-feathers above bronze-green margined blue on inner webs and becoming bluish-black towards tips, lateral tail-feathers greenish-blue broadly tipped bluish-white; underside of tail bluish-white; underwing-stripe absent; bill greyish; iris brown; legs greyish-brown.

21 specimens: wing 125–135 (127.9) mm, tail 143–182 (158.5) mm, exp. cul. 12–14 (12.8) mm, tars. 16–19 (16.7) mm.

ADULT FEMALE Crown and nape blackish-brown; forehead and periophthalmic region yellowish-buff with faint red margins to some feathers; face, throat, breast and sides of neck dusky buff-yellow suffused brownish-orange and feathers tipped dull orange-brown; abdomen, vent and under tail-coverts pale blue with some red on centre of abdomen; rump and upper tail-coverts bright turquoise-blue; duller red restricted to lesser and innermost median wing-coverts; underwing-coverts, bend of wing, primary-coverts and outer webs of primaries greyish-blue; underwing-stripe present.

6 specimens: wing 121–126 (124.7) mm, tail 144–168 (161.5) mm, exp. cul. 12–14 (12.6) mm, tars. 16-18 (16.6) mm.

JUVENILES Resemble adult female; young males have some emerald-green on face and breast, with both the blackish crown and red wing-coverts noticeably darker; underwing-stripe present.

DISTRIBUTION Northeastern Australia, where recorded in central and southern Queensland, from about the Tropic of Capricorn,

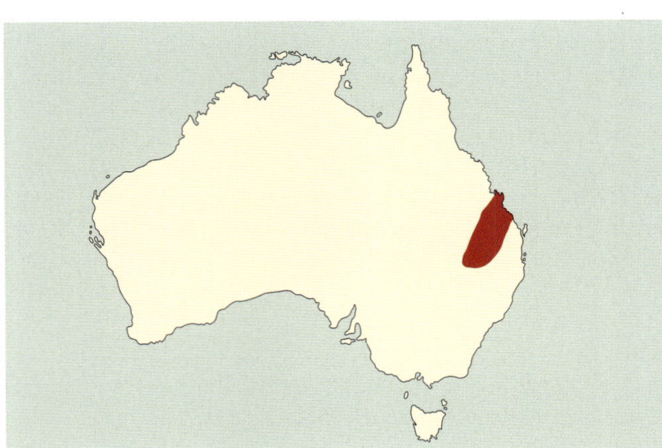

inland from near Rockhampton to the upper Comet and Nogoa Rivers, south to the Goondiwindi district and the upper Brisbane River valley, occasionally ranging near to Brisbane (Storr 1984). Westernmost records include the upper Dawson River, the Roma and St George districts, and the Warrego River, where a pair was collected in 1847 by Thomas Wall, a member of the Kennedy Expedition. Another western record from near Augathella, on the upper Nive River, may refer to the Turquoise Parrot *Neophema pulchella* (in Higgins 1999). Records from the Inverell and Casino districts, in northeastern New South Wales, have been discounted (Olsen 2007).

John Gilbert, who discovered the species on the Darling Downs in 1844, noted in his diary for 17 June 1845 that he again met with 'the new *Platycercus* of the Downs', on the Mitchell River near the base of Cape York Peninsula, and this gave rise to claims that a northern, presumably isolated population occurred on the Peninsula. What we know of the three closely allied anthill-nesting species points to allopatry being a key feature of their ranges, and I have no doubt that, despite his competence as an observer, Gilbert misidentified Golden-shouldered Parrots *Psephotellus chrysopterygius*.

STATUS It is a tragedy that arguably the most beautiful of Australian parrots has been lost. When he collected the first specimen, an adult male, on 17 May 1844, between Tummaville and Yandilla Stations, on the Condamine River, south-central Queensland, John Gilbert was so impressed by its beauty that he wrote to John Gould requesting that this new species be named after him:

> *… almost the first Bird shot is a totally new Parrot of the size and form of* Platycercus haematogaster *but without exception the most beautiful of the whole tribe I have ever seen in Australia; the mingling of the beautiful shades of green is its most conspicuous and beautiful character. If you have not already honoured my poor name in your works, I know of no species that wd delight me more to see Gilberti attached to than this beautiful bird.*

Gould did not name the species after Gilbert, but recognised the beauty of the adult male when he gave the common name 'Beautiful Parrot' to the species in his *The Birds of Australia*, where two males were illustrated because he had not received a specimen of the adult female. Indeed there are very few specimens of adult females among the many specimens in museums throughout the world. Although a targeting of adult males by collectors was likely, it is possible that a debilitating sex imbalance in populations was responsible for very poor recruitment.

Records of the eggs of Paradise Parrots being taken in the St George district, southern Queensland, in the 1960s, proved to be a hoax and, despite reports of sightings as late as the 1990s, I am convinced that the Paradise Parrot is extinct and probably has been so for about 70 years. The history of its decline and eventual disappearance has been documented meticulously by Penny Olsen in her *Glimpses of Paradise*: *The Quest for the Beautiful Parakeet* (National Library of Australia 2007). The last reliable records were sightings by Eric Zillman near Wallaville, in the Burnett River valley, southeastern Queensland, in 1938, and the last confirmed record was on 17 March 1922, when C. H. Jerrard photographed a pair at the nest at Manar Station in the upper Burnett River valley (in Olsen 2007). The last sighting in the same district by Jerrard was on 13 November 1927 when he observed a lone male. I concur with Olsen in discounting a claim by Florence Irby that a party of five birds was seen near Casino, far northeastern New South Wales, on 21 November 1926, during a period of drought and extensive bushfires (in Cayley 1938). Cayley also quotes records from the Inverell district, in northern New South Wales, but again these should be discounted.

The patchy distribution noted by Gilbert indicates that prior to European settlement, Paradise Parrots probably were declining. In the 1880s, they were locally common, though generally scarce, and some birds were shipped to London, where they were highly prized by aviculturists. However, by the early 1900s shipments to London had ceased, and there was a noticeable absence of birds from areas where they formerly occurred. The fall in numbers was quite dramatic, for as early as 1915, the species was listed as 'missing' by A. J. Campbell, who asked C. Barnard if it was still to be found in the Duaringa district, central Queensland. Barnard replied that he had not seen a single bird since the 1902 drought. During the next two years, Barnard searched old haunts, but failed to find any parrots. He attributed their disappearance to a loss of seeding grasses resulting from two years without a wet season, followed immediately by the 1902 drought and widespread heavy stocking.

In 1918, Alec Chisholm took up the search and launched a press campaign for information on the species. His efforts were rewarded when, in December 1921, C. H. Jerrard wrote from the upper Burnett River district to say that he had seen a pair of Paradise Parrots. In March 1922, Jerrard photographed a pair at their nest in a terrestrial termitarium, and his sightings in that district continued until 1927. In the 1960s, Chisholm conducted another press campaign in the hope of finding parrots, but without success. The publicity generated much enthusiasm, which was manifested in numerous, obviously false records, most of which could be attributed to misidentification of Red-vented Bluebonnets *Northiella haematogaster*. Reports came from localities as far afield as Broken Hill, New South Wales, Hebel, southern Queensland, and Cape York Peninsula, but none could be verified.

Rumours about the continued existence of Paradise Parrots have persisted to the present time, and claims have been made that a few pairs are held in captivity and some have been shipped illegally to overseas aviculturists. Lendon (1979) maintained that convincing evidence of the existence of the species was provided by a colour transparency of a male taken at an undisclosed locality some time in the 1960s. I have not seen this photograph, which was subjected to extreme secrecy, but I place no more credence in its authenticity than in other unsubstantiated claims emanating from similarly unreliable sources, often involving persons with a vested interest in promoting Golden-shouldered Parrot *Psephotellus chrysopterygius* × Mulga Parrot *Clarkona varia* hybrids as true Paradise Parrots. In September 1974, I visited the home of a Sydney aviculturist to see his pair of Paradise Parrots, and the birds obviously were *P. chrysopterygius* × *Clarkona varia* hybrids, but the owner genuinely believed them to be Paradise Parrots and had paid a very high price for them.

Plate 9
Paradise Parrot *Psephotellus pulcherrimus*
UPPER AND CENTRE adult ♂ LOWER adult ♀

F KNIGHT 2008

Olsen documents a number of reported sightings between the 1960s and 1990s together with unsuccessful searches undertaken in the same period, and here I mention a couple of reports with which I became involved, though I place no credence in any of them. Kiernan (1993) claimed that in central Queensland, between late February and mid April 1990, five Paradise Parrots regularly were seen in the company of Pale-headed Rosellas *Platycercus adscitus* at a station homestead and he personally observed the parrots in mid April. Wildlife authorities organised an extensive search in this area, but no Paradise Parrots were seen and the possibility of misidentification could not be discounted. John McCloy claimed that in June 1993, approximately 5 km west of Yelarbon, southeastern Queensland, he saw a pair of Paradise Parrots at the side of the road (pers. comm. 1998). Other reports of unconfirmed sightings made in the 1990s have come from various localities in south-central Queensland and northeastern New South Wales. I suspect that misidentification of Bluebonnets or Mulga Parrots in districts where formerly they were scarce or unknown could be responsible for some reported sightings of Paradise Parrots.

The rapid disappearance of these parrots between the 1880s and 1920s is indicative of a declining species being unable to withstand the impact of a sudden change in its environment. Olsen points out that settlement of the upland grasslands in southern Queensland occurred at about the time of discovery of the Paradise Parrot, and the resulting development of pastoral practices with overgrazing and altered fire regimes exacerbated by drought brought about fundamental changes to groundcover vegetation, particularly native grasses. I am strongly of the opinion that, like the closely allied Golden-shouldered Parrot and Hooded Parrot *Psephotellus dissimilis*, Paradise Parrots fed primarily on the small seeds of native grasses, and changes to groundcover vegetation were the principal cause of their decline and eventual disappearance. I would point out that other specialist seedeaters, notably the Squatter Pigeon *Geophaps scripta*, Turquoise Parrot *Neophema pulchella*, Star Finch *Neochmia ruficauda* and Black-throated Finch *Poephila cincta*, have been affected similarly and have declined dramatically in southern Queensland, the two finches having disappeared entirely from that part of their range. In his recollections of encounters with Paradise Parrots, Eric Zillman referred to these vegetation changes and linked them to those affecting Golden-shouldered Parrots (in Debus 2013):

> *The introduction of beef cattle, with its regime of burning off to promote green feed for stock and other changes to the land, I believe, sounded the death knell for the ground- feeding, mound-nesting Parrot ... I concluded that the introduction of fire had made a vast change to the vegetation, compared with when I observed the Paradise Parrot near Wallaville in the 1930s. Fire had greatly increased the regeneration of native species (eg. shrubs, eucalypts and taller rank grasses) that were not present before the graziers' burning regime. Having visited Cape York Peninsula seven or eight times (including camp-outs of up to a month) between 1969 and 2004, I feel that the demise of the Paradise Parrot and its preferred habitat can be likened to the situation of the endangered Golden-shouldered Parrot* Psephotus chrysopterygius *and its habitat of today.*

The rapid spread of prickly pear *Opuntia vulgaris* also may have contributed to the changes in groundcover vegetation. Secondary pressures take on increased significance when impacting on a species already threatened by a primary pressure, and Olsen refers to secondary pressures that may have contributed to the extirpation of Paradise Parrots. I have already mentioned a possible debilitating sex imbalance in local populations, as reflected in the very few specimens of females in museum collections, and this may have resulted in poor recruitment. Another possible cause of poor recruitment is a widespread loss of nesting sites because of the destruction of termite mounds to obtain material for the floors of settlers' cottages and the surfaces of tennis courts. Zillman describes the apparent destruction of a nest during these operations:

> *In 1938, my family and other workers were levelling termite mounds for the material that then formed the base for tennis courts. We hired a two-ton truck and dug up 13 loads of a dozen mounds each (or about 150 mounds for one tennis court). After lopping off the top of a mound, I found a chamber containing four eggs. We pondered the owner's identity, and I did not link the termite mound to the Parrots, or finally identify the species of parrot (by its official name) I and my father had been seeing, until I consulted a copy of A. J. Leach's field guide,* An Australian Bird Book *(Leach 1926). That was when we realised for certain what the birds were. The existence of the birds had been known to people in the area, but they were simply known locally as the 'anthill parrots'. At that time, we did not know that the parrots were so rare.*

Predation of nests by foxes and feral cats could have impacted on populations in later years, and may have been responsible for a higher mortality of females. Trapping for the live-bird market may have contributed to the decline of some populations, especially in the vicinity of Brisbane, but I suspect that its impact was local in extent.

HABITATS Paradise Parrots occurred mainly in broad river valleys, where they frequented open forests or eucalypt woodlands with an understorey of annual and perennial grasses. In May 1965, Alan Lendon visited the upper Burnett River valley, in southern Queensland, where a nesting pair was photographed by Jerrard in 1922, and he found the locality to be in wooded undulating country with numerous terrestrial termitaria (see Lendon 1979).

HABITS What we know of the habits of Paradise Parrots comes from the writings of early naturalists. Chisholm was enthralled by the beauty of the birds and noted that the bright red wing-coverts and abdomen of the male were particularly conspicuous in the field. In *Birds and Green Places* (1929), he wrote:

> *For there was the male parrot, the gorgeous red of his underparts gleaming in the midday sun. He was sitting quietly on a dry branch of the tree, and did not appear to be much concerned when I cruised about in order to glimpse his red shoulders, black cap and red forehead. Indeed, the only effect of the intrusion was to cause him to cease a sweet, plaintive piping, not unlike that of a young bird, and perhaps to sit up a trifle straighter.*
>
> *. . .They rose as I approached too closely, and there was a flash of glowing colours as both flew off across a dry creek. Following, I got within twelve yards of the pair, and remained there for about fifteen minutes, admiring the beautiful colour-scheme of the male and the ease with which both birds stripped the grasses.*

The confiding behaviour of these parrots is a feature of a recollection by Keith Williams of his encounter with a pair at a dam in sclerophyll forest, near Ipswich, southern Queensland, in the 1920s; he wrote (*Queensland Ornithological Society Newsletter*, November 1977):

> *While we were occupied with our shrimping activity, we saw a pair of small parrots flying towards the shallow end of the dam and alight on the ground some little distance from the water's edge. We immediately realised that they were quite different from any parrot we had seen before. From our position across the dam, we observed that they were sprightly little birds with heads alertly raised and they approached the water with short intermittent runs. The birds did not go to the edge of the main body of water but stopped and drank from the small puddles in the hoofprints left by the watering stock.*
>
> *We decided to try to get closer to them so we crossed the concrete overflow and then crawled along a trench that bordered the western side of the dam. As we raised our heads above the grassed rim of the trench, we were about 6 m from the birds so we lay quietly for a long time and watched them. The afternoon sun was behind and we had a perfect view of them. They seemed unoccupied except for an occasional preening and a spasmodic run from one spot to another. Now and then they dipped their beaks quickly into the muddy puddles, but no bathing took place.*
>
> *Rising cautiously from our prone position we were surprised when the birds, though they showed awareness of our presence, did not fly. After climbing through the barbed wire fence we moved quietly closer to them and were only 3–4 m from them when they flew. They did not go very far before they settled on the shallowly sloping hillside and appeared to be picking up seed from the short grass. We followed them quietly and again were able to get close to them. This was repeated a few times before they crossed the ridge and then flew across Deebing Creek, which runs nearby and were lost to view.*
>
> *We hurried home and consulted a bird book owned by my Uncle and agreed that we had seen a pair of Paradise Parrots. It was exciting to see such beautiful birds but we were not to know until some time later how rare they were.*
>
> *We watched for them each time we were in the area but they were never seen again.*

A confiding behaviour was recalled also by Eric Zillman in his recollection of encounters with a pair of Paradise Parrots near Wallaville, in the Burnett River valley:

> *It was just the one pair of parrots each time we went out, and we always saw them. I remember the colours, the red flashing wings and brownish back as the male took off in front of us. They were always on the ground, feeding, when we found them. Sometimes we went out dipping once a month, and they were there each time, at the far end of the property. The Parrots favoured short grasses in sparsely wooded country. They were not shy; we could approach to within about 10 metres of them before they flushed and flew off.*

The flight was said to be rather swift and with slight undulation, so presumably was similar to the flight of the Golden-shouldered Parrot *Psephotellus chrysopterygius*. Referring to birds in flight, Zillman recalls that 'They looked mainly brown but sometimes you'd catch a red wing flash' (Barnes and Zillman 2007). This description by Zillman of birds in flight reinforces the ease of misidentifying Bluebonnets *Northiella haematogaster* or female Mulga Parrots *Clarkona varia*.

CALLS When located by Chisholm, an adult male was giving a 'sweet, plaintive piping, not unlike that of a young bird'. When attacked by two Noisy Miners *Manorina melanocephala*, a feeding pair was heard to emit a short, sharp musical *queek* (Chisholm 1929).

DIET AND FEEDING It seems that Paradise Parrots fed almost exclusively on seeds of native grasses, procured on the ground. Chisholm (1929) noted that the method of feeding was to seize a grass stem near the base and move the bill along it to strip the seeds, which then were picked up from the ground. If the stem was thick, the bird would clamber on and bend it down with its weight. Seeds of herbaceous plants, vegetable matter and insects and their larvae probably were included also in the diet.

BREEDING Nesting was recorded in September, December and March. The nest was in a chamber at the end of a tunnel, all excavated by the birds in a terrestrial termitarium. I reject claims that they sometimes nested in tunnels excavated in the banks of watercourses or in tree-hollows (see Cayley 1938). No nesting material was used, the three to five eggs being laid on a mat of crumbled earth. Olsen (2007) points out that, as in nests of the Golden-shouldered Parrot *Psephotellus chrysopterygius* and the Hooded Parrot *P. dissimilis*, nests of the Paradise Parrot probably were co-inhabited by larvae of a unique *Trisyntopa* housekeeping moth which fed on the excreta of nestlings, and these moths would have followed the parrot to extinction.

Jerrard noted the behaviour of the pair of parrots at the nest he was photographing (in Chisholm 1922). On 21 January 1922, the birds recommenced excavation of a hole that they had visited previously. The date of laying was not determined, but by 4 March, the female was brooding. Only the female incubated and the male was never seen to enter the tunnel. At times he came to the entrance, peered in and uttered soft chirping notes until the female emerged. Then they flew off together, presumably to feed. He accompanied her back to the nest and she entered almost immediately. This nest failed and was opened up by Jerrard to reveal five addled eggs. The entrance tunnel was approximately 40 mm in diameter and about 23 cm in length. It entered the nesting chamber at the top and to one side. The chamber was roughly circular, about 38 cm to 45 cm in diameter and up to 20 cm high. The floor of the chamber was below ground-level.

EGGS The eggs are rounded to broad-elliptical and with slight gloss. A clutch of five eggs was taken near Duaringa, central Queensland, on 10 December 1896, and these eggs measure 21.2 (20.3–21.8) ×17.6 (17.3–18.0) mm (MV BE01414).

GENUS *LATHAMUS* Lesson

Lathamus Lesson, *Traite d'Orn*., livr. 3, 1830, p. 205. Substitute name for *Nanodes* Vigors and Horsfield preoccupied. Type of *Nanodes* Vigors and Horsfield, by subsequent designation, is *Psittacus discolor* Shaw ie. White.

This monotypic genus has been associated variably with the lorikeets, the fig parrots and the playcercine or 'rosella' parrots. Certain anatomical features, including structure of the skull, pelvis and legs, the rounded cere, the small, non-protruding bill, the presence of a superficial left carotid, and pterylosis or feather-tract pattern, all point to a relationship with *Platycercus, Psephotus* and allied genera. Biochemical analyses corroborate the inclusion of *Lathamus* in this core group of platycercine genera (Christidis *et al.* 1991). The presence of a variable underwing-stripe and red under tail-coverts is further evidence of platycercine affinities, while the crude 'brush' tongue and long pointed wings, characteristics so well marked in the lorikeets, presumably are the results of convergence due to similar modes of life.

The genus occurs in southeastern Australia, where its breeding range is restricted to Tasmania.

ENDANGERED

Swift Parrot

Lathamus discolor (Shaw)

Psittacus Discolor Shaw, *in* White, *Journ. Voy. New South Wales*, 1790, pl. 49 (vicinity of Port Jackson, New South Wales).

DESCRIPTION Length 25 cm. Weight males 46–76 g, females 45–72 g.
ADULT MALE General plumage bright green, lighter and more yellowish on underparts; forehead, throat and foreparts of cheeks red; lores and borders of red facial areas yellow; turquoise-blue suffusion on ear-coverts; crown dark blue; some birds have few red feathers on underparts; vent and under tail-coverts dull red, feathers variably tipped yellow and pale green; underwing-coverts, bend of wing and lesser wing-coverts rich red; outer median and secondary wing-coverts greenish-blue; primary coverts and outer webs of primaries violet-blue, latter narrowly margined pale yellow; inner webs of tertials scarlet; central tail-feathers above dull brownish-red, tipped dark blue; lateral feathers dull dark blue, margined brownish-red; underside of tail dusky grey; underwing-stripe variable; bill brownish-horn; iris pale yellow to orange-yellow; legs dusky brown.
15 specimens: wing 116–128 (121.1) mm, tail 102–127 (116.8) mm, exp. cul. 12–14 (12.7) mm, tars. 14–17 (15.1) mm.
ADULT FEMALE Similar to, but slightly duller than adult male; less red on face and bisected by yellow at lores; red markings on underparts confined to flanks and not extending to abdomen or breast; underwing-stripe variable.
10 specimens: wing 118–125 (120.2) mm, tail 105–122 (114.8) mm, exp. cul. 12–14 (12.5) mm, tars. 14–16 (15.0) mm.
JUVENILES Duller than adult female; less orange-red on face and smaller, duller blue patch on forecrown; less red on bend of wing; under tail-coverts dull yellow suffused pale pink-red towards bases and on inner webs, and shaft-streaked dull bluish-grey; pale underwing-stripe present; bill darker brown; iris brown.

DISTRIBUTION Southeastern Australia, including Tasmania, larger islands in Bass Strait, and from southeastern Queensland through eastern New South Wales to Victoria and southeastern South Australia. Breeding occurs only in Tasmania, mostly in eastern coastal and subcoastal areas south to Southport Lagoon, and to a lesser extent in the northwest and possibly in the midlands.

STATUS In the late 1970s, reports of fewer numbers of Swift Parrots in both the breeding and wintering ranges had given rise to concerns about the status of the species and highlighted a lack of information about its ecology. In response to these concerns, a survey was conducted in the breeding range during the 1987–1988 breeding season, and this revealed a surprisingly low estimated breeding population of only 1320 pairs. Brown (1989) points out that of all Tasmanian endemic bird species only the critically endangered Orange-bellied Parrot *Neophema chrysogaster* has a smaller population.

Recent studies have revealed that numbers are declining at an alarming rate, and the species is severely threatened by loss of habitat coupled with high predation at nests (Heinsohn *et al.* 2015). This is yet another example of a secondary adverse pressure becoming more significant when coupled with a primary pressure. In both the breeding and wintering ranges, favoured nesting and foraging trees have been felled in logging operations or in landclearance for agriculture. Brown estimates that within the breeding range conservatively no more than a third of the original forests of blue gums *Eucalyptus globulus* now remain, and these are continuing to be lost, with clearfelling for woodchips being the major threat. In the main wintering range, from central Victoria to south-central New South Wales, approximately 75 per cent of the original box-

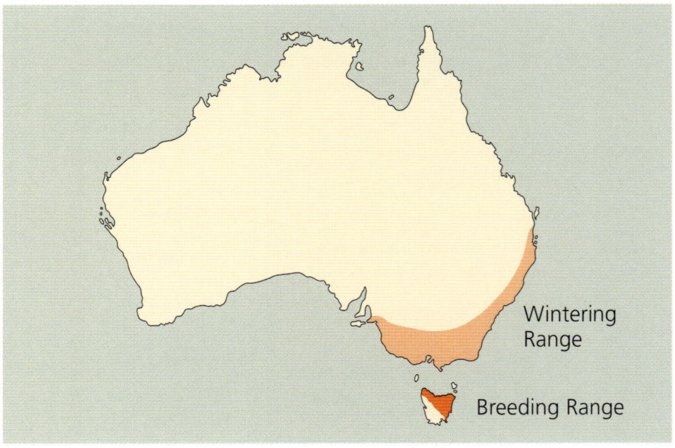

Plate 10
Swift Parrot *Lathamus discolor*
UPPER juvenile LOWER adult ♂

F. KNIGHT '15

ironbark woodlands have been lost through mining, agricultural and timber harvesting practices, but in remaining stands in reserves there continues to be intense cutting for poles, fenceposts and firewood.

A monitoring of nests of Swift Parrots for the 2010–2011 to 2013–2014 breeding seasons with motion-activated cameras revealed severe, though highly variable rates of predation on eggs, chicks and sitting females by introduced Sugar Gliders *Petaurus brevipes* (Stojanovic *et al.* 2014; Heinsohn *et al.* 2015). Predation of eggs and adult females ranged from 100 per cent at some breeding sites on mainland Tasmania to zero predation on Maria and Bruny Islands where Sugar Gliders are absent, and the intense predation pressure equated to an overall annual mortality rate of 50.9 per cent for nesting females across Tasmania, though nest survival analyses indicated that this is likely to be an underestimate of true female mortality. Of particular concern is that the high nesting success rate occurring at times on Maria and Bruny Islands appears to be insufficient to buffer the population decline. It was revealed also that on the Tasmanian mainland survival of nests is a function of mature forest cover in the surrounding landscape, with the likelihood of predation decreasing with increasing forest cover. The use of population viability analyses demonstrates that the current population of Swift Parrots, estimated at 2500 birds, is likely to decrease by 78.8 to 94.7 per cent over only three generations of 12 to 18 years (Saunders and Heinsohn 2008; Heinsohn *et al.* 2015).

The Swift Parrot is listed as critically endangered under the *Australian Environment Protection and Biodiversity Conservation Act 1999.*

HABITATS In eastern Tasmania, the breeding distribution of Swift Parrots coincides with the natural range of blue gums *Eucalyptus globulus*, a principal food tree, and in northwestern Tasmania the key feature is the presence of swamp gums *E. ovata*, another important food source, and planted blue gums (Webb *et al.* 2012). Dry sclerophyll woodland dominated by these two eucalypts is the favoured habitat, but birds can be encountered in more open country, including remnant stands of woodland or scattered blue gums in farmlands, and in the 1960s, when undoubtedly they were more plentiful, I observed them in single trees in suburban parks and gardens in and around Hobart. In November–December 2004 to 2006, surveys of nests were undertaken at four sites in southeastern Tasmania, where the forest comprised a range of tree age-classes from young regrowth of less than 50 years to senescent or dead trees at more than 200 years, and the main forest types at these sites were (Webb *et al.* 2012):

(i) Fern Tree – dry messmate stringybark *Eucalyptus obliqua* interspersed with dry white peppermint *E. pulchella*,
(ii) Maria Island – wet and dry forest of messmate stringybark with blue gums and white gums *E. viminalis* subdominant or locally dominant in some patches,
(iii) Bruny Island – dry forest or woodland of white peppermint with blue gums subdominant or locally dominant in some patches,
(iv) Meehan Range – dry white peppermint-blue gum forest and woodland with dry forest or woodland of silver peppermint *E. tenuiramis* and black peppermint *E. amygdalina* common in the surrounding area.

Approximately 10 000 km² of eucalypt forest in eastern Tasmania can be utilised for breeding (Saunders and Heinsohn 2008; Webb *et al.* 2014). After nesting has finished and the flowering of blue gums has passed its peak, these parrots disperse throughout much of Tasmania and can be found in most wooded habitats, with a preference for dry sclerophyll forests or woodlands, but less frequently in wet sclerophyll forests and rarely in temperate rainforests or subalpine habitats. At these times they are seen also in exotic trees, street trees in towns or cities, and in orchards or vineyards (Hindwood and Sharland 1964; Ridpath and Moreau 1966).

The mainland overwintering range is approximately one million square kilometres, and again there is an association with favoured food trees (Heinsohn *et al.* 2015). In Victoria the bulk of the wintering population is found in the box-ironbark forests of central regions, where red ironbark *Eucalyptus sideroxylon* and yellow gum *E. leucoxylon* are prevalent. The distribution is centred on these box-ironbark forests, particularly where red ironbarks are well represented, and the parrots often visit town parks or occur sporadically elsewhere in dry forests or woodlands, and wooded farmlands, but seldom are seen in treeless areas, rainforests, wet sclerophyll forests or areas where the annual rainfall is less than 400 mm (Emison *et al.* 1987). During the compilation of atlas records in Victoria, between January 1973 and June 1986, there were 229 records, and a reporting rate of only 4 per cent was achieved in the range. Reporting on the spatial distribution of Swift Parrots over two successive winters in the box-ironbark forests of central Victoria, MacNally and Horrocks (2000) suggest that their presence may be linked to the intensity of flowering of golden wattle *Acacia pycnantha* and the density of aggressive honeyeaters, especially Noisy Miners *Manorina melanocephala*, and there was no apparent relationship with the intensity of eucalypt flowering. These patterns were difficult to explain because the parrots seem to depend on eucalypt nectar, with few records of feeding on racemes of golden wattle, and the positive correlation with honeyeaters is difficult to reconcile with the lack of correspondence with eucalypt flowering. In the course of these investigations, it was found that in some years the parrots utilise forest remnants as small as 10 ha, so demonstrating the importance of protecting all remnant patches. In southern Victoria, I have seen flocks of Swift Parrots feeding in flowering eucalypts surrounding sportsfields in the outer suburbs of Melbourne and alongside the Melbourne to Geelong freeway.

In New South Wales, Swift Parrots frequent most types of open forests and woodlands in coastal districts and on the central-western and southwestern slopes of the Great Dividing Range, but numbers are influenced by seasonal conditions and can vary markedly from year to year (Cooper *et al.* 2016). On the central-western and southwestern slopes foraging habitat is available in all years, but north of the central coast seasonal variation is more pronounced, though this area is an important drought refuge. From April to October in the years 2001 to 2005, field studies of the utilisation of wintering habitat by Swift Parrots were undertaken in New South Wales at 35 sites on the mid-north, central and southern coasts and at 18 sites on the central-western and southwestern slopes of the Great Dividing Range (Saunders and Heinsohn 2008). During the course of these studies a cumulative total of 2149 parrots was recorded at the sites, which were among scattered individual trees within a rural or urban environment, in remnant patches of forest or woodland less than 10 000 ha in size and surrounded by rural or urban environments, and in continuous stands of forest or woodland exceeding 10 000 ha in size. Within these three main landscape types potential foraging habitats were defined as areas with previous Swift Parrot records or containing known or potential food trees. In coastal habitats 69 per cent of foraging sites were among scattered trees, 17 per cent were in remnant patches of forest or woodland, and 14 per cent were in continuous forest or woodland, and in the western slopes habitats 89 per cent were in remnant patches and all remaining sites were among scattered trees, but in both coastal and western slopes habitats the greatest abundance of parrots occurred within remnant patches of forest or woodland, emphasising the importance of this landscape type.

MOVEMENTS Swift Parrots arrive in Tasmania during September, remain to breed and then return to the mainland during March and April. Birds have been recorded on the mainland in all months, though extremely rarely between late October and the end of February, with occasional individuals seen at this time presumed to be stragglers taking advantage of a profuse flowering of local eucalypts, though I suspect that they could be old or unhealthy individuals incapable of undertaking the flight to Tasmania. Similarly, there are reports of birds overwintering in Tasmania. Thomas (1970) reported that throughout the winter of 1970, parties of up to 11 parrots were observed in blue gums *Eucalyptus globulus* at Lutana, a suburb of Hobart. Brown (1989) notes that in 1985, many birds overwintered in the Hobart area, and in 1986 up to 12 birds were recorded around Hobart between May and July, but in 1988 the only record for June and July was of a single bird seen in Hobart on 8 June, so it would seem that 1988 was one of those years when there was virtually complete migration across Bass Strait.

In mid December to early January, as the flowering of blue gums comes to an end in the nesting areas in eastern Tasmania, Swift Parrots commence to disperse in a westerly direction, taking advantage of the flowering of other eucalypts through central and western Tasmania. Departure for the mainland starts in late February, though it can be variable, and in some years may be delayed until late March or early April. Brown points out that available evidence indicates that migration is over western Bass Strait, and the 300 km wide water crossing is made by small parties of 10 to 20 birds in direct daytime flight from coast to coast, probably in about five hours at flight speeds of up to 80 km per hour. At what appeared to be a large pre-migratory assemblage of up to 500 birds at a roost located at Tullah, northwestern Tasmania, there was a protracted morning departure between about 0620 hours and 0920 hours, with a peak at approximately 0800 hours, and most birds fed in the flowering roosting trees before leaving, but even with such a late morning departure there would be sufficient daylight for post-arrival feeding on the mainland.

Earliest arrivals are detected in southern Victoria at the end of February, but numbers build up during March, and by April the parrots are fairly widespread throughout Victoria and are moving north into southern New South Wales. The relatively dry valleys of the Tambo and Snowy Rivers, in southeastern Victoria, may be corridors for birds moving to and from wintering areas in New South Wales (Emison *et al*. 1987). Brown points out that an analysis of wintering records suggests that in most years more than 90 per cent of the population overwinters in Victoria and New South Wales, and the paucity of records from southeastern South Australia indicates that the species is basically a latitudinal migrant, in general moving north and south rather than east and west. Nevertheless, the parrots do wander nomadically in search of food and at times there are large-scale movements away from the core wintering range. Such movements probably are influenced by available food supplies. Spectacular irruptions occur on occasions, and birds may be seen at localities from which they have been absent for many years. Hindwood and McGill (1958) noted that during the early 1900s the species was recorded commonly in the vicinity of Sydney, New South Wales, but subsequently seldom was observed until the autumn and winter of 1936, when considerable numbers came to flowering eucalypts in northern suburbs, and then again in June and July 1958 there was a spectacular invasion, with flocks, sometimes containing up to 2000 birds, being reported from many suburbs. Similarly in 2002, when severe drought conditions prevailed, there was a massive influx of birds in the Central Coast, where there was an incidental observation of mass roosting by 650 birds, equating to some two-thirds of the estimated total population (in Saunders and Heinsohn 2008).

Return to Tasmania commences in late August to early September, with the parrots again becoming widespread in Tasmania after mid September, but the exodus from Victoria continues until October and occasionally into early November.

Gartrell and Jones (2012) note that there is no evidence of synchronous physiological or behavioural changes in Swift Parrots prior to migration, and this may reflect the weakly synchronised migration or that the fairly short-distance migration requires minimal physiological preparation.

HABITS In habits and general behaviour, Swift Parrots resemble lorikeets and mixed flocks occasionally are seen feeding in flowering trees. Many observers experience difficulty identifying Swift Parrots in the field. The most reliable distinguishing feature is the call, which differs markedly from the screeching notes emitted by lorikeets. In flight, the red underwings, together with the fine, pointed tail, are diagnostic. As no Australian lorikeet species has red under tail-coverts, this marking is a further aid to identification.

Saunders and Heinsohn (2008) report that during field studies undertaken at winter foraging sites in New South Wales, Swift Parrots were observed in flocks of up to 200 birds, with a mean size of 25 birds, in coastal habitats and up to 60 birds, with a mean size of 20 birds, in habitats on the western slopes of the Great Dividing Range. In Tasmania the flocks can be quite large, particularly at pre-migratory roosts in the northwest. Brown (1989) points out that even during the breeding season the flocking behaviour is strong, with birds tending to congregate to feed, and an analysis of feeding groups seen in December showed that 57 per cent were in groups of five or more and 33 per cent in flocks of 10 or more. They are arboreal, spending much of the time feeding among the topmost branches of flowering eucalypts, and coming to the ground only rarely to drink or to examine seeds and fallen blossoms. They are noisy birds and their presence is readily detected. While feeding, they are oblivious to the approach of an intruder, and may be watched from underneath the feeding tree. They are always on the move, clambering about amidst the foliage, frequently hanging upside down to get at blossoms, and darting from tree to tree. Brown notes that when moving from one tree to the next, they seldom alight amongst the outer foliage but prefer to fly into the tree to land on a thick inner branch and then literally run out to flowers at extremities of the outer branches. I have seen a feeding flock suddenly leave a tree for no apparent reason, fly above the forest canopy for a few minutes, and then return to the same tree to resume feeding as if there had been no interruption. Hindwood (1939) noted that after feeding they often rest on topmost leafless branches of tall trees. Communal nighttime roosts are occupied by the parrots, and feeding during the early morning and the evening is in nearby trees, but during the day flocks travel much farther afield to forage.

The Swift Parrot is aptly named, for its strong, direct flight is exceedingly swift, the wingbeats producing an audible 'whirring' sound. Flocks move in the air with remarkable precision. Collisions in flight with manmade structures, notably power lines, tennis courts and windows often result in death or serious injury.

CALLS Brown (1989) identifies five calls given by adults, and all are quite different from the screeching call-notes of lorikeets. A very soft *chit* is repeated at intervals of 20 to 30 seconds by birds to maintain contact with each other, usually when intently feeding in the canopy of a eucalypt. The familiar alarm call is a loud series of sharp, penetrating *kik-kik-kik* notes repeated up to 20 or 30 times in ten seconds, and is given by birds when taking flight and in the early stages of sustained flight, or by alarmed and excited birds while perched, often in an alert posture prior to take off, or by birds flying from one tree to another. A series of musical warbling notes, which may be given only by males, can be heard at any time of the year, but is most prevalent in association with courtship activities soon after arrival in the breeding areas; single birds can be seen uttering these singing notes while perched alone on a branch inside the canopy, but the call is heard most frequently amongst groups of birds rather than from individuals. Females soliciting feeding or being fed by males give a disyllabic or occasionally trisyllabic high-pitched call, with the first note audibly stronger and higher. On one occasion, birds were heard to emit a very soft, unusual *churr* as they settled at a nighttime roost. Strident, rasping begging calls are given by juveniles and recently fledged birds when soliciting parental feeding.

DIET AND FEEDING Lorikeets rely heavily on pollen rather than nectar for food, and their specialised, brush-tipped tongues are well adapted for sweeping up pollen, but Swift Parrots are primarily nectar-feeders, and the less specialised, brush-tipped tongue is more suited to lapping or siphoning nectar (Brown 1989). They are voracious feeders, and there are numerous reports of the plumage of feeding birds becoming completely matted with nectar. While nesting, these parrots are dependent on nectar from the flowers of blue gums *Eucalyptus globulus* and swamp gums *E. ovata*, the former being a source of copious quantities of nectar. An analysis of all feeding records during the period of August 1987 to January 1988 showed that at 78 per cent of feeding sites blue gums were the food source, and when combined with swamp gums these two eucalypts made up 91 per cent of all feeding records. A comparison of feeding rates in three species of eucalypts highlights the higher nectar flow in flowers of blue gums; in flowering white gums *Eucalyptus viminalis* birds were seen to feed at a rate of 3 flowers per second, occasionally slowing to 2 flowers per second, while in swamp gums the rate was 4.5 flowers per second, but in blue gums it was almost exactly one flower per second, and one bird was seen to spend as much as 11 seconds feeding at one flower, obviously utilising a particularly rich source of nectar. After blue gums have ceased flowering, Swift Parrots have been observed feeding at flowers of other eucalypts, taking fruits or berries from both native and cultivated trees or shrubs, and eating unripe seeds of native grasses as well as insects and their larvae.

In the mainland wintering range, flowering eucalypts again are the major food source, though there are indications that psyllids and their scaly coverings or lerps feature prominently in the diet, either as a subsidiary food or as a replacement food in years when there is poor flowering of eucalypts. Brown notes that in Tasmania, females appear to feed regularly on psyllids and lerps just prior to egg-laying, but at other times the insects are only a secondary food. In Victoria, the parrots feed mainly in winter-flowering plants, especially red ironbarks *Eucalyptus sideroxylon* and ornamental trees or shrubs (Emison *et al.* 1987). In the Bendigo district, central Victoria, they are attracted to flowering red ironbarks and yellow gums *Eucalyptus leucoxylon*.

Saunders and Heinsohn (2008) report that during surveys of winter foraging sites in New South Wales, Swift Parrots were observed feeding in known food trees, including swamp mahogany *Eucalyptus robusta*, forest red gum *E. tereticornis*, blackbutt *E. pilularis*, spotted gum *Corymbia maculata* and red bloodwood *C. gummifera* in coastal habitats and red ironbark *Eucalyptus sideroxylon*, yellow box *E. melliodora*, white box *E. albens* and grey box *E. microcarpa* in habitats on the western slopes of the Great Dividing Range. Two additional tree species identified as being potentially important food sources were gum-topped box *Eucalyptus mollucana* and rough-barked angophora *Angophora floribunda*, though the latter was recorded only as a source of lerps during drought conditions. During the course of the surveys, nectar and lerps comprised similar proportions of foraging observations, but lerps seemed to be more important in coastal habitats. The presence of other nectar-feeding species impacted on the likely occurrence of Swift Parrots at foraging sites. In coastal habitats, a presence of non-aggressive Little Lorikeets *Parvipsitta pusilla* and Noisy Friarbirds *Philemon corniculatus* indicated an increased likelihood of Swift Parrots occurring and the presence of aggressive Rainbow Lorikeets *Trichoglossus haematodus* indicated a decreased likelihood, while in habitats on the western slopes of the Great Dividing Range, a presence of non-aggressive Little Lorikeets, Fuscous Honeyeaters *Lichenostomus fuscus* and Dusky Woodswallows *Artamus cyanopterus* and aggressive Red Wattlebirds *Anthochaera carnunculata* indicated an increased likelihood, and the presence of aggressive Noisy Miners *Manorina melanocephala* indicated a decreased likelihood.

In Victoria and southern New South Wales, I have watched Swift Parrots, often in the company of Musk Lorikeets *Glossopsitta concinna* and *Lichenostomus* honeyeaters, feeding in flowering sugar gums *Eucalyptus cladocalyx* planted as windbreaks in farmlands and along roadsides. Hoskin (1991) notes that in the suburbs of Sydney, New South Wales, they have been seen feeding in flowering swamp mahogany *Eucalyptus robusta*, and were observed sipping nectar from inflorescences of saw banksias *Banksia serrata* and swamp banksias *B. robur*. Edwards (1921) recalled that near Tathra, on the south coast of New South Wales, they were seen clinging to flowering spikes of unidentified *Xanthorrhoea* grass-trees.

Lea and Gray (1935) reported that numerous seeds were found in the stomach of a bird collected near Hobart, Tasmania, while birds taken near Adelaide, South Australia, had in their crops and stomachs numerous small caterpillars, psyllids and lerps, seeds, some vegetable matter and sand or grit. Brown (1989) notes that in the stomachs of eight birds accidentally killed in the Hobart area there were insects and their larvae, particulate and amorphous vegetative matter, gravel, plastic fragments, and feathers, the last presumably being ingested while preening, but it was pointed out that the likely main food, namely nectar, would not be present in the stomach.

BREEDING Field studies indicate that provided both birds survive the winter, adults remain paired from year to year, and it is presumed that courtship and pair-maintenance activities occur mostly in late winter just prior to departure from the mainland, for pairs commence nesting activities almost immediately after arrival in the breeding areas (Brown 1989). At one location near Hobart, on 11 September, a pair was found to be occupying a nesting hollow only three weeks after the first arrivals had been recorded. If not already paired, first-year birds undertake courtship and pair formation activities, often involving aerial chases and squabbling between rival males, before searching for nesting sites, and this usually delays the commencement of nesting until October or early November. The male adopts a lead role in selection of a nesting hollow, with the female following closely behind. Both birds arrive in a tree and, in turn, hurriedly scamper along each secondary limb, inspecting every crack, knothole or broken branch, often hanging upside down while peering into possible crevices, before moving on to repeat the search in the next tree. Not always will the hollow first selected be used eventually for nesting, and one pair was seen to spend three days scratching and excavating inside a hollow before abandoning it.

The nest is in a hollow limb or hole in a living or dead eucalypt, and the close proximity of many nests to each other reflects the gregarious behaviour of these parrots being retained throughout the breeding season. Brown notes that nests most often are located in well drained, dry sclerophyll woodland on steep hillsides or ridges in areas where bushfires have resulted in an abundance of hollows in dead stags on living trees or in numerous dead trees. Surveys for nests were undertaken at sites in southeastern Tasmania during November–December in the 2004 to 2006 breeding seasons, and a total of 128 nest-hollows was described from 117 nesting trees (Webb *et al.* 2012). The trunk diameter at breast height of nesting trees was 33 cm to 202 cm, with a mean of 105 cm, and tree height was 10 m to 45 m, with a mean of 23 m. Nesting trees were characterised by having a trunk diameter at breast height exceeding 80 cm, having five or more potential hollows, having dead limbs penetrating the crown, and showing clear signs of senescence. The relative probability of a tree being used as a nesting tree increased with the number of potential hollows and with the trunk diameter at breast height. Of the 128 nest-hollows, 90, or 70 per cent were in branches and 38, or 30 per cent were in trunks, with 87 entrances, or 68 per cent being at knotholes or branch stubs, 21 entrances, or 16 per cent being at fissures and 20, or 16 per cent being at spouts. During the surveys there was dramatic variability in food availability, so the parrots moved to breed wherever food was most available, and they did not reuse nesting sites in successive years when food availability declined, thus enabling them to exploit a variable food resource for successful breeding (Stojanovic *et al.* 2015).

Prior to egg-laying, both male and female make regular visits to the selected hollow, spending much time scraping and chewing at the inside walls, and the female spends increasingly longer periods inside the hollow, leading to fulltime occupancy just prior to laying of the first egg. A clutch normally comprises four or five eggs, which are laid on a bed of decayed wood dust at the bottom of the hollow. All of 20 clutches of eggs taken on Bruny Island, southeastern Tasmania, between 1926 and 1937, are of four or five eggs, with 4.4 eggs being the average clutch size (ANWC Collection).

Only the female incubates, and she is fed by the male at intervals of three to five hours. At times the sitting female leaves the nest in response to calling from the approaching male, and both fly to a nearby tree, but on other occasions he comes to the nest, peers into the hollow and calls to coax her to leave. Regurgitated food is passed from the male to the female at up to 15 times on each visit, and is followed by copulation if laying is still in progress. Immediately after feeding, the female returns to the nest, while the male often perches quietly near the nest before flying off. Hatching appears to be asynchronous, for at one nest there was a lapse of six days between departure from the nest of the first and last chicks, and this suggests that incubation commences with laying of the second or third egg. Newly hatched chicks are brooded by the female for at least the first two weeks, during which time feeding visits by the male increase to a frequency of one every one to two hours. Both parents then share feeding duties, making visits to the nest at an average interval of 1.5 hours when the chicks are four weeks old, but at intervals of less than an hour as the chicks grow older. At three to four weeks old, the chicks clamber up to the hollow entrance to be fed by the incoming adults, and they leave the nest approximately six weeks after hatching. For the first few days, fledglings remain near the nest, often perching immobile for hours at a time or calling loudly if adults appear in the area. Eventually they can be seen begging for food while following adults in flight through the treetops.

There is no evidence of second broods being reared, but it is possible that pairs may renest after failure of their first attempts, and this would explain the appearance of fledglings very late in the season.

EGGS The eggs are broad-elliptical and with a slightly glossy surface. On Bruny Island, southeastern Tasmania, on 20 November 1936, at a height of approximately 7 m in a living eucalypt, in a 60 cm deep hollow with an entrance at the side of a dry limb, a clutch of five eggs had been laid on an old nest of Tree Martins *Petrochelidon nigricans*, and these eggs measure 24.5 (24.2–24.7) × 20.1 (19.8–20.5) mm (ANWC E07232). Average measurements of 88 eggs from 20 clutches of either four or five eggs are 24.7 (22.0–27.0) × 20.3 (19.0–22.0) mm (Hindwood and Sharland 1964).

GENUS *Eunymphicus* Peters

Eunymphicus Peters, *Check-list of Birds of the World*, **3**, 1937, p. 269 (New name for *Nymphicus* of authors, not of Wagler, *Monogr. Psittac.*, 1832, p. 490; see *Cat. Bds Brit. Mus.* **20**, p. 590).

Although there has been general agreement that this genus is a member of the platycercine group, there has been some uncertainty about its nearest alliances. Peters (1937) expressed the view that it is related to *Cyanoramphus*, but Cain (1955) proposed linking it with *Purpureicephalus.* Molecular analyses have confirmed the traditional view that it is most closely related to *Cyanoramphus*, with *Prosopeia* a sister group to this clade, which in turn appears to be allied to *Platycercus* (Boon *et al*. 2008).

A peculiar coronal, non-erectile 'crest' of elongated feathers is the most conspicuous morphological feature of the midsized parrots belonging to this genus. As noted by Peters, the lores are less densely feathered than in *Cyanoramphus*, and there is a distinct notch in the upper mandible. The sexes are alike, and juveniles are like adults, but with less extensive, duller facial markings and a pale bill.

Eunymphicus is restricted to New Caledonia and to Ouvéa, in the nearby Loyalty Islands, in the southwest Pacific Ocean. The two allopatric and morphologically different populations formerly were treated as subspecies, but molecular analyses indicate that there is clear genetic separation which, together with some behavioural isolation, warrants specific differentiation (Boon *et al*. 2008).

VULNERABLE

Horned Parakeet

Eunymphicus cornutus (Gmelin)

Psittacus cornutus Gmelin, *Syst. Nat.*, **1**, part 1, 1788, p. 327 (New Caledonia).

DESCRIPTION Length 32 cm. Weight approximately 130–140 g.
ADULTS General plumage bright green, more yellowish on underparts; rump greenish-yellow; ear-coverts and hindneck yellow, variably tinged green on hindneck; forehead and forecrown red; face black; 'crest' of two elongated feathers black tipped red; primaries and outer secondaries blue-black, on outer webs strongly suffused deep violet blue and narrowly margined paler blue; central tail-feathers above green becoming violet-blue towards tips, lateral tail-feathers violet-blue tipped pale blue at tips of outer webs; bill pale bluish-grey becoming black towards tip; iris orange-red; legs dark grey.
10 males: wing 154–171 (162.2) mm, tail 139–175 (163.8) mm, exp. cul. 19–21 (20.4) mm, tars. 19–22 (21.2) mm.
6 females: wing 151–161 (156.7) mm, tail 150–166 (156.2) mm, exp. cul. 17–21 (18.7) mm, tars. 20–22 (20.5) mm.
JUVENILES Facial markings greyish and less extensive; feathers of forehead and forecrown black tipped red; ear-coverts pale green; hindneck olive-green; bill horn-coloured; iris brownish.

DISTRIBUTION Restricted to Grande Terre, New Caledonia.

STATUS As early as the 1940s, concern was expressed about declining numbers of Horned Parakeets on New Caledonia by Warner (1947), who, during 1944 and 1945, found them to be fairly common in more inaccessible areas, but he remarked that numbers must have declined drastically since 1882, when they were reported from all forested areas. Vincent (1967) claimed that numbers were low enough to cause concern. In December 1976, during a visit to New Caledonia on behalf of the Australian Museum, Anthony Stokes found *cornutus* to be fairly common in the Rivière Bleue Forest Reserve, but it was rarely seen elsewhere, and he suggested that protection afforded by the Reserve was a key element in long-term survival of the parrots, though the capture of live birds for pets will continue to be a threat (pers. comm. 1977). Bregulla (1993) confirmed that by the 1960s and 1970s these parakeets were relatively uncommon, but were known to frequent particular regions. Between July and November 1998, surveys were undertaken at nine sites in humid forest, and Horned Parakeets were recorded at five of these sites, more frequently in southern regions, where they were common up to 1100 m at the Reserve Speciale de Faune et de Flore de la Nodela and up to 700 m at Rivière Pujémia, but rare or uncommon up to 450 m at Néoua and at Rivière Bleue Forest Reserve, and rare or uncommon also up to 1200 m at Mount Ignambi in the north (Ekstrom *et al.* 2002). Population estimates made at this time were of 1000 to 3000 birds and 720 pairs, but a more recent study, using distance sampling density data, records and ecological niche modelling, indicates that the parrots are more widely distributed and more common than previously believed, producing a roughly estimated population of possibly 8000 to 9000 birds, equivalent to 5300 to 6000 mature individuals (in Birdlife International 2016). During surveys undertaken in 2003 to 2006, the parakeets were recorded on 57 per cent of massifs in the northern province and 42 per cent of massifs in the southern province, being locally common in central sections of the mountain chain, with numbers having remained stable at Rivière Bleue Forest Reserve during the past 20 years (in Birdlife International 2016).

Only 28 per cent of primary vegetation remains on New Caledonia, and loss of habitat is identified as the principal threat to populations of Horned Parakeets. In remnant patches of rainforest, degradation by feral pigs and introduced Rusa Deer *Rusa timorensis* poses an additional pressure and selective logging removes potential nesting trees. Predation of nests by feral cats

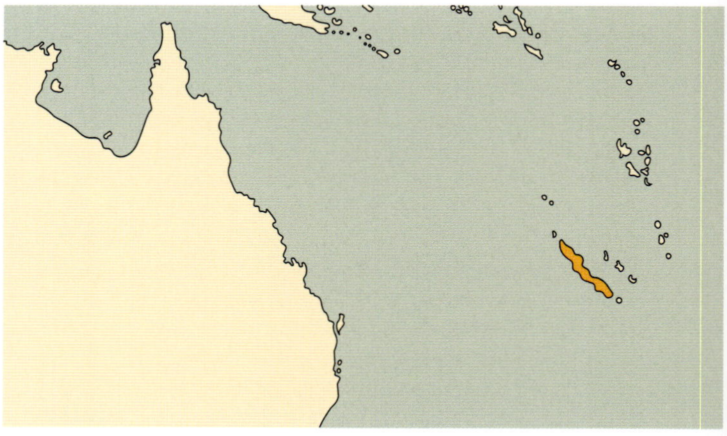

and rats is likely to impact on recruitment, though it has not been quantitatively assessed. There is little documentation of trapping or nest-robbing, though the parakeets are held in captivity on New Caledonia and are sought locally by collectors for the live-bird trade, but nesting occurs mostly in remote areas so poaching is unlikely to be a major threat (in Birdlife International 2016). Psittacine circoviral disease (beak and feather disease) has been detected in Red-fronted Parakeets *Cyanoramphus novaezelandiae* on New Caledonia, so is perceived as a potential threat to Horned Parakeets. These parakeets are protected by New Caledonian law, but safeguarding habitat remains the most important requirement. Only 4 per cent of New Caledonia currently is set aside for conservation of fauna and flora, and many protected areas do not have strict bans on mining, so establishment of a network of nature reserves and wildlife corridors would assist in preserving preferred forest habitat and would give the parakeets an important link between isolated tracts of forest (Legault *et al.* 2011).

Eunymphicus cornutus is listed on Appendix I to the Convention on International Trade in Endangered Species of Wild Fauna and Flora (CITES).

HABITATS From studies of large-scale habitat selection by three species of parrots in New Caledonia, undertaken between 2002 and 2010, it was concluded that Horned Parakeets used rainforest more frequently than expected based on its restricted availability along the central mountain chain (Legault *et al.* 2011). Forest cover proved to be the most influential factor, with a strong preference for areas with 80 per cent to 100 per cent forest cover and large tracts of forest exceeding 1000 km². They favour forests on metamorphic or volcanic soils, and this may be the reason why they are more common in central regions and patchily distributed in the south. Also, preliminary research into their diet indicates that their food plants are mainly associated with valley forest. When these parakeets were encountered outside forest, the average distance to the nearest forest edge was only 66 m, and they avoided shrublands and *Melaleuca*-dominated savanna. Rainfall was the second most important influential factor, with preference shown for areas receiving 1500 mm to 2000 mm of rainfall per year. They selected mid montane altitudes between 400 m and 600 m, though altitude was the least important influential factor and they did frequent montane forests, but their avoidance of altitudes below 200 m could be due, at least in part, to the very widespread clearance of lowland forests. Despite these preferences, birds have been observed crossing open scrub between tracts of forest, and flocks have been seen feeding in open savanna or in valleys in close proximity to rural dwellings (in Birdlife International 2016).

HABITS Often located by their characteristic call-notes, Horned Parakeets are noisy, active birds usually encountered in pairs or small parties flying above the treetops or feeding in the mid to upper stages of forest trees, where they scurry among the foliage in search of berries and fruits. Warner (1947) recalled that pairs or groups of four or more typically were encountered in *Agathis-Araucaria* forests, and on one occasion a flock of 10 birds was observed feeding in a small tree about 3 m high. They were shy and difficult to approach, invariably flying off when disturbed to alight in the tops of tall trees at a distance of 100 m or more, and there they perched quietly. If disturbed again they would fly off silently, flocks dispersing in pairs. At night they roosted in the treetops or in hollows, and were astir about an hour before dawn. When a group moved off to feed, one bird usually took the lead, and the swift flight was slightly undulating.

CALLS Dutson (2011) describes the calls as a repeated trumpeting *khoo* or *kho-khoot*, together with softer chattering and a range of shrieks and chuckles.

DIET AND FEEDING The diet comprises fruits, berries and seeds procured in trees and bushes, with the flesh and seeds of cultivated papaya *Carica papaya* a favoured food. Macmillan spent a couple of hours watching a pair feed on the berries of vines and seeds of various trees and shrubs, especially seeds of the 'penubre tree' and, while feeding, the two birds occasionally displayed to each other, bowing several times, ruffling up their feathers and calling intermittently (in Warner 1947). Crop and stomach contents from four birds collected by Macmillan comprised remains of flowers, large milky seed kernels, and yellow fruit-pulp with melon-like seeds.

BREEDING Breeding commonly is recorded during October to December, and every nest found by Macmillan was in a living 'teak' tree *Metrosideros laurifolius* (in Warner 1947). Between 2002 and 2008, observations were undertaken at Grandes Fougères Park, where five nests were found in tree hollows, and Rivière Bleue Forest Reserve where six nests were found on the ground in crevices between rocks (Theuerkauf *et al.* 2009). Clutches comprised two to four eggs, but in the 2003–2004 breeding season five eggs were found in one tree hollow regularly visited by two pairs, and on one occasion two females were seen peacefully sharing the cavity while the two males were chasing each other at the hollow entrance. As the two pairs used this nest at the same time and the clutch size was larger than in any other recorded clutch, it is unlikely that all five eggs were laid by only one female. There was always only one female in the nest during incubation and chick-rearing, but it was not possible to determine whether both females took turns in the nest or if only one female incubated the eggs and fed the chicks. Based on differences in behaviour of the two pairs, it was suspected that both continued to attend the nest, but at no other time were they seen together in the nest at the same time.

EGGS Measurements of five eggs are given as 26.4 (25.4–28.0) × 21.7 (20.3–23.6) mm (Schöwetter 1964).

Ouvéa Parakeet

Eunymphicus uvaeensis (Layard and Layard)

ENDANGERED

Nymphicus uvaeensis E. L. and E. L. C. Layard, *Proc. Zool. Soc. London*, 1882, p.408, pl. 26, F. 2 (Ouvéa, Loyalty Islands).

DESCRIPTION Length 32 cm.
ADULTS General plumage bright green, slightly paler on underparts; rump yellowish-green; ear-coverts and hindneck green; red restricted to centre of forehead; face blackish-green; 'crest' of six elongated green feathers; primaries and outer secondaries blue-black, on outer webs strongly suffused deep violet blue and narrowly margined paler blue; central tail-feathers

above green becoming violet-blue towards tips, lateral tail-feathers violet-blue tipped pale blue at tips of outer webs; bill pale bluish-grey becoming black towards tip; iris orange-red; legs dark grey.
8 males: wing 151–169 (161.3) mm, tail 136–158 (151.0) mm, exp. cul. 19–22 (20.9) mm, tars. 19–22 (20.5) mm.
13 females: wing 149–167 (155.0) mm, tail 127–155 (146.8) mm, exp. cul. 18–22 (19.3) mm, tars. 18–21 (19.5) mm.
JUVENILES Like adults, but generally duller and with less red on centre of forehead; bill horn-coloured; iris brownish.

DISTRIBUTION Confined to Ouvéa in the Loyalty Islands, New Caledonia.

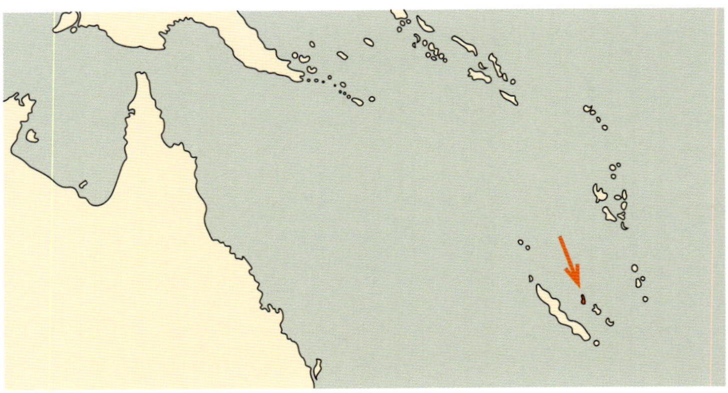

STATUS On Ouvéa Island, with a total area of 132 km², Ouvéa Parakeets occur mainly in a northern tract of forest totalling some 2000 ha in area. Between 27 November and 10 December 1993, a population count using fixed distance line transects was undertaken, the transects totalling 134.5 km of walk tracks and roads through all possible habitats, and of the 73 presumably different parakeets that were observed, 65 were in the north of the island, five were in the centre and three were in the south, giving an estimated population of 617 birds with a minimum of 274 and a maximum of 617, and at that time another 100 to 200 birds were held in captivity, mostly in New Caledonia (Robinet *et al.* 1995; Robinet *et al.* 1996). Loss of habitat again was identified as the principal threat, with a large reduction in the extent of forest having taken place to make way for coconut plantations and subsistence crops. The remaining forest was further fragmented through clearance for agriculture. Mindful of the vulnerability of populations on small islands to introduced predators, a study was undertaken between July 1995 and October 1996 to assess the impact of predation by Pacific Rats *Rattus exulans* on Ouvéa and to investigate the feasibility of translocating parakeets to nearby Lifou Island, which is much larger in size and is densely forested (Robinet *et al.* 1998). A rat trapping program confirmed the presence of Pacific Rats on Ouvéa and the presence of both Pacific Rats and Black Rats *Rattus rattus* on Lifou, with differences in levels of predation of quail eggs placed in natural tree hollows during the main bird breeding season reflecting a significantly higher predation by Black Rats on Lifou, so making that island unsuitable for translocation. Other predators on Ouvéa include feral cats, which are known to take nestlings but are not common, and Pacific Boas *Candoia bibronii*, which have been found in nests during the breeding season but are rare, so overall predators do not appear to pose a major threat to the parakeets, although Brown Goshawks *Accipiter fasciatus* are said to take many juveniles. There is a long tradition on Ouvéa of taking nestlings for pets, and this often entailed chopping into hollows to gain access to the chicks, so making the sites unsuitable for future nesting. Poaching levels increased when collectors of rare birds became interested in the parakeets, and it seems that between September 1992 and February 1993 more than 50 chicks were taken from nests, with the majority being illegally exported from the island (Robinet *et al.* 1995).

In an effort to save the Ouvéa Parakeets, the Association for the Protection of the Ouvéa Parakeet (ASPO) was established in 1993, and it was set up as a non-governmental organisation of mostly local members to ensure that its proposed conservation measures were accepted by the island community (Barré *et al.* 2010). A long-term study of the biology and ecology of the parakeets was initiated as a basis for two recovery plans that were prepared with local residents for acceptance by local authorities and the Loyalty Islands administration. The recovery plans recommended increasing law enforcement to tackle illegal trade, mitigating habitat degradation and restoring suitable habitat, monitoring nesting sites, testing artificial nest-hollows, preventing the introduction of Black Rats, locating and destroying nests of Honey Bees in the wild, and improving public awareness of the value of the parakeets. The public awareness campaign has been particularly successful, and is credited with bringing about an elimination of nest poaching, which has been the most important factor in achieving a substantial increase in the population. A continuous presence of local guides effectively prevented nest poaching, and many people on the island now appreciate the parakeet, which has become the island emblem and has been included in the logo of the Ouvéa Council. As a consequence of these actions, the mean density of parakeets on Ouvéa increased threefold from 10 birds per km² in 1993 to 34 birds per km² in 2009, equating to a population increase of 34 per cent over three generations, and the present total population is estimated at 3100 to 3200 birds (Barré *et al.* 2010; in Birdlife International 2016). The bulk of the population is still in the north of the island, but numbers are increasing in the south (Dutson 2011). Conservation efforts will be restricted to Ouvéa as a translocation of birds to Lifou Island is no longer seen as a viable option because of the presence of Black Rats on that island and because two previous attempts failed. In about 1925 approximately 100 birds were taken to Lifou, but were said to have returned immediately to Ouvéa, though the authenticity of that claim has been questioned. In 1963, a second attempt made by the New Caledonian Forestry Service involved the capture of 15 birds at the beginning of September and subsequent release by an employee of the Agricultural Service at a site on Lifou chosen because of the presence of dense forest with nearby cleared areas containing fruit trees, but a follow-up search in 1981 failed to locate any birds.

HABITATS Ouvéa Parakeets are very much birds of old-growth rainforest, where nest-sites are available in mature trees, but they do forage in nearby second-growth forest and in fruit or vegetable gardens, especially where papaya is cultivated. Comprising trees 10–15 m in height, the principal tract of old-growth forest is in the north of the island, where it is less than 2500 ha in extent in an area of some 10 km in length and an average 1 km

Plate 11
UPPER Horned Parakeet *Eunymphicus cornutus* (adult)
LOWER Ouvéa Parakeet *Eunymphicus uvaeensis* (adult)

F. KNIGHT '11

in width (Robinet *et al.* 1995). During studies of habitat use undertaken during the 1994 and 1995 breeding seasons, the observed proportions of usage were equally highest in tall forest, Melanesian fallow and recent fallow, with the next highest usage in medium forest, and little usage in old fallow and domestic gardens and no usage at all in coconut plantations or coastal vegetation (Robinet *et al.* 2003).

HABITS Ouvéa Parakeets are noisy, active birds easily located by their distinctive calls, and they usually are encountered in pairs or small parties flying above the treetops or feeding in the mid to upper stages of forest trees, where they scurry among the foliage in search of berries and fruits. During studies of activity patterns and habitat use by radio-tagged birds undertaken during July to December 1994 and again in July to December 1995, it was determined that four juvenile parakeets had an average home range of 2.34 ha, and estimated home ranges of adults ranged from 1.21 to 2.8 ha, with the 2.27 ha mean home ranges of males being larger than the 1.41 ha mean home ranges of females (Robinet *et al.* 2003). The parakeets were almost exclusively arboreal, spending most of their time in the shaded canopy, where 71.5 per cent of observations were made, or in the unshaded canopy, where 17.7 per cent of observations were made, and in the understorey, where 10.6 per cent of observations were made, but only 0.2 per cent of observations were of birds on the ground. Similarly, most records were of birds at 5 m to 7 m above the ground, where 55.7 per cent of observations were made, or above 7 m, where 24.1 per cent of observations were made, but there was less use at 3 m to 5 m, where 14.3 per cent of observations were made. On average, 47 per cent of time was spent on foraging activities, with two peak periods being in the late morning, between 0900 and 1100 hours, and mid-afternoon, between 1400 and 1600 hours. During the middle of the day, most time was spent roosting, preening and participating in nesting activities, including incubation and allofeeding, while during the first and last hours of the day vocalisation and some agonistic interactions were the main activities, though at these times other infrequent activities could have remained undetected. Although foraging was the main recorded behaviour, it accounted for less than half of the total activity of the parakeets.

CALLS The contact call is described as a oft-repeated loud nasal, trumpeting *khoot*, with a screeching alarm call given when flushed and a range of other shrieks and chuckles (Dutson 2011).

DIET AND FEEDING Studies of the diet and feeding behaviour were undertaken between July and December in 1994 and again in 1995, and it was found that the phenology of food plants was marked by a long fruiting season, leading to a relative abundance of food throughout the year, especially during the breeding season. The average proportion of food plants with fruits was 70.9 per cent, with some plants, including *Ficus* spp., papaya, and the vines *Maesa novocaledonica*, red almond *Rhamnella vitiensis* and small passionfruit *Passiflora suberosa* bearing fruit almost throughout the year while some others, such as *Elattostachys apetala*, show great variation in their fruiting season between years and between individual plants. Most plants have a fruiting season during the hot months of October to March, leading to a greater abundance of food during this period, followed by a decrease in April. During the studies, 770 foraging observations involved 23 different food plants, and seven of these were involved in more than two-thirds of observations, with *Ficus* spp. alone representing more than 30 per cent of all observations. The most strongly selected food plants were those with fruit all year round, and *Ficus* spp., papaya, *Rhamnella vitiensis*, canary beech *Polyalthia nitidissima*, brush bloodwood *Baloghia inophylla* and burny vine *Trophis scandens* were significantly preferred by the parakeets. *Ficus* fruits are a major food in all months, with a maximum from May to October, before and at the beginning of the breeding season. Papaya is eaten mostly during the second half of the breeding season, from November to January, with seeds of *Trophis scandens*, *Baloghia inophylla* and *Harpullia neocaledonica* then eaten successively from January to July. Apart from the flesh of papaya, seeds ranging in size from 1.5 mm to 45.0 mm were eaten almost exclusively, and there was no evidence of insects being taken, although an exceptional incursion near to the ground may have been for this purpose.

BREEDING Much of what is known of the nesting behaviour of these parakeets comes from studies undertaken between June 1994 and February 1996, when nesting was recorded from early August until late January (Robinet and Salas 1999). From 16 active nests and 24 clutches followed during the study, the first egg was laid on 30 July and the last chick fledged on 24 January, with a first peak of egg-laying occurring in August to early September, followed by a second peak in early October, and some laying, including second clutches, in November–December, but no second clutches were recorded for birds which first laid during the second peak time. The four recorded second clutches were produced by two marked pairs and an unmarked pair, and renesting was not related to nest failure, as all of the four first clutches of these pairs were successful. Nests were in hollows in the trunks of living trees, and all 71 active or former nests identified during the study were at an average height of 183 cm above the ground in trees of only five species, with 37 being in *Syzygium pseudopinnatum*, 27 in red condoo *Mimusops elengii*, five in *Dysoxylum rufescens* and only one in kwila *Intsia bijuga*. Hollow entrances averaged 11.5 cm in diameter, and egg-chambers were at an average depth of 97.8 cm. Two pairs were found nesting at the same time only 30 m apart without any evidence of agonistic behaviour, and this suggested that only the nest itself was defended as a breeding territory. There appeared to be a strong fidelity to nesting sites, and three identified pairs used the same nests in successive years. Preparation of the nest commenced some two to seven weeks before laying, during which time the hollow was cleaned out and at the end a roughly circular 10–15 cm deep depression was excavated in a lining of powdered wood, presumably chewed from walls of the hollow.

Of the 24 clutches monitored in the study, 22 comprised three eggs and two comprised two eggs. The second egg was laid at an average 24 hours after the first, and intervals of 24 to 72 hours elapsed between laying of the second and third eggs. Behavioural observations suggested that incubation by the female commenced after laying of the second egg, and it lasted 19 to 22 days. Hatching of the first and second chicks occurred at the same interval as between laying of the first and second eggs, but the third chick hatched an average 48 hours after the second. Newly hatched chicks were covered with thin white down, and at 10 days the eyes started to open, becoming fully opened at 14 days, when the white down was replaced by dark grey down. Also at 14 days, feather quills appeared on the wings and tail, and the wing and tail feathers were fully developed at about 32 days. At about 37 days red showed on the forehead and the elongated 'crest' feathers appeared. Fledging took place at an average 43 days after hatching.

Parental behaviour was monitored at three nests and, while incubating and brooding the chicks during 15 to 20 days after hatching, the female was fed by the male three to six times during the day. In response to his characteristic soft 'trumpet' calls, the female left the nest swiftly and alighted on a branch some 10 to 30 m from the nest, where she was joined by the male. The female solicited feeding by presenting her head held slightly back, with her bill open, and emitting a loud *clok* note. With intermittent pauses, the male spent an average 10 minutes feeding regurgitated crop contents to the female, and then she returned to the nest. Time spent in the nest by the female decreased from 73 per cent during the first 10 days after hatching to 57 per cent in days 11 to 20, when she continued to be fed by the male but also foraged increasingly by herself. After about 20 days, the chicks were fed by both parents, with an average interval of 145 minutes between feeding visits, and the female spent only 20 per cent of the time in the nest, with absences lasting an average 70 minutes. After leaving the nest, the young birds were fed by the male, and possibly by the female, for several weeks. Some three to seven months after fledging, the bill of young birds progressively turned bluish-grey and black, and the iris became orange-red.

From the 70 eggs laid in the course of this study, 63 chicks hatched. Of the 54 hatchlings for which data were available, only 34 reached fledging age, giving a fledging rate of 63 per cent. Of these 34 fledglings, nine were captured by local people, producing an overall breeding success rate of only 42 per cent, although this increases to 57 per cent if human trapping is excluded. On average, each nesting pair produced 2.9 eggs per clutch and fledged 1.65 young, from which only 0.75 was still alive 30 days after fledging, representing a 74 per cent loss of initial investment. Of the 24 clutches monitored in the study, five failed before fledging, due to human disturbance, screw-worm infestation, or presumed predation. In broods that partially failed within the first eight days, six out of seven deaths were caused by starvation of the third sibling. After fledging, four out of the six recorded deaths were caused by raptor attack, presumably by Brown Goshawks *Accipiter fasciatus*. A third major cause of loss was poaching of nestlings by islanders for pets, and this accounted for the loss of nine chicks from four nests.

EGGS Robinet and Salas (1999) list 27.1 × 22.5 mm as the average measurements of three eggs. A damaged egg laid in captivity is in the Natural History Museum at Tring, UK, and it measures 31.2(?) × 19.7 mm (Harrison and Holyoak 1970).

GENUS *Cyanoramphus* Bonaparte

Cyanoramphus Bonaparte, *Revue Mag. Zool.*, (2) **6**, 1854, p. 153. Type, by subsequent designation. *Psittacus pacificus* 'Forster' = *Conurus phaeton* Des Murs 1845 = *Psittacus erythronotus* Kuhl 1820 (G. R. Gray, *Cat. Gen. Subgen. Bds*, 1855, p. 86).

An absence of elongated 'crest' feathers readily differentiates this genus from closely allied *Eunymphicus*. The species are small to midsized, stocky parrots with long, graduated tails. The tarsi are decidedly longer than in other platycercine genera and there is no notch in the upper mandible. Smith (1975) pointed out that there is a unique tube-like, bony overgrowth of the ear-canal, which isolates *Cyanoramphus* from other genera. Molecular analyses confirm its inclusion in the 'core platycercine' group (Joseph *et al*. 2012).

Sexual dimorphism is slight, the female being smaller with a smaller head and narrower upper mandible. The underwing-stripe is prominent in juveniles and adult females, but slight or absent in adult males.

It seems that *Cyanoramphus* parakeets are still actively speciating, and they are capable of successful interspecific hybridisation, especially where one of two species is numerically dominant, resulting in a scarcity of prospective mates, and this is exacerbated where habitats have been modified by human activity (in Kearvell *et al*. 2003). Triggs and Daugherty (1996) warn that unnaturally high levels of hybridisation caused by human modification of habitats can pose a serious threat to a rare species because genetic swamping of that rarer species can lead to extinction of the natural genotype.

With an apparent centre of origin in New Zealand, *Cyanoramphus* parakeets are believed to have dispersed across ocean barriers to many islands in the South Pacific Ocean, with their historical range extending from the Society Islands south to subantarctic Macquarie Island and from Lord Howe Island east to the Chatham Islands (in Boon *et al*. 2001). This range has contracted following the extinction of populations in the Society Islands and on Lord Howe and Macquarie Islands, and habitat degradation, exacerbated by predation by introduced mammals, particularly rats and mustelids, is impacting severely on many remaining populations, including those in mainland New Zealand.

Red-fronted Parakeet

Cyanoramphus novaezelandiae (Sparrman)

Psittacus Novae Zelandiae Sparrman, *Mus. Carls.*, fasc. 2, 1787, no. xxviii and pl. (New Zealand, i.e. Dusky Sound, South Island).

OTHER NAMES Red-fronted Parrot or Kakariki, Red-crowned Parakeet; New Caledonian Parakeet (New Caledonia); Tasman Parrot, Green Parrot (Norfolk Island).

DESCRIPTION Length 30 cm. Weight 100 g.

ADULT MALE General plumage bright green, lighter and more yellowish on underparts; forehead, crown. lores and patch on ear-coverts behind eye rich red; red patch on each side of rump; underwing-coverts yellowish-green; primary-coverts violet-blue; outer webs of outer primaries violet-blue, shading to green towards tips; tail above dark green with fine greenish-yellow margins to outer webs of outermost feathers; underside of tail dusky olive; no underwing-stripe; upper

mandible silver-grey, becoming black at tip and along cutting edge, lower mandible grey-black; iris orange-red; legs greyish-brown.
18 specimens: wing 138–150 (145.5) mm, tail 138–183 (154.5) mm, exp. cul. 20–23 (21.3) mm, tars. 21–25 (23.4) mm.
ADULT FEMALE Like male but smaller in size, and with smaller, narrower bill; less red on ear-coverts behind eye.
15 specimens: wing 128–137 (134.1) mm, tail 136–169 (145.1) mm, exp. cul. 17–19 (17.4) mm, tars. 21–23 (22.2) mm.
JUVENILES Resemble adults, but with less extensive red markings on head; only few red feathers on each side of rump; shorter tail; underwing-stripe present; bill pale pink-white in fledglings, but soon changing to colour of adults; iris pale brown.

DISTRIBUTION New Zealand and outlying islands, from the Kermadec Islands south to the Chatham Islands, the Antipodes Islands and the Auckland Islands, and also New Caledonia and Norfolk Island; formerly on Lord Howe and Macquarie Islands.

SUBSPECIES Populations of *C. novaezelandiae* on more remote islands, including Norfolk Island and New Caledonia, have been isolated for a long time, so it is not surprising that molecular analyses have shown these to be highly divergent, and both have been accorded specific status under the phylogenetic species concept (Boon *et al.* 2001). I have stated repeatedly my reasons for not recognising specific differentiation for these and other island populations, pointing out that morphological differences are very slight, and I am not convinced that, despite long periods of isolation, they would retain reproductive integrity, a key requirement for specific status under the biological species concept. Furthermore, I hold reservations about the thoroughness of the molecular analyses, and it is unfortunate, though understandable, that these analyses did not include material from extinct forms, two of which could be important in determining the affinities of birds on Norfolk Island and New Caledonia. Birds formerly occurring on Lord Howe Island resembled those on Norfolk Island in size, but shared plumage features with New Caledonian birds. Also large in size were birds from Macquarie Island, but these resembled the nominate subspecies in plumage colouration. Also, I would point out that misidentification of some specimens used in the molecular analyses has been acknowledged. These are the same factors that prompted del Hoyo and Collar (2014) to recognise all morphologically differentiated forms as subspecies, and I favour that arrangement.

At this time similarly plumaged subspecies in New Zealand and on the outlying islands, and in New Caledonia are not endangered, so are not included in this text.

CRITICALLY ENDANGERED

1. *Cyanoramphus novaezelandiae cookii* (G. R. Gray)
Platycercus cookii G. R. Gray, *List Bds Brit. Mus.*, 1859, Psittacidae, p. 13 (New Zealand, error = Norfolk Island).
This subspecies, the description of which is given above, is confined to Norfolk Island and possibly adjacent Philip Island.

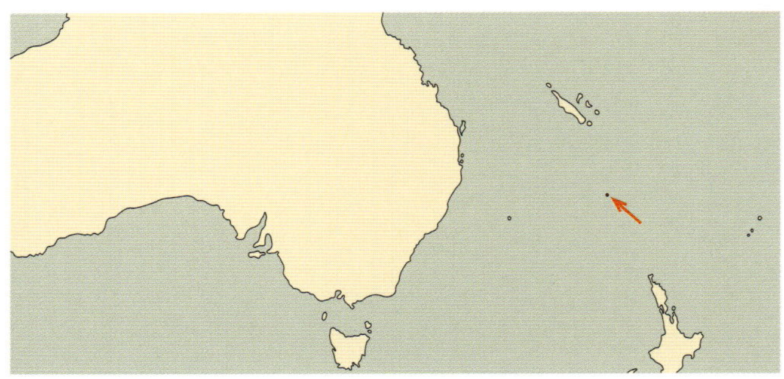

EXTINCT

2. *Cyanoramphus novaezelandiae subflavescens* Salvadori
Cyanoramphus subflavescens Salvadori, *Ann. Mag. Nat. Hist.*, (6) **7**, 1891, p. 68 (Lord Howe Island).
ADULT MALE Similar to *cookii*, but general plumage more yellowish, especially on cheeks and upperparts; red markings on head noticeably less extensive.
1 specimen (UKNHM 1881.5.14812, incorrectly labelled as female): wing 149 mm, tail 162 mm, exp. cul. 22 mm, tars. 22 mm.
ADULT FEMALE Like male, but smaller and with smaller, narrower bill.
1 specimen (UKNHM 1881.5.1.4813, incorrectly labelled as male): wing 147 mm, tail 145 mm, exp. cul. 19 mm, tars. 21 mm.
This subspecies formerly occurred on Lord Howe Island; it is now extinct.

I have examined carefully these two specimens and must agree with Salvadori (1891), who suspected that the labels had been transposed.

Plate 12
Red-fronted Parakeet *Cyanoramphus novaezelandiae*
UPPER *C. n. subflavescens* (adult ♂)
LOWER *C. n. cookii* (adult ♂)

F. KNIGHT 2011

3. *Cyanoramphus novaezelandiae erythrotis* (Wagler)
Psittacus erythrotis Wagler, *Abh. bayer, Akad. Wiss., Math.-Phys.* K1., **1**, 1832, p. 426 (Macquarie Islands, ex Vigors, *Zool. Journ.*, **1**, 1825, p. 529, suppl. pl. 1).
ADULT MALE Like *cookii*, but outer webs of primaries paler blue and washed green.
2 specimens: (not sexed, but obviously males by measurement): wing 145 and 145 mm, tail 128 and 152 mm, exp. cul. 18 and 20 mm, tars. 24 and 25 mm.
ADULT FEMALE Similar to male, but smaller and with smaller, narrower bill.
1 specimen: (not sexed, but obviously female by measurement): wing 142 mm, tail 158 mm, exp. cul. 15 mm, tars. 21 mm.
This subspecies formerly occurred on Macquarie Island, but is now extinct. Oliver (1955) suggested that extant *hochstetteri* from the Antipodes Islands should be synonymised with *erythrotis*; apparently the differences are slight and may be encompassed within the range of individual variation.

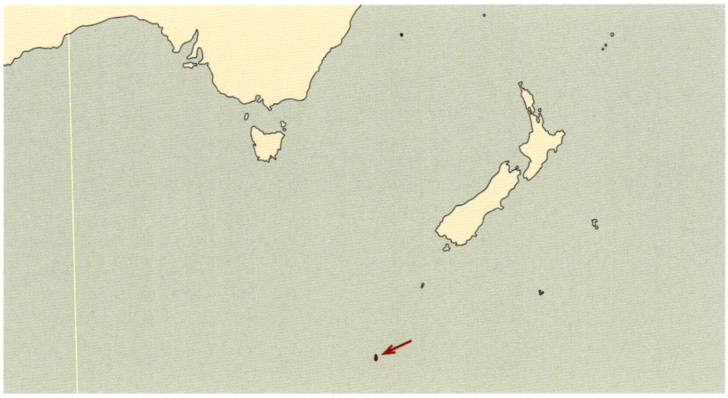

STATUS On Norfolk Island, the Red-fronted Parakeet, or Green Parrot as it is known locally, is extremely endangered. It is a classic example of an island population declining under a combination of pressures, notably loss of habitat, competition from introduced species, predation by rats and feral cats, and disease. When Basset Hull visited the island in October 1908, the parrots were already rare and a special protection order had been issued by the then Chief Magistrate. Smithers and Disney (1969) recalled that during their stay of two weeks in late November 1968, parrots were seen only in Palm Glen, in the National Park, where the resident population was estimated to be only a few pairs, the low numbers being due largely to competition with the introduced Crimson Rosella *Platycercus elegans*.

In 1976, an outbreak of psittacine circoviral disease (parrot beak and feather disease) was detected in Crimson Rosellas, causing very high mortality, and in February 1977 sightings of an apparently diseased Red-fronted Parakeet were reported. In 1995, this disease was found again in wild and captive Red-fronted Parakeets, causing the death of two captive females, and now is thought to be widespread in the wild population (Fortescue *et al.* 1999).

Surveys undertaken between May 1977 and December 1978 revealed that the population on Norfolk Island was from a known minimum of 17 to an estimated maximum of 30 birds, and the parrots were dependent on native forest in and around the National Park. Surveys conducted again in 1981–1982 found that the estimated population had declined to less than 30 birds, so in 1983 an intensive recovery program was initiated, with the dual objectives of establishing a captive breeding population and of affording extra protection to the wild population (Hicks and Greenwood 1989). Attention was focused on finding and protecting nests so that the rate of fledging success could be increased. Only one wild nest was known to be active between 1983 and 1985, but with an improving searching proficiency fieldworkers were able to find a new nest in 1986, two more in 1987, six new ones in 1988, and nine more in 1989. Provision of nestboxes of varying sizes proved ineffective in overcoming the shortage of natural nesting sites, for the boxes were accepted only by competitors, especially Common Starlings *Sturnus vulgaris*, and none was occupied by the parrots. Although clutches of up to six or eight eggs often were laid, only a few chicks fledged from each nest, and it was known that one female had laid four clutches over a period of eight months, but had produced only three fledglings. Smaller chicks in large broods usually perished, so they were removed for handrearing or transferred to other nests with small broods, and both techniques proved successful. Predation was a major cause of nesting failure, with eggs, chicks and even brooding females being taken by rats, so baiting stations were set up in the vicinity of each nest, and modifications were made to thwart access by rats to the hollows. Species competing for nesting hollows, especially Crimson Rosellas *Platycercus elegans* and Common Starlings, also were removed. Between November 1987 and August 1989, rosellas were evicted from hollows on 12 occasions, and in two instances Red-fronted Parakeets subsequently renested in the hollows. Finally, there was intensification of the trapping of feral cats around all active nests. A long-term objective was to translocate some parrots to nearby predator-free Philip island, where vegetation is recovering slowly following the removal of introduced grazing mammals. This major effort of intensive management seemed to achieve outstanding success, for the estimated population in 1991 of at least 40 birds in the wild and 13 in captivity had increased by June 1999 to a wild population thought to number approximately 160 birds (Fortescue *et al.* 1999). I am unsure that such high numbers were present, but success of the program was evident.

In October 2007, I revisited Norfolk Island and during one week conducted the same early morning transects through the National Park that alerted me to the critically low numbers in 1977–1978. On these transects in 2007 I recorded parrots on only two occasions, a sighting of one pair on one morning and a sighting of a party of four birds on another morning, so it seemed that the population had again declined dramatically. My concerns increased when I observed extremely high numbers of Crimson Rosellas throughout the Park and evidence that the rodent control program had long since been abandoned, for virtually all traps were in varying stages of disrepair. My fears were confirmed in 2013, when studies commissioned by Birdlife Australia, The Nature Conservancy and Island Conservation Groups revealed that numbers had collapsed to between 46 and 92 birds, with a critical shortage of females resulting in possibly only 11 breeding pairs (*Australian Birdlife*, **2**(4):12, 2013). In response to this

---

Plate 13
Red-fronted Parakeet *Cyanoramphus novaezelandiae*
UPPER *C. n. erythrotis* (adult ♀)
LOWER *C. n. erythrotis* (adult ♂)

F. KNIGHT
'10

alarming finding, protection measures were initiated quickly, so that in early 2014 approximately 40 predator-proof nest sites were scattered throughout the Park, some showing signs of nesting activity, and rat baiting and cat trapping efforts were resumed. I am firmly of the opinion that the Norfolk Island Green Parrot ranks with the Orange-bellied Parrot *Neophema chrysogaster* and Coxen's Fig Parrot *Cyclopsitta coxeni* as the most endangered of Australian parrots and, if required protection measures are not fully maintained at all times, it will be added to the appalling list of extinct birds from Norfolk Island.

Red-fronted Parakeets presumably were forest birds also on Lord Howe Island, though it was claimed that they visited settlement gardens and caused damage to crops. In 1869, E. S. Hill accompanied a judicial party on a visit to Lord Howe Island and while there carried out some observations on birds. His published notes of 1870 include this remark:

> *The paraquet also was a nuisance to the cultivators, once appearing in flocks; now I saw but a solitary pair in their rapid flight through the foliage and recognised them only by their peculiar noise.*

This was the last reported sighting and, when visiting the island in October 1907, Basset Hull was informed by a Mrs Nichols that some years earlier the parrots were very plentiful, but gradually were exterminated because of the damage they caused in fruit gardens (Basset Hull 1910).

On treeless Macquarie Island, these parrots were terrestrial, living and nesting among tussocks, particularly along the seaboard (in Falla 1937). A Russian expedition, under the command of Fabian Gottlieb von Bellinghausen, visited Macquarie Island in 1820, and collected 20 specimens of Red-fronted Parakeets. They also obtained a living bird from a sealer in exchange for three bottles of rum. During the next decade, the parrots were mentioned in several reports and live birds were frequently brought back to Sydney, where they were in demand as cagebirds. In 1877, John Inches Thomson, of Port Chalmers, New Zealand, and a party were wrecked on the island in the schooner *Beneleugh* and, in his account of the adventures of the party published in *Voyages and Wanderings in Far-off Seas and Lands* (1912), Thomson recalled that he '.....shot some paroquets, and occasionally we were successful in knocking them over with stones......there appeared to be great numbers of them'. When John Scott, from the University of Otago, New Zealand, visited the island in 1880, the parrots 'occurred in great numbers round the shore'. That was the last report of their existence. A. Hamilton, also from Otago University, visited the island in 1891 and found that the parrots had disappeared completely (in Taylor 1979). Extinction of both the Red-fronted Parakeet and the Macquarie Island Landrail *Gallirallus philippensis macquariensis* coincided with a sudden large increase in the numbers of feral cats and introduced Wekas *Gallirallus australis*, apparently in response to the introduction of rabbits, so ending a scarcity of food in winter which presumably had limited severely the numbers of these predators (Taylor 1979). Rabbits, Wekas and rats have been removed from Macquarie Island, so with assistance from New Zealand wildlife authorities it may be possible to restore a population of Red-fronted Parakeets on Macquarie Island by translocating birds from treeless Antipodes Islands.

I do not share the view that recognition of the Norfolk Island population as a separate species enhances conservation priorities, for that endangered population is deserving of the highest conservation effort irrespective of its taxonomic status. *C. n. cookii* is listed as endangered under the Australian *Environment Protection and Biodiversity Conservation Act 1999*, and is included in Appendix I to the Convention on International Trade in Endangered Species of Wild Fauna and Flora (CITES). Other subspecies are included in the listing of *Cyanoramphus novaezelandiae* in Appendix I.

HABITATS Even as early as the early 1900s, there was a close association between Red-fronted Parakeets and remaining areas of native vegetation on Norfolk Island. When visiting the island in October–November 1908, Basset Hull (1910) found that the parrots were almost entirely restricted to gullies running up towards the crest of Mount Pitt, and these same gullies are still favoured by the birds. The 410 ha Norfolk Island National Park, centred on Mount Pitt, together with some adjoining areas, contains the last extensive tract of native forest on Norfolk Island and it is there that the parrots usually are observed. At times they visit other parts of the island, showing a preference for scattered patches of native vegetation; I have never seen them in cleared areas around the Administration Centre at Kingston.

HABITS On Norfolk Island, Red-fronted Parakeets are active in the early morning, when pairs or small parties may be seen flying to and from favoured feeding areas. Once they have settled to feed, these parrots are extremely difficult to detect, so well does their green plumage blend with the foliage, and their presence is betrayed only by the continuous clicking of their mandibles, the steady stream of debris falling to the ground below and the occasional emission of their characteristic call-notes. They are mainly arboreal, but will come to the ground, where they forage for fallen seeds by scratching about in the earth and debris with fowl-like sideways movements of their long legs. They are not shy and usually will allow a close approach. By mid-morning, feeding has ceased and the birds move to a sunny, but not exposed perch near the top of a forest tree and here they spend some time preening. During the remainder of the day, they are quiet and inconspicuous.

It is my impression that in Norfolk Island National Park, pairs occupy defined territories, at least during the breeding season. There were localities to which I could go each morning and be fairly confident of seeing a pair of parrots. I have seen birds fly to the very tops of tall exposed Norfolk Island pines *Araucaria heterophylla* and call loudly for several minutes, an action that appeared to be for territorial advertisement.

On 2 December 1978, I was watching two parrots feeding in wild olive trees *Olea africana* when one was seen to bite off a leaf, chew it and then agitatedly rub the macerated pulp into its fluffed plumage. On 8 December 1978, I saw two parrots in another olive tree and both were rubbing into their plumage macerated bark, which they had stripped from a branch.

The rather swift, buoyant flight is only slightly undulating and comprises rapid, shallow wingbeats interspersed with glides on outspread, slightly decurved wings. These parrots are strong fliers, twisting and turning through the trees with remarkable ease. When alighting, they spread their tails.

CALLS The characteristic call-notes cannot be mistaken for those of any other species. In flight or when alarmed, the parrots give a prolonged, repetitive *kek-kek-kek-kek-kek*, and this also is the call that I have heard from birds perched atop pine trees. While feeding, they utter an occasional subdued *kek-kik. . . kek-kik. . . kek-kik* or *kek-kik-kek. . . kek-kik-kek*.

DIET AND FEEDING On Norfolk Island, I have seen Red-fronted Parakeets taking seeds and fruits from a variety of native and introduced trees or shrubs. They are particularly fond of the blossoms and fruits of bloodwoods *Baloghia lucida* and blossoms of whiteoaks *Lagunaria patersonia*. I have watched them feeding on unripe berries of lantana *Lantana camara*, flowers and fruits of introduced guava *Psidium guajava* and native guava *Rhodomyrtus psidioiods*, fruits of wild tobacco *Solanum mauritianum* and seeds of wild olive *Olea africana*. They eat emergent leaf shoots of Norfolk Island pines *Araucaria heterophylla* and use their robust bills to prise seeds from opening cones. On 25 October 2007, at 0720 hours, in the National Park, I watched four parrots spend 10 minutes feeding on leaf petioles of kawakawa *Macropiper excelsum*, and they favoured the petioles of new leaves. On the rare occasions that I have seen these parrots on the ground, they have been foraging for fallen seeds of wild olive.

When visiting Macquarie Island in 1891, A. Hamilton was informed by sealers that the parrots formerly frequented the seaboard, where they searched among stranded seaweed for crustaceans and other small animals (in Oliver 1955).

BREEDING Hicks and Greenwood (1989) note that on Norfolk Island, egg-laying has been recorded in all months, and some pairs breed more than once in a year. One marked female laid a clutch of five infertile eggs in December, followed by another clutch of six eggs in February, resulting in the death of the sole nestling, and then another clutch in March, but before the last of three chicks had fledged from this nest in June she had commenced a clutch of six eggs in another nest. Most nests found between 1983 and 1989 were in hollows in living trees, mainly in ironwoods *Nestegis apetala*, bloodwoods *Baloghia lucida*, Norfolk Island pines *Araucaria heterophylla* or tea-trees. Entrance holes usually were within 2 m of the ground, with hollows averaging about a metre in depth, and nesting chambers varied from 20 cm to 40 cm in diameter. One unsuccessful nest was found in the root mass of a fallen pine, and another inside the crown of a dead tree-fern. On Macquarie Island, nests were under tussocks or in burrows in the ground (in Falla 1937).

A clutch of one to eight eggs is laid on decayed wood dust at the bottom of the hollow, and incubation by the female apparently commences after laying of the second or third egg. Also at nests on Norfolk Island, it was observed that during the incubation period of approximately 21 days, the sitting female is fed by the male when she emerges from the nest. Feeding of the nestlings is undertaken mainly by the female, although on occasions the male may take over entirely the feeding of nestlings while the female moves to another nest to commence laying a new clutch. Young birds leave the nest some seven weeks after hatching, but for a further two to three weeks are dependent on their parents and remain near to the nest.

Nesting success is low on Norfolk Island, with the smallest chicks in large broods usually perishing in the nest, and predation by rats of eggs, chicks and even brooding females remains a major cause of failure. Fifteen chicks fledged from eight breeding attempts monitored in 1987–1988, and 21 fledged from 14 attempts monitored in 1988–1989 (Hicks and Greenwood 1989). While on Norfolk Island in October 2007, I was told that in the 2006 breeding season only 10 chicks were banded and fledged, but improved success was achieved in 2007 when 20 chicks were banded and fledged between February and June, with another three nests failing in late winter possibly due to cold conditions.

In the *Sydney Journal* of 4 February 1822, Thomas Raine lists the birds known to sealers on Macquarie Island and notes that '... All the small birds, even the parrot, make their nests underground, so that the declivity of the mountains appears like a rabbit warren' (in Falla 1937). In 1880, John Scott from Otago University visited the island and, in his subsequent account of that visit, he notes that the parrots were 'nesting under tussocks' (in Falla 1937). On similarly treeless Antipodes Islands, Red-fronted Parakeets nest in burrows at the base of tussocks and ferns (Heather and Robertson 2015).

EGGS The eggs are rounded to broad-elliptical and without gloss. A set of two eggs of *C. n. cookii*, presumably an incomplete clutch, was taken at Steel's Point, Norfolk Island, on 12 October 1892, and these eggs measure 26.7 × 22.4 mm and 26.8 × 22.1 mm (AM O.50770). Taken on Norfolk Island in October 1909, and presumed to be from the same clutch of seven eggs, are a set of two eggs measuring 27.9 × 23.4 mm and 27.1 × 23.0 mm (ANWC E04369) and a set of three eggs measuring 27.7 (27.0–28.3) × 21.9 (21.5–22.5) mm (ANWC E04370).

A single egg taken at Mount Pitt, Norfolk Island, on 28 February 1909, is significantly larger and almost certainly is an egg of the Crimson Rosella *Platycercus elegans* (NMV BE01434).

Chatham Islands Parakeet

Cyanoramphus forbesi Rothschild

ENDANGERED

Cyanorhamphus forbesi Rothschild, *Proc. Zool. Soc. London*, 1893, p. 529 (Chatham Islands).

OTHER NAME Forbes' Parakeet.

DESCRIPTION Length 28 cm. Weight 70–80 g.
ADULTS General plumage bright green, lighter and more yellowish on underparts; sides of face bright emerald-green; crimson frontal band not extending to eyes; crown golden yellow; crimson patch on each side of rump; underwing-coverts yellowish-green; outer webs of primaries greenish-blue, shading to green towards tips; tail above dark green with fine greenish-yellow margins to outer webs of outermost feathers; underside of tail dusky olive; no underwing-stripe; upper mandible pale bluish-grey, becoming dark grey at tip and along cutting edge, lower mandible grey-black; iris orange-red; legs greyish-brown.
6 males: wing 121–131 (128.2) mm, tail 115–152 (130.5) mm, exp. cul. 13–15 (14.4) mm, tars. 19–22 (20.5) mm.
4 females: wing 121–126 (122.5) mm, tail 108–135 (123,5) mm, exp. cul. 13–14 (13.3) mm, tars. 20–21 (20.5) mm.
3 unsexed: wing 122–129 (126.0) mm, tail 128–158 (139.7) mm,

exp. cul. 15–16 (15.7) mm, tars. 21–24 (22.7) mm.
JUVENILES Like adults but with shorter tail; iris pale brown.

DISTRIBUTION Restricted to the Chatham Islands Group, New Zealand, where now resident only on Mangere and Little Mangere Islands; wandering yellow-crowned birds occasionally recorded on other islands in the Group probably are hybrids from Mangere Island.

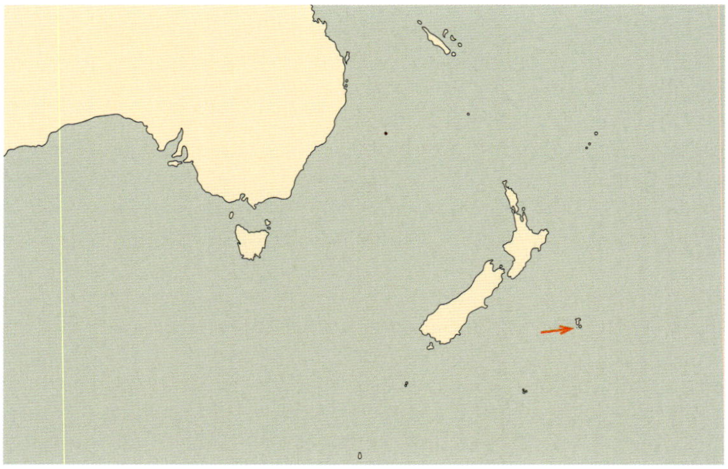

STATUS Triggs and Daugherty (1996) point out that Chatham Islands Parakeets are the most critically endangered of New Zealand parakeets. Phenotypical or 'pure' Chatham Islands Parakeets survive only as a very small population in remnant forest on Little Mangere Island and as a low proportion, estimated in 1996 at 50 to 120 birds, of 'yellow-crowned' birds on Mangere Island, where hybridisation with Red-fronted Parakeets *Cyanoramphus novaezelandiae* is a very serious threat (Birdlife International 2016). Prior to the early 1900s, when Mangere Island was totally forested, Chatham Islands Parakeets were dominant on both Mangere and Little Mangere Islands, but in the early 1900s virtually all forest cover was cleared on Mangere Island to make way for agriculture and cats were introduced, eventually bringing about the loss of both parakeets. Fleming (1939) noted that during a visit to the Chatham Islands between 28 November 1937 and 24 January 1938, Chatham Islands Parakeets were confined to Little Mangere Island, where they were quite common, and the population was estimated very conservatively at 100 birds. After the disappearance of cats from Mangere Island in the 1950s and a cessation of farming in 1968, when the island was made a Fauna and Flora Reserve, both Red-fronted and Chatham Islands Parakeets returned. Taylor (1975) reported that during visits made in 1970 and 1973 the small patch of forest on Little Mangere Island was inhabited by only about 10 pairs of Chatham Islands Parakeets and constituted their last stronghold because on man-modified Mangere Island there was widespread hybridisation with Red-fronted Parakeets. In 1970, there were about 60 parakeets on Mangere Island, of which only 8 per cent were identified as Chatham Islands Parakeets, with 32 per cent being Red-fronted Parakeets, and the remaining 60 per cent easily recognised as hybrids. In 1973, Taylor estimated the total population of parakeets on Mangere Island at about 100 birds, of which only 6 per cent were identified as Chatham Islands Parakeets and the remainder comprising equal numbers of Red-fronted Parakeets and hybrids. Extreme habitat modification was presumed to be a major factor contributing to hybridisation, together with the large imbalance in numbers between the two species in the early stages of recolonisation of Mangere Island.

In 1976, wildlife authorities commenced culling Red-fronted Parakeets and hybrids on Mangere Island in an effort to protect the Chatham Islands Parakeets, and revegetation programs were started. The objective is to maintain numbers of Red-fronted Parakeets at less than about 10 per cent of the total parakeet population. A massive tree-planting program is being undertaken and nestboxes are being provided (Heather and Robertson 2015). Chatham Islands Parakeets have benefited from these measures, allowing their numbers to increase to a few hundred, but hybridisation continues to be a serious threat, and continuing monitoring is required to ensure that levels of hybridisation are kept low enough to allow increased positive assortative mating. Data from molecular analyses show that, to date, these conservation measures have effectively helped to preserve the genetic integrity of Chatham Islands Parakeets (Boon *et al.* 2000).

The Chatham Islands Parakeet is included in Appendix I of the Convention on International Trade in Endangered Species of Wild Fauna and Flora (CITES).

HABITATS A preference by Chatham Islands Parakeets for dense, unbroken forest and scrub, and a poor adaptation to more open vegetation, was noted by Taylor on Little Mangere Island, where they were seen only in the remnant patch of forest at the summit, and scattered patches of scrub and herbs on the northern and eastern cliffs were frequented only by a resident pair of Red-fronted Parakeets and a few others that visited from Mangere Island. This difference in habitat preferences presumably was a major factor in ecological separation of the two species on Mangere Island. When large-scale deforestation left only a small remnant area of about 4 ha of forest, Chatham Islands Parakeets declined from being the dominant species to near extirpation, and during the early stages of revegetation Red-fronted Parakeets have continued to be advantaged.

Nixon (1994) recalls that on Mangere Island, during winter 1979 and summer 1980, the two species differed markedly in habitat use during winter, when Red-fronted Parakeets were present in all vegetation types, but were particularly common in grassland, and Chatham Islands Parakeets almost always were encountered in or around taller vegetation. However, during the summer breeding season, there was a shift in habitat use, mainly by Red-fronted Parakeets, so that it became very similar for the two species.

Hopefully, revegetation of Mangere Island will in the long-term redress the imbalance in favour of Red-fronted Parakeets, so restoring the major isolating mechanism preserving the reproductive integrity of both species.

HABITS Taylor (1985) notes that Chatham Islands Parakeets are seen most commonly in pairs or singly, and larger groups, other than probable family parties, are rarely recorded. Observations on both Little Mangere and Mangere Islands suggest that pairs are sedentary and attached to their nesting territories throughout the year. During the breeding season on Little Mangere Island, pairs are strongly territorial, defending their territories while calling loudly, and females are less aggressive though they do fly with their mates in aerial chases of intruding birds. This territorial defence cannot be maintained in open areas, and presumably that is another factor in facilitating hybridisation with Red-fronted Parakeets on Mangere Island.

Nixon (1994) notes that on Mangere Island, during winter 1979 and summer 1980, differences in feeding and social behaviour were observed between Chatham Islands Parakeets and Red-fronted Parakeets or hybrids. Mixed groups of both species and hybrids were observed, but Chatham Islands Parakeets often were encountered in pairs, whereas Red-fronted Parakeets more often formed larger flocks. Also, there was a difference in comparative approachability of the two species, with Chatham Islands Parakeets being seen in flight less often than Red-fronted Parakeets. Differences in perching heights reflected, in part, the general habitat preferences of the two species, with Chatham Islands Parakeets showing little seasonal variation in perch height, and those seen on the ground invariably were near to scrub or forest, but in winter, when Red-fronted Parakeets occurred mostly in open habitats, 61 per cent of observations were of birds on the ground, compared with 21 per cent of observations in summer.

CALLS In flight these parakeets emit a rapid, high-pitched *ki-ki-ki-ki-ki*, but when feeding they are silent or utter a chatter and babble (Heather and Robertson 2015).

DIET AND FEEDING Taylor (1985) reports that more than 100 observations of foraging birds on Little Mangere Island between 1970 and 1976 indicated that in October to November the main foods were invertebrates, comprising more than 40 per cent, and flowers making up 35 per cent, with seeds making up 18 per cent. During March to May, more leaves and berries were eaten, but invertebrates remained important. Caterpillars and scale insects were taken in the dense canopy of forest and scrub, and birds commonly were seen on the forest floor searching for caterpillars dislodged from the canopy.

Nixon (1994) reports that on Mangere Island, between September 1976 and December 1980, observations of foraging by Chatham Islands Parakeets were recorded and analyses made of crop contents from Red-fronted Parakeets and hybrids killed by wildlife authorities in their culling program. It was found that, although there was a greater frequency of insects in the crops of *forbesi*-like hybrids, these hybrids and phenotypical *forbesi* have taken advantage of abundant food sources in grassland and scrub country adjacent to their favoured habitat in the remnant forest. They appear to have readily included non-forest foods in their diet, taking more seeds and leaves so that their diet has become more like that of Red-fronted Parakeets, and mixed groups of all three parakeets were seen feeding together. Records of food items taken by Chatham Islands Parakeets and hybrids in winter 1979 and in summer 1980 are listed in Table 3.

Food items	Winter 1979		Summer 1980	
	C. forbesi	Hybrids	*C. forbesi*	Hybrids
grass leaf	15	1		
Olearia traversi leaves	6	1	3	1
Olearia traversi seeds			5	1
Olearia traversi flowers			1	1
Olearia traversi bark	3			
Acaena seeds		2		
Plagianthus twigs	3		1	
Plagianthus bark	1		1	
Plagianthus flowers			1	
Plagianthus fruits/seeds			3	
Plagianthus leaves			2	
Hebe leaves	1			
Muehlenbeckia australis seeds	1		1	
Disphyma flowers	1			
Sonchus grandifolius seeds			6	1
Sonchus grandifolius flowers			1	1
Phormium tenax fruits			1	2
Urtica australis seeds			1	
Parietaria debilis leaves			1	
pseudococcid insects	3	1		
regurgatory feeding			3	

Table 3 Recorded feeding observations of Chatham Islands Parakeets and hybrids on Mangere Island in winter 1979 and summer 1980 (after Nixon 1994).

BREEDING Taylor (1985) notes that the breeding season is extended, eggs being laid at any time between October and March, and nests are in hollows in living or dead trees. In November 1976, on Little Mangere Island, a freshly prepared nest was found inside a dead tree trunk, and it was lined with powdered wood. Also on Little Mangere Island, numbers of immature birds were seen with their parents in December and recently fledged juveniles have been seen until late May (Fleming 1939; Taylor 1985).

EGGS Measurements of two eggs are given as 25.4 × 21.6 mm and 26.7 × 21.6 mm (in Schönwetter 1964).

Orange-fronted Parakeet

Cyanoramphus malherbi Souancé

ENDANGERED

Cyanoramphus Malherbi Souancé, *Rev. et Mag. Zool.* (2), **9**, 1857, p. 98 (Locality unknown = South Island, New Zealand.)

OTHER NAME Malherbe's Parakeet.

DESCRIPTION Differentiating Orange-fronted Parakeets from Yellow-crowned Parakeets *Cyanoramphus auriceps* in the field can be quite difficult (Kearvell *et al.* 2014). The only reliable distinguishing features are the orange frontal band and orange patches on each side of the rump, and detection of both are dependent on obtaining a good, uninterrupted sighting of the birds.

Length 20 cm. Weight males 40–52 g, females 30–41 g.

ADULTS General plumage green, only slightly paler on underparts; orange frontal band, becoming paler towards eyes; crown pale yellow; orange patch on each side of rump; outer webs of primaries violet-blue; bill pale bluish-grey becoming dark grey towards tip; iris orange-red; legs brown.

7 males: wing 95–114 (105.7) mm, tail 101–119 (111.1) mm, exp. cul. 11–15 (13.3) mm, tars. 18–19 (18.4) mm.

9 females: wing 97–110 (102.4) mm, tail 107–119 (111.0) mm, exp. cul. 11–13 (11.7) mm, tars. 17–20 (18.3) mm.

JUVENILES Like adults, but less extensive dull yellow on crown and duller orange frontal band; iris pale brown.

DISTRIBUTION New Zealand, where formerly occurred throughout North and South Islands, but now confined to South Island, where recorded only from South Hurunui, Poulter and Hawdon Rivers valleys in inland Canterbury (Heather and Robertson 2015).

Historically, it may have occurred on the Auckland Islands, where Orange-fronted Parakeet mitochondrial DNA haplotypes have been found in both Red-fronted and Yellow-crowned Parakeets (see Rawlence *et al.* 2015). Captive-bred birds introduced in South Island to Chalky Island, Fiordland, and Maud and Blumine Islands in Marlborough Sounds, and in North Island to Mayor Island, in Bay of Plenty.

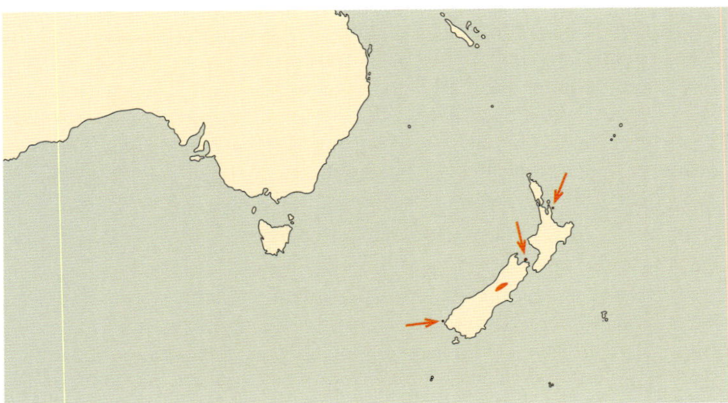

GENERAL NOTES Holyoak (1974) suggested that the Orange-fronted Parakeet is merely a colour morph of the common Yellow-crowned Parakeet *Cyanoramphus auriceps*, pointing out that there appear to be no consistent structural differences, and colour differences could be due to small changes in carotenoid pigmentation, which probably is under simple genetic control. Nixon (1981) found no significant differences between museum specimens, whereas the same techniques distinguished both forms from the Red-fronted Parakeet *Cyanoramphus novaezelandiae*. Nixon did caution that finding live Orange-fronted Parakeets provided an opportunity to learn more by field observations and captive breeding and, should specific status be confirmed, a detailed ecological study would be warranted. This uncertainty surrounding taxonomic status of the Orange-fronted Parakeet persisted, and after reviewing the evidence, including captive cross-breeding experiments, Taylor (1998) reasserted his earlier conclusion that it is a colour morph. Recognition of specific status was proposed when research based on mitochondrial DNA, assortative pairing, bill morphology, vocalisation, and comparative ecology strongly indicated that the Orange-fronted Parakeet is a distinct species (Kearvell *et al.* 2003).

STATUS Recognition of specific status for the Orange-fronted Parakeet focused attention on its rarity in the wild, and the need for immediate conservation measures. In the late 1800s, these parakeets were said to be 'fairly represented' in large flocks of parakeets that came to the Canterbury Plains, in the South Island, though they were 'nowhere as common' as Yellow-crowned or Red-fronted Parakeets, and their former occurrence in North Island is confirmed by museum specimens with reliable locality data (in Kearvell *et al.* 2003). In the late 1890s, they were said to be the dominant species in some South Island districts, and museum specimens indicate that they were fairly common in hills around Nelson, also in South Island. Extensive landclearance and the introduction of mammalian predators, especially mustelids and rats, brought about dramatic declines in populations of all parakeets and, being the least numerous, Orange-fronted Parakeets would have been particularly vulnerable. The first warning about the decline of Orange-fronted Parakeets in particular seems to have come from Phillpotts (1919) who noted:

> *The three species of* Cyanoramphus *which were once so common in Otago are now seldom seen or heard in the small forests.* Cyanoramphus malherbi *Souancé, which was never so abundant as the other two, is in all probability extinct ....*

Sightings became less frequent, and an absence of records after 1965 gave rise to speculation that these parakeets could be extinct. In September 1980, Orange-fronted Parakeets were 'rediscovered' near the confluence of the Hope and Kiwi Rivers in Lake Sumner Forest Park, North Canterbury, where they occurred in a ratio of about one to 12 with Yellow-crowned Parakeets, and in early February 1984 birds were seen in the company of larger numbers of Yellow-crowned Parakeets in the Hawdon River valley, in Arthur's Pass National Park (in Taylor *et al.* 1986; Read and McClelland 1984). Small populations subsequently were found in the South Branch Hurunui River valley, in Lake Sumner Forest Park, and the Poulter River valley, also in Arthur's Pass National Park, North Canterbury. In 1999–2000, irruptions of Black Rats *Rattus rattus* in two successive summers brought about a collapse in the total population from a possible 500–700 birds to an estimated 100–200 birds (in Birdlife International 2016).

Kearvell and Farley (2016) report that data from recent surveys of wild populations undertaken in the three river valleys in South Island reveal that there is an excess of males in the populations. In the Hawdon River valley, in the pre-breeding season, the proportion of males ranged between 0.574 and 0.586 for the three seasons from 2009 to 2011, with only one count in the pre-breeding season in 2009 not differing significantly from a 0.5 ratio. For three counts taken during the breeding season, the ratios in favour of males increased to a range of 0.678 to 0.740, but this is not surprising because it is at a time when most females spend much of their time incubating. Limited data suggest that prior to declines in the wild populations, caused largely by introduced mammalian predators, the adult sex ratio may have been nearer to parity, so any male-skewed sex ratio in the surviving wild population poses an additional threat.

Successful translocations of captive-bred birds to predator-free Chalky Island, in Fiordland, commenced in December 2005, with a total of 45 birds being released there between 2005 and 2007, and breeding brought about an increase in the population to an estimated 100 to 200 birds in early 2011 (in Birdlife International 2016). More recently, there has been an apparent decline, accompanied by an increase in numbers of Yellow-crowned Parakeets, and in 2013 it was suspected that some 50 to 150 adult Orange-fronted Parakeets were present. Translocations to predator-free Maud Island, in Marlborough Sounds, South Island, between 2007 and 2009, involved the release of 62 captive-bred birds, and successful breeding resulted in an estimated population of 60 to 100 birds in early 2011, but in early 2012 there were thought to be only 30 to 50 mature birds on the island. Between December 2009 and April 2013, translocations to Mayor Island, in the Bay of Plenty, North Island, involved the release of 95 captive-bred birds, but little subsequent monitoring was undertaken,

---

Plate 14
UPPER Orange-fronted Parakeet *Cyanoramphus malherbi* (adult ♂)
LOWER Chatham Islands Parakeet *Cyanoramphus forbesi* (adult ♂)

F. KNIGHT 10

though in 2013 the total population was thought to be 50 to 100 mature individuals. Translocations were carried out also on Blumine Island, in Marlborough Sounds, where 61 captive-bred birds were released in 2011–2012, and nesting was confirmed in 2013, when 50 to 100 birds were thought to be present (in Birdlife International 2016). In addition to these translocations, wildlife authorities are implementing predator control measures, involving trapping and baiting programs in South Island valleys where wild populations survive. The total population probably is 500 to 600 birds, with fewer than 250 birds surviving in the wild in valleys in South Island (Heather and Robertson 2015; Kearvell and van Hal 2016).

Symptoms consistent with psittacine circoviral disease (beak and feather disease) have been noted in birds on Maud Island, and antibodies for the disease have been detected in birds from Maud Island and in Red-fronted Parakeets used as foster parents at the captive-breeding facility.

HABITATS Although recorded from near sea level up to subalpine scrublands, recent sightings have been mainly of birds in, or at the margins of *Nothofagus* forest below 900 m. Crouchley noted that in the Hope-Kiwi Rivers area, Lake Sumner Forest Park, nesting birds were found only in forest dominated by tall red beech *Nothofagus fusca* between 600 m and 900m, where they seemed to prefer areas bordering on forest dominated by mountain beech *N. solandri*, and this habitat was shared with Yellow-crowned Parakeets *Cyanoramphus auriceps* (in Taylor 1998). In the upper reaches of the south branch of the Hurunui River, in Lake Sumner Forest Park, South Island, in the summer of 1998–1999, at a study site 7 km long and 600 m wide along the valley floor at an altitude of 750 to 900 m, sympatric Orange-fronted and Yellow-crowned Parakeets were surveyed, with particular note made of habitat use, behaviour and diet (Kearvell *et al.* 2002). At this study site the forest canopy had a mean height of 21 m, and was dominated by red beech, with silver beech *Nothofagus menziesii* and mountain beech also common. The generally open understorey was dominated by silver beech, but broadleaf *Griselinia littoralis*, mountain toa toa *Phyllocladus alpinus*, *Pseudopanax* spp. and *Coprosma* spp. also were common, and native and introduced grasses and herbs were present in clearings. Both parakeets spent most of their time in the uppermost 20 per cent of the forest stratum, and both favoured the shaded canopy, using small branches and twigs for most of their activities. Both were observed in beech trees far more frequently than in any other plant species, and both targeted a very similar size range of trees. Orange-fronted Parakeets appeared to use the ground and low-growing shrubs more frequently than Yellow-crowned Parakeets, and it was suspected that this could make the Orange-fronted Parakeets more susceptible to predation.

Between October 2008 and May 2010, studies of habitat use by translocated Orange-fronted Parakeets were undertaken on Maud Island, and Ortiz-Catedral (2012) reports the findings of those studies. Observations were made along the track network on the island, and also in accessible coastal areas during low tide, with 29 'parakeet plots' being identified by making each tree in which a bird was observed the centre of a 25 m² plot. Each time a parakeet was observed, a record was made of its activity, such as perching, calling or foraging. To assess habitat features selected by the parakeets, 23 'random plots' were selected by using a table of random numbers for a list of 250 accessible points established along the entire track system across all vegetation types. It was found that 'parakeet plots' had higher frequencies of the highest category of canopy cover (76–100%) than 'random plots', and the mean height of the highest tree within 'parakeet plots' was significantly higher than that of 'random plots'. There was no significant difference in density of stems with a diameter at breast height of greater than 20 cm, but the mean density of stems with a diameter at breast height less than 20 cm was significantly higher in 'parakeet plots'. The lowest category of understorey cover (0–25%) was significantly more frequent in 'parakeet plots', and 'parakeet plots' had the lowest category of groundcover vegetation (0–25%) significantly more frequently than 'random plots'. Of the 29 'parakeet plots', 22 were located in regenerating scrub of manuka *Leptospermum scoparium* and kanuka *Kunzea ericoides*, four were in plantations of *Pinus radiata*, and three were in coastal forest dominated by nikau *Rhopalostilis sapida*, kohekohe *Disoxylum spectabile*, titoki *Alectryon excelsus*, kawakawa *Macropiper excelsum* and mahoe *Melycitus ramiflorus*. It was concluded that in this study parakeets were recorded in areas of forest with a greater frequency of high canopy cover, low understorey and low groundcover vegetation, suggesting an active use of ecotones of available habitat types.

HABITS Orange-fronted Parakeets usually are encountered singly or in pairs, though in autumn and winter they may associate in small flocks, and almost invariably they will be in the company of larger numbers of Yellow-crowned Parakeets *Cyanoramphus auriceps*. When feeding or resting in the treetops, both species can be difficult to observe, so well does their plumage blend with the foliage. In Lake Sumner Forest Park, South Island, along the upper reaches of the south branch of the Hurunui River, comparative ecological studies of both parakeets were undertaken in the spring and summer of 1998–1999 (Kearvell *et al.* 2002). Orange-fronted Parakeets were observed calling less frequently than Yellow-crowned Parakeets, but more frequently were seen performing comforting activities, which included bathing, preening, drinking and daytime roosting. Comfort activities were performed almost exclusively on small branches, and both species often left their feeding stations to undertake such activities. Feeding was the dominant daytime activity, accounting for 61 per cent of observations for Orange-fronted Parakeets and 59 per cent of observations for Yellow-crowned Parakeets.

Ortiz-Catedral (2009) reports that behavioural observations were made on translocated Orange-fronted Parakeets on Maud Island, in Marlborough Sounds, South Island, between March 2007 and January 2009, and again the most commonly observed behaviour was feeding, with each feeding bout lasting from one to 52 minutes, with a mean of five minutes. Of 153 feeding bouts, 132 were of single birds, 17 were of pairs, two were of trios and two were of groups of four birds (Ortiz-Catedral and Brunton 2009). Bouts of preening and resting also were common, with the former lasting from one to 32 minutes, with a mean of four minutes, and the latter lasting one to 10 minutes, with a mean of two minutes. Bouts of calling lasted one to 15 minutes, with a mean of three minutes, and birds moved about for periods of one to 15 minutes, with a mean of four minutes. Sleeping was the most infrequently observed behaviour, with bouts lasting one to 13 minutes, with a mean of six minutes.

Read and McClelland (1984) recall that in early February 1984, in the Hawdon River valley, in Arthur's Pass National Park, South Island, up to three Orange-fronted Parakeets and three Yellow-crowned Parakeets were observed following and feeding with flocks of Yellowheads *Mohoua ochrocephala*, and also associating with them in mixed foraging flocks were Warblers *Gerygone igata*,

Tomtits *Petroica macrocephala*, Fantails *Rhipidura fuliginosa* and Brown Creepers *Finschia novaeseelandiae*.

CALLS A rapid, high-pitched chattering *ki-ki-ki-ki-ki* usually is given in flight, and is the most frequently heard call. Although there are significant differences from the flight call of Yellow-crowned Parakeets in the number of syllables per unit time, the calls are similar, making it very difficult to distinguish the two species by vocalisation (Kearvell and Briskie 2003). During comparative ecological studies undertaken in the valley of the upper south branch of the Hurunui River, in Lake Sumner Forest Park, South Island, in the spring and summer of 1998–1999, calling by both species was restricted mostly to upper levels of the beech forest, with 77 per cent of calls from Orange-fronted Parakeets and 87 per cent of calls from Yellow-crowned Parakeets coming from the shaded canopy and upper understorey combined, while no calling was heard from the ground (Kearvell *et al.* 2002).

DIET AND FEEDING Foraging observations for both Orange-fronted and Yellow-crowned Parakeets were recorded at a study site in the upper reaches of the south branch of the Hurunui River, in Lake Sumner Forest Park, South Island, during the spring and summer of 1998–1999, which was a beech mast year, so large quantities of beech seeds were available from about late January (Kearvell *et al.* 2002). In summer the diet of both species consisted almost entirely of seeds, with seeds of mountain beech *Nothofagus solandri* comprising 65 per cent of seed source observations for Orange-fronted Parakeets and 51 per cent of seed source observations for Yellow-crowned Parakeets. Both species adopted similar foraging techniques when taking invertebrates, which were a significant component of the diet for both species, especially in spring, with the main source being bark of red beech *N. fusca*, and it featured in 53 per cent of source observations for Orange-fronted Parakeets and 58 per cent of source observations for Yellow-crowned Parakeets. Both species took small larvae of Tortricidae leaf rollers from the undersides of leaves in the subcanopy of silver beech *N. menziesii*, and also much larger leaf roller larvae and unidentified lepidopteran caterpillars from leaves of red beech. From early December aphids and Tineidae leaf miners were commonly taken, particularly in heavily infested red beech trees. Mainly from mountain beech, both species took Margarodidae honeydew aphids, and they also consumed larvae of Coccidae scale insects. Flowers of all three beech species were an important component of the diet in spring, but at no time were any parakeets observed feeding on the flowers of other plants. Both parakeets were seen on the ground browsing on herbs and ferns, including blue star creeper *Pratia angulata*, white star creeper *P. perpusilla*, mountain myrrh *Oreomyrrhis colensoi*, Lyall's parahebe *Veronica lyallii*, maniototo button daisy *Leptinella maniototo* and little hard fern *Blechnum penna-marina*.

Ortiz-Catedral and Brunton (2009) report that between March 2007 and January 2009, opportunistic observations on the diet and foraging of translocated Orange-fronted Parakeets were undertaken on Maud Island, in Marlborough Sounds, South Island. The parakeets foraged between 3 m and 8 m above the ground, eating mostly material from 14 plant species, which featured in 96 per cent of observations, and three introduced plants featured in 10 per cent of these observations. Most food items were fruits, which made up 61.4 per cent of total items, and leaves, making up 12.4 per cent of items, with flowers making up only 3.9 per cent and invertebrates just 3.3 per cent of total items. Fruits and leaves of mahoe *Melicytus ramiflorus* were a major component of the diet. Plant species and the food items ingested by the parrots are listed in Table 4. Also consumed were bark and sticks of tutu *Coriaria arborea*, manuka *Leptospermum scoparium*, mahoe, akiroha *Olearia paniculata*, pine *Pinus radiata*, karo *Pittosporum* sp. and whauwhaupaku *Pseudopanax arboreus*. There was one observation of birds taking grit at a sand bank. The very low proportion of invertebrates contrasted with foraging observations of wild parakeets in Lake Sumner Forest Park, and it was suggested that this may be related to the different composition of vegetation at the two study sites, the low parakeet population density during the time of the study on Maud Island and methodological restrictions during the survey on the island.

Plant species (*introduced species)	Food items	Proportion of diet (feeding bouts in brackets)
Melycitus ramiflorus mahoe	fruits, leaves, flowers	43.70 (52)
Aristotelia serrata makomako	fruits, leaves	13.44 (16)
Coprosma robusta karamu	fruits	8.40 (10)
Leptospermum scoparium manuka	fruits	7.56 (9)
Carpodeus serratus putaputaweta	fruits, leaves	5.88 (7)
*Cytisus palmensis** tree lucerne	flowers, leaves	5.04 (6)
Pseudopanax arboreus whauwhaupaku	fruits	5.04 (6)
*Acer pseudoplatanus** sycamore	fruits	3.36 (4)
Alectryon excelsus titoki	fruits	1.68 (2)
Hebe stricta koromiko	flowers	1.68 (2)
*Pinus radiata** radiata pine	leaves	1.68 (2)
Dodonea viscosa akeake	leaves	0.84 (1)
Disoxylum spectabile kohekohe	flowers	0.84 (1)
Pittosporum sp. karo	fruits	0.84 (1)

Table 4. Food plants and food items recorded in foraging observations of Orange-fronted Parakeets on Maud Island, 2007–2009 (after Ortiz-Catedral and Brunton 2009).

BREEDING Kearvell and Steeves (2015) report that because mixed pairs of *Cyanoramphus* parakeets have been reported throughout the genus, and because Orange-fronted Parakeets are sympatric with Yellow-crowned Parakeets in all three river valleys where wild populations occur, the level of assortative mating between these two species was investigated during 1999 to 2011. Of 355 nests recorded during the investigation, 351 were pure pairings and only four were mixed pairings. With one exception, the ratio of Orange-fronted Parakeets to Yellow-crowned Parakeets encountered during annual surveys ranged between zero and 0.78. These results indicate that the two parakeets exhibit assortative mating, even when the Orange-fronted Parakeet is outnumbered by Yellow-crowned Parakeets.

Nesting by Orange-fronted Parakeets can take place at almost any time of the year, but egg-laying occurs mostly during December to April, and favoured nesting sites are cavities in the branches or trunks of trees, especially holes in mature or old beech trees and in standing dead trees (Heather and Robertson 2015). In the summer of 1998–1999, during comparative ecological studies of Orange-fronted and Yellow-crowned Parakeets in the valley of the south branch of the Hurunui River, in Lake Sumner Forest Park, characteristics of selected nest sites were recorded, and not unexpectedly it was found that the two parakeets selected similar sites in cavities in beech trees (Kearvell 2002). There were no detectable differences in the height of nests nor in the choice of tree size, as measured by trunk diameter at breast height, but data on beech tree species indicated that Yellow-crowned Parakeets used silver beech *Nothofagus menziesii* significantly more frequently and red beech *N. fusca* less frequently, when compared to the almost exclusive use of red beech by Orange-fronted Parakeets. Overall, it was found that both parakeets chose large, old mature trees, predominantly red and silver beech, of large diameter, averaging taller than the mean canopy, and most nests were accessed through knot holes in trunks. Also, it was found that they preferred a hole facing the sun and definitely away from the shaded, colder side of the tree.

Between 2001 and 2012, nests of Orange-fronted Parakeets and Yellow-crowned Parakeets were studied in all three South Island valleys with remnant populations of Orange-fronted Parakeets (Kearvell and van Hal 2016). A total of 139 nests of Orange-fronted Parakeets were found and, apart from 28 that were in standing dead trees, all were in living beech trees, with 94 in red beech, 13 in mountain beech *Nothofagus solandri* and three in silver beech. The mean nest height was 10.87 m, with a range of 2 m to 21 m, and the mean nesting tree height was 20.6 m, with a mean diameter at breast height of 79.6 cm and a range of 30 cm to 141 cm. A strong 83.7 per cent majority of nests were located in tree trunks, and only 16.3 per cent were in branches. Access to 87.2 per cent of nests was through knot holes, and to only 12.8 per cent was through slits. Entrance holes varied in size from the smallest, at 20 × 40 mm to the largest at 250 × 150 mm, and both of these were knot holes. The smallest single dimension recorded was a width of only 20 mm. The deepest egg-chamber was 100 cm down from the nest entrance, and seven chambers were level with the entrance. Northeast-facing and south-facing nests were in approximately equal proportions. In this study also there were no differences between the two parakeets in their choice of nesting tree height, height of nest entrance above the ground or diameter at breast height of the nesting tree. Again in this study there were limited differences in nest site selection between the two parakeets, and the large degree in overlap in nest site characteristics was well demonstrated by observations that about 10 per cent of nests were used by both species at varying times. There were significant differences in the selection of nest tree species, with the number of nests of Orange-fronted Parakeets in standing dead trees being twice the number of nests of Yellow-crowned Parakeets, and there was a 16.8 per cent selection of silver birch for nesting by Yellow-crowned Parakeets, but only a 2.1 per cent selection of these trees by Orange-fronted Parakeets. Also, Orange-fronted Parakeets tended to use smaller cavities, though there was much overlap.

Kearvell (2013) reports that in the Poulter River valley, in Arthur's Pass National Park, South Island, a nesting hollow that had been used successfully by both Orange-fronted and Yellow-crowned Parakeets was taken over by Common Starlings *Sturnus vulgaris*, so raising concerns that these aggressive competitors may pose a local threat, especially on the predator-free islands where translocated populations of these endangered parrots are being established.

A much wider selection of nesting sites was found on Maud Island, in Marlborough Sounds, South Island, where studies of nesting behaviour by translocated, captive-bred birds were undertaken from March 2007 to January 2009 (Ortiz-Catedral *et al.* 2009). During spring, summer and autumn, nests were found in holes in a dead branch of a kohekohe *Disoxylum spectabile* and in dead stems of tree-ferns *Cyathea medullaris*, in vacant nesting burrows excavated in soil banks by Sacred Kingfishers *Todiramphus sanctus*, and in a hole at ground level formed by the roots of a fallen tutu *Coriaria arborea*, though this last site collapsed due to heavy rain. Copulations were observed on 27 April and 27 November, and the earliest recorded date for the commencement of egg-laying was 25 April, when the first egg in a clutch of five was laid synchronously in two nests. For the last nest found during the study, it was estimated that egg-laying commenced on 20 January. During egg-laying, males visited sitting females a mean 0.4 times per hour for feeding and a mean 0.43 times per hour on non-feeding visits. Apart from two instances, sitting females left the nest only when their partners arrived in the area, staying outside the nest for a mean 17.5 minutes, with a range of one to 90 minutes. Between visits by males, sitting females spent a mean 60.3 minutes inside the nest, with a range of 10 to 132 minutes. In one nest with a clutch of five eggs, chicks hatched after incubation by the female lasting approximately 22 days, and the similar size and degree of feather down development indicated low hatching asynchrony. The chicks were fed by both parents, and the surviving three chicks fledged 43 days after hatching. Recently fledged juveniles were awkward fliers and remained in the vicinity of the nest for about four weeks, often perching quietly until the parents returned to feed them. Some eight weeks after fledging, two banded siblings were observed foraging with a banded non-parent adult about 700 m from their nest, and approximately 16 weeks after fledging a banded juvenile was seen to be chased away by its parents, who were prospecting for a new nest. Banded birds with a known age of 7.2 months were members of breeding pairs.

EGGS Measurements of 24.0 × 19.0 mm are given by Heather and Robertson (2015), and Schönwetter lists 23.2 × 18.0 mm as the measurements of a single egg in the Vienna Museum Collection.

Black-fronted Parakeet

Cyanoramphus zealandicus (Latham)

Psittacus zealandicus Latham, *Index Orn.*, **1**, 1790, p. 102 (New Zealand, error = Society Islands).

OTHER NAME Tahiti Parakeet.

DESCRIPTION Length 25 cm.

ADULTS General plumage dull green, paler and somewhat bluish on underparts; forehead purple-black, becoming grey-black on forecrown; lores and narrow stripe behind eyes scarlet; broad superciliary band and ear-coverts to beneath eyes bright emerald green; lower back brownish-red becoming dark red on rump; upper wing-coverts and inner secondaries olive-green; primaries and outer secondaries dusky grey, on outer webs broadly margined violet-blue; central tail-feathers dull green, on inner webs suffused dull blue, lateral tail-feathers dull greenish-blue,

Plate 15
UPPER Black-fronted Parakeet *Cyanoramphus zealandicus* (adult)
LOWER Raiatea Parakeet *Cyanoramphus ulietanus* (adult)

F. KNIGHT '09

underside of tail dusky brown; bill pale bluish-grey becoming black towards tip; legs greyish-brown.
2 unsexed: wing 136 mm and 145 mm, tail 135 mm and 144 mm, exp. cul. 15 mm and 19 mm, tars. 25 mm and 25 mm.
JUVENILES (Rothschild 1907): Forehead dull bluish-black; head brownish; greyish-green underparts; brown markings on back; eye-stripe and rump markings chestnut-red.

DISTRIBUTION Formerly occurred on Tahiti, in the Society Islands.

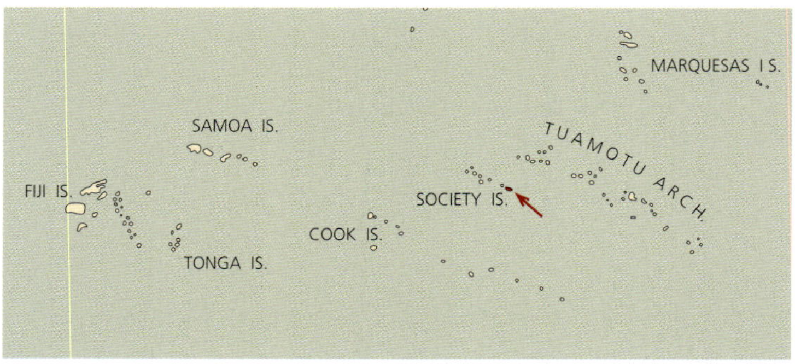

GENERAL NOTES Greenway (1967) noted that the first specimens of the Black-fronted Parakeet probably were obtained by naturalists on the first voyage of Captain James Cook, for in the British Museum there is an unpublished drawing of the bird by Sydney Parkinson, who was on the first voyage, and it is plate 8, titled ' no. 5, Green Peroquit Otahite, Aá'. Possibly two additional specimens were collected by Forster and Ellis on Cook's second voyage (Hume and Walters 2012). A specimen was obtained by G. Amadis, who stayed in Tahiti in 1842 as a member of the expedition of the French frigate *Vénus* under the command of Admiral A. A. Dupetit-Thouars, and the rather worn tail-feathers suggest that the bird may have been held in a cage for some time (Voisin *et al.* 1995).

The Black-fronted Parakeet disappeared about the mid to late 1800s, and the last-known specimens were taken by Lt M. J. de Marolles in 1844. Marolles is said to have shot three birds at Port Phaeton on the isthmus of Taravao, and at that time the parakeets were rare, being found only on the isthmus and in the mountains (in Hume and Walters 2012). Marolles saw only about four or five parakeets in total, and was told by local residents that they lived in big trees on inaccessible scarps and deep valleys. Greenway reported being told by Anthony Curtiss, a former resident of Tahiti, that apart from the name 'á-á', there is no longer any tradition of the occurrence of a green and red parrot on the island, and a careful search by members of the Whitney South Sea Expedition in the 1920s was unsuccessful. Hume and Walters suggest that as Tahitians greatly prized green and red parrot feathers brought from Tonga, it is possible that past excessive hunting may have been at least partly responsible for the disappearance of the Black-fronted Parakeet. I strongly suspect that, in common with other extinct birds from South Pacific islands, the introduction of mammalian predators, especially rats and cats, was a major factor in its extirpation.

AVAILABLE SPECIMENS In the Liverpool Museum, UK, there are two ex Derby Museum specimens, one of which is thought to have come from the collection of Sir Joseph Banks, so may be Latham's type (Greenway 1967). There is another specimen in the Natural History Museum at Tring, UK. The specimen collected by G. Amadis is in the Muséum d'Histoire Naturelle de Perpignan, France (Voisin *et al.* 1995). One of the birds taken by Lt. Marolles is in the Muséum National d'Histoire Naturelle, Paris, but the whereabouts of other specimens is unknown. In the Muséum de Genève, Switzerland, there is a complete skeleton that was purchased on 7 September 1842, and is thought to have come from one of Cook's voyages (Schauenberg 1969).

Raiatea Parakeet

Cyanoramphus ulietanus (Gmelin)

Psittacus ulietanus Gmelin, *Syst. Nat.*, **1**, pt 1, 1788, p. 328 (Ulietea =Raiatea, Society Islands).

OTHER NAME Society Parakeet.

DESCRIPTION Length 25 cm.
ADULTS Head blackish-brown, paler towards neck; mantle and back olive-brown, feathers finely edged darker brown; wings and wing-coverts brown; rump and upper tail-coverts brownish-red; breast to abdomen olive-yellow, feathers very finely edged pale greyish-brown; thighs and under tail-coverts dull olive-brown; underwing-coverts and outer webs of primaries and secondaries grey-mauve; central tail-feathers above brownish-olive tinged grey-mauve, lateral feathers mauve-grey; bill grey; legs grey-brown.
2 unsexed: wing 130 mm and 145 mm, tail 132 mm and 133 mm, exp. cul. 19 mm and 19 mm, tars. 22 mm and 25 mm.
JUVENILES Undescribed.

DISTRIBUTION Known only from Raiatea, in the Society Islands.

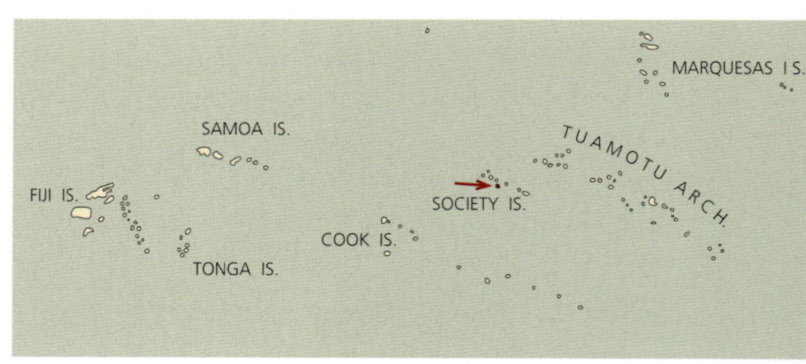

STATUS The Raiatea Parakeet is known from only two specimens which are believed to have been collected by Georg Forster in 1773 or 1774 during Captain James Cook's second voyage. Greenway (1967) pointed out that it is unlikely that any specimens were collected on Raiatea or neighbouring islands during Cook's first voyage, in September 1768, 'as a result of difficulty with the natives', so it is presumed that both specimens were collected on the second voyage. The parakeets presumably became extinct soon after their discovery, and nothing is known of their habits.

SPECIMENS AVAILABLE One specimen is in the Natural History Museum at Tring, UK (UKNHM 21.R.28.1.84), and the second specimen, which was purchased at the sale of the Leverian Museum in London in 1806, is in the Naturhistorisches Museum, Vienna, Austria (NMW 50.687).

TRIBE **PEZOPORINI** Bonaparte

Molecular analyses have confirmed that *Pezoporus* is not related to the superficially similar Budgerigar *Melopsittacus undulatus*, but together with *Neopsephotus* and *Neophema* forms a clade that is sister to the 'core platycercines', and I am adopting the recommendation that the two groups be accorded tribal rank within the subfamily Platycercinae (see Joseph *et al*. 2012).

GENUS ***Pezoporus*** Illiger

Pezoporus Illiger, *Prodromus*, 1811, p. 201. Type by monotypy, *Psittacus formosus* Latham 1790 (= *Psittacus wallicus* Kerr), not of Scopoli 1769, indeterminable *fide* Salvadori, 1891, *Cat. Bds Brit. Mus.*, **20**, p. 605.

The two midsized, terrestrial parrots belonging to this genus are characterised by a highly distinctive, strongly mottled plumage which provides excellent camouflage amidst groundcover vegetation. Traditionally, they have been differentiated generically because of morphological differences, particularly wing and tail structure, but molecular analyses indicate that they are congeneric. The Ground Parrot *P. wallicus* is a slim bird with a long, sharply graduated tail comprising narrow, pointed feathers, while the wings are short and rounded. There is a conspicuous narrowing of the anterior of the cranium, the tarsi are long, and the claws are long but only slightly curved. Also with proportionately long tarsi, the Night Parrot *P. occidentalis* has a short, slightly graduated tail comprising broader, pointed feathers, and the proportionately long wings are pointed. The head is broad without any narrowing of the anterior of the cranium, and lines of short, hair-like feathers are present both above and below the large, fleshy naked cere. The curved claws are very short. Obvious sexual dimorphism is absent in both species, and juveniles resemble adults, with yellowish-white underwing-stripes being present in adults and juveniles of both sexes.

The genus is widespread throughout much of Australia, including Tasmania, but has not been recorded from the tropical north.

Ground Parrot

Pezoporus wallicus (Kerr)

Psittacus wallicus Kerr, *Anim. Kingd.*, **1**, pt 2, 1792, p. 581 (New South Wales = region of Port Jackson, *apud* Mathews, 1917). New name for *Psittacus formosus* Latham 1790, not of Scopoli 1769.

DESCRIPTION Length 30 cm. Weight males 68–83 g, females 59–79 g.

ADULT MALE General plumage rich green, mottled black and yellow; narrow orange-red frontal band; crown and nape green with prominent black shaft-streaking to feathers; cheeks, ear-coverts and throat to breast green with little or no black shaft-streaking to feathers, which are faintly margined dusky black; mantle to upper tail-coverts and upper wing-coverts green, all feathers broadly centred black and with yellow transverse barring; abdomen to flanks and under tail-coverts greenish-yellow transversely barred brownish-black; underwing-coverts dull green tinged bluish at bend of wing; flight feathers above dark green on outer webs, brown on inner webs with pale yellow spot on primaries and two pale yellow spots on secondaries forming conspicuous underwing-stripes; central tail-feathers above green lightly barred yellow on outer webs, below brownish; lateral tail-feathers greenish-yellow transversely barred brownish-black; bill brownish-grey; iris pale yellow; legs pink-brown.

13 specimens: wing 118–139 (128.6) mm, tail 161–195 (177.6) mm, exp. cul. 12–15 (13.7) mm, tars. 25–27 (25.9) mm.

ADULT FEMALE Like male, but usually with black shaft-streaking to feathers of cheeks, ear-coverts and throat to breast.

12 specimens: wing 120–138 (128.9) mm, tail 158–200 (177.4) mm, exp. cul. 13–15 (13.6) mm, tars. 24–27 (25.9) mm.

JUVENILES Differ from adults in having general plumage duller yellowish-olive, particularly on lower underparts; no orange-red frontal band; black shaft-streaking to feathers of cheeks and ear-coverts; feathers of breast marked black and yellow as on underparts; shorter tail; bill pink-horn in very young birds; iris brown.

DISTRIBUTION Coastal regions of southwestern and southeastern Australia, including Tasmania and some offshore islands.

SUBSPECIES Molecular analyses indicate an ancient divergence more than two million years ago between populations of the Ground Parrot in southeastern and southwestern Australia, and it was cautiously suggested that Western Ground Parrots be recognised as a separate species for conservation prioritisation, planning and management purposes (Murphy *et al*. 2010). I do not doubt that these populations have been isolated for a long time, but morphological differences are almost negligible, and I very much doubt that that they maintain reproductive integrity, a requirement for specific status under the biological species concept. Consequently, I prefer to differentiate the two populations subspecifically, an arrangement adopted also by del Hoyo and Collar (2014). Also, I am opposed to conservation needs being used to support specific differentiation, and must emphasise strongly that the conservation of all vulnerable or endangered populations should be assessed independently of whether those populations are species or subspecies. If the Western Ground Parrot is recognised as a separate species then I suggest that to reflect intrageneric affinities the Night Parrot *Pezoporus occidentalis* should be differentiated subgenerically.

1. *Pezoporus wallicus wallicus* (Kerr)

The nominate subspecies, the description of which is given above, occurs in coastal southeastern Australia, from southeastern Queensland south to extreme southeastern South Australia, and also in Tasmania and some offshore islands.

In southeastern Queensland, the Ground Parrot is confined to coastal sandplains from the west coast of Fraser Island and near Maryborough south to the Caloundra district, and inland to the Wolvi and Beerwah districts, but formerly it ranged north to Woodgate and south to near Brisbane (Storr 1984; McFarland 1991c). In easternmost New South Wales, it is known to occur in scattered pockets of suitable habitat, mostly in reserves along coastal and adjacent plateau areas from the Byron Bay and Evans Head districts south to the Hastings River estuary or possibly

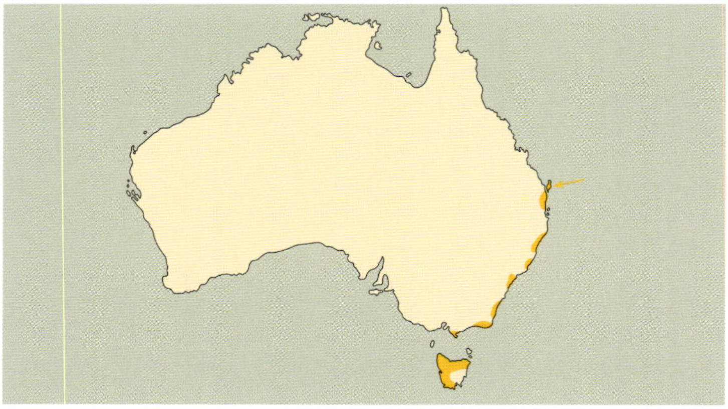

Myall Lakes, and from near Royal National Park, Barren Grounds Nature Reserve, and the Budderoo Plateau south to the Victorian border at Nadgee Nature Reserve. Formerly, it was widespread along most of the coastal plain, including areas now occupied by major urban centres. Similarly in coastal and near coastal Victoria, the range has become fragmented, with records now coming from east Gippsland between the New South Wales border and Wilson's Promontory, west of Port Phillip Bay in the vicinity of Lake Connewarre and Torquay, the Carlisle River drainage basin in the southwestern Otway Ranges, and at Discovery Bay in the far southwest (Meredith and Jaremovic 1990; in Higgins 1999). The species formerly ranged through coastal and near coastal southeastern South Australia, north to the Adelaide Plains and the southernmost Mount Lofty Ranges, though by the 1940s it apparently was restricted to the extreme southeast between Port MacDonnell and the Victorian border.

The Ground Parrot is widespread in Tasmania, having been recorded from both coastal and inland localities, mostly in the western half. It formerly occurred on Flinders Island in Bass Strait, but I can find no evidence of past or present occurrences on King Island, and doubts have been expressed about early records from the Kent Group (see Jones *et al.* 1970). It is present on Hunter Island, off the northwestern coast (Pinner and Bird 1974).

CRITICALLY ENDANGERED

2. *Pezoporus wallicus flaviventris* North
Pezoporus flaviventris North, *Aust. Mus. Special Cat.*, no. 1, 3, pt 2, 1911, p. 175 (King George Sound, Western Australia).
ADULTS Like *wallicus*, but abdomen brighter yellow with indistinct, interrupted barring of dusky black.
2 males: wing 135 mm and 138 mm, tail 164 mm and 176 mm, exp. cul. 14 mm and 15 mm, tars. 26 mm and 30 mm.

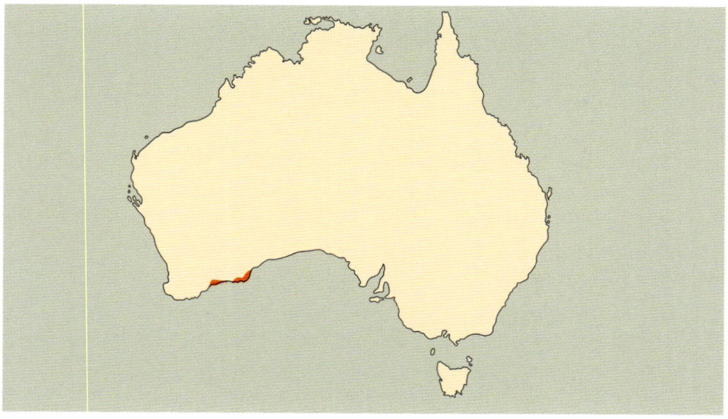

3 females: wing 118–133 (127.0) mm, tail 145–176 (165.3) mm, exp. cul. 13–15 (14.3) mm, tars. 26–28 (27.3) mm.
This poorly differentiated subspecies is restricted to coastal southwestern Western Australia, where formerly it ranged north of Perth and possibly to Geraldton. It currently is known only from Fitzgerald River and Cape Arid National Parks and Nuytsland Nature Reserve on the southern coast.

STATUS Ground Parrots have suffered greatly from destruction of their specialised habitats, the result of extensive urban and agricultural development along the seaboard, and formerly they were plentiful in areas now occupied by the major cities of Sydney, Newcastle, Wollongong, Melbourne and Adelaide. They continue to be threatened because of low numbers, and more importantly because the loss of habitat means that their range is discontinuous with marked fragmentation of the genetic pool available to most populations. In New South Wales and Victoria, some protected pockets of ideal habitat, with a past history of frequent burning or grazing, have not been recolonised, probably because alienation of surrounding lands precludes any immigration of birds.

It is conceded that estimates of a total population of 5500 birds in eastern Australia are based on very little recent information, but fragmentation of the specialised habitat and its vulnerability to wildfires pose a continuing threat to surviving populations (Garnett *et al.* 2011). Mindful that species with small populations often have reduced genetic variability, making them susceptible to higher risk of inbreeding, randomly amplified DNA fingerprinting was used to determine levels of genetic diversity in populations of Ground Parrots at three breeding sites in southeastern Queensland (Chan *et al.* 2008). One site in Cooloola National Park was approximately 50 km distant from the other two sites in Noosa National Park, with non-heath bushland and major urban infrastructure in the intervening area, and at Noosa National Park the two sites were connected by a 7 km corridor of non-heath vegetation. Blood samples were taken from 20 parrots captured at the sites and, despite the small sample size, there was sufficient evidence to suggest that the low levels of genetic variation in the sampled populations at least equal those found in the lower end of diversity for endangered species. These results indicate that the vulnerable status of Ground Parrots in eastern Australia may be justified, and caution needs to be exercised to ensure that genetic diversity is not reduced further through loss of habitat.

Southwestern Tasmania undoubtedly remains a stronghold, and inclusion of large tracts of suitable habitat in the South West National Park provides excellent safeguards for the future maintenance of strong populations in that part of the state. Conversely, Ground Parrots are scarce along coastal Victoria, where densities of between 0.5 birds per 10 ha and 2.6 birds per 10 ha were estimated at four sites in 1979–1980, and the statewide population was estimated at approximately 600 breeding adults (Meredith and Isles 1980). Meredith and Jaremovic (1990) note that in 1983 at Croajingolong National Park, a major reserved tract of habitat in eastern Victoria, fire reduced the local population to between 75 and 85 birds, or 25 per cent of the pre-fire population, but the heathlands regenerated well and there was no long-term threat to the survival of Ground Parrots in the park. Farnes (2007) reports that in far southwestern Victoria, the parrots are restricted to the Long Swamp area of Discovery Bay Coastal Park, where an encroachment of native and introduced plants in dried-out sections has reduced the area of *Baumea* sedgelands, the favoured habitat, and a small number of parrots was seen on each of three visits in 2005. There are no recent reports from neighbouring southeastern South Australia, so Ground Parrots are presumed to be extinct in the state (South Australian Ornithological Association Inc. 2008).

The present status of Ground Parrots in New South Wales is uncertain, and long-term population studies are required to determine their distribution, abundance and population trends (Cooper *et al.* 2016). The New South Wales Bird Atlassers compiled 1085 atlas records up to December 2006, and these are in just 54 of 2853, or 1.9 per cent of 10-minute blocks, and there was a strong decline in annual reporting rates between 1986 and 2006 (Cooper *et al.* 2016). Loss of coastal habitat to urbanisation and agricultural development has been very extensive in New South Wales, and there is no doubt that this has brought about a massive decline of Ground Parrots in the state. Remaining populations are in scattered remnants of habitat, most of which are protected in national parks and reserves, but their fragmentation severely restricts or precludes altogether any interchange of birds between populations, so increasing the likelihood of inbreeding, especially in small populations, and has prevented any recolonisation of some areas recovering after fires. Baker (1997) estimates the population in New South Wales at about 2000 birds, including 100 birds at Barren Grounds Nature Reserve and Budderoo National Park, less than 550 birds at Jervis Bay, 300 birds at Nadgee Nature Reserve and 100 birds at Ben Boyd National Park, with smaller numbers at other locations. At Barren Grounds Nature Reserve, density estimates have varied between 1.3 parrots per 10 ha in 1982–1983 up to 4.0 birds per 10 ha after the fledging of young in 1987, but in early August 2001 a count of only 29 birds in an 80 ha survey area suggests that there has been a sudden decline in numbers and this is cause for concern (in Higgins 1999; Neilly 2001).

McFarland (1991c) points out that a numerically large population exists within the restricted range in southeastern Queensland, where a majority of the estimated 2900 to 4900 birds occur in four reserved tracts of suitable habitat. Greg Roberts tells me that in recent years there has been a dramatic decline in some populations in southeastern Queensland, and a formerly strong population of as many as 300 birds on the Sunshine Coast has declined to fewer than 40 birds in four scattered areas of wallum heath extending from Marcus Beach in the north to Marcoola in the south (pers. comm. 2015).

Of most concern is the endangered population along the south coast of Western Australia. Since 1900, numbers of Ground Parrots have declined by some 60 per cent and recently the population was estimated at less than 140 birds, with the majority confined to Cape Arid National Park, where a wildfire in January 2011 may have brought about a further reduction (Bondin *et al.* 2011). Birds have not been recorded in Fitzgerald River National Park since 2012, though authorities remain hopeful that some birds may still be present there. In November 2015, a devastating wildfire destroyed vegetation in almost 90 per cent of Cape Arid National Park, leaving only two unburned pockets of habitat, where an unknown number of Ground Parrots survived. More recently, birds have been seen and heard in adjoining Nuytsland Nature Reserve for the first time since 2006, but at the time of writing the total number of parrots in the wild population was not known. A multi-faceted recovery program has been established and, in addition to the implementation of fire management strategies for maintenance of habitat suitability, baiting for foxes *Vulpes vulpes* and feral cats is being carried out in the national parks and a captive breeding program has been set up with a view to reintroducing parrots to parts of the former range.

The Western Ground Parrot *P. w. flaviventris* is listed as critically endangered under the *Australian Environment Protection and Biodiversity Conservation Act 1999*, and is afforded special protection under state legislation. *Pezoporus wallicus* is listed on Appendix I of the Convention on International Trade in Endangered Species of Wild Fauna and Flora (CITES).

HABITATS Ground Parrots inhabit coastal and contiguous plateau heathlands or sedgelands, as well as estuarine flats, swamps and, in some localities, grasslands or pastures. Primarily, they are birds of heathlands or sedgelands, where vegetation communities vary regionally, and their occasional presence in other habitats often is a consequence of fires or seasonal shortages of preferred foods.

At Cooloola National Park, a stronghold in southeastern Queensland, the preferred habitat is subtropical closed graminoid-heathlands extending inland from the coastal dune system as expansive plains or in a patchy pattern (McFarland 1988a). The 70 to 100 per cent foliage cover is from 0.5 m to 2.0 m in height, and is dominated by swamp grasstrees *Xanthorrhoea fulva,* together with spreading rope-rush *Empodisma minus,* conesticks *Petrophile shirleyae*, hakeas, banksias and *Leptospermum* species, with occasional emergent plants including wallum banksia *Banksia aemula,* broad-leaved paperbark *Melaleuca quinquinervia* and broad-leaved white mahogany *Eucalyptus umbra*. Within this heathland, the parrots appear to use all parts of the dry and wet microhabitats, but seldom use sedgelands or shrub and tree dominated areas (McFarland 1991a). Shifts between dry and wet areas correspond to changes in seed availability and accessibility and may be linked also to climate. Gosper (1995) reports similar seasonal shifts between wet and dry areas in graminoid-heathlands on coastal sandplains in Bundjalung National Park, northeastern New South Wales. On a low sand ridge, between 60 m and 70 m in width, a dry diverse shrub community grades from low, relatively sparse heath averaging less than 0.7 m in height and 10 to 30 per cent foliage cover to dense closed heath averaging 1.3 m in height and 80 per cent foliage cover, with characteristic shrubs including wallum banksia *Banksia aemula*, juniper wattle *Acacia ulicifolia, Persoonia, Leptospermum* and *Melaleuca* species, and the epacrids *Woollsia pungens* and green five corners *Styphelia viridis*. In adjacent low-lying, seasonally inundated wet heathland, a uniformly low, closed graminoid community is characterised by sedges and small shrubs, with dominant species including zig-zag bog-rush *Schoenus brevifolius, Sporadanthus interruptus*, rusty banksia *Banksia oblongifolia*, heath banksia *B. ericifolia* and egg-and-bacon plant *Dillwynia floribunda* together with scattered xanthorrhoeas and taller broad-leaved paperbarks. In this wet heathland, standing water to a depth of 25 cm is present following heavy wet-season rainfall in some years.

At Barren Grounds Nature Reserve, at approximately 300 m on the escarpment behind the Illawarra coastal plain, in southern New South Wales, these parrots frequent temperate heathlands, with a marked preference for low, rather sparse heath along ridges. I have noticed that they move out of areas where long protection from fire has resulted in growth of tall dense heath with a predominance of woody perennials, mainly *Leptospermum* and *Hakea* species. Conversely, they come into areas of regenerating heath a few years after it has been burned, and it is obvious that optimum habitat is heathland three to eight or 10 years after burning. Vegetation height and floristic diversity seem to be of critical importance to the birds, optimum habitat being heath 0.5 m to 1.5 m in height and with a high species diversity that includes sedges, rushes and epacrids, with scattered taller woody perennials such as dwarf eucalypts, banksias and paperbarks. I have encountered very few parrots in any extensive areas of damp matted vegetation, though this is where buttongrass *Gymnoschoenus sphaerocephalus*, a favoured food plant, is most prevalent.

In coastal Victoria, peak densities of Ground Parrots are recorded in heathlands 15 years after burning, and apparently there then comes a rapid population decline so that heathlands with a post-fire age exceeding about 20 years do not support parrots (Meredith *et al.* 1984). Also, there is an absence of parrots from heathlands where three or more fires have occurred at intervals of less than four to six years. Using these data, Meredith and Jaremovic (1990) recommend that in east Gippsland, eastern Victoria, 1000 ha of graminoid heathlands should be burned every two years over a 20 year cycle and 30 ha of diverse shrub heathlands should be burned every two years over a 16 year cycle, but no active management is required in sedgelands. Bryant (1994) notes that during the statewide survey conducted in Tasmania between October 1989 and December 1990, Ground Parrots were recorded in heathlands of all ages exceeding one year after burning, but most suitable were those with post-fire ages of four to 13 years.

At a study site in Fitzgerald National Park, on the south coast of Western Australia, between 1996 and 2004, a monitoring of calling frequencies was undertaken to determine the habitat preferences of Ground Parrots in relation to fire (Burbidge *et al.* 2007). Calling frequency and presumably population levels increased as the vegetation age increased from 39 to 43 years after fire, and it was concluded that the parrots do not need fire to maintain habitat suitability in the scale of 40 to 45 years after fire. However, it was found also that the parrots used recently burned vegetation provided that it was near to older vegetation. In September 2008, at Cape Arid National Park, a radio-tracked bird appeared to prefer six-year-old vegetation for feeding, but returned to vegetation more than 40 years old for roosting, and it was presumed that nesting also occurred within the older vegetation (Comer *et al.* 2009). In recent years, at Cape Arid National Park, other birds have been observed similarly utilising vegetation of differing ages after fire, so again there are indications that Ground Parrots may need a mosaic of vegetation of differing post-fire ages along the southern coast of Western Australia.

MOVEMENTS Recaptures of banded birds at Barren Grounds Nature Reserve, southern New South Wales, where birds have been recaptured up to eight years after banding, and at Cooloola National Park, southeastern Queensland, where a bird was recaptured two years and five months after banding, indicate that adults are sedentary. Juveniles also remain in the natal areas for up to three months, but thereafter disappear, suggesting either high mortality or more probably a strong autumn dispersal. Further evidence of autumn dispersals of young birds was obtained at Cooloola National Park, where no juveniles were retrapped more than four months after being banded as nestlings, and increased numbers of unbanded birds caught during autumn were suspected to be first-year birds moving through the area (McFarland 1991d).

It is during these autumn dispersals of first-year birds or when all birds are forced from areas by fire, that sightings are more likely to be made in suboptimal habitats and at times well outside areas of normal occurrence. There are instances of birds moving considerable distances. Sightings made in alpine heathlands in the Victorian alps were between 150 km and 220 km from the nearest known breeding areas, and birds that appear each year in sedgelands in the Gippsland Lakes district, eastern Victoria, must have flown at least 80 km from breeding sites in the Marlo region (Meredith *et al.* 1984). Meredith (1984) points out that over the years there have been more than 20 records from areas that normally do not support Ground Parrots, and all have been made during the post-breeding period of January to March. In southwestern Australia, for the first time since 2006, birds were seen and heard in Nuytsland Nature Reserve after wildfire destroyed much of their habitat in adjoining Cape Arid National Park in November 2015, and this may represent a post-fire dispersal.

HABITS Ground Parrots are shy, elusive birds. To see them, it is necessary to visit an area where they are known to occur and methodically tramp through the vegetation in the hope of flushing one or two birds. Rarely are they detected on the ground before taking flight. With an audible flutter, a parrot will rise up from the vegetation about 5 m to 10 m ahead and then fly 30 m or more before dropping back into cover. If followed immediately, it can be flushed a second or even a third time, but thereafter the flights become longer. Ground Parrots rely largely on their plumage colouration to escape detection. At Barren Grounds Nature Reserve, southern New South Wales, a bird flushed by another member of our survey team alighted about 5 m in front of me, and by standing perfectly still, I was able to watch it walk to the base of a dwarf *Banksia* bush; with its back towards me and head turned sharply to the left, it stood motionless and watched me for nearly 10 minutes, its cryptic colouration blending superbly with the vegetation, but as soon as I moved, it ran off and rose into the air about 10 m ahead.

The misconception about these parrots being crepuscular or mostly nocturnal arose from observations of active, flying birds just before dawn and after dusk, together with early reports of captive birds sitting still during the day but becoming active after sunset. Also, the other two terrestrial parrots, the Night Parrot *Pezoporus occidentalis* and the Kakapo *Strigops habroptila*, are known to be nocturnal, so there may have been an expectation that the Ground Parrot would be active at night. Recent field investigations have established that Ground Parrots forage throughout the day, with peak periods of activity in early and mid morning and late afternoon, but voluntary flying and vocalisations are restricted to periods of between 10 and 40 minutes before sunrise and after sunset (Meredith *et al.* 1984; McFarland 1991a). At Cooloola National Park, southeastern Queensland, a home range averaging 9.2 ha was occupied by each bird, with adults using smaller areas than subadults, probably reflecting the occupancy of better feeding sites by long-term residents. Also, it was found that adult males occupied smaller ranges than did adult females, so suggesting that males may be dominant over females, with their smaller ranges being centred on the best food

Plate 16
Western Ground Parrot *Pezoporus wallicus flaviventris* (adult)

F. KNIGHT '15

supplies. All nighttime roosts were in dry heathlands, which were well away from summer foraging sites in wet heathland, but in or near to feeding areas used at other times.

Ground Parrots run quickly and, because of their long legs, they lack the waddling gait so characteristic of most parrots. When disturbed or pursued, they crouch low and run through the vegetation, frequently pausing and standing fully erect to look about. Though capable of perching, they seldom do so, but readily will clamber across the tops of matted vegetation to get at seedheads or climb adeptly amidst the foliage of shrubs to peck at flowers or fruits. There is an early record of a bird alighting on the top of a three-railed wooden fence, while at Bundjalung National Park, in northeastern New South Wales, birds were seen to alight in shrubs and then perch or move about at between 60 cm and 90 cm above the ground (in North 1911; Gosper 1995). Norman Favaloro watched a parrot climb a *Xanthorrhoea* spike to feed on the seeds (pers. comm. 1967).

Courtney (1997) points out that a competent climbing and perching capability was evident also in captive birds and, when climbing about, their long claws seemed to act as extensions to the toes, helping to prevent their feet from falling through between the stems. Their long legs enabled them to step up on to tussocks with minimal effort and, at times when feeding on the ground, they would reach up above their heads with one foot to pull down seeding grasses. They were ambidextrous, using either foot to hold entire seedheads on the ground while extracting seeds. During daily water-bathing, these birds squatted deeply in the water, dunked the head and vigorously flapped their wings in actions similar to those of bathing rosellas. When sunbathing, they would lie stretched out on one side, with the uppermost wing fully spread over the legs. Apart from courtship feeding, no physical contact was seen to occur between the three captive individuals. At night they roosted well apart, usually on the ground though occasionally on a low branch. Apparent ritualised fighting took place between two three-year-old males; facing each other and from a distance of at least 0.5 m, these two birds walked quickly towards each other, leaping into the air at each other but, when close, avoiding bodily contact at the last moment by rolling onto their sides with the wings extended vertically, and then gliding past each other with their undersides almost touching before immediately rolling back into normal horizontal flying to descend and alight.

The swift, erratic flight is most characteristic and consists of rapid, shallow wingbeats alternated by gliding on motionless, downcurved wings. The tail is held in a slightly uptilted position, similar to that of a pheasant in flight, and with each series of wingbeats the parrot turns its body, thus producing a 'zig-zag' flight path. Ground Parrots usually fly low to the ground, but Condon (1942) saw one bird rise to a height of more than 15 m. In flight, the yellowish-white underwing-stripe and the yellow outer tail feathers are conspicuous. Contrary to most reports, these parrots are strong fliers. Furthermore, the ease and buoyancy of their flight suggest that they are quite capable of flying long distances. At Barren Grounds Nature Reserve, southern New South Wales, a bird that I flushed from a nest flew through a small stand of trees and circled back behind me, covering a distance of almost 0.5 km. When alighting, a parrot seems to drop suddenly into the vegetation; just before it disappears into cover, there is a quick fluttering of wings, which checks speed and enables the bird to drop at a more acute angle, giving the impression of a headlong, tumbling dive.

CALLS Ground Parrots regularly call after dusk and before dawn, but rarely at other times. Several birds may call simultaneously, but usually one call is followed by a response, and numbers of birds in an area can be estimated by listening to these calls. In Fitzgerald River National Park, on the south coast of Western Australia, during a frequency of calling monitoring undertaken between 1996 and 2000, calling in the morning started about an hour before sunrise and continued for an average of 36 minutes, while in the evening calling commenced about 25 minutes after sunset and continued for approximately 30 minutes, and there was a tendency for calling to continue later on cloudy mornings and begin earlier on cloudy evenings (Burbidge *et al.* 2007).

The thin, high-pitched call is of three or four clear measured bell-like notes, *tee . . . tee . . . stit* or *tee . . . tee. . . tee. . . stit*, followed occasionally by a number of sharper notes on an ascending scale, terminating with a drawn-out note, *tee . . . tee . . . tee . . . tee---ee*. McFarland (1991b) recognises four variations of the level or ascending components of these calls, which carry well on a still morning or evening, but are scarcely audible when strong, blustery winds are blowing. At Cooloola National Park, southeastern Queensland, three other calls were heard only during the breeding season; at dusk, a frog-like croaking note was given by males approaching incubating females at the nest, a repetitious *teek* was emitted by juveniles while following an adult, and nestlings gave a harsh *kaarr* call when disturbed. At Barren Grounds Nature Reserve, southern New South Wales, we recorded the same *kaarr* calls from disturbed nestlings, but these nestlings also uttered a mellow *cheep-cheep*, and a parent was heard to announce its arrival at the nest by emitting three or four loud, whistle-like notes.

Courtney (1997) points out that the juvenile food-begging call does not develop in chicks until they are about to fledge, and it is a pure-sounding, regularly repeated quavering whistle, which remains unchanged with increasing age until discarded at independence. Recently fledged juveniles in captivity were heard to warble softly and constantly for long periods, and captive adult males uttered a frog-like 'croak-click' or occasionally a rapid succession of throaty 'clacking' noises. Two other calls rarely given by captive adults were an extraordinarily long soft, thin whistle of unvarying pitch ending with about three brief whistles of the same pitch, and a brief disyllabic 'squeak'.

DIET AND FEEDING The diet comprises mainly seeds, small fruits or berries, buds, blossoms and leaf shoots of heathland or swampland herbs and grasses. These are picked up from the ground or taken from plants by reaching up from the ground to pull down seedheads and flowers or by clambering on top of matted vegetation and amidst the foliage of taller shrubs. At Barren Grounds Nature Reserve, southern New South Wales, I have watched a feeding bird stand on plant stems to force seedheads down to the ground and then pass its bill along the stems and over the seedheads, thus spilling seeds to the ground. Insects and their larvae have been found in crop samples, but may have been ingested accidentally.

McFarland (1991a) reports that at Cooloola National Park, southeastern Queensland, parrots took seeds from 40 plant species, with 19 of the known species being from dicotyledonous plants, primarily of Fabaceae, Epacridaceae, Rutaceae and Proteaceae, and 15 from monocotyledonous plants, mostly of Cyperaceae, Restionaceae and Poaceae. Important food plants producing or retaining seeds throughout most or all of the year included curly sedge *Caustis recurvata*, variable smoke-bush *Conospermum taxifolium*, dwarf spider oak *Grevillea leiophylla,* coastal pseudanthus *Pseudanthus orientalis* and *Sprengelia sprengelioides*, while other more seasonal food plants of importance were flat-stemmed cord-rush *Restio complanatus*, bog-rush *Schoenus paludosus,* Chaffy scale-rush *Lepyrodia scariosa*, and *Epacris* species. Sizes of seeds taken ranged from 0.6 mm to 7.0 mm, and seeds not recorded in crop samples were those of plants with very small or very large seeds or with seeds enclosed in hard, woody fruits. Insect items recorded in crop samples were coleopteran larvae averaging 4 mm in length, hemipterans averaging 1.1 mm in length and egg sacs, but these were rare and probably were ingested accidentally. Non-seed vegetative material and sand grains were present in several crops.

Gosper (1995) notes that in Bundjalung National Park, in northeastern New South Wales, there appears to be an association between Ground Parrots and snow wreath *Woollsia pungens*, the flowers of which seem to be a favoured food, for birds frequently were seen feeding in or underneath flowering shrubs, and often were flushed from the shrubs. Alan Rayward told me that near Byron Bay, also on the far north coast of New South Wales, parrots were observed feeding on seeds of *Paspalum* and *Compressum* grasses; they moved about quite quickly on the ground, pulling down grass stems as they walked and occasionally stopping to scratch in the soil with both feet and then probe with the bill, as if trying to dig up a root or germinating seed (*in litt.* 1974). At Barren Grounds Nature Reserve, southern New South Wales, I have observed parrots feeding on seeds of slender rope-rush *Leptocarpus tenax* and buttongrass *Gymnoschoenus sphaerocephalus*. At Mallacoota Inlet, easternmost Victoria, Norman Favaloro watched a bird feeding on *Xanthorrhoea* seeds (pers. comm. 1967). At Croajingolong National Park, also in eastern Victoria, seeds and other plant material, mostly glumes and anthers, with five coleopteran larvae, presumably ingested accidentally, were present in crop samples, and of identified seeds those of zig-zag bog-rush *Schoenus brevifololius* were present in 30 per cent of samples, of *Cassytha* species in 18 per cent of samples, of pale twig-rush *Baumea acuta* and *Lepidosperma* in 5 per cent of samples, of *Gonocarpus* in 4 per cent of samples, and of *Restio complanatus* and *Gahnia* species in about 1 per cent of samples (Meredith *et al.*1984). Condon (1942) noted that near Port MacDonnell, in extreme southeastern South Australia, parrots were seen in an area where the only available food appeared to be the blackish seeds of bare twig-rush *Baumea juncea*, and this same plant dominates in sedge-heaths frequented by the birds at Long Swamp, in neighbouring far southwestern Victoria.

Bryant (1994) points out that in Tasmania the bulk of the diet probably comprises seeds of plants from the Restionaceae, and a list of suggested food plants include 13 species of dicotyledons and 12 species of monocotyledons. In western Tasmania, birds have been observed taking seeds of branching scale-rush *Lepyrodia tasmanica*, river rose *Bauera rubioides*, swamp boronia *Boronia parviflora* and buttongrass *Gymnoschoenus sphaerocephaus* (Brown 1979). At Cleland, northern Tasmania, parrots were observed under almost leafless briar bushes, probably *Rosa rubiginosa*, feeding on fallen fruits (Fletcher 1908).

BREEDING McFarland (1988b) points out that there is a distinct trend towards earlier nesting with decreasing latitude, and egg-laying in southeastern Queensland commences up to three months earlier than in Victoria or Tasmania. In southeastern Queensland, eggs have been recorded during July to November, and during research undertaken at Cooloola National Park, between 1986 and 1988, of the 31 nests examined, 17 contained clutches known or inferred to have been laid in September, 11 to have been laid in August, and three in October (McFarland 1988b, 1991b). On the north coast of New South Wales, nests with eggs or chicks normally are recorded during September to December, but near Byron Bay, in the second week of March, Rayward found a nest containing three eggs (*in litt.* 1974). On the south coast of New South Wales and in coastal Victoria, nests with eggs or chicks have been recorded during September to December, while in Tasmania nests with eggs have not been recorded earlier than October, and nests with chicks have been found as late as February (McFarland 1988b).

Generally situated near the base of a tussock or stunted bush and well hidden by overhanging vegetation, the nest is in a shallow excavation in the soil normally well lined with chewed stems and leaves. In a wet situation, the nest may be on a raised platform in the middle of a tussock. I am sceptical of the claim made by Webber (1948) that a nest was found in a fallen hollow log. McFarland (1991b) notes that at Cooloola National Park, southeastern Queensland, all nests were of similar construction, being in a domed cavity formed by the birds cutting away thin stems of sedges, rushes or small herbs at the centre of a sward of spreading rope-rush *Empodisma minus*, swamp grasstree *Xanthorrhoea fulva* or rusty banksia *Banksia oblongifolia*. The cut stems were nipped into pieces between 5 mm and 30 mm in length, which lined a scrape in the sand to build up a nest platform. Nests were almost totally obscured by overhanging vegetation, which was significantly higher than that in the immediate surrounds, and there were only one or two access tunnels, openings to which were in no preferred direction. Nests were well spaced, the density averaging 1.0 to 1.3 nests per 10 ha.

Chris Sonter describes a nest that he found at Barren Grounds Nature Reserve, southern New South Wales, on 19 October 1975 (*in litt.* 1978). It was a cup-shaped structure placed on the ground at the base of a buttongrass tussock and the overhanging grass enclosed an oblong-shaped chamber around the nest; this chamber was approximately 75 cm in length, 25 cm high and about 20 cm wide. The nest was approximately 2 cm deep and had an overall diameter of about 18 cm; it was constructed from strips that the bird had chewed from vegetation within the chamber and the uniform length of these was about 9 cm. When found, the nest contained four eggs.

At Evans Head, on the far north coast of New South Wales, a clutch of four eggs was taken on 13 September 1976, and the nest is described as follows (ANWC E04364):

> *Nest well concealed within a clump of Restiad known as* Lepyrodia interrupta*, 2 feet height, which was continuous & merged with similar surrounding densely matted clumps, growing on boggy ground, with 1½ inches of water in places. The site was 300 yards down from the start of the water meadow & 25 yards from the eastern edge of the tract, 75 yards wide at this point. Nest, a shallow bowl, neatly constructed of thin living sections of stalks of* Lepyrodia interrupta*, bitten by the parrot from the interior of the clump to form a chamber 8 inches diameter × 9 inches in height. A sloping tunnel-way, 17 inches in length, led to this cavity & the nest, from an entrance 5 × 4 inches diameter.*
>
> *Diameter of nest: 6 inches*

Diameter of egg-cavity: 5 inches
Depth of egg-cavity: 2¾ inches

Near Noosa Heads, southeastern Queensland, a clutch of four eggs was taken on 27 September 1980, and these eggs were laid on bare sand (ANWC E04365):

> *Nest: none made. This was a most unusual situation. The bird had laid four eggs on bare sand; no hollow, nor any lining material whatsoever. The eggs were beneath the centre of a thick low shrub, some 1.3 of a metre in from the outer edge of the shrub. Viewed from directly overhead the eggs could just be seen through the foliage of the shrub. The sitting bird flushed immediately after the rope passed over the shrub. The site was well out in an open plain of low (0.5–0.8 metre) shrubs.*

McFarland (1991b) notes that at Cooloola National Park, eggs were laid asynchronously over two or three days, and most clutches were of three or four eggs. Incubation was undertaken only by the female, and feeding of the sitting female, presumably by the male, took place at least three times during the day – in the early morning, late morning and late afternoon or at dusk. During the day, the male walked to the nest, but at dusk flew in with the final feed, and he gave the frog-like call when approaching the nest, at least at the dusk visit. At 21 days after laying, asynchronous hatching occurred over at least two days, but females continued to brood the newly hatched chicks continuously for four days, and then only at night for two days, after which feeding visits to the nest apparently were made only by males. Chicks seemed to be fed in the early morning, mid morning, at midday or in the early afternoon, and in the late afternoon or at dusk. Although capable of running from the nest at 18 days after hatching, chicks normally remained in the nest for 24 days, and then remained in the near vicinity for at least three weeks. When capable of flying, the juveniles, while uttering their *teek* calls, commenced following an adult. Of 63 eggs in 19 nests, 66.7 per cent hatched, and the fledging success rate of 57.1 per cent was equivalent to an average of 1.9 chicks fledging from each nest.

At Barren Grounds Nature Reserve, in November 1967, Peter Fullagar and I found a nest containing four newly hatched chicks, which were covered in very dense greyish-black down, obviously for protection against frequent exposure to cold, damp conditions. Our observations at this nest provided data on parental attendance similar to that recorded by McFarland at Cooloola National Park. The chicks were fed before dawn, in mid-morning at about 1000 hours and in the evening at about 1900 hours. Only one adult was seen to visit the nest, but we do not know whether it was the same bird each time. It flew in and alighted approximately 5 m from the nest (JMF) or at the nest (PJF). Approach to the nest was through a tunnel on the western side. We noticed that each time the chicks were handled for weighing, they emitted their begging cries whenever we placed our hand in the tunnel, but remained silent when approached from other directions. For the first two weeks, the nestlings were fed in the nest, but thereafter they moved out into the surrounding vegetation to meet the incoming parent. When an estimated 21 days old, they would leave the nest at the slightest disturbance, and one bird was absent for two days. A few days later, they had vacated the nest and were found sheltering in nearby tussocks; at this stage, they were fully feathered, but could not fly, and their tails were only about 50 mm in length.

Courtney (1997) notes that the pink bills of captive nestlings commenced to darken at about 58 days, and gradually became the grey of adults at six months. Change in colour of the cere from pink to cream-white and then grey occurred over the same six-month period. The dark brown irides gradually paled, and at six months had become pale brown with underlying white. At about 100 days, these juveniles had acquired a small patch of red feathers immediately above the centre of the cere, and within 20 days these red feathers had spread outward to form the complete frontal band. Courtship displays and courtship feeding were first observed when the birds were almost three years old. In typical platycercine manner the male stood upright and lifted the 'shoulders' well away from the body to display the bluish undersides of the carpal joints, while the tips of the wings remained touching the back; there was no fanning of the long tail, which was pressed firmly against the ground. While in this posture, the bird stood still for many seconds or walked slowly in small circles, calling frequently with a series of soft whistles. Occasionally while perching on a branch, a male was seen to thrust his body up and down vigorously several times by bending and straightening his legs, and emitting the same soft whistles. One presumed male was seen to make preliminary head-bobs and then, with a gentle head-bobbing, pass regurgitated food into the slightly upturned bill of a presumed female.

EGGS The eggs are rounded to broad-elliptical with slight gloss. A clutch of four eggs of *P. w. wallicus* was taken on 13 September 1976, at Evans Head, on the far north coast of New South Wales, and these eggs measure 26.8 (26.3–27.6) × 21.7 (21.3–22.0) mm (ANWC E04364). Another clutch of four eggs of *P. w. wallicus* was taken on 27 September 1980, near Noosa Heads, southeastern Queensland, and these eggs measure 26.2 (25.5–27.2) × 22.2 (21.8–22.8) mm (ANWC E04365). A clutch of three eggs of *P. w. flaviventris* was taken on 20 November 1913, at Wilsons Inlet, on the south coast of Western Australia, and these eggs measure 27.2 (27.0–27.7) × 22.3 (22.2–22.5) mm (MV BE1440).

Night Parrot

Pezoporus occidentalis (Gould)

ENDANGERED

Geopsittacus occidentalis Gould, *Proc. Zool. Soc. London*, 1861, p. 100 (West Australia = 8 miles south-east of Mount Farmer, Western Australia, after Wilson, *Emu*, **37**, 1937, p. 80).

DESCRIPTION Length 23 cm. Weight unsexed (live bird) 104 g. ADULTS General plumage colouration yellowish green, mottled dark brown, black and yellow; crown and nape green, feathers shaft-streaked black; discernible cheek-patches and ear-coverts yellowish-green, becoming more yellowish towards throat, where feathers centrally edged subterminally brownish-black; back, wing-coverts and rump yellowish-green with dark brown bases to feathers and pale brownish-black markings in form of arrowhead spots and transverse barring; primaries, secondaries and primary-coverts dark brown, edged olive-green towards bases, and on inner webs concealed pale yellow spots forming underwing-stripe; underwing-coverts pale green; foreneck, breast and flanks

yellowish-green with black and yellow barring; abdomen and under tail-coverts rich yellow; central tail-feathers dark brown with yellow indentations and green margins, lateral tail-feathers dark brown barred yellow; bill greyish-brown; iris black (Gould); legs brown.
4 males: wing 142–150 (147.0) mm, tail 92–113 (103.0) mm, exp. cul. 13–14 (13.3) mm, tars. 20–23 (21.8) mm.
3 females: wing 136–145 (137.3) mm, tail 95–106 (98.6) mm, exp. cul. 12–13 (12.3) mm, tars. 22–23 (22.3) mm.
12 unsexed: wing 132–153 (143.8) mm, tail 88–115 (98.7) mm, exp. cul. 10–14 (12.5) mm, tars. 20–24 (22.2) mm.
JUVENILES Undescribed; young birds said to be dull and very plain, with some yellow on throat and neck (Burgoin, in Wilson 1937).

DISTRIBUTION The Night Parrot has been reported from widely scattered localities in the arid interior of Australia. Of the 24 known specimens, no fewer than 19 were collected in the north of South Australia, and 16 of these were taken by F. W. Andrews, collector with the South Australian Museum (see Forshaw *et al.* 1976). Of three specimens, including the type, taken in Western Australia, there is some doubt about the provenance of the specimen said to have been collected by J. T. Cockerell. The remaining two specimens were obtained recently in western Queensland (Boles *et al.* 1994; McDougall *et al.* 2009). There are unconfirmed sight records from all mainland states and the Northern Territory, and Blyth (1996) cautions that inclusion of reasonably convincing sightings from Western Australia and the southern Northern Territory undermines any delineation of geographical centres of distribution.

There are scattered reports from the interior of Western Australia, north to the arid southern sector of the Kimberley division, where there have been four unconfirmed records, and south to southern salt lakes in the vicinity of Lakes Varley and King, with westernmost records coming from localities between long. 118°E and long. 120°E, including Ethel Creek, Lake Austin, and the designated type locality near Mount Farmer in the upper Murchison River region (Blyth 1996; Johnstone and Storr 1998). In the Northern Territory, most reports have come from the southern range country west from Harts Range and north to Barrow Creek and Mount Doreen homestead, but McKean (1985) reports possible sightings to the north of Keep River National Park and at nearby Kildurk Station, in the far northwest near the Western Australia border. There is another unconfirmed report from Waterloo Station, also in northwestern Northern Territory, where stockmen claimed to have seen about 12 birds during September 1970 (Wolfe pers comm. 1970). An egg, suspected of being of a Night Parrot, was found in the Tanami Desert, west of Tennant Creek, in the early 1980s, but subsequently has proved to be of a quail (see Higgins 1999).

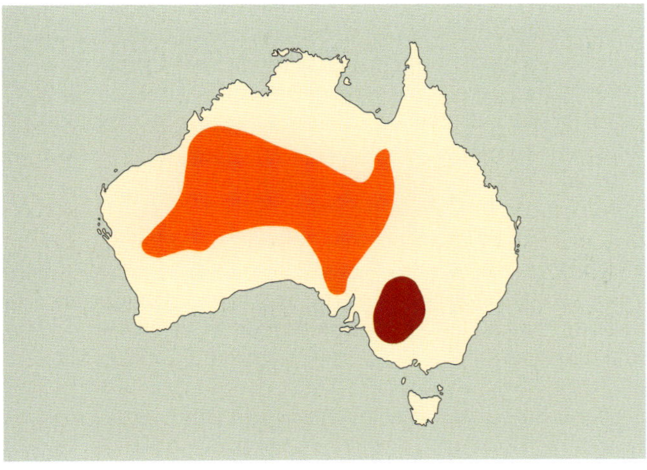

In South Australia, records have come from Brown's Well, in the east, from the Gawler Ranges, the Lake Eyre basin and to the south of Oodnadatta, in the north, and from the vicinity of Cooper Creek, in the far northeast. It was in the vicinity of Cooper Creek, in the far northeast, where the first specimen was collected in 1845, a sighting was made in early June 1979, and a sighting of three birds was made in late July 1986. In 1867, a Night Parrot survived the long sea voyage to Britain, where it lived for a short time at London Zoo, and Sclater (1867) reported that this bird was sent by Ferdinand Müller, who wrote: 'This peculiar parrot was presented to me by Mr Ryan, on whose sheep station, on the Gawler Ranges west of Spencer Gulf, it was obtained'. Early reports from northwestern Victoria date from the late 1800s and were quite vague, but more credible records dating from the early 1900s include sightings made approximately 13 km north of Bellbird Bore and about 20 km south of Cowangie, while additional sightings were made during the 1950s at Ross's Spring, 13 km south of Tutye (Emison *et al.* 1987).

The only record from New South Wales dates from August 1897, when a Mr Kelly shot a bird and found a nest among '... a kind of porcupine grass' on a saltbush plain approximately 65 km from Oxley, in the southwest (Kershaw 1943). In western Queensland, there have been scattered records from near Cloncurry south to near Boulia, where a road-killed specimen was obtained in October 1990, in Diamantina National Park, where a dead bird was collected in 2006, and the far south from northwest of Birdsville, at the eastern edge of the Simpson Desert, southeast to Nappa Merrie Station, with doubtful reports coming from west of Quilpie and some 120 km west of Charleville (in Higgins 1999; McDougall *et al.* 2009).

GENERAL NOTES In 1861, when describing and naming this species, John Gould wrote:

> *I have had in my possession for many years the skin of a Parrakeet, which was sent to me direct from Perth, in Western Australia, and which differs, in my opinion, both generically and specifically from every known species. In general appearance, and especially in its colouring, it resembles the* Pezoporus formosus*; but on carefully comparing it with that species, some remarkable differences are apparent.*

Gould's failure to perceive immediately the differences between the Night and Ground Parrots is all the more apparent because we now know that, in November 1847, the first Night Parrot specimen was sent to him by Charles Sturt. This specimen was collected by John McDouall Stuart on 15 October 1845, near Lake Goyder, in the Cooper Creek area of far northeastern South Australia, was identified incorrectly as a Ground Parrot, forwarded to Lord Derby and subsequently lodged in the Liverpool Museum, UK (see Forshaw *et al.* 1976). Gould never realised the error he had made and the misidentification remained undetected until July 1928, when Gregory Mathews examined the specimen and amended the label. It is interesting to note that Mathews, usually so meticulous in these matters, did not appreciate the significance of this specimen and made no mention of it in his subsequent writings.

STATUS Over the years, the Night Parrot has remained the most mysterious of Australian birds, and after the 1880s, when most specimens were collected, it was not until September 1912 that another specimen was collected at Nichol Spring, Western Australia, but because of poor preparation it rotted away and the remains were lost (see Wilson 1937). Unconfirmed sightings at irregular intervals were the only evidence of continued existence of the parrots, and then in early June 1979 four birds were seen on Cooper Creek, due east of Lake Perigundi, in far northeastern South Australia (Parker 1980). Subsequently, unconfirmed sightings continued to be reported, though less regularly, from widely scattered parts of the range, with more recent reports coming from western Queensland, where 'rediscovery' of the species started in mid October 1990 with recovery of a desiccated road-killed specimen 36 km north of Boulia (Boles *et al.* 1994). Attention continued to be focused on western Queensland, where seven reliable sightings of Night Parrots occurred in the Cloncurry district from early 1992 to mid 1993 (Garnett *et al.* 1993). In mid September 2006, a headless, desiccated specimen was found by the side of an access road in Diamantina National Park, and detached feathers in a wire barb on a fence indicated that the bird may have hit the fence in flight, or alternatively it could have been killed by a predator (McDougall *et al.* 2009). In May 2013, at an undisclosed locality in western Queensland, the first photographs of a living Night Parrot were obtained by John Young, and at 1900 hours on 4 April 2015, at another then undisclosed locality in western Queensland, a Night Parrot was captured in a mist net by Steven Murphy and the Recovery Team. The Night Parrot had indeed been 'rediscovered'!

There is insufficient information to determine the present status of the Night Parrot and, despite the many reported sightings, it is not possible to ascertain whether there has been a change in the population size. The species may be endangered as a consequence of the degradation of habitats by grazing of domestic or feral mammals and by altered fire regimes, together with predation by foxes or feral cats, for these factors have been implicated in the extinction of native mammals in the arid zone. Conversely, the parrots may always have been scarce, with irregular irruptions bringing about local upsurges in numbers in response to propitious conditions. The collection by Andrews of up to 16 specimens in the regions of the Gawler Ranges and Lake Eyre, northern South Australia, during the 1870s, has prompted speculation that there was a local irruption at that time. Similarly, the birds were said to be particularly common in the Alice Springs district, southern Northern Territory, in 1892, when 'many were brought in by cats' (in North 1896). Local irruptions may have contributed also to the apparent presence of reasonable numbers at extremities of the range at various times.

Predation by feral cats has been identified as a serious threat. In the early 1900s, articles about birds in northwestern Victoria were contributed to national and regional newspapers by a writer using the pseudonym 'Mallee Bird'. Menkhorst and Ryan (2015) identify this writer as Charles H. McLennan, a bushman and well-regarded ornithologist who worked on pastoral runs established among terminal lakes of the Wimmera River in the Big Desert of northwestern Victoria, and his notes, supplemented with information from a local Aboriginal man named Jowley, are some of the earliest published accounts of the life history of Night Parrots. In a contribution to *The Australasian* of Saturday 4 April 1908, 'Mallee Bird' blamed predation by feral cats for the disappearance of Night Parrots in northwestern Victoria:

> *One of the most interesting of the birds that haunts the spinnifex tracts in Sunset country is the now rare night parrot, whose plaintive whistling note heard in the still of the night makes one in a sense of his solitude. They are unfortunately becoming extinct. Some years ago hundreds of domestic cats were turned loose in the Mallee, in the hope that they would kill off the rabbits. They evidently preferred a bird diet and the night parrots, occupying in daytime old rabbit burrows, fell easy prey to them. The ever increasing fox is another enemy,and when I find feathers or remains of the night parrot there are generally fox or cat traces in the loose soil. On the wing the night parrot takes short, quick flights, dropping suddenly like a snipe. According to Jowley, an old aboriginal of this district, they live in a tussock of turpentine grass and lay from four to six eggs, rather round in shape.*

To date there is no evidence that cats are taking parrots at the study site in western Queensland, and it is suggested that maintenance of good spinifex cover gives the parrots effective refuge in the dense, prickly clumps (Murphy *et al.* 2015). I see the major threat to be loss or degradation of habitat by wildfire and excessive grazing by stock or feral herbivores, for this deprives Night Parrots of food and exposes them to predation. However, I would not downplay the significance of predation, especially by feral cats, as a major threat, for there were reports of many Night Parrots being killed by feral cats at Alice Springs in 1892, and the spread of feral cats was put forward as the cause of disappearance of the parrots from northwestern Victoria.

The study site in central-western Queensland is in Pullen Pullen Nature Reserve, and management will be focused on protection of the 'rediscovered' population of Night Parrots, with a feral cat control program being implemented. To afford extra protection an exclusion zone has been set up around the Reserve, and this prohibits entry by unauthorised persons.

Previously, I noted that a paucity of authenticated records may not be an accurate reflection of the scarcity of Night Parrots, but probably is due to the difficulty of finding extremely secretive, nocturnal birds in country with little or no habitation. Indeed, I suggested that these parrots may not be rare, but present in low numbers over a wide range, and a similar view was expressed by Schodde and Mason (1980). Despite confirmation of the continued existence of birds in southwestern Queensland and a continuation of unconfirmed sightings at widely scattered localities elsewhere, I have become less optimistic about the status of the Night Parrot. There is tangible evidence of the existence of birds only in western Queensland, and it seems that the loss or degradation of habitat and predation have brought about significant declines in overall numbers, with the species having been extirpated from some regions, including northwestern Victoria, western New South Wales and much of central Australia, where pastoral activities have been prominent. Consequently, I must concur with the assessment made by an expert committee that the Night Parrot is suffering a continuing decline (in Garnett *et al.* 2011).

The Night Parrot is listed as endangered under the Australian *Environment Protection and Biodiversity Conservation Act 1999*,

Plate 17
Night Parrot *Pezoporus occidentalis* (adult)

F. KNIGHT '09

and is listed on Appendix I to the Convention on International Trade in Endangered Species of Wild Fauna and Flora (CITES).

HABITATS A close association between Night Parrots and *Triodia-Plectrachne* grasslands features frequently in historical accounts of encounters with the birds. Many sightings have been made in these hummock grassland sandplains, usually characterised by widely spaced large clumps of spinifex with much intervening bare ground and without trees or with scattered eucalypts, acacias or cassias. Recalling his experiences with these parrots in the north of South Australia, Andrews (1883) referred to their close association with spinifex grasslands, but commented also on his collecting specimens at Cooper Creek, where the birds were observed to 'conceal themselves during the day in thick patches of shrubby samphire on the salt flats'. Parker (1980) suggests that these grasslands are visited seasonally by the parrots to take advantage of widespread seeding after rains, but chenopod-dominated flats associated with lake systems seem to be the preferred habitat during dry seasons. In early June 1979, also on Cooper Creek, due east of Lake Perigundi, in the far northeast of South Australia, four birds were flushed from *Sclerolaena intricata* and then went down in dense *Muehlenbeckia* lignum (Parker *in litt.* 1979).

Blyth (1996) notes that in Western Australia, most of the 25 reports of Night Parrots, dating from between the 1930s and 1995, are from areas consistent with the parrots being inhabitants of samphire and spinifex country in the arid zone, but there are unexpected sightings from chenopod-dominated lands near southern salt lake systems. Davis and Metcalf (2008) report the sighting of two, or possibly three parrots at dusk on 12 April 2005, at Minga Well, in the Pilbara region, where habitats ranged from samphire flats surrounding the Fortescue Marsh in the south to *Triodia*-covered foothills of the Chichester Ranges in the north, and in extensive belts of mulga *Acacia aneura* woodland some 300 m north of the spinifex-chenopod ecotone. The well was in an area degraded and open from intense stock pressure.

Menkhorst and Isles (1981) report that at Ross's Spring, in northwestern Victoria, Night Parrots were seen in open herbaceous flats surrounding the small permanent freshwater spring fed by groundwater and run-off from the surrounding dunes. Extending to the southwest was another flat area supporting mallee scrubland featuring red mallee *Eucalyptus oleosa*, green mallee *E. viridis* and white mallee *E. dumosa* with an understorey dominated by spinifex *Triodia irritans*, mallee heath *Brachyloma ericoides*, mallee pine *Callitris verrucosa* and acacias. All sightings were made in mallee scrubland with spinifex in the understorey.

The desiccated road-killed specimen obtained north of Boulia, in western Queensland, in mid October 1990, was beside the road in an area of low, sparse *Astrebla* grasses, *Calotis* burr-daisy and chenopods, with some areas of bare ground, but spinifex was not apparent in the immediate vicinity, and the only trees were along a watercourse several kilometres in the distance, with little standing water being present for some distance in either direction along the road, except possibly for bores or dams (Boles *et al.* 1990). It was not possible to ascertain whether this bird had been killed at the place where it was found or had been transported on the front of a vehicle for an undetermined distance before falling by the roadside. Between early 1992 and mid 1993, within the Mount Isa uplands, in northwestern Queensland, seven sightings were made in grasslands and low open woodlands, with or without spinifex *Triodia modesta*, on land systems ranging from black-soil plains to gravel-encrusted slopes, and no single plant species characterised these sightings (Garnett *et al.* 1993). In Diamantina National Park, the second desiccated specimen was found on a low dissected tableland near the head of a small creek bordered by sparse shrubland of gidgee *Acacia cambagei*, crimson turkey-bush *Eremophila latrobei* and desert cassia *Senna artemisioides*, with the nearest spinifex grasslands being about 4 km to the south and some 12 km to the north (McDougall *et al.* 2009). Also in western Queensland, the undisclosed site where the first photographs were taken was 'in a massive island of dense spinifex, isolated from similar habitats by a sea of gibber plain' (see *Australian Birdlife*, **2**(3): 26–28, 2013).

A good insight into habitat utilisation was obtained from radio tracking of the parrot captured by the recovery team at Pullen Pullen Nature Reserve in central-western Queensland (Murphy *et al.* 2015). The landscape comprised more-or-less discrete patches of dense spinifex separated by bare areas that act as natural fire-breaks, and there was no evidence of fires in satellite and aerial imagery dating back to 1951. During the day the radio-tracked bird roosted in burrows in clumps of spinifex, and on one occasion it was located roosting in one of just five clumps of spinifex on an open stony plain. At night it made regular and predictable movements into non-spinifex habitats, including chenopod-dominated systems on stony plains and perennial grasslands on alluvial deposits. In an area dominated by a sparse cover of gidyea burrs *Sclerolaena* sp., another bird was seen sitting near the head of a drainage line cut away from the side of a track by a grader.

MOVEMENTS It has been assumed that Night Parrots are nomadic. Andrews (1883) claimed that numbers fluctuated in response to seasonal conditions, with a profuse seeding of spinifex in response to rainfall resulting in the parrots being locally common, but in dry years little seed was available and they were scarce or absent altogether. There is evidence that the parrots obtain moisture from succulent vegetation, so the suggestion that movements are made in search of surface water may not be correct and, while food supplies are adequate, Night Parrots may be sedentary in areas of favourable habitat.

HABITS A large, dark eye is conspicuous in the first photographs of a live Night Parrot, and shows clearly that these parrots are nocturnal, a feature fully confirmed in the daily routine of the radio-tracked bird at the study site in Pullen Pullen Nature Reserve, central-western Queensland. Nighttime activities commenced at about 1845 hours when a few calls were exchanged with another Night Parrot that evidently was roosting nearby, and then the radio-tracked bird flew off on one of its regular and predictable nightly movements of up to 7 km to foraging areas. It returned only when the very first light appeared on the eastern horizon, and retreated into its daytime roost in a burrow in a spinifex clump. There was a period of 12 days when the bird could not be located, and then it was relocated only 600 m from where it had been roosting. An encounter with another Night Parrot is documented as follows:

> *At 10 pm one evening we happened to slowly drive past another Night Parrot on the side of a track, lying prone as if to avoid detection. We had time to turn the vehicle around and pull up with this bird in our headlights and proceeded to watch it through binoculars. It hopped to the side of the track, turned and then seemed to hop-run, a little faster this time, back to its original spot. It then turned again momentarily before flying off low into the darkness. We knew it wasn't 'our' bird*

*because there was no signal through the radio receiver and we couldn't see the short, shiny aerial of the transmitter. Later inspection of the site showed it to be an area dominated by a sparse cover of Gidyea Burrs (*Sclerolaena *sp.) and that the bird was sitting near the head of a drainage line cut away from the track by a grader. The Gidyea Burrs growing in that drainage were 31 per cent more succulent than those growing on the open plain nearby, leading us to speculate that the bird was there to chew on the leaves for moisture. This might help explain why we haven't been able to detect Night Parrots drinking at water points using cameras. Interestingly, the specimen collected in 1990 from the roadside north of Boulia also had Gidyea Burrs in its feathers.*

Monitoring activities of the radio-tracked bird confirmed habits reported in early accounts. Andrews (1883) noted that on one occasion a parrot was caught in a hut, where it may have been attracted by the light, and was placed in a box with a handful of dry grass; next morning the bird could not be seen because it had gathered the grass into a heap and had drawn out the inside stems to form a hole in which it was hiding. There were reports of Night Parrots sheltering in rocky caves and an elderly Aboriginal from northwestern Australia told Stephen Davies that they will burrow into sandy soil (Davies pers. comm. 1975).

In *The Mildura Cultivator* of Saturday 23 August 1913 Charles H. McLennan, writing under the pseudonym 'Mallee Bird', described the habits of Night Parrots in northwestern Victoria:

This bird, as its name implies, is nocturnal in its habits. The tall dense spinnifex grass is where this strange bird makes its home and often throughout the night it utters a plaintive whistling cry. Between thirty and forty years ago they were found in the spinnifex grass on the sand ridges around Pine Plains. The last I saw of the birds was some eight years ago, between Sunset Country and Koorlong. Of late years the bird has become somewhat of a rara avis, *mainly through the liberation in the mallee of many hundreds of domestic cats – to combat the rabbits ... Being terrestrial in their habits they are a difficult bird to flush in the daytime without the aid of a good dog. So far I have not heard of the nest being found, but the information that I received from the old men aboriginals was that the birds build their nest in the dense spinnifex grass.*

Walton recalls that at Ross's Spring, also in northwestern Victoria, between 1950 and 1959, these parrots were never seen voluntarily moving about during daytime, but sheltered in small tunnels or burrows under spinifex tussocks, from which they would run out and scatter like quail when disturbed (in Menkhorst and Isles 1981). They preferred to escape by running across the ground, although they could fly quite well, but if approached cautiously they were fairly quiet, often allowing an approach to within 10 m before flying a short distance to realight on the ground and then run away. When flushed by a dog, some birds were seen to perch on the lower branches of a mallee tree, less than a metre above the ground. Another report of perching birds comes from south of Cloncurry, northwestern Queensland, in early May 1993, at about 1730 hours, shortly before sundown, when two parrots were seen sitting low in a bush adjacent to the road (in Garnett *et al.* 1993).

In early June 1979, due east of Lake Perigundi, in far northeastern South Australia, parrots were flushed twice from underneath *Sclerolaena* bushes (Parker *in litt.* 1979). A single parrot rose up approximately 30 m in front of approaching camels, and later persons walking through the area disturbed three birds from the base of a bush. The first bird flew only 4 m or 5 m, low to the ground, before pitching into the base of dense *Muehlenbeckia* lignum, whereas the other three parrots flew much farther, at a height of 3 m to 4 m, and then also dropped down into dense lignum. Intensive searching failed to relocate any of the birds.

Most reports are of Night Parrots being seen singly or in pairs, though Andrews recalled that on occasions parties of up to eight or more were seen coming to water well after nightfall. At Ross's Spring, in northwestern Victoria, a party of five birds was the largest group observed by Walton.

The flight is said to be erratic, but without undulation, and the birds seldom rise more than a couple of metres above the groundcover vegetation. It has been claimed also that when flushed in daytime, these parrots are slow and uncertain in their movements, generally flying only a short distance to the shelter of tussocks or bushes. Whitlock (1924) recalled being told by Aborigines at Hermannsburg Mission, southern Northern Territory, that in daytime these parrots were poor fliers and could be run down and caught. Keartland flushed a parrot from a clump of spinifex during the day and noticed that it did not drop suddenly when alighting, as does the Ground Parrot *Pezoporus wallicus*, but glided down and then moved off along the ground at right angles (in North 1898). Schodde and Mason (1980) note that the disproportionately long, pointed wings and short tail are consistent with a capability of long-distance flying, and flight at nighttime apparently is strong, because Andrews claimed that birds would fly up to 8 km to water. On one occasion at the study site in Pullen Pullen Nature Reserve, a parrot flushed from its roosting burrow flew 90 m at a height of some 2 m, in much the same manner as a Ground Parrot, before turning sharply and dropping into a slightly larger run of spinifex (Murphy *et al.* 2015).

CALLS The earliest record of vocalisation comes from Andrews, who reported (in Campbell 1901):

The name given to this bird by the aborigines is 'Myrrumbing', from the supposed resemblance of their whistling note to the sound of that word. They have also a very peculiar croaking note of alarm whilst at the water, which much resembles the loud croak of a frog.

Of the captive bird at London Zoo, Murie (1868) reported 'that its voice was a double note, harsh and loud'. While in flight, a parrot flushed by Keartland was heard to give a short, sharp note, repeated several times (in North 1898). In northwestern Victoria, the parrots were said to utter 'a plaintive whistling cry', and McGilp (1931) quoted a claim made by John McDonald, of Cootanoorina Station, south of Oodnadatta, South Australia, that when coming to water at night, the parrots frequently uttered a low, disyllabic whistle. Vocalisations reported by Bourgoin were of a flushed bird giving a squeak as if it had been hurt, a far-carrying, mournful and prolonged whistle from birds flying to and from water, and a 'squawk' from fledglings begging for food (in Wilson 1937).

At the study site in Pullen Pullen Reserve, central-western Queensland, between August 2013 and February 2015, more than 4500 Night Parrot calls were recorded, and listening for the calls using automated recorders with follow-up computer scanning seems to be a practical and repeatable survey method

(Murphy *et al.* 2015). Normally there was a brief exchange of calls just prior to nightfall and then again just before dawn, but for three months after a period of above-average rainfall the parrots were quite vocal, and up to 200 calls were recorded during the night, with a peak in calling just after nightfall, followed by a quiet period of a few hours before calling picked up again at about midnight and continued until dawn. It was speculated that this time of intense vocalisation may have been associated with breeding. The 'frog-like' call described by Andrews was confirmed as sounding like *grriet* given as a single note or sometimes repeated, and was heard mainly as a short-range contact call, but was emitted also when birds were alarmed. A *ding-ding* call was heard during the first hour after sunset, and was likened to the tinkling notes of the Bell Miner *Manorina melanophrys*.

DIET AND FEEDING In early reports it was claimed that Night Parrots feed on seeds of grasses and herbaceous plants, with seeds of *Triodia* spinifex being a favoured food. To the south of Cloncurry, northwestern Queensland, between early 1992 and mid 1993, birds were observed pecking at the ground, so presumably they were feeding on small seeds, and near to one site seeds of purple nineawn *Enneapogon purpurascens* were found to be abundant (Garnett *et al.* 1993). Seeds of two *Sclerolaena* species were removed from amidst feathers of the desiccated carcass found beside the road north of Boulia, western Queensland, in mid October 1990, and it seems that the leaves of these plants may be an important source of moisture in times of drought (Boles *et al.* 1994; Murphy *et al.* 2015). The crop of the specimen collected by Bourgoin at Nichol Spring, Western Australia, was filled with spinifex seeds and seeds of 'limestone herbage' (in Wilson 1937). Murie (1868) noted that the captive bird at London Zoo showed a marked preference for greenfoods.

Schodde and Mason (1980) point out that an association with chenopodiaceous vegetation, an apparent preference for greenfoods in captivity and the habit of feeding out in fields and claypans at night suggest that Night Parrots may take the nutritious fruits and green shoots of chenopods. Furthermore, the bluntly rounded bill may be adapted to picking up seeds from the ground and to masticating succulent plant stems or fruits. A caking of earth on the inside of the upper mandible of some specimens suggests that the bill is used also for digging in the soil.

Andrews (1883) observed that, after drinking and briefly fluffing out their feathers, the parrots flew off to feed on spinifex seeds, returning to the waterhole two or three times during the night. Davis and Metcalf (2008) recall that at Minga Well, in the Pilbara region of Western Australia, at 1815 hours on 12 April 2005, three birds came to the well, but only two remained drinking during the observation period, having run forward after alighting some 1.5 m from the water's edge.

BREEDING Garnett notes that pairs were seen in an area to the south of Cloncurry, in northwestern Queensland, during April to June, a time when many seed-eating birds breed in tropical Queensland, and this could indicate that in this area Night Parrots also breed at that time (in Garnett *et al.* 1993).

At Pullen Pullen Nature Reserve, central-western Queensland, on 24 April 2016, a nest containing two eggs was found at the base of a clump of spinifex. It was a shallow scraped depression in the sand, approximately 150 mm in diameter and lined with short lengths of what appeared to be spinifex stems (Murphy *in litt.* and photograph 2016). Unfortunately, this nest was predated.

John McDonald of Cootanoorina Station, south of Oodnadatta, northern South Australia, described a nest from which he had taken four chicks (in McGilp 1931). It was in a clump of spinifex, the entrance being through a tunnel starting low down on the side and almost under the clump; at the end of the tunnel there was an enlarged cavity in which the birds had placed some small sticks. McDonald had seen three or four nests, one in a 'samphire bush', and the entrance was always through a tunnel into the centre of a tussock or bush, where a platform of small sticks was placed between the central stems; in no instance was a nest actually placed on the ground. McDonald claimed that the four white eggs were about the same size (ca. 33.0 × 23.0 mm) as those of the Crested Pigeon *Ocyphaps lophotes,* but I would expect them to be significantly smaller and approximately of the same measurements as eggs of the closely allied Ground Parrot *Pezoporus wallicus*.

Julian Ford was informed by an employee of Jeedamya Station that in April 1937, in breakaway country near Ghost Rocks and the southeastern shore of Lake Ballard, southeastern Western Australia, he and the manager found an adult and five young birds in a nest under a large spinifex bush on a spinifex flat (in Storr 1986).

Ives (1971) described a possible nest found on 14 July 1970, approximately 16 km northeast of Balfour Downs homestead, in the Pilbara region of Western Australia. It was in a large dormant spinifex tussock, a little over a metre in diameter and a metre in height. From ground-level, at the edge of the tussock, a tunnel, approximately 80 mm in diameter, ascended slightly where the spinifex stems were densest and then descended to ground-level at the centre of the tussock, where it terminated in a chamber about 25 cm in diameter. The floor of this chamber was actually on the ground, which had been scraped to a depth of 12 mm and densely lined with 12 cm long terminal sections of spinifex leaves. In a longitudinal arrangement, similar leaves were scattered along the floor of the tunnel. These leaves were well frayed and clearly had been chewed from a tussock, but not the tussock in which the nest was situated. Tracks in the soil, at the approach to the tunnel, suggested that the nest was being built or had been completed recently. On the following day, some 100 km to the northeast, a fleeting glimpse was obtained of four green parrots flushed from among spinifex, and the low swift flight coupled with the failure of a thorough search to relocate them, strongly suggested that they were Night Parrots.

On 6 August 1966, a pair of parrots and a nest were located by a patrol officer and his Aboriginal assistant at a small water-filled rockhole, known locally as Takutju, in the Lake Auld district at the southern edge of the Great Sandy Desert, Western Australia; this nest was in a large spinifex tussock about 20 m from the rockhole (Stewart *in litt.* 1969). There is a vague report of a nest with five eggs having been found in northwestern Victoria (Campbell 1901).

EGGS Mathews (1948) describes three eggs said to have been laid in captivity in 1862. He accepted the authenticity of this clutch simply because the scientific name and the date were inscribed on each egg and because the eggs were in the collection of Henry Munt, 'a most reliable man' who collected only white eggs. Invariably, the authenticity of this clutch has been rejected by subsequent authors, and I am of the same opinion. The eggs were oval-shaped, measured 25.2 (25.0–25.5) × 19.6 (19.2–20.0) mm, and presumably now are in the Natural History Museum at Tring, UK.

GENUS ***Neophema*** Salvadori

Neophema Salvadori, *Cat. Bds Brit. Mus.*, **20**, 1891, pp. xvi, 539, 569. Type, by original designation, *Psittacus pulchellus* Shaw.

Molecular analyses have confirmed that the small, ground-frequenting parrots belonging to this genus, together with closely allied *Neopsephotus,* form a clade with *Pezoporus* that is sister to the 'core platycercines' (Joseph *et al*. 2012). In *Neophema* parrots the tail is graduated, but the central feathers and next two pairs are of equal length. The upper mandible is not distinctly notched, and the white underwing-stripe is variable. Within this genus there are two species groupings, which are differentiated by patterns of plumage colouration and sexual dimorphism, and I follow Schodde (1997) in treating these groupings as subgenera.

The genus is widely distributed through central and southern Australia, including Tasmania, islands in Bass Strait and offshore islands. The migratory Orange-bellied Parrot *N. chrysogaster* is critically endangered, but other species are secure.

SUBGENUS ***Neonanodes*** Mathews

Neonanodes Mathews, *Novit. Zool.*, **18**, 1912, p. 279. Type, by monotypy, *Psittacus chrysogaster* Latham.

A prominent blue frontal band and absence of red are features of the plumage colouration in the four species belonging to this subgenus. Sexual dimorphism is very slight, and the pale underwing-stripe is variable. Juveniles are duller than adults, usually lacking the blue frontal band, and have yellowish lower mandibles, which soon change to the color of adults.

This subgenus is widespread through southeastern and southwestern Australia, including Tasmania, islands in Bass Strait and offshore islands.

CRITICALLY ENDANGERED

Orange-bellied Parrot

Neophema chrysogaster (Latham)

Psittacus chrysogaster Latham, *Index Orn*., 1, 1790, p. 97 (no locality = Adventure Bay, Tasmania, *apud* Schodde 1997).

OTHER NAME Orange-bellied Grass-Parakeet.

DESCRIPTION Other *Neophema* species often show orange abdominal markings, so care must be exercised when identifying Orange-bellied Parrots in the field. The most reliable distinguishing features for field identification are the bright grass-green upperparts and the unique 'buzzing' alarm call.
Length 22 cm. Weight 41–51 g.
ADULT MALE Entire upperparts, including crown and nape, bright grass-green; broad frontal band deep blue, bordered above by a faint pale blue line; sides of head bright green, merging into yellowish-green on lores, throat and breast; abdomen greenish-yellow, merging into yellowish-green on flanks and bright yellow on vent and under tail-coverts; variable orange patch on centre of abdomen; bend of wing, carpal edge and outer secondary-coverts deep violet-blue; inner secondary-coverts green, on outer webs washed greenish-blue; primary-coverts and outer webs of primaries dark violet-blue, latter narrowly margined pale yellow; underwing-coverts violet-blue; central tail-feathers above green, becoming greenish-blue towards tips, lateral feathers greenish-blue, becoming yellow towards tips; underside of tail yellow; underwing-stripe absent; bill greyish-black; iris dark brown; legs brownish-grey.
11 specimens: wing 107–113 (110.9) mm, tail 95–110 (103.0) mm. exp. cul. 10–12 (10.6) mm, tars. 15–17 (15.7) mm.
ADULT FEMALE Generally duller than male, particularly on upperparts where there are scattered dull green feathers; narrower, slightly paler blue frontal band with upper bordering pale blue line replaced by indistinctive paler edging; less orange on centre of abdomen; underwing-stripe variable.
5 specimens: wing 106–108 (107.4) mm, tail 97–104 (99.8) mm, exp. cul. 10–11 (10.2) mm, tars. 15–16 (15.8) mm.
JUVENILES Noticeably duller than adult female, though upperparts still decidedly brighter green than in similar species; very faint dull blue edging to some frontal feathers, but disappearing with wear; smaller, ill-defined orange abdominal patch; underwing-stripe present; females duller than males and with broader underwing-stripe.

DISTRIBUTION Tasmania and some offshore islands, some islands in Bass Strait, and coastal areas of extreme southeastern mainland Australia from southeastern South Australia east to central Victoria, or extralimitally to eastern Victoria or possibly southeastern New South Wales.

Stephenson (1991) points out that the breeding range is confined to a narrow coastal strip of southwestern Tasmania, from Macquarie Harbour south to South West Cape, and most pairs nest within 20 km of Melaleuca, near Port Davey. Elsewhere in the state, Orange-bellied Parrots now are recorded mainly in western coastal districts and on King Island, but in past years occasional reports have come from other localities. In November 1977, a bird was observed in high country near Margate, and in the summer of 1976–1977 there was an unconfirmed record from

the southern shores of Lake Pedder (Newman *in litt.* 1978). Brown and Wilson (1980) note that there have been a few records from the southeast, where formerly the species was more widespread. There are only sporadic records from the northeast, including sightings at Cape Portland in September 1977 and at Musselroe Bay in September 1978 (in Higgins 1999). Green (1969) points out that the species does not occur on islands in the Furneaux Group, in Bass Strait, but Littler (1910) mentioned a sighting of birds on Ninth Island, off the northeastern coast, in 1909.

In Victoria, the wintering range extends along the coast east to Westernport Bay, and occasionally to Corner Inlet or Jack Smith's Lake. In southeastern South Australia, these parrots overwinter in coastal districts west and north to about Lake Alexandrina and the Coorong, rarely reaching Adelaide, where two birds were seen at the ICI Saltworks in early February 1978, and Fleurieu Peninsula. Extralimitally, they have occurred as vagrants to Spencer Gulf, where a single bird was seen near Port Augusta during August–September 1992 (Klau and Langdon 1994). Another report from west of St Vincent Gulf dates from late December 1886, when birds were seen in the Edithburgh district, southernmost Yorke Peninsula, but this could be attributable to misidentification because it seems most unlikely that Orange-bellied Parrots would have been there at that time of the year (see Jarman 1965).

The former occurrence of a remnant or possibly isolated population in the Sydney region, New South Wales, certainly is anomalous, and has been subjected to much discussion (Cooper *et al.* 2016). Morris and McGill (1980) suggest that the birds may have been escaped cagebirds, but this seems unlikely for the species always has been extremely rare in captivity. Early Sydney records date from a brief period between the late 1880s and 1907, and five specimens were collected. North (1911) reported that three birds were taken at Long Bay, in Sydney Harbour, while at nearby Middle Head, J. A. Thorpe collected a male and female that were 'breeding in a low hollow stump'. I could locate only one specimen, which almost certainly is one of the two collected by George Masters at Long Bay, and there is no doubt about its identity (B1499, Macleay Museum, University of Sydney). The nesting pair said to have been collected at Middle Head apparently have been lost, so the breeding record cannot be verified. A second specimen from New South Wales is much more interesting because it is labelled as a 'male in moult', but is a juvenile and, according to the label, it was collected at Thredbo on 15 January 1917 (ANWC B41281). It comes from the D'Ombrain Collection, but unfortunately we have no additional data. Mid January certainly would be very early for a juvenile from Tasmania to reach the mainland and come as far north as Thredbo.

North was informed that in the late 1880s, these parrots were numerous about Penshurst and Blacktown, now densely populated suburbs of Sydney, and in 1907 two were captured at Riverstone. Althofer (1934) claimed to have seen a single bird near Wellington, but I concur with McGill (1960) in attributing this record to misidentification of the Turquoise Parrot *Neophema pulchella*, and it is likely that birds recorded about Penshurst, Blacktown and Riverstone also were Turquoise Parrots. Two recent records are of a bird observed at Comerong Island in the Shoalhaven River estuary, on the south coast, in June 1988, and a juvenile seen and photographed at Shellharbour and Maroubra in September 2003. It is possible that in the 1800s to early 1900s, when Orange-bellied Parrots were more numerous, the mainland wintering range was more extensive, with birds regularly overwintering along the south coast of New South Wales, occasionally ranging north to around Sydney or reaching subcoastal districts. The two recent records suggest that it is their very low numbers that now normally restrict occurrences to a much reduced wintering range.

STATUS Between 1838 and the early 1900s, there were reports of Orange-bellied Parrots being abundant, suggesting that irruptions may have occurred in certain areas or throughout the range, with birds moving as far afield as the Sydney region. However, none of these reports can be confirmed and I suspect that references to 'flocks of thousands' could be exaggerated or could refer to mixed species aggregations involving mainly Blue-winged Parrots *Neophema chrysostoma*. Undoubtedly, Orange-bellied Parrots were more numerous and widespread in the early 1900s, but the decline seems to have commenced at about that time. Between 1838 and 1886, they were found to be common in the Actaeon Islands, at the mouth of the D'Entrecastaux Channel, southeastern Tasmania, and were recorded occasionally around Hobart and New Norfolk, but by the early 1920s had disappeared from all of these localities. Jarman (1965) pointed out that the early history of these parrots in Victoria is poorly documented, but much of what now is known of their status in the wintering range comes from annual counts that have been undertaken there and in neighbouring southeastern South Australia since 1979, and these counts are the basis for estimates of the total population. Between 1979 and 1997, the minimum numbers of parrots seen at wintering sites in coastal Victoria averaged 58.2, but there was a very significant decline from a count of 100 in July 1983 to 32 in July 1991 (in Higgins 1999). It seems that during the 1920s, Orange-bellied Parrots were fairly common in parts of southeastern South Australia, where flocks of 20 to 30 were encountered regularly, usually late in the year, but numbers decreased thereafter and smaller flocks were recorded less frequently. However, McKechnie recalled that in 1940 there arrived at the Adelaide bird markets about 40 juveniles said to have been trapped near the lower Murray River lakes or at the Coorong (in Jarman 1965). Between 1979 and 1997, counts of overwintering birds in southeastern South Australia, averaged only 9.6 birds, and varied from nil in each July of 1987, 1991, 1992 and 1997 to 28 in July 1989, and in July 2005 the only record was of a single bird seen in the Coorong National Park (in Higgins 1999; *Wingspan*, **15**(4): 6, 2005). In later years, nearly all sightings have been in or near the Coorong. In May 2007, five birds were seen on Younghusband Peninsula, in the Coorong, and these possibly included four birds seen earlier in the month on an island in the Coorong (Rogers 2010). Very few, if any birds have reached southeastern South Australia in recent years (Menkhorst pers. comm. 2015).

Concern for the plight of Orange-bellied Parrots was expressed by Mathews (1917), but indications of their rarity did not become apparent until Jarman summarised historical and recent reports. In 1978, attention was focused on their status in Victoria with the announcement that a major petrochemical plant would be established at Point Wilson, the best-known wintering locality. This prompted a study of the parrots, and in the winter of 1978, when a co-ordinated survey was carried out in known and potential wintering areas along coastal Victoria

Plate 18
Orange-bellied Parrot *Neophema chrysogaster*
UPPER adult ♂ LOWER juvenile

F KNIGHT 2009

in an attempt to obtain an estimate of the total population, between 83 and 97 birds were located (Loyn pers comm. 1979). In August 1979, a meeting of representatives of government and non-government agencies selected the Orange-bellied Parrot as a species deserving of a comprehensive conservation program, and that provided the impetus for co-operative efforts to protect the birds and their habitats. A significant achievement in 1984 was publication of the *Orange-bellied Parrot Recovery Plan*, the first national recovery plan for a single species in Australia, and in 1991 there followed publication of the *Orange-bellied Parrot Recovery Plan: Management Phase*. Largely in fulfilment of recommendations in these plans, wintering habitat was managed in Victoria and was protected within a reserve on King Island. Research was undertaken to enhance saltmarsh vegetation through propagation of favoured food plants, especially oak-leaf goosefoot *Chenopodium glaucum*. A monitoring of population trends was maintained through continuation of the annual winter counts, and sightings of marked birds were appraised to determine survival rates and migratory patterns. A captive breeding program was established at Taroona in Tasmania and at Healesville Sanctuary in Victoria. Despite setbacks with mortality from psittacine circoviral disease (parrot beak and feather disease), successful breeding was accomplished, and at regular intervals captive-bred birds were released to augment the wild population.

Loss of wintering habitat was identified as a significant factor in the decline of the wild population, so high priority was given to expanding and improving the vitality of saltmarsh vegetation at the main wintering sites. Competition from other species, notably introduced Goldfinches *Carduelis carduelis* and Greenfinches *Carduelis chloris*, for favoured food plants is significant in some parts of the wintering and migrating ranges, and human disturbance of beachfront habitats also is a problem at some sites, particularly in southeastern South Australia. The known breeding range is entirely protected within the World Heritage Area of southwestern Tasmania, but competition for nesting sites from Common Starlings *Sturnis vulgaris* was identified as a potential problem if their numbers increase locally, so nestboxes were provided to counteract this potential competition and to increase the nesting success rate (Brown pers. comm. 1999). Another potential problem in the breeding range is possible hybridisation with Blue-winged Parrots; photographic evidence of an interspecific pairing was obtained at Melaleuca in late December 2010, of a male Orange-bellied Parrot feeding a female Blue-winged Parrot (Joseph pers. comm. 2011). I suspect that any progeny from interspecific pairings would not be fertile.

Recovery criteria set out in the *Orange-bellied Parrot Recovery Plan: Management Phase* (1991) included the establishment within five years of a stable wild population of 250 birds supplemented by annual releases of at least 10 captive-bred birds, and within 10 years to have a self-sustaining wild population of 400 birds with sufficient secure wintering habitat and food supplied. Unfortunately, achieving these objectives has proved elusive and, despite the best efforts of all involved with the program, the decline in numbers has continued.

Findings of a long-term capture-mark-recapture study using capture histories of 848 known-age wild parrots indicate that maximum survival occurs in the second year of life, but declines thereafter (Holdsworth *et al.* 2011). Survival rates for both adults and juveniles varied considerably across the 20 years of the study, but there was no evidence of a decline in survival. However, between 2000 and 2008 there was an annual decline of 12 per cent in numbers of adult birds observed each year at the breeding grounds, so current survival rates do not appear to be a factor inhibiting population growth. It was suggested that the observed decline is more likely to result from a decline in female participation in breeding resulting in a decrease in the recruitment of juveniles. It seems that the decline in female participation in breeding is due to a serious sex imbalance in favour of males in the wild population.

Although precise numbers of Orange-bellied Parrots surviving in the wild may not be known, numbers seen during counts at wintering sites and observed in summer at the only known breeding location in the Port Davey–Melaleuca area of southwestern Tasmania indicate that it is less than 50 birds, and possibly less than 30 birds. Approximately 300 birds are held in captivity, and at regular intervals captive-bred birds have been released at Melaleuca to augment the wild population. In the 2007–2008 summer 80 captive-bred birds were released into the wild, and in November 2013, when just 17 birds returned to Melaleuca, 24 captive-bred birds were released there, but in November 2015 only 23 birds returned to Melaleuca, so releases of captive-bred birds are not achieving the desired results. The annual survival of wild adults in 2015 was 0.57, which is lower than the average over the previous five years, but similar to the long-term average, and the annual survival of juveniles was 0.12, which is lower than the average over the previous five years and the period 1996–2005, most likely reflecting an outbreak of psittacine beak and feather disease at Melaleuca in 2014–2015.

I am not hopeful that with the current program the wild population of Orange-bellied Parrots can be saved. Maintenance of that population, even in the short term, will depend on regular releases of captive-bred birds, but I am not confident that these releases can sustain the population in the long term, and a new strategy needs to be considered. Maintaining, or preferably increasing the captive population should be a priority, for that population provides the best safeguard against extinction while a reassessment of long-term actions is formulated. Expert advice could be sought on the possibility of eventually having a non-migratory captive population from which birds could be translocated to areas of suitable habitat with all-year food availability.

The Orange-bellied Parrot is listed as critically endangered under the Australian *Environment Protection and Biodiversity Conservation Act 1999*, and is afforded special protection under state legislation. Also, it is listed on Appendix I to the Convention of International Trade in Endangered Species of Wild Fauna and Flora (CITES).

HABITATS Orange-bellied Parrots occur almost exclusively in coastal and subcoastal areas, but are less restricted to littoral habitats than are Rock Parrots *Neophema petrophila*. Stephenson (1991) points out that in southwestern Tasmania their breeding habitat lies to the west of the central mountains and plateaus, and is a mosiac of temperate forest, rainforest and extensive moorland plains intersected by wooded creeks and rivers. This vegetation mosaic is determined by complex interactions of weather, fire, soils and drainage. Of primary importance as foraging habitats are the moorlands, where fire regulates the height and density of the vegetation which is dominated by sedges and heaths, including buttongrass *Gymnoschoenus sphaerocephalus* and principal food plants, notably branching scale-rush *Sporadanthus tasmanicus,* flat-stemmed cord-rush *Eurychorda complanatus,* lemon-scented boronia *Boronia citriodora,* swamp boronia *B. parviflora,* tiny alpine everlasting *Helichrysum pumilum* and tiny flannel flower *Actinotus bellidioides*. These plants fruit most profusely and are most accessible to the parrots at between three and five years after burning. Other important food plants occur amidst taller vegetation bordering creeks, and roosting sites are in scattered *Melaleuca* and *Leptospermum* thickets. Adjacent forests are dominated by mature Smithton peppermints *Eucalyptus nitida,* favoured for nesting, with some swamp gums *E. ovata* occurring along rivers or larger creeks, but extensive tracts of temperate rainforest are avoided by the parrots.

While migrating along the west coast of Tasmania, Orange-bellied Parrots frequent littoral heathlands and low scrublands behind the foredunes as well as sheltered grassy areas among rocky foreshores. Green (1984) recalls that at Ordnance Point, on the northwestern coast of Tasmania, during a biological survey conducted in March–April 1981, two birds were observed resting and feeding amongst shingles on the shore of a sheltered bay. On islands in the Hunter Group, in Bass Strait, parrots have been recorded in saltmarshes in the broad intertidal zones and in fringing *Melaleuca-Leptospermum* scrublands, as well as in scrublands or recently burned heathlands on stabilised sand-dunes, and beyond the dunes in pastures with remnant stands of eucalypt woodland. On King Island, migrating birds frequent pastures infested with *Juncus* rushes and dandelions, but may be seen also in coastal sand-dunes or fringing groves of melaleucas, while at the Sea Elephant River estuary they favour the extensive saltmarshes which extend up to 5 km inland (in Jarman 1965; Stephenson 1991).

In their mainland wintering range along coastal Victoria to southeastern South Australia, Orange-bellied Parrots occur predominantly in low-lying, sheltered coastal areas, with more than 95 per cent of records being made within 5 km of the seaboard, the exceptions being at large lake systems such as Lake Connewarre, southeast of Geelong on the Bellarine Peninsula, Victoria, and Lake Alexandrina, near the Coorong, South Australia (Ehmke *et al.* 2008). Though differences in dependence on saltmarshes are evident in habitat preferences in eastern and western sectors of the wintering range, a number of features, in addition to availability of food plants, are common to favoured habitats. An affinity between the parrots and water has been evident in detailed habitat assessments. At the 30 foraging sites recorded in 2006–2007, the mean distance to water was 40 m, and birds routinely foraged at well-sheltered sites surrounded by water, including small islands and 'spits' around the edges of tidal estuaries, rivers, coastal wetlands, inundated saline pastures, drains and sewage lagoons. In saltmarshes they favoured areas with narrow avenues of bare ground dissecting expanses of beaded glasswort interspersed with taller rushes, sedges or sea-blite *Suaeda australis*, and in rough, weedy pastures they selected sites where taller rushes and clumps of grasses or herbs gave good protection from cold winds and cover from predators. It has been suggested that changing climate may be forcing the parrots to forage in moister areas of saltmarshes and coastal wetlands or pastures. Less rainfall deprives saltmarshes of freshwater, and this may reduce seed availability or palatability for the parrots, so explaining the recent shift away from feeding in drier saltmarsh communities to feeding in the brackish-freshwater interface, where the parrots are feeding in the more luxurient, freshest areas of their saltmarsh habitat.

In the 1980s, my encounters with Orange-bellied Parrots were at a traditional site on the western shore of Port Phillip Bay, approximately 50 km southwest of Melbourne, where in those years good numbers of parrots frequented an area that included natural dry saltmarsh, improved pasture and artificial lagoons on the southern edge of a vast sewage farm, while coastal features included a muddy tidal lagoon backed by flooded saltmarsh and sheltered by two mobile sandy spits which support dune and saltmarsh vegetation (see Loyn *et al.* 1986). This area, which encompasses the Western Treatment Plant and The Spit Nature Conservation Reserve, is the only location in Victoria where Orange-bellied Parrots were located during winter counts conducted in 2013, 2014 and 2015 (Menkhorst *in litt.* 2015). At this site, I observed birds feeding among halophytic plants on tidal flats and then at high tide retreating to nearby pasture paddocks. No parrots have been seen at Lake Connewarre in recent years, but in the past they frequented saltmarsh at the edge of the lake and on islands in the lake, and the most favoured stands appeared to be those that were established as the delta islands expanded by sedimentation (in Starks *et al.* 1992). The saltmarshes and islands of Swan Bay, in southernmost Port Phillip Bay, also were traditional wintering habitats, and on Swan Island the parrots regularly fed and rested on the golf course, but again these sites have not been occupied in recent years. Farnes (2007) reports that in far southwestern Victoria, there are records from the late 1980s and early 1990s of small flocks feeding in dunes, scrubs, saltmarshes and pastures along the coastal fringe of Discovery Bay and in the vicinity of Lake Yambuk, where six birds were seen in May 2005, with records coming also from saltmarshes west of the Glenelg River and near the mouth of the Fitzroy River.

In southeastern South Australia, it seems that more use is made of coastal dune systems, where substantial beds of sea rocket *Cakile maritima* and sea spurge *Euphorbia paralias* are prevalent above the high tide mark, and food plants such as bidgee-widgee *Acaena novaezelandiae* are present in the dune swales. Stephenson notes that at Carpenter's Rocks, near Port Macdonnell, the parrots have been seen along the beach and occupying a nearby roost dominated by coast wattle *Acacia sophorae* in association with coast beard-heath *Leucopogon parviflorus* and coastal daisybush *Olearia axillaris*. In early June 1985, at Lake Eliza, I encountered five Orange-bellied Parrots in the company of small parties of Elegant Parrots *Neophema elegans* in lakeshore marshland and open grassland with scattered low shrubs. Eckert (1990) recalls that in mid July 1981 and early June 1983, near Lake Alexandrina, a few Orange-bellied Parrots were seen with an occasional Rock Parrot *Neophema petrophila* and mixed flocks of Blue-winged Parrots *N. chrysostoma* and Elegant Parrots feeding in paddocks of sunflower stubble and in a standing sunflower crop. Klau and Langdon (1994) report that at Chinaman's Creek Research Station, approximately 23 km south of Port Augusta, on 3 August 1992, a single Orange-bellied Parrot was seen with a flock of Blue-winged Parrots and several Rock Parrots feeding amidst samphire, chenopods and scattered *Cakile* plants behind *Nitraria* bushes above the high tide mark.

MOVEMENTS Orange-bellied Parrots are migratory, undertaking regular seasonal movements between their breeding range in southwestern Tasmania and wintering grounds along the mainland coast from southeastern South Australia to central or eastern Victoria. Very occasionally, a few birds will remain in Tasmania or on King Island throughout the winter, and a lone bird is known to have remained on the wintering grounds near Point Wilson, in Port Phillip Bay, Victoria, during the summer of 1978–1979. Departure from the breeding range occurs in two phases by differing age groups, with adults leaving in late February to early

March, and post-fledging flocks of juveniles usually departing between mid March and late April (Brown and Wilson 1984). Unpaired adults and probably unsuccessful breeders are the first birds to leave in February, and it is possible that some pairs may leave as early as mid December following nest failures.

Northward migration takes place up the west coast of Tasmania, through the western Bass Strait islands, and across to the mainland coast, with a majority of birds apparently passing through King Island, where adults commence arriving in mid March followed by juveniles in April. Birds regularly stop over on King Island to feed in saltmarshes or pastures, and in some years small numbers may overwinter there; in the winter of 1959, when parrots did not arrive at the usual wintering sites in Victoria, 75 birds were recorded on King Island (in Starks *et al.* 1992). First arrivals at mainland wintering sites usually are recorded during the last week of March, though occasional records as early as the beginning of February suggest that at times the first birds may bypass King Island to fly diagonally across Bass Strait from northwestern Tasmania to Victoria. Between 1983 and 1997, the earliest recorded dates for birds at any of three major wintering sites in Victoria varied from 2 February in 1988 and 1991 to 10 May in 1987 (in Higgins 1999). Most early arrivals in Victoria are of single birds or small parties, and only after early April are groups of 10 or more birds recorded. At the main wintering site along the western shore of Port Phillip Bay, Victoria, numbers slowly build up to reach maximum levels in late July to mid August. Arrivals in southeastern South Australia usually are later, with the first birds normally being seen in April or early May, but there can be marked variations from year to year; in 1988 an unusually early record was made on 29 February, whereas in the following year the first bird was not seen until 4 June (Starks 1992). At Carpenter's Rocks, near Port Macdonnell in southeastern South Australia, first arrivals usually are seen in mid July (Starks *et al.* 1992).

Stephenson (1991) points out that departure from mainland wintering sites commences in September, and the last birds usually have left by November. Between 1983 and 1997, the latest recorded dates for birds at either of two major wintering sites in Victoria varied from 9 September in 1990 to 7 November in 1987 (in Higgins 1999). Return to the breeding range is undertaken more rapidly than the northward migration, and records from King Island suggest that returning birds pass over the island or make only brief stopovers; between 1958 and 1964, McGarvie obtained only a single record late in the year, when a group of six parrots was seen on 30 October 1961 (in Jarman 1965). An examination of specimens in the South Australian Museum reveals that most were collected from flocks encountered in the Robe district, on the coast of southeastern South Australia, during late October to early November, so suggesting that this locality may be a departure point for flights across Bass Strait. Farther south, at Carpenter's Rocks, near Port Macdonnell, between 1979 and 1997, departure usually was in late August or early September (in Higgins 1999). Brown and Wilson (1984) recall that in late September 1980, four birds were seen flying out to sea at Point Lonsdale, at the entrance to Port Phillip Bay, Victoria, and Littler (1910) reported that in late September 1909, a pair and then six parrots were seen to come to Ninth Island, off the northeastern coast of Tasmania, rest briefly and then fly off towards the Tasmanian coast. Earliest arrivals at the breeding range are in late September, and maximum numbers are present by mid November (Brown and Wilson 1984). The shortest interval between sightings of a banded bird on the mainland and in the breeding range is four days (in Higgins 1999).

Starks points out that sightings of banded birds suggest a fidelity to wintering sites; of 14 banded birds seen in more than one non-breeding season, only one changed sites (in Higgins 1999). Another bird banded as an adult in early 1992 was observed on Swan Island, western Port Phillip Bay, in 1992, 1993 and 1994, also demonstrating a high degree of wintering site fidelity, and in 1995 possible confusion in identifying band colours cast some doubt on a probable first record of a parrot moving between major wintering sites in the same season.

I suspect that, like other neophemas, Orange-bellied Parrots probably travel extensively at night while on migration. At 2300 hours on an April night in 1905, one parrot struck the lighthouse at Cape Wickham, northernmost King Island. Not unexpectedly, all recoveries of banded birds have been at the place of banding in southwestern Tasmania, because that is where all trapping takes place, but sightings of colour-banded birds have confirmed trans Bass Strait migration by wild birds and by captive-bred birds released into the wild (see Loyn *et al.* 2005). The colour-banded juvenile observed and photographed at Maroubra, New South Wales, in September 2003 was a nestling when banded in the previous summer at Melaleuca, southwestern Tasmania (see *Wingspan*, **13**(4): 41, 2003).

HABITS Much of what we know of the habits of Orange-bellied Parrots comes from writings and reports dating from times when the birds were far more numerous. At their wintering grounds or during migration, these parrots occasionally were seen in flocks, sometimes comprising up to 50 or more birds but normally they were well dispersed as small groups of 10 or less individuals. Large groups tended to remain separated from other species, but single birds, pairs or small parties often fed in the company of other neophemas, Goldfinches *Carduelis carduelis*, Greenfinches *C. chloris* or House Sparrows *Passer domesticus*. While field studies were undertaken between 1978 and 1980, at the main wintering site on the western shore of Port Phillip Bay, Victoria, it was noted that the parrots occupied a nighttime roost in extensive 1 m to 3 m high shrublands of shrubby glasswort *Tecticornia arbuscula* in flooded saltmarsh and within 3 km of all feeding areas (Loyn *et al.* 1986). Departure of small groups from this roost occurred from about 40 minutes before dawn to a few minutes after dawn, and the birds usually gathered in larger groups to feed in one or more areas. Feeding was almost continuous throughout the day, being interrupted only if the birds were disturbed or when they paused to drink, bathe or move to new feeding areas, these latter activities occupying less than half an hour in mid winter but longer intervals during the warmer weather of autumn and spring. Flights to new feeding areas usually took place within two hours of midday, or at other times if the birds were disturbed. Visits to drinking sites were made immediately after leaving the nighttime roost, in mid morning or in the early afternoon, and it appeared that each bird drank only once per day, though it is probable that moisture was obtained from succulent vegetation or more probably from puddles and freshwater channels within the feeding areas. In pairs or groups of up to10, the parrots came to a drinking site, where they sat quietly nearby for up to 10 minutes before approaching the water and, unless the presence of several puddles allowed simultaneous drinking, they usually drank one at a time. Return to the nighttime roost, usually in larger groups than at departure, occurred from 40 minutes before sunset to 10 minutes after sunset, and the birds remained quiet after settling down for the night.

When these parrots are feeding on the ground or amidst low shrubby vegetation, their plumage blends well with the surrounds, often making detection difficult. They are rather timid, seldom allowing a close approach, and when disturbed they rise high into the air, calling loudly, before dropping down again to move away just above ground or water level, soon disappearing from sight. Stephenson (1991) notes that after returning to the breeding area in southwestern Tasmania, these parrots congregate in communal roosts in dense heath and 2 m to 3 m high *Melaleuca* copses along certain creeks. Soon after sunrise, pairs leave these roosts to go to their nesting sites, and the day is spent preparing the hollows for laying, interspersed with bouts of foraging.

Orange-bellied Parrots are capable of flying long distances. Their strong, rapid flight resembles that of the Rock Parrot *Neophema petrophila* in having a jerky side-to-side rocking motion, and is undertaken with rapid wingbeats interspersed with brief glides on downcurved wings, producing a slightly undulating pattern. When disturbed, these parrots characteristically rise high into the air before again dropping down and flying off at a low level.

CALLS The normal contact call, often given in flight or occasionally by a perched bird, is a monosyllabic sharp *tzit* repeated at intervals of one to three seconds, and Stephenson (1991) points out that typically it is given at the apex of the undulating flight pattern. When alarmed, the parrots emit a sharp *chitter-chitter . . . chitter-chitter*, repeated so rapidly as to produce an overall 'buzzing' effect, and this is a valuable aid to identification. Feeding may be accompanied by soft warbling notes.

DIET AND FEEDING The diet comprises seeds, fruits, berries and flowers of sedges, herbaceous and halophytic plants, and grasses procured mostly on or near the ground and amidst the foliage of lower shrubs. Brown and Wilson (1984) report that in southwestern Tasmania, Orange-bellied Parrots have been observed to hold food plants down on the ground while extracting seeds, to clamber through thick clumps of sedges and hold many stems in one or both feet while taking seeds or fruits, to energetically climb over the tops of boronias or other tall heath plants while stripping flowers or ripening seeds, and to flutter up from the ground to grasp stalks of buttongrass *Gymnoschoenus sphaerocephalus* and then balance with the aid of flapping wings while taking the seeds. Upon arrival at the breeding grounds in spring, the parrots feed mainly on seeds of branching scale-rush *Sporadanthus tasmanicus* and flat-stemmed cord-rush *Eurychorda complanatus* in sedgelands aged between five and 12 years after burning, and then from mid November to January the seeds of lemon-scented boronia *Boronia citriodora,* swamp boronia *B. parviflora,* tiny alpine everlasting *Helichrysum pumilum* and tiny flannel flower *Actinotus bellidioides* are the main foods for nesting pairs (Stephenson 1991). From mid January, the parrots continue to feed from these plants, and also move into taller vegetation along creeks, where lemon-scented boronia and buttongrass are the key food plants. Towards the end of the breeding season, birds can be seen in littoral areas feeding on seeds of bidgee-widgee *Acaena novaezelandiae* and occasionally on seeds of coast tussock-grass *Poa poiformis*. Brown and Wilson (1984) note that feeding rates recorded during the breeding season varied from 15 to 30 pecks per minute, and in the early part of the season females fed at up to three times the rate of males.

During field investigations undertaken between 1978 and 1980 at the main wintering site on the western shore of Port Phillip Bay, Victoria, these parrots were seen to feed mainly by biting on ripe or developing seeds and fruits from low herbs and shrubs, mostly chenopods, and salty tissue surrounding seeds of many saltmarsh plants usually was discarded (Loyn *et al.* 1986). Feeding was on the ground or amidst the foliage of taller shrubs, especially shrubby glasswort *Tecticornia arbuscula*, and fallen seeds or fruits were taken from the ground only when other food sources were scarce. Food preferences changed with seasonal availability, and utilisation of food plants varied according to the timing of their fruiting or seeding or to when the plants were submerged by rainfall, spring tides or sewage irrigation. Most food plants flowered in late summer and carried seed through to early winter, so ample food was available to arriving parrots, with beaded glasswort *Sarcocoria quinqueflora*, sea-blite *Suaeda australis* and oak-leaf goosefoot *Chenopodium glaucum* being favoured food plants at this time. Food became progressively scarcer through the winter, and in July to early August the parrots fed on a wide range of plants, with much variation from year to year. Yugovic (1984) points out that grey glasswort *Tecticornia halocnemoides* is an important food source from late April to mid August, being utilised early in the season mainly when preferred feeding areas in wet saltmarsh or filtration paddocks are flooded by tides or irrigation, but from July to mid August it is considered to be particularly important during a period of food scarcity. Other foods taken at this time include flowers of wild turnip *Rapistrum rugosum*, fallen seeds of beaded glasswort and oak-leaf goosefoot, and seeds of curled dockweed *Rumex crispus*, buckshorn plantain *Plantago coronopus*, southern sea-heath *Frankenia pauciflora* and hastate orache *Atriplex hastata*. In mid August, good supplies of food become available from winter or spring flowering plants, and seeds or flowers of shrubby glasswort and seeds of capeweed *Arctotheca calendula* are principal foods until the parrots depart for the breeding range. Feeding rates vary according to food plants, with 24 pecks per minute being recorded when the birds are feeding on shrubby glasswort, 30 pecks per minute on grey glasswort, and 37 pecks per minute on sea-blite.

In southeastern South Australia, most reports have been of birds in coastal sand-dunes feeding on sea rocket *Cakile maritima* or bidgee-widgee *Acaena novaezelandiae*, or in pastures feeding on capeweed *Arctotheca calendula*. Eckert (1990) reports that at Tolderol Game Reserve, southeastern South Australia, in early June 1983, a few Orange-bellied Parrots were seen with an occasional Rock Parrot *Neophema petrophila* and mixed flocks of Blue-winged Parrots *N. chrysostoma* and Elegant Parrots *N. elegans* taking seeds from ripening seedheads of commercially grown sunflowers *Helianthus annuus*. Klau and Langdon (1994) recall that on 3 August 1992, at Chinaman's Creek Research Station, about 23 km south of Port Augusta, South Australia, a lone Orange-bellied Parrot was observed with several Rock Parrots and a flock of more than 60 Blue-winged Parrots feeding on coast groundsel *Senecio pinnatifolius*. At 1600 hours on 4 June 1985, at Lake Eliza, southeastern South Australia, I watched five Orange-bellied Parrots with pairs and small parties of Elegant Parrots feeding on the ground and apparently taking fallen seeds of nut grass *Cyperus rotundus*.

At Werribee, Victoria, Len Robinson picked up a freshly killed Orange-bellied Parrot, and in its stomach were seeds, including those of annual sea-blite *Suaeda maritima* and grey saltbush *Atriplex cinerea* (in Jarman 1965).

BREEDING Stephenson (1991) points out that in early October many of the first birds to return to breeding areas in southwestern Tasmania already are paired, and each morning, soon after sunrise, these pairs fly from communal roosts to the nesting trees, where part of the day is spent by females cleaning out and occupying their hollows interspersed with bouts of foraging. Some pairing occurs after arrival, with chasing and tail-fanning by males and courtship feeding being observed. On one occasion, the male of a pair that had been feeding together for about an hour was seen to preen the throat of the female, utter a sharp note while raising slightly his fanned tail, and spread his wings momentarily while facing the female before again preening her, a sequence that was repeated five times before both birds flew off (Brown and Wilson 1980).

The nest is in a hollow limb or hole in a living or dead eucalypt, with preference being shown for Smithton peppermints *Eucalyptus nitida*. Brown and Wilson (1980) report that of eight nests, one was in a dead swamp gum *Eucalyptus ovata* and the remaining seven were in living peppermints at heights of between 4 m and 25 m, with five being in holes in trunks and the remaining three in broken limbs (in Higgins 1999). Nest entrances were through ends of broken limbs, a hole on the upper side of a branch, an elongated split in a branch, or knotholes in trunks, with rounded entrance holes having diameters of 6 cm to 10 cm, the hole on the upper side of a limb being 9 cm × 18 cm, and the elongated split in a branch being 89 cm long and 7 cm wide. Depths of three nesting hollows were between 46 cm and 60 cm, and the internal diameter of one hollow was 10.5 cm. Hinsby (1947) noted that at Macquarie Harbour, southwestern Tasmania, one nest was used for six consecutive years.

At Melaleuca, nestboxes are being used increasingly by nesting pairs, and in the 2015–2016 breeding season nine nests were in nestboxes and two nests were in tree hollows. Two of the occupied nestboxes were in a lot of 10 nestboxes set up on poles, and three chicks fledged from one of these nests.

Nesting commences in early to mid November, and the female occupies the hollow for several days prior to egg-laying. Eggs are laid at intervals of two days, and Holdsworth reports that for one clutch the first egg was laid at 1100 hours and the second two days later at about 1500 hours (in Higgins 1999). Of 13 clutches, three were of three eggs, two were of four eggs, seven were of five eggs, and one was of 6 eggs, to give a mean clutch size of 4.5 eggs. Incubation by the female commences with laying of the first egg and lasts between 21 and 24 days. The sitting female is fed by the male at intervals of two to three hours. He flies up to 5 km from feeding areas to the nesting tree to call the female from the nest, either from a nearby perch or by poking his head into the entrance hole, and both then fly from the nest, the female usually following the male, to a favoured perch within 100 m of the nest. If laying is still in progress, copulation will precede his feeding the female by regurgitation up to 40 times; the female then returns to the nest and, after resting briefly, the male flies back to resume foraging out on the sedgelands. For the first 10 days after hatching, nestlings are brooded by the female, and she feeds them with food brought to the nest by the male. After 10 days, the female ceases daytime brooding, and joins with the male in feeding of the chicks. At two nests, where both parents were feeding chicks, 20 successive feeding visits were made at mean intervals of 59 minutes and 78 minutes (in Higgins 1999). Older chicks come to the nest entrance to be fed. Some four to five weeks after hatching, young birds, often with traces of down still adhering to their feathers, leave the nest, and recently fledged youngsters have been seen being fed by adults. Juveniles band together in foraging flocks, and their departure on northward migration takes place later than that of adults.

Holdsworth reports that in nestboxes set up at the study site at Melaleuca, southwestern Tasmania, in the 1996–1997 breeding season, 49 chicks hatched from 52 eggs to give a hatching success rate of 94 per cent, and 37 young birds fledged for a fledging success rate of 71 per cent, equalling 3.08 youngsters from each nest (in Higgins 1999). Although female breeding participation was high in the 2015–2016 breeding season, and the number of nests was consistent with recent previous seasons, nesting productivity was disappointing, with the average clutch size being a low 3.2 eggs. The hatching success was 52 per cent and the fledging success was 47 per cent to give an overall breeding success rate of only 24 per cent.

EGGS I have doubts about the authenticity of most eggs in museum collections. In the Tasmanian Museum, Hobart, there are clutches of eggs of both the Orange-bellied Parrot and the Blue-winged Parrot *Neophema chrysostoma*, collected by Swindells, Brent and Harrison at Bothwell and Melton Mowbray, southeastern Tasmania, in 1898 and 1899, but the possibility of misidentification cannot be dismissed. The clutch of four eggs of *Neophema chrysogaster* taken by Swindells at Bothwell measures 21.2 (20.3–22.5) × 17.5 (17.2–18.0) mm.

Also in the Tasmanian Museum there are two clutches, one of three eggs catalogued as coming from the Brent Collection and another of two eggs catalogued as originating from Victoria, with the collector identified as C. French and the date of collection as October 1897. However, all five eggs have the same Brent Collection set marking and date of collection (7 November 1899), so apparently are from the same clutch. It is such anomalies that give rise to my doubts.

In the South Australian Museum, Adelaide, there is a clutch of five eggs laid in captivity at Adelaide Zoo during December 1944 and presumably identification is reliable; these slightly glossy eggs are rounded to broad-elliptical and measure 21.6 (20.4–22.6) × 17.7 (17.1–18.0) mm. In the Campbell and White Collections, Melbourne, there are eggs from near Adelaide, South Australia, and obviously all are from aviary birds, though this is acknowledged only by Campbell.

Measurements of 22.0 (21.8–22.1) × 18.2 (17.6–18.6) mm are of five infertile eggs laid by captive birds in the present captive breeding program, and measurements of 22.8 (21.6–24.8) × 18.8 (17.2–19.9) mm are of 20 failed eggs from nests of wild birds at Melaleuca, southwestern Tasmania (Holdsworth *in litt.* 1999).

SUBFAMILY **LORIINAE** Selby

Parrots belonging to this subfamily are categorised in three readily discernible groups, which together were identified in molecular analyses as comprising a subclade within the Old World parrots – the brush-tongued lories and lorikeets, the somewhat specialised fig parrots, and the ubiquitous Budgerigar *Melopsittacus undulatus* (Wright *et al.* 2008). Each group here is accorded tribal status.

TRIBE **LORIINI** Selby

Parrots in this tribe are known collectively as lories or lorikeets. They are a readily recognisable, discrete group of nectarivorous parrots confined to the Indo-Pacific region, from Henderson and the Marquesas Islands, central-east Pacific Ocean, west to Mindanao, southern Philippine Islands, and Nusa Tenggara, Indonesia, and molecular analyses suggest that they originated in New Guinea about 10 million years ago (Schweizer *et al.* 2015). They have tight, glossy plumage and most species are brilliantly coloured, with green, red and blue predominating. They are birds of the treetops, rarely coming near the ground, and moving in noisy flocks from one stand of flowering trees to the next. They have anatomical modifications associated with the collection and ingestion of pollen, nectar and soft fruits, which make up their diet. The bill is relatively elongated and laterally compressed, while the tongue is tipped with elongated papillae, forming a 'brush-like' appendage. Steinbacher (1934) pointed out that the ventriculus or gizzard is weak and not muscular, while compound glands are arranged linearly along the walls of the proventriculus.

GENUS ***Charmosyna*** Wagler

Charmosyna Wagler, *Abh. k. Bayer. Akad. Wiss., Math.-Phys. Kl.*, **1**, 1832, p. 493. Type, by monotypy, *C. papuensis* = *Psittacus papuensis* Gmelin = *Psittacus papou* Scopoli.

Charmosyna lorikeets are small to midsized parrots with a slim body form and a sharply pointed, strongly graduated tail, with very elongated central feathers forming conspicuous streamers in one species. The bill is fine and pointed, and the unfeathered cere is prominent. Shaft-streaked feathers on the crown of some species are suggestive of the strongly shaft-streaked, erectile crown feathers characterising *Vini* species, and Amadon (1942) treated *Charmosyna* as a subgenus of *Vini*. Sexual dimorphism is slight or absent, and juveniles generally are duller than adults.

Distribution is centred in the New Guinea region, where a number of species occur sympatrically, but extends east to New Caledonia and the Fijian Islands and west to Maluku Province, Indonesia. In all parts of the range, these lorikeets are very difficult to observe amidst the forest canopy, especially the smaller species, and even in areas where they are known or suspected to be present in reasonable numbers there can be a marked scarcity of records, so contributing to difficulties in estimating population levels.

CRITICALLY ENDANGERED

Blue-fronted Lorikeet

Charmosyna toxopei (Siebers)

Hypocharmosyna toxopei Siebers, *Treubia*, **7**, suppl., 1930, p. 252, pl. 4 (Buru).

DESCRIPTION Length 16 cm.
ADULTS General plumage green, more yellowish on underparts and greater underwing-coverts; forehead green, forecrown blue, paler and less extensive in females; chin and throat greenish-yellow; yellow band across undersides of secondaries, more pronounced in females; tail above green narrowly tipped dull yellow, underside dusky yellow and basally marked orange-red; bill orange-red; iris yellow-orange; legs red-orange.
2 males: wing 85 mm and 90 mm, tail 73 mm and 73 mm, exp. cul. 13 mm and 13 mm, tars. 11 mm and 12 mm.
3 females: wing 83–88 (85.3) mm, tail 65–77 (70.3) mm, exp. cul. 12–12 (12.0) mm, tars. 11–12 (11.3) mm.
JUVENILES General plumage colouration darker and duller; chin and throat more greenish; well-defined yellow band on undersides of flight feathers.

DISTRIBUTION Confined to the island of Buru, South Maluku Province, Indonesia.

STATUS Difficulties in determining the status of small green lorikeets in forested habitats is amply demonstrated by uncertainties about the status of Blue-fronted Lorikeets on Buru, for it seems that they could be locally common though targeted searches repeatedly have been unsuccessful. Only the original specimens have been collected, and these were seven birds captured alive with 'bird lime' on the west side of Lake Rana and brought to L. J. Toxopeus, the collector with the 'Boeroe-Expeditie 1921–1922' (in Siebers 1930). There were no further records until Smiet (1985) identified as Blue-fronted Lorikeets the small

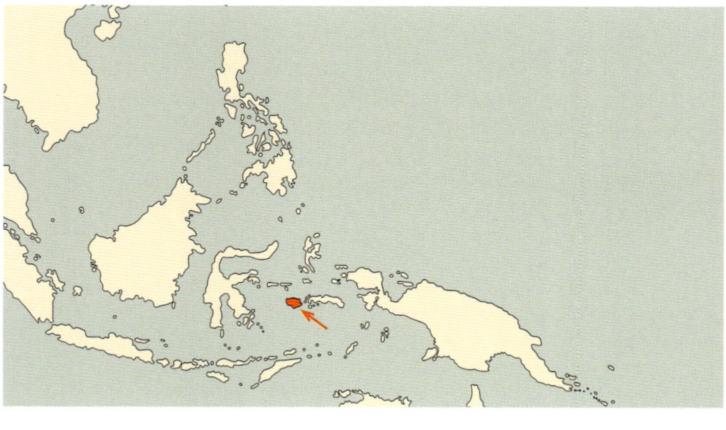

lorikeets observed in November 1980, when occasional pairs and more frequent small groups of up to 10 birds were found to be quite common in plantations, primary forest and secondary growth. An unconfirmed report in 1989, together with reports by local villagers in the 1990s, reports of birds being trapped in 1998 and a sighting of three small lorikeets in 2006 probably refer to this species (in Birdlife International 2016). Robson records confirmed sightings made in November 2014 when at least two birds initially, and a lone bird two days later, were seen feeding in the same tree, and one bird was photographed (in Birdlife International 2016).

Although montane forests remain largely undisturbed, most coastal lowland forest has been cleared, and much of the forest in northern parts of the island has been selectively logged or degraded by shifting agriculture. Also, there is at least one fairly large-scale logging operation on Buru. Because of this loss of lowland forest and the paucity of confirmed records, it is suspected that the total population of Blue-fronted Lorikeets is small, probably numbering 50 to 250 mature birds, and is declining (in Birdlife International 2016).

HABITATS The original specimens were collected in lower montane forest between 850 m and 1000 m, and the birds seen and photographed in November 2014 were at 1300 m. There is one aural record from a coconut plantation at or near sea-level, and recent observations of small lorikeets, presumed to be this species, have come from plantations and selectively logged primary and secondary forest at about 600 m. These records, together with anecdotal information provided by local villagers, suggest that Blue-fronted Lorikeets occur principally in lower montane forest, but in some years or in seasonal movements they come down into coastal lowlands (in Birdlife International 2016).

MOVEMENTS Information provided by local people suggests that these lorikeets come down to the lowlands in the hot months of March–April and August to November (in Birdlife International 2016). This is somewhat surprising because I would have expected them to avoid the humid lowlands at these times.

HABITS Very little is known of the habits of Blue-fronted Lorikeets. They have been seen singly and in pairs, but apparently they more commonly associate in groups of up to 10 birds. Toxopeus described the flight as being direct, though not particularly swift, and the tail is spread to reveal the orange-red markings on the underside (in Siebers 1930).

CALLS The call has been described as a very shrill *ti…ti…ti…ti… ti-ti-ti* (in Siebers 1930).

DIET AND FEEDING These lorikeets have been observed and photographed feeding at blossoms in flowering trees, where undoubtedly they gather nectar and pollen.

BREEDING There are no nesting records.

Plate 19
UPPER Blue-fronted Lorikeet *Charmosyna toxopei* (adult ♂)
LOWER New Caledonian Lorikeet *Charmosyna diadema* (adult ♀)

New Caledonian Lorikeet

Charmosyna diadema (J. Verreaux and Des Murs)

Psitteuteles diadema J. Verreaux and Des Murs, *Rev. et Mag. Zool.* (2), **12** 1860, p. 390 (New Caledonia).

DESCRIPTION Length 19 cm.
ADULT MALE Undescribed.
ADULT FEMALE General plumage green, paler and more yellowish on forehead, lores and underparts; crown violet-blue; underwing-coverts green; vent red; tail above green tipped yellow, undersides yellow marked red and black at bases of lateral feathers; bill orange-red; legs orange.
1 specimen (type): wing 91 mm, tail 77 mm, exp. cul. damaged; tars. 16 mm.

DISTRIBUTION Known only from Grande Terre, New Caledonia.

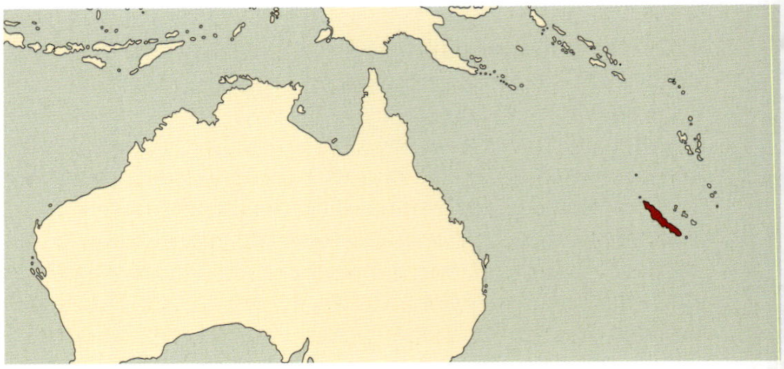

STATUS Two females collected at unknown localities in 1859 are the only specimens of the New Caledonian Lorikeet, and one of these is the type, but the second specimen has been lost. Another specimen was collected in 1913 in 'forests behind Oubatche', which corresponds to Mount Ignambi, but it was not preserved (in Birdlife International 2016). There have been no confirmed reports since the collection of these specimens, and no unconfirmed reports since 1976. Also, several months of specific searching in 1998 were unsuccessful, so if these lorikeets still survive the population is assumed to be extremely low, possibly numbering fewer than 50 mature birds.

Although montane humid forests are not threatened, other habitats in the lowlands have almost disappeared so if there was any dependence on these habitats by the lorikeets the detrimental impact would have been severe. It is likely that introduced mammalian predators, particularly rats, would have contributed to a decline, and the suggestion has been made that a particular susceptibility to predation by rats may have brought about the almost total disappearance of these lorikeets by the time Europeans arrived in New Caledonia to record their presence (Ekstrom *et al.* 2002). It is possible that disease, particularly avian malaria, also was a contributing factor, as it has been implicated in the demise of birds elsewhere in the South Pacific region.

HABITATS Some anecdotal information on possible habitat preferences was included in unconfirmed records given to Tony Stokes when he undertook fieldwork in New Caledonia in November–December 1976, on behalf of the Australian Museum, Sydney (Stokes, pers. comm. 1979). When shown a coloured plate

F. KNIGHT 10

of *C. diadema*, a resident with an interest in local natural history told Stokes that he had seen one bird in low scrubland near Yaté Lake, in the south of the island, many years ago, probably in the 1920s. A senior forestry officer familiar with local birds told Stokes that his first record of this species was in 1953 to 1954, on the La Foa to Canala road, when a pair was seen flying from rainforest to open *Melaleuca* woodland, where they alighted in a low tree. All that he could remember about the appearance of these birds was that they were predominantly greenish with some yellow on the abdomen. This observer recorded the lorikeets again on 3 June 1976, to the west of Mount Panie, also in the rainforest-*Melaleuca* woodland ecotone, when his attention was attracted by a call differing slightly from that of the common Rainbow Lorikeet *Trichoglossus haematodus*, and looking up he saw two small green parrots dart quickly from a tree and fly overhead.

HABITS Nothing is known of the habits of these lorikeets, but presumably they were birds of the forest canopy where they foraged in flowering trees for nectar and pollen. The resident who saw a bird at Yaté Lake told Stokes that it was silent during the five minutes that it was under observation.

SPECIMEN AVAILABLE The only available specimen is the type (MNHN 762A) held in the Museum National d'Histoire Naturelle, Paris.

CRITICALLY ENDANGERED

Red-throated Lorikeet

Charmosyna amabilis (Ramsay)

Trichoglossus (*Glossopsitta*) *amabilis* Ramsay, *Proc. Linn. Soc. New South Wales*, **1**, 1875, p. 30 (Fiji).

DESCRIPTION Length 18 cm.
ADULTS General plumage green, paler on underparts and underwing-coverts; ear-coverts shaft-streaked bluish-green; lores, cheeks and throat red, bordered below by a yellow band; thighs dark red; tail above green tipped yellow, below dull yellow with black and red markings at bases of lateral feathers; bill orange; iris yellow; legs orange.
10 males wing 94–100 (96.3) mm, tail 69–79 (72.8) mm, exp. cul. 10–11 (10.8) mm, tars. 12–13 (12.5) mm.
8 females: wing 91–96 (93.9) mm, tail 68–80 (74.0) mm, exp. cul. 10–11 (10.8) mm, tars. 12–13 (12.4) mm.
JUVENILES Yellow band on throat only faintly indicated; thighs dull greyish-mauve faintly tinged red.

DISTRIBUTION Recorded from Viti Levu, Vanua Levu, Ovalau and Taveuni, in the Fiji Islands.

STATUS Watling (2001) points out that the Red-throated Lorikeet is very rare, and predation by Black Rats *Rattus rattus* may be responsible for its rarity and continuing decline. Its distribution appears to be quite patchy, and records have become much fewer in recent years, the decline apparently having been quite dramatic since the 1970s, when it was reported to be locally common at some localities on both Viti Levu and Taveuni (see Holyoak 1979). The last record by observers familiar with these lorikeets was in 1993, but a sighting at Mount Tomaniivi on Viti Levu in 2001 is supported by detailed field notes (in Birdlife International 2016). Nearly all recent records on Viti Levu have been in the Mount Tomaniivi area, though in 1998 two birds were seen in the Nausori highlands, but in 2001–2002 searches were undertaken in the central highlands, where birds have been seen most frequently since the 1970s, and during 49 days in the field no birds were seen or heard (in Birdlife International 2016; Swinnerton and Maljkovic 2002). Continuing surveys on Viti Levu also have been unsuccessful. Watling notes that there are unconfirmed records from the 1980s and 1990s from lowland areas of Ovalau, uplands on Taveuni and from the Natewa peninsula on Vanua Levu (in Birdlife International 2016). Dick Watling confirms that there have been no records on Viti Levu since 1993, so the lorikeets may no longer occur there, and better prospects for finding birds may be on Taveuni (*in litt.* 2016).

In view of failures to locate birds during specific searches, the population is assumed to be very low, and possibly may be fewer than 50 birds. Deforestation and predation by introduced mammals, especially Black Rats, have been identified as likely causes of declining numbers of Red-throated Lorikeets. Lowland and hill forests have been, or are being cleared in much of Fiji, and in highland areas an ongoing expansion of logging operations with associated roadworks is likely to increase the density of rats and the Indian Brown Mongoose *Herpestes fuscus* in any stands of forest supporting remnant populations of lorikeets. Survival of such remnant populations within the Tomaniivi Nature Reserve or in any other reserves within the range will be dependent on effective predator control programs.

HABITATS Watling (2001) notes that Red-throated Lorikeets are found in mature forest, with an exceptional observation in mangroves reported from Ovalau. Gorman (1975) reported that during observational surveys undertaken on Viti Levu between November 1970 and May 1973, these lorikeets were observed in the lowlands at about 150 m, and in the highlands between 780 m and 970 m at three sites – the Nabukulevu area, the Nadarivatu Plateau and Mount Tomaniivi; habitats at these sites were described as follows:

(i) Nabukulevu area: three-storey forest with the main closed canopy at 20 m, emergents up to 25 m, and the dense lower canopy at 15 m; the dominant tree species were Pacific kauri *Agathis vitiensis*, *Calophyllum vitiense*, *Myristica castaniifolia* and kauvula *Endospermum macrophyllum*, and

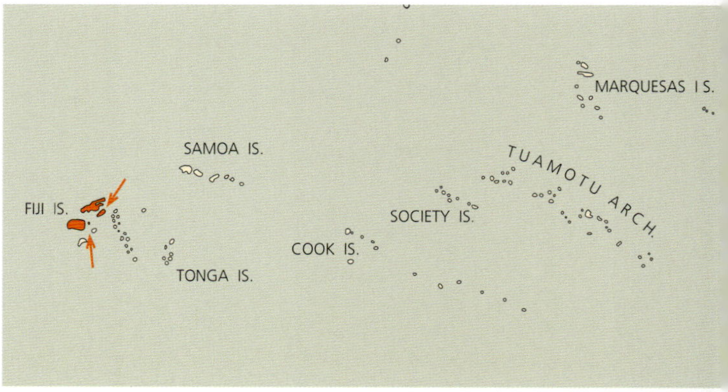

Plate 20
Red-throated Lorikeet *Charmosyna amabilis* (adults)

F KNIGHT '10

shrubs included *Gironniera celtidifolia*, *Parinari glaberrima* and *Dillenia biflora*,

(ii) Nadarivatu Plateau: *Agathis vitiensis* and *Endospermum macrophyllum* were emergent above a closed canopy of broad-leaved species such as *Parinari insularum*, *Myristica* spp., Java cedar *Bischofia javanica*, sasawira *Dysoxylum richii* and *Gonystylus punctatus*,

(iii) Mount Tomaniivi: a two-canopied forest on lower slopes of the mountain, with the top canopy at 20 m and dominant tree species being *Calophyllum neo-ebudicum*, *Myristica* spp., *Endospermum macrophyllum*, *Agathis vitiensis* and *Bischofia javanica*.

Holyoak (1979) recalled that in June–July 1973, also on Viti Levu, these lorikeets were found to be not uncommon on a forested ridge near Waisa, Vunidawa, at 200–250 m, and at about 800 m on a ridge near Naitaradamu, while on Taveuni they were seen or heard on five days in rainforest at 510–1000 m.

HABITS The very little information that we have on habits comes from observations recorded in the 1970s. Holyoak (1979) recalled that on Viti Levu and Taveuni sightings were of a group of five birds, a pair, at least two together and of single birds, which probably were with others that remained undetected. Being predominantly green, they were very inconspicuous and usually kept to the canopy of tall trees. They flew about much in the treetops, the flight being 'quick and direct with rapid wingbeats'. Watling (2001) notes that the rapid wingbeats produce a fluttering effect.

CALLS Watling describes the call as a brief, high-pitched monosyllabic squeak given while feeding and in flight when it is louder and more resonant.

DIET AND FEEDING At 700 m on Taveuni, Holyoak (1979) observed two birds actively feeding in the canopy of a tall tree, often hanging upside-down to reach flowers. Several times they used their opened bills to split the 13 mm long corolla tubes of the white flowers, and then poked the tongue about the base of each corolla tube. At least two Collared Lories *Phigys solitarius* and four Wattled Honeyeaters *Foulehaio carunculata* were feeding at flowers of this tree.

BREEDING There are no nesting records.

GENUS *Vini* Lesson

Vini Lesson, *Illustr. de Zool.*, 1831, pl. 28. Type, by monotypy, *Vini coccinea* Lesson = *Psittacula kuhlii* Vigors.

Amadon (1942) differentiated *Vini* and *Charmosyna* subgenerically, but Steadman and Zarriello (1987) linked *Vini* with monotypic *Phigys* from the Fijian Islands and noted that although it is distinct osteologically from other genera of parrots, including *Charmosyna*, the species within *Vini* are rather uniform osteologically, with differences in size and occasionally in proportions being the only diagnostic features at the specific level. Lorikeets belonging to this genus are noticeably stockier in body form than *Charmosyna* species and have short, rounded tails. Erectile feathers of the crown are prominently shaft-streaked. There is no sexual dimorphism, but in the two predominantly blue species – *V. peruviana* and *V. ultramarina* – juveniles differ markedly from adults.

Apart from the more widespread Blue-crowned Lorikeet *V. australis*, which is secure in some parts of its range, all species are at risk. They occur on isolated islands in the South Pacific region, and fossil material reveals that current and now extinct species were more widely distributed at the time of human arrivals some 2000 years ago (see Steadman 1985; Steadman and Zarriello 1987). Two extinct species – *V. sinotoi* and *V. vidivici* – are known from bones collected in Polynesian archaeological sites in the Marquesas and Society Islands (Steadman and Zarriello 1987). McCormack and Künzlé (1996) point out that in the 1990s the Blue Lorikeet *V. peruviana* survived on five islands compared with more than a dozen islands in the 1940s, the Ultramarine Lorikeet *V. ultramarina* survived on only one island compared with three islands in the 1970s, and the Rimatara Lorikeet *V. kuhlii* occurred only on Rimatara Island after having been extirpated in the Cook Islands.

Although now known to occur on more than five islands, the Blue Lorikeet certainly has disappeared from other islands in its former range. The introduction of rats, especially *Rattus rattus*, has been associated with the decline of these lorikeets, and exploitation of *V. kuhlii* for its red feathers to be used in ceremonial head-adornments probably brought about its extirpation in the Cook Islands. Existing species continue to be threatened by destruction of habitat and the introduction of predators. Disease also may pose a risk, for Holyoak (1973a) warned that avian malaria, spread by alien mosquitoes imported in airliners, has joined the destruction of forest, the introduction of cats and goats, and local persecution to endanger endemic Polynesian landbirds.

Long-term survival of the extremely beautiful *Vini* lorikeets will depend on concerted efforts to protect remaining populations.

ENDANGERED

Rimatara Lorikeet

Vini kuhlii (Vigors)

Psittacula Kuhlii Vigors, *Zool. Journ.*, **1**, 1824, p. 412, pl. 16 (Toohooteterooha Island. A day's sail from Otaheite = Rimatara, Austral Islands).

OTHER NAME Kuhl's Lory.

DESCRIPTION Length 19 cm. Weight 55 g.

ADULTS Crown green, feathers prominently shaft-streaked brighter yellowish-green; nape mauve-black, feathers broadly shaft-streaked purple-blue; upperparts green, more yellowish on mantle and becoming bright yellowish-green on lower back to upper tail-coverts; outermost primaries greyish-black edged mauve-blue on outer webs, inner primaries and outer secondaries greyish-black broadly edged green on outer webs; underwing-coverts green; lores, throat and cheeks red, becoming dark purple

on abdomen and thighs; under tail-coverts greenish-yellow; tail above scarlet tipped green, below red at base and broadly tipped olive-green; bill orange; iris red; legs dark orange-brown.
10 males: wing 120–136 (129.6) mm, tail 64–75 (69.7) mm, exp. cul. 11–12 (11.3) mm, tars. 64–75 (69.7) mm.
4 females: wing 124–134 (128.8) mm, tail 67–72 (69.8) mm, exp. cul. 11–12 (11.3) mm, tars. 18–19 (18.3) mm.
JUVENILES Red underparts variably suffused dusky mauve-black; bill dusky brown; iris brown.

DISTRIBUTION Endemic to Rimatara, in the Austral Islands, French Polynesia, but introduced to Teraina, Tabueran and Kiritimati, in the Line Islands, Kiribati, and birds translocated from Rimatara to Atiu, in the Cook Islands.

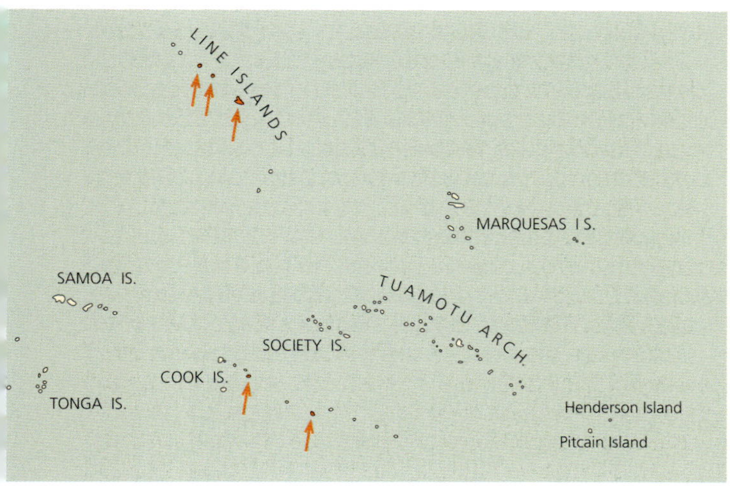

STATUS Lieberman and McCormack (2008) point out that Rimatara Lorikeets historically inhabited several islands in the southern Cook Islands and nearby Rimatara in French Polynesia, but since about 1800 they have survived only on Rimatara, having been intensely hunted in the Cook Islands for the bright red feathers to be used in ceremonial head-adornments and in trade. Surveys undertaken on Rimatara between 1992 and 2004 revealed an estimated population of 750–900 birds to be fairly stable, and this stability was due largely to protection measures put in place in about 1900 by Queen Tamaeva III, who prohibited the exporting, exploiting or harming of the lorikeets in any way. Also contributing to their survival on the island was an abundance of food resulting from intensive horticulture and agriculture on the island and probably most importantly an absence of Black Rats *Rattus rattus*. Seitre (2015) reports that during a visit to Rimatara he found the lorikeets to be very common and conspicuous.

When Thomas Street visited Washington Island, now Teraina in the Line Islands, Kiribati, with a United States survey expedition in 1874, he met a party of Polynesian voyagers who had come from islands to the south to harvest coconuts, and they had with them a number of Rimatara Lorikeets as pets, so in this way the species probably was introduced to Teraina and Tabuaeran, the Line Islands sometime before the 1800s. In 1957, six birds were released on Kiritimati, also in the Line Islands. In the 1990s, in the Line Islands, a small population was surviving on Tabuaeran despite the presence of *Rattus rattus*, and a good population of more than 1000 birds survived on Teraina, where there were no perceived threats (Watling 1995). On Kiritimati, where the more arid climate was not considered conducive to establishment of a permanent population, a few birds were said to be surviving in 1999, but in 2008 it was suggested that the lorikeets were common in plantations to the north of the village of London (in Birdlife International 2016).

The translocation in 2007 of birds from Rimatara to Atiu, in the southern Cook Islands, appears to have been successful, for successful breeding has been recorded and within 15 months of the translocation birds were seen on the neighbouring island of Mitiaro, some 50 km from Atiu (Lieberman and McCormack 2008). Heptonstall (2011) reports that in May–June 2010, an observer point survey was undertaken to estimate numbers on Atiu, and counts from two days of observations were 104 and 111 lorikeets, with a presence of juveniles confirming that successful breeding continues to occur. To safeguard nests from aggressive Common Mynas *Acridotheres tristis*, a successful program was undertaken to eradicate mynas, but a consequence of their removal was a dramatic upsurge in damage to coconut palms by coconut stick-insects *Graeffea crouanii* so consideration is being given to reintroducing mynas to the island (McCormack *in litt.* 2016).

Watling (1995) points out that in the Line Islands, Rimatara Lorikeets have been liberated on Kiritimati several times in the relatively recent past but, despite the absence of *Rattus rattus*, the arid, unpredictable climate with periodic extremely severe droughts probably precludes any long-term successful establishment of a resident population. In February 1993, Watling undertook a three-day survey on Tabuaeran, where *Rattus rattus* is present, and he found a small population of perhaps fewer than 50 birds on the northwestern island of the atoll. If this population still survives, it important to know how it has managed to do so for at least 70 years because, as noted by Watling, it may have important implications for the conservation of *Vini* lorikeets on islands where Black Rats are present. Also in February 1993, on the basis of transect counts in all parts of Teraina, Watling estimated the population of lorikeets to be between 1000 and 1600 birds, and it appears to be secure provided that Black Rats are not accidentally introduced.

In summary, it is estimated that at least 1000 birds are present on Rimatara and Atiu, with another 1000 birds probably present in the Line Islands, Kiribati, and although populations are thought to be stable on Rimatara and Teraina, and increasing on Atiu, invasive predators are likely to be causing a decline on Tabuaeran and Kiritimati, so the overall trend is suspected to be a slow to moderate, continuing decline. I am somewhat more optimistic about the long-term survival of these lorikeets provided that populations can be safeguarded against rat predation.

HABITATS McCormack and Künzlé (1996) recall that in August 1992, when an observer-point survey was undertaken on Rimatara, the highest density of 2.2 lorikeets per hectare was recorded in mixed horticultural woodlands, which cover some 32 per cent of the island and in which are planted many introduced food plants and coconut palms, while densities in other habitats were 1.4 birds per hectare in village gardens, 0.9 birds per hectare in mixed fernlands, grasslands and plantations of introduced white albizia *Paraserianthes falcataria* in the centre of the island, 0.8 birds per hectare in coastal coconut plantations, and 0.6 birds per hectare in indigenous forests and scrublands along an ancient raised reef around the northwestern half of the island, but no birds were present in swamplands. Seitre (2015) reports that he encountered birds all over the island, including in the centre of town and in gardens, where they were easily seen in low vegetation. Lieberman and McCormack (2008) note that during the 15 months following their release on Aitu, the lorikeets were seen feeding in areas of cultivation over the entire island.

Watling (1995) reports that on Teraina, in the Line Islands, Kiribati, these lorikeets utilise all terrestrial habitats, and commonly are seen flying over villages and visiting trees in village gardens.

HABITS Seitre (2015) notes that on Rimatara these lorikeets were normally encountered in small groups, and this was more usual than with other *Vini* species. Parties of two or more pairs were quite common, and at times up to 10 or 12 birds were seen feeding in the same tree, often with obvious social interactions, and pairs with up to three dependent fledglings also were observed. They usually were quite noisy, especially in flight, when alighting in the treetops, in interactions between parents and their offspring or in interactions between pairs or small groups, but feeding often was undertaken in silence. On some days there was an inexplicable absence of calling, and prolonged periods would pass without any birds being heard. The flight is swift and direct, the 'whirring' of wingbeats being audible as the birds pass overhead.

CALLS The call is described as a shrill croak or screech (Pratt *et al.* 1987).

DIET AND FEEDING The diet comprises nectar and pollen, gathered primarily from coconut blossoms and banana flowers. McCormack and Künzlé (1996) report that on Rimatara these lorikeets were observed feeding in 12 species of flowering plants, taking mostly nectar, though also licking the leaf-stalks of some plants, and they appeared to extract seeds from the fruits of Pacific ironwoods *Casuarina equisetifolia*. The food plants with frequencies of feeding observations are listed in Table 5. Local people reported that the birds also fed frequently at the flowers of cultivated coffee plants *Coffea arabica*. Other foraging records are of birds feeding at the flowers of tree hibiscus *Hibiscus tiliaceus* and the rotting leaf-borders of fish-poison trees *Barringtonia asiatica* (Seitre and Seitre 1991).

BREEDING Bruner (1972) noted that on Rimatara the breeding season is reported to be during January to March, and nesting is restricted to coconut palms, presumably in a hollow in the trunk or in a hole in a rotting coconut still adhering to the tree. McCormack and Künzlé (1996) report that local residents believed that nests were in holes in fish-poison trees *Barringtonia asiatica*, *Pisonia grandis* trees and white albizias *Paraserianthes falcataria*, although it eventuated that none of the informants had actually seen a nest. A pair of lorikeets was seen burrowing into the end of a rotten branch of a tree hibiscus *Hibiscus tiliaceus*, but the activity subsequently was abandoned. Watling (1995) reports being informed that on Tabuaeran, in the Line Islands, Kiribati, a nest had been found in the rotten trunk of a *Pandanus* tree, and this suggested that the small population of lorikeets on the island

Food plants	Frequency of feeding observations
Inga ynga inga (introduced)	33
Casuarina equisetifolia Pacific ironwood	18
Paraserianthes falcataria white albizia (introduced)	18
Musa spp. banana (introduced)	9
Mangifera indica mango (introduced)	8
Ceiba pentandra kapok (introduced)	7
Cocos nucifera coconut palm	6
Erythrina variegata Indian coral-tree (introduced)	4
Hibiscus rosa-sinensis Chinese hibiscus (introduced)	3
Adenanthera pavonina red-bead tree (introduced)	3
Syzygium jambos rose apple (introduced)	3
Albizia lebbeck siris tree (introduced)	2

Table 5. Food trees with frequency of feeding observations recorded on Rimatara in August 1992 (after McCormack and Künzlé 1996).

may have survived because nesting sites were away from the stands of coconut palms favoured by rats.

Seitre (2015) reports that on Rimatara pairs with dependent juveniles were observed in August. On Atiu, in the southern Cook Islands, in August 2008, two chicks were observed and photographed as they emerged from one of two nests in the top of dead, hollow trunks of coconut palms (in Lieberman and McCormack 2008). A day after first being seen at the hollow entrance, the first chick emerged, only to be attacked by Common Mynas *Acridotheres tristis*, but the lories prevailed, and a second chick fledged two days later. In the Bernice P. Bishop Museum, Honolulu, there are two specimens in juvenal plumage that were collected on Tabuaeran, in the Line Islands, on 28 July 1922 (BBM 4784 and 4786).

Some details of nesting in captivity have been reported (Lee 1935; Patten 1947). No more than two eggs were laid, and during incubation the male spent much time in the nestbox. Only one chick fledged, and it left the nest about seven weeks after hatching. At a little over six months, the upper mandible and legs changed to orange.

Henderson Island Lorikeet

Vini stepheni (North)

Calliptilus ? stepheni North, *Rec. Austr. Mus.*, **7**, 1908, p. 20 (Henderson Island).

OTHER NAME Stephen's Lory.

DESCRIPTION Length 19 cm. Weight male 55 g, females 42–51 g (Graves 1992).

ADULTS Forehead, crown and nape to wings and upper back rich green, becoming greenish-yellow on lower back to flanks and tail-coverts; erectile feathers on crown prominently shaft-streaked paler green; lores, cheeks and ear-coverts to sides of abdomen red with purple at centre of abdomen and on thighs; variable band of green and purple across breast; underwing-coverts green marked red; tail greenish-yellow; bill deep orange; iris yellow-orange; legs orange.

Plate 21
UPPER Rimatara Lorikeet *Vini kuhlii* (adult)
LOWER Henderson Island Lorikeet *Vini stepheni* (adult)

F. KNIGHT '08

4 males: wing 126–133 (128.5) mm, tail 83–93 (87.8) mm, exp. cul. 11–12 (11.5) mm, tars. 16–18 (16.8) mm.
4 females: Wing 124–127 (125.8) mm, tail 85–91 (87.8) mm, exp. cul. 11–12 (11.8) mm, tars. 16–18 (17.0) mm.
JUVENILES Lores red, but underparts green with purple suffusion on throat and abdomen, and small red patch on centre of abdomen; tail darker green; bill brown; iris pale brown; legs orange-brown.

DISTRIBUTION Confined to Henderson Island, in the Pitcairn Group, eastern Polynesia.

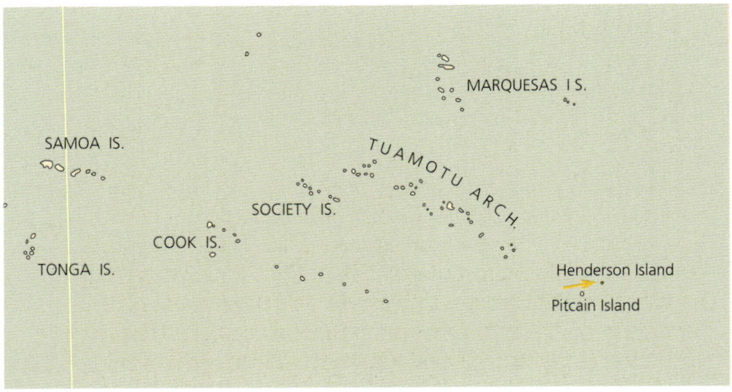

STATUS No estimate of the population of Henderson Island Lorikeets has been made since counts were made by Gary Graves during a visit to the island in May 1987 (Graves 1992). He estimated their local abundance by recording the number of birds seen along the beach and interior forest trails during a continuous census, taking into account the location of flocks and the destination of flying birds. Because lorikeets are attracted from the interior to coconut groves along the beach, it was difficult to estimate the size of the population, but a density of 0.2 to 0.5 birds per hectare was speculated and that equated to an estimated population of 720 to 1820 individuals. Because Henderson Island is remote, without human habitation and relatively undisturbed, it is assumed that the population has remained stable, and the lorikeets seem to have adapted to the presence of Pacific Rats *Rattus exulans*, which were introduced by the Polynesians (in Birdlife International 2016). In August 2011, operations were undertaken to eradicate Pacific Rats. Graves notes that predation on the eggs, nestlings and adults of landbirds on Henderson Island currently is limited to the Pacific Rat and a large terrestrial hermit crab, but he warns that despite the great isolation of the island, accidental introduction of predators, disease vectors and diseases probably is inevitable because of unauthorised landings of yachts, and he recorded undisputable evidence of past landings.

HABITATS Graves (1992) notes that these lorikeets occur along the beach, where they are attracted to coconut groves, and throughout the low, almost impenetrable forest that covers the elevated interior.

HABITS Graves (1992) recalls that during his visit in May 1987, Henderson Island Lorikeets were conspicuous in coconut palms along the beachfront, and frequently were observed in small parties of three to five birds, possibly family groups, flying 20–40 m above the forest canopy. Despite their brilliant plumage colouration, they were difficult to locate once they alighted amidst foliage, and feeding birds often were detected by their calls. They made relatively large daily movements.

CALLS Graves notes that soft, twittering calls were emitted by feeding birds, but flight calls were louder and could be heard at a distance of 100 m away from the surf.

DIET AND FEEDING Trevelyan (1995) reports that during field studies of foraging ecology undertaken on Henderson Island between January and March 1992, Henderson Island Lorikeets were found to be generalists in their feeding habits, consuming nectar, pollen and fruits from a wide range of food plants from beaches to the inland plateau. Insect larvae also were a major component of the diet, but these were from a very specific source, namely lepidopteran larvae taken from the sporangia of *Phymatosorus* ferns. Feeding occurred at all stratifications, from canopy level in coconut palms to ground-level shrubs. Two main sources of nectar and pollen were *Scaevola sericea*, a creeping herb, and *Timonius polygamus*, a shrub growing up to 5 m in height which produces solitary female flowers and smaller male flowers in cymes, with each cyme at any one time having a number of buds and flowers at different stages of development. The lorikeets fed almost exclusively at male flowers. A comparison of *Timonius* flowers which had been plucked by the lorikeets with non-visited flowers revealed that both pollen and nectar had been consumed, and after examining visited and non-visited *Scaevola* flowers it was assumed that nectar was the main food taken by the birds. *Timonius* and *Scaevola* flowers have very different nectar yields, with the latter containing larger volumes of more concentrated nectar, but they grow at much lower densities than *Timonius* flowers, so are a less predictable nectar resource. The lorikeets fed on the two flower types at different frequencies, and it was hypothesised that the net rate of energy gained from feeding in a similar sized patch of the two plant species was comparable. Although nectar volumes in *Timonius* flowers decreased from early mornings until late afternoons, the lorikeets did not show a corresponding daily pattern of foraging behaviour.

Other food plants visited for nectar and pollen were *Cyclophyllum* sp., *Xylosma suaveolens*, *Thespesia populnea*, *Cordia subcordata*, *Psydrax* sp., *Senecio* sp., coconut palms *Cocos nucifera* and *Pandanus tectorius*. Fruits of *Eugenia rariflora*, *Nesoluma st-johnianum*, *Guettardia speciosa* and *Timonius polygamus* were eaten, and there was one observation of lorikeets extracting 'juice' from *Caesalpinia* leaves. During spot checks carried out at the study site, 48 per cent of all observations were of birds feeding at *Scaevola* flowers, with 28 per cent being of birds taking insect larvae, almost exclusively lepidopteran larvae from sporangia of *Phymatosorus* ferns, 22 per cent feeding at *Timonius* flowers, and 2 per cent taking *Eugenia* fruits. Trevelyan concludes that Henderson Island Lorikeets have a quite generalist diet compared with other *Vini* species, and their ecology is interlinked with several plant and at least one insect species found on Henderson Island.

BREEDING There are no nesting records. A male with enlarged gonads was collected in April (AMNH 192976), and Graves notes that, of five specimens collected in May, one female had a greatly enlarged ovary, but the others had quiescent gonads. Amadon (1942) concluded that the post juvenal moult begins on the head, and the red underparts are acquired gradually.

VULNERABLE

Blue Lorikeet

Vini peruviana (P. L. S. Müller)

Psittacus peruvianus P. L. S. Müller, *Natursyst.*, suppl., 1776, p. 80 (Peru, error = Tahiti).

OTHER NAME Tahitian Lory.

DESCRIPTION Length 18 cm. Weight 31–34 g.
ADULTS General plumage dark purple-blue; erectile feathers of crown prominently shaft-streaked paler mauve-blue; ear-coverts, throat and upper breast white; primaries and outer secondaries dusky black, on outer webs broadly margined dark purple-blue; underwing-coverts dark purple-blue; tail above dark purple-blue, below dusky black; bill orange; iris brownish-yellow; legs orange.
6 males: wing 108–116 (112.5) mm, tail 66–74 (68.2) mm, exp. cul. 10–11 (10.5) mm, tars. 14–16 (14.8) mm.
6 females: wing 107–116 (110.8) mm, tail 65–69 (66.5) mm, exp. cul. 9–11 (10.5) mm, tars. 14–15 (14.7) mm.
JUVENILES Entire underparts dark grey-blue with greyish-white markings on chin; bill black; iris dark brown; legs dark orange-brown.

DISTRIBUTION Recorded from the Society Islands and northern atolls in the Tuamotu Archipelago, French Polynesia, and introduced to Aitutaki in the Southern Cook Islands.

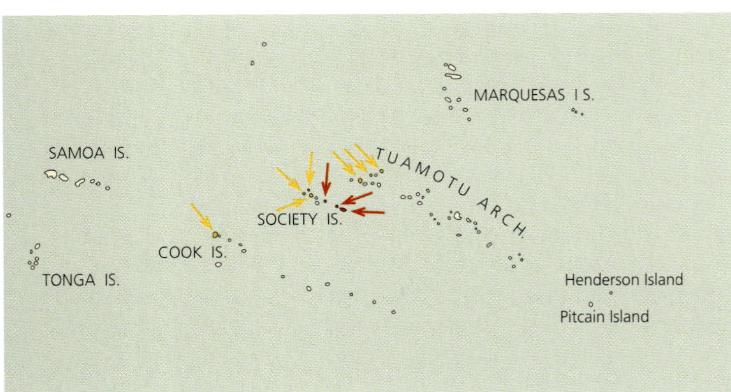

STATUS Blue Lorikeets are widely, though unevenly distributed on islands in southeastern Polynesia, but have disappeared from several islands within their range. In the Society Islands, Wilson (1907) suspected that as early as 1904 they no longer nested on Tahiti, and presumably soon after that time they became extinct there and on nearby Moorea. In the early 1900s they were found to be fairly common on Bora Bora, also in the Society Islands, and on Rangiroa in the Tuamotu Archipelago, and in the early 1920s a number of specimens were collected on Bora Bora by members of the Whitney South Sea Expedition from the American Museum of Natural History, but they certainly had disappeared from Bora Bora well before my visit in 1971. Holyoak and Thibault (1984) suggest that disappearance of these lorikeets from much of the Society Islands correlates with spread of the introduced Swamp Harrier *Circus approximans*, but I suspect that the spread of rats and cats is more likely to have been the cause, and rats were very common on Bora Bora at the time of my visit. In the Society Islands, populations still occur on Motu One and Mauae, where up to 250 and 350–400 pairs respectively were estimated to be present in the early 1970s, and on Maupihaa, where previously thought to be extinct but again recorded in 1999 (in Birdlife International 2016).

Seitre (2015) notes that on islets or motus along the southern shores of Rangiroa, in the Tuamotu Archipelago, there are two populations of Blue Lorikeets estimated at about 1000 birds in total, and this population appears to be stable, or possibly has increased slightly since a previous visit made in the 1980s. Recent estimates of populations occurring elsewhere in the Tuamotu Archipelago include 100 birds on Kaukura, 500 on Arutua, 200 on Apataki and 50 on Tikehau, while on Tiamanu Motu in Apataki atoll an estimate of 300 birds made in 1989 may be lower than an estimate made 10 years earlier (in Birdlife International 2016).

McCormack and Künzlé (1996) point out that studies of fossils and middens on Aitutaki, in the Southern Cook Islands, have shown that Rimatara Lorikeets *Vini kuhlii* formerly occurred there, but there are no remnants of Blue Lorikeets and the two species would not have coexisted on the small island, so it can be concluded that *V. peruviana* was introduced after the extirpation of *V. kuhlii*. This introduced population on Aitutaki probably is the best monitored population, and as early as 1899 these lorikeets were the only landbirds found on the island by Townsend, who commented that they were common pets of the local people (in Townsend and Wetmore 1919). Wilson (1993) reports that, during a visit to Aitutaki between 14 and 18 February 1991 to make preliminary observations on the distribution, abundance and habitat preferences of the lorikeets,136–142 birds were seen or heard while walking or cycling along roads and tracks on most parts of the island and, although it was not possible to make an accurate estimate of the total population from these figures, it was suggested that there must be at least several hundred birds on the island.

In February 2010, Cyclone Pat caused a loss of more than 50 per cent of the population, including the loss of virtually all juveniles, but Jennings (2011) reports that a post-cyclone study using distance sampling revealed that the surviving population comprised an estimated 1448 birds, and there is every chance that numbers again will increase.

When recent survey data are combined, the total population is estimated at 1500 to 7000 mature birds, but a slow to moderate decline is suspected (in Birdlife International 2016).

HABITATS Throughout much of their range, Blue Lorikeets frequent littoral groves of coconut palms, and to a lesser degree are found in banana plantations or village gardens. Seitre (2015) notes that near Blue Lagoon Tourist Village, on the southern shores of Rangiroa, the lorikeets are plentiful on islets well planted with coconut palms, but to the south and east, where the small islets lack coconut palms there are no lorikeets. Wilson (1993) reports that these lorikeets showed a wide but clumped distribution on Aitutaki, in the Southern Cook Islands and, apart from one sighting of a single bird in a megaherb-grassland area, all 132 observations were made in wooded areas, with the preferred habitat being coconut plantations, where 18.2 per cent of observations were made, while an additional 34 per cent of observations were in plantations or gardens containing both coconuts and bananas, but all of the 11.4 per cent of observations in extensive coastal forests of hibiscus *Hibiscus tiliaceus* were in areas with an admixture of coconut palms, suggesting that this habitat generally was avoided, and only 5.3 per cent of observations were in banana plantations. In plantations or gardens containing both coconuts and bananas, it seemed that the lorikeets made greater use of coconut palms than of banana trees.

HABITS Blue Lorikeets usually are encountered in pairs or flocks of up to 10 birds (Pratt *et al.* 1987). Wilson (1993) recalls that in February 1991, on Aitutaki, in the Southern Cook Islands, 57 of 63 observations were of single birds, accounting for 18 observations, pairs, accounting for 16 observations, or parties of up to four birds, accounting for 23 observations, and each of groups of five and seven birds were seen on only three occasions. They are not shy, and Jennings (2011) recalls that on Aitutaki, while she sat quietly on a verandah, birds would feed in flowering plants right in front of her. Seitre (2015) notes similar confiding behaviour at Rangiroa, in the Tuamotu Archipelago, where the lorikeets were conspicuous and noisy as they flew about in and between coconut palms, feeding within 5 m of observers and showing inquisitive behaviour, even approaching quite close in response to imitations of their calls. Their flight is rapid with quick, short wingbeats (Pratt *et al.* 1987).

CALLS The call is described as a hissing, rolling, high-pitched screech, usually disyllabic, and resembling *schee-schee* (Pratt *et al.* 1987). Taylor (1984) reports that on Aitutaki, these lorikeets seemed to be especially vocal in the evenings.

DIET AND FEEDING The diet consists of pollen, nectar and soft fruits, with flowering coconut palms and banana trees being common food sources, but I suspect that insect larvae also are eaten. Child (1981) reports that on Aitutaki, in August–September 1980, birds were seen feeding in flowering coconut palms and mango trees *Mangifera indica*. Wilson (1993) reports 11 foraging observations made on Aitutaki in February 1991, when a flock of six birds was seen feeding on mango flowers, a group of three birds was seen eating fruits of nano *Morina citrifolia*, and two birds were seen nipping through the bases of large unidentified red flowers as if gathering nectar. Jennings (2011) recalls seeing birds use this same technique to take nectar from tubular *Hibiscus* flowers, and one lorikeet was seen to visit 10 or more flowers in a single *Hibiscus* tree. Also on Aitutaki, it was noticed that after the passage of Cyclone Pat in February 2010, the birds moved away from their primary food sources of coconut, banana, mango and kapok flowers in favour of smaller, more herbaceous plants which commenced to re-flower relatively quickly.

BREEDING Bruner (1972) noted that breeding occurs in the May to July dry season, and the nest is in a hollow in a tree or hole in a rotting coconut still attached to the tree. The claim made by Bruner that nests are constructed from grass and sticks and placed in a tree almost certainly is not correct, and may refer to occasional use by the lorikeets of such nests initially built by mynas, finches or other birds, though this also seems unlikely.

Low (1985) gives details of nesting in captivity. Clutches comprised two eggs, and both parents participated in incubation and care of the chicks, with the male assuming a prominent role in both tasks. Incubation normally commenced with laying of the first egg and lasted 25 days. On one occasion the two chicks in a brood hatched four days apart, and on another occasion both hatched on the same day, but usually the interval between hatching of siblings was two days. Most nestlings were removed for handrearing, but one chick reared by its parents left the nest 61 days after hatching.

EGGS An egg laid in captivity and held in the Natural History Museum at Tring, UK, is rounded and measures 19.4 × 17.2 mm (Harrison and Holyoak 1970). Low (1985) lists a smaller 17.0 × 15.0 mm as approximate measurements of eggs laid in captivity.

ENDANGERED

Ultramarine Lorikeet

Vini ultramarina (Kuhl)

Psittacus ultramarinus Kuhl, *Nova Acta Acad. Caes. Leop. Carol.*, **10**, 1820, p. 49 ('Consp. Psittacorum'). (New Holland, error = Marquesas Islands.)

DESCRIPTION Length 18 cm. Weight 35 g.
ADULTS Forehead rich blue; crown and occiput mauve-blue, erectile feathers prominently shaft-streaked paler blue; upperparts dull blue, paler on rump; lores and ear-coverts white, remaining underparts white mottled dark blue; dark mauve-blue band across breast; underwing-coverts dull blue; thighs and under tail-coverts mauve-blue; tail pale blue tipped whitish; bill orange tipped black; iris yellow-orange; legs orange.
6 males: wing 118–127 (123.3) mm, tail 71–80 (75.5) mm, exp. cul. 11–12 (11.7) mm, tars. 15–16 (15.5) mm.
6 females: wing 113–124 (117.2) mm, tail 70–78 (74.5) mm, exp. cul. 11–12 (11.7) mm, tars. 15–17 (15.5) mm.
JUVENILES Underparts dark blue, paler on sides of breast and flanks; lores and ear-coverts variably marked greyish-white; bill black; iris dark brown; legs orange-brown.

DISTRIBUTION Recorded from Ua Pou, Nuku Hiva and Ua Huka in the Marquesas Islands, French Polynesia, and translocated to Fatu Hiva, also in the Marquesas Islands, but apparently now surviving only on Ua Huka.

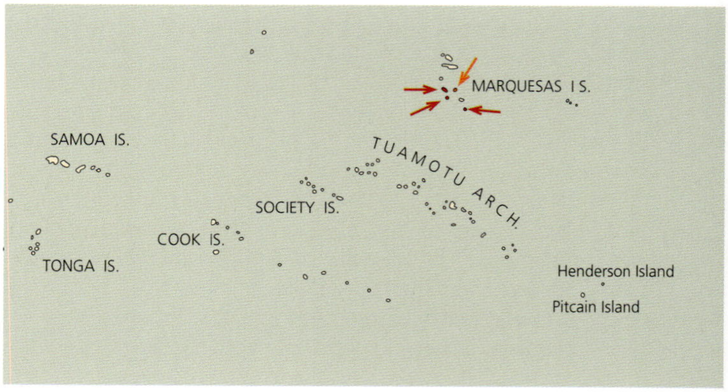

Plate 22
UPPER Blue Lorikeet *Vini peruviana* (adult)
LOWER Ultramarine Lorikeet *Vini ultramarina* (adult)

F. KNIGHT 10

STATUS In the 1970s, Ultramarine Lorikeets were reported to be widespread and fairly common on Ua Pou, with an estimated population of 250–350 pairs, fairly common on Ua Huka, with an estimated population of 200–250 pairs, and scarce and localised on Nuku Hiva, with a population of 40–150 individuals (Holyoak and Thibault 1984). In July 1977, on Ua Pou they were seen commonly near the coast and up on the highest ridges, and they were locally abundant in the valley above Hakahetau village (Montgomery *et al.* 1980). Subsequent searches in the late 1980s revealed that the lorikeets were declining on Ua Pou and may have been extirpated on Nuku Hiva, but on Ua Huka there was a thriving population, which reportedly originated from two captive birds translocated from Ua Pou in the early 1940s (in Birdlife International 2016). No birds were found on Nuku Hiva in 1990 or in 2004, so that population also was considered to be extinct, and at least three birds seen there in 1998 were thought to be vagrants from Ua Huka rather than members of a relictual population. In 1991, a survey confirmed that the effective range was restricted to one island, Ua Huka (Kuehler *et al.* 1997). In response to this finding, it was decided to establish a satellite population, so between 1992 and 1994, in three translocations, 29 birds were taken from Ua Huku to the island of Fatu Hiva. Initially, these translocations were successful, with 51 birds being counted on Fatu Hiva in 1997, but by 2000, when Black Rats *Rattus rattus* had become established, fewer than 10 birds were seen, and in 2007 the population was considered to be extinct (in Birdlife International 2016). On Ua Huka, the population was estimated at about 1300 birds in 1991, increasing to an estimated 1763–2987 in 2004, a level at which it has remained relatively stable, with 2094 individuals reported in 2009.

The assumption that predation by Black Rats has been the principal cause of the extirpation of Ultramarine Lorikeets from some islands in their historic range was confirmed by the disappearance of the translocated population on Fatu Hiva following the establishment of rats. Habitat degradation probably has been another contributing factor because all islands have been devastated by very high levels of grazing and fire, with much of the original dry forest being replaced by grassland, and upland forests have suffered extensive damage. Caroline Blanvillain from the Société d'Ornithologie de Polynésie confirms that these lorikeets now are restricted to Ua Huka, where trained dogs are being used to detect the presence of rats in incoming cargo (*in litt.* 2016). Based on these latest counts from Ua Huka the total population is estimated at 1000–2499 individuals, with the overall trend likely to be a slow and ongoing decline, but if Black Rats reach Ua Huka decline of the only remaining population will be rapid and severe (Birdlife International 2016).

The Ultramarine Lorikeet is listed on Appendix I of the Convention on International Trade in Endangered Species of Wild Fauna and Flora (CITES).

HABITATS During surveys undertaken on Ua Huku in 1991, highest counts of birds at 11.8–22.0 birds per hour were made in mixed forest-plantations dominated by coconut palms, hibiscus *Hibiscus tilaceus*, banana trees, mango trees, with a wide variety of other, less numerous trees, including breadfruit, banyan figs *Ficus benghalensis*, screw-pines *Pandanus tectorius*, mape *Inocarpus fagifer*, kava *Pometia pinnata*, kapok *Ceibe pentandra*, aeho *Miscanthus floridulus*, bamboo, ahia *Syzygium malaccense*, tamanu *Calophyllum inophyllum*, motoi *Cananga odorata*, papaya, ihi *Portulaca lutea* and noni *Morinda citrifolia*. Fairly high counts were made also in gardens and villages, where many of the same trees were present, but at lower densities, and there were low counts in drier forest or more open woodland on slopes where there were lower hibiscus trees, some guavas *Psidium guajava* and sheoaks *Casuarina equisetifolia* (Kuehler *et al.* 1997). Approximately half of the 78 km² of the island was assessed to be suitable habitat for the lorikeets. On Ua Pou, in late July 1977, they were encountered near the coast, feeding in coconut palms and *Erythrina* trees, and also up on high ridges in the interior (Montgomery *et al.* 1980).

Food plants	Food items
Calophyllum inophyllum tamanu	flowers/nectar
Carica papaya papaya	fruits
Casuarina equisetifolia sheoak	fruits
Citrus sinensis cultivated oranges	flowers/nectar
Cocos nucifera coconut palm	flowers/nectar
Coffea arabica coffee	flowers/nectar
Cordia subcordata sea trumpet	flowers/nectar
Coroza oleifera swampy palm	fruits
Eryhtrina sp. coral tree	flowers/nectar
Eugenia malaccense	flowers/nectar
Guettarda speciosa zebra wood	flowers/nectar
Hibiscus tiliaceus beach gardenia	flowers/nectar
Inga edulis icecream bean	flowers/nectar
Mangifera indica mango	fruits, flowers/nectar
Morinda citrifolia noni	flowers/nectar
Musa sp. banana	flowers/nectar
Pisonia grandis pisonia	flowers/nectar
Pometia pinnata kava	fruits
Psidium guajava guava	fruits
Syzygium malaccense Malay apple	flowers/nectar
Tectona grandis teak	flowers/nectar
Spondias dulcis ambarella	flowers/nectar

Table 6. Food plants and food items from those plants observed being eaten by Ultramarine Lorikeets during surveys on Ua Haku in 1991–1994 (after Kuehler *et al.* 1997).

HABITS Ultramarine Lorikeets usually are encountered in pairs or less commonly in small flocks, and they are very active, continuously fluttering from tree to tree and seldom resting for long periods in any one tree (Holyoak 1975). On Ua Pou, in late July 1977, raucous groups of up to a dozen birds typically were seen in rapid flight (Montgomery *et al.* 1980). Seitre (2015) reports that they are less confiding than other *Vini* lorikeets, being noisy and conspicuous in flight, but they tend to become very quiet after alighting, often continuing to feed silently in village gardens despite the presence of passers-by, but if approached too closely they immediately take flight. Janβen (2010) also found these lorikeets to be fairly shy, and feeding birds continually moved around to the far side of a tree when under observation. Holyoak described the flight as being swift and direct, with rapid wingbeats, but when flying down mountain slopes the birds will make short glides with the down-curved wings partially closed towards the body. They generally fly at or below treetop level, though on long flights they may spiral up to considerable heights.

CALLS The contact call is described as a sharp, piercing whistle transcribed as *iiiii*, and occasionally a loud, high-pitched screech or a disyllabic *to-weet* call are repeated several times (Pratt *et al.* 1987).

DIET AND FEEDING The diet consists of nectar, pollen, fruits, berries, buds and insects and their larvae. Holyoak (1975) recorded 10 observations of birds feeding at flowers, three of birds eating flower buds, two of birds eating fruits, and two of birds taking grubs and large adult hemipterans from leaves and branches. Thibault (1973) reported that on Ua Huka they were observed feeding mainly on pollen and nectar from coconut and banana flowers, and fruits of cultivated breadfruit and mango trees, and when mangoes were ripe flocks came to villages to feed in garden trees. Food plants and items seen to be eaten by these lorikeets during surveys undertaken on Ua Huku between 1991 and 1994 are listed in Table 6. Crop contents from two birds collected on Ua Pou comprised nectar, pollen, fragments of flowers and remains of large grubs (Holyoak 1975).

BREEDING According to Bruner (1972) nesting takes place during June to August, but a male in breeding condition (AMNH 195162) was collected in September, and Janβen (2010) reports that on Ua Huku a nest was found in February, and behaviour of the parents indicated that this nest probably contained chicks. Nests are in hollows in trees or possibly in a hole in a rotting coconut still attached to the tree, and the nest found by Janβen was in the top of a rotting protuberance near the base of a tree trunk.

Tavistock (1939) recorded some details of a successful breeding in captivity. The clutch comprised two eggs, but only one was laid in the nest. The incubation period was not determined, but both parents incubated and cared for the chick. The young bird left the nest about eight weeks after hatching.

EGGS Two eggs laid in captivity and held in the Natural History Museum at Tring, UK, measure 22.6 × 18.4 mm and 22.5 × 18.7 mm (Harrison and Holyoak 1970).

GENUS ***Lorius*** Vigors

Lorius Vigors, *Zool. Journ.,* **2,** 1825, p. 400. Type, by original designation, *Psittacus Domicella* Linnaeus.

Parrots belonging to this genus are midsized, stocky birds with short, slightly rounded and markedly broad tails. The bill is broader and less pointed than in other lories. Sexual dimorphism is absent, and juveniles resemble adults, but have pointed tips to the central tail-feathers.

Lorius is distributed from the eastern Solomon Islands west to New Guinea, including the western Papuan Islands, and the Moluccas, in Maluku Province, Indonesia. In eastern parts of this range, *Lorius* lories are secure, particularly in New Guinea where Black-capped Lories *L. lory* are widespread and common, but the two species occurring in the Moluccas have declined in numbers because of habitat loss and capture for the live-bird market.

ENDANGERED

Purple-naped Lory

Lorius domicella (Linnaeus)

Psittacus Domicella Linnaeus, *Syst. Nat.*, edn 10, **1**, 1758, p. 100 (Asia = East Indies *ex* Edwards, pl. 171 = Ambon).

DESCRIPTION Length 28 cm. Weight 235 g.
ADULTS General plumage red, darker on back; forehead, lores, crown and occiput black; nape black, feathers variably shaft-streaked violet or purple; variable yellow markings on sides of upper breast, sometimes forming incomplete pectoral band; thighs violet; wings and wing-coverts green, on bend of wing variably marked pale blue; underwing-coverts blue; broad yellow band across undersides of flight feathers; tail red broadly tipped deep brownish-red; bill orange-red; iris reddish-brown to orange; legs dark grey.
10 males: wing 160–173 (165.9) mm, tail 89–105 (96.6) mm, exp. cul. 25–27 (26.2) mm, tars. 22–25 (23.1) mm.
3 females: wing 157–163 (160.6) mm, tail 92–97 (94.3) mm, exp. cul. 25–26 (25.3) mm, tars. 23–25 (23.6) mm.
JUVENILES More extensive deeper purple shaft-streaking to feathers on nape; yellow markings on breast more extensive and usually forming a pectoral band; greater underwing-coverts blue margined black; tail faintly tipped blue; bill orange-brown.

DISTRIBUTION Occurs on Seram and possibly Ambon, southern Moluccas, Indonesia; introduced to Buru, southern Moluccas, but no longer present.

STATUS Widespread and intense trapping for the live-bird market together with ongoing deforestation brought about by logging, oil drilling and hydroelectric power schemes are suspected to be causing a rapid and continuing decline in numbers of Purple-naped Lories (Birdlife International 2016). On Seram most recent records have come from Manusela National Park, where these lories are said to be scarce in submontane forest between 400 m and 900 m, and from the Wae Fufa catchment in the east, where they are said to be fairly common on ridges between 900 m and

1050 m (in Birdlife International 2016). Bowler and Taylor (1989) report that when observations were made in Manusela National Park, between July and September 1987 as part of ornithological research conducted under the auspices of the Operation Raleigh Seram Expedition, Purple-naped Lories were encountered at a rate of only 0.7 birds per hour of observation. They seem always to have occurred at low densities because they were found to be uncommon during the course of the Second Freiburg Moluccan Expedition in 1911–1912 (in Stresemann 1914). In 2006 they were absent at all sites where they had been found to be fairly common in 1994, and widespread trapping may have removed them from most of Seram by 2009. Bishop (1992) points out that on Ambon there is a large human population and an extremely limited area of undisturbed forest, so a lack of recent reports of these lories is not surprising. The only evidence of the continued presence of Purple-naped Lories on Ambon is unconfirmed local reports that they survive above Hila (Birdlife International 2016).

In 2000, the total population was estimated at 1500–7000 mature birds, but in view of the apparent scarcity of these lorikeets and a lack of recent records, together with subsequent habitat loss and ongoing extensive trapping, numbers are suspected to have declined such that the population now is best estimated at 1000–2499 mature birds (Birdlife International).

HABITATS Purple-naped Lories frequent hill and submontane primary or secondary forest, particularly along ridges, within a fairly restricted altitudinal range of 300 m to 1100 m, being more common between 600 m and 1000 m.

In a recent study they were not seen in logged forest, suggesting an intolerance of forest degradation (Birdlife International 2016). Smiet (1985) reports that on Seram, in March 1980, on three occasions, single birds were observed in montane primary forest in an area where Blue-eared Lories *Eos semilarvata* also were present.

HABITS Coates and Bishop (1997) note that these lories usually are seen singly or in pairs, and their habits presumably are similar to the habits of other *Lorius* species, but there is very little recorded information. They are noisy and conspicuous in flight above or through the forest, but less so when in the canopy or subcanopy of flowering trees, where they are easily overlooked while resting or feeding quietly amidst the foliage. They are shy, seldom allowing a close approach, and this probably is a consequence of intense trapping or nest-robbing to secure birds as pets.

CALLS The only recorded information on vocalisation is the comment made by Smiet (1985) noting that the call is melodious.

DIET AND FEEDING As with other *Lorius* species, the diet undoubtedly consists of nectar, pollen and fruits or berries procured in the treetops. These lories are said to feed on fruits of an apparently thinly scattered *Pandanus* (Juniper and Parr 1998).

BREEDING I know of no records from the wild. Spence (1955) recorded some details of a successful breeding in captivity. Two eggs were laid, and were incubated only by the female. One chick hatched on the 24th day, and the second chick hatched on the 26th day. Only one chick was reared, and it left the nest three months after hatching. At an age of about four months its bill gradually changed to orange.

EGGS Recorded measurements of eight eggs, presumably laid in captivity, are 32.0 (30.5–33.6) × 25.5 (25.0–26.6) mm (Schönwetter 1964).

GENUS *Eos* Wagler

Eos Wagler, *Abh. k. Bayer. Akad. Wiss., Math.-Phys. Kl.*, **1**, 1832, p. 494. Type, by subsequent designation, *E. indica* (Gmelin) = *Psittacus histrio* P. L. S. Müller (G. R. Gray, *List Gen. Bds*, 1840, p. 52).

Eos lories are midsized parrots with a short rounded tail, and the plumage colouration is red with black and blue markings, but without green. There is no sexual dimorphism, but juveniles show dusky margins to the body feathers, have pointed instead of rounded tips to the tail-feathers, and orange-brown to black bills.

In the Indonesian Archipelago, *Eos* lories occur on islands in Teluk Cenderawasih, West Papua, and in the Moluccan, Kai and Tanimbar Islands. Two species – *E. bornea* and *E. semilarvata* – occur on Seram, where they are segregated altitudinally, but other species replace each other geographically.

ENDANGERED

Red and Blue Lory

Eos histrio (P. L. S. Müller)

Psittacus histrio P. L. S. Müller, *Natursyst.*, suppl., 1776, p. 76 (East Indies = Sangihe Islands).

DESCRIPTION Length 31 cm. Weight males 210–217 g, females 191–197 g.

ADULTS General plumage bright red, darker on rump; broad purple-blue band on hindcrown; mauve-blue band from eyes through ear-coverts and sides of neck to meet deep purple-blue of mantle and upper back; scapulars black; primaries and secondaries margined black on outer webs and tipped black; upper wing-coverts red tipped black; broad blue band across breast; thighs black; under tail-coverts red washed blue; tail above reddish-purple, below red; bill orange-red; iris red; legs grey.
8 males: wing 161–181 (167.8) mm, tail 111–131 (121.5) mm, exp. cul. 20–22 (20.9) mm, tars. 19–23 (21.2) mm.

Plate 23
Purple-naped Lory *Lorius domicella* (adults)

F. KNIGHT '16

6 females: wing 160–168 (164.0) mm, tail 108–123 (116.0) mm, exp. cul. 20–23 (21.5) mm, tars. 20–23 (21.2) mm.
JUVENILES (AMNH 616914) Blue of crown extending to nape and below eyes; feathers of head and breast variably margined dusky blue; less defined blue pectoral band; thighs dull mauve-blue; bill orange-brown.

DISTRIBUTION Recorded on the Sangihe and Talaud Islands, and possibly Nanusa Island, in the Molucca Sea, Indonesia.

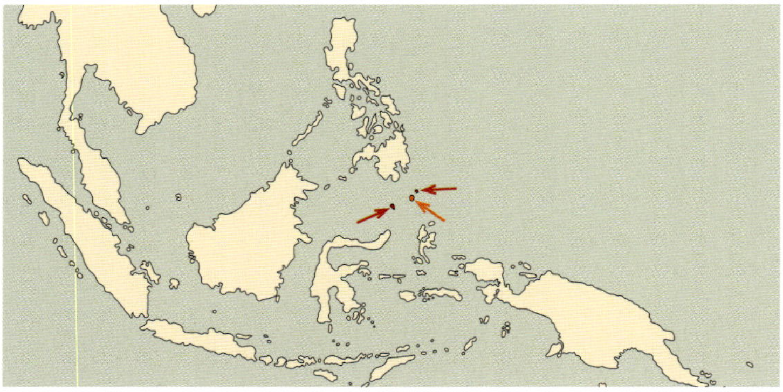

SUBSPECIES
1. *E. h. histrio* (P. L. S. Müller)
The nominate subspecies, as described above, has been recorded on Sangihe, Siau and Ruang Islands, in the Sangihe Group, but it may be extinct.

2. *E. h. talautensis* Meyer and Wiglesworth
Eos histrio talautensis Meyer and Wiglesworth, *Journ. f. Orn.*, **42**, 1894, p. 240 (Talaut Islands).
ADULTS Less black on wing-coverts and flight feathers.
10 males: wing 160–168 (164.2) mm, tail 106–133 (119.2) mm, exp. cul. 20–21 (20.4) mm, tars. 21–23 (22.1) mm.
8 females: wing 161–173 (165.4) mm, tail 117–136 (125.6) mm, exp. cul. 19–20 (19.6) mm, tars. 21–22 (21.5) mm.
Recorded from the Talaud Islands, where now almost exclusively confined to Karakelang; escaped cagebirds recorded on Sangihe Island.

3. *E. h. challengeri* Salvadori
Eos challengeri Salvadori, *Cat. Bds Brit. Mus.*, **20**, 1891, p. 18 (in key), p. 22 (Miangas Island).
ADULTS Blue band on breast less extensive and variably intermixed red; blue band from eyes does not extend to meet blue mantle; smaller size.
2 males: wing 155 mm and 156 mm, tail 101 mm and 107 mm, exp. cul. 19 mm and 20 mm, tars. 19 mm and 22 mm.
1 female: wing 152 mm, tail 98 mm, exp. cul. 19 mm, tars. 23 mm.
Known only from the type series collected on Miangas, in the Nanusa Islands, and now apparently extinct.

STATUS As early as the late 1800s, Red and Blue Lories were uncommon on Sangihe Island, where they had retreated into the mountainous interior because of the spread of coconut plantations around the coast, but were very common in the Talaud Group, where flocks were seen flying from one island to the next to roost (in Meyer and Wiglesworth 1898). In 1978, Murray Bruce found the situation on Sangihe Island and the Talaud Group to be much the same as it was in the late 1800s (in White and Bruce 1986). A very different situation confronted David Bishop when he visited the islands in mid May 1986, for his observations indicated that the lories were no longer present on Sangihe Island, which had been almost totally converted to coconut and nutmeg plantations, and he failed to find birds on Salibabu, in the Talaud Islands during three days of intense searching, though it was possible that a small population survived in remaining forest at the western end of the island (Bishop 1992). Bishop noted also that little or no forest exists on Siau, in the Sangihe Islands, and both Ruang, also in the Sangihe Group, and Miangas, in the Nanusa Islands, are so tiny as to be of dubious practical conservation value even if the lories survived there. On Karakelang, the largest island in the Talaud Group, Red and Blue Lories were found to be widespread, occurring in small numbers in coastal coconut groves and more commonly in forest. Riley (2002) reports that no birds were seen on Sangihe Island during a survey undertaken in 1998–1999, so the endemic population was considered to be extinct, and it was suspected that most, if not all birds seen near the village of Talawid in October 1995 were escaped, or more likely released cagebirds brought to Sangihe from the Talaud Islands.

The Red and Blue Lory is yet another Indonesian parrot seriously endangered by the combination of deforestation and intense trapping for the live-bird market. At present, it is confined to the Talaud Islands, where it survives almost exclusively on Karakelang, and in 1999 research suggested that 1000–2000 birds were being taken from this island each year (Birdlife International 2016). Based on data collected on Karakelang Island in 1996, the population estimate was of 9400–24 150 lories, with population densities of 26.7–65.9 birds per km² in undisturbed forest and 9.1–24.9 birds per km² in agricultural smallholdings (Lambert 1997). Riley (2003) reports that during a visit to Karakelang Island in 1999, the population densities were found to have remained stable, or declined slightly, with 15.2–28.1 birds per km² in primary forest and 4.8–19.3 birds per km² in secondary habitats, equivalent to a total population of 8230 to 21 400 birds, and small numbers of these lories were seen on neighbouring Salibabu Island, but the population was small and apparently restricted to primary forest in the central hills. This estimated total population of 8230–21 400 birds equates roughly to 5500 to 14 000 mature individuals (Birdlife International 2016). Riley points out that although 350 km² of primary forest is protected on Karakelang Island, with 250 km² set aside as a wildlife reserve, management is absent and the forest continues to be threatened by agricultural encroachment, illegal logging and fire. Because of this forest degradation and continuing trapping, the population is suspected to be declining rapidly (Birdlife International 2016).

The Red and Blue Lory is listed on Appendix I of the Convention on International Trade in Endangered Species of Wild Fauna and Flora (CITES).

HABITATS Riley (2003) points out that on Karakelang Island, Red and Blue Lories do forage in secondary habitats, and frequently are encountered in coconut plantations, but they are much more common in forest, where cavities in large trees are available for nesting.

Plate 24
Red and Blue Lory *Eos histrio talautensis* (adults)

F. KNIGHT 11

HABITS Red and Blue Lories are noisy, conspicuous birds, especially when in flight above the forest canopy. They usually are encountered in pairs or small parties, more often during regular flights between resting and feeding areas. Larger numbers congregate at nighttime roosts, which may be on offshore islands, and at these roosts the birds are very vocal. The flight is swift and direct.

CALLS Short, harsh, somewhat chattering screeches are given in flight, and may be heard before the birds are seen, but feeding birds normally are quiet.

DIET AND FEEDING These lories visit fruiting or flowering trees, including coconut palms, to feed on nectar, pollen or soft fruits, and they will feed on cultivated fruits in village gardens. In addition to coconut palms, feeding has been recorded in *Ficus*, *Canarium* and *Syzygium* trees, and insects and their larvae also are included in the diet (in Juniper and Parr 1998).

BREEDING Local residents suggest that nesting commences in April–May, with fledging taking place during June and July, but trappers also claim that the breeding season coincides with the main fruiting season in November–December (in Juniper and Parr 1998). Nests are in cavities in tall trees, and the clutch is said to be one or two eggs (Birdlife International 2016; in Meyer and Wiglesworth 1898).

Sweeney (1998) gives some details of breeding in captivity, including the development of chicks artificially incubated and reared. A clutch comprised two eggs, laid approximately 48 hours apart, and incubated by the female for 25 to 26 days. Newly hatched chicks were extensively covered with long greyish-white natal down, and the bill and claws were black. The eyes commenced to open at day 10, and usually were fully opened by day 14, when pin-feathers appeared through the skin. By day 20 secondary down was present over much of the body, and by day 30 pin-feathers were present on the head and wings. By day 40 feathers had emerged from the sheaths on the head and abdomen, and both wing-feathers and tail-feathers were emerging. By day 60 the chicks were well feathered over much of the body, and by day 70, when almost completely feathered, they were feeding independently. Birds were capable of breeding at two years.

EGGS Sweeney (1998) records measurements of 26.3 (25.1–27.1) × 22.4 (21.8–22.7) mm for eight eggs laid in captivity.

TRIBE **CYCLOPSITTACINI** Salvadori

The specialised arboreal fig parrots belong to this tribe, and their inclusion with lories and lorikeets in the same subclade, as determined by molecular analyses, is not surprising for there are morphological, behavioural and vocal similarities. They are small, stocky parrots with an extremely short, wedge-shaped tail that does not exceed half the length of the wing. The proportionately large, broad bill is deeper than it is long, with a pronounced notch in the upper mandible, and the culmen is prominently ridged. The tongue is smooth. Schodde (1997) points out that an osteological feature of the skull is the completely ossified orbital ring free from zygomatic processes.

Sexual dimorphism is pronounced in most forms, and strong underwing-stripes are present in adults and juveniles of both sexes.

Distribution is confined to the New Guinea region and northeastern Australia.

GENUS ***Cyclopsitta*** Reichenbach

Cyclopsitta Reichenbach, *Vollst. Naturgesch*., 2 *Vögel*, 1 *Avium Syst. nat*., 1850, tab. 1, p. xxxii. Type *apud* Storr = *Psittacula diophthalma* Hombron and Jacquinot.

Holyoak (1970) advocated incorporation of the New Guinea genus *Psittaculirostris* in this genus, but I prefer to retain separation of the two genera. The fig parrots in *Cyclopsitta* are appreciably smaller than *Psittaculirostris* species and there is no elongation of the ear-coverts. The cere is naked and prominent, not feathered as in *Psittaculirostris*. Sexual dimorphism is present in most forms, but not in *C. coxeni*, and young birds normally resemble the adult female.

Cyclopsitta fig parrots occur in the New Guinea region, including the Aru Islands, and in northeastern Australia.

My preference has been to retain Coxen's Fig Parrot as a subspecies of the widespread, polytypic Double-eyed Fig Parrot *C. diophthalma*, though recognising that, of the three subspecies confined to the three major tracts of tropical rainforest along the northeastern coast of Australia, it most approaches the degree of differentiation typical of a species. It is treated as a separate species by del Hoyo and Collar (2014), so that arrangement is adopted here.

CRITICALLY ENDANGERED

Coxen's Fig Parrot

Cyclopsitta coxeni Gould

Cyclopsitta coxeni Gould, *Proc. Zool. Soc. London*, 1867, p. 182 ("... scrub on the east coast", = near Brisbane, Queensland).

DESCRIPTION Length 15 cm.

ADULTS General plumage colouration green, slightly more yellowish on underparts; centre of forehead blue, with green bases to some feathers showing through; some reddish feathers on lores and sides of forehead at base of upper mandible; bright yellowish-green in front of and above eyes; lower cheeks to ear-coverts red, bordered below by variable band of mauve-blue; flanks and sides of breast bright yellow; underwing-coverts yellowish-green slightly mottled dull greenish-blue; outer webs of primaries and outer secondaries deep blue, inner webs dull brownish-grey with two prominent cream-yellow patches forming broad underwing-stripes; inner webs of innermost tertials and

secondary-coverts orange-red; tail dark green; bill pale grey at base, becoming dark grey towards tip; iris dark brown; legs greyish-green.
8 males: wing 93–100 (95.6) mm, tail 42–47 (43.8) mm, exp. cul. 14–16 (15.6) mm, tars. 13–16 (14.2) mm.
4 females: wing 92–94 (92.8) mm, tail 42–46 (43.8) mm, exp. cul. 15–16 (15.5) mm, tars. 14–15 (14.8) mm.
JUVENILES (?, AM A1176) Ear-coverts orange-red becoming yellowish-orange anteriorly; lower cheeks green with reddish-brown bases to some feathers showing through; lores and sides of forehead green; base of lower mandible horn coloured.

DISTRIBUTION Within a restricted range in northeastern Australia, Coxen's Fig Parrot is very sporadically dispersed in coastal and contiguous upland areas from the Maryborough, or possibly Rockhampton district, southeastern Queensland, south to the Macleay River valley, northeastern New South Wales.

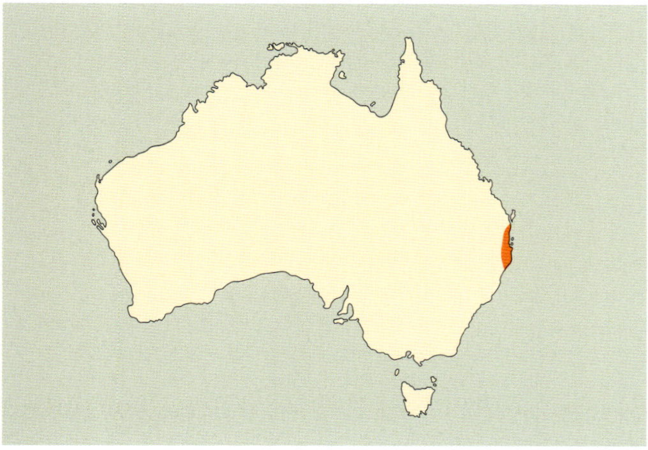

STATUS Widespread landclearance and logging of rainforest trees in the 1890s have left only small, very fragmented patches of subtropical rainforest, the preferred habitat of Coxen's Fig Parrots, and it is this loss of habitat that is responsible for the near extinction of these parrots. Recent targeted searches have been unsuccessful, and infrequent unconfirmed sightings make up the only evidence of their continued survival, so photographic confirmation is required. Gynther (2013) cites credible records from Fraser Glen, on the Sunshine Coast in southeastern Queensland, in January 2013, and from the Cambridge Plateau area of Richmond Range National Park, in far northeastern New South Wales, in September 2013, as being among an irregular, but continuing trickle of sightings and call records providing good support for the continued survival of these fig parrots, but he does emphasise the importance of obtaining photographs or sound recordings. There have been no recent records from the southern sector of the range, south of the Richmond River in northeastern New South Wales. With low reliability, the total population of Coxen's Fig Parrot has been estimated at 100 birds and with even lower reliability, this was estimated to be divided into four subpopulations of a maximum of 50 individuals each (Garnett *et al.* 2011). A low rate of sightings has persisted for several decades, so the population presumably has been very small for a long time.

Under state legislation the Coxen's Fig Parrot is categorised as endangered, and is afforded special protection. It is listed as endangered also under the Australian *Environment Protection and Biodiversity Conservation Act 1999* and is listed on Appendix I to the Convention on International Trade in Endangered Species of Wild Fauna and Flora (CITES).

In north Queensland, the closely related, or possibly conspecific Red-browed Fig Parrot *C. diophthalma macleayana* and Marshall's Fig Parrot *C. d. marshalli* are common within their restricted ranges.

HABITATS Holmes (1990) points out that lowland subtropical rainforest is the prime habitat for Coxen's Fig Parrots, but small local populations now inhabit mainly dry rainforest and cool subtropical rainforest, particularly in foothills of the McPherson and Great Dividing Ranges. Schodde and Mason (1996) note that all confirmed native food trees are elements of lower altitude rainforest, both complex and simple and dry, so such forest is the required habitat, but dependence on the fruiting of different food trees may necessitate access to a range of complex forest types. My only encounter with these fig parrots occurred more than 60 years ago, in October 1955, just north of Uki, in the Murwillumbah district, far northeastern New South Wales, when I had a fleeting, but good view of three birds as they flew out from a stand of loquat trees in cultivated farmland into a nearby narrow band of depauperate gallery forest along a small watercourse. Schodde and Mason point out that the continuing presence of these fig parrots in the Murwillumbah district indicates that sufficient habitat remains to maintain a diminished population, but it is unknown whether this habitat is capable of supporting a breeding population or is suitable only for foraging groups visiting from neighbouring forests in the Border Ranges region.

MOVEMENTS Holmes (1990) notes that Coxen's Fig Parrots apparently undertake regular seasonal movements. Schodde and Mason (1996) point out that recent records from the Murwillumbah district, in far northeastern New South Wales, are consistent with the claim that these fig parrots are seasonally nomadic throughout their range, breeding in isolated pairs in major stands of rainforest during the summer months of October to March, and congregating in small groups to forage more widely through pockets of lowland rainforest during March to October. Chisholm (1924) recalled that in May 1922, he was informed by Charles Benfer, an observer resident in the Blackall Ranges, southeastern Queensland, that at that time only a few birds were present, but the majority were expected to return in August, presumably after wintering in the north. Holmes notes that a similar pattern prevails at the Sarabah Range, on the western Lamington Plateau, southeastern Queensland, where fig parrots have been seen only during August to February.

HABITS The little information that we have indicates that the habits of Coxen's Fig Parrots do not differ from habits of the other fig parrots in north Queensland. Difficulties in detecting these small green parrots while they are feeding in rainforest trees are highlighted in recollections by Neville Cayley of his early encounters with the birds, as quoted in his *Australian Parrots: Their Habits in Field and Aviary*, published in 1938:

> *When a boy and resident in Casino, Richmond River, New South Wales, I shot one of these birds in mistake. I had fired at a Fruit Pigeon in a tall fig tree and judge my astonishment when down fell a small parrot along with the pigeon. My father identified it as a Fig Parrot, the only name the species was known by in that district. . . . Visiting the same brush,*

> *which was quite close to the town, we were fortunate enough to locate these lorilets on several occasions. They were feeding in the fig-trees, usually in company with pigeons, fig-birds and other fruit-eating species. Difficult to locate, the only indication one had that any birds were feeding in the trees was the quantity of falling fruit dislodged by them. It is a neck-breaking task locating brightly coloured birds feeding among the tree-tops, and extremely difficult to identify them without the aid of field-glasses. On several visits to the far northern rivers of New South Wales during later years attempts to locate these birds proved fruitless; certainly much of the virgin scrub had disappeared, but there were large areas still untouched by axe or fire. Only the older residents remember the species; they saw small flocks occasionally, especially when felling and clearing the big scrubs.*

Most of the recent unconfirmed sightings of Coxen's Fig Parrots have been of single birds or pairs, but there is at least one report of a party of four birds (in Garnett *et al.* 2011). I suspect that in earlier times, when they were more numerous small flocks would have been common outside the breeding season, and these flocks would have congregated at communal nighttime roosts, associations that I have recorded for Marshall's Fig Parrot *C. diophthalma* on Cape York Peninsula.

Fig parrots are very 'lorikeet-like' in many of their actions. When moving along a branch, they crouch low, body aligned parallel to the branch, and progress by grasping the branch with the inward-turned feet, one over the other.

The flight of fig parrots is quite swift. After each series of wingbeats, the wings are withdrawn towards the body, but so rapid is this action that it causes no noticeable undulation in flight. The parrots fly in a straight line, generally above the forest canopy, and do not twist and turn through the trees.

CALLS I doubt that the calls of Coxen's Fig Parrot would differ from the calls of other *Cyclopsitta* fig parrots that I have heard in north Queensland and in Papua New Guinea. The contact call is a sharp, penetrating *tseet*, repeated two or three times in quick succession, each note being clear and quite distinct. This call is given during flight and just after alighting or prior to departure. The alarm call is a shrill rolling screech, while a softer screech sometimes is given in flight, particularly during the evening return flight to roost. There are reports of feeding birds keeping up a soft chattering while feeding, but this has not been my experience and I have found fig parrots to be particularly quiet while feeding. At Iron Range, Cape York Peninsula, I found that Marshall's Fig Parrots *C. diophthalma* would respond to an imitation of their call-notes by calling loudly while repeatedly circling back to a tree near to where I was calling, and one male fluttered down to the lowest branches only a metre or so above my head.

DIET AND FEEDING Although fig parrots eat mostly seeds from ripe or near-ripe fruits of native figs, they have been recorded taking a variety of native and cultivated fruits and berries as well as nectar or pollen, and I suspect that insects and their larvae also feature prominently in the diet, especially during the breeding season. They are methodical feeders, returning to the same tree, or even to the same fruit, until the seed supply is exhausted. Observations of feeding Marshall's Fig Parrots at Iron Range National Park, Cape York Peninsula, north Queensland, lead me to suspect that insects and their larvae are an important food item for fig parrots. At Iron Range, I noticed that birds were returning time after time to certain trees of hairy fig *Ficus hispida* and feeding on specific fruits, but leaving other fruits that appeared to be at the same stage of ripeness. Subsequent investigation revealed that many fruits were parasitised by fig wasps (Agoanidae), and I strongly suspect that it was these parasitised fruits that the birds were selecting.

Holmes (1990) notes that the principal food trees recorded for Coxen's Fig Parrot are Moreton Bay fig *Ficus macrophylla* and Watkin's fig *F. watkinsiana*, with less frequently recorded native food plants being rusty fig *F. rubiginosa*, white fig *F. virens*, deciduous fig *F. superba*, sandpaper fig *F. fraseri*, blue quandong *Elaeocarpus grandis*, sour cherry *Syzygium corynanthum* and probably bollygum *Litsea reticulata*. Nectar or pollen is obtained occasionally from flowers of eucalypts or melaleucas. Holmes points out also that an interesting feature of the foraging behaviour of these fig parrots is their tendency to favour certain trees; at the Sarabah Range, on the western Lamington Plateau, southeastern Queensland, a rusty fig tree was visited by birds on successive days for at least a week in September and October, and another tree for three successive days in January. Also, at the Conondale Range, southeastern Queensland, birds were seen coming to the same Moreton Bay fig tree in October–November of successive years. I suspect that figs parasitised by fig wasps were attracting the parrots to specific trees. Cultivated exotic plants appear to be utilised as food sources when native foods are scarce, and this probably is during the cooler winter months between March and October.

Recorded exotic food plants include edible fig *Ficus carica*, loquat *Eriobotrya japonica* and late cotoneaster *Cotoneaster lacteus*. In northeastern New South Wales, Coxen's Fig Parrots have been observed in the company of Crimson Rosellas *Platycercus elegans* feeding on fruits of blue quandong, and with Scaly-breasted Lorikeets *Trichoglossus chlorolepidotus* and Musk Lorikeets *Glossopsitta concinna* feeding at the flowers of silky oaks *Grevillea robusta*, and at Bunya National Park, southeastern Queensland, they were with Little Lorikeets *Parvipsitta pusilla* feeding in what appeared to be a forest plum *Davidsonia pruriens* (in Lendon 1979).

BREEDING Nesting presumably takes place between August and December, during late spring and early summer, but I know of no authentic breeding records for Coxen's Fig Parrot, and have given my reasons for rejecting a record from Urangan, near Maryborough, southeastern Queensland (see Forshaw 1979). Also, I strongly doubt the reliability of another record from the Maryborough district, where at Mount Poppel a clutch of three eggs was taken on 15 September 1908 from a nest in a small hole in the limb of a tree at approximately 6 m above the ground, but measurements of these eggs strongly suggest that they may be those of the Little Lorikeet *Parvipsitta pusilla*.

Nests of *Cyclopsitta* fig parrots found in north Queensland, Australia, and in New Guinea have been in cavities excavated by the birds in a rotten stump or tree trunk, in a dead limb of a living tree or, on one occasion, in the base of an epiphyte growing on the trunk of a rainforest tree. There is no reason to suspect that

Plate 25
Coxen's Fig Parrot *Cyclopsitta coxeni*
UPPER adult CENTRE TOP juvenile CENTRE BOTTOM adult
LOWER adult

F. KNIGHT 2009

nests of Coxen's Fig Parrots would not be similarly excavated by the birds. I suspect that as with the other species, a normal clutch would comprise two or three eggs, and incubation by the female would last approximately 18 to 20 days, with fledging occurring some 40 days after hatching.

EGGS In Museum Victoria, Melbourne, there is a clutch of three eggs, registered as eggs of Coxen's Fig Parrot, taken at Mount Poppel, south of Maryborough, southeastern Queensland, on 15 September 1908, and these measure 20.3 (20.0–20.7) × 16.2 (16.1–16.2) mm but, in my opinion, these eggs are too small and probably are eggs of the Little Lorikeet *Parvipsitta pusilla*.

SUBFAMILY **AGAPORNITHINAE** Salvin

Included in this subfamily are the African *Agapornis* lovebirds, the Papuo-Malesian hanging parrots and monotypic *Bolbopsittacus* from the Philippine Islands, and molecular analyses indicate that the subfamily is sister to Loriinae. All species are small parrots with very short, broad and slightly rounded tails, with modifications in bill structure reflecting foraging behaviour.

GENUS ***Loriculus*** Blyth

Loriculus Blyth, *Journ. As. Soc. Bengal*, **19**, 1850, p. 236. Type, by monotypy, *Psittacus galgulus* Linnaeus.

Arboreal *Loriculus* parrots popularly are known as 'hanging parrots' because of their strange roosting habit of hanging upside down. They have small, finely pointed bills, well suited to probing into blossoms to feed on nectar, and the tail is so short that it sometimes is hidden by the tail-coverts.

The genus is distributed from eastern New Guinea and the Bismarck Archipelago west to the Indian subcontinent, including Sri Lanka. Two species occur in Sulawesi, in the Indonesian Archipelago, but elsewhere the species replace each other geographically.

EXTINCT

Philippine Hanging Parrot

Loriculus philippensis (P. L. S. Müller)

Psittacus philippensis Müller, *Natursyst.*, suppl., 1776, p. 80 (Philippines, *ie* Luzon, *ex* Buffon).

OTHER NAME Colasisi.

DESCRIPTION Length 14 cm. Weight 32–40 g.

ADULT MALE General plumage colouration bright green, slightly paler and more yellowish on underparts and darker on upper wing-coverts; forehead and forecrown red; hindcrown to upper back golden yellow with orange band across nape; red patch on throat to upper breast, and bordered yellow-orange at sides of throat; rump and upper tail-coverts bright red; pale blue on sides of rump; primaries and outer secondaries greyish-black, on outer webs broadly margined dark green; undersides of wings greenish-blue; tail above dark green, below greenish-blue; bill orange-red; iris brown; legs orange.

7 specimens: wing 96–104 (100.0) mm, tail 44–51 (43.1) mm, exp. cul. 14–15 (14.4) mm, tars. 11–14 (13.0) mm.

ADULT FEMALE Hindcrown to hindneck dull golden yellow; mantle and upper back green suffused orange-yellow; no red patch on throat; forecheeks and at base of bill suffused blue.

8 specimens: wing 95–104 (99.8) mm, tail 45–56 (49.6) mm, exp. cul. 12–15 (13.4) mm, tars. 12–14 (13.4) mm.

JUVENILES Like adult female, but little or no red on forehead.

DISTRIBUTION Philippine Islands, including the Sulu Archipelago.

SUBSPECIES

1. *L. p. chrysonotus* Sclater

Loriculus chrysonotus Sclater, *Ibis*, 1872, p. 324, pl. 11 (Cebu, Philippine Islands).

This subspecies is described above, and it formerly occurred on Cebu, central Philippines.

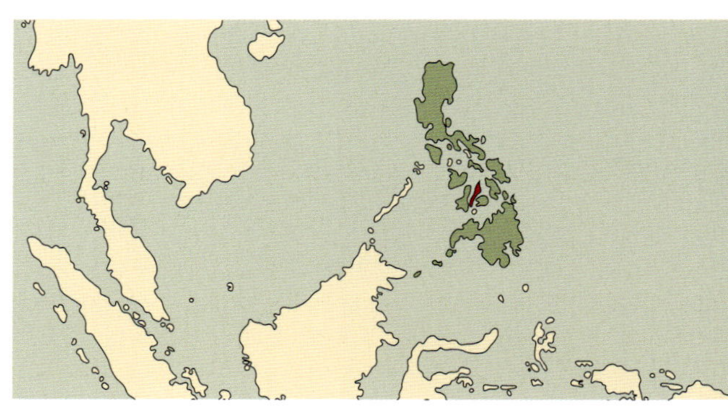

2. *L. p. siquijorensis* Steere

EXTINCT

Loriculus Siquijorensis Steere, *List Bds Mamms Steere Exped.*, 1890, p. 6 (Siquijor, Philippine Islands).

ADULT MALE Red patch on forehead and forecrown, but hindcrown to mantle and upper back green; small red patch on throat.

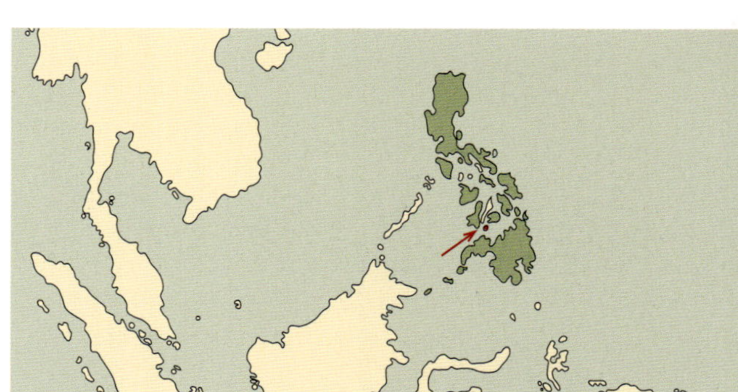

5 specimens: wing 97–100 (98.6) mm, tail 41–50 (45.8) mm, exp. cul. 14–15 (14.8) mm, tars. 13–14 (13.2) mm.
ADULT FEMALE Like male, but no red throat patch; sides of forehead, lores, cheeks and chin to upper throat suffused pale blue. 6 specimens: wing 98–104 (101.3) mm, tail 45–51 (48.2) mm, exp. cul. 13–15 (13.3) mm, tars. 13–15 (14.0) mm.
JUVENILES Similar to adult female, but little or no red on forehead. This subspecies formerly occurred on Siquijor, central Philippines.

3. Other subspecies of the polytypic Philippine Hanging Parrot are distributed throughout the Philippine Islands, including the Sulu Archipelago.

STATUS Philippine Hanging Parrots are widely distributed throughout much of the Philippine Archipelago and, although the population trend appears to be decreasing because of extensive deforestation, the decline is not considered to be sufficiently rapid to render the species eligible for vulnerable status (Birdlife International 2016). While not disagreeing with that assessment, I would point out that declines are uneven with some island populations more at risk than others, particularly on smaller islands. In the early 1990s, it was suggested that these heavily-trapped parrots must be either scarce or rapidly declining on Bohol Island in the central Philippines (Brooks *et al.* 1995). The apparent extinction of two distinctive subspecies is further cause for concern.

Philippine Hanging Parrots are popular cagebirds and are traded freely between islands, so occasional records from both Cebu and Siquijor Islands usually are attributed to escaped cagebirds, and there are no confirmed records of the endemic subspecies. Birds of other subspecies are sold as cagebirds in Cebu City, so there are likely to be numerous escaped birds throughout the island (Kennedy *et al.* 2000).

It seems that *chrysonotus* persisted in some numbers on Cebu Island until 1929, when live birds became available in Britain and the United States, but there have been no subsequent confirmed records. In the early 2000s, only seven small, fragmented patches of degraded forest survived on Cebu, and between 2001 and 2004 mist-netting, line transects, point counts and interviews with local persons were used in surveys to monitor bird populations in three of these forest patches (Paguntalan and Jakosalem 2008). During these surveys a single hanging parrot was observed in Tabunan Forest, where escaped cagebirds previously had been reported, and, because of a light yellow, instead of orange-yellow on the head and back, this bird also may have been an escaped cagebird. There was a report by local villagers of a hanging parrot fitting the description of *chrysonotus* being seen in forest near Nug-as village in August 2004, but this has not been verified.

In the early 1950s, Dioscoro Rabor suspected that *siquijorensis* could be extinct when survey parties failed to find any birds, and his suspicions were strengthened when an adult male collected on the island in December 1954 proved to be an example of the quite different *L. p. regulus*, which presumably was an escaped cagebird (in Rand and Rabor 1960). On Siquijor Island, in the early 1990s, four remaining patches of forest covered a total of only 781 ha, and three of these patches were surveyed between 27 and 30 August 1991 (Evans *et al.* 1993). No hanging parrots were recorded, and again recent reports have been referred to escaped cagebirds of other subspecies.

HABITATS Philippine Hanging Parrots frequent all forest types and forest patches, even urban gardens, up to 1000 m, with occasional records up to the limits of mossy forest at about 2500 m (Kennedy *et al.* 2000). They regularly visit coconut plantations to feed in flowering palms. Between 400 m and 800 m, at Tabunan Forest, Cebu, where a likely escaped cagebird was observed during surveys undertaken between 2001 and 2004, the closed-canopy forest with trees exceeding 25 m in height covered an area of at least 1.85 km² and formed a thin, segmented strip of forest on a steep limestone hillside, with closed-canopy areas of less than 0.1 km² (Paguntalan and Jakosalem 2008). In this forest the vegetation was dominated by *Ficus*, *Homolanthus*, *Syzygium*, *Macaranga*, *Leeia*, *Dillenia*, *Leucosyke* and *Mangifera* trees, together with *Sterculia philippinensis* and *Diospyros philippensis*. The gradient changed drastically from a gentle 11–34 per cent at the interface between the forest and abandoned farmlands to more than 40–50 per cent at the forest edge, with active and abandoned farmland within the forest and in the middle of the forest gradually dividing the forest stand into two main blocks. On Siquijor Island, in the early 1990s, probable escaped cagebirds were reported at the only locality where some primary forest remains (Evans *et al.* 1993).

HABITS Philippine Hanging Parrots usually are encountered singly, in pairs or less commonly in small parties. They spend much of the day feeding in the middle to upper stages of flowering trees, where they can be difficult to detect amidst the foliage, though invariably their presence will be indicated by the continual calling and movement of leaves as they actively clamber along branches to get at blossoms or fruits. They have been seen nearer to the ground feeding in flowering or fruiting bushes, and they will join with other frugivores, including *Coracina* cuckoo-shrikes, Coletos *Sarcops calvus* and *Aplonis* glossy starlings, in mixed foraging assemblages. Though rather shy when disturbed in the forest, they can be surprisingly bold when feeding in coconut palms or urban gardens.

The swift flight is characteristically undulating and, when accompanied by calling, is a valuable aid to field identification.

CALLS The contact call is described as a distinctive sharp twitter emitted as a single note or repeated rapidly up to three times to resemble *twick … twick … twick*, or given as a prolonged staccato series especially when leaving a perch to fly some distance (Kennedy *et al.* 2000). Also uttered is a rapidly repeated disyllabic *seep … seep*.

DIET AND FEEDING The diet comprises nectar, blossoms and soft fruits, and these parrots are attracted to flowering coconut palms and banana trees. Also, they have learned to take fermenting coconut nectar being harvested by villagers in joints of bamboo fitted over the cut ends of flowering stalks, and at times the parrots will become intoxicated. They will join with other frugivores to feed on *Ficus* fruits (Rand and Rabor 1960).

BREEDING Little is known of the nesting habits. On Bohol Island, central Philippines, in April–May, birds with enlarged gonads were collected, and three nestlings were taken from a hole 12 m above the ground in a dead tree standing at the edge of a small clearing in primary forest (Rand and Rabor 1960).

Low (1980) gives details of a successful breeding in captivity. The clutch comprised three eggs, the first two being laid on successive days, and the third egg was laid two days after the second. Incubation lasted 20 days, and only the female brooded, but some six to nine days after the chicks had hatched the male commenced roosting in the nestbox at night. The young birds fledged approximated five weeks after hatching.

EGGS Harrison and Holyoak (1970) give 15.5 × 13.7 mm as measurements of a single egg of *L. p. chrysonotus* laid in captivity, but I suspect that identification may be doubtful because the measurements are significantly smaller than the measurements of 18.7 (18.4–19.0) × 16.4 (16.2–16.7) mm for three eggs of *L. p. worcesteri* also laid in captivity.

ENDANGERED

Wallace's Hanging Parrot

Loriculus flosculus Wallace

Loriculus flosculus Wallace, *Proc. Zool. Soc. London*, 1863 (1864), p. 488 (Flores).

DESCRIPTION Length 12 cm.
ADULT MALE General plumage green, noticeably lighter on underparts; nape to upper mantle suffused brownish-orange; elongated red patch on throat; rump and upper tail-coverts crimson; undersides of wings greenish-blue; tail above green tipped paler green, on outer feathers the pale tips washed orange-red, underside of tail greenish-blue; bill orange-red; iris orange; legs orange.
ADULT FEMALE Little or no red on throat; iris brown (Wallace).
1 unsexed (type): wing 79 mm, tail 33 mm, exp. cul. 11 mm, tars. 12 mm.
JUVENILES Similar to female, but with duller bill and legs (Coates and Bishop 1997).

DISTRIBUTION Recorded only from Flores and neighbouring Rinca Island, East Nusa Tenggara, Indonesia.

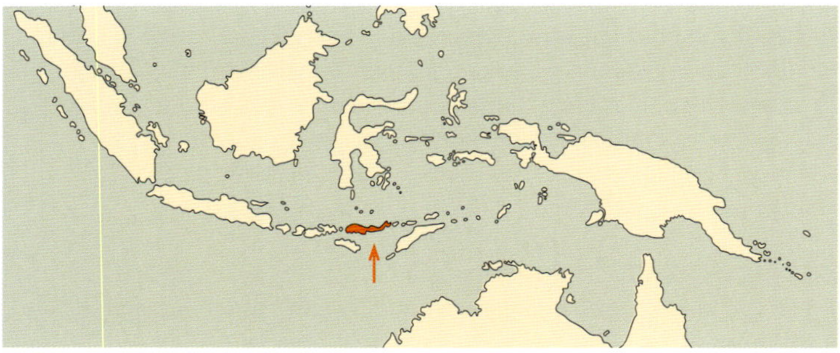

STATUS For more than a century after its discovery, Wallace's Hanging Parrot remained a mysterious bird, about which almost nothing was known, and confirmation of its continued existence on Flores did not come until May 1978, when Father Erwin Schmutz located birds in rainforest at 1035 m and subsequently elsewhere at lower elevations (Bishop *in litt.* 1987). During a survey undertaken in July to September 1993, these hanging parrots were found only at Tanjung Kerita Mese, western Flores, where they were common, and in early November 1994, a small flock was recorded at another locality in western Flores (Butchart *et al.* 1996). Although possibly locally common at some localities, with a density of about 10 birds per hectare reported at Tanjung Kerita Mese, they remain generally scarce and are threatened by habitat loss resulting from extensive deforestation for agriculture, with the remnant fragmented forest further impacted by firewood collection, commercial logging and timber extraction for housing construction to meet the needs of an expanding human population (Birdlife International 2016). No semi-evergreen forest below 1000 m is included within gazetted protected areas. Birds were first recorded on neighbouring Rinca Island, in Komodo National Park, in late April 2003, and nesting activities were observed there in April 2006 (Imansyah *et al.* 2008).

Based on an assessment of known records, descriptions of abundance, and range size, the current population is estimated at 2500–9999 mature birds, with rapid declines suspected if conservation measures are not implemented in the near future (Birdlife International 2016).

HABITATS At Tanjung Kerita Mese, western Flores, in July–September 1993, Wallace's Hanging Parrots were recorded between 450 m and 980 m, in primary semi-evergreen rainforest with a few deciduous emergents and a canopy height of 35–40 m, and the majority of records were in a narrow altitudinal band from 850 m to 980 m, particularly where large *Ficus* trees were prevalent, with the parrots appearing to be rare below 850 m (Butchart *et al.* 1996). At another locality in western Flores, in early November 1994, a small flock was recorded in a narrow corridor of semi-evergreen forest (in Butchart *et al.* 1996).

In late April 2003, in the south of Rinca Island, birds were observed in coastal moist deciduous forest, which was dominated by *Pterospermum javanicum*, a tree that can reach 25 m in height, and in April 2006 nesting activities were observed at the same location (Imansyah *et al.* 2008). These records from outside the known altitudinal range and preferred habitat on Flores suggest that these hanging parrots may have broader habitat preferences, so could be more widespread and less threatened, but this requires further investigation (Birdlife International 2016). Coates and Bishop (1997) include moist deciduous forest on limestone and eucalypt forest as habitats also frequented by Wallace's Hanging Parrots, though I suspect that eucalypt forest may be visited merely to feed in flowering trees.

HABITS Wallace's Hanging Parrots usually are encountered in pairs or small flocks of up to 10, or occasionally 20 birds. They are noisy and conspicuous when in flight above the forest canopy but, despite their intermittent calling while feeding in the treetops, they can be difficult to detect as they clamber among the foliage in search of fruits or blossoms. At Tanjung Kerita Mese, western Flores, in July–September 1993, all sightings were of birds flying above the forest canopy or feeding in the canopy or subcanopy down to a minimum height of 10 m, but the much earlier sighting at Paku, western Flores, by Father Schmutz of a bird perched only 2.5 m above the ground suggests that at times these parrots may feed in the mid-storey (Butchart *et al.* 1996). Also

---

Plate 26
UPPER LEFT Philippine Hanging Parrot
Loriculus philippensis chrysonotus (adult ♂)
UPPER CENTRE Philippine Hanging Parrot
Loriculus philippensis siquijorensis (adult ♂)
UPPER RIGHT Philippine Hanging Parrot
Loriculus philippensis siquijorensis (adult ♀)
LOWER LEFT Wallace's Hanging Parrot
Loriculus flosculus (adult ♂)
LOWER RIGHT Wallace's Hanging Parrot
Loriculus flosculus (adult ♀)

F. KNIGHT '10

at Tanjung Kerita Mese, a bird was seen in an apparent display flight, weaving in and out of the forest canopy while uttering a distinctive call, and birds regularly were seen drooping and flicking their wings, or wing-shivering to highlight the red rump and upper tail-coverts, an action that I suspect could be a contact or territorial signal. The flight is swift and direct with an audible 'whirring' of the wingbeats.

CALLS At Tanjung Kerita Mese, western Flores, these hanging parrots were heard calling continuously in flight and while feeding (Butchart *et al.* 1996). The flight call is described as a sharp, screeching *strrt...strrt*, which is louder and more low-pitched than the call given while perched, and one bird was seen to perform an apparent display flight while giving a slightly hoarse *chi-chi-chi-chi-chi* call.

DIET AND FEEDING At Tanjung Kerita Mese, most foraging observations were of birds feeding in *Ficus* trees, and less commonly they were seen feeding on berries, flowers and buds in other trees (Butchart *et al.* 1996). Local dispersal of the parrots appeared to be related closely to the distribution of fruiting fig trees, and any dependence on a single food source could add to their vulnerability.

BREEDING In mid April 2006, in the south of Rinca Island, a nest was found in a hollow approximately 10 cm wide and at a height of about 15 m in a dead branch of a *Terminalia catappa* tree (Imansyah *et al.* 2008). The attendant male and female were observed alternating in nesting activities and, while one bird was inside the hollow, the partner waited outside. The time spent inside the hollow by each bird was approximately 10–15 minutes. The nesting tree was near to several large *Ficus* trees, but the birds were not observed feeding on the fruits. Nothing further is known of the nesting habits.

EGGS Schönwetter gives 19.7 (19.5–20.0) × 16.0 (15.5–16.3) mm as measurements of three eggs, but these almost certainly are eggs of the Green Hanging Parrot *L. exilis*, the confusion being due to the practice of treating *L. flosculus* and *L. exilis* as conspecific.

SUBFAMILY **PSITTACULINAE** Vigors

Molecular analyses have identified a sister relationship between parrots in this subfamily and the platycercine parrots, and within this subfamily the analyses have recovered two clades, together with the *Micropsitta* pygmy parrots as a separate lineage not embedded within either of these two clades (in Joseph *et al.* 2012). The two clades are differentiated at tribal level, and *Micropsitta* also is given tribal rank. The collective term 'coral-billed parrots' has been applied to parrots belonging to the two clades, and refers to the smooth-surfaced yellow to red bill of most species. Another conspicuous external feature is their pericyclic irides.

TRIBE **PSITTACULINI** Vigors

A smooth-surfaced, usually pale-coloured bill and a contrastingly coloured rump or underwing-coverts are two prominent plumage features characterising parrots in this tribe. The bill lacks a pronounced culmen ridge, and in some genera is red in colour. The pericyclic irides are pale-coloured or brown, and white irides seem to be correlated with greenish legs (Smith 1975). The inconspicuous cere is partly or fully feathered. Smith pointed out that the courtship display is complex, and in most genera sexual dimorphism is pronounced. Juveniles may resemble the adult female or they may exhibit a distinctive juvenile plumage with young males usually acquiring the adult female plumage before attaining adult male plumage. There is no underwing-stripe in juveniles or adults.

This tribe is widely distributed from the Mascarene Islands and Africa through India and South-east Asia to Australia.

GENUS ***Prioniturus*** Wagler

Prioniturus Wagler, *Abh. K. Bayer. Akad. Wiss., Math.-Phys. Kl.*, **1**, 1832, p. 490. Type, by monotypy, *Psittacus platurus* Vieillot.

As their name implies, midsized, stocky parrots belonging to this genus are characterised by a conspicuous elongation of central feathers of the short squarish tail in the form of bare shafts terminating in spatules. There is an indistinct notch in the upper mandible of the proportionately robust bill, and the cere is closely bordered by feathers, only the nares being unfeathered. Sexual dimorphism is pronounced in some species, but slight in others. Juveniles resemble or are duller than the adult female and lack the elongated central tail-feathers.

Prioniturus is distributed from the Philippine Islands to Sulawesi, including adjacent islands, and Buru in Maluku Province, Indonesia. Like most forest birds in the Philippine and Indonesian Archipelagos, they are adversely affected by loss of habitat, and two species are endangered, but all are very poorly known and I suspect that better information would reveal that additional species are at risk.

ENDANGERED

Green Racquet-tail

Prioniturus luconensis Steere

Prioniturus Luconensis Steere, *List Bds Mamms Steere Exped.*, 1890, p. 6 (Marinduque and Luzon).

DESCRIPTION Length 29 cm.

ADULT MALE General plumage yellowish-green, paler on head and mantle and slightly darker on upper wing-coverts and more yellowish on underparts, particularly under tail-coverts; primaries and outer secondaries dusky grey-black suffused green on outer webs, and on three outermost primaries slightly suffused dull blue towards tips of outer webs; white margins to inner edges of secondaries; underwing-coverts yellowish-green; undersides of flight feathers and tail-feathers greenish-blue; upperside of tail

green, and spatules at tips of central bare shafts black variably suffused blue; bill pale bluish-grey becoming horn-coloured towards tip; iris dark brown; legs bluish-grey.
8 specimens: wing 143–150 (147.0) mm, tail 125–155 (142.5) mm, exp. cul. 16–19 (17.3) mm, tars. 16–18 (16.8) mm.
ADULT FEMALE Green of general plumage less yellowish, slightly darker; shorter bare shafts of central tail-feathers.
8 specimens: wing 148–158 (151.9) mm, tail 114–134 (125.6) mm, exp. cul. 16–19 (17.3) mm, tars. 16–18 (16.9) mm.
JUVENILES Like adults, but central tail-feathers tipped blue and with pointed tips, but no elongated bare shafts.

DISTRIBUTION Confined to Luzon and Marinduque, northern Philippine Islands.

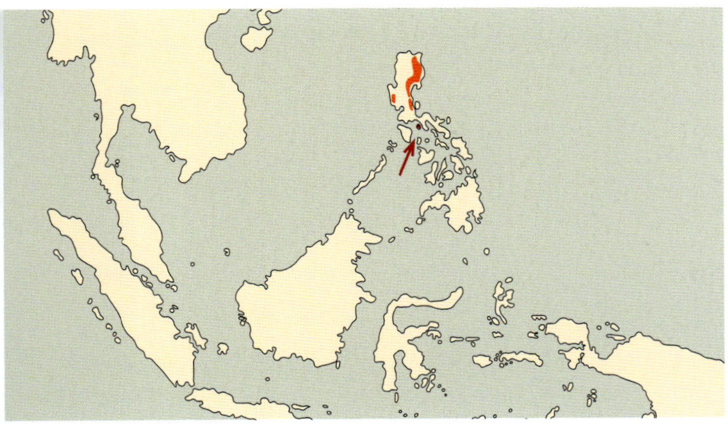

STATUS Green Racquet-tails formerly were widespread and locally common throughout northern and central Luzon and on Marinduque, but there have been no recent records from Marinduque, and on Luzon the population seems to be declining rapidly, having all but disappeared from the southern Sierra Madre Mountains, and virtually all recent records have come from the northern Sierra Madre Mountains (Kennedy *et al.* 2000; Birdlife International 2016). Since about 2000, these racquet-tails have been recorded at only seven out of 29 historic sites and, although possibly a function of poor coverage, this may reflect a real contraction of the range (Española *et al.* 2013). Between December 2009 and September 2010, distance sampling was used along nearly 500 km of line transects at 14 sites in surveys across Luzon for 25 species of parrots, pigeons and hornbills, which are large-bodied forest frugivores, and during these surveys there were 21 encounters with Green Racquet-tails at only four sites, with just a single bird recorded in the largest reserve (Española *et al.* 2013). From these surveys population estimates were of 246 (42–1434) birds for Northern Sierra Madre Natural Park and 174 (80–380) for Subic Bay Forest Reserve and Bataan Natural Park. These data suggest that there are fewer than 250 mature birds in each of the two remaining subpopulations, and imply that the estimated total population of 1500 to 3800 birds could include fewer than 2500 mature individuals (Birdlife International 2016). Following on from these surveys, it was recommended that the greatest conservation concern must be directed at these racquet-tails.

Reasons for the apparent rapid decline in numbers of Green Racquet-tails are not readily understood, for the birds seem to be absent from some areas of potentially suitable habitat, and there is little demand for them in the live-bird market (see Kennedy *et al.* 2000). However, they have disappeared from a number of localities, largely due to deforestation and the fragmentation of remnant forest (Española *et al.* 2013). Loss of forest habitat has been massive, and presumably is the cause of their probable extinction on Marinduque, where only 3 per cent cover remained in 1988 (Birdlife International 2016). On Luzon, forest cover in the Sierra Madre Mountains has declined by 83 per cent since the 1930s, and most remaining areas are subject to logging concessions, which will involve additional degradation from road-building operations with consequent upsurges in illegal clearing. These racquet-tails now may occur in only two locations, with protection afforded only in the Subic Bay area. At Quezon National Park, where they were regular in the 1980s, Green Racquet-tails have been replaced by the larger Montane Racquet-tail *Prioniturus montanus*, so indicating that they may be disadvantaged also by competition from larger frugivores.

HABITATS Green Racquet-tails frequent lowland and foothill forests from 300 m up to 1000 m, and on Marinduque they were recorded above 1000 m (Birdlife International 2016). At times they may be encountered in secondary-growth or heavily degraded forest, and occasionally venture out to feed in fruiting trees in open areas or cultivation adjacent to forest margins.

HABITS In pairs or small groups, arboreal Green Racquet-tails are noisy and conspicuous when in flight above the forest canopy, but are easily overlooked while feeding quietly amidst the treetops where their green plumage colouration blends with the foliage. They are particularly active in the late afternoon. The swift, direct flight is undertaken with rapid wingbeats.

CALLS Vocalisation comprises an assortment of screeches and gurgling or chiming notes, including a harsh *aaaak*, a whinnying *we-li-li*, a ringing *linnng*, and a disyllabic *yuur-witt* with the final note rising sharply (Kennedy *et al.* 2000).

DIET AND FEEDING I am not aware of any feeding records, but the diet presumably comprises fruits, berries, and probably flowers and leaf-buds procured in the forest canopy and mid stages. Bananas and banana flowers are said to be favoured foods, and there is an early report of birds coming to cornfields to feed on both flowers and grain (see Hachisuka 1934).

BREEDING Nests are in hollows in the trunks or major branches of forest trees, but nothing is known of the nesting habits. Young birds were recorded in May (Whitehead 1899).

CRITICALLY ENDANGERED

Sulu Racquet-tail

Prioniturus verticalis Sharpe

Prioniturus verticalis Sharpe, *Bull. Brit. Orn. Club*, **3**, 1893, p. 10 (Tawi Tawi, Sibutu and Bongao).

DESCRIPTION Length 30 cm.
ADULT MALE General plumage green, noticeably darker and duller on back and upper wing-coverts; mantle and underparts more yellowish-green; forecrown bright blue with central red spot; primaries suffused blue on outer webs and on inner webs towards

tips; underwing-coverts green; undersides of flight feathers and tail-feathers greenish-blue; upperside of tail green tipped black on outer feathers, and spatules at tips of central bare shafts black variably suffused blue; bill pale bluish-grey becoming horn-coloured towards tip; iris dark brown; legs bluish-grey.
7 specimens: wing 163–185 (174.6) mm, tail 125–146 (136.0) mm, exp. cul. 21–22 (21.6) mm, tars. 18–20 (18.9) mm.
ADULT FEMALE Forecrown green sometimes lightly suffused blue, but without red spot; shorter bare shafts of central tail-feathers.
4 specimens: wing 171–176 (174.0) mm, tail 119–127 (122.0) mm, esp. cul. 21–24 (22.3) mm, tars. 18–21 (19.5) mm.
JUVENILES Like adults, but central tail-feathers tipped blue and with pointed tips, but no elongated bare shafts.

DISTRIBUTION Endemic in the southwestern Sulu Archipelago, southernmost Philippine Islands, where historically recorded on six islands from Tawi Tawi south to Tumindao, but now probably occurring only on Tawi Tawi and possibly Sibutu (Birdlife International 2016).

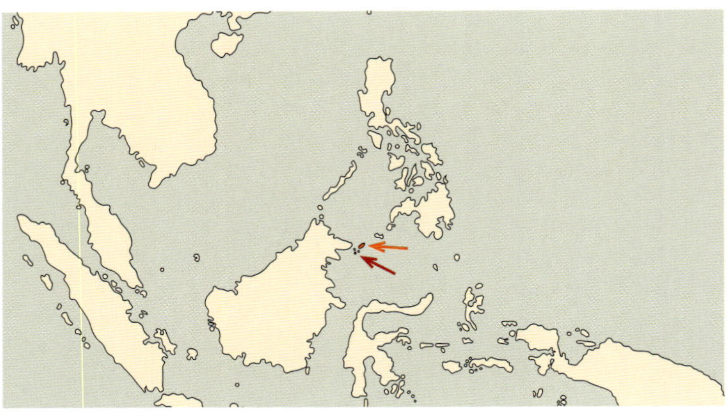

STATUS In the early 1900s, the Sulu Racquet-tail was reported to be locally common, particularly on Tawi Tawi, but by the 1970s it had declined dramatically because of widespread deforestation (Birdlife International 2016). In September–October 1971, occasional sightings by Discoro Rabor on Tawi Tawi and Sibutu were of pairs or single birds (In duPont and Rabor 1973b). There have been no records from Tumindao and Manuk Manka for more than 80 years, although this may reflect a lack of search effort in that time, and the species is regarded as probably extinct on Bongao and Sanga Sanga. It may persist on Sibutu, but in recent times reliable reports have come only from Tawi Tawi, where in the early 1990s very small numbers were known to be present at three sites. Clearance of the remaining forest is continuing on Tawi Tawi, and these racquet-tails are becoming increasingly scarce, with local people reporting that they are the least encountered of parrot species on the island. In early 2008, during a visit of one week, there was no response to the playing of taped calls, and during seven days in the field in early 2010 there was only one aural record. During a visit of five days in January 2012, three birds were seen, but security restrictions prevented searches in all but a small area, and it was acknowledged that the parrots could be more numerous in areas not visited.

The total population previously was estimated at fewer than 1000 birds, but more recent observations indicate that it now could be less than 250 mature individuals (Birdlife International 2016). Lambert (1993b) points out that during a visit to Tawi Tawi in late September 1991 no racquet-tails were observed far away from good forest, so they would be threatened by deforestation, and this is exacerbated by the targeting of larger birds by hunters armed with high-powered rifles, with these racquet-tails being easy targets because of their tameness.

HABITATS In the early 1900s, Sulu Racquet-tails were found to be abundant in mangroves on Tawi Tawi, where pairs occasionally were observed also in forested areas, and in forested parts of Sibutu and Sanga Sanga (McGregor 1909). In September–October 1971, Discoro Rabor occasionally observed pairs in high flight above densely forested areas on the larger islands, and single birds or pairs sometimes were flushed from fruiting trees in dense patches of remnant forest in the interior, especially on Sibutu and Tawi Tawi (in du Pont and Rabor 1973b). Lambert (1993b) notes that in late September 1991, small numbers were observed in forest and at the margins of forest in the northern part of Tawi Tawi, and a nest was found in a grove of palm trees in an agricultural area near to forest, but no birds were seen in agricultural areas away from forest.

HABITS Most recent reports of Sulu Racquet-tails are of sightings of single birds or pairs, and this probably reflects their scarcity rather than a non-flocking behaviour. They usually are seen flying high above the forest canopy, their screeching call-notes making them highly conspicuous, but while feeding in the upper stages of fruiting trees they are quiet and easily overlooked. Sometimes in the company of pigeons or other parrots, they climb actively among the branches, their predominantly green foliage blending well with the foliage. Lambert (1993b) notes that they are very tame. They are particularly active soon after sunrise, seeming to delight in dashing through the trees and wheeling about high overhead, all to the accompaniment of much screeching. The flight is swift and direct, with deliberate wingbeats.

CALLS Recorded calls include harsh, raspy *aaaaack* notes and a disyllabic *lee-aaack* with emphasis on the second syllable and sounding like a rusty gate (in Kennedy *et al.* 2000).

DIET AND FEEDING Sulu Racquet-tails are forest frugivores, and fruits procured in the upper stages of forest trees are the principal food items.

BREEDING Lambert (1993b) reports that on Tawi Tawi, in late September 1991, in a grove of palms in an agricultural area near to forest, a female was observed at a nest-hole in a large palm tree with a broken top. Nothing further is known of the nesting habits.

Plate 27
UPPER Green Racquet-tail *Prioniturus luconensis* (adult ♂)
LOWER Sulu Racquet-tail *Prioniturus verticalis* (adult ♂)

F. KNIGHT '16

AFRO-ASIAN DISTRIBUTION

The Afro-Asian Distribution, as determined for this book, extends from the Malay Archipelago, Indochina, China and the Indian subcontinent to Africa, Madagascar and the western Indian Ocean islands. Cacatuoidea is represented only by a self-sustaining population of introduced Yellow-crested Cockatoos in Hong Kong and, although all three families in Psittacoidea are represented, generic diversification is low, with parrots being a less prominent component of the avifauna than in Australasia or the Neotropics. *Psittacula* parrots are particularly widespread, especially on the Indian subcontinent and in Southeast Asia, and in Africa most species belong to just two genera – *Poicephalus* and *Agapornis*.

Earliest historical accounts of parrots as pets are referrable to species from the Indian subcontinent and North Africa. It is possible that the Rose-ringed Parakeet *Psittacula krameri* from North Africa was known to the ancient Egyptians, though there appear to be no records in their writings or art. Ctesias, a Grecian slave who became court physician to Artaxerxes II in 401 BC, gave a fairly accurate description of the Plum-headed Parakeet *Psittacula cyanocephala* and wrote romantically of the bird's ability to speak the language of its native India and the claim that it could be taught to learn Greek. It was probably Alexander the Great who introduced to Europe tame parrots from the Far East, and Alexandrine Parakeet, the English name for *Psittacula eupatria*, honours the warrior king. Aristotle almost certainly based his descriptions of parrots on birds brought back by the triumphant armies of his pupil – Alexander. Parrots, presumably *Psittacula* species from North Africa and the Indian subcontinent, were well known to the ancient Romans, and 'talking' birds were status symbols among the noble classes.

INDIAN SUBCONTINENT AND SOUTHEAST ASIA

In common with other forest-dependent species, parrots have been adversely affected by habitat loss in Southeast Asia and especially in Indochina, where deforestation has been particularly widespread, but an ability to utilise manmade habitats has prevented any species from becoming endangered. Jeyarajasingam notes that forest now covers about 40 per cent of peninsular Malaysia and the loss of lowland and hill dipterocarp as well as peat swamp forests is a key factor in the decline of many forest-dwelling birds (in Jeyarajasingam and Pearson 2012). Throughout Southeast Asia, dipterocarp forests have been heavily exploited for timber. Deforestation has brought about moderately rapid declines in populations of Long-tailed Parakeets *Psittacula longicauda*, particularly on islands, but they are able to utilise manmade habitats, including plantations of oil palms *Elaeis guineensis*. In Indochina, where parrots are prevalent in dry dipterocarp forest, the Blue-rumped Parrot *Psittinus cyanurus*, Alexandrine Parakeet *Psittacula eupatria* and Blossom-headed Parakeet *P. roseata* are scarce to locally common at extremities of their ranges (Robson 2000). The Blossom-headed Parakeet seems to be scarce to uncommon also in other parts of its range, in Nepal, northeastern India and Bangladesh to southeastern China (Rasmussen and Anderton 2005; MacKinnon and Phillips 2000). A long tradition of protection of all animals probably has helped parrots and other birds to withstand to varying degrees the impact of widespread deforestation on the Indian subcontinent, where there has been overexploitation of forests for fuelwood, timber and fodder, with a prevention of regeneration by overgrazing, and clearance for agriculture, especially shifting cultivation, plantations and urbanisation (Grimmett *et al.* 1999). In the northeastern highlands, Grey-headed Parakeets *Psittacula finschii* are uncommon in hill forest and cultivation up to 2100 m, but elsewhere other species remain common. This is particularly true of the Rose-ringed Parakeet *Psittacula krameri* which is a serious pest in agricultural areas, and has proved to be highly invasive where introduced in other regions, at times posing a threat to native parrots.

AFRICA

Parrots are poorly represented in Africa. There is a concentration of species in the central tropical belt, but I suspect that radiation has been hindered by a prevalence of other groups in habitats where parrots would be expected to be prominent. During fieldwork undertaken in East Africa, I found that hornbills and turacos are the conspicuous, common large-bodied avian frugivores, while in drier forests and open woodlands there is a strong preponderance of seed-eating pigeons and doves, and in drier woodlands and grasslands, favoured by *Agapornis* lovebirds, seed-eating grassfinches and sparrows are particularly plentiful. Nowhere did I observe an abundance of parrots, and that contrasted markedly with my experiences in Australia.

Of the four genera occurring in Africa, *Psittacus* and *Poicephalus* are the more distinctive in that their relationship is with New World parrots in a lineage that diversified some 60 to 65 million years ago, corresponding to the initial separation of Australia and New Guinea from East Antarctica (Wright *et al.* 2008). The two *Psittacus* species are confined to tropical forests of Central and West Africa, where there has been extensive deforestation, and the adverse impact on populations from habitat loss has been compounded by excessive trapping for the live-bird market, so that the Grey Parrot *P. erithacus* and the Timneh Parrot *P. timneh* now are listed as vulnerable. *Poicephalus* is widespread in sub-Saharan Africa and, although nowhere particularly abundant, most species are fairly common. A notable exception is the endangered Cape Parrot *Poicephalus robustus*, which is the most southerly distributed of African parrots and is a habitat specialist confined to Afromontane forests at 1000 m to 1700 m in eastern Republic of South Africa. Within these forests, now reduced to patches ranging in size from less than a hectare to 1000–1500 hectares or more, the parrots are dependent on mature *Podocarpus* trees for food and nesting sites. Because of past exploitation of these trees for timber and present exploitation by landholders for poles, there is a shortage of midsized to large trees, so remnant stands of forest are depauperate with little variation in tree species and a generally young age structure. Little natural regeneration to climax forest cannot be expected for many decades, even with the most optimistic forecast and immediate action.

Affinities of *Agapornis* and *Psittacula*, the remaining African genera, are with the psittaculine parrots of Asia, and their colonisation of Africa obviously is of much more recent occurrence. Indeed, the African representatives of *Psittacula* normally are differentiated only subspecifically from the widespread Rose-ringed Parakeet *P. krameri* of Asia, though it is possible that more focused studies may show that recognition of African populations as a separate species could be warranted. In North Africa, the Rose-ringed Parakeet is common throughout most of its range.

Grouped with the *Loriculus* hanging parrots and monotypic *Bolbopsittacus* from Asia in the subfamily Agapornithinae, the *Agapornis* lovebirds are widely distributed in sub-Saharan Africa, where the closely related species are almost totally allopatric, and in Madagascar. An unusual behavioural feature shared with the hanging parrots is the carrying of nesting material tucked under their feathers. Based largely on behavioural and morphological charactersistics, two well-differentiated species groups, one intermediate species – *A. roseicollis* – and one, little-known aberrant species – *A. swindernianus* – have been identified (Moreau 1948; Dilger 1960). In the more 'primitive' group, comprising the Grey-headed Lovebird *A. canus*, Red-faced Lovebird *A. pullarius* and Black-winged Lovebird *A. taranta*, sexual dimorphism is present, there are no naked eyerings, and many small pieces of nesting material are carried by females thrust amidst feathers of the entire body. In the second group, comprising Fischer's Lovebird *A. fischeri*, Yellow-collared Lovebird *A. personatus*, and the very closely allied, possibly conspecific Nyasa Lovebird *A. lilianae* and Black-cheeked Lovebird *A. nigrigenis*, sexual dimorphism is lacking, there are prominent naked eyerings, and large pieces of nesting material are carried by females in their bills. The Peach-faced Lovebird *A. roseicollis* lacks sexual dimorphism and a naked eyering, and variably sized nesting material is carried by females tucked under feathers of the rump and lower back. The little-known, arboreal Black-collared Lovebird *A. swindernianus* also lacks sexual dimorphism and has an inconspicuous feathered eyering, but little is known of its nesting habits. The Black-collared Lovebird is confined to tropical forests in Central and West Africa where, in common with the *Psittacus* parrots, it has been adversely affected by widespread deforestation and now is generally scarce and declining, especially in West Africa where apparently it has been extirpated in Liberia (Gatter 1997). In East Africa, Fischer's Lovebird, the Nyasa Lovebird and the Black-cheeked Lovebird occupy quite restricted ranges and are threatened by intense trapping for the live-bird market. The Black-cheeked Lovebird is particularly vulnerable because it has quite narrow habitat preferences, being closely associated with mopane (*Colophospermum*) woodland, moving into *Baikiaea* woodland in the wet season, and the total population has been estimated at 10 000 birds. Other species remain locally to generally common, though the Red-faced Lovebird possibly is declining in West Africa.

MADAGASCAR, COMOROS ISLANDS AND SEYCHELLES ARCHIPELAGO

Distinctiveness of the avifauna of Madagascar rivals that of the mammals, and attesting to that distinctiveness are the two *Coracopsis* vasa parrots. Long considered an 'old' or 'ancient' group with doubtful affinities, *Coracopsis* has been linked by molecular analyses to monotypic *Psittrichas*, another relict form endemic to the midmontane forests of New Guinea, with the common ancestor having occurred in Australasia, and it is suggested that dispersal from Australasia to Madagascar some 40 million years ago could have been facilitated by volcanic islands in the south Indian Ocean possibly serving as stepping stones (Schweizer *et al.* 2010).

Ekstrom (2013) points out that, despite widespread deforestation, both *Coracopsis* species remain common on Madagascar. The Greater Vasa Parrot *C. vasa* is less dependent on native forest than the Lesser Vasa Parrot *C. nigra* and, although originally a forest bird, it can be found in all habitats where there are trees suitable for nesting, including secondary forest, plantations, other wooded areas and farmlands. Hunting and trapping occurs, mainly to protect crops, but there is no evidence that this adversely affects populations, and it occurs in many protected areas.

In the Comoros islands, it is fairly common on Grande Comore but scarce on Anjouan and Mohéli. In Madagascar, the Lesser Vasa Parrot is more forest-restricted than the Greater Vasa Parrot and seldom is encountered far from forest, but it has a large range and the population apparently is stable. It is captured for pets, trade or food, but this may not pose a significant threat, and the ongoing loss of forest probably is more significant. Conversely, the distinctive subspecies on the Comoros islands is rare in remnant stands of degraded forest.

A vulnerable *Coracopsis* population on Praslin, in the Seychelles Archipelago, has been treated as indistinguishable from populations of *C. nigra* on the Comoros islands or as a separate species – the Seychelles Black Parrot *C. barklyi*. In the late 1960s, it was estimated that only 30–50 birds survived, but subsequently there has been a steady increase in numbers, particularly between 1985 and 1997 when a 40 per cent increase brought the estimated population up to 200–300 individuals (Rocamora and Laboudallon 2013). Since 1999, when the population was estimated at 300–400 birds, the abundance has fluctuated, and in 2008 an estimated 654 birds were present on Praslin and six to 10 birds on neighbouring Curieuse.

At lower altitudes on Praslin, these parrots are best known from native forests, but they occur also in river valleys with tall trees and mixed forests with palms and introduced broadleaved trees, and they commonly feed in cultivated or abandoned orchards, other agricultural lands, or in gardens of hotels and in residential areas.

Very similar to, and possibly conspecific with the extant Alexandrine Parakeet *Psittacula eupatria* from Asia, the Seychelles Parakeet *P. wardi* formerly occurred on Mahé, Silhouette and possibly Praslin, in the Seychelles Archipelago. It disappeared sometime after 1893, when the last specimen was collected on Mahé. In 1866, Edward Newton failed to find any birds on Mahé, but did observe them on Silhouette, and was told that they had almost been exterminated by ruthless shooting because of the damage that they did to maize crops (Newton 1867). A claim was made that the clearing of forests to make way for coconut plantations also contributed to their extinction.

MASCARENE ISLANDS

In a scholarly review of available evidence, Cheke and Hume (2008) have recreated a fascinating description of undisturbed ecosystems on the Mascarene Islands, where together with the Dodo *Raphus cucullatus* and Solitaire *Pezophaps solitaria* there occurred a remarkable array of parrots. Arrival of the first European travellers in the late 1500s and early 1600s had an immediate and devastating impact on these ecosystems, with widespread deforestation, excessive hunting and the introduction of competitors and predators eventually bringing about appalling levels of extinction. From a few museum specimens, fossil remains and contemporary accounts by early travellers, six endemic parrots have been described, but only the Mauritius Parakeet *Psittacula eques* now survives. Hume (2007) notes that because of the paucity of fossil parrot remains, the number of species may well have exceeded the number currently recognised. A detailed comparative analysis of fossil skeletal material indicates that affinities of these parrots are within the Psittaculini, a wide-ranging tribe occurring mainly in the Indian subcontinent and Southeast-Asia to Australasia, and colonisation of the Mascarene Islands occurred by island-hopping from India during times of low sea levels. It has been suggested that extant *Psittacula eupatria* may be the founding species for all *Psittacula* parrots that occur or did occur on islands in the Indian Ocean, and among these parrots there is a southwards progression in the loss of characters of *P. eupatria*, including reduction in size (in Hume 2007). Molecular phylogenetic analysis suggests that the extinct Mascarene species experienced recent divergences within their clades, implying that the Indian Ocean islands have played a key role in the evolutionary radiation of *Psittacula* species (Jackson *et al.* 2015). Extinct *P. wardi* from the Seychelles Archipelago groups with *P. eupatria* from Asia and diverged 3.83 million years ago, whereas extinct *P. exsul* and extant *P. eques* from the Mascarene Islands group with *P. krameri*, which is native to Asia and Africa. *P. exsul* is ancestral to the *P. eques* lineage, having diverged approximately 3.82 million years ago, with subsequent divergence within the *P. eques* lineage, between the extinct population on Réunion and the extant population on Mauritius, occurring about 0.61 million years ago.

Mauritius Parakeet

Recovery of the Mauritius Parakeet from the brink of extinction certainly is a major conservation success, and is a fitting reward for teams of dedicated fieldworkers who have undertaken an intense rehabilitation program since the 1980s, when fewer than 20 birds were known to survive in the wild and extinction seemed imminent. A sudden and dramatic fall in numbers in the 1970s coincided with the clearing of a large area of foraging habitat to make way for a forestry plantation, and in late December 1979 a cyclone caused severe damage in all areas of remnant forest. A disease outbreak also may have contributed to the sudden decline. Some or all of these factors probably were implicated, with the loss of habitat being particularly severe.

In 1996 only 5 per cent of the island supported native vegetation, with remnant stands of upland forest, the favoured habitat, continuing to be impacted by cyclones, the spread of invasive introduced plants and disturbance by introduced pigs and Rusa Deer *Cervus timorensis*. The surviving parrots were confined to the southwest of the island, where they occupied about 40 km^2 of upland native forest centred on Black River Gorges National Park.

Known threats to Mauritius Parakeets were addressed in an intensive conservation program commenced in 1973, and early attention was given to improving the low level of recruitment. A scarcity of nesting sites in remaining large native trees was compounded by strong competition from other cavity-nesting species, and some nests were susceptible to predation, particularly from Black Rats *Rattus rattus* and Crab-eating Macaques *Macaca fascicularis*, which took eggs and chicks or at times killed sitting females. Repairs and improvements were made to known and prospective nesting sites, including the fitting of predator-deterrent collars to the trunks of nesting trees, and inspection holes were made into nest chambers so that the hatching and development of chicks could be monitored. If monitoring indicated that a nest was likely to fail, the eggs or chicks were removed for artificial incubation and handrearing. The handreared chicks were used to establish a captive breeding program or were released back to the wild. In addition to safeguarding protected nest-sites against competitors and predators, artificial nestboxes were provided, and eventually these were used by a strong majority of nesting pairs. With these measures spectacular success has been achieved in boosting the level of recruitment, and release into the wild of captive-bred birds has augmented the population significantly. This increase in numbers has highlighted the shortage of food in the small remnant of native forest, much of which is degraded by exotic plants, so supplementary feeding has become a key element of the conservation program. Psittacine beak and feather disease is present in the population and continues to pose a serious threat, but strict hygiene measures and protocols have been introduced to ensure that management does not facilitate spread of the disease. The population now approaches 700 birds, but threats remain and longterm survival of these parakeets is dependent on maintenance of intensive conservation actions.

Subspecifically differentiated from the Mauritius Parakeet were 'Perroquets verts ayant un collier noir' that formerly occurred on Réunion Island and are known only from a single specimen and from illustrations, but it is not known whether these illustrations were based on living birds or preserved specimens. Hume (2007) notes that, considering the available habitat on Réunion and survival of the Mauritius Parakeet, it is surprising that the parakeets died out very early in Réunion's post-human history. The last reference to green and grey parrots on Réunion was by Joseph-François Charpentier de Cossigny in 1732, and it indicates that both probably were hunted to extinction:

> *The woods are full of parrots, either completely grey or completely green. They were eaten a lot formerly, the grey especially, but both are always lean and very tough whatever sauce one puts on them* (translation by Hume).

Mascarene Grey Parakeet

Holyoak (1973b) described the Mascarene Grey Parakeet *Psittacula bensoni* from subfossil material obtained on Le Pouce Mountain, Mauritius, in the early 1900s, and identified it as a second, smaller *Lophopsittacus* species. After re-examining the material, Hume (2007) determined that the species clearly is derived from *Psittacula* stock and is similar to the extant Alexandrine Parakeet *P. eupatria*, but larger and more robust in some elements. It also appears to have been atypical in colouration, being all grey instead of green or partially green as common to all other known *Psittacula* species. Hume notes that Mascarene Grey Parrots were particularly targeted by hunters and, being very tame, they were easily caught, as reported by Willem van West-Zanen, who visited Mauritius in 1602. The only known drawing of these parrots is in an engraved scenic drawing depicting parrot-catching, which is included in the published account of his visit, and first encounters by his crew with parrots are described as follows:

> *... some of the people went bird hunting. They could grab as many birds as they wished and could catch them by hand. It was an entertaining sight to see. The grey parrots are especially tame and if one is caught and made to cry out, soon hundreds of the birds fly around ones' ears, which were then hit to the ground with little sticks. Also just as tame are the pigeons and turtle doves, that let themselves be caught easily ...* (translation by Hume).

A similarly depressing account is given by Admiral Steven van der Hagen, who visited Mauritius in 1606 and 1607, and again he specifically mentions the ease of killing 'grey parrots':

> *During all our time there, we lived on turtles, dodos, pigeons, grey parrots and other game, which we caught in the woods with our hands. Besides their usefulness to us, there was also much amusement to be got from them. Sometimes when we caught a grey parrot, we made it call out, and at once hundreds more came flying around, and we were able to kill them with sticks* (translation in Barnwell 1948).

Hume points out that despite this persecution, Mascarene Grey Parakeets appear to have remained reasonably common until the 1750s, but the population must have crashed shortly afterwards because the last mention of them is in the account by Charpentier de Cossigny in 1764. Subfossil remains have not been found on Réunion and, in the absence of any evidence of their being different, both the 'completely grey' and 'completely green' parrots' last mentioned by Charpentier de Cossigny in 1732 are suspected to have been *P. bensoni* and *P. eques* in keeping with a pattern of occurrences of other species on both islands (see Hume 2007).

Newton's Parakeet

Newton's Parakeet *Psittacula exsul* is known from two specimens and subfossil material collected on Rodrigues Island. It seems that there were green and blue morphs, and the green morph may have had red wing 'epaulettes' as are present in *P. eupatria* and other Asiatic species, but both specimens are of the blue morph and, as would be expected, they lack red on the wing-coverts. François Leguat reported that these parakeets were abundant during his stay on Rodrigues in 1691–1692:

> *There are abundance of green and blew Parrets, they are of a midling and equal bigness; when they are young, their Flesh is as good as young Pigeons.*

Although in decline from the 1760s, when there was severe deforestation on the island, these parrots survived until the 1870s, when the last records were made, and the second specimen was collected in 1875.

Leguat's Parrot

The second endemic parrot on Rodrigues was Leguat's Parrot *Necropsittacus rodericanus*, which was described from a subfossil partial bill by Alphonse Milne-Edwards in 1867, and presumably it is the green parrot referred to by Leguat. Subfossil remains of at least four birds, including one partial associated skeleton, have been collected from the Plaine Corail, Rodrigues (in Hume 2007). Hume points out that *Necropsittacus* had proportionately the largest head and jaws of any of the Mascarene parrots, and in life may have been somewhat similar to the extant Great-billed Parrot *Tanygnathus megalorynchos*, but with an even larger head and longer tail. The plumage colouration almost certainly was uniformly green. The only detailed account of living birds was made by Julien Tafforet in 1726:

> *The parrots are of three kinds and in numbers. The largest are larger than a pigeon, and have a very long tail, the head large as well as the beak* [N. rodericanus]. *They mostly come on the islet which is to the south of the island, where they eat a small black seed, which produces a shrub whose leaves have the smell of the orange-tree, and come to the mainland to drink water. The second species is slightly smaller and more beautiful, because they have green plumage like the preceding, a little more blue, and above the wings a little red as well as their beak* [Psittacula exsul ♂]. *The third species is small and altogether green, and the beak black* [P. exsul ♀].

The islet referred to by Tafforet is Ile Gombrani, and he refers also to the bois de buis *Fernelia buxufolia* being a food plant for both Leguat's Parrot and Newton's Parakeet (in Hume 2007). Both parrots had become rare by 1761, when Abbé Alexandre Pingré visited Rodrigues to view the transit of Venus, and he regretted their scarcity because they were so good to eat. Pingré noted also that large numbers of fires were lit by tortoise-hunters to clear vegetation, so deforestation as well as excessive hunting probably had reduced populations of both parrots by this time. The last mention of Leguat's Parrot in life is in Pingré's account, so it must have died out soon after, presumably due to deforestation, excessive hunting and probable predation of eggs and chicks by rats (Hume and Walters 2012).

Rothschild (1907) ascribed to *Necropsittacus* a parrot from Réunion and described by Dubois in 1674:

> *Green parrots the same size* [presumably as Psittacula eques] *with head, upper parts of the wings, and tail the colour of fire* (translation by Hume).

There is no other basis for the occurrence of a red and green parrot on Réunion, and Hume points out that, if it did exist, it may have been a *Psittacula* derivative because the description corresponds well with the colouration of *P. eupatria*, except for the red tail.

Mascarene Parrot

The Mascarene Parrot *Mascarinus mascarin* formerly occurred on Réunion and possibly also on Mauritius, and is known from two museum specimens, both of which are from captivity and are in poor condition. It was first mentioned by a visitor to the island in 1671–1672 (Dubois 1674):

> *Parrots a little bigger than pigeons, with plumage the colour of squirrel fur, a black hood on the head, the beak very large and the colour of fire.*

Reference to the plumage being 'the colour of squirrel fur' has given rise to a claim that the brown plumage of the two specimens 'is clearly an artefact of fading due to light exposure, from grey to brown' (Hume and van Grouw 2014). All of the early drawings depict a brown-plumaged bird, and that has become the orthodox image, which is adopted in this book. Carl Wilhelm Hahn supposedly saw the last living bird in the menagerie of the King of Bavaria in 1834, and this date consistently has been quoted as the date of extinction (see Greenway 1967). Hume (2007) points out that, although not conclusive, it is very doubtful that Hahn saw a living Mascarene Parrot as late as 1834, if at all, and the 1834 date of extinction is unfounded. The last mention of living birds on Réunion was in the early 1700s, and captive birds were alive in Paris during the 1780s, but there are no records indicating that they survived after this time, and the species almost certainly was extinct by 1800 (Hume and Walters 2012). Almost nothing is known about these parrots in the wild, or of reasons for their extinction.

Broad-billed Parrot

Known from subfossil remains collected on Mauritius and from contemporary drawings, the Broad-billed Parrot *Lophopsittacus mauritianus* certainly is the most distinctive of the extinct Mascarene parrots. Dutch travellers to the island in 1598 and 1601 categorised it separately from other parrots, and referred to it as the 'Indian raven', presumably because of its raucous call. There are subfossil remains derived from at least 28 individuals, and Hume (2007) notes that measurements of these remains indicate that these parrots were large with sexual dimorphism in size greater than in any extant parrot, total lengths of males being 55–65 cm and of females being 45–55 cm. There was a strong sexual difference also in size of the massive bill, with bills of males being approximately 20 per cent larger than bills of females. These features are readily apparent in the only known drawing of living birds, a pen-and-ink sketch in the 1601 journal of the Dutch East India Company ship *Gelderland* and attributed to Joris Laerle. This sketch depicts two large-headed parrots with a massive bill attached to a flattened skull, and at the base of the bill there is a tuft of feathers forming a distinctive frontal 'crest', a feature not known in any other Mascarene parrot. Hume points out that ridges on the frontal region of the fossilised skull indicate that the 'crest' was attached firmly to the cranium and was not erectile. While the sketch shows well the appearance of living birds there is no information on plumage colouration, and there are differing descriptions in early writings. Reference to colouration is included in the first account, which stems from the voyage of Admiral Jacob van Neck in 1598 (in Hume 2007):

> *Is a bird which we called the Indian Crow, more than twice as big as the parroquets, of two or three colours.*

A slightly more detailed description is included in an account by Johann Christian Hoffman and dating from 1673–1675:

> *There are also geese, flamingos, three species of pigeon of various colours, mottled and green perroquets, red crows with recurved beaks and blue heads, which fly with difficulty and have received from the Dutch the name of 'Indian crow'* (translation by Hume).

In an account of his stay on Mauritius in the years 1666–1669, Johannes Pretorius wrote in detail about the native fauna and provided the only description of keeping now-extinct bird species in captivity. Hume and Winters (2015) point out that during the 17th century parrots regularly were captured for the bird trade, and often were given as gifts or sold by the Dutch, so it is most likely that they were maintained in captive conditions on Mauritius prior to exportation. Referring to the Broad-billed Parrot, Pretorius wrote:

> *The Indian ravens are very beautifully coloured. They cannot fly and are not often found. This kind is a very bad tempered bird. When captive it refuses to eat. It would prefer to die rather than to live in captivity.*

Differences in these descriptions of plumage colouration are reflected in reconstructed drawings by several artists. The first coloured reconstruction was by Henrik Grönvold in *Extinct Birds* (Rothschild 1907), and he depicted an all-dark blue parrot presumably based on the text by Rothschild:

> *... Except a few further osseous remains, mostly collected by Sir Edward Newton, nothing more of importance was found relating to this bird till Professor Schlegel discovered in the Library of Utrecht the manuscript journal kept during the voyage to Mauritius in A.D. 1601–1602 of Wolphart Hermanszoon, in which among other items of natural history there is a sketch of* Lophopsittacus *from life, and the statement that it was wholly of a grey-blue colour.*

This same all-blue colouration was adopted in an illustration of two birds in *The Dodo and Kindred Birds, or the Extinct Birds of the Mascarene Islands* (Hachisuka 1953), and it has been widely accepted, even being reproduced on a postage stamp from Mauritius. Hume (2003) examined the *Gilderland* journal, and points out that no text other than a brief description of the Dodo is found in any of the bird illustrations, so the all-blue plumage colouration must be disregarded. I concur also with the suggestion made by Hume that the distinctively drawn, but peculiar facial mask is not composed of bare skin, but is feathered and may represent a different colour. Based on the description by Hoffman, a reconstruction by Ria Winters depicts the parrot with a red bill, pale blue face, mauve-blue nape and neck, and the remainder of the body, including the tail, is red (in Hume and

The only known drawing of living Broad-billed Parrots *Lophopsittacus mauritianus* is this pen-and-ink sketch in the 1601 journal of the Dutch East India Company ship *Gelderland*, and attributed to Joris Laerle.

Winters 2015). I agree that the bill probably was red, at least in adult males, but would suggest that a more likely interpretation of Hoffman's description would be a general plumage colouration of more subdued reddish-brown with a pale bluish-grey face, not dissimilar from the plumage colouration of *Mascarinus mascarin*, and that pattern is adopted in the frontispiece (p. ii) depicting reconstructions of the three extinct Mascarene parrots known only from subfossil material.

Although the ecology of the Broad-billed Parrot is unknown, Hume (2007) suggests that some insight into its habits may be gained by looking at species with morphological similarities, particularly the large-billed Hyacinth Macaw *Anodorhynchus hyacinthinus* and Palm Cockatoo *Probosciger aterrimus*. Like those species, Broad-billed Parrots may have used their massive bills to feed on the hard nuts or seeds of palms and other trees or shrubs. Palms and *Pandanus* screw-pines formerly dominated the lowland areas on Mauritius, where prodigious quantities of their fruits would have accumulated on the ground to be eaten by the parrots or by the *Cylindraspis* giant tortoises. Undigested endocarps voided by the tortoises also may have been eaten by the parrots in the same manner as the macaws eat the undigested endocarps voided by cattle. The assumption that these parrots were flightless comes from large birds with proportionately short, rounded wings being depicted in Laerle's sketch, but Hume (2007) notes that obscured pencil outlines underneath the finished ink lines show that the wing to body length is not particularly short and also more clearly defined is a large alula, which is an adaptation to prevent stalling in slow-flying birds. There is reduction in the sternal keel, but not to the extent of preventing flight. These features tend to confirm the statement made by Hoffman that the parrots could 'fly with difficulty'. There is no basis at all for the claim made by Hachisuka (1953) that these parrots were nocturnal and, as noted by Hume (2007), we can conclude from the meagre available information that observers found them to be active during daylight hours. With respect to habitat preferences, it can be noted that the few records are from the drier, leeward side of the island, though this may be an artefact of that region being more easily accessible for observers and hunters.

Hume (2007) notes that the large Broad-billed Parrots would have been a ready source of food for early mariners and, because of their confiding nature together with poor flight, they would have been caught quite easily. Nests in large tree hollows, or possibly among rocks on the ground, would have been highly vulnerable to predation by introduced rats and monkeys. Hume points out also that, although the last mention of these parrots is in the account by Hoffman dating from 1673–1675, it is possible that they persisted for some time after that date because by then the Dutch presence on Mauritius was quite limited. There is no record of a specimen, either alive or dead, ever being taken out of Mauritius.

An erroneous claim by Rothschild (1907) that the Broad-billed Parrot was '...wholly of a grey-blue colour' was responsible for this plumage colouration being adopted by artists in reconstruction drawings, including this illustration from *The Dodo and Kindred Birds*, or *Extinct Birds of the Mascarene Islands* (Hachisuka 1953). There is no basis for the claim, and in contemporary writings the parrots are referred to as being 'of two or three colours'.

SUPERFAMILY
PSITTACOIDEA Rafinesque-Schmaltz

Apart from cockatoos and the *Strigops-Nestor* parrots from New Zealand, all parrots are included in this superfamily, and a major radiation is reflected in a systematic arrangement recognising three groupings at family level rank (Joseph *et al.* 2012). Two of these families are represented in Africa and the Indian Ocean Islands.

FAMILY **PSITTACIDAE** Rafinesque-Schmaltz

In molecular analyses African *Psittacus* and *Poicephalus* have been recovered as sister to Neotropical parrots, and this Afro-Neotropical group making up the family Psittacidae is sister to all other parrots in the superfamily, and may be worthy of differentiation as another superfamily. It is hypothesised that a common ancestor lived in Antarctica and became separated from the Australasian lineages when Antarctica began to split from Australia in the late Eocene or early Oligocene, approximately 35 million years ago (Schweizer *et al.* 2010). Africa and the Neotropics subsequently were colonised from Antarctica, and parrots in these two regions now are differentiated at subfamily rank (see Joseph *et al.* 2012).

SUBFAMILY **PSITTACINAE** Rafinesque-Schmaltz

Molecular analyses have indicated a close relationship between *Psittacus* and *Poicephalus*, the two African genera included in this subfamily. It is suggested that in the middle to late Miocene, when a major change in vegetation brought about a reduction in rainforests, with corresponding increases in areas of open vegetation in Australia, Africa and South America, there emerged several dry-adapted lineages, and *Poicephalus*, which includes several dry-adapted species, split from *Psittacus*, a bird of tropical lowland forests (Schweizer *et al.* 2010).

Because they are very popular as cagebirds, the *Psittacus* parrots have been excessively exploited in many parts of tropical Africa, where deforestation has resulted in significant habitat loss, so both species are listed as vulnerable. Conversely, *Poicephalus* parrots remain common throughout much of sub-Saharan Africa, and only the Cape Parrot *P. robustus* in southernmost Africa is threatened by loss of habitat.

GENUS ***Poicephalus*** Swainson

Poicephalus Swainson, *Classif. Bds*, **2**, 1837, p. 301. Type, by subsequent designation, *P. senegalensis* (L.) Swainson = *Psittacus senegalus* Linnaeus (G. R. Gray, *List Gen. Bds*, 1840, p. 52).

Parrots belonging to this genus are midsized to large, stocky birds with short, squarish tails and proportionately stout, heavy bills. There is no bare facial patch, but the unfeathered cere is prominent. A small post-orbital process and a straight, rather short squamosal are osteological features of the skull. Sexual dimorphism is variable, being prominent in some species but slight or absent in others, and juveniles are duller than adults.

The genus is restricted to, and widespread in sub-Saharan Africa.

ENDANGERED

Cape Parrot

Poicephalus robustus (Gmelin)

Psittacus robustus Gmelin, *Syst. Nat.*, **1**, pt 1, 1788, p. 344 (No locality = South Africa).

OTHER NAME Brown-headed Parrot if conspecific with *P. suahelicus*.

DESCRIPTION Length 33 cm. Weight 310–400 g.
ADULT MALE Plumage variable; head and neck dull olive-yellow to greenish-brown with darker dusky-brown centres to feathers producing scalloped appearance; lores blackish-brown; pink-red frontal marking sometimes present; scapulars and upper wing-coverts greenish-black broadly edged green; primaries and secondaries black, on outer webs finely margined dull green; rump and underparts green variably tinged dull blue; thighs, bend of wing, carpal edge and outer underwing-coverts orange-red; tail blackish-brown; bill horn-coloured; bare eyering grey-white; iris dark brown; legs bluish-grey.
11 specimens: wing 204–223 (213.1) mm, tail 81–97 (89.3) mm, exp. cul. 31–37 (34.1) mm, tars. 21–25 (22.0) mm.
ADULT FEMALE Like male, but well-defined pink-red frontal band often extending to forecrown.

6 specimens: wing 201–218 (208.5) mm, tail 83–93 (88.3) mm, exp. cul. 32–35 (33.5) mm, tars. 21–23 (22.3) mm.
JUVENILES No orange-red on thighs, bend of wing, carpal edge or outer underwing-coverts; head and neck brownish-olive; orange-pink frontal band; tail dark green.

DISTRIBUTION Eastern Republic of South Africa, where disjunctly distributed from the Alice district, Eastern Cape Province, to the Karkloof region of KwaZulu-Natal, with a small relict population some 800 km farther north at about lat. 23° in Soutpansberg, Limpopo Province.

SUBSPECIES Taxonomic status of the endangered Cape Parrot has been the focus of much debate. It often is treated as conspecific with the more common Brown-headed Parrot *P. suahelicus* from south-central and west Africa, from which it is differentiated by head colour, and that arrangement is adopted by del Hoyo and Collar (2014) with the comment that justification of species status is hampered by inadequate evidence and argument. Results of multilocus DNA analyses identified *P. robustus* as genetically distinct from the two forms of *P. suahelicus* and supported recommendations to elevate the Cape Parrot to species status based on morphological, ecological and behavioural assessments (Coetzer *et al.* 2015). That arrangement is adopted here.

STATUS Perrin (2013) points out that numbers of Cape Parrots have declined greatly in the past 50 years, particularly in Eastern Cape Province, probably because of the loss of habitat and nesting sites, depletion of their natural food resources, illegal capture and trade, and persecution as agricultural pests. Occurrences are discontinuous throughout the range, being restricted to localised forest patches, and there has been a decline in sightings in many districts where previously these parrots were common. There were no reliable estimates of numbers, though reports in the early 1900s referred to sightings of large flocks and birds feeding in great numbers. Commencing in April 1997, a national census has been undertaken annually with groups of two or more observers assigned to count parrots in montane forest patches where the birds had been recorded in the past five years, and then counts were extended to forest patches where the parrots were expected to occur (Perrin 2013). Observers positioned themselves at vantage points to record the presence and behaviour of the parrots, and observations were made during the first three to five hours after sunrise and the last three hours before sunset. Densities were estimated initially at Hlabeni and Ingeli State Forests, in southern KwaZulu-Natal, and later were extended to other parts of the range to estimate the total population. Estimates of the total population based on these counts, questionnaires and personal communications showed numbers to be exceedingly low, even in preferred habitat, and the annual censuses have been important in monitoring population trends. In 2001, optimistic crude estimates of numbers suggested that there were 200 birds in Eastern Cape Province, 100 in the Transkei, 150 in south-central KwaZulu-Natal and fewer than 50 birds in the northern population in Limpopo, giving a tentative total of 500 parrots. In 2013, at least 225 volunteers were posted at 84 localities in the three provinces and, despite poor weather conditions, two counts yielded 1182 and 1317 parrots, with the maximum number of birds seen in each of the areas suggesting that there were at least 1402 parrots in the wild at that time (in Downs *et al.* 2013).

Perrin points out that distributional data for Cape Parrots show that the population is contracting in its range and numbers probably are too low to recolonise peripheral patches of suitable forest habitats. Although the parrots show good dispersal capabilities, the nature of the landscape surrounding these forest fragments, together with a probable need for social learning in determining food availability, further reduces the likelihood and efficacy of recolonisation. To improve access to natural food resources, new techniques using vocalisation playbacks are being used to attract parrots to depleted forest patches that local populations are unable to investigate effectively for potential feeding sites (in Downs *et al.* 2013).

Habitat preservation is perceived as being of critical importance in ensuring the long-term survival of Cape Parrots and, because the birds are mobile, a large mosaic of forested areas supporting a strong population is required (in Perrin 2013). The parrots show little inclination to use artificial nestboxes, so preservation of old, hollow-bearing trees for nesting is of paramount importance. When nutritionally stressed, the parrots will utilise alternative food sources, including other native fruiting trees and pecan nuts, so providing these sources can be beneficial when there is a shortage of *Podocarpus* fruits. Persecution of these parrots as agricultural pests and capture for the live-bird trade are secondary pressures that have the potential to impact very severely on local populations, so must be eliminated. There may be a role for captive-breeding to establish a 'back-up' population from which birds can be released to augment local wild populations, but in any captive-breeding program stringent regimes must be put in place to minimise disease outbreaks, particularly with psittacine beak and feather disease, which has been identified as a serious threat to wild populations (in Downs *et al.* 2013). Three years of data collection in Eastern Cape Province suggest that severe outbreaks of psittacine beak and feather disease in the wild population could be linked to a food resource bottleneck between January and March each year.

In response to concerns over the presence of psittacine beak and feather disease in populations in Eastern Cape Province and KwaZulu-Natal, a systematic monitoring of the isolated population

Plate 28
Cape Parrot *Poicephalus robustus*
FOREGROUND adult ♂ BACKGROUND adult ♀

F. KNIGHT '16

in Limpopo Province was initiated in 2012, and early assessments suggest that this population could comprise fewer than 100 birds with the largest single flock containing 30 individuals (in Downs *et al.* 2013). No clinical signs of the disease have been observed, but presence of the virus in the population has been confirmed from blood and feather samples. In addition to the monitoring, an educational program has been established to engage school children with the conservation of parrots and other forest birds. As part of the program, children participate in the planting of native trees around schools. At certain times of the year, Cape Parrots feed in and around villages, so in the future these trees may well provide resources for Cape Parrots. Planting native trees and the provision of nestboxes are facets of conservation programs being implemented in Eastern Cape and KwaZulu-Natal Provinces.

HABITATS Cape Parrots are habitat specialists, being dependent for food and nesting on Afromontane forests dominated by yellowwoods *Podocarpus* spp. at 1000 to 1700 m (in Perrin 2013). Along the mountains or escarpments from Eastern Cape Province, including the Transkei, to the midlands of KwaZulu-Natal, these forests occur in patches ranging in size from less than a hectare to 1000–1500 hectares or more. Characterised by heavy summer mist, they are coarse-grained with a predominance of shade-intolerant species, and are dominated by lemonwood *Xymalos monospora*, narrow-leaved yellowwood *Podocarpus falcatus*, Henkel's yellowwood *P. henkelii* and broad-leaved yellowwood *P. latifolius*. The Transkei mist-belt forests are dominated by narrow-leaved yellowwood, sometimes as an emergent, red pear *Scolopia mundii*, broad-leaved yellowwood, lemonwood, and Cape beech *Rapanea melanophloeos*. They differ from the Eastern Cape mist-belt forests in the larger numbers of narrow-leaved and broad-leaved yellowwoods and smaller numbers of Henkel's yellowwoods. In the Amatola Mountains, near the southern extremity of the range, mist-belt forests vary from tall forest to scrub forest along a gradient from cool, mountain slopes with heavy summer mists to lowland areas, and comprise two subtypes. Along the Amatola escarpment, mid-altitude forests are relatively large and species-rich, with a dominance of emergent narrow-leaved yellowwoods. Conversely, scarp forests along the lower east–west quartzite ridges between the Zuurberg Mountains and King William's Town, are a complex of scattered, small forest patches, in which the most prominent canopy trees are narrow-leaved yellowwoods, red currants *Rhus chirindensis* and white ironwoods *Vepris lanceolata* (in Perrin 2013).

A survey of two forest patches in the KwaZulu-Natal midlands, where Cape Parrots are present, showed them to be generally depauperate, with little variation in tree species composition. There is a general lack of midsized trees, probably the result of past exploitation for timber and present exploitation by rural landholders for poles. Perrin points out that because forests in key parts of the range of Cape Parrots have been reduced in area, and because of their young age structure, there can be little natural regeneration to climax forest for many decades, even with the most optimistic forecast and immediate action.

MOVEMENTS Cape Parrots are locally mobile, with both short-distance daily movements and long-distance seasonal movements being undertaken in search of food. Local, daily movements of 10–20 km are made along regular flight paths between nesting or roosting sites and foraging areas, and movements of 100 km or more are undertaken in certain seasons when there are food shortages. Skead (1964) reported that in Eastern Cape Province the parrots occupied traditional roosting and nesting sites in the high mountain forests of the Stutterheim, Amatola and Katberg Ranges, from where they made foraging flights in varying distances either over short distances to nearby bush country and scattered forest patches or over longer distances of 60–90 km to the coastal bush. Flocks may be seen daily at one place for a time and then be absent for months.

HABITS Much of what is known of the habits of Cape Parrots comes from field studies undertaken in Hlabeni State Forest and the Weza-Ngele Forest Complex, in southern KwaZulu-Natal (in Perrin 2013). Flocks of up to 20 or more birds occupy communal nighttime roosts, usually in the canopy of emergent yellowwood trees. Daily activity commences at sunrise, when individual birds spend time preening and sunbathing while perched on exposed branches, and these actions include scratching and stretching to the accompaniment of soft chirping, all interspersed with climbing movements and interactions with other members of the flock. Allopreening takes place between all individuals and probably serves a social function to maintain group cohesiveness. Dispersal from the roost to feeding sites usually is by small parties of two to five birds, but at times larger flocks my leave as a unit, and bouts of foraging persist until late morning. Cape Parrots are highly arboreal, coming to the ground only to drink, and they are adept at climbing among the branches to take fruits, their actions being slow and deliberate as they walk forward along a stout limb, placing one foot in front of the other or stretching up to grasp a higher branch in the bill before lifting one foot at a time up to the next level. In the middle of the day, the parrots sleep or rest in the canopy of tall trees, and when not at rest they often interact with each other, typically chasing in short flights, tussling with bills or lunging at each other in mock combat, one bird sometimes dropping down to hang upside-down from its perch. Feeding recommences in the mid to late afternoon and at dusk the birds return to the communal roost. Weather conditions may influence the pattern of daily activities, with the birds being more active on cool, misty days.

Interspecific actions occur mostly with African Olive Pigeons *Columba arquatrix*, which have similar habits, but there is little interaction with other avian frugivores.

The flight is strong and purposeful, with rapid wingbeats featuring little movement above the horizontal. The birds often fly as pairs or in larger groups, with the group breaking up into subgroups of up to five individuals, which move in different directions before regrouping. In swift, direct flight between forest patches, single birds, pairs or small flocks typically fly high while calling loudly, and at other times they can be seen wheeling and swerving about high above the forest canopy. Birds commonly fly out from a perch to circle overhead before returning to the perch, and if disturbed they will take flight to dart through the trees while screeching loudly.

CALLS Perrin (2013) notes that Cape Parrots have a distinctive vocal repertoire, and most frequently heard calls comprise five separate elements described as *tzu-wee...zu-wee...zeu-wee... zz-keek* and a nasal *zeek*, the first four elements being identified by an initial syllable ascending in frequency, followed by a double-harmonic trailing syllable. A series of *screet...screet...screet* shrieks is emitted in flight, and when disturbed the birds give a rapid rasping call with up to 12 identifiable harmonics. Pairs at the nest usually are quiet, but infrequent chirps may be heard, and adults or chicks threatened in the nest give an aggressive succession of *zeeks* that becomes a growl as the threat increases. Duets performed by pairs, probably mated pairs, show controlled synchrony, and probably function to maintain pair bonds, locate the position of partners and maintain territories.

DIET AND FEEDING Cape Parrots feed primarily on *Podocarpus* seeds, which make up approximately 70 per cent of their diet, and it is mainly when these are not available that other foods are eaten. The heavy, robust bill is well adapted to extracting seeds from the fruits, which are taken while unripe and inaccessible to most other frugivores. A fruit is held in the foot while the bill is used to peel away the skin and flesh, which are discarded and the seed kernel then eaten. Fruits of narrow-leaved yellowwoods *Podocarpus falcatus* are favoured because the large size provides a high nutritional return for less effort, and the trees are high-yielding over extended fruiting periods. Climatic factors influence fruiting patterns, at times bringing about poor yields or even crop failures, and then the parrots will utilise alternative food sources within and outside the forests. They regularly take the green seed pods of introduced black wattle *Acacia mearnsii*, which is grown commercially, and will raid crops of pecans, apples, pears, plums, apricots and peaches (in Perrin 2013). Usually during August to December, they feed on dead flower heads of Natal sugarbush *Protea caffra*, which grows in grassland mosaics surrounding the forest patches, and these may be an important food item when forest fruits are scarce.

The usual foraging pattern is for these parrots to feed in single-species flocks in *Podocarpus* trees for extended periods, rarely moving to other fruiting trees. During February and March, when feeding in white stinkwoods *Celtis africana*, they regularly form mixed-species foraging flocks by associating with African Olive Pigeons *Columba arquatrix*, Red-winged Starlings *Onychognathus morio* and Purple-crested Turacos *Gallirex porphyreolophus*.

Drinking is observed most frequently during the morning, and at the study site in Hlabeni State Forest a favoured drinking site used regularly throughout the year was where a perennial stream flowed over an overhanging rock with water dripping into a small shallow pool (in Perrin 2013). A lone bird, usually a male, flew down first and then, depending on the size of the flock, one to five birds came to drink while others remained in nearby shrubs. Each bout of drinking was brief, with birds flying down to drink for no longer than one minute before returning to perch in the shrubs.

BREEDING Perrin (2013) points out that the breeding success of Cape Parrots is determined by the availability of very old trees, especially those showing signs of senescence, which provide nesting sites, and these trees are increasingly scarce. Young parrots are fed almost exclusively on seeds of narrow-leaved yellowwoods, which results in a restricted breeding season, though erratic fruiting patterns can at times extend breeding episodes. Socio-sexual behaviour is exhibited by pairs as early as April or March, and this includes courtship display featuring the 'archangel display' with the orange-red underwings prominently displayed by lifting the wings above the back, together with courtship feeding and copulation. Courtship prior to copulation involves several behaviours, including 'switch-sidling' displays, wing displays, allopreening and head-bobbing, with intermittent 'archangel displays' accompanied by vocal duetting, and the male feeds regurgitated food to the female with exaggerated head movements. Copulation is a highly ritualised behaviour pattern, and can occur well before and long after egg-laying to maintain the pair bond.

Nesting occurs during September to December, but has been recorded also in May (in Perrin 2013). Apart from one nest in a living black wattle *Acacia melanoxylon*, all recorded nests were in natural cavities or snags in dead emergent canopy trees, and 71 per cent of these were in *Podocarpus* trees, which remain standing long after death and often have cavities in the trunk where branches have broken away. At heights of 6–12 m above the ground, nesting cavities were in main trunks, large branches or at the tops of dead trees, and one nest probably was an old nesting cavity of woodpeckers. At study sites in Ingeli and Hlabeni State Forests, in KwaZulu-Natal, three nest-sites were occupied in consecutive years. One nesting cavity at a height of 10 m in a 25 m tall east-facing dead tree was examined when the tree fell in strong winds. This cavity was 170 × 75 mm and the chamber was 300 × 300 × 660 mm in extent, with the only lining being woodchips and flakes. The parrots had excavated the entrance hole and soft, decomposed heartwood, presumably to widen or deepen the chamber.

A normal clutch comprises two to five, usually three or four eggs, and incubation by the female lasts 28 to 30 days. Newly hatched chicks are covered in very short white down which changes to golden yellow at three weeks. Both parents feed the nestlings in response to their soliciting by chirping continuously, but the female spends more time in the nest than does the male. The parents are cautious and vigilant when flying to and from the nest, and they usually give loud squawks when approaching the nest. Both enter the cavity to feed the chicks but, when well-grown, the chicks will appear at the nest entrance to be fed. The male spends variable time at the nest and in the vicinity of the nest, but roosts elsewhere. The chicks fledge at 55 to 79 days after hatching, and remain to be fed by the parents for a further few months, after which they mix with adult flocks. Females probably breed for the first time at age two or older, and erratic breeding is responsible for a low overall recruitment rate.

EGGS Measurements of eggs are listed as 30.4–39.2 (34.1) × 26.0–30.2 (27.9) mm (in Perrin 2013).

FAMILY **PSITTACULIDAE** Vigors

Of two major subgroupings that have been identified within Psittaculidae, the so-called 'coral-billed' parrots make up a more Afro-Asian radiation, with the platycercines or 'broadtailed parrots', together with the lories and their allies, making up a predominantly Australo-Papuan radiation. The subgroupings are differentiated as subfamilies.

SUBFAMILY **PSITTACULINAE** Vigors

Molecular analyses have identified a sister relationship between parrots in this subfamily and the Australo-Papuan platycercine parrots, and within this subfamily the analyses have recovered two clades, which are differentiated at tribal level (Joseph *et al.* 2012). The collective term 'coral-billed parrots' has been applied to parrots belonging to the two clades, and refers to the smooth-surfaced yellow to red bill of most species. Another conspicuous external feature is their pericyclic irides.

TRIBE **PSITTACULINI** Vigors

Two prominent plumage features characterising parrots in this tribe are a smooth-surfaced, usually pale-coloured bill and a contrastingly coloured rump or underwing-coverts. The bill lacks a pronounced culmen ridge, and in some genera is red in colour. The pericyclic irides are pale-coloured or brown, and white irides seem to be correlated with greenish legs (Smith 1975). The inconspicuous cere is partly or fully feathered. Smith pointed out that the courtship display is complex, and in most genera sexual dimorphism is pronounced. Juveniles may resemble the adult female or they may exhibit a distinctive juvenile plumage, with young males usually acquiring the adult female plumage before attaining adult male plumage. There is no underwing-stripe in juveniles or adults.

This tribe is widely distributed from the Mascarene Islands and Africa through India and Southeast Asia to Australasia.

GENUS ***Psittacula*** Cuvier

Psittacula Cuvier, *Leçons d'Anat. Comp.*, **1**, 1800, table at end. Type, by subsequent designation, *Psittacus alexandri* Linnaeus (Mathews, *Bds Austr.*, **6**, 1917, p.169).

A strongly graduated tail, in which the feathers, particularly the elongated central pair, are narrow and attenuated is the most conspicuous morphological feature of midsized parrots belonging to this genus. The bill is heavy and robust with a distinct notch in the upper mandible. Thompson (1900) pointed out that in the skull there is a wide interruption to the orbital ring and a rather large postfrontal process arching downwards.

Psittacula is very widely distributed from the Mascarene Islands and Africa to Southeast Asia, and one species – *P. krameri* – is the most widely distributed of all parrots. An interpretation of findings from molecular analyses indicates that divergence of *Psittacula* from other parrots occurred at some point between the Late Miocene and Early Pliocene (3.4–9.7 mya), and may have been prompted by formation of the Tibetan plateau and Himalayan mountains, which took place during the Late Miocene (5–24 mya) (Groombridge *et al.* 2004). Divergence of *Psittacula* into three main lineages was a comparatively early event. The occurrence of *Psittacula* in the Seychelles Archipelago and the Mascarene Islands is in accordance with the high proportion of biota found across the western Indian Ocean islands showing affinities with Asia rather than Africa, and low sea levels over the previous 10 million years may have facilitated radiation from India by 'island hopping' (see Warren *et al.* 2010). Molecular analyses indicate that extinct *P. wardi* and the closely related, extant *P. eupatria* form one of two separate clades within a very widespread Asian lineage, with *P. wardi* being the first divergence (Kundu *et al.* 2012).

EXTINCT

Seychelles Parakeet

Psittacula wardi (E. Newton)

Palaeornis wardi E. Newton, *Proc. Zool. Soc. London*, 1867, p. 346. (Seychelles)

DESCRIPTION Length 41 cm.

ADULT MALE General plumage green, slightly paler and more yellowish on underparts; occiput, nape and sides of neck suffused blue; obscure black line from feathered cere to eye; broad black band across lower cheeks and continuing as narrow collar across hindneck; deep maroon patch on upper wing-coverts; elongated central tail-feathers above blue tipped yellow and becoming green towards bases, lateral tail-feathers green and underside of tail yellow; bill red; iris probably yellow; legs grey.

2 specimens: wing 204 mm and 208 mm, tail 184 mm and 187 mm, exp. cul. 33 mm and 34 mm, tars. 22 mm and damaged.

ADULT FEMALE Like male, but no broad black band across lower cheeks.

6 specimens: wing 182–204 (194.5) mm, tail 200–261

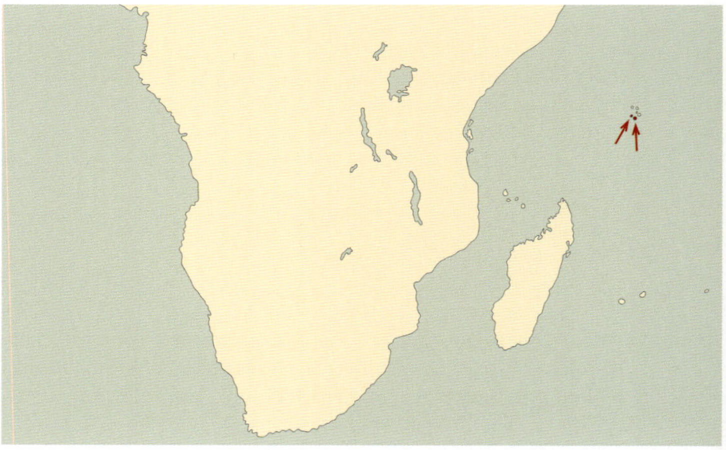

Plate 29
Seychelles Parakeet *Psittacula wardi*
UPPER adult ♂ LOWER adult ♀

F. KNIGHT '11

(228.0) mm, exp. cul. 29–34 (31.7) mm, tars. 20–22 (21.0) mm.
JUVENILES Resemble adult female, but with shorter central tail-feathers.

DISTRIBUTION Formerly occurred on Mahé, Silhouette and possibly Praslin, in the Seychelles.

SUBSPECIES The Seychelles Parakeet obviously is very closely allied to, and at times is considered to be conspecific with the Alexandrine Parakeet *P. eupatria*, which is widespread in the Indian subcontinent and Southeast Asia. It differs by lacking the rose-pink collar on the hindneck, and by its smaller size and shorter but broader wings and tail. Hume (2007) points out that an x-radiograph of a female specimen revealed that it is larger than female *P. eupatria* in the cranium, rostrum, mandible, ulna and tibiotarsus, but smaller in tarsometatarsal and carpometacarpal length. Hume suggests that these characters warrant specific status for *P. wardi*, and that arrangement is adopted here.

STATUS Seychelles Parakeets disappeared sometime after 1893, when the last specimen was collected on Mahé, and 1906, when M. J. Nicoll visited the islands and subsequently made no mention of seeing them in his account of the visit (in Greenway 1967). A 'considerable number' were found in 1811, but they were rare by 1867, and in 1870 a few specimens were sent to Cambridge University. In June 1881, H. M. Warry collected two specimens, one of which is in Liverpool Museum and the other is in the American Museum of Natural History, New York. In 1866, E. Newton failed to find any birds on Mahé, but did observe them on Silhouette, and was told that they had almost been exterminated by ruthless shooting because of the damage that they did to maize crops (Newton 1867). He mentioned hearsay reports that the parrots occurred also on Praslin, but if so they probably disappeared at an earlier date. It was claimed that the clearing of forest to make way for coconut plantations, together with indiscriminate shooting and trapping, brought about their extinction (Newton and Newton 1876).

HABITATS Newton (1867) reported that at 190 to 220 m on Silhouette these parakeets were encountered along the edge of forest adjoining a maize field.

HABITS Virtually nothing was recorded on the habits of Seychelles Parakeets, but they presumably resembled the habits of closely allied Alexandrine Parakeets *Psittacula eupatria*, which usually associate in parties or small flocks in forests or most wooded habitats, and make conspicuous daily flights, sometimes at a considerable height, between communal nighttime roosts and feeding areas. Newton (1867) recalled that the parakeets seen on Silhouette were wary and could not be approached, presumably because of the constant persecution.

SPECIMENS AVAILABLE Hume (2007) notes that there are 10 specimens, and according to Greenway (1967) these are held in the Natural History Museum at Tring and the Museum of Zoology at Cambridge University, UK, in the Museum National d'Histoire Naturelle, Paris, France, and in the American Museum of Natural History, New York, and the Museum of Comparative Zoology at Harvard University, USA.

Mauritius Parakeet

Psittacula eques (Boddaert)

ENDANGERED

Psittacus eques Boddaert, *Table Planches Enluminées*, 1783, p. 13 (Based on 'Perruche à collier, d'l'sle Bourbon', Daubenton *Planches Enluminées*, pl. 215).

OTHER NAME Echo Parakeet.

DESCRIPTION Length 42 cm. Weight 167–193 g.
ADULT MALE General plumage rich green, slightly more yellowish on underparts; underwing-coverts greenish-yellow; narrow black line from cere to eyes; broad black band across lower cheeks and continuing as black line on to sides of neck, where bordered underneath by narrow rose-pink band forming interrupted collar; occiput and nape strongly suffused blue; tail green above, dusky yellow below; upper mandible dark red, lower mandible black; iris pale yellow; legs grey.
8 specimens: wing 177–190 (184.3) mm, tail 164–200 (185.9) mm, exp. cul. 22–24 (23.5) mm, tars. 19–22 (20.1) mm.
ADULT FEMALE No blue on occiput; rose-pink replaced by yellowish-green collar on hindneck; dark green band across lower cheeks; central tail-feathers washed blue; bill entirely black.
5 specimens: wing 175–182 (180.0) mm, tail 162–186 (170.2), mm, exp. cul. 21–23 (22.4) mm, tars. 20–22 (20.8) mm.
JUVENILES Like adult female; bill coral-pink in very young birds of both sexes; iris pale brown.

DISTRIBUTION Confined to Mauritius, and formerly occurred on Réunion, Mascarene Islands.

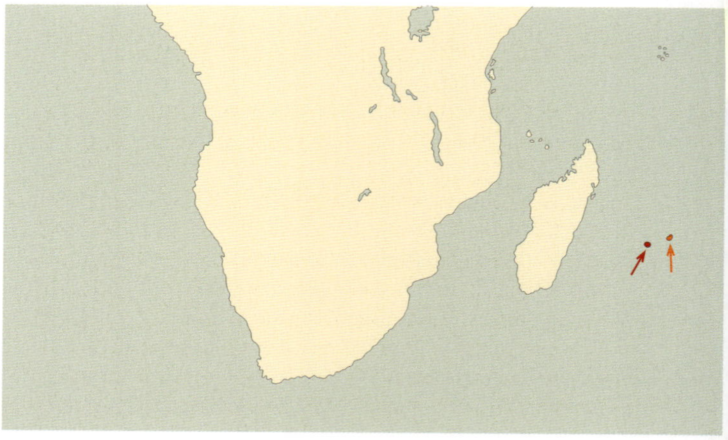

SUBSPECIES Peters (1937) treated this form as a subspecies of the Rose-ringed Parakeet *Psittacula krameri*, but I fully concur with the now widely accepted view that morphological and ecological differences are sufficient to warrant its being considered a separate species. It is a larger, heavier bird, with a distinctly shorter, broader tail and sexual dimorphism in bill colouration.

Plate 30
Mauritius Parakeet *Psittacula eques echo*
LEFT adult ♂ RIGHT adult ♀

F KNIGHT 09

Less obvious is the relationship between the Mauritius Parakeet and 'Perroquets verts ayant un collier noir', which formerly occurred on Réunion Island and are known only from a specimen of uncertain provenance and from illustrations, but it is not known whether these illustrations were based on living birds or preserved specimens. Hume (2007) notes that the type illustration by Martinet in Buffon's *Histoire Naturelle des Oiseaux* is a poor rendition, but in a far superior drawing by Georges Barraband in Levaillant's *Histoire Naturelle des Perroquets* few, if any distinct plumage differences from the Mauritius Parakeet are apparent.

Levaillant captioned Barraband's drawing of an adult male 'Perruche à double collier', and it shows clearly a rose-pink collar encircling the neck, which certainly is a distinct plumage difference, for in adult males from Mauritius the rose-pink bands on sides of the neck do not extend to form a collar encircling the hindneck. A *Psittacula* specimen without a tail is held at the Royal Museums of Scotland (NMS Z 1929.186.2), Edinburgh, where it was received before 1819 as part of the collection of Louis Dufresne, and on Dufresne's specimen label it is identified a *'La Peruche à double collier. Levt'*. The entry in Dufresne's catalogue for this specimen reads *'La Perruche à double collier. Levt. Pl. 39.'*, referring to Barraband's drawing. A comparison of this specimen with specimens of adult males from Mauritius revealed that it differs in having a thin rose-pink collar encircling the hindneck, whereas in Mauritius birds the rose-pink is restricted to crescent-shaped markings on each side of the neck (Jones *et al.* 2016). I concur with the finding that this specimen is from Réunion, and it should be designated as a neotype of *Psittacus eques* Boddaert.

Molecular analyses of toepad material from the Edinburgh specimen revealed a low, but detectable nucleotide divergence between it and specimens of the Mauritius Parakeet, suggesting that populations on Réunion and Mauritius were divergent only at a subspecific level (Jackson *et al.* 2015). I accept that subspecific differentiation is appropriate for populations of ring-necked parakeets on Mauritius and Réunion.

1. *Psittacula eques echo* (Newton and Newton)
Palaeornis echo A. Newton and E. Newton, *Ibis*, 1876, p. 284, pl. 6 (Mauritius).
This subspecies, as described above, occurs on Mauritius.

2. *Psittacula eques eques* (Boddaert)
ADULT MALE Differs from *echo* by having rose-pink markings on sides of the neck continuing to form a narrow collar encircling the hindneck; underparts slightly darker green; slightly larger size. 1 specimen (neotype, in Jones *et al.* 2016): wing 193 mm, tail missing, exp. cul. 24.5 mm, tars. 19.5 mm.
ADULT FEMALE Undescribed.
Formerly occurred on Réunion; now extinct.

STATUS In the early 1900s, concern already was being expressed about the status of the Mauritius Parakeet, but in the 1970s a sudden and dramatic fall in numbers coincided with clearing of a large area of foraging habitat in *Sideroxylon-Helichrysum* scrubland to make way for a forestry plantation, and in late December 1979 a cyclone caused severe damage in all areas of remnant forest (Jones 1980). A disease outbreak also may have contributed to the sudden decline. While some or all of these factors probably were implicated, a rapid fall in numbers could be expected following prolonged declines in recruitment levels and a consequential ageing of the population brought about by long-standing adverse pressures of habitat loss or degradation and high levels of competition and predation at nests. In the 1970s and 1980s few pairs bred and, of those that did lay eggs, few succeeded in successfully rearing young so that fewer than 20 birds were known to survive in the wild and extinction seemed imminent (Jones *et al.* 2013). The loss of habitat has been particularly severe, and in 1996 only 5 per cent of the island supported native vegetation, with remnant stands of upland forest, the favoured habitat, continuing to be impacted by cyclones, the spread of invasive introduced plants and disturbance by introduced pigs and Rusa Deer *Cervus timorensis*. The parrots were confined to the southwest of the island, where they occupied about 40 km² of upland native forest centred on Black River Gorges National Park.

An intensive conservation program commenced in 1973 has focused on addressing each of the known threats to these parrots, and early attention was given to improving the low level of recruitment. There was an obvious scarcity of suitable nesting cavities in large native trees, and the parakeets often chose sites that were falling apart, too shallow, or would be flooded during the heavy summer rains. There was strong competition for available sites, with other cavity-nesting species, including White-tailed Tropicbirds *Phaeton lepturus* and introduced Common Mynas *Acridotheres tristis*, taking over hollows in which the parakeets had commenced nesting. Also, some nests were susceptible to predation, particularly from Black Rats *Rattus rattus* and Crab-eating Macaques *Macaca fascicularis*, which took eggs and chicks or at times killed sitting females. Repairs and improvements were made to known and prospective nesting sites, including the fitting of predator-deterrent collars to the trunks of nesting trees, and inspection holes were made into nest chambers so that the hatching and development of chicks could be monitored. Even when nesting in the improved sites, some pairs continued to have problems, with threats to chicks coming from attacks by blood-sucking fly larvae or inadequate feeding by the parents. Attacks by fly larvae were overcome by replacing the nest substrate with wood shavings treated with insecticide, and a protocol was developed for removing for artificial incubation or handrearing eggs or chicks from nests that looked likely to fail. These handreared chicks were used to establish a captive breeding program or were released back to the wild. The provision of nest-boxes initially was unsuccessful, but subsequently birds released from the captive breeding program used them more readily, eventually being followed by wild pairs, and during the 2009–2010 breeding season 78 per cent of nesting attempts occurred in artificial nestboxes (Richards 2010). The fitting of predator-deterrent collars to the trunks of nesting trees has been highly effective in eliminating predation by rats, and natural hollows have been deepened and baffles fitted in nestboxes to prevent the macaques from reaching eggs or chicks. Occupation of nest sites by Honey Bees *Apis mellifera* and Yellow Wasps *Polistes hebraeus* continues to be a problem, and some success has been achieved by attaching black plastic to the underside of the baffle on nestboxes, so deterring the bees from building a comb.

Spectacular success has been achieved in boosting the level of recruitment, and release into the wild of captive-bred birds has augmented the population significantly so that following a record fledging of 134 chicks in the 2009–2010 breeding season the wild population in 2010 was boosted to an estimated 500 birds (Richards 2010). This increase in numbers has highlighted the shortage of food in the small remnant of native forest, much of which is degraded by exotic plants, so supplementary feeding has become a key element of the conservation program. Before

The upper figure in this drawing by François-Nicolas Martinet, dated 1760, is the earliest coloured drawing of *P. e. eques* from Réunion, and shows the complete rose-pink collar.

Dated 1801, and reproduced in *Histoire Naturelle des Perroquets* by François Levaillant, this fine drawing of *P. e. eques* by Jacques Barraband also shows the complete rose-pink collar.

the introduction of supplementary feeding, food shortages during the breeding season were a major factor limiting population growth, but breeding pairs now are able to meet their nutritional requirements for successful nesting. By watching captive-bred birds feeding at supplementary feeding stations, wild pairs have learnt to take supplementary foods, so that now more than 50 per cent of the population feed on these supplementary foods. The population continues to increase, and Carl Jones tells me that at the time of writing it is approaching 700 birds (*in litt.* 2016).

Psittacine beak and feather disease is present in the population and continues to pose a serious threat. Richards (2010) notes that an isolated case was recorded in 1996, but in 2004 there was a significant outbreak, prompting an intensive screening program involving the collection of blood samples for testing. In five years, a total of 552 parrots were tested for the active virus, and this revealed that the disease is widespread with more than 30 per cent of sampled birds having encountered the virus. It was found that young birds less than two years old are most affected and, although virulence of the disease varies from year to year, 40 to 50 per cent of fledglings are lost to it and associated infections. On a more positive note, screening has shown that 29 birds recovered from the disease, though it is not known whether these birds remain carriers with a potential for passing the disease to their offspring. Strict hygiene measures and protocols have been introduced to ensure that management does not facilitate spread of the disease, and investigations are being undertaken to determine whether the gathering of groups of birds at feeding stations increases the spread of infection.

In response to numbers of Mauritius Parakeets nearing the maximum carrying capacity within Black River Gorges National Park, birds have been released in the mountains of eastern Mauritius to establish a second population (Jones *in litt.* 2016). The prospect of reintroducing birds to Réunion also has been raised, and introduction to the Seychelles has been mentioned as another possibility (in Birdlife International 2016).

Recovery of the Mauritius Parakeet has been extremely rewarding, and plaudits must go to the dedicated personnel involved in the management program. However, threats remain, so long-term survival of these parakeets is dependent on maintenance of intensive conservation actions, and at this time the endangered status is warranted.

Hume (2007) notes that, considering the extent of available habitat on Réunion and the survival of these parakeets in much diminished habitat on Mauritius, it is surprising that they died out very early after settlement of Réunion. The last reference to green and grey parrots on Réunion was by Joseph-François Charpentier de Cossigny in 1732, and it indicates that both probably were hunted to extinction:

> *The woods are full of parrots, either completely grey or completely green. They were eaten a lot formerly, the grey especially, but both are always lean and very tough whatever sauce one puts on them* (translation by Hume).

Supporting a suspicion that *P. eques* may have survived on Réunion until the late 1700s is a description by Oustalet in 1896 of an unlabelled pencil drawing from about 1770, and this description notes '.....encircled by a black collar to the front, red at the back, widening at the nape'.

The Mauritius Parakeet is listed on Appendix I of the Convention on International Trade in Endangered Species of Wild Fauna and Flora (CITES).

HABITATS Jones (1980) points out that Mauritius Parakeets are almost totally dependent on native vegetation, which now occurs on less than 2 per cent of Mauritius, and their last stronghold is in the southwest, in the upland forests of Black River Gorges National Park where some of the largest mature trees survive. They regularly venture into intermediate and lowland forest or scrub to feed, with different areas being utilised seasonally.

HABITS Early reports are of these parakeets being encountered singly, in pairs or small parties, but now that numbers are much higher and the population is concentrated in Black River Gorges National Park larger flocks are observed more frequently, particularly at or near feeding stations. Pairs are readily identified within the groups and pair maintenance behaviour, including allopreening and courtship feeding, can be observed throughout the year. Peak periods of activity are during the morning, and especially again in the middle to late afternoon, when there are conspicuous flights to and from feeding areas, accompanied by much calling. Unlike the introduced Rose-ringed Parakeet *P. krameri*, these parakeets are strictly arboreal, and their plumage blends well with the foliage as they clamber amongst the foliage in search of food. During the middle of the day small groups may be seen resting in the tops of large, emergent trees. They are confiding, and normally will allow a close approach, but they react with alarm at the appearance of macaques, and have been observed mobbing a Mauritius Kestrel *Falco punctatus*. For approximately an hour before sunset pre-roosting flights are particularly evident, with groups of parrots having been observed calling excitedly while circling repeatedly above the forest canopy and pausing intermittently atop the emergent branches of tall trees (Jones *in litt.* 1984). Nighttime roosts usually are in stands of mature trees with dense foliage, at sheltered locations on hillsides or in ravines, and the parrots select perches well in and near to the main trunk. The flight is more laboured than that of Rose-ringed Parakeets, and with noticeably slower wingbeats.

CALLS Jones (1987) notes that Mauritius Parakeets call throughout the year, but are more vocal during the breeding season. Bouts of calling are more frequent in the early morning and especially in the evening before roosting. A variety of calls are given, and all are quite different from those given by the more vocal Rose-ringed Parakeets. The normal contact call is a low, nasal *chaa-choa...chaa-choa*, emitted at a rate of approximately two notes per second, while the alarm or excitement call is a higher-pitched *chee-chee-chee-chee* at the rate of three or four notes per second. While perched, both sexes utter a prolonged creaking *yowl*, which is complex with numerous harmonics, and it is emitted also in courtship (Horne 1987). When suddenly disturbed or frightened, a bird may give a short, sharp *ark*, and perched birds of all ages and both sexes often whistle quietly or chatter and make a wide variety of sounds (in Jones *et al.* 2013). Also, there are reports of birds whistling melodiously from perches at the tops of trees. On two occasions Jones heard a deep, quiet *purring* call from an adult female just after alighting in a tree.

DIET AND FEEDING Native plants are the primary food source, and from these Mauritius Parakeets take fruits, seeds, flowers, buds, leaves, emergent shoots, twigs, bark and sap, with different plants being targeted at different times of the year. In November to March they were seen feeding in *Sideroxylon-Helichrysum* scrub and in April the seeds of Mauritian ochna *Ochna mauritiana* seemed to be the main food (Temple 1978). Jones (1987) reports that of 95 feeding episodes, 53 per cent were on fruits, 31 per cent on leaves, 12 per cent on flowers and 4 per cent on bark or sap. Temple recorded 239 feeding observations in more than 30 plant species, and featuring in more than 25 per cent of these observations were tamanu *Calophyllum parviflorum*, myrrh *Canarium paniculatum*, brittlewood *Nuxia verticillata*, Mauritian toad tree *Tabernaemontana mauritiana*, ordeal tree *Erythrospermum monticolum*, bois colophane rouge *Protium obustifolium*, jackal berry *Diospyros mespiliformis*, great red milkwood *Mimusops maxima* and bois makak *M. petiolaris*, together with *Eugenia* myrtles and *Labourdonnaisia* milkberries. Between 1993 and 1995, during two breeding seasons, 80 per cent of feeding observations were in five native tree species – *Tabernaemontana persicariifolia*, bois de perroquet *Cordemoya integrifolia*, bois dur *Securinega durissima* and black ebony *Diospyros tessellaria* (Jones *et al.* 2013). *Tabernaemontana* leaves and shoots are important in October to January, followed by flowers which become available from January and then fruits are targeted during March to September. Fruits of *Canarium paniculatum*, *Cordemoya integrifolia*, *Ficus reflexa*, *Syzygium glomeratum* and introduced privet *Ligustrum robustum* are taken in October to December, and seeds of *Securinega* and exotic star fruit *Averrhoa carambola* together with *Canarium* leaves and shoots are eaten during January to April. Other regular food items include fruits of *Calophyllum parviflorum*, *C. eputamen*, haronga *Harungana madagascariensis*, bois maigre *Nuxia verticillata*, *Gaertnera* spp. and exotic *Litsea monopetala* and black sage *Cordia currasavica* (Jones *et al.* 2013). In addition to star fruit, other introduced plants utilised for food include strawberry guava *Psidium cattleianum* and wild apple *Solanum auriculatum*.

Feeding is arboreal, the parakeets often hanging upside-down from a branch to remove fruits, flowers or leaves with the bill, and these then are held in the foot to be eaten. When feeding on the leaves of Mauritian toad trees, the parakeets typically scoop out the mesophyll or leafy centre with the lower mandible before discarding the petiole and midrib (in Perrin 2013). Many leaves are partly eaten, being discarded after only partial scooping of the mesophyll and, in a sample of 152 leaves, 49 were bitten off and dropped, 34 were partly eaten, and in 69 most of the mesophyll had been eaten. Similarly, several fruits may be sampled and dropped before one is eaten.

Jones points out that many of the native food plants now are scarce, and food shortages have been implicated in decline of the population. More leaves are eaten in winter, when there is a lower availability of fruits and seeds. There is overlap between the diet of these parakeets and diets of the Pink Pigeon *Streptopelia mayeri*, the Mauritius Bulbul *Hypsipetes olivaceus* and especially the Mauritius Fruit-bat *Pteropus niger*. The Crab-eating Macaque *Macaca fascicularis* is a particularly destructive food competitor because it frequently pulls away unripened fruits.

BREEDING Late September to early October appears to be the peak period for egg-laying, though there are estimates of laying as early as the beginning of August and as late as mid November (Jones *in litt.* 1984). Recorded natural nests have been in cavities in large, living native trees, especially in *Canarium*, *Mimusops* and *Labourdonnaisia* trees, at 6 m to 10 m above the ground, but nowadays nesting largely takes place in nestboxes. For 81 nests in trees, the mean height above the ground was 7.7 m, with a range of 1.5–13.7 m, and for 72 nesting hollows the mean depth was 81 cm, with a range of 16–290 cm, with most being at 30–40 cm (Jones *et al.* 2013). The entrance hole usually is a vertical crack or elliptical aperture from 5.5–29.0 cm in width and 6.0–45 cm in height, with the mean for 71 entrances being 10 × 18 cm (in Jones *et al.* 2013). Nesting hollows inspected during a survey carried out by Stanley and Barbara Temple were sited so that exposure to the southeasterly trade winds was avoided (in Jones *in litt.* 1984).

Defence of the nest site by both sexes commences before egg-laying, but subsequently is mainly by the male. In pre-copulatory behaviour typical of *Psittacula* parakeets, the male wipes his bill along a branch and then in a slow walk moves towards the female to preen the feathers of her nape. To solicit copulation the female squats on the branch in a horizontal position. The male mounts on the back of the female and copulation is accompanied by a ritualised bobbing of his head. Copulation may last for up to five minutes, during which time the male may change sides frequently. After copulation, the male may feed the female, and both preen before flying off to the accompaniment of shrill *cheeek* calls (in Perrin 2013).

A normal clutch comprises two to four eggs, laid at intervals of one to two days, with incubation by the female commencing after laying of the first or second egg and lasting 22 to 23 days. While incubating, the female is fed by the male four or five times during the day or about once every four hours, typically leaving the nest to be fed on a nearby branch, but sometimes she will be fed inside the hollow, especially during cold, wet weather. For the first two weeks after hatching the chicks are brooded by the female, but thereafter are left unattended for much of the day. At a nest under observation by Stanley and Barbara Temple, an adult visited at intervals averaging 79 minutes, and remained in the nest for an average of 11 minutes to feed the chicks (in Jones *in litt.* 1984). The chicks develop slowly, with darkish feather tracts visible on the back and primary quills just beginning to show at five days. At 10 days, the feather tracts are more obvious with down tips breaking through, the eyes appear as slits as they commence to open, and the legs and feet change from pinkish to pale grey. At 15 days, when the eyes are almost fully opened, there is a fine covering of greenish-grey down on the upperparts, flanks and crop, an emergence of the secondary quills and appearance of feather tracts on the crown. A complete covering of greenish-grey down is present at 20 days, when the wing quills are well developed and a few green coverts appear on the upper wings. At 30 days feathered tracts are present on the underparts and rump, the wing and tail feathers are emerging and much of the upperparts is feathered. At 40 days all contour feathers are well developed, as are the lengthening wing and tail feathers, but down is retained on the lower back, rump and flanks. At 50 days, the chicks are quite active, flapping their wings and coming to the entrance hole. Fledging occurs at 50–69 days, with a mean of 57.3 days for 64 nests (Jones *et al.* 2013). During the survey carried out between 1973 and 1975 by Stanley and Barbara Temple, fledging dates were recorded between mid November and mid January, with most young birds leaving the nest in mid to late December (in Jones *in litt.* 1984). Young birds remain with their parents for some time, and immatures being fed by adults have been observed as late as March.

The monitoring of nest sites, with required intervention to safeguard eggs or chicks, has increased the nesting success rate. Richards (2010) reports that in the 2009–2010 breeding season eggs were laid in 74 of 78 monitored nests, and the proportion of females fledging at least one chick rose to 82 per cent. More than 220 eggs were laid, producing 156 chicks, and an increase in the number of young females successfully fledging chicks was most encouraging. Chicks fledged from 11 nests where the female was three years or younger. Conversely, breeding attempts in non-monitored nests were either unsuccessful or of very limited success and only three unringed fledglings from these nests were seen during the three years prior to the 2009–2010 breeding season. Highly successful breeding in the 2009–2010 season, including the survival of all four chicks in several nests, more than compensated for mortality resulting from the serious outbreak of psittacine beak and feather disease in that year.

Attendance at a nest by up to five or six additional adult males has been recorded. Apparently these extra males were evident only after the chicks had hatched and it was assumed that they were helpers. However, their attempts to feed the sitting female and the chicks were rebuffed and their presence seemed to be disruptive, so the likelihood is that their presence is merely a consequence of an overall low population density with a higher preponderance of sexually active males. Support for this explanation has come with at least one report of nest failure because of aggressive interference from an additional adult male.

EGGS The eggs are rounded, and measurements of 38 eggs in nests in the wild are 32.3 (29.4–34.5) × 26.1 (23.0–27.6) mm (Jones *et al.* 2013).

Newton's Parakeet

Psittacula exsul (A. Newton)

EXTINCT

Palaeornis exsul A. Newton, *Ibis*, 1872, p. 33 (Rodrigues).

OTHER NAME Rodrigues Parakeet.

DESCRIPTION Length 40 cm.
Presumably green and blue morphs existed, and mention is made in an early account by Julien Tafforet of a red alar patch being present in the adult male of the green morph. Only two specimens survive, and both are of the blue morph.
ADULT MALE General plumage greenish-blue with a greyish cast, paler on underparts; head noticeably darker and without greyish suffusion; fine black line from cere to eyes; chin black; broad black stripe across lower cheeks to sides of neck, then becoming narrow and continuing up to nape; lower back and rump paler, lighter blue; primaries deep greenish-blue; tail dark greenish-blue above, greyish underside; upper mandible red, lower mandible black; iris yellow (Vandorous); legs grey.

1 specimen: wing 199 mm, tail 207 mm (abraded); exp. cul. 25 mm, tars. 22 mm.
ADULT FEMALE Very faint black line on forehead; black stripes not extending beyond sides of neck; crown suffused grey; upper mandible black.
1 specimen (type): wing 192 mm, tail 212 mm, exp. cul. 24 mm, tars. 23 mm.
JUVENILES Undescribed.

DISTRIBUTION Formerly occurred on Rodrigues, in the Mascarene Islands.

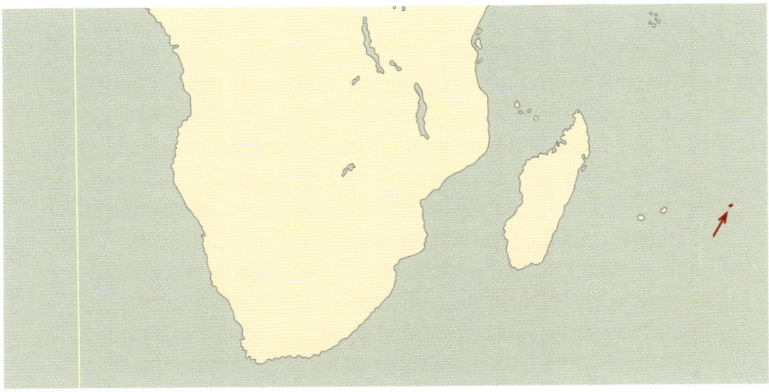

STATUS The last records of Newton's Parakeet were made in the 1870s, and presumably it became extinct at about that time. Only two specimens were collected, the first being a female, which was collected in 1871 by George Jenner, the then magistrate on the island. This specimen was preserved in alcohol and given to Edward Newton, a colonial administrator on Mauritius, who in turn sent it to his brother, and it was used by Alfred Newton to describe the species (Newton and Newton 1876). The second specimen, a male, was shot by a local resident William Vandorous on 14 August 1875 and given to the assistant colonial secretary William Caldwell, who in turn forwarded it to Edward Newton. Caldwell remarked that he had seen several birds, but could not get near one. Henry Slater, a naturalist who stayed on the island for three months in 1874, reported to Newton that on 30 September in that year he saw a single bird in forest towards the southwestern end of the island (in Newton 1875). It has been postulated that the male collected by Vandorous may have been the same bird seen by Slater in the previous year.

Hume (2007) documents accounts from early visitors to Rodrigues and notes that although Newton's Parakeets survived until the 1870s, they were in decline from the 1760s. François Leguat reported that they were abundant at the time of his stay on the island between 1691 and 1692:

> *There are abundance of green and blew Parrets, they are of a middling and equal bigness; when they are young, their Flesh is as good as young Pigeons. ....Hunting and fishing were so easie to us, that it took away from the Pleasure. We often delighted ourselves in teaching the Parrots to speak, there being vast numbers of them. We carried one to Maurice Isle, which talked French and Flemish.*

A live bird was received by the naturalist Philibert Commersen on Mauritius during the 1770s, where it was described as a long-tailed, greyish-blue parrot with a black collar (in Oustalet 1897). Hume notes that the parrots were still common when Julien Tafforet was on Rodrigues in 1726, but had become rare by the time of a visit in 1761 by Abbé Alexandre Pingré, a French astronomer who had come to monitor the transit of Venus, and he referred to their continued presence on the southern islets (translation by Hume):

> *On the 19th at Isle Mombrani, the multitude of grey terns on our side served exactly as a parasol; they fly about our heads, in the manner more or less to ease the heat of the sun. In an additional premium to this there were tropic birds and their eggs. There are also some frigates, some tratras, some perruches.*

After the visit by Pingré, there was severe deforestation on the island, and an increase in numbers of free-roaming livestock. In 1843, a government surveyor Thomas Corby was sent to Rodrigues to ascertain the suitability of the land to support cattle, and he noted that the western side of the island, although severely deforested, contained the best stands of palms and *Pandanus* screw-pines. Corby also referred to the presence of many wild bullocks, pigs, great flights of guinea fowl and green parrots, indicating that Newton's Parakeets remained fairly numerous, but they had become extremely scarce by 1871, when the first specimen was received by Alfred Newton. There were no records after the second specimen was collected in 1875, and Hume suggests that the last few survivors may have been wiped out by a devastating series of cyclones in the following year.

HABITATS Newton's Parakeets presumably frequented native forest, and extensive destruction of this habitat was a major factor in their decline and subsequent extinction.

HABITS Leguat commented on the partiality of Newton's Parakeets to the nuts of bois d'olive *Cassine orientale* and made mention of the bois du buis *Fernelia buxifolia* being a food tree for Newton's Parakeets and for Leguat's Parrot *Necropsittacus rodericanus* (in Hume 2007). Nothing is known of habits of Newton's Parakeets, though they probably were similar to the habits of Mauritius Parakeets.

SPECIMENS AVAILABLE Both specimens (18/Psi/67/h/1 ♀ type and 18/Psi/67/h/2 ♂) are held in the Museum of Zoology at Cambridge University, UK.

Plate 31
Newton's Parakeet *Psittacula exsul*
IN FLIGHT adult ♂ CENTRE adult ♂ RIGHT adult ♀

F. KNIGHT '09

GENUS ***Mascarinus*** Lesson

Mascarinus Lesson, *Traité d'Orn.*, livr. 3, 1830, p. 188. Type, by tautonymy, *Mascarinus madagascarensis* Lesson = *Psittacus mascarin* Linnaeus.

The extinct species belonging to this monotypic genus was a midsized parrot with a large red bill and a moderately long, broadly rounded tail. Traditionally, it has been associated with the Psittaculini and, apart from plumage colouration, it resembles the *Tanygnathus* parrots from Southeast Asia. Alternatively, it has at times been linked with *Coracopsis* from Madagascar and the Comoros Archipelago, probably because of a similarity in the brown plumage colouration, and it has been noted that *Coracopsis* and *Mascarinus* are the only parrots that naturally lack psittacin in their plumage (in Hume and van Grouw 2014). An extraordinary finding from molecular analyses that a cytochrome *b* sequence from mitochondrial DNA of *Mascarinus* is embedded in *Coracopsis* has been questioned on the basis that the mitochondrial DNA was extracted from the damaged type specimen (MNHN 211) and alternative hypotheses concerning the placement of *Mascarinus* were not considered (see Kundu *et al.* 2012; Joseph *et al.* 2012). Molecular analyses of DNA extracted from the only other specimen (NMW 50.688) indicates that the previously obtained cytochrome *b* sequence probably is an artificial composite of partial sequences from two other parrot species and that *Mascarinus* is indeed part of the *Psittacula* diversification, placed close to *P. eupatria* and *P. wardi* (Podsiadlowski *et al.* 2017). I strongly support this finding, and am of the opinion that all three extinct monotypic genera from the Mascarene Islands – *Mascarinus*, *Necropsittacus* and *Lophopsittacus* – can be placed in Psittaculini.

Mascarinus formerly occurred on Réunion, and possibly on Mauritius, in the Mascarene Islands.

EXTINCT

Mascarene Parrot

Mascarinus mascarin (Linnaeus)

Psittacus mascarin Linnaeus, *Mantissa*, 1771, p. 524 ('Mascarina').

DESCRIPTION Length 35 cm.

ADULTS General plumage brown tinged greyish, distinctly paler and slightly yellowish on underparts and rump to upper tail-coverts; forehead and lores to forecheeks and throat black, forming a distinctive facial mask; remainder of head greyish-lilac; tail dark brown with white at base of lateral feathers; bill red; legs brownish.

1 unsexed (type): wing 211 mm, tail 152 mm, exp. cul. 32 mm, tars. 24 mm.

I unsexed: wing (damaged), tail 144 mm, exp. cul. 36 mm, tars. 22 mm.

JUVENILES Undescribed.

The only two specimens presumably were captive birds and are in poor condition, the type having been damaged badly by sulphur fumigations in 1792–1793 and the second specimen has irregular white feathering. Despite the poor condition, both specimens are brown, although faded. All of the early illustrations, including the first coloured illustration by François-Nicolas Martinet in 1779, depict a brown bird, and that has become the orthodox image. Hume and van Grouw (2014) argue that live birds were grey, not brown, and the pale brown colouration of the specimens 'is clearly an artefact of fading due to light exposure, from grey to brown'. They cite two written descriptions of live birds to support this claim. In a book published in 1674, Sieur Dubois documented his travels to the Mascarene Islands in 1671–1672, and he gave the best description (translation by Hume):

> *Parrots a little bigger than pigeons, with plumage the colour of squirrel fur, a black hood on the head, the beak very large and the colour of fire.*

Plate 32
Mascarene Parrot *Mascarinus mascarin* (adults)

From personal observation, Hume (2007) interprets 'the colour of squirrel fur' as a variable dark blackish grey/brown. In 1760, the following description of a fresh specimen was given by Mathurin Brisson (translation by Hume):

> *.....head and neck ash-grey; back, wings, rump and underparts very dark ash-grey; tail dark ash-grey, with lateral feathers white at base; bare skin around eye region, bill, bare skin around upper bill and iris bright red; legs and feet pale flesh; claws grey-brown.*

I have much difficulty in acknowledging that all of the early illustrations depicting the body plumage as brown are wrong. While it is quite probable that over time the two specimens have faded due to light exposure, it is more likely that the fading merely involved a loss of intensity in brown plumage and not a change from grey to brown. It is probable that the two most prominent early illustrations, namely by Martinet in 1779 and by Jacques Barraband in 1805, were drawn from the type specimen or from another specimen of a captive bird, but at that early time it is highly unlikely that the specimens would have changed colour from grey to brown due to light exposure. This then raises the possibility that the colouration change was due to poor husbandry, particularly food deficiency in captivity. Clipped primaries on the Vienna Museum specimen leave little doubt that it was a cagebird, and the irregular white feathers well may be a consequence of poor husbandry, though it is not uncommon in wild birds, but I doubt that other specimens would have consistently changed colour from grey to brown as a consequence of captive conditions. Whether the Mascarene Parrot was brown or grey seems to be unresolved, but we have adopted the so-called orthodox image of a brown bird because that is the plumage colouration of the two specimens and also is the colouration consistently depicted in early illustrations.

DISTRIBUTION Formerly occurred on Réunion, in the Mascarene Islands. From a quote of 'russet parrots' being seen on Mauritius in the early 1600s, the suggestion has been made that the Mascarene Parrot may have occurred also on Mauritius, but Hume (2007) points out that, although Réunion and Mauritius share a number of genera, including parrots, the former occurrence

F KNIGHT '11

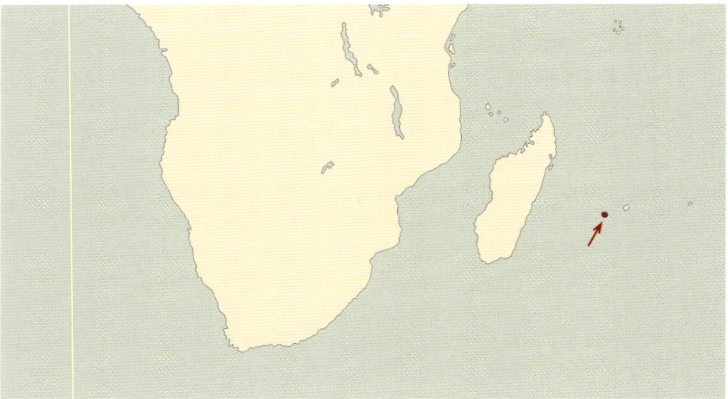

A brown plumage colouration is depicted in this fine drawing of *Mascarinus mascarin* by Jacques Barraband, dated 1801, and reproduced in *Histoire Naturelle des Perroquets* by François Levaillant.

of *Mascarinus* on both islands can be confirmed only by the discovery of fossil remains on Mauritius.

STATUS According to Greenway (1967) the last known living Mascarene Parrot was in the garden of King Maximilian of Bavaria in 1834, but Hume notes that this is extremely unlikely, and the species almost certainly was extinct by about 1800 (in Hume and van Grouw 2014). The last mention of these parrots on Réunion was made by Jean Feuilley in 1704 (translation by Hume):

> *There are several sorts of parrot, of different sizes and colours. Some are the size of a hen, grey, the beak red; others the same colour the size of a pigeon, and yet others, smaller are green. There are great quantities, especially in the Sainte-Suzanne area and on the mountainsides. They are very good to eat, especially when they are fat, which is from the month of June until the month of September, because at that time the trees produce a certain wild seed that these birds eat.*

Hume and Walters (2012) note that these parrots appear to have survived longer than other parrots from Réunion, and at least three or four captive birds were alive in Europe around the latter half of the 18th century. Mention of live birds being seen in Paris was made by P. Mauduyt in 1784 (translation by Hume):

> *The Mascarin is found at Ile Bourbon; I have seen several alive in Paris, they were rather gentle birds; they had in their favour only that the red beak contrasted agreeably with the dark backgroundof their plumage; they had not learnt to talk.*

In 1805, Levaillant commented that they were very rare and he had seen three specimens, one of which was in Mauduyt's possession, so apparently no live birds survived in captivity in Paris at that time.

HABITATS Nothing is known of the habitat preferences, other than the comment made by Feuilley that these and other parrots were particularly abundant in the Sainte-Suzanne area, which is in the northeast of the island, and on mountainsides.

HABITS We have no information on the habits of Mascarene Parrots, apart from the remark made by Jean Feuilley that they and other parrots became fat when feeding on certain seeds during June to September.

SPECIMENS AVAILABLE The type specimen (MNHN 211) is in the Museum National d'Histoire Naturelle, Paris, and another specimen (NMW 50.688) is in the Naturhistorisches Museum, Vienna.

NEOTROPICAL DISTRIBUTION

NEOTROPICAL DISTRIBUTION

For the purposes of this book, the Neotropical Distribution encompasses the Neotropical Faunal Region and southern sector of the Nearctic Region, extending from the United States of America south through Central and South America and including the Caribbean Islands. Although characterised by a lack of diversity, with all species belonging to a single subfamily (Arinae), parrots are very well represented in the Neotropical Distribution, where are found some of the most familiar groups, including the large, spectacular *Anodorhynchus* and *Ara* macaws, the short-tailed, stolid *Amazona* parrots, collectively known as 'amazons', and the slim, narrow-tailed *Aratinga*, *Psittacara* and *Pyrrhura* parakeets. The dominant geographical component is the South American continent, where most species occur, and here the distribution of parrots is influenced strongly by two major topographical features – the Andes and the Amazon River with its tributaries. The influence of the Andes is reflected in the association of some species with one or more of the three forested zones: tropical (up to 1000 m), subtropical (1000 to 2500 m), or temperate (2500 m up to the tree-line at about 3500 m), and altitudinal limits to the distribution of parrots are more pronounced in South America than elsewhere in the worldwide range.

NORTH AMERICA

In any discussion about extinct and endangered Neotropical parrots, attention promptly focuses on loss of the Carolina Parakeet *Conuropsis carolinensis*, the only parrot endemic to the United States. Its extirpation, together with that of the Passenger Pigeon *Ectopistes migratorius*, has become legendary, with blame almost universally attributed to persecution by early colonists. While it is difficult to disagree with the assessment that excessive and wanton hunting was a major factor in disappearance of the Passenger Pigeon, the role of hunting in extinction of the Carolina Parakeet seems to be less straightforward. In his excellent book, *The Carolina Parakeet: Glimpses of a Vanished Bird*, and in his Foreword to this book, Noel Snyder points out that the disappearance of these parakeets may well have been due largely to other factors. That the parakeets were shot in large numbers, both for sport and to protect crops, is evident from written accounts by Audubon (1831) and others, but whether persecution was sufficiently intense or widespread to bring about their extirpation remains doubtful. Such doubts were expressed by Daniel McKinley, who painstakingly chronicled the decline of Carolina Parakeets throughout much of their range, and he raised the possibility of disease being a significant factor in their demise (see McKinley 1980). Audubon took live parakeets to Britain, and they were popular cagebirds in Europe, so they were captured for the live-bird market, but it is likely that this impacted significantly on populations only as the decline in numbers accelerated. Of course, habitat alienation or destruction almost certainly impacted adversely on populations but, as pointed out by Snyder, it was localised and adequate potential nest-sites and food sources seem to have remained available. In summary, it can be said that a lack of reliable evidence in early accounts prevents extinction of the Carolina Parakeet being attributed to any single cause, but these same accounts demonstrate forcefully that the species faced formidable adverse pressures associated with the spread of European settlement. A combination of loss of forested habitats and intensive persecution could have brought about a rapid decline in much of the range, and disease may have played an increasingly significant role as numbers reached low levels.

A second parrot species – the Thick-billed Parrot *Rhynchopsitta pachyrhyncha* – disappeared from the southwestern United States at about the same time as the Carolina Parakeet was lost in the east, but it survives south of the border in western Mexico, and very occasionally a few birds wander north into the pine forests of Arizona and New Mexico. As with the Carolina Parakeet, causes of the disappearance of Thick-billed Parrots from this northernmost part of their range are poorly known, again because no field studies were undertaken while they occurred in the region. Early reports of these parrots in Arizona, where they were locally common in pine forests of the Chiricahua Mountains, consistently refer to their being shot in numbers and, when interviewed by Noel Snyder and colleagues, elderly residents familiar with the birds in bygone decades confirmed that they were subjected to heavy hunting pressure, especially from hungry miners and woodsmen (in Snyder *et al.* 1989). When interviewed before his death in 1976, Ralph Morrow, a lifelong resident of the Chiricahua Mountains, and an Arizona Game and Fish Department agent for many years, believed that hunting pressure was sufficient to have brought about their extirpation. He admitted to having shot many Thick-billed Parrots in the early 1900s, and knew of others who had done the same. At that time many persons were 'living off the land', and their depredations undoubtedly impacted on local wildlife. Wild Turkeys *Meleagris gallopavo* vanished from the region at about this time, as did Bighorn Sheep *Ovis canadensis*, Elk *Cervus canadensis* and Pronghorn Antelope *Antilocapra americana*, all surely because of hunting, and it seems reasonable that Thick-billed Parrots succumbed to the same pressure. The mammals and turkeys have been reintroduced, but reintroducing Thick-billed Parrots proved more difficult and, although encouraging, the only attempt did not succeed. To some extent the parrots may have suffered also from fairly extensive cutting of the pine forests, which was associated mainly with early mining activities and has resulted in the forests of Arizona and New Mexico now being almost uniformly mature with logging operations being limited in extent. South of the border, logging of the mountain forests has been much more intensive and widespread, so restricting very significantly any northward movements of Thick-billed Parrots into the southwestern United States.

CENTRAL AMERICA

From northern Mexico south to southern Panama, widespread deforestation, coupled with hunting and capture for the live-bird market, both domestically and internationally, has impacted very severely on parrot populations. In northern Mexico, Thick-billed Parrots *Rhynchopsitta pachyrhyncha* have lost both foraging and nesting habitat from extensive logging of mountain pine forests, and for the closely allied Maroon-fronted Parrot *R. terrisi* this logging has deprived them of essential foraging habitat, a threat that takes on increased importance when the logged forests are near to their nesting sites in cliff-faces. Increased frequency and intensity of forest fires is causing passive deforestation, for replanting with native species is occurring only on small parcels of privately owned lands, and naturally regenerating areas usually become covered with dense oak-chaparral vegetation, which is of no value to the parrots (in Snyder *et al.* 2000).

It has been estimated that by 1995 nearly 99.5 per cent of *Pinus-Quercus* forest in the Sierra Madre Occidental had been at least partially logged, and only 22 km² of high-elevation old-growth forest, the preferred nesting habitat for Thick-billed Parrots, remained intact (in Snyder *et al.* 1999). In some districts, the disappearance or near disappearance of these parrots and loss of the Imperial Woodpecker *Campephilus imperialis*, a sympatric species also closely associated with highland pine-oak forests, coincided with the almost complete logging of local pine forests. Because they do not 'talk', Thick-billed Parrots are not popular pets in Mexico, but increasing demands in other countries resulted in an upsurge in trapping for the export market, and this reached a peak in the mid 1980s, when there were large numbers of confiscations in the United States (in Snyder *et al.* 2000).

Conversely, *Amazona* parrots are highly popular as cagebirds, both domestically and internationally, and excessive exploitation for the live-bird market is a secondary pressure that has contributed very significantly to rapid declines in populations threatened by landclearance. Species strongly threatened by this combination of habitat loss and excessive exploitation are the Green-cheeked Amazon *Amazona viridigenalis* and Lilac-crowned Amazon *A. finschi* from northern to central Mexico, and the Yellow-headed Amazon *A. oratrix* from central Mexico to northwestern Honduras. A legacy of intense trafficking in these three species is the establishment of feral populations in southwestern United States, particularly in the Los Angeles urban district, where it has been suggested that 25 per cent of the world population of Green-cheeked Amazons now occurs.

Popularity as cagebirds has contributed to very significant declines in populations of the large *Ara* macaws in Central America. Although all species occurring in Central America occur also in South America, there often is subspecific differentiation. Howell and Webb (1995) point out that in Mexico the Military Macaw *Ara militaris* has been widely extirpated by the pet trade, and the Scarlet Macaw *A. macao* may vanish completely from some regions within a few years. Jones (2004) points out that the total population of Scarlet Macaws in Belize may not exceed 150 birds, with their range being much reduced since the 1970s, and Gallardo (2014) notes that illegal widespread trafficking continues in Honduras, where it is likely that there are more of these macaws in cages than in the wild.

Of particular concern is the endangered Great Green Macaw *A. ambiguus*, which is generally uncommon to very rare throughout its entire range and is intolerant of habitat disturbance. It has retreated in response to a relentless destruction of lowland forests, disappearing altogether from much of its historical range and remaining fairly common only in very few remote, undisturbed areas, mostly in Central America. Gallardo (2014) points out that the largest remaining populations in Central America likely occur at the northern limit of the range, in easternmost Honduras, and serious efforts must be made to curtail further habitat destruction and poaching for the cagebird market. It is fairly numerous also in Nicaragua, where populations in the Bosawas Reserve and in extensive tracts of lowland forest within the vast Indio de Maíz Biological Reserve may constitute the second largest global subpopulation (in Birdlife International 2016). Conversely, in the Caribbean lowlands and foothills of Costa Rica, where formerly it was widespread, the Great Green Macaw has been forced to near extinction by destruction of approximately 90 per cent of its forested habitat, and in the 1990s, the remnant population was estimated at less than 200 birds (Chassot and Arias 2002). Field research with conservation objectives is being undertaken in northern Costa Rica, where nests are being monitored to determine productivity and recruitment levels, while adults are being fitted with radio transmitters to monitor survival rates and migration movements. Also in northern Costa Rica a public awareness campaign has been initiated to generate support among local communities for conservation efforts. The Ara Project is a non-profit organisation set up in Costa Rica to breed both Great Green Macaws and Scarlet Macaws in captivity for the sole purpose of release into the wild, and in 2011 the first Great Green Macaws were released at Manzanillo, in the extreme southeast (Taylor 2011).

Pyrrhura parakeets are widely distributed in South America, where some endangered species occur in very restricted ranges, mostly in upland forests, but two species are found in the far south of Central America. The Sulphur-winged Parakeet *P. hoffmanni* is generally fairly common in closed primary forest on ridges and partly wooded hillsides in southern Costa Rica and western Panama, but the Azuero Parakeet *P. eisenmanni* is restricted to the southwestern Azuero Peninsula, south-central Panama, where only a single population occurs predominantly in Cerro Hoya National Park, and the total range is considered to be less than 700 km². Long-term survival of this population is dependent on adequate protection of remaining stands of forest.

SOUTH AMERICA

In any consideration of threats to parrots in South America, attention usually is focused on accelerating deforestation in the Amazon River basin, and that focus of attention is warranted because the long-term consequences of destruction of the Amazonian rainforests could be far-reaching. Many parrots occur in these rainforests, and there is no doubt that this expanding loss or degradation of rainforest habitat poses a serious risk to species confined to that region. Three Amazonian species – the Golden Parakeet *Guaruba guarouba*, Santarém Parakeet *Pyrrhura amazonum* and Diademed Amazon *Amazona diadema*, are listed as endangered, but I suspect that in future years other species will be added to the list.

The adverse impact of landclearance on parrot populations has been more devastating in some other regions, where species with restricted ranges or with highly specialised habitat requirements already have been lost or are in danger of disappearing. It seems that only two species are extinct, though another species – Spix's Macaw *Cyanopsitta spixii* – may survive only in captivity. All available evidence indicates that Glaucous Macaws *Anodorhynchus glaucus*, which formerly occurred in the southeast, between lats 27°S and 31°S and centred on middle reaches of the Río Paraguay, Río Paraná and Río Uruguay, have been extinct since the early to mid 1900s but it is difficult to determine with any certainty those factors which brought about their extinction. Available evidence indicates that extinction probably was caused by a combination of habitat destruction and excessive exploitation for food and the pet trade. Lightly wooded savanna with palm groves, the preferred habitat, is easily burned and converted to grazing land, so presumably this habitat transformation occurred throughout centuries of intense cattle grazing in the region. Loss of the Sinú Parakeet *Pyrrhura subandina*, known only from the Río Sinú valley in northwestern Colombia, typifies the vulnerability of species with very restricted ranges. Despite extensive searches, the Sinú Parakeet has not been recorded since 1949, and all forest habitat apparently has been lost at the four sites where specimens were obtained.

Closely allied to the Glaucous Macaw, and apparently occurring only as two colonies in northern Bahia, central-eastern Brazil, the Indigo Macaw *Anodorhynchus leari* remained a mysterious bird for more than a century, being known only from captive birds until 1978 when birds finally were located on the Raso da Catarino, in northern Bahia. Habitat degradation is likely to have brought about initial declines, though hunting for food and feathers probably were contributing factors, and there was local persecution because of damage to maize crops. With its specialised feeding and nesting requirements, the species is particularly vulnerable to habitat interference, and poor regeneration of *Syagrus* palms, its primary food source, due to livestock overgrazing has been identified as a continuing threat. Persecution by farmers because of depredations on peanut crops contributed to a decline in numbers of the Red-fronted Macaw *Ara rubrogenys*, which is restricted to a small range in central-southern Bolivia where, at the beginning of the 1990s, it was estimated that 40 per cent of the habitat had been destroyed and deforestation was continuing, with regeneration of groundcover vegetation being seriously impaired by overgrazing. Like the Indigo Macaw, this species nests in crevices in cliff-faces, so the availability of food resources in proximity to nesting sites is essential to the maintenance of satisfactory recruitment levels. A second endemic species in Bolivia, the Blue-throated Macaw *Ara glaucogularis* remained a mysterious bird for more than a century after its original description, being known only from a few museum specimens of uncertain provenance, and there was much speculation about its relationship to the familiar Blue and Yellow Macaw *Ara ararauna*. That it was a separate species only became evident in the 1970s, when small numbers regularly were found in the company of much greater numbers of Blue and Yellow Macaws in the Llanos de Mojos, an isolated expanse of flat savanna grassland, at 200 m to 300 m altitude, on the upper Río Mamoré system, in Beni, where a healthy population of palms is the single outstanding ecological requirement. Populations of the Great Green Macaw *Ara ambiguus* generally are less common than in Central America, and the southernmost, subspecifically differentiated population survives in western Ecuador as two critically endangered subpopulations in the northwest and southwest. The primary subpopulation, which could be as low as 30 to 40 birds, is in the northwest where persistent hunting for food poses a continuing threat. Only very small, relict populations survive in southwestern Ecuador, where deforestation has been particularly extensive, and total numbers are critically low. Deforestation in western Ecuador has been responsible for very significant declines also in numbers of endemic Grey-cheeked Parakeets *Brotogeris pyrrhoptera* and Lilacine Amazons *Amazona lilacina*. Continued illegal trapping for the live-bird trade, together with persecution and habitat destruction or fragmentation, have been identified as causes of a very rapid population decline in populations of the Grey-cheeked Parakeet that has occurred during the past 10 years, but the rate of decline is expected to be slower during the next decade. Numbers of Lilacine Amazons similarly have declined substantially as a result of deforestation and persecution, and the total population is assumed to be in continuing decline because of ongoing habitat destruction and unsustainable levels of exploitation.

Probably no species more highlights the plight of parrots confined to upland forests than does the Indigo-winged Parrot *Hapalopsittaca fuertesi*. Discovered in 1911, when specimens were collected in montane cloud forest on slopes of the Cordillera Central, in western Colombia, it was 'lost' until 2002 when a flock of 14 birds was found in a small remnant stand of montane forest in the same region. Protecting some remnant stands of montane forest in reserves and the provision of nestboxes to overcome an acute shortage of mature trees with hollows are key elements of a concerted conservation effort for these parrots, and that has slowed or stopped the decline, but total numbers remain very low, being estimated at only 160 birds. Formerly widespread and common in all three Andean cordilleras in western Colombia and neighbouring northwestern Ecuador, Yellow-eared Parrots *Ognorhynchus icterotis* suffered a catastrophic decline in numbers during the mid 1900s, and were brought to near extinction primarily by large-scale deforestation and localised persecution. Available evidence indicates that loss of their montane habitat was primarily responsible for near extinction of these parrots in Colombia and probable extinction in northwestern Ecuador. In addition to the large-scale destruction of stands of wax palms, on which the parrots are so dependent for nesting sites, the loss of elfin forest deprived populations of crucial seasonal food sources. These threats are being addressed in a very successful conservation program, which features extensive reafforestation and widespread planting of wax palms. Another species occurring in montane forests and closely associated with wax palms, the vulnerable Golden-plumed Parakeet *Leptosittaca branickii*, is benefiting from a similarly oriented conservation program in the Colombian sector of its range.

In addition to the Santarém Parakeet, other *Pyrrhura* species are seriously threatened by habitat loss or degradation in their restricted ranges, which often are in upland forests on slopes of the Andes or of isolated massifs. Within its restricted range along the western slope of the Cordillera Central, in northeastern Colombia to the Venezuela border, the Perijá Parakeet *Pyrrhura caeruleiceps* has lost some 70 per cent of its original habitat, and only 20 per cent of its current area of occupancy is protected in reserves, so the loss of habitat is expected to continue. Habitat destruction and poaching for the pet trade have been identified as the principal threats to the surviving population. In northeastern Colombia, the Santa Marta Parakeet *Pyrrhura viridicata* is restricted to the Sierra Nevada de Santa Marta, where it is seriously threatened by loss of habitat. Extensive clearance of native forest to make way for *Pinus* and *Eucalyptus* plantations has left only 15 per cent of the original vegetation cover remaining in the Sierra Nevada de Santa Marta. A nature reserve has been established to protect part of the range, and within this reserve much effort is directed at the eradication of exotic trees, especially the removal of *Pinus* saplings, and nestboxes are being provided to compensate for the loss of mature native trees, though there is competition for these nestboxes from the more aggressive Scarlet-fronted Parakeets *Psittacara wagleri*. On western slopes of the Andes, in southwestern Ecuador, numbers of the recently discovered El Oro Parakeet *Pyrrhura orcesi* are quite low and the highland forest habitat is fragmented as a consequence of widespread deforestation. Tracts of land at and around the type locality at Buenaventura, in El Oro Province, were purchased and this eventually brought about establishment of the Buenaventura Nature Reserve, where conservation efforts have been concentrated. An extensive reafforestation of cleared areas is being undertaken in the reserve and artificial nestboxes are being set up to overcome a shortage of natural nest-sites, while a community awareness program is being conducted, especially among local school children. DNA analyses of blood samples are being used to determine gene flow in the population as an indication of potential inbreeding and loss of genetic diversity within isolated subpopulations, because it is suspected that both threats may be accentuated by the communal breeding system, with few individuals reproducing each year and the likelihood of closely related individuals pairing with each other. In northern Brazil, two *Pyrrhura* species – the Maroon-faced Parakeet *P. pfrimeri* and the Grey-breasted Parakeet *P. griseipectus* – are restricted to small tracts of dry forest on the slopes of isolated massifs, and both are seriously threatened by habitat loss and degradation. Known only from between the Serra Geral massif and the Rio Paraná, in northeastern Goiás and southeastern Tocantins, the Maroon-faced Parakeet is endangered because within its extremely small and severely fragmented range patches of dry forest are being cleared for pasture or selectively logged to meet a high demand for durable timbers used for fence poles. Grey-breasted Parakeets now are known to survive in only three localities in Ceará, and dramatic declines suffered by the extremely small population are continuing. Programs are being implemented to address the high level of habitat loss and to counteract a continuing illegal poaching for the live-bird trade. At least 11 private reserves are being set up, and in a large-scale education and public awareness campaign the Grey-breasted Parakeet is being promoted as a flagship species, with ecotourism enterprises being fostered as a means of creating alternative sources of income for local residents.

Another region of prominent endangerment is in coastal Brazil to northeastern Argentina, where the Atlantic forest has been fragmented by landclearance for urban development, agriculture and plantations. Among parrots occurring in these forests are two *Touit* parrotlets – the Golden-tailed Parrotlet *T. surdus* and the Brown-backed Parrotlet *T. melanonotus* – about which very little is known because of their secretive habits and the difficulty of observing such small, predominantly green parrots in the forest canopy. Indeed virtually nothing was known of the habits of Brown-backed Parrotlets until birds visited urban gardens at Ubatuba, in coastal São Paulo state, between 2009 and 2011, so giving observers an opportunity to learn something of their habits. Lower montane evergreen forest between 500 m and 1200 m are favoured by these parrotlets, which now are distributed through the fragmented habitat as subpopulations, each of which comprises less than 1000 individuals, and a continuing moderate decline is likely because of habitat loss and degradation. Because of this extensive deforestation in coastal Brazil, coupled with poaching for the cagebird market, Red-browed Amazons *Amazona rhodocorytha* have disappeared entirely from parts of their range, and elsewhere are very locally dispersed in fragmented forest remnants. Illegal trafficking poses a serious threat in many parts of the range, and apparently is the overriding threat in Espirito Santo. Conflict with agriculture also is a problem, and in some districts these parrots are considered to be pests in papaya, coffee and cocoa plantations. Both threats probably will continue, so an ongoing rapid decline is likely. The same threats of widespread deforestation, poaching of nestlings for the pet trade and local persecution as a pest in croplands have brought about a dramatic decline in numbers of Vinaceous-breasted Amazons *Amazona vinacea*, and they have been extirpated in parts of their range, especially in Argentina and Paraguay. Although common to abundant in the 19th century, they now are rare throughout their entire range, with highest numbers surviving in southeastern Brazil, where there is a close association with *Araucaria* forests, and destruction of these forests has been identified as the principal reason for the decline. At a forested property in São Paulo state, southeastern Brazil, a facility has been established for rehabilitating confiscated captive parrots for release into the wild, and Vinaceous-breasted Amazons are included in the program.

WEST INDIES

Ara and possibly *Anodorhynchus* macaws, *Amazona* parrots and *Psittacara* parakeets occurred in the West Indies at the time of 'discovery' by Europeans, and in the writings of early voyagers are references to all being widespread and common. Most prominent among these writings are accounts given by two French Dominican friars – Jean Baptiste Du Tertre and Jean-Baptiste Labat. Du Tertre was in the West Indies from 1640 until 1658. In January 1694, Labat arrived at Martinique, and he travelled widely throughout the West Indies until returning to Europe in 1706. Both were botanists, and obviously were competent observers, their accounts of wildlife, plants and native peoples often being accompanied by their drawings in multi-volume works published mostly after their return to Europe. *Histoire générale des Antilles habitées par les Français* by Du Tertre was published in four volumes between 1667 and 1671, and there were two editions of *Nouveau voyage aux iles de l'Amerique* by Labat, the first published in six volumes in 1722 and the second in eight volumes in 1742.

In 1667, Du Tertre noted (translation in Wiley and Kirwan 2013):

> *These birds are so dissimilar according to the grounds where they procure their food, that every island has its Parrots, its Aras, and its Parroquets, different in size of body, in tone of voice, and in tints of the plumage.*

Determining with any degree of certainty the number of species that may have occurred originally in the West Indies is not possible, for many of the records are highly doubtful, and there is little or no evidence to support the existence of some species that have been described in the literature. Parrots were hunted for food and were kept as pets by Amerindians and subsequently by European settlers, so it is highly likely that live birds were traded between islands, and possibly were brought from the mainland, resulting in some species becoming established on islands where they had not occurred previously. I suspect that a total of as many as 60 endemic species being present originally, as suggested by Williams and Steadman (2001), is excessive, but the present occurrence of 12 endemic species probably does represent a significant loss of species from the region.

It has been suggested that up to 15 species of macaws may have occurred, with most Greater Antillean and several Lesser Antillean islands supporting one or more indigenous species (Williams and Steadman 2001). Only the Cuban Macaw *Ara tricolor*, which was extirpated in about the 1860s, is known from museum specimens, and the recovery of subfossil bones is evidence of macaws being present on Puerto Rico, the Virgin Islands, where birds probably were transported from Puerto Rico, and in the Guadeloupe Archipelago. Wiley and Kirwan (2013) point out that occurrences on other islands are based on casual accounts by travellers, colonists and the few naturalists who passed through, or resided in the islands during the 17th and 18th centuries, but some of these early reports have been compiled in such a way by authors of world or regional faunas as to make identification of species and their locations difficult if not impossible. Adding to the confusion is the regular transportation of parrots, including macaws, between islands and probably from the South American mainland by the native Amerindians and by early European settlers. No fewer than seven species have been named from accounts written by visitors in the 18th and 19th centuries, prompting Hume and Walters (2012) to comment that:

> *The confusion surrounding the original diversity of the Antillean macaws has been extraordinary, and no other group of extinct birds has aroused such overzealous misinterpretation. The evidence on which the majority of species have been described is, at best, inadequate, and based on the vaguest of accounts.*

I concur with this critical assessment, and would point out that there is no evidence of more than one macaw having occurred on Cuba, by far the largest island in the West Indies, so it seems unlikely that more than one species would have occurred on any other island.

From a left tibiotarsus of an immature bird found in kitchen middens at Concordia, near Southwest Cape on St Croix, in the Virgin Islands, Wetmore (1937) described and named *Ara autochthones*. This species subsequently was named the Puerto Rican Macaw because, in April 1987, several associated bones evidently from a single bird were found at an archaeological site in south-central Puerto Rico, but Olson and Maíz López (2008) caution that, although it is assumed to have been endemic to the West Indies, prehistoric interisland transport of parrots by humans renders impossible any interpreting of its natural distribution. I am sure that it is more likely to have occurred naturally on Puerto Rico and transported to the Virgin Islands. It was larger than the Cuban Macaw, but nothing more is known of its appearance.

Diego Álvarez Chanca, a physician who accompanied Columbus on his second voyage, made the first reference to macaws occurring on Guadeloupe, in the Lesser Antilles, when in a letter of 1494 he reported taking '*two parrots, very large and quite different from the parrots we had before seen*' from the houses of Carib inhabitants (in Wiley and Kirwan 2013). A quite detailed account of red macaws in Guadeloupe was compiled in 1654 by Du Tertre, who wrote (translation in Clark 1905a):

> *We have in Guadeloupe three of the parrot kind, viz:- Macaws, Parrots and Parrakeets, each different from those which inhabit the neighbouring islands; for each has its parrots different from those of the others in size, voice and colour. The Macaw is the largest of all the parrot tribe; for although the parrots of Guadeloupe are larger than all other parrots, both of the islands and of the main land, the Macaws are a third larger than they.*
>
> *The head, neck, underparts, and back are flame colour. The wings are a mixture of yellow, azure, and scarlet. The tail is wholly red, and a foot and a half long. The natives hold the feathers of the tail in great esteem; they stick them in their hair, and pass them through the lobe of the ear and septum of the nose to serve as moustaches, and consider themselves much more genteel and worthy of the admirations of Europeans.*
>
> *The bird lives on berries, and on the fruit of certain trees, but principally on the apples of the manchioneel, which is a powerful and caustic poison to other animals. It is the prettiest sight in the world to see ten or a dozen macaws in a green tree. Their voice is loud and piercing, and they always cry when flying. If one imitates their cry, they stop short. They have a grave and dignified demeanour, and so far from being alarmed by many shots fired under a tree where they are perched, they gaze at their companions who fall dead on the ground without being disturbed at all, so that one may fire five or six times into the same tree without their appearing to be frightened.*
>
> *The natives make use of a stratagem to take them alive; they watch for a chance to find them on the ground, eating the fruit which has fallen from the trees, when they approach quietly under cover of the trees, then all at once run forward, clapping their hands and filling the air with cries capable not only of astounding the birds, but of terrifying the boldest. Then the poor birds, surprised and distracted, as if struck with a thunderbolt, lose the use of their wings, and, making virtue of necessity, throw themselves on their backs and assume the defensive with the weapons nature has given them – their beaks and claws – with which they defend themselves so bravely that not one of the natives dares to put his hand on them. One of the natives brings a big stick which he lays across the belly of the bird, who seizes it with beak and claws; but while he is occupied in biting it, the native ties him so adroitly to the stick that he can do with him anything that he wishes.*
>
> *The flesh of this bird is very tough, and considered by many unwholesome, and even poisonous. I never had any ill effects from it, although we inhabitants often eat it.*

In a later account, dated 1667, Du Tertre adds information about breeding (translation in Clark 1905a):

> *The male and female are inseparable companions, and it is rare that one is seen singly. When they wish to breed (which they do once or twice a year) they make a hole with their beaks in the stump of a large tree, and construct a nest with feathers from their own bodies. They lay two eggs, the size of those of a partridge.*

Accompanying this later account is a crude drawing depicting one of these macaws sitting on a branch below three perched *Amazona* parrots.

Attributed to François-Nicolas Martinet, and varying slightly in hand-coloured reproductions in copies of *Planches Eluminées d'Histoire Naturelle* (1765–1780), this drawing of a red macaw with an entirely red tail closely resembles descriptions of the Guadeloupe Macaw *Ara guadeloupensis* given in 1654 by Jean-Baptiste Du Tertre and in 1742 by Jean-Baptiste Labat.

A very similar description of the macaw on Guadeloupe was given in 1742 by Labat, who wrote (translation in Clark 1905a):

> *The Macaw, which I place at the head of the parrots, is the largest of all the parrot tribe, either in these islands or on the mainland. It is the size of a full grown fowl. The feathers of the head, neck, back, and underparts are flame colour; the wings are a mixture of blue, yellow, and red; the tail, which is from fifteen to twenty inches in length, is wholly red. The head and beak are very large, and it walks gravely; it talks very well, if it is taught when young; its voice is strong and distinct; it is amiable and kind, and allows itself to be caressed.*
>
> *There are Macaws, Parrots and Parrakeets in each of our islands, and it is easy to tell from their plumage from which island they have come. Those from Guadeloupe are generally larger than the others, but the parrakeets are smaller.*

Hume and Walters (2012) point out that Labat's important comment that parrots from each island were recognisably different eliminates the possibility that the extinct macaws in the West Indies were escapees of South American species.

Plate 12 in D'Aubenton's *Planches eluminées d'histoire naturelle* (1765–1780) is a drawing by François-Nicolas Martinet depicting a red macaw with an entirely red tail and closely resembling the descriptions given by Du Tertre and Labat, prompting Hume and Walters to suggest that at least one specimen of the Guadeloupe Macaw must have reached Europe. They suggest also that it evidently was a close relative of the Scarlet Macaw *Ara macao*, and possibly was a derivative of it.

Based on descriptions in accounts from Du Tertre and Labat and on the drawing by Martinet, *Ara guadeloupensis* was described by Clark (1905a) as being apparently similar to *A. macao*, but smaller and with the tail wholly red. In addition to Guadeloupe, Martinique and possibly Dominica were included in the range by Clark, presumably because 'These three islands collectively show affinities to the Greater Antilles and to South America, at the same time having genera and species peculiar to themselves'. A terminal phalanx with a size and morphology supporting its representing an *Ara* macaw was found during excavations undertaken in the spring of 2013 and 2014, at a late Pleistocene fossil deposit on Marie-Galante, in the Guadeloupe Archipelago, and this provides credible evidence for the existence of an extinct, possibly endemic macaw (Gala and Lenoble 2015).

Less credible is evidence for the former existence of macaws on other islands. Three species are listed as having occurred on Jamaica, but Wiley and Kirwan (2013) point out that identifications in many original accounts are muddled, and evidence is good for only one species. Although Jamaica is a large island and is one of only two of the Antilles that supports two endemic *Amazona* species, the occurrence of only one species would be in keeping with my assertion that no island in the West Indies supported more than one species, and I fully concur with the comment made by Hume and Walters (2012) that unsubstantiated claims of three or even four macaws having occurred sympatrically on Jamaica epitomises the over-zealous passion often associated with extinct birds. Events leading to naming of the Red-headed Macaw *Ara erythrocephalus* by Rothschild (1905) well demonstrate this approach. In his *The Birds of Jamaica* (1847) Phillip Gosse acknowledged assistance given by Richard Hill, a resident and reliable observer, and with reference to macaws he wrote:

> *A letter just received from Mr Hill, who kindly assisted my inquiries on the subject, says:- I have ascertained with unquestionable certainty that Macaws are occasionally, if not constantly, denizens of our mountain forests. They are found exclusively in the central mountains westward of the island, and are observed on the skirt of the partially cleared country, at an elevation of 2500 or 3000 feet above the sea. They have been surprised in small companies feeding on the full-eared maize, while the grain was soft, milky and sweet, and the very husk was sugary. Every description I have received of them makes the species to be the* Ara militaris, *the Great Green Macaw of Mexico. The head is spoken of as red; the neck, shoulders and underparts of a light and lively green; the greater wing-coverts and quills blue; and the tail scarlet and blue on the upper surface, with the under plumage both of the wings and tail, a mass of intense orange yellow.*

Hill did not observe the birds, and the hearsay description, which could have been widely inaccurate or may have been of introduced parrots or even of local *Amazona* parrots, was accepted by Rothchild as referring to a new species.

More credible is evidence for the occurrence on Jamaica of the Yellow-headed Macaw *Ara gossei*, also named by Rothschild (1905), and known from the description of a lost specimen shot in about 1765 by a Mr Odell and from a contemporary drawing. The specimen was seen by a Dr Robinson, who sent a description to Gosse, who in turn wrote:

> *If this not be the Tricolor of Le Vaillant* [Cuban Macaw], *which is the only macaw I am aware of marked with a yellow nape, it*

is probably undescribed. The two descriptions do not, certainly, agree exactly; yet still I cannot but think the bird seen by Robinson, whose description I give below, to be this very rare species. Of the present specimen the Doctor says 'This bird I saw stuffed. The legs and tail were wanting. It seemed less than the common Red and Blue Macaw [introduced Scarlet Macaw]. *By what I can judge from this sample, this bird has never been figured or described. Sir Henry Moore, the late Lieutenant Governor, often assured me that the Jamaican Macaw was very different from any he had ever seen. The subject now before us was shot in the mountains of Hanover parish, about ten miles east of Lucea, by Mr Odell'.*

Gosse was uncertain whether this macaw differed from the Cuban Macaw, and Clark (1905d) also was unsure, though suggesting that it possibly was a closely related species or subspecies. Noting that the overall plumage pattern is very similar to that of the Cuban Macaw, Olson and Maíz López (2008) suggest that it may well represent an example of that species that had been taken to Jamaica. When naming it as a new species, Rothschild did so on the basis that the description was different from that of *A. tricolor*. Hume and Walters (2012) note that a painting of a probable Scarlet Macaw, painted on Jamaica in 1765, confirms that this mainland species at least had been introduced to the island at that time, but there seems to be reasonable evidence that at least one other red macaw, along with *A. tricolor*, once occurred in the West Indies. Indeed, it seems that all three macaws, for which there is credible evidence of a former occurrence in the West Indies, namely *A. tricolor*, *A. guadeloupensis* and *A. gossei*, were so similarly plumaged, small to midsized birds that they could have been a single species transported between islands or, together with macaws which possibly occurred on other islands, may have been closely related and shared a similar red and yellow plumage pattern, so indicating a possible derivative association with the Scarlet Macaw *Ara macao*.

I concur with Hume and Walters in regarding other macaws described as having occurred in the West Indies as being doubtful or invalid, and this is especially so for the Violet Macaw *Anodorhynchus purpurascens* described by Rothschild (1905) without any credible evidence and said to have occurred on Guadeloupe. Only the genus *Ara* is known from the West Indies, and it is highly unlikely that *Anodorhynchus*, which is confined to the South American mainland, reached the West Indies unless transported there.

Psittacara parakeets that closely resemble, and obviously are closely related to species occurring in Central America, remain widely distributed and common on many islands in the West Indies, and historical accounts indicate that they formerly occurred on additional islands. The Puerto Rican Parakeet *Psittacara maugei* from Puerto Rico and nearby islands is the only extinct species known from museum specimens. Although apparently widespread and fairly common on Puerto Rico when the first two specimens were collected in 1797, they disappeared sometime before 1835, when C. Moritz, a collector for the Zoological Museum at Berlin, failed to find them during a stay of four months. They evidently were present in reasonable numbers on nearby Mona Island in about 1875, when wings were collected by Claudio Bloch, a Danish physician, but must have been almost extinct by 1892 when the last specimen was collected by Wilmot Brown. Wiley (1991) refers to the widespread environmental degradation on Puerto Rico, and notes that extinction of the Puerto Rican Parakeet is not remarkable in view of the near-complete loss

This drawing of the Yellow-headed Macaw *Ara gossei* by John Smit (in Rothschild 1907) is based on the description of a lost specimen and a drawing made in Jamaica in about 1765.

of habitat. Persecution of the parakeets because of their depredations on agricultural crops is mentioned in early accounts, and it seems that there was a deliberate effort to eradicate the birds. Olson (2015) suggests that disease also may have contributed to the demise of these parakeets and, in addition to domestic poultry with their set of diseases, exotic parrots had been brought to Puerto Rico since the earliest times of Spanish colonisation.

The second extinct *Psittacara* species is the Guadeloupe Parakeet *P. labati* described by Rothschild (1905) from the following brief description given by Labat in 1742 (translation in Hume and Walters 2012):

Those of Guadeloupe are about the size of a blackbird, entirely green, except a few small red feathers, which they have on their head. Their bill is white. They are very gentle, loving, and learn to speak easily.

An earlier reference was made in 1667 by Du Tertre, who wrote of 'Macaws, Parrots and Parrakeets' occurring on Guadeloupe. I agree with Hume and Walters in stating that there is little reason to doubt the former occurrence of parakeets on Guadeloupe, but their specific status is unclear. Bond (1956) suggested that they may have been captive Cuban Parakeets *Psittacara euops*, but transporting live parakeets to Guadeloupe from distant Cuba seems unlikely when birds would have been available on nearer Puerto Rico and Hispaniola. Marien and Koopman (1955) were convinced that small, predominantly green parakeets formerly occurred on Guadeloupe, and may have been related to the Green Parakeet *P. holochlorus* from Central America but, because of the distance between Guadeloupe and Central America, *P. labati* should be treated as a separate species. A reconstructed drawing by John Gerrard Keulemans (in Rothschild 1907) depicts an all-green parakeet with a single red patch on the crown, but from the description given by Labat it seems more likely that markings on the head were in the form of variably scattered small red feathers, as is found in some other *Psittacara* species. These parakeets were not present on Guadeloupe in the early 1800s, so presumably disappeared sometime between Labat's visit in the 1690s and then, but reasons for their extinction are not known (in Hume and Walters 2012).

Much less convincing are references to parakeets formerly occurring on other islands, including Barbados, Martinique and Dominica. Hume and Walters (2012) point out that Frederick Albion Ober, who collected specimens on Dominica from 1876 to 1878 and again in 1880, refers to the occurrence of a parakeet '*about the size of our Northern Carolina Parakeet, but more robust*', but he did not collect a specimen. In a brief discussion about evidence for the existence of a parakeet on Dominica, Clark (1905b) notes '*There appears to be no description extant of the species which formerly inhabited Dominica, but it seems to have been exterminated at an early date*'. Brown-throated Parakeets *Eupsittula pertinax* occur in the Virgin Islands, where

This reconstruction drawing by John Gerrard Keulemans of the Guadeloupe Parakeet *Psittacara labati* (in Rothschild 1907) is based on a brief description given in 1742, but the pattern of red markings on the head may have been quite different.

they were introduced sometime prior to 1860, date of the earliest specimen that I could locate, and more recently they have been recorded on other islands, including Guadeloupe, Martinique and Dominica (see Raffaele *et al.* 1998). I strongly suspect that early accounts and a possible drawing of parakeets formerly occurring on Martinique can be referred to introduced Brown-throated Parakeets, which probably were introduced also to Barbados, and it is not improbable that they were introduced to Dominica.

Based solely on plumage features and body size, *Amazona* parrots occurring in the West Indies have been divided into two main groups – five extant midsized species found in the Greater Antilles and four extant, together with three extinct, larger species in the Lesser Antilles (Snyder *et al.* 1987). Based on colours of the forehead, chin, forewing and abdomen, Lack (1976) postulated that the White-fronted Amazon *A. albifrons* from Central America was ancestral to the Cuban Amazon *A. leucocephala*, which in turn gave rise to the Yellow-billed Amazon *A. collaria* in Jamaica and the Hispaniolan Amazon *A. ventralis*, with the Puerto Rican Amazon *A. vittata* resulting from the colonisation and differentiation of *A. ventralis* and the Black-billed Amazon *A. agilis* originating in Jamaica from a separate invasion by *A. albifrons* via the Honduran-Nicaraguan bulge. An analysis examining 18 characters of colour and size for more than 300 sexed specimens of all Greater Antillean species, plus *A. albifrons* and the Yellow-lored Amazon *A. xantholora* from Central America, corroborated the basic pattern proposed by Lack and revealed three subgroups (in Snyder *et al.* 1987):

(i) *A. albifrons* and *A. xantholora* from Central America,
(ii) *A. leucocephala* from Cuba and neighbouring islands, *A. ventralis* from Hispaniola and *A. collaria* from Jamaica, and
(iii) *A. agilis* from Jamaica and *A. vittata* from Puerto Rico.

Bond (1978) suggested that amazons in the Lesser Antilles may have been derived from two or more separate invasions from South America. Among these larger amazons, it was postulated that similarities in plumage colouration indicated a close relationship between the endangered Imperial Amazon *A. imperialis* from Dominica and the extinct Guadeloupe Amazon *A. violacea*, which is not known from specimens, and it was noted that both the Red-necked Amazon *A. arausiaca* from Dominica and the St Lucia Amazon *A. versicolor* share several plumage characteristics with the poorly described, extinct Martinique Amazon *A. martinicana*, which also is not known from specimens (Snyder *et al.* 1987). The St Vincent Amazon *A. guildingii* is well differentiated by its highly distinctive plumage colouration. Apart from some semblance in plumage colouration between the Imperial Amazon and the endangered Violet-breasted Amazon *A. vinacea* from eastern Brazil, no extant South American species shows plumage features similar to those displayed by Lesser Antillean species, but it was suggested that the Blue-cheeked Amazon *A. defresniana* from Guyana and Surinam may be the closest mainland relative (Snyder *et al.* 1987). Russello and Amato (2004) report that consistent with these hypotheses and not refuting the suggestion that *A. agilis* originated from a separate invasion are molecular analyses indicating that the Greater Antilles amazons are closely related and Central American *A. albifrons* is basal to the Greater Antilles species, except *A. agilis*. Conversely, molecular analyses indicate that Lesser Antillean amazons are not closely related to each other, and the various species are allied with different species or species groups in South America.

Established in the late 1960s, an intensive conservation program to save the critically endangered Puerto Rican Amazon *A. vittata* has attracted widespread attention, and slowly, but surely is achieving its objectives. Widespread deforestation and persecution were primary causes of a dramatic decline in numbers of these amazons on Puerto Rico, and of their extirpation on nearby Culebra and Vieques Islands sometime before 1900. A thorough census of the wild population was undertaken in 1968, and only 24 birds were located, all in the eastern part of the island in virtually the only remaining old-growth forests in the Luquillo Mountains. By 1971, the count had dropped even further to 16 birds, but then the decline was arrested, with the population fluctuating for some time between 15 and 20 individuals. The surviving population had declined to only 13 birds when the conservation program commenced. Two approaches adopted in the program were to undertake intensive field studies of the wild population with a view to implementing a recovery strategy and to establish a captive-breeding program with an objective of releasing captive-bred birds into the wild population. During the course of field studies in the Luquillo forests it was found that very few tree hollows were suitable for nesting and a number of pairs were unable to breed because of an inability to locate good sites. This scarcity of sites was addressed by providing artificial nest-sites, and modification of these artificial sites eventually overcame high levels of predation of eggs and chicks by invasive Pearly-eyed Thrashers *Margarops fuscatus*, a major cause of nesting failure. Parasitism by warble flies *Philornis pici* also was responsible for losses of some nestlings. Predation of both adults and fledglings by very common Red-tailed Hawks *Buteo jamaicensis* was another pressure restricting recovery of the population. Implementation of protection measures arrested the decline, but nesting success remained low and numbers increased very slowly. As a safeguard and to provide birds for reintroduction to the wild, a captive breeding population was set up in the 1970s, and this has been successful, enabling captive-produced chicks to be fostered into wild nests and captive-bred birds to be released into the wild. Success with these measures has resulted in new wild populations becoming established in reserves in other parts of Puerto Rico, and at the time of writing more than 100 Puerto Rican Amazons are being monitored in the wild, with nearly 400 additional birds held in captivity.

On Dominica, in the Lesser Antilles, a conservation effort is being maintained to safeguard surviving populations of the endangered Imperial Amazon *A. imperialis*, which largely replaces the Red-necked Amazon *A. arausiaca* in montane and elfin forest above 600 m. In the mid 1970s it was suggested that the total population comprised 150 birds, and a strong decline in numbers was attributed to the loss and degradation of mountain forest, primarily because of conversion to plantations and damage by hurricanes, the latter being particularly severe in 1979 when Hurricane David brought about widespread destruction. Hunting also was identified as a continuing threat. So severe was the impact of Hurricane David that after its passage it was suspected that as few as 50 birds survived in the Morne Diablotin area in the north of the island, and the small southern population in Morne Trois Pitons National Park had been extirpated. It seems that recovery was slow because in 1993 the population comprised only 80–100 birds. Since its near extinction by Hurricane David, recovery of the Imperial Amazon has been the subject of intense field research and a concerted conservation effort, and a very significant achievement in January 2000 was the establishment of Morne Diablotin National Park, encompassing 5300 ha of pristine mountain forest and a stronghold of these amazons. In December 2000 a small population was located in the vicinity of Morne Trois Pitons National Park. In the past these parrots were shot for food or in the hope of obtaining injured birds for the illicit live-bird market, but this has been reduced considerably, if not eliminated, as a result of successful awareness and education programs emphasising the importance of the species as the national bird and the need to protect it. Investigations are being undertaken to address conflict with agriculture, particularly in citrus orchards near to the national parks and forestry reserves. Competition for nest-sites from the more numerous Red-throated Amazon could increase as lowland forest is lost and contact between the two species becomes more frequent, and habitat limitations may restrict increases in numbers. Despite these concerns, the long-term outlook is encouraging, and there is an expectation that the slow improvement in status of the Imperial Amazon will continue.

Reference is made in accounts from early observers to parrots, almost certainly *Amazona* species, formerly occurring on Guadeloupe and Martinique, also in the Lesser Antilles, but their existence is not supported by specimens. Evidence for the existence of an amazon on Guadeloupe is convincing, with quite detailed accounts accompanied with crude drawings being given by Du Tertre and Labat. The first description was given in 1664 by Du Tertre, who wrote (translation in Clark 1905c):

> *The Parrot of Guadeloupe is almost as large as a fowl. The beak and the eye are bordered with carnation. All the feathers of the head, neck, and underparts are of a violet colour, mixed with a little green and black, and changeable like the throat of a pigeon. All the upper part of the back is brownish green. The long quills are black, the others yellow, green and red, and it has on the wing-coverts two rosettes of a rose colour. When it erects the feathers of its neck, it makes a beautiful ruff about its head, which it seems to admire, as a peacock its tail. It has a strong voice, talks very distinctly, and learns quickly if taken young. It lives on the wild fruits which grow in the forests, except that it does not eat the manchioneel. Cotton seed intoxicates it, and affects it as wine does a man, and for that reason they eat it with great eagerness. The flavour of its flesh is excellent, but changeable, according to the kind of food. If it eats cashew nuts, the flesh has an agreeable flavour of garlic; if 'bois des inde' it has a flavour of cloves and cinnamon; if on bitter fruits, it becomes bitter like gall. If it feeds on genips, the flesh becomes wholly black, but that does not prevent it from having a very fine flavour. When it feeds on guavas it is at its best, and then the French commit great havoc among them.*

In 1667, Du Tertre added some brief details on breeding (translation in Clark 1905b):

> *We had two which built their nest a hundred paces from our house in a large tree. The male and female sat alternately, and came one after the other to feed at the house, where they brought their young when they were large enough to leave the nest.*

A brief description was given in 1742 by Labat, who wrote (translation in Clark 1905c):

> *The Parrots of these islands are distinguishable from those of the mainland of Guinea* (= Guiana?) *by their different*

This reconstruction drawing of the Guadeloupe Amazon *Amazona violacea* by John Gerrard Keulemans (in Rothschild 1907) is based on descriptions given in the 17th and 18th centuries, but there are doubts about its specific differentiation from the Imperial Amazon *A. imperialis*, which still occurs on nearby Dominica.

This reconstruction drawing of the Martinique Amazon *Amazona martinicana* does not match the description from 1742 of 'the parrot of Martinique' by Jean-Baptiste Labat, who referred to its similarity to the Red-necked Amazon *A. arausiaca*, which still occurs on nearby Dominica.

> *plumage; those of Guadeloupe are a little smaller than the Macaws. The head, neck, and underparts are slaty, with a few green and black feathers; the back is wholly green, the wings green, yellow, and red.*

Based on the descriptions from Du Tertre and Labat, the species formerly occurring on Guadeloupe was named *Amazona violacea* by Gmelin, and a restoration drawing was made by John Gerrard Keulemans (in Rothschild 1907), but doubts persist about its specific differentiation from the Imperial Amazon of Dominica. An ulna from an archaeological site on Marie Galante, in the Guadeloupe Archipelago, initially was assigned to the Guadeloupe Macaw *Ara guadeloupensis,* but subsequently was found to be perfectly intermediate in size and robustness between two individuals of *Amazona imperialis*, and a tibiotarsus found at the same site was referred to *A. violacea* because of its similarity to *A. imperialis* (in Olson and Maíz López 2008). These findings prompted Olson and Maíz López to assert that *A. violacea* probably is the same as *A. imperialis*. The cause or causes of extirpation of amazons on Guadeloupe are not known, but a clue was given by Comte de Buffon in 1779, when he noted (translation in Clark 1905c):

> *We have never seen this parrot, and it is not found in Cayenne. It is even very rare in Guadeloupe today, for none of the inhabitants have given us any information concerning it; but that is not extraordinary, for since the islands have been inhabited, the number of parrots has greatly diminished, and Du Tertre remarks in particular of this one that the French colonists wage a terrible war on it in the season when it is especially fat and succulent.*

It certainly is possible that the parrots were hunted to extinction by early colonists.

Also depicted in a restoration drawing by Keulemans (in Rothschild 1907) is the Martinique Amazon *Amazona martinicana*, named by Clark (1905c) and based on a description of 1742 from Labat, who referred to its similarity to the Red-necked Amazon *Amazona arausiaca* from Dominica when he wrote (translation in Hume and Walters 2012):

> *The Parrot is too common a bird for me to stop to give a description of it........*[The parrot] *of Dominica has some red feathers in its wings, in its tail and under the throat. That of Martinique resembles it, except that the top of the head is slate colour with a little red.*

Hume and Walters (2012) point out that, although its former occurrence on Martinique is not supported by subfossil remains, there is no reason to doubt the existence of this species, but the drawing by Keulemans does not match Labat's description, and it may have differed from *A. arausiaca* only in having some red feathers on the head. It was not recorded after the early 1700s, but nothing is known of the causes of its extinction.

SUPERFAMILY
PSITTACOIDEA Rafinesque-Schmaltz

Apart from cockatoos and the *Strigops-Nestor* parrots from New Zealand, all parrots are included in this superfamily, and a major radiation is reflected in a systematic arrangement recognising three groupings at family level rank (Joseph *et al.* 2012). Only one of these families is represented in the Neotropics.

FAMILY **PSITTACIDAE** Rafinesque-Schmaltz

In molecular analyses Neotropical parrots have been recovered as sister to African *Psittacus* and *Poicephalus*, and this Afro-Neotropical group making up the family Psittacidae is sister to all other parrots in the superfamily, and may be worthy of differentiation as another superfamily. It is hypothesised that a common ancestor lived in Antarctica and became separated from the Australasian lineages when Antarctica began to split from Australia in the late Eocene or early Oligocene, approximately 35 million years ago (Schweizer *et al.* 2010). The Neotropics and Africa subsequently were colonised from Antarctica, and parrots in these two regions now are differentiated at subfamily rank (see Joseph *et al.* 2012).

SUBFAMILY **ARINAE** G. R. Gray

A widely adopted arrangement for Neotropical parrots has been to recognise at least two groupings, one of primarily long-tailed species such as the macaws, conures and allies, and the other of primarily short-tailed species such as amazons and allies, but molecular analyses have identified four primary lineages, and these lineages here are recognised as distinct tribes in an arrangement proposed by Schodde *et al.* (2013). The singularly distinctive *Forpus* parrotlets alone comprise the Forpini, and the status of only one species – the Yellow-faced Parrotlet *F. xanthops*, is of concern. Restricted to arid woodlands and cacti-dominated desert scrublands at mid montane elevations in northwestern Peru, this species has been subjected to excessive exploitation for the live-bird trade and possibly is threatened by habitat degradation from overgrazing by goats, so is given vulnerable status (in Birdlife International 2016).

TRIBE **ARINI** G. R. Gray

Some of the most familiar of Neotropical parrots are among the midsized to large, long-tailed, sexually monomorphic species included in this tribe. The bill is proportionately large and strong, and the cere may be feathered or unfeathered. Anatomical features include a deeply defined, strongly muscled temporal fossa, an open, square shaped auditory meatus, an orbital ring completely or almost completely ossified by extension of the prefrontal, and a well-developed uropygial gland (Schodde *et al.* 2013). Because of their size and spectacular plumage colouration, the large macaws have a long history of exploitation for the live-bird trade, and their tail-feathers are used widely by native peoples for ceremonial dress. This exploitation, coupled with widespread destruction of their forest or woodland habitats, has brought about strong declines in numbers and the extirpation of some populations. Similarly affected are some of the more specialised *Psittacara*, *Aratinga* and *Pyrrhura* parakeets, with species restricted to small ranges or occurring in threatened montane habitats being most at risk.

This tribe is widespread throughout Central and South America, and formerly was represented by the now extinct Carolina Parakeet *Conuropsis carolinensis* in the eastern United States.

GENUS ***Anodorhynchus*** Spix

Anodorhynchus Spix, *Av. Bras.*, **1**, 1824, p. 47, pl. 11. Type, by monotypy, *Anodorhynchus maximiliani* Spix = *Psittacus hyacinthinus* Latham.

Highly distinctive, predominantly blue macaws belonging to this genus are among the most spectacular of all parrots. They are very large birds with a massive bill and long, graduated tail, the bill being well adapted to splitting open very hard palm nuts, which are their principal food. The lores and face are feathered, but there is a prominent bare eyering and variable bare lappets at the base of the lower mandible. The orbital ring is incomplete, and there are other notable osteological features in the skull, chiefly in the auditory region. The sexes are alike, and young birds resemble adults.

Yamashita and Valle (1993) point out that the three recognised species clearly are monophyletic, and the first split probably was between *Anodorhynchus hyacinthinus* and *A. leari-A. glaucus*. The *A. leari-A. glaucus* group became isolated in marginal, semiarid eastern habitats, and later differentiated into the northern *A. leari* and the southern *A. glaucus*.

This genus is confined to tropical and subtropical South America, where it is or was represented by the widespread and locally common Hyacinth Macaw *A. hyacinthinus*, the endangered Indigo Macaw *A. leari* in a restricted range in northeastern Brazil and by the extinct Glaucous Macaw *A. glaucus* in the south. The Guadeloupe Violet Macaw *A. purpurascens* was based on a written description of 'le gros Perroquet de la Guadeloupe' by an early voyager to Guadeloupe, in the Lesser Antilles, but Wiley and Kirwan (2013) point out that at best, the evidence for its existence is weak.

ENDANGERED

Indigo Macaw

Anodorhynchus leari Bonaparte

Anodorhynchus leari Bonaparte, *Naumannia*, **6**, 1856, *Consp. Psitt.*, in Beilage no. 1. New name for *Macrocercus hyacinthinus* Lear, not *Psittacus hyacinthinus* Latham (no locality, presumed to have come from Brazil).

OTHER NAME Lear's Macaw.

DESCRIPTION Length 75 cm. Weight 940–950 g.
ADULTS General plumage colouration violet-blue, paler and slightly tinged greenish on head and neck; dusky suffusion on throat to upper breast; underside of tail dusky grey; yellow 'tear-drop' shaped lappet at each side of lower mandible, but not extending beneath lower mandible; bill grey-black; prominent bare eyering yellow; iris dark brown; legs dark grey.
4 males: wing 374–391 (384.8) mm, tail 354–371 (362.0) mm, exp. cul. (abnormal growth) 66–73 (68.5) mm, tars. 34–36 (35.0) mm.
1 female: wing 389 mm, tail 343 mm, exp. cul. (abnormal growth) 65 mm, tars. 40 mm.
2 unsexed: wing 396 mm and 406 mm, tail 376 mm and 401 mm, exp. cul. (abnormal growth) 67 mm and 71 mm, tars. 38 mm and 41 mm.
JUVENILES Like adults, but bare eyering and facial lappets paler yellow; shorter tail.

DISTRIBUTION Northern Bahia, central-eastern Brazil, where two colonies are at Toca Velha and Serra Branca, south of the Raso da Catarina plateau.

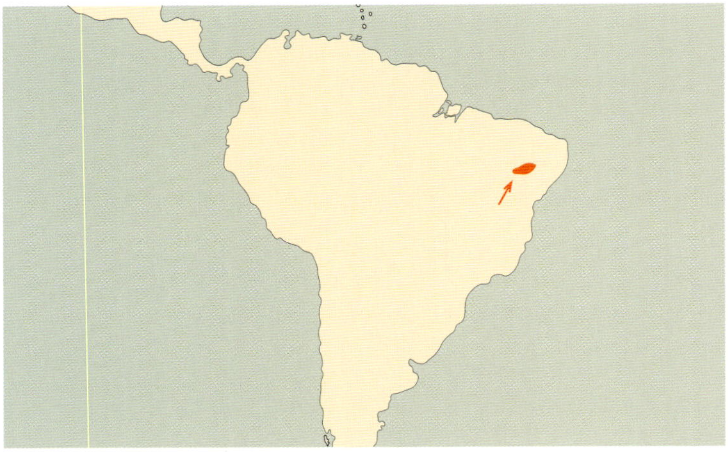

Plate 33
Indigo Macaw *Anodorhynchus leari* (adults)

STATUS Although known since the 1850s, when first painted from a captive bird by Edward Lear and subsequently formally described by Charles Lucien Bonaparte in 1858, the Indigo Macaw remained a mysterious bird for more than a century. As it was known only from captive birds, Voous (1965) was prompted to suggest that it may be a Hyacinth Macaw *Anodorhynchus hyacinthinus* × Glaucous Macaw *A. glaucus* hybrid. European and American zoos occasionally received single birds in consignments of Hyacinth Macaws, but the provenance of these smaller macaws remained unknown. Acting on information provided by local residents, including former trappers, and after three earlier unsuccessful searches, Helmut Sick finally located the species in December 1978, when a flock of about 20 birds was encountered on the Raso da Catarino, in northern Bahia (*in litt.* 1979). It now is known from Toca Velha and Serra Branca, two sites south of the Raso da Catarina plateau. In June 1995 about 20 birds were counted in a nighttime roost at another site some 200 km to the east, and initially it was presumed that these roosting birds represented a distinct eastern subpopulation, but now is considered to refer to birds from the known populations that have followed stands of fruiting *Syagrus* palms (Birdlife International 2016).

Although illicit trapping of adults and taking of nestlings for the live-bird trade certainly has been a serious threat to the remnant populations, I doubt that it was significant in reducing numbers to dangerously low levels. These macaws always have been extremely rare in captivity, with very few individuals being offered, and this suggests that habitat degradation is more likely to have brought about initial declines. Hunting for food and feathers probably contributed also to these initial declines. With its specialised feeding and nesting requirements, the species is particularly vulnerable to habitat interference, and poor regeneration of *Syagrus* palms, its primary food source, due to livestock overgrazing has been identified as a continuing threat (Snyder *et al.* 2000). Reporting the results of fieldwork undertaken from 12 to 30 July 1983, Yamashita (1987) recorded only two localised colonies in the studied area and, assuming that the remaining 40 per cent of the known range consisted entirely of optimum habitat, he estimated the total population to comprise far less than 200 birds. It seems that numbers remained low until the mid 1990s, when a rapid increase commenced, and while this may have in part reflected improved survey methods, it can be attributed to intensive conservation efforts (Birdlife International 2016). In 2001, it was reported that a recent census indicated that the population comprised 246 individuals, representing a substantial increase from the 170–180 birds of a few years earlier (Gilardi 2001). Population estimates of 400–500 birds were made in 2004, increasing to 630 in 2006 and to 960 in 2008, with the latest estimate of 1123 birds in 2010 (in Birdlife International 2016). Although the population is growing, it is likely that it includes a high proportion of subadults, and there are thought to be at least 258 mature birds.

F. KNIGHT '09

A multifaceted conservation program is being undertaken, with strong efforts being made to combat the ever present poaching of adults and chicks for the live-bird market. That some poaching continues is evidenced by the arrival of two chicks at the Los Pozos pet market in Santa Cruz city, Bolivia, in November 2004 (Herrera and Hennessey 2007). Gilardi (2001) reports that the landowner, on whose property critical breeding and foraging habitat is sited, is eager to protect the macaws from poachers and has been extremely helpful in facilitating habitat restoration. An important part of that restoration is the establishment of plantations of *Syagrus* palms, while simultaneously testing various techniques for managing palm habitat, such as fire, grazing, planting seeds and seedlings, and the possible value of transplanting juvenile palms. The macaws do take corn, particularly at times of food shortages, and that causes conflicts with farmers, so protocols to compensate farmers for crop losses have been initiated. Safeguarding nesting areas is given high priority, and that has been boosted by the acquisition of 1450 ha for the Canudos Biological Station (in Birdlife International 2016). Former trappers are among persons now employed to provide year-round protection of the birds and their nesting sites, and their contribution has been effective in combating poaching. Williams (2003) notes that these guards take great pleasure from being outdoors, and all are very proud of their role in the conservation of Indigo Macaws, with effectiveness of their efforts being reflected in the absence of any evidence of poaching during the 2002 breeding season. Williams notes also that educating the local people about the plight of these macaws and the efforts being made to save them has generated a lot of pride within local communities and has raised awareness that nesting sites are protected at all times. Essential to success of the conservation program is a better understanding of the biology of these macaws, and research projects are underway. Gilardi (2001) reports that birds are being photographed from hides set up at feeding stations to determine if individuals can be identified by differences in bill markings, as this would enable field researchers to study the basic natural history of the species by calculating population size as well as monitoring the rate of recovery and detecting potential problems such as low juvenile survival. Gilardi notes also it is proposed to evaluate potential sites throughout the historic range for the release of confiscated or captive-bred birds (in Birdlife International 2016).

Part of the range is within the Estação Ecológica do Raso da Catarina, a federal reserve, and the species is protected under Brazilian law. All *Anodorhynchus* macaws are listed on Appendix I of the Convention on International Trade in Endangered Species of Wild Fauna and Flora (CITES).

HABITATS Yamashita (1987) points out that the range of the Indigo Macaw is within the 'caatinga' region of Brazil at an altitude of about 380 m to 800 m. The vegetation community in this region is semiarid to arid scrubland with a sparse groundcover of few grasses and dominated by cacti, stands of *Syagrus* palms, and deciduous, often spiny trees and bushes remaining leafless for many months and with characteristically pale grey bare branches. Rainfall is low and unpredictable, and there can be prolonged periods of drought. Raso da Catarina, the last known stronghold of the species, is a strongly fissured plateau near the confluence of Rios Sapão and São Francisco, and here the macaws are dependent on cliff-faces in rugged gorges for nesting and roosting sites.

HABITS From observations carried out on the Raso da Catarina, during 12 to 30 July 1983, Yamashita (1987) provides the first account of the habits of Indigo Macaws. He found the birds to be very shy, which contrasts with later reports of their being confiding, with birds seen and photographed feeding in palm trees alongside village houses while children played on the ground below, and I suspect that this change in behaviour could be attributed to diminished hunting (see Hartley 1998). Yamashita notes that there seemed to be an elaborate social organisation within each flock using a roosting site. Roosts were in burrows created by weathering in the top two-thirds of cliff-faces, which varied in height from 30 m to 60 m. Burrows often were within 0.5 m of each other, and the fairly narrow entrance tunnels allowed passage of only one bird at a time, but up to four macaws occupied a burrow. Some birds roosted on shallow shelves or by clinging to the cliff-face. Departure from the roosting cliffs was before dawn, and at dusk birds returned from their feeding grounds. Just after sunset two or three macaws returned to the roosting site, flying in over the canyon while calling loudly, and then they perched quietly in the tallest tree, habitually biting off the tips of topmost branches, so making regularly used perching trees identifiable by the broken branch tips. After about 10 minutes these advance 'scouts' began to call loudly, and soon the remainder of the flock approached. Sizes of flocks varied, and groups of 22, 23, 19 and 23 birds were recorded on successive days at the same canyon. Incoming birds passed over the canyon, calling all the while, then descended to alight near the 'scouts', and all sat quietly as darkness fell. A resumption of calling signalled a short flight by all birds to the cliff-face, where they paused briefly in silence before entering the burrows. The 'scout' birds roosted in an adjoining canyon, and it was they that made pre-dawn flights over the main roosting site, again to the accompaniment of loud screaming, which brought all birds out from their burrows to form a wheeling flock, that soon flew from sight. Daily flights to and from feeding areas often were over long distances, and on four occasions a feeding flock was observed to be dispersed in small subgroups of two or three birds in palm trees spaced 5 m to 30 m apart. A 'sentinel' bird always kept watch while the flock was feeding, and birds alternated their watching and feeding activities.

Williams (2003) reports that a similar pattern of daily activity was observed during the 2002 breeding season. Peak levels of activity were in the early morning and in the evening, with the middle of the day spent in a few trees, where groups of about 20 birds would gather to shelter from the heat. Even before sunrise calls were heard from inside the nesting burrows, and at dawn pairs emerged to join others in flight above the cliffs, where they called and interacted with each other before flying purposefully towards their feeding sites some 12 km away. Aerial displays were seen in the evening, when returning birds would fly around the cliffs, often diving down onto each other, but it could not be determined whether this was play or territorial defence.

The flight is strong and graceful, the macaws being well able to fly long distances between roosting or nesting sites and feeding areas. Old (2001) comments that they fly like swallows, swooping and gliding with ease. In flight, these macaws usually travel in pairs, one to the side of, and slightly behind its partner.

CALLS Cries resembling *greee-ah* and *ara-ara…trrahra* are among reported croaking and screeching calls (Juniper and Parr 1998). All call-notes are less raucous and more high-pitched than calls of the more familiar Hyacinth Macaw *Anodorhynchus hyacinthinus*.

DIET AND FEEDING Indigo Macaws feed almost exclusively on the nuts of licuri palms *Syagrus coronata*, which make up more than 90 per cent of their diet, so their long-term survival is very much dependent on protection of this food source. It is claimed that each macaw eats up to 350 nuts per day. Williams (2003) notes that normally a pair will alight among the fronds of a palm tree and then climb down towards the raceme of fruits. Each bird will bite off a stem bearing eight to 20 or more fruits and either remain sitting with it in the palm tree or fly to a nearby tree to then work its way through each fruit. The fibrous pericarp is stripped quickly and dropped to the ground below, and then, after being rotated in the bill, the hard nut is split neatly into two halves from which the endosperm is scooped out and eaten before the next fruit is grasped. Williams observed that a macaw often passes its bill along a palm frond, possibly to remove accumulated residue.

Yamashita (1987) describes similar feeding techniques, and notes that the macaws will come to the ground in search of fallen palm nuts. After grasping one of these nuts, a bird flies to a nearby palm tree, where the nut is held in the foot while being mandibulated to remove the pericarpal skin, and then the hard nut is easily opened with two perfect transverse cuts before extracting the endosperm. Yamashita points out that when lands are cleared for subsistence farming or free range cattle and goat farming, licuri palms are left because in the dry season they are an important food source for cattle that eat the racemes and young fronds. Although the macaws prefer mature palm fruits and the cattle prefer green fruits, many farmers believe that the birds compete with their cattle for food.

Other recorded foods include *Dioclea* fruits, as well as fruits of pinhão *Jatropha pohliana*, imbu *Spondias tuberosa* and baraúna *Schinopsis brasiliensis*, *Melanoxylon* seeds and *Agave* flowers (in Birdlife International 2016). At times of drought, or when there is a poor fruiting of palms, these macaws will attack corn crops and that is another source of conflict with farmers.

BREEDING Nesting takes place in the early wet season, from February to May, and nests are in cavities formed by weathering of sandstone cliff-faces. Virtually all that is known of the breeding biology comes from field studies being undertaken at the Canudos Biological Station (see Pacifico de Assis 2010). Of 10 examined nesting cavities, five were single, narrow and deep tunnels and five were galleries varying in size and without a recognisable pattern. Depths ranged from 2 m to 18 m, with egg chambers located anywhere along these depths, and some openings were sufficiently large for entry by researchers. Nesting cavities are completely dark and very damp. Observations indicated that pairs use the same nesting cavities from year to year, but confirmation awaits genetic analyses of blood sample from chicks. The two or three eggs are laid on the sandy floor of the chamber, and some chewing of the substrate by the parents was indicated by marks around the nest, but in two successful nests eggs were laid on a smooth rock surface. It was noted that in broods of three nestlings the youngest often is significantly smaller and weaker than its older siblings. During two years of research the death from natural causes of six chicks was documented, the causes being falls from a nest, possible but unconfirmed predation, and a lack of parental care. The entrance to a cavity from which two nestlings fell was repaired with a mix of concrete and sandstone from the cliff-face, and in another cavity sandstone rocks were used to fill a deep hole between the entrance and the nest chamber, and both repaired cavities were reused for nesting. There were instances of nesting cavities being abandoned after occupation by bees, and one nest that had been successful since the 1980s was not used after 2006 because of a hive established 5 m in from the entrance, but after removal of the bees in 2009 it was reused for nesting and three chicks fledged. In response to good rains in 2002, there was a successful breeding season, with 25 chicks fledging at one nesting site (Williams 2003).

Little additional information is available from the few accounts of breeding in captivity, but the traditional hand-rearing of chicks indicates that the nestling period probably is about three months and young birds become independent soon after fledging. Williams reports that fledged birds often are seen with their parents at a feeding site and they are very clumsy, initially showing no interest in palm fruits, but over time they progress to the point of stripping the pericarp to get at the nut, though it seemed that some time would be taken in being able to split open a nut.

EGGS An egg laid by a bird at the London Zoo is held in the collection at the Natural History Museum, Tring, UK, and measures 57.0 × 38.4 mm (Harrison and Holyoak 1970).

EXTINCT

Glaucous Macaw

Anodorhynchus glaucus (Vieillot)

Macrocercus glaucus Vieillot, *Nouv. Dict. Hist. Nat.*, **2**, 1816, p. 259. (South America between lat 27° and 30°S.... on the banks of the Paraná and Uruguay Rivers, *ex* Azara, **1**, p. 402).

DESCRIPTION Length 72 cm.

ADULTS General plumage greenish-blue, more greenish on underparts; head and neck prominently tinged greyish-green; throat dark greyish-brown, paler suffusion on cheeks and upper breast; tail above greenish-blue, below dark grey; undersides of wings dark grey; bill grey-black; bare eyering yellow, and paler yellow 'tear-drop' shaped lappet at each side of lower mandible, but not extending underneath mandible; iris dark brown; legs dark grey.

5 males: wing 360–373 (365.0) mm, tail 361–375 (370.6) mm, exp. cul. 64–68 (66.8) mm, tars. 34–40 (36.6) mm.

1 female: wing 353 mm, tail 352 mm, exp. cul. 66 mm, tars. 36 mm.

3 unsexed: wing 352–364 (360.0) mm, tail 340–381 (354.3) mm, exp. cul. 63–68 (65.3) mm, tars. 35–36 (35.6) mm.

JUVENILES Undescribed.

DISTRIBUTION Formerly restricted to a region between lats 27°S and 31°S and centred on middle reaches of the Río Paraguay, Río Paraná and Río Uruguay in southeastern Paraguay, northern Argentina, in Corrientes, probably Misiones and possibly Chaco, Entre Ríos and Santa Fé, and in southeastern Brazil, in Rio Grande do Sul and possibly Santa Catarina and Paraná; probably occurred also in Artigas, northwestern Uruguay.

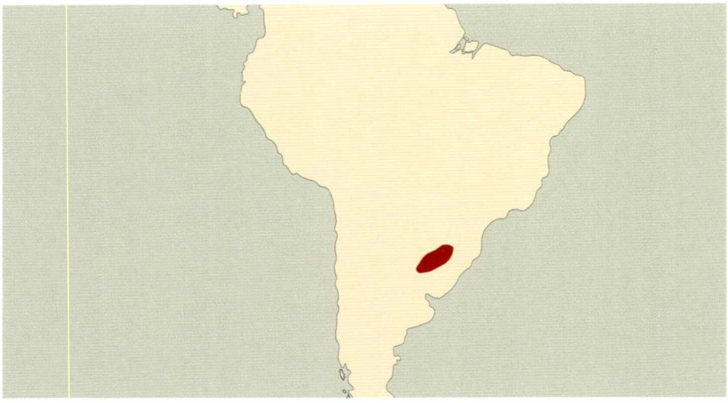

STATUS The Glaucous Macaw is extinct, and all available evidence indicates that it has been so since the early to mid 1900s. Apparently specimens were not taken later than 1860, and there are doubts about identification of the last recorded living specimen. Orfila (1936) referred to a bird held during the 1930s in the Buenos Aires Zoological Gardens and thought to have come from Brazil. A photograph of this bird cannot be positively identified, and the likelihood is that a bird brought from Brazil was an Indigo Macaw *Anodorhynchus leari*. Ridgely (1981) points out that if this bird was *A. leari*, then the last confirmed live specimen of *A. glaucus* was another captive bird seen by Jean Delacour at the Jardin d'Acclimation in Paris from 1895 to 1905. I give no credence at all to a claim that there was a Glaucous Macaw in Australia for many years, and eventually it passed into the collection of Edward Hallstrom (in Low 1980).

Because historical reports comprise mainly brief, vague accounts from early naturalists of encounters with these macaws, it is difficult to determine with any certainty those factors which brought about extinction of the species. Yamashita and Valle (1993) point out that original accounts referring to Glaucous Macaws as being very common are typical reactions to encounters with sedentary, conspicuous populations of large, blue macaws, and it is quite likely that when first reported by naturalists the relict population already was extremely local and declining. Also, it should be noted that the restricted natural range of *Butia* palms, the presumed staple food source, would have supported only a relatively small, localised population of large macaws.

It has been suggested that population declines occurred independently of human activities, but available evidence indicates that extinction was caused by a combination of habitat destruction and excessive exploitation for food and the pet trade. Lightly wooded savanna with palm groves is easily burned and converted to grazing land, so presumably this habitat transformation occurred throughout centuries of intense cattle grazing in the region. In southeastern Brazil, virtually all *Butia* palm groves had been destroyed by the early to mid 1900s, and in northern Argentina the last extensive stands of palms were converted to croplands between 1940 and 1960. Surveys undertaken during 2003 in most of the historical range of the Glaucous Macaw revealed that remnant sparse, low-density stands of *Butia* palms would not support a viable population of macaws.

Attributed to José Sánchez Labrador, a Jesuit missionary in Argentina and Paraguay during the mid 1700s, is a report that the missionaries organised the taking of macaw chicks from nests for use as pets. In the 19th century, live birds apparently were sent to European zoos. Nests in most riverbank cliff-faces would have been easily accessible, and there are early claims of riverboat travellers shooting macaws for food. Such depredations on both adults and chicks could have extirpated entire colonies and, coupled with widespread habitat destruction, brought about extinction of the species.

Although presumed to be extinct, the Glaucous Macaw remains listed on Appendix I of the Convention on International Trade in Endangered Species of Wild Fauna and Flora (CITES).

HABITATS In accordance with known habitat requirements of the closely related Indigo Macaw *Anodorhynchus leari*, it can be assumed that Glaucous Macaws also were dependent on an availability of cliff-faces for nesting or roosting sites and access to palm groves for feeding. Most historical records mention an association with riverine habitats, where nesting and roosting sites were present in riverbank cliff-faces, but there is evidence that the birds also frequented lightly-timbered savanna away from rivers, and here they may have nested in tree hollows. Unconfirmed reports of possibly the last sightings of these macaws in the wild come from near Corrientes city, northeastern Argentina, where, between 1900 and 1910, several birds were seen in gallery forest along the Río Riachuelo, and before 1926, when birds were seen sitting in trees in forest surrounding the city.

HABITS Reports from some early naturalists indicate that these macaws were gregarious, but we have virtually no information on their habits. Presumably, they resembled the closely allied Indigo Macaw *Anodorhynchus leari* in undertaking conspicuous daily flights between cliff-face nesting or roosting sites and feeding areas in nearby or distant palm groves.

CALLS Undescribed.

FEEDING Yamashita and Valle (1993) point out that within the range of the Glaucous Macaw the only colonial palm species with extractable endosperm free of lignin is the yatay palm *Butia yatay*, and confirmation that these palm nuts were the staple food is found in at least one account from an early naturalist.

BREEDING In the late 18th century, Félix de Azara saw a number of pairs along the Río Paraná, and noted that they nested not only in hollows in the trunks of trees, but more frequently in holes '.....made in the vertical banks of the Paraná and Uruguay Rivers' (in Orfila 1936). In December 1823, Freidrick Sellow, a botanist, journeyed from Guaíba to Caçapava de Sul, in southern Rio Grande do Sul, extreme southeastern Brazil, and the translation of an entry in his journal reads '*A blue macaw nests here in holes in the rock cliffs.*' Belton (1984) correctly points out that Sellow's record can be assumed to pertain to the Glaucous Macaw.

SPECIMENS AVAILABLE Hume and Walters (2012) note that specimens are held in the Rijksmuseum van Natuurlijke Historie, Leiden, Netherlands, the Natural History Museums at Liverpool and Tring, UK, and the National Natural History Museum, Washington, USA.

Plate 34
Glaucous Macaw *Anodorhynchus glaucus* (adults)

F KNIGHT 10

GENUS *Ara* Lacépède

Ara Lacépède, *Tableaux Ois.*, 1799, p. 1. Type, by subsequent designation, *Psittacus macao* Linnaeus (Ridgway, *Bull. U.S. Nat. Mus.*, no. 50, pt 7, 1916, p. 119).

Macaws belonging to this genus are large to very large parrots with proportionately large bills and long, graduated tails. The most conspicuous external feature is a wholly or partly bare face, sometimes with widely spaced lines of diminutive feathers. The orbital ring is complete, and its posterior portion is somewhat flattened (see Hargrave 1970). Of particular interest are three sibling species or 'species pairs', each comprising two very similarly plumaged species.

The sexes are alike, and young birds resemble adults.

This genus is widespread in Central and South America, and formerly occurred in the West Indies. The writings of early voyagers indicate that, in addition to *Ara tricolor* in Cuba, macaws formerly occurred elsewhere in the West Indies, but supporting evidence comes only from Guadeloupe and from St Croix, in the Virgin Islands. Species from other islands have been named, but their existence cannot be established with certainty.

ENDANGERED

Great Green Macaw

Ara ambiguus (Bechstein)

Psittacus ambiguus Bechstein, in Latham's *Allgem. Uebers. Vög.*, **4**, Th. 1, 1811, p. 65 (South America, *ex* Levaillant, *Hist. Nat. Perroquets*, **1**, p. 16, pl. 6 = northwestern Colombia by designation of Chapman, *Am. Mus. Novit.*, no. 205, 1925, p. 4).

OTHER NAME Buffon's Macaw.

DESCRIPTION Length 85 cm. Weight 1080–1300 g.
ADULTS General plumage yellowish-green; forehead and feathered lines on bare lores scarlet; bordering bare face, throat tinged olive-brown; lower back to tail-coverts pale blue; flight feathers above dull blue, below dusky yellow; tail above orange-red broadly tipped dull blue, below olive-yellow; bill dark grey tipped paler; bare face pink-white (deep pink when excited) with fine lines of olive-brown feathers; iris dull yellow; legs grey.
12 males: wing 356–422 (391.6) mm, tail 330–468 (399.1) mm, exp. cul. 65–81 (71.6) mm, tars. 34–41 (37.3) mm.
12 females: wing 372–407 (392.2) mm, tail 385–459 (426.2) mm, exp. cul. 64–74 (68.5) mm, tars. 34–39 (36.2) mm.
JUVENILES Duller than adults, particularly on underparts; indistinct yellowish margins to scapulars and inner secondaries; shorter central tail-feathers tipped dull yellow; iris grey.

DISTRIBUTION Central America and northwestern South America from easternmost Honduras to northwestern Colombia, and in western Ecuador.

SUBSPECIES

1 *Ara ambiguus ambiguus* (Bechstein)
The nominate subspecies, as described above, is distributed from the Caribbean lowlands of eastern Honduras and easternmost Nicaragua to Costa Rica and Panama, mostly on the Caribbean slope, to northwestern Colombia, east to the Río Sinú, Córdoba, and south to the Serranía del Baudó, Chocó.

2 *Ara ambiguus guayaquilensis* Chapman
Ara ambigua guayaquilensis Chapman, *Am. Mus. Novit.*, no. 205, 1925, p. 2 (Cerro Bajo Verde, Chongón Hills, 20 miles northwest of Guayaquil, Ecuador).
ADULTS similar to *ambiguus*, but with a smaller, narrower bill; undersides of tail and flight feathers more greenish.
No specimens examined.

Restricted to western Ecuador, where apparently occurring in two isolated populations; in the north in Esmeraldas and western Imbabura, and in the south in the Chongón Hills, near Guayaquil, Guayas.

This poorly differentiated subspecies is variable, particularly in colour of the underparts, wings and tail, and a specimen from El Placer, Esmeraldas, northwestern Ecuador, showing an approach to the Military Macaw *A. militaris* in its darker green colouration prompted a suggestion that not only is *guayaquilensis* synonymous with *ambiguus*, but *A. ambiguus* and *A. militaris* may be conspecific (Fjeldså *et al.* 1987). I concur with Ridgely and Greenfield (2001a) in stressing the need for additional information before this possibility can be resolved, but would suggest that differences between *A. ambiguus* and *A. militaris*, although subtle, are sufficiently consistent to warrant recognition of separate species. Furthermore, the presence of two large, similarly plumaged green macaws conforms to the 'species pairs' pattern evident in the two red and the two blue and yellow *Ara* species. Also, I would point out that, if hybridisation between *A. ambiguus* and *A. militaris* does occur in northwestern Ecuador, it may be a fairly recent phenomenon and could be a consequence of very small local populations of one or both species. Genetic swamping through hybridisation can occur when local populations decline to such low numbers as to become unstable, and certainly there are indications that in parts of northwestern South America both *A. ambiguus* and *A. militaris* are present only in very small numbers.

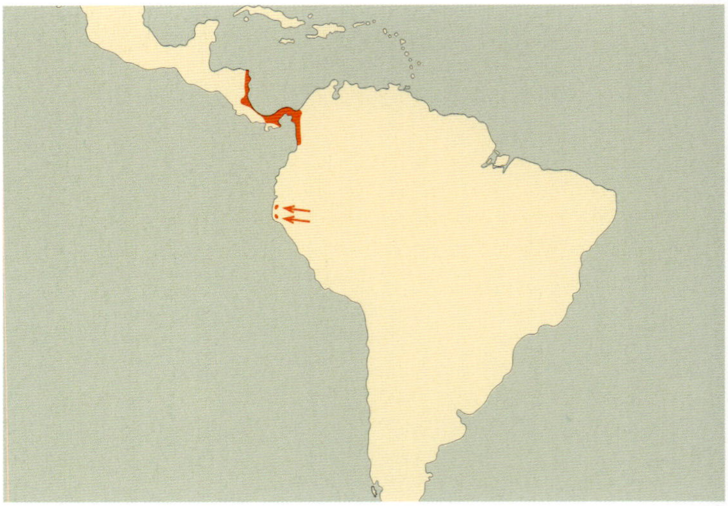

STATUS Generally uncommon to very rare, the Great Green Macaw is intolerant of habitat disturbance and has retreated in response to a relentless destruction of lowland forests, disappearing altogether from much of its historical range and remaining fairly common only in very few remote, undisturbed areas. At the northern limit of its range, in easternmost Honduras, it is an uncommon resident, with a local stronghold in the Río Platano Biosphere Reserve. It is significantly more numerous in Nicaragua, where populations in the Bosawas Reserve and in extensive tracts of lowland forest within the vast Indio de Maíz Biological Reserve may constitute the second largest global subpopulation (in Birdlife International 2016). Formerly widespread throughout the Caribbean lowlands and foothills of Costa Rica, the Great Green Macaw has been forced to near extinction by destruction of approximately 90 per cent of its forested habitat. In the 1990s, the remnant Costa Rican population was estimated at less than 200 birds with about 35 breeding pairs holding on precariously in the far northeast, mostly from the Río Sarapiquí east to the Río San Juan, and long-term survival was dependent on maintaining a viable corridor of suitable habitat connecting with the Indio de Maíz Biological Reserve in neighbouring southern Nicaragua (Chassot and Arias 2002). Similarly in Panama, where it is more widespread and numerous than other macaws, habitat destruction and persecution have extirpated populations in some accessible districts, while elsewhere it is becoming increasingly scarce or locally dispersed, and even in Darién, easternmost Panama, where it remains widespread and locally common, numbers are declining in all but the most remote regions (Ridgely and Gwynn 1989). Collar (1997) confirms that it is fairly common in the Darién Biosphere Reserve, and this protected area appears to be a major stronghold. There are few records from northwestern Colombia, but Ridgely suspected that in the remote Serranía del Baudó, Chocó, good numbers could be present in the fairly extensive tracts of forest (*in litt.* 1977). Conversely, to the north in the Río Atrato valley, it has been virtually extirpated from much of the local range by large-scale landclearance for agriculture (Rodríguez-Mahecha and Hernández-Camacho 2002).

Prior to the late1940s, when large-scale deforestation commenced, these macaws probably were distributed throughout much of lowland western Ecuador, but remnant populations in the northwest and southwest now are critically endangered, and the primary population, which could be as low as 30 to 40 birds, is in the northwest, where records compiled during the 1980s and 1990s include a sighting of a flock of eight birds in western Esmeraldas (Ridgely and Greenfield 2001a; in Birdlife International 2016). Persistent hunting for food poses a continuing threat to birds in the northwest. Only very small, relict populations survive in southwestern Ecuador, where deforestation has been particularly extensive, and total numbers are critically low. Nesting still takes place in the Bosque Protector Cerro Blanco, Guayas, where improved protection may enable the extremely small population to survive, but Waugh suspected that in 1995 as few as nine birds may have been present in this reserve. López-Lanús and Sócola (2000) note that outside the Bosque Protector Cerro Blanco and adjacent areas the only nearby record was from some 30 km to the west, at Hacienda González, where five birds were seen flying together, and this was the same locality where the species was first encountered in July 1996. A few birds may persist also in the coastal cordillera east of Manglaralto, western Guayas, but details are sketchy, and at Machalilla National Park, southwestern Manabí, local residents claimed that the species disappeared several decades before initiation in 1990–1991 of the first major fieldwork (Ridgely and Greenfield 2001a).

In 2009, the population in southern Nicaragua and northern Costa Rica was estimated at 1530 birds, and to the south, in Darién, southern Panama, and neighbouring northern Colombia, the population recently was estimated at less than 1700 mature birds or less than 2500 in total (in Birdlife International 2016). Only eight birds were found during a census conducted in 2010 in Esmeraldas and in the Cordillera Costera Chongón-Colonche, Guayas, western Ecuador, and the critically endangered population is suspected to be as low as 30 to 40 birds, with the majority being in Esmeraldas and very small numbers surviving in the Cordillera Costera Chongón-Colonche (in Birdlife International 2016). Recent estimates suggest that the global population is less than 2500 mature birds, or less than 3700 with juveniles and immatures included. These estimates are higher than those made in the 1990s, but it is not known whether this reflects a real increase in numbers or improved survey techniques.

Field research with conservation objectives is being undertaken in northern Costa Rica, where nests are being monitored to determine productivity and recruitment levels, while adults are being fitted with radio transmitters to monitor survival rates and migration movements. Also in northern Costa Rica a public awareness campaign has been initiated to generate support among local communities for conservation efforts. The Ara Project is a non-profit organisation set up in Costa Rica to breed both Great Green Macaws and Scarlet Macaws *Ara macao* in captivity for the sole purpose of release into the wild, and in 2011 the first Great Green Macaws were released at Manzanillo, in the extreme southeast (Taylor 2011). In the Bosque Protector Cerro Blanco, in western Guayas, southwestern Ecuador, in a program to protect the small remnant population of these macaws initiated by the Fundacion Pro-Bosque in 1993, focus is on locating and protecting nest sites to prevent the taking of chicks for the national pet trade, anti-poaching patrols by park guards being a priority, and native trees used by the macaws for food and nesting feature in habitat restoration programs.

The Great Green Macaw is listed on Appendix I of the Convention on International Trade in Endangered Species of Wild Fauna and Flora (CITES).

HABITATS Though primarily a bird of humid lowland forests, the Great Green Macaw ranges out also into partially cleared areas, at least to feed, and regularly is found in densely forested foothills up to 1000 m, or locally to 1500 m in Darién, eastern Panama (Ridgely 1981). Throughout Central America there is a close association with the almendro tree *Dipteryx panamensis*, which is a principal food source and provides suitable nesting hollows, so the breeding range is restricted almost exclusively to almendro forests. Howell and Webb (1995) note that in eastern Honduras this species usually is encountered in humid evergreen forest or adjacent areas. Likewise in Nicaragua, it is seen mainly in lowland forests on the Caribbean slopes, but at times is observed flying over adjacent *Pinus* savanna. In Costa Rica, it is confined mainly to the Caribbean lowlands and foothills, locally up to about 600 m or higher, and prefers the canopy of humid lowland forest, though often coming to feed in remnant *Dipteryx* trees in pastures or semiopen country (Stiles and Skutch 1989). In Panama, it favors humid forest, especially in hilly areas (Ridgely and Gwynn 1989). Hilty and Brown (1986) note that in northwestern Colombia, it frequents humid lowland forest and intervening tracts of semiopen country, with a preference being shown for wetter areas than are favoured by the Military Macaw *Ara militaris*.

In western Ecuador, these macaws frequent the canopy and borders of both humid and deciduous forest in the lowlands and foothills up to about 800 m (Ridgely and Greenfield 2001a).

In the southwest, they are recorded mostly in the Bosque Protector Cerro Blanco, a 6000 ha private reserve at the eastern extremity of the Cordillera Costera Chongón-Colonche, and here the three plant community types are tropical dry forest at 50–300 m elevation, riparian forest at 150–350 m elevation and moist tropical forest at 300–400 m elevation (Berg *et al.* 2007). Between 150 m and 350 m, forests are dominated by second-growth deciduous vegetation ranging from 50 to 80 years in age, and pigío *Cavanillesia platanifolia* is the dominant tree species in a discontinuous canopy reaching 35 m in height. At Hacienda González, approximately 25 km northwest of the Bosque Protector Cerro Blanco and also in the Cordillera Costera Chongón-Colonche, in July 1996, birds were encountered primarily in a band of *Cavanillesia* woodland (Pople *et al.* 1997).

MOVEMENTS Seasonal movements and irregular local wandering, apparently in response to changes in food availability, have been reported from some regions. When more numerous and widespread in Costa Rica these macaws would appear in numbers at a locality and remain for a few weeks or months before departing, their presence usually coinciding with the fruiting of *Dipteryx* trees (see Stiles and Skutch 1989). Radio-telemetry studies undertaken in northeastern Costa Rica have revealed a post-breeding dispersal from the lowlands to higher elevation forests in mountain ranges to the west, as well as northwards to drier forests in southern Nicaragua (Chassot and Arias 2002). At Playa de Oro, northern Esmeraldas, northwestern Ecuador, these macaws occur in most districts only as foraging visitors, with breeding seemingly restricted to a narrow zone between 150 m and 300 m (in Ridgely and Greenfield 2001a).

HABITS Great Green Macaws seem to be less gregarious than other large macaws, normally being encountered in pairs or small groups of up to eight birds, and seldom in larger flocks or singly, though there is a report of non-breeding flocks of 50 or more birds forming when flock attendants come together from huge areas (in Birdlife International 2016). Generally they keep to the uppermost branches of tall trees, where their green plumage blends well with the foliage, and their presence is betrayed by occasional squawks or by a steady stream of discarded food remnants falling to the ground. Conversely, they are conspicuous and noisy in the air, and will fly long distances, sometimes over open country, to reach feeding areas. Slud (1964) described seeing birds flying about well below canopy height in a partial clearing and remarked that this was while he was close nearby. Ridgely and Gwynn (1989) also comment on these macaws being quite tame, and note that feeding birds often allow a remarkably close approach, but Huber (1933) recalled that in the Eden district, northeastern Nicaragua, they were very wary, and I suspect that this may have been a consequence of local persecution.

During field studies undertaken between June 1999 and May 2000, in the Bosque Protector Cerro Blanco, western Guayas, southwestern Ecuador, Great Green Macaws were observed in every month and on 71 different days, representing 65 per cent of census days (Berg *et al.* 2007). Sightings were of single birds and groups of up to nine birds, the groups being more prevalent during the non-breeding season. The amount of time macaws were in view varied, with an increase at the beginning of the December to May wet season, and there was a period of 27 days in November 1999 and again in March 2000 when no macaws were seen. While being observed, the macaws spent 75 per cent of the time perching, 9 per cent foraging, 6 per cent preening, 5 per cent flying, 3 per cent allofeeding, and 2 per cent in unidentified activities.

These macaws are strong fliers, and long-distance flights usually are high in the air, with the fairly swift, direct flight comprising steady, somewhat leisurely, shallow wingbeats.

CALLS Extremely loud and heard from afar, the raucous calls include shouts, squawks and a growling *aa-aahrk* or *aowrk* (Stiles and Skutch 1989).

DIET AND FEEDING Recorded food items include seeds, nuts, fruits, flowers, bulbs, roots and bark, and these are procured in the treetops. Slud (1964) pointed out that with their powerful bills, these macaws can open the toughest-shelled nuts, such as those of the monkey pot *Lecythis costaricensis*. Seeds of almendro *Dipteryx panamensis* constitute a major proportion of the diet throughout Central America, the birds often travelling long distances in search of fruiting trees, and the targeting of these trees in past logging operations would have impacted severely on local populations. Felling almendro trees now is prohibited in Costa Rica. Field studies undertaken in northeastern Costa Rica have identified *Sacoglottis trichogyna* as a favored food source when *Dipteryx* seeds become scarce or unavailable.

López-Lanús and Sócola (2000) give details of observations made in 1994 and 1997 on the foraging habits of nesting pairs in Bosque Protector Cerro Blanco, western Guayas, southwestern Ecuador, where feeding tended to be in one place until the supply was exhausted, and the changing of foraging areas every three or four days was reflected in different directions followed by pairs when leaving the nest. Succulent bulbs of the arboreal orchids *Epidendrum collare* and *Encyclia apera* were an important component of the diet, making up 71 per cent of the diet, and consumption was consistent throughout the study periods. The macaws tested these bulbs by biting into them, and dehydrated or dead bulbs were discarded. Consumption of the bulbs lasted between 10 seconds and five minutes, with a maximum intake of up to four per minute, and several bulbs would be taken from the same plant or the birds would go searching from plant to plant. Seeds of amarillo *Centrolobium ochroxylum* were taken during the entire 1997 study period, with an apparent tendency towards ingestion immediately after heavy rain, and also after heavy falls of rain the macaws drank water retained at the base of *Tillandsia* leaves. Seeds, flowers and bark fragments from pigío *Cavanillesia platanifolia* were an important component of the diet, and an overlap in the consumption of flowers and seeds was due to differing times of flowering of different trees. Bark was taken only from pigío trees, and on two occasions a macaw was seen extracting and chewing strips of bark about 3 cm wide by 40 cm long, but more often birds were seen scraping the bark with their mandibles and then using the tip of the upper mandible to scoop up the powdered bark. López-Lanús and Sócola speculate that bark and powdered bark were consumed

Plate 35
Great Green Macaw *Ara ambiguus* (adult)

F. KNIGHT '15

as detoxicants. Combined consumption of flowers, seeds and bark from *Cavanillesia platanifolia* exceeded the intake of seeds of *Centrolobium ochroxylum*, but seeds of *Centrolobium ochroxylum* were taken in greater quantities than seeds of *Cavanillesia platanifolia*, except during October–November when preferences were reversed. Other items taken in smaller quantities included seeds of balsa *Odroma pyramidalis*, beldaco *Pseudobombax guayasense*, bototillo *Cochlospermum vitifolium* and *Ipomea* sp., together with fruits of pechiche *Vitex gigantea* and inflorescences of *Cecropia litoralis*.

Another foraging study was undertaken in the Bosque Protector Cerro Blanco between June 1999 to May 2000, when these macaws were recorded feeding more on seeds rather than either pulp or mesocarps, and they utilised at least eight species of food plants (Berg *et al.* 2007). Most time was spent feeding on amarillo fruits, fruits and flowers of *Terminalia valverdeae*, and feeding in pigío trees, where they took fruits, buds, flowers, leaves and bark. Based on earlier unpublished feeding records and anecdotal accounts from local residents it was determined that the macaws typically pry open exocarps with their bills and consume only the seeds, but they do consume the entire fruits of *Cecropia litoralis* and *Vitex gigantea*.

BREEDING At Guápiles, northeastern Costa Rica, a female collected on the last day of February had in the oviduct an egg ready to be laid (Carriker 1910). López-Lanús and Sócola (2000) note that in or near the Bosque Protector Cerro Blanco, Guayas, southwestern Ecuador, chicks were present in a nest found in mid August, and incubation apparently had just commenced in another nest found in mid July, with fledging of the single chick occurring in mid November, so the breeding season presumably was in the dry season, between June and November.

Chassot and Arias (2002) report that radio-telemetry studies undertaken in northeastern Costa Rica have revealed that breeding pairs occupy large, non-overlapping home ranges. During the seven years of study, 41 nests were found in natural hollows in living trees, with 88 per cent being in almendro trees *Dipteryx panamensis*, but no more than 19 nests were active in a year, and during one breeding season only 13 of 29 hollows were occupied. Between 1996 and 1999, a monitoring of 13 adults fitted with radio-transmitters showed that breeding pairs generally do not occupy the same nesting sites in consecutive years. During five years, 33 young birds fledged from 18 successful nesting attempts at 15 sites, and this represents a success rate of 1.83 fledglings per nest. A monitoring of 23 fledglings from 12 nests revealed that 15 survived until commencement of the next breeding season, when juveniles separate from adults, and this represents a first-year survival rate of 1.25 juveniles per nest.

Berg and Horstman (1996) report that on 14 August 1994 an occupied nest was found in the Bosque Protector Cerro Blanco, Guayas, southwestern Ecuador. This nest was in a natural hollow in a dead, but standing pigío tree *Cavanillesia platanifolia.* On 6 September two nestlings were seen inside the hollow, and were seen again on 7 and 29 September. The nest was vacated on 1 October. In April 1995, a pair of macaws, possibly the same nesting pair, returned to the site, and on several occasions were observed entering the hollow. However, in late July the termite-infested upper section of the tree broke away, leaving the hollow exposed, and the site was abandoned.

López-Lanús and Sócola record data obtained from observations at a nest found in the Bosque Protector Cerro Blanco in mid August 1994 and at another nest found near to the reserve in mid July 1997. At the first nest, the parents normally were absent for three periods in the day, returning each time to feed the two chicks. The first period was 30 to 45 minutes after dawn, with their absence lasting 20 to 30 minutes, and a few hours later the parents again left to gather food. In the afternoon, they generally were absent for a prolonged period. Cleaning out of the nesting hollow by both parents was first observed 26 days before fledging of the second chick, and was recorded daily or on every other day. Using bill and feet, the parents cleared the floor of the hollow and removed debris by using the feet to eject it behind and below the body. From approximately eight days before fledging of the chicks, the parents were absent for increasingly longer periods during the day and at night roosted away from the area, returning to the nest at sunrise. A period of 29 days elapsed between appearance of the first chick at the nest entrance and fledging of the second chick. Practice flying by the chicks was observed at the nest entrance seven days before fledging, and an interval of two days occurred between first flights of the two chicks. These first flights were decisive, prolonged and steady, each youngster following a parent away from the nesting area, and the first fledgling did not return to the area during the next two days when return visits were made by both parents to feed the second chick in the nest. The second nest was in an exposed, vertical hollow some 20 m above the ground in the upper crown of a lone pigío tree *Cavanillesia platanifolia.* On 13 July, both parents were observed feeding at a distance of 600 m from the nesting tree, but from the next day until 4 September, when she presumably was incubating and then brooding the young nestling, the female remained in or near to the nest at all times. Three or four times during the day, each for periods of two to six minutes, she left the nest and flew to a nearby tree, never more than 60 m away, to be fed regurgitated food by the male, and normally she also ate two to four orchid bulbs. Absences by the male to gather food lasted between 20 minutes and three hours. After 4 September, when the chick would have been well advanced, the female accompanied the male to gather food, and both were absent for prolonged periods of up to three hours. The minimum foraging area of this pair was estimated at about 2000 ha. As the chick grew older it was left unattended for increasingly longer periods. On 12 October, while the parents were absent, the nestling successfully repulsed an attack by a Collared Forest Falcon *Micrastur semitorquatus*. In response to a presence of raptors the parents spent more time at the nest defending the chick, and during mid to late October up to 10 attacks by the falcons occurred on some days, so forcing both parents to remain at the nest throughout the day. On a couple of occasions the male was seen to feed regurgitated food to the chick, apparently despite not being able to leave the nest to gather food. On 20 September, the parents were seen cleaning out the nesting hollow, using the same techniques observed in 1994 at the first nest, and this continued daily or on every other day until usurpation of the hollow by a pair of falcons. Feeding of the nestling initially was by the female, but subsequently the male also was seen to enter the hollow to feed the chick. On 14 October the chick was detected flapping its wings while inside the hollow, and from this date there was a more frequent repetition of these actions. On 31 October, for the first time the chick was seen poking its head out of the nest entrance. Attacks on both parents and chick by Collared Forest Falcons became intense on 6 November, the parents being prevented for two days from coming to the nest to feed the chick. Up to 20 repeated attacks eventually resulted in the hollow being taken over by the falcons after the chick had been ejected and fell to the ground. It was rescued and, five days later when it could fly, was reunited with its parents. While the chick was held in the rescue centre, the parents remained in the area, roosting at night in another pigío tree some 200 m from the nesting tree. After being reunited with its parents, the young bird was seen to be left behind when accompanying the adults, and on several occasions the parents had to circle back to reach it and again guide it in the direction they were flying, always with increased vocalisation from both adults and youngster. During the first week, the fledgling was clumsy when alighting, often colliding with branches, and for the first two weeks it always remained with the parents, being fed by them until at least 1 December. After two weeks, the young bird remained alone in trees near to the nest for increasingly longer periods while the parents left to gather food, and apparently the adults called to locate the waiting youngster. At nighttime, the young bird roosted away from the adults in another pigío tree, where it sat amidst thicker branches nearer to the centre of the crown. Gradually, the trio spent increasingly longer periods away from the nesting area, being absent for up to two or three days at a time.

EGGS Elliptical; a single egg in the Dresden Museum Collection measures 55.0 × 46.0 mm (Schönwetter 1964). Gale (1983) gives 55.0 × 45.0 mm and 46.0 × 38.0 mm as the measurements of two eggs laid in captivity.

CRITICALLY ENDANGERED

Blue-throated Macaw

Ara glaucogularis Dabbene

Ara glaucogularis Dabbene, *Hornero,* **2**, 1921, p. 225. New name for *Ara caninde* of authors, not *Sittace caninde* Wagler. (Paraguay = Santa Cruz de la Sierra, Santa Cruz, Bolivia).

DESCRIPTION Length 85cm. Weight 750 g.
ADULTS Entire upperparts, including forehead and forecrown, blue; outer webs of primaries and outer secondaries darker violet-blue; throat and broad band extending up side of neck to ear-coverts greenish-blue; breast to abdomen and underwing-coverts yellow; under tail-coverts blue; undersides of flight feathers yellow; tail above blue, below dusky yellow; bill grey-black; bare cere and face white with broad greenish-blue feathered lines across face; iris yellow; legs dark grey.
1 male: wing 330 mm, tail 422 mm, exp. cul. 47 mm, tars. 31 mm.
3 unsexed: wing 350–363 (358.3) mm, tail 453–480 (462.3) mm, exp. cul. 50–56 (53.0) mm, tars. 31–32 (31.3) mm.
JUVENILES Like adults, but darker blue on throat; shorter tail; iris grey.

DISTRIBUTION Now known only from Beni, north-central Bolivia, where a small remnant population survives as two subpopulations north and south of Trinidad, in the Llanos de Mojos, east of the upper Río Mamoré, and another small population in the Municipal Protected Area of the Great Tectonic Lakes of Exaltación; formerly occurred also in neighbouring Santa Cruz. Historical records from northernmost Paraguay and northern Argentina, in Salta, probably are attributable to misidentification of the Blue and Yellow Macaw *Ara ararauna*.

STATUS For more than a century after its original description, the Blue-throated Macaw remained a mysterious bird known only from a few museum specimens of uncertain provenance, and there was much speculation about its relationship to the familiar Blue and Yellow Macaw *Ara ararauna*. That it was a separate species only became evident in the 1970s, when a number of live birds were exported from Bolivia and, in response to his inquiries, Robert Ridgely was informed by local bird-dealers that in the drainage of the upper Río Mamoré small numbers of Blue-throated Macaws regularly were found in the company of much greater numbers of Blue and Yellow Macaws (*in litt.* 1980). This information ultimately proved to be accurate. In 2016, another small population comprising at least 10 adults was located in the recently established Municipal Protected Area of the Great Tectonic Lakes of Exaltación.

Despite a complete lack of information on the distribution and status of the species, no fewer than 60 Blue-throated Macaws were exported from Bolivia in 1980. Hesse and Duffield (2000) note that the highest estimated figure for international trade between 1981 and 1992 is 390 birds, with a majority being exported before 1984, when Bolivia prohibited the export of live animals. However, these estimates are likely to be too low and, based on discrepancies between documented and undocumented exports, a figure of 1200 birds is suggested by Yamashita and Machado de Barros (1997). In the light of what we now know about the very low numbers within a quite restricted range, excessive exploitation for the live-bird trade was largely responsible for near extinction of the wild population.

Yamashita and Machado de Barros note that the restricted distribution and specialised habitat requirements fit a pattern common to almost all medium-sized macaws. This appears to be so, for reports of occurrences in Paraguay and northern Argentina can be attributed to misinterpretation of early references to *A. ararauna*, but Hesse and Duffield point out that interviews with ex-trappers suggest that historically the range was much broader in Bolivia. Currently, there are no Blue-throated Macaws in southwestern Beni, where prior to the advent of intense trapping in the 1970s, they were reported to be present in the vicinity of Santa Rosa, a locality some 225 km west of the now known range.

Surveys and field studies commenced soon after formal rediscovery of the species in August 1992, when the first observations of wild birds were made in the region that had been identified by local bird-dealers (Jordan and Munn 1993). During this and subsequent searches very few birds were located, and it became apparent that total numbers were extremely low. Yamashita and Machado de Barros recall that during surveys across 2000 km², undertaken in 1993–1994, only 54 birds were located and, extrapolating from these counts, the total wild population was estimated to comprise 200 birds. Hesse and Duffield report that consistently low counts of between 21 and 36 birds were made during surveys carried out between 1993 and 1998, and extrapolation of the count of 36 individuals to the total area of suitable habitat gave a total population estimate of 120 macaws. Despite limitations in census techniques employed during these surveys and the unreliability of extrapolating counts to the total area of suitable habitat when the species is known to be unevenly distributed, the results were sufficiently consistent to confirm that only a very small remnant population survived in the wild, and the species was critically endangered.

At the time of their rediscovery, Blue-throated Macaws occurred exclusively on privately owned lands devoted principally to cattle ranching, so participation by landholders was essential in implementing both short-term protection measures and longer-term conservation strategies. Hesse and Duffield point out that cattle grazing has been the dominant landuse practice for centuries, and there is little evidence of its impacting negatively on the ecological requirements of these macaws, although trampling or grazing by stock and regular dry-season burning by landholders could affect adversely the regeneration and growth of young palms. Climatic conditions can impact on nesting success, and Kyle (2006) reports nest failures at a time of severe drought with accompanying destructive fires and again when torrential rains flooded nesting hollows and caused the collapse of nesting trees. Secondary pressures impacting adversely on a population become increasingly more significant when the population is low in numbers, and pressures further threatening these macaws were identified. There was competition for nest-sites from other cavity-nesting species, including the larger macaws, and some nests were lost to predation, parental neglect at times of food shortages or abandonment following occupancy by bees, and deaths of nestlings from infestations by mites or *Philornis* botfly larvae were recorded. Also, these and other macaws were hunted for feathers and wings to be used in ceremonial adornments. Since the 1990s, intense efforts have been made to counter these threats as part of a combined national and international conservation program largely under the auspices of Fundacíon Armonía, a non-governmental organisation based in Santa Cruz, and fieldworkers from this organisation, assisted by volunteers and funding from international bodies, have been undertaking field investigations of the biology of Blue-throated Macaws.

In 2007, at a farm site in Santa Ana de Yacuma, west of the Río Mamoré, 70 macaws were found at a non-breeding roost and, with funding assistance from international agencies, the farm was acquired to be set aside as the 5000 ha Barba Azul Nature Reserve. A census undertaken in this reserve in 2011 revealed that it provides protection for at least 100 birds, which represents about 25 per cent of the entire population. With additional land acquisitions, this reserve has been expanded to approximately 11 000 ha, and it provides protection for a significant proportion of the entire population of Blue-throated Macaws. Through community workshops and meetings, liaison with landowners has been set up to promote management practices that minimise adverse impacts on the macaws in farmlands. Indigenous people are being persuaded to use artificial feathers in ceremonial adornments, and a strong educational program has been initiated in local schools to foster a pride in having these rare birds as an integral part of the region. As part of the field investigations, nests are being monitored in an effort to improve productivity. Artificial nestboxes are being set up, and it has been found that these are more likely to be occupied when positioned near to known nest-sites. Monitoring of nests also enables intervention to be made when the survival of chicks is threatened, and at times supplementary feeding proves beneficial, particularly if the youngest chick in a brood of three is seen to be weakening. Although now seen as being of lower significance, poaching remains of concern, and the vigilance of law enforcement agencies is supported. The captive population of Blue-throated Macaws undoubtedly exceeds the wild population, which currently is estimated at approximately 300 birds, and there is the potential for releasing captive-bred birds into the wild.

The Blue-throated Macaw is listed on Appendix I of the Convention on International Trade in Endangered Species of Wild Fauna and Flora (CITES).

HABITATS Blue-throated Macaws are restricted to the Llanos de Mojos, an isolated expanse of flat savanna grassland, at 200 m to 300 m altitude, on the upper Río Mamoré system, and bordered by Amazonian rainforest to the north, deciduous woodland to the east, the Chaco to the south, and humid Andean foothills to the west. Yamashita and Machado de Barros (1997) point out that within this region the macaws occur only in seasonally flooded savanna with natural or artificial elevated islands of forest dominated by motacú palms *Attalea phalerata* with semideciduous emergent trees, including *Enterolobium cortisiliqun*, *Tabebuia heptaphilla* and Panama trees *Sterculia striata*, and also in gallery woodland along watercourses, though only where motacú palms represent more than 60 per cent of component tree species. They favour tall trees and areas with continuous fruit production from motacú palms, the most important food source, but are absent from swamps dominated by groves of moriché palms *Mauritia flexuosa*, a habitat favoured by the sympatric Blue and Yellow Macaw *Ara ararauna*. A very strong association between Blue-throated Macaws and motacú palms, with the birds not occurring in areas where these palms are not prevalent, prompts Hesse and Duffield (2000) to emphasise that a presence of forest islands or fragments containing a healthy population of palms is the single outstanding ecological requirement of this critically endangered species.

HABITS Unless observed in good light, when the blue throat can be seen, Blue-throated Macaws can be difficult to distinguish from the sympatric Blue and Yellow Macaw *Ara ararauna*, but a smaller size, slimmer build and longer tail give them a different appearance. These macaws normally are encountered in pairs or small parties of up to five or six birds, but this could be a consequence of low numbers and, when more numerous, they may have associated in larger flocks, and there is a report of 70 birds being seen at a dry season roost site (in Birdlife International 2016). Yamashita and Machado de Barros (1997) recall that of 112 records made during surveys undertaken in 1993–1994, 43 were of pairs, 68 were of family groups, and the remaining single

---

Plate 36
Blue-throated Macaw *Ara glaucogularis* (adults)

F. KNIGHT 10

record was of a flock of 15 birds. Like other large macaws, they are conspicuous in the air, especially during early morning and late afternoon flights between nighttime roosts and feeding areas, or when perched on projecting, often leafless branches of tall trees. Such perches, particularly atop tall *Tabebuia* trees, are habitually used for preening and bill-wiping, the latter often leaving on the branch telltale deposits of matted fruit-pulp scraped from sides of the bill. Jordan and Munn (1993) report that between 0630 and 0830 hours a pair was observed preening and allopreening on a treetop perch above a living *Acromonia* palm in which a nesting cavity apparently was being excavated, and this suggests to me that paired adults probably are territorial, at least during the breeding season, and territories are centred on nesting sites.

These macaws are rather wary, and normally will not allow a close approach. Occasionally they are seen in the company of Blue and Yellow Macaws, and seeing mixed groups in flight prompted Duffield and Hesse (1997) to speculate that the two species may come together at communal nighttime roosts.

The direct, fairly swift flight is characterised by regular, strong wingbeats. When flying in the company of Blue and Yellow Macaws, the slimmer body shape and longer tail of this species are quite apparent.

CALLS In common with other large macaws, Blue-throated Macaws are noisy, and the loud, raucous call-notes often are heard well before the birds are seen. Jordan and Munn (1993) point out that all call-notes are distinctly higher in frequency and of a different quality than those of the Blue and Yellow Macaw, and these differences can be detected at a distance of hundreds of metres. The typical loud call follows a distinctively alternating pattern, and there is a distinctive rolling introduction to the flight call (in Birdlife International 2016).

DIET AND FEEDING Differences in dietary preferences of Blue-throated Macaws and Blue and Yellow Macaws are reflected in differences in the size and shape of bills of the two similar species. With its more massive bill, the Blue and Yellow Macaw is capable of breaking open palm nuts to extract the kernels, whereas the smaller, finer bill of the Blue-throated Macaw is suited to peeling away the soft pulp or mesocarp of ripe palm fruits. Yamashita and Machado de Barros (1997) point out that mesocarp from ripe fruits of motacú palms *Attalea phalerata* is the most important food item for this species, and characteristically narrow marks made on fruits by the chisel edge of the fine lower mandible facilitate identification in the field of fruits attacked by the macaws. Other food items recorded by Yamashita and Machado de Barros are mesocarps and inflorescences from *Acrocomia*, *Syagrus* and *Astrocaryum* palms, leaf stems and unripe seeds from *Hura crepitans*, flower petals from *Cochlospermum hybiscoides*, unripe seeds of *Sapindus saponaria* and leaf stems from *Genipa americana*, but all except the unripe *Hura* seeds represented only a small proportion of the food intake. Jordan and Munn (1993) recall that on occasions these macaws were seen to open and then drink juice from very immature fruits of *Attalea* and *Acrocomia* palms.

BREEDING What is known of the breeding biology comes mostly from studies undertaken from early August to late March during five consecutive breeding seasons between 2007–2008 and 2011–2012, when it was found that egg-laying generally was concentrated at the end of the dry season, in September–October, in the northern subpopulation and during the beginning of the wet season in the southern subpopulation, resulting in a prolonged laying interval from August to May (Berkunsky *et al.* 2014). During the study period, 64 individual birds were identified in the study area, and they included 16 distinct pairs that laid at least one egg. The number of active breeding pairs in a given area either remained constant or decreased, depending on the site. Of 31 monitored nesting attempts, 19 were in natural tree cavities, six were in wooden nestboxes, and one was in a PVC nestbox. The natural tree cavities included 11 in dead *Attalea* palms, one in a dead *Acrocomia* palm, one in a Panama tree *Sterculia apetala* and, of the remaining six in living hardwood trees, four were in ajo-ajo *Gallesia integrifolia* and two were in vilca *Anaedanthera colubrina*. Measurements of the nesting trees and nesting cavities are listed in Table 7. Most cavities in hardwood trees and nestboxes were reused by the macaws at least once during the study period, but cavities in dead palms were never reused before the trees fell. All pairs that occupied nestboxes had nested in the same tree or in a tree within a few metres from where the pair had nested in previous seasons. A high nest-site fidelity was evident in monitored pairs reusing the same nestboxes and cavities in hardwood trees. Daily observations at nests were concentrated in the morning, from sunrise to 1000 hours, and again in the afternoon, between 1500 hours and dusk. At monitored nests, inspections were made once a week during incubation, daily or every second day during the first three weeks after hatching, and then twice a week until fledging. A nest was deemed successful if it produced at least one fledgling. Clutches of one to three eggs were recorded, and the average clutch size for 29 nests was 2.53 eggs. Both instances of clutch replacement were in nests that failed during incubation, but no eggs from these second clutches hatched. Incubation by the female lasted 25 days, during which time she spent most of the time inside the nesting cavity or perched nearby. Rarely was a male seen to enter the nesting cavity, but he often perched nearby. A mean hatching success of 72 per cent was recorded for 23 nests, with hatching failure being the major cause of egg losses. All eggs hatched in 12 nests, while one egg failed to hatch in nine nests, and no eggs hatched in two nests. In the 21 nests where hatching occurred, recorded dates of hatching of the first chick were between 3 May and 15 January, with the mean date at 23 October. Fledging occurred approximately 85 days after hatching. All nestlings successfully fledged in monitored nests, but this probably was attributable to the intensive conservation management program. Parents appeared to attend to their fledglings for an extended period, possibly through the next breeding season, and in some

Tree and cavity characteristics	Sample size	Measurements
Diameter of tree trunk at breast height (cm)	7	30–107 (65.7)
Diameter of tree trunk at nest entrance (cm)	4	52–62 (57.5)
Height of nest entrance (m)	17	2.5–14 (8.6)
Maximum diameter of nest entrance (cm)	8	13–33 (24.1)
Minimum diameter of nest entrance (cm)	8	9–33 (17.9)
Depth of nest cavity (cm)	11	2–110 (52.3)
Internal diameter of nest cavity (cm)	6	26–33 (28.8)

Table 7. Measurements of nesting trees and nesting cavities at study sites in 2007–2008 to 2011–2012 (after Berkunsky *et al.* 2014).

years no breeding behaviour was shown by breeding pairs seen to be accompanied by juveniles of the previous year. Nesting attempts and nesting success varied between years, with up to 10 pairs nesting in some years, but in other years only two attempts were made. During the study period, no new adult pairs were recruited into the breeding population, and it was suggested that this lack of recruitment of breeding pairs could be a consequence of a low survival rate of pre-breeding birds or an extremely low density impeding effective pair formation.

On 11 October 2011, parental feeding of two large chicks was observed at a nest in a cavity in the trunk of a *Attalea* palm standing 25 m in from the periphery of a palm-dominated 'island' measuring 300 × 250 m (Pederson and Pederson 2002). The parents spent much time sitting together in a tall tree some 50 m from the nest and overlooking the entire area or among fronds of a palm adjacent to the nesting tree. Departure from the area by the adults for foraging was accompanied by much calling, and they were absent for two hours each time. They returned silently, alighting in the same tall tree away from the nest, and then proceeded either directly to the nesting tree or to adjacent palm fronds above the nest, where they always rested before the female climbed down to the nest entrance. Before entering the nest, the female pre-regurgitated a number of times and then spent 20 to 30 minutes feeding and staying with the chicks in the nest. While the female was in the nest, the male usually remained in the adjacent palm fronds above the nest. After emerging from the nest, the female joined the male and both flew away. The chicks were fed at intervals of three to four hours throughout the day. A pair of Blue and Yellow Macaws *Ara ararauna* came to the nesting palm, spent about five minutes climbing up and down the trunk, looked down at the chicks and chewed at the nest entrance before being chased away by the parents.

EGGS A single egg laid in captivity measures 45.0 x 35.0 mm (Arndt *in litt.* 1986).

ENDANGERED

Red-fronted Macaw

Ara rubrogenys Lafresnaye

Ara rubro-genys Lafresnaye, *Rev. Zool.*, 1847, p. 65. (Bolivia.)

DESCRIPTION Length 60 cm. Weight 450–650 g.
ADULTS General plumage pale olive-green; forehead, crown and ear-coverts orange-red; thighs red; lesser wing-coverts, bend of wing, carpal edge and lesser underwing-coverts orange-red; greater underwing-coverts and undersides of flight feathers olive-yellow; outer webs of primaries and outer secondaries blue, darker on primary-coverts; tail above olive suffused blue and broadly tipped blue, below olive-yellow; bill grey-black; bare face pink with fine lines of minute olive-brown feathers; iris orange; legs dark grey.
9 males: wing 295–314 (303.8) mm, tail 275–351 (328.7) mm, exp. cul. 47–52 (49.3) mm, tars. 24–28 (26.3) mm.
5 females: wing 299–310 (305.6) mm, tail 305–336 (323.8) mm, exp. cul. 46–51 (48.4) mm, tars. 26–30 (27.4) mm.
JUVENILES Little or no orange-red on forehead or crown; lesser wing-coverts, bend of wing and carpal edge green; lesser underwing-coverts paler orange; thighs green tinged orange; shorter tail; iris brown.

DISTRIBUTION East slope of Andes in central-southern Bolivia, from southern Cochabamba and western Santa Cruz south to eastern Potosí and northern Chuquisaca, mainly in valleys of Ríos Grande, Mizque, Caine and Pilcomayo.

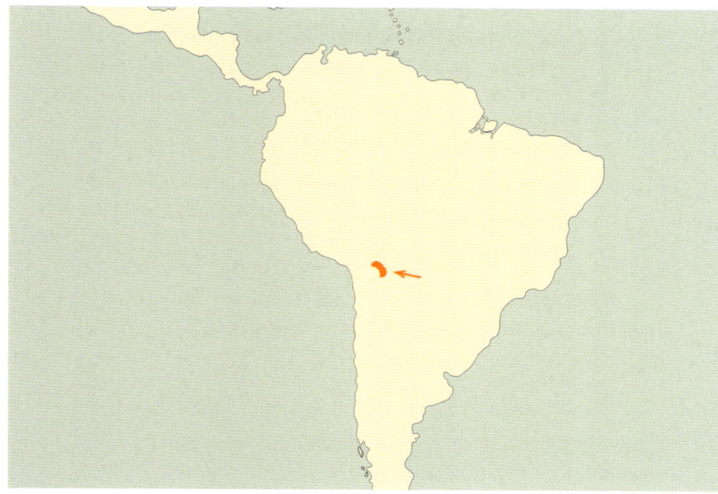

STATUS Within a restricted range, Red-fronted Macaws are threatened by habitat loss, persecution as crop pests and capture for the live-bird market. At the beginning of the 1990s, it was estimated that 40 per cent of the habitat had been destroyed and deforestation was continuing, with regeneration of groundcover vegetation being seriously impaired by overgrazing (in Pitter and Christiansen 1995). In addition to serious degradation of the habitat from intense grazing, especially by goats, firewood cutting and charcoal production have targeted several important food trees, and resulting food shortages can be expected to bring about local increases in attacks on crops. There is an urgent need to protect remnant native vegetation in river valleys and to replant native trees. Damage to crops and consequent persecution by farmers also need to be addressed. The macaws attack maize and peanuts or groundnuts crops, but it appears that damage levels may vary locally. Along the Río Caine, northern Potosí, while field studies were being carried out in October–November 1990, no persecution by farmers was observed, and at that time threats to local populations were considered to be minimal (Boussekey *et al.* 1991). Conversely, along the Río Chico, northern Chuqusiaca, the macaws greatly favoured maize and were considered to be serious pests by local farmers, who repeatedly chased the birds from ripening crops (Pitter and Christiansen 1995). Tameness makes these macaws particularly vulnerable to shooting, and Kyle (2005a) reports that stories of crop-raiding macaws being shot are not uncommon.

Prior to 1983, when this species was included on CITES Appendix I, and before 1986, when Bolivia prohibited the export of all wildlife, several hundreds of these macaws reputedly were trapped annually during the 1970s and early 1980s for export to the United States, Europe and Japan. Undoubtedly, this high level of trapping impacted adversely on the total population within its very restricted range, and this is confirmed by local reports of declining numbers. Pitter and Christiansen (1995) note that a trapper from near Vallegrande and along the Río Chico claimed that the local population of macaws had declined during the previous 12 years because of habitat loss and excessive trapping. A similar claim is reported by Lanning (1991), who was told by local residents that within a period of five to 10 years the number of macaws that could be seen 'on a good day' in one particular valley had fallen from 200 or more to just 30 or 40. For the most part the implementation of protective measures seems to have been effective in safeguarding local populations from poaching, but there have been claims that some illegal capture of both adults and nestlings is continuing (Pitter and Christiansen 1995). In 2005, there were reports of trappers arriving in trucks during the previous 10 years to 'harvest parrots', and the use of large nets baited with maize or corn enabled them to take up to 50 or 100 macaws in a single day (in Kyle 2005a). In the 12 month period between August 2004 and July 2005, 26 Red-fronted Macaws passed through the Los Pozos pet market, in Santa Cruz city, and it is suspected that this represents only a small proportion of the parrot trade in Bolivia (Herrera and Hennessey 2007).

Attesting to difficulties in determining the status of Red-fronted Macaws are the widely varying population estimates. After surveys were carried out between late December 1981 and early March 1982, Lanning (1991) estimated the total population to be 3000 to 5000 birds. During field studies undertaken in October–November 1990, in a 200 ha study site extending 2 km along the Río Caine valley, northern Potosí, the local population was estimated to be at least 60 birds, and all were feeding almost exclusively in a 30 ha area of cultivated land, at a density of two birds per hectare (Boussekey *et al.* 1991). Pitter and Christiansen (1995) report that Red-fronted Macaws seemed to be common in most parts of the range visited between September 1991 and March 1992 and, from extrapolating counts of 2.5 to 4 birds along a kilometre stretch of river to an assessment of feeding areas in all river valleys within the range, they estimated the total population to be between 2000 and 4000 birds. However, Collar (1997) warns that assessing the total population is difficult, and only 555 to 626 birds were counted in the main part of the range in 1991, suggesting that possibly less than 1000 macaws survive, and long-term prospects for the species are doubtful, with landclearance, persecution by farmers, and trapping for the live-bird trade identified as principal threats. An attempt to cover the entire range was made in a survey undertaken in 2011, when 130 pairs were counted, with 67 to 86 pairs breeding, and 545 non-breeding birds were aggregated at seven localities during the breeding season, so giving a total population of 805 birds (in Birdlife International 2016). This survey also revealed that five of 28 breeding locations known from the previous five years were unoccupied in 2011. Based on all survey counts, the total population is estimated at 1000–4000 birds, roughly equating to 670–2700 mature individuals, and an ongoing moderate decline is suspected because of habitat loss, persecution as a crop pest and continued illegal trade (Birdlife International 2016).

The Red-fronted Macaw is listed on Appendix I of the Convention on International Trade in Endangered Species of Wild Fauna and Flora (CITES).

HABITATS In semiarid intermontane river valleys, mostly between 1100 m and 2500 m, Red-fronted Macaws frequent temperate and subtropical thorn woodland, dry deciduous or semideciduous forest, and cacti-dominated scrubland with scattered *Schinus* and *Acacia* trees. Pitter and Christiansen (1995) point out that in areas where they and the Military Macaw *Ara militaris* both occur there is detectable habitat differentiation, with Red-fronted Macaws favouring arid habitats and the Military Macaw keeping to humid areas.

At approximately 2000 m, along the Río Caine in one of the semiarid intermontane valleys of the Cordillera Real, in northern Potosí, studies of these macaws were undertaken in a valley running roughly northwest to southeast for more than two kilometres along the river and with a width of about one kilometre; this valley was enclosed by an even relief, and was subdivided along its length by low ridges (Boussekey *et al.* 1991). Lying between the 300 mm and 600 mm annual isohyets, the valley supported xerophytic vegetation comprising mainly cacti, bromeliads, spiny bushes, the most widespread of which was caranda *Prosopis kuntzei*, and scattered trees, the most common of which was the evergreen false wild pepper *Schinus molle* with a maximum height of about 7 m. Apparently referring to this same area, identified as being around Sucusuma, northern Potosí, Pitter and Christiansen (1995) further describe the vegetation as being temperate and subtropical, mainly deciduous dry forest and woodland with patches of semideciduous forest in the river valleys, and common trees, mostly between 7 m and 12 m tall, included acacias, *Prosopis kuntzei* and *P. chilensis*, rosewood *Tipuana speciosa*, *Schinus molle*, kebrako *Aspidosperma quebracho-blanco*, *Loxopterygium grisebachii* and *Schinopsis haenkeana*, together with *Cereus* cacti and many other cacti, such as *Browningia caineana, Yungasocereus inquisivensis, Optunia cochabambensis, Harrisia tetracantha, Cleistocactus parviflorus* and *Neoraimondia herzogiana*. Logging for firewood and charcoal production, landclearing for agriculture, and grazing by cattle and goats resulted in a mosaic of agricultural lands, scrubby areas, and heavily degraded forest with a majority of trees less than 30 years old. Located some 90 km from Sucusuma was another study site in a valley along the Río Chico, a tributary of the Río Grande in northern Chuquisaca, where the vegetation also was temperate and subtropical woodland and dry forest with abundant cacti, and common trees again were acacias and *Prosopis* species, *Loxopterygium grisebachii, Schinopsis haenkeana, Schinus molle* and *Aspidosperma quebracho-blanco* (Christiansen and Pitter 1993a). Also at this site, many ravines running perpendicular to the river were filled with tall vegetation, while most on the eastern side of the river were surrounded by high cliffs, which contained crevices used by the macaws for nesting, but vegetation on mountain slopes bordering the narrow riverbed was heavily degraded by cutting for firewood, charcoal production and agriculture, with grazing by livestock, mainly goats, preventing regrowth.

HABITS Spectacular amidst their stark, dry surroundings, Red-fronted Macaws are noisy, sociable and fairly tame, regularly feeding to within 100 m of farm dwellings or following behind ploughs at distances of less than 30 m, and normally allowing an approach to within 20 m before taking flight. Much of what we know of their habits comes from surveys carried out in much of the known range between late December 1981 and early March 1982 (Lanning 1991), and from field studies undertaken in the vicinity of Sucusma, on the Río Caine, northern Potosí, in October–November 1990 and between late September and late October 1991 (Boussekey *et al.* 1991; Pitter and Christiansen

Plate 37
Red-fronted Macaw *Ara rubrogenys* (adults)

F. KNIGHT 14

1995), and some 90 km away, in an area along the Río Chico, 2 km to 25 km south of Puente Arce, northern Chuquisaca (Pitter and Christiansen 1995). A surprisingly high count of 58 sightings of single birds was recorded by Lanning during 305 hours of observation time in a wide section of the range, and other counts were 101 sightings of pairs and 45 sightings of flocks of from three to 12 birds, with the two largest flocks of 10 and 12 macaws being presumed roosting flights seen in the late afternoon and evening. In October–November 1990, in the Sucusuma district, single birds were seen rarely, comprising less than 4 per cent of all counts, while the mean number in a flying flock was 5.8 birds, in a perched flock was 4.8 birds, in a feeding flock was 13.9 birds, and in a drinking flock was 7.0 birds, to give a mean number of 7.2 birds in a flock for all activities. These counts show that flock sizes varied with activity. Perched birds were in small groups, with 30 per cent of observations being of pairs and 15 per cent being of two pairs, and flying birds normally were seen also in small flocks, with 42 per cent of observations being of a single pair and 17 per cent being of two pairs, although flights of flocks of up to 30 birds did occur, particularly towards the end of the day. Conversely, feeding clearly was a more sociable activity, with larger foraging flocks consistently being observed, especially in cultivated lands. Four major movements of birds occurred daily, commencing between 0530 hours and 0700 hours when macaws arrived at feeding areas from cliff-faces to the south and southeast, with return flights to these cliffs taking place between 0700 hours and 1000 hours. Between 1400 hours and 1730 hours, the macaws came back to feed a second time, and between 1700 hours and 1830 hours they again departed towards cliffs in the south and southeast. Between late September and late October 1991, in the same district, single birds rarely were seen, and again variable flock sizes were associated with different activities, with a mean flock size of 45 birds for nighttime roosting far exceeding the next highest mean flock size of 19 birds recorded for feeding. In flight, a mean flock size of five birds was recorded for departure from nighttime roosts, while for other flights the mean flock size was four birds, and for arrival at feeding areas the mean flock size was six birds. The mean flock size for drinking was five birds, and for resting in trees during the hot midday it was seven birds. Nighttime roosting was not monitored at the study site along the Río Chico, in northern Chuquisaca, where mean flock sizes for all activities except drinking were significantly lower than in the Sucusuma district.

In the Sucusuma district, between late September and late October 1991, nighttime roosting for the local population of approximately 100 macaws normally was in two 50 m to 200 m high vertical cliff-faces rising on both sides of the river, the birds sitting on ledges protected from rain and wind by overhanging rocks, but on two occasions small groups of macaws were found roosting in trees near the feeding areas. Numbers of macaws roosting in each cliff-face varied from night to night, and these roosting sites were used also by Blue-crowned Parakeets *Psittacara acuticaudatus*, Mitred Parakeets *P. mitratus*, Cliff Parakeets *Myiopsitta luchsi* and Turquoise-fronted Amazons *Amazona aestiva*. Departure of macaws from these roosts commenced after first light, but before the sun rose over the mountain tops, and birds continued to leave in groups over a period of 30 minutes. At Sucusuma, birds usually flew from nighttime roosts directly to the feeding grounds, but along the Río Chico they often first flew to trees in which they perched briefly or sometimes for up to several hours before continuing on to the feeding grounds, and while in these trees they generally were quite restless, moving about among the branches, flying from tree to tree while calling loudly, or preening, playing and squabbling. Trees selected for perching always were in exposed, prominent positions, such as on top of a ridge or on the riverbank, so offering full, uninterrupted views of surroundings, and in regularly used trees chewing and constant rubbing of bills resulted in a clearly visible wearing down of the branches with tips being broken off.

Feeding normally occurred between 0600 hours and 1000 hours and again from 1400 hours to 1900 hours, though along the Río Chico the two periods were not as well separated as at Sucusuma. After feeding, the macaws went to the riverbed to drink from small ponds or small, less turbulent streams, and at Sucusuma drinking appeared to be restricted to two periods, whereas along the Río Chico it occurred throughout the day. During the hot midday hours, the macaws rested amidst the foliage of taller shady trees, mainly *Schinopsis, Tipuana, Loxopterygium, Aspidosperma* and *Acacia* trees growing in quiet valleys or gorges near to the feeding areas. At Sucusuma, resting periods usually lasted from about 0900 hours to 1400 hours or 1600 hours, but along the Río Chico there was no defined resting period and birds could be seen resting at any time of the day. Resting birds sometimes interacted animatedly with each other, and on one occasion a flock of 15 macaws, perched in trees beside the river, was seen to indulge in playful antics, with birds fluttering from perch to perch, some with wings opened wide while hanging upside down from the thinnest branches, and others bickering noisily.

These macaws are strong fliers, and their normal flight speed has been estimated at 60 km per hour. In the Sucusuma district, northern Potosí, pairs were seen to show no obvious difficulty or to make no apparent adjustment to their line of flight when crossing the Río Caine against a very violent sandstorm (Boussekey *et al.* 1991). They can fly at very great height, especially when near cliffs, and then will descend with a series of long glides. Attesting to their strong flight is the description by Kyle (2005) of a spectacular aerial display:

> *......Not only did I find Red-fronted Macaw nests in the surrounding cliff faces, I also got to see pairs of Red-fronts put on a spectacular flying display. At one point in the afternoon I found myself in the mouth of a small canyon perched between two large, blood red sandstone outcrops. The structure of the canyon effectively created a wind tunnel by channelling strong gales through it. I watched in awe as several macaws faced the oncoming wind headfirst with effortless skill, hanging like kites over the canyon. At times they shot rapidly upwards, caught in a strong gust of wind, only to twist and dive downwards to resume their almost static hovering position above the canyon mouth. The narrow, pointed wings of Red-fronts are ideally suited to the windy conditions found in these desert-like valleys. I get the impression they are flying not to actually go somewhere but rather for the joy of flight itself.*

CALLS Given regularly in flight, while perched or at rest, and frequently while feeding, the most commonly heard vocalisation is a short, rather shrill and very ringing parakeet-like call, which is more high-pitched than would be expected from a large macaw (Boussekey *et al.* 1991). This same call is emitted also by juveniles, although less powerfully and more high-pitched. The frequency and intensity of this call increases with heightened social excitement, notably when perched birds are about to fly off or when about to alight on the ground. A lower, raucous *raa-aah*, closely resembling the call of other large macaws, is given when alarmed. On one occasion a perched pair was observed to call in duet for about 10 minutes, the notes being rather melodious, but continuous and monotonous, and accompanied by a light vibrato. Other harsh squealing and shrieking calls have been likened to the sounds of excited domestic pigs (Juniper and Parr 1998). Lanning (1991) points out that nestlings emit a monotonous *nyah* that is more high-pitched and nasal than the calls of adults.

DIET AND FEEDING Seeds and fruits of native plants are eaten when seasonally available, but cultivated peanuts or groundnuts and maize are major components of the diet, and flocks of Red-fronted Macaws can cause significant damage to crops of ripening maize. Lanning (1991) records 12 observations of macaws feeding on fruits of *Jatropha hieronymi*, while two birds were seen feeding on maize and one was observed nibbling on fruits on the top of a large *Cereus* cactus. Pitter and Christiansen (1995) report that along the Río Chico, northern Chuquisaca, during January and February, the birds fed mainly on maize and *Jatropha* fruits, with the latter continuing to be the chief food until May. From April to October, the main foods were *Loxopterygium, Aspidosperma, Tipuana* and *Schinopsis* seeds, with a new maize crop also becoming available at this time, and from October to January the birds fed mainly on fruits of mistol *Ziziphus mistol*. Among native fruits there seemed to be a preference for larger items, such as *Jatropha* fruits, over smaller *Celtis* or *Cnidoscolus* fruits, and the macaws seemed to concentrate on one abundant fruit for a certain period, only occasionally selecting other, less common foods. In *Jatropha* trees with abundant fruits there rarely were any aggressive interactions between feeding macaws, which often fed within 20 cm to 50 cm of each other. Also there was much variability in feeding intensity in these trees; at times the feeding was energetic and focused, with birds spending long periods in one part of the tree, but at other times they adopted a seemingly disinterested approach, moving about in the tree and dropping fruits to the ground or taking a long time to eat a single item. When all three seeds in each fruit were eaten, the mean handling time between detachment from the stem and consumption of the last seed was 51 seconds per fruit, and on one occasion 13 birds were seen to eat 89 fruits in 50 minutes, each bird on average eating seven fruits during the period. Also along the Río Chico, when feeding on maize, the macaws were seen to detach the entire cob or take seeds from an attached cob while balancing on the stalk, and on one occasion 12 birds ate 17 cobs in 36 minutes, spending an average 20 minutes at each cob.

In the Sucusuma district, northern Potosí, fruiting native trees seemed to be present only in the very few remnants of semideciduous vegetation in the valleys, and from late September to November, when field studies were undertaken, virtually no fruits were available in deciduous vegetation on the hillsides, so the macaws foraged almost exclusively in cultivated fields for peanuts and seeds of *Cenchrus* weeds. In October, at the start of the rainy season, peanuts are sown, and flocks of macaws follow behind the plough to pick up any nuts from last season that are turned up in the tilled soil (Boussekey *et al.* 1991). Germinating nuts also are dug up by the birds, and in June attempts are made to pull ripe nuts from the soil. In the period between harvest and sowing of a new crop, the macaws systematically search for any unharvested nuts on the surface of fallow fields. At least four hours of each day are spent walking with a characteristically graceless, heavy gait back and forth across fields, foraging in a systematic, uninterrupted manner, the long tail contributing to awkwardness and progress is impeded by numerous large clods of soil turned up by the plough. Close observations of a dozen foraging birds revealed that they took an average 100 steps per minute, each step being of about 5 cm, so representing a distance of 300 m covered in an hour. While walking, the birds picked up food items with chicken-like pecking actions, and digging or scratching actions averaged seven per minute. Most food items appeared to be small enough to be swallowed directly or nearly so, but larger items, such as a whole peanut, were picked up with the bill and then transferred to one foot to be held while being opened with the bill. Also in the Sucusuma district, birds occasionally were seen feeding on pods of *Prosopis chilensis* and *P. kuntzei*.

BREEDING Commencement of the breeding season in October–November apparently coincides with start of the rains. In the Río Caine valley, northern Potosí, during field studies undertaken in October–November 1990, copulation was observed on the mornings of 30 and 31 October; this was a very noisy performance, with the male vigorously tossing his head, quickly circling round the female, feeding her by regurgitation, and then mating with her on the ground, the two birds standing side-by-side (Boussekey *et al.* 1991). Also at this time, the presence of young birds estimated to be between six and eight months old suggested that hatching took place between March and May, with egg-laying being from February to April, though early clutches may have been laid in November–December.

Nests are in crevices or holes in cliff-faces or rarely in cavities in palm trees. Lanning (1991) notes that along the Río Mizque drainage system seven active nests were found in crevices in sheer or overhanging parts of cliffs composed of sandstone or conglomerates, and entrances faced northeast, east, southeast, southwest and west. These nests were between 2 m and 20 m down from the top of the cliff and between 10 m and 48 m up from the base. The two closest nests were 200 m apart on separate cliff-faces, and two other nests were 500 m apart on separate cliff-faces. Two active nests of Blue-fronted Amazons *Amazona aestiva* and six active nests of Mitred Parakeets *Psittacaraa mitrata* were on the same cliff-faces and within 100 m of nests of the macaws, while an active nest of Cliff Parakeets *Myiopsitta luchsi* was 30 m below a macaw nest. Christiansen and Pitter (1993a) report that along the Río Chico, in northern Chuquisaca, in a 15 m high cliff-face rising from the riverbed, an active nest was 12 m up in a crevice facing the river and, slanting slightly to one side, this crevice was 1.2 m long and 30 cm high.

In April 2011, some 30 km distant from the nearest cliff-nesting colony, in the Área Natural de Manejo Integrado El Palmer, Chuquisaca, three active nests were found in dead pasopaya palms *Parajubaea torallyi*, and five pairs of Red-fronted Macaws were observed (Rojas *et al.* 2014). At 2585 m, on a southwesterly oriented, densely forested steep slope, one nest was at a height of 8 m in a palm with a trunk diameter of 30 cm at a height of 1.8 m and, with a diameter of approximately 30 cm, the nest hole clearly had been enlarged by the birds. At 2696 m, on a west-southwesterly oriented, palm-dense steep slope, in dead palms with trunks approximately 32–35 cm in diameter, one nest was at a height of 14 m and another was at a height of 20 m. Another nest was in a dead palm in a stand of palms some 5 km from the two nests. Dead palms are scarce in this area, and regeneration is being restricted by human activities, so it is not expected that numbers of pairs nesting in palm trees would be high. Despite their protected status, these endangered palms are being cut down, so posing a threat to this population of macaws.

Along the Río Chico, prospecting for nest-sites and territorial behaviour towards other macaws as well as Mitred Parakeets and Cliff Parakeets were observed at least a month before the presumed start of incubation at the beginning of February (Christiansen and Pitter 1993a). Selection of a nest-site involved much cautious inspection of the cavity, with male and female taking turns to enter the hole, and together they sat at the entrance to demonstrate occupancy. They also flew repeatedly back and forth between the entrance and the top of the cliff, sitting on the cliff-face while repeatedly 'cooing' for a few minutes and alternately quietly looking around, eventually becoming less restless and sitting calmly at the entrance for longer periods. On 15 occasions during this pre-laying period, when other macaws were seen near a nest under observation, the occupying pair responded on four occasions by flying from a nearby tree to the nest entrance, while on another three occasions intruders were chased off by the male or by the pair, and once the pair came out to the cavity entrance, but on seven occasions no reaction was seen from the pair sitting at the nest entrance. On 29 occasions during the period of incubation, the arrival of other macaws near the nest elicited an aggressive response, mostly involving the male flying towards the intruder and supplanting it, sometimes gaping at it with opened bill, while on eight occasions the male or the pair appeared at the nest entrance in response to vocalisations from approaching macaws, but on 11 occasions there was no response from the pair to the presence of other macaws.

Incubation is undertaken only by the female, though at night the male often roosts with his sitting mate inside the nesting cavity. Along the Río Mizque drainage, Lanning recorded activity patterns at nests with incubating females, and at all three nests males roosted at night inside the cavity. Each pair copulated once during the day. At two nests, the sitting female left the nest seven times during the day, and at the other nest the female left four times, with periods of absence for all three females lasting three to 15 minutes, during which time each was fed by the male, usually while perched on a limb below and in sight of the nest. At these three nests, one or both parents were inside the cavity during 95 per cent of the three days, and both were in sight of the nest at all times. Christiansen and Pitter (1993a) report that at a nest in a cliff-face along the Río Chico, during four days of observation, the male was not seen to roost at night inside the cavity, but arrived shortly after sunrise, usually to feed the female and then they copulated. The female quickly returned to the nest and the male flew off to feed, being absent for periods of 34 to 59 minutes between 0630 hours and 0810 hours. The male then returned to again feed the female, and during the middle of the day he rested in the nest or in a nearby tree. In the afternoon, between 1400 hours and 1700 hours, the male foraged away from the nest for periods of 59 to 127 minutes before again returning to feed the female, and then at night he flew off to roost, sometimes accompanying other macaws. The female never left the nesting area, except for a few short flights of less than five minutes duration.

Along the Río Mizque drainage system, Lanning recorded parental activity during five days of observation at three nests containing chicks. Two pairs roosted at night in their nests, but the third pair arrived at the nest an hour after dawn and departed 20 minutes before dusk. One or both parents entered the nest two to six times during the day, each visit lasting between 14 and 189 minutes. Both parents fed the chicks, and at least two chicks were audible in each of two nests, with 'gurgling' sounds heard while they were being fed. One or both parents were inside the nesting cavity during 39 per cent of the five days, both were outside the nest but in the immediate vicinity during 6 per cent of the five days, and both were away from the area for the remaining 55 per cent of five days.

Christiansen and Pitter (1993b) report that in the vicinity of Sucusuma, along the Río Caine, in northern Potosí, and along the Río Chico, in northern Chuquisaca, juveniles remained with their parents during the first year, and were fed by both parents at any time of the day, although about half of the observed feedings occurred immediately or shortly after the parents had been feeding in the early morning or late afternoon. Juveniles were more active than adults, spending much more time chewing on bark, branches, twigs or leaves. When eight to 12 months old, juveniles were observed to forage efficiently. Of 26 observations of family groups made in October–November 1990, in the Río Caine valley, 24 were of pairs with a single youngster, and only single observations were of a pair with two and a pair with three offspring, so indicating that a majority of pairs rear only a single young bird each year, and this probably is attributable to a scarcity of available food resources (Boussekey *et al.* 1991).

Cuban Macaw

Ara tricolor Bechstein

EXTINCT

Ara tricolor Bechstein, in Latham's *Allgem. Uebers. Vög.*, **4**, Th. 1, 1811, p. 64, pl. 1. (South America, *ex* Levaillant, *Hist. Nat. Perroquets*, **1**, p. 13, pl. 5 = Cuba.)

DESCRIPTION Length 50 cm.
ADULTS General plumage red, paler and more scarlet on underparts; crown tinged yellow and merging into bright yellow on nape and hindneck; upper back brownish-red with dull green margins to feathers; lesser wing-coverts brown edged red; rump and lower back pale blue; outer webs of primaries and secondaries dark blue; ear-coverts and throat orange-red; undersides of tail and flight feathers brownish-red; tail above red tipped blue; bare face probably white; bill dark brown-grey, paler at tip; iris yellow; legs brownish.
2 females: wing 275 mm and 279 mm, tail 215 mm and 232 mm, exp. cul. 44 mm and 44 mm, tars. 27 mm and 30 mm.
7 unsexed: wing 276–290 (281.6) mm, tail 230–291 (269.0) mm, exp. cul. 42–46 (44.1) mm, tars. 26–30 (27.7) mm.
JUVENILES Undescribed.

DISTRIBUTION Formerly Cuba, probably including Isla de Pinos, and possibly Hispaniola; now extinct.

Plate 38
Cuban Macaw *Ara tricolor* (adults)

F KNIGHT '09

SUBSPECIES Claiming that Cuba was ornithologically unknown until 1822, so any species described before that date could not have come from Cuba, and that on Hispaniola there occurred a macaw that differed from the Cuban species only in having a white instead of ochraceous bare cere and face and a slightly smaller bill, Wetherbee (1985) transferred the name *Ara tricolor* from the extinct Cuban Macaw to an also extinct macaw from Hispaniola, and renamed the Cuban species *Ara cubensis.* I concur totally with Walters (1995), who rejects this rearrangement, pointing out that there are species from Cuba described as early as 1758 and differences in colour of the bare facial skin can be attributed to physiological factors, particularly the flushing associated with excitement or sexual activity. Also, I found no differences in bill size from comparisons between specimens of known provenance. Consequently, *Ara tricolor* is retained for the Cuban Macaw.

STATUS Noting that several dates have been proposed for extinction of the Cuban Macaw, including about 1850, about 1864, about 1884, about 1885, between 1864 and 1885, and the end of the 19th century, Wiley and Kirwan (2013) point out that the last reliable reports were those by Johannes Gundlach from Ciénaga de Zapata, western Cuba, in the early 1850s, by Father Antonio Perpeña in central and southwest Cuba in 1867, and possibly the secondhand record by Walter Zappey for about 1864 on Isla de Pinos. Barbour (1923) reported the collection of specimens by Gundlach:

> *Gundlach collected a number of birds from the last band which came regularly to feed in a small group of paraiso trees in the yard or batey of the colonia at Zarabanda. These trees are still standing, and I have talked with an aged planter who was with Gundlach when he shot his last pair.*

The secondhand record of the last pair known on Isla de Pinos being shot at San Francisco de la Vega, near the Ciénaga de Lanier, was reported as follows (in Bangs and Zappey 1905):

> *It has been supposed that perhaps the Cuban Macaw still lingered in the Isle of Pines. Unfortunately this is not so. The last pair known in the island was shot at La Vega, near the Ciénga, about the year 1864, and none have been seen since. This information was furnished by the man on whose plantation they were shot.*

Cuban Macaws undoubtedly featured in the widespread harvesting of animals for food, pets and inter-island trade that predated European arrival. Accounts from the earliest explorers strongly suggest that they were kept commonly as pets, and they were traded internationally almost from the time of the voyages of Christopher Columbus. They were given to the kings of Spain, and over the centuries thousands were exported to Europe. Many of the existing museum specimens had been captive birds in European 'menageries', and it is known that they were very popular as ornamental birds in Sweden in the 1700s (in Wiley and Kirwan 2013). In addition to being captured for trade, these macaws were hunted for food and for feathers to be used in ceremonial adornments, prompting several writers to suggest that hunting was the main, or even the only reason for their extinction. Referring to their being killed for food and taken as nestlings for pets, Greenway (1967) comments that there is no other known reason for their extirpation. The means of obtaining nestlings was particularly detrimental because it involved the felling of nesting trees. Nests were located by observing the adults and, when the chicks were well advanced, they were obtained by cutting down the palm tree. Not only would this have resulted in the death of some chicks, it deprived the birds of suitable nesting trees. Wiley and Kirwan point out that after many years of such selective destruction, the macaws would have been left with few, and possibly only suboptimal, nest sites. There is little doubt that habitat destruction also contributed to the extinction of these macaws, and there are early reports of their becoming scarce as native vegetation was cleared for cultivation (in Wiley and Kirwan 2013). It is likely that conflict with farmers occurred locally as they raided crops, especially corn, and this would have resulted in their persecution. As local populations declined to low numbers they would have been particularly vulnerable to such persecution and also to natural forces. Barbour (1923) reported being told that no macaws were seen in western Pinar del Río, western Cuba, after the great hurricane of 1844. Wiley and Kirwan comment that even if that storm did not eliminate the last macaws in western Cuba, the devastating hurricanes of October 1846 and August 1856 could have further decimated small remnant and fragmented areas of habitat so severely as to cause extirpation of some macaw populations.

HABITATS Wiley and Kirwan (2013) note that Cuban Macaws mostly were found in lowland savanna, including coastal regions, and interior forests at low elevations. They frequented open terrain with scattered trees, especially palm savanna. The habitat around Las Lomas de Rompe, southwestern Cuba, where Perpiña encountered macaws in 1867, formerly consisted of extensive rainforest-like gallery forests, but now only a few degraded stands of forest remain, the natural vegetation having been replaced by treeless marshy meadows and moist savannas. In western Cuba, the area around Guane, in Pinar del Río, where macaws were reported to have disappeared after a severe hurricane in 1844, the flatlands and gently rolling hill country originally were characterised by mixed palm-pine woodlands with a loose canopy layer, in which the palms *Colpothrinax wrightii* and *Acoelorraphe wrightii*, together with tropical pines *Pinus tropicalis* and Caribbean pines *P. caribaea*, were dominant (in Wiley and Kirwan 2013). On Isla de Pinos, in the area where the last known pair on the island was reported to have been shot, there has been little change since the mid 1800s, with mangroves at the southern fringe of the Ciénaga de Lanier, and a variety of seasonally inundated habitats, including semi-deciduous forest, marshy grasslands, hardwood and palm hummocks, and palm savanna in the interior (in Wiley and Kirwan 2013).

HABITS There is very little information on the habits of Cuban Macaws. They were said to have occurred in pairs and family groups, though the report from Barbour (1923) of birds being collected from 'the last band' suggests that small flocks may have assembled to feed. The apparent ease with which these macaws were captured or killed suggests that, like other Caribbean macaws, they were confiding, with little fear of humans, and this would have contributed to their demise.

CALLS Gundlach likened the loud call to that of the macaws of Central America (in Wiley and Kirwan 2013).

DIET AND FEEDING Recorded food items include fruits, seeds of introduced Chinaberry trees *Melia azedarach*, tender shoots and buds. Olson and Suárez (2008) suggest that the dorso-ventrally compressed skull, which in other macaws is an adaptation for the attachment of strong muscles, indicates that these macaws fed on extremely hard-shelled palm nuts, and their distribution may have been inextricably linked with this food source.

BREEDING Local people told Gundlach that nests were in cavities in palm trees, but nothing further is known of the breeding behaviour (in Wiley and Kirwan 2013).

SPECIMENS AVAILABLE Wiley and Kirwan (2013) document 19 specimens held in 15 museums or institutions in Cuba, Europe and the United States.

GENUS *Cyanopsitta* Bonaparte

Cyanopsitta Bonaparte, *Rev. Mag. Zool*, **6**, 1854, p. 149. Type, by monotypy *Sittace spixii* Wagler.

The sole small blue macaw belongs to this monotypic genus. It is a midsized parrot with a long, graduated tail. Only the lores to around the eyes is bare, and there are no feathered lines. There is no sexual dimorphism, and young birds resemble adults.

Although a predominantly blue plumage colouration is shared with the large *Anodorhynchus* macaws, molecular analyses indicate that *Cyanopsitta* is more closely allied to the similarly midsized *Primolius* and *Orthopsittaca* macaws (see Wright *et al.* 2008; Kirchman *et al.* 2012).

CRITICALLY ENDANGERED

Spix's Macaw

Cyanopsitta spixii (Wagler)

Sittace Spixii Wagler, *Abh. K. Bayer. Akad. Wiss., Math.-Phys. Kl.*, **1**, 1832, p. 675. Based on *Arara hyacinthinus* Spix, not of Latham, type from Joazeiro, Río São Francisco, Brazil.

DESCRIPTION Length 56 cm.
ADULTS General plumage blue, darker on back, wings and upperside of tail; forehead, cheeks and ear-coverts grey tinged blue, becoming greyish-blue on remainder of head and neck; slight greenish-blue tinge on breast and abdomen; underside of wings and underside of tail dark grey; bill grey-black; bare lores to around eyes grey; iris yellow; legs dark brownish-grey.
3 males: wing 261–299 (274.7) mm, tail 313–355 (335.0) mm, exp. cul. 31–33 (32.3) mm, tars. 23–26 (24.3) mm.
6 females: wing 247–286 (273.0) mm, tail 265–378 (322.8) mm, exp. cul. 31–35 (32.7) mm, tars. 22–27 (24.3) mm.
JUVENILES Like adults, but bare lores to around eyes pale grey; bill grey-black with broad white stripe along culmen; shorter tail; iris brown.

DISTRIBUTION In northeastern Brazil there are records from southernmost Maranhão and extreme northeastern Goiás east to northern Bahia and extreme southwestern Pernambuco. The last known wild population disappeared from the Rio São Francisco valley, northern Bahia, and Juniper and Yamashita (1991) suggest that, in historical times at least, Spix's Macaw has occurred only in this valley, and many, if not all records from elsewhere stem from escaped captive birds or more likely the misidentification of other species. Olmos (1993) reports that a pair of macaws fitting the description of this species was observed at Serra Branca, southeastern Piauí, in March–April 1975, and a villager from Angical, also in southeastern Piauí, claimed to have observed a pair of small, pale blue macaws in early 1991, but surveys undertaken in the same area proved fruitless.

STATUS Although Spix's Macaw was first described by the German naturalist Georg Marcgrave in 1638, when he was working in Pernambuco, northeastern Brazil, the first specimen was not collected until 1819, when another German naturalist, Johann Baptist von Spix, collected what he thought was a Hyacinth Macaw *Anodorhynchus hyacinthinus* in the Curaçá district, approximately 90 km northeast of Joazeiro, northernmost Bahia, and it was in this same district that the then last known wild bird, a male paired with a female Blue-winged Macaw *Primolius maracana*, disappeared in 2000. In

June 2016, a lone bird was seen and photographed in riparian woodland near Curaçá, but the origin of this bird is uncertain, and it may have been released from captivity in response to local awareness of increased anti-poaching patrols and warnings against trapping.

The decline of Spix's Macaw culminated with the capture of adults and removal of nestlings for the live-bird trade being responsible for extirpation of the only known small remnant population. There is a dearth of reliable field data, but it seems that almost from the time of its discovery Spix's Macaw suffered significant habitat degradation. An almost total loss of preferred habitat probably brought about a major decline in numbers, for in 1985, when only five birds were found in the Curaçá district, only three fragmented stands of degraded *Tabebuia* dominated gallery woodland totalling no more than 30 km² appeared to remain in the area, and this remnant was seriously threatened by overgrazing, which prevented almost all regeneration, and by harvesting for fuelwood (Juniper 1991; Collar 1997).

The plight of Spix's Macaw has been publicised very widely, even forming the storyline for two successful animated movies featuring Blu, a male smuggled to the United States, and taken to Rio de Janeiro to mate with a female. The history of trafficking in these rare parrots is indeed a sorry saga, and has been well documented by Juniper (2002). Trapping and export certainly was occurring in the 1960s, because in 1964 I saw and photographed a lone bird in the collection at Busch Gardens in Florida, USA. In the 1970s and 1980s, as numbers in the wild fell to extremely low levels, the population at Curaçá was specifically targeted by trappers and nest-robbers, and their activities are well documented by Juniper (2002). At least 23 birds, or possibly as many as 40, were taken from the population, and in 1984, by misleading a protective landholder, trappers were able to take seven adults and five nestlings. In 1985 another adult was shot, evidently with the intention of injuring it for capture, and in 1986 efforts to catch two of the three surviving birds at the nesting hollow brought about the failure of two breeding attempts (Roth *in litt.* 1987). By 1988 all birds had disappeared, apparently taken by trappers, but in 1990 a sole survivor was located, and for the next decade regularly was seen in the company of a Blue-winged Macaw. Tests carried out on discarded feathers collected under a roosting tree confirmed that the surviving bird was a male, so in 1995 a captive female was released in the area. Although this female met up with the male, a pair bond was not established, possibly because of a pre-established bond with the Blue-winged Macaw, and soonafter the released female disappeared. With disappearance of the male in 2000, the wild population was presumed to have been lost. After 2000, intensive searches were undertaken in the Curaçá district and, as part of a concerted public awareness campaign, posters were distributed asking local people to report any sightings of these macaws, so it seems unlikely that a bird would have remained undetected in the area until 2016. If the bird seen in June 2016 was released from captivity, its survival could augur well for success of any reintroduction of captive-bred birds into the wild.

The history of these macaws in captivity also is well documented by Juniper (2002). Successful breeding was enhanced by initial cooperation between holders of captive pairs and, when necessary, birds were moved from one breeding facility to another to achieve suitable pairings. In 2000, the total number of publicly declared birds in captivity was 60, of which 54 were captive-bred, and in 2012 the official captive population totalled 80 birds, with about another 13 individuals in private ownership (Birdlife International 2004, 2016). Under the auspices of the Brazilian government a Permanent Committee for the Recovery of Spix's Macaw was established and cooperation between the holders of birds resulted in annual increases in the captive population, with artificial insemination of females contributing to these successes. The Committee was succeeded by the Working Group for the recovery of Spix's Macaw, which is responsible for coordinating the captive breeding program, and the objectives are to set up on-site breeding and reintroduction facilities in areas in Bahia that have been purchased for this purpose and where the natural vegetation will be restored after the removal of livestock (in Birdlife International 2016).

Spix's Macaw is listed on Appendix I of the Convention on the Conservation of Endangered Species of Wild fauna and Flora (CITES).

HABITATS In the Curaçá district, northernmost Bahia, where the last wild birds were observed along Melância Creek, the dominant vegetation community extending up to 100 km either side of the Rio São Francisco was dry caatinga scrubland characterised by a predominance of Euphorbiaceae, but Spix's Macaw strongly favoured gallery woodland of caraiba trees *Tabebuia caraiba* bordering seasonal watercourses (Roth *in litt.* 1987). Juniper and Yamashita (1991) note that evidence which they gathered revealed the former presence of these macaws only in remnant patches of *T. caraiba* woodlands and, during a week-long period in July 1990, the lone wild bird consistently was seen in the vicinity of Melância Creek, where caraiba trees grew. Dead crowns of these trees were strongly favoured by the bird, and a preference was shown for particular perches. It was suggested that these observations, together with the discovery of a second site in very similar habitat, where birds had occurred recently, indicate that caraiba gallery woodlands are a critical habitat because of the availability of nesting sites. It was reported also that during a survey undertaken in June–July 1990, only three good patches of these gallery woodlands were found, and there was no regeneration. Two factors were identified as seeming to be important determinants of the distribution of caraiba gallery woodlands. The presence of seasonally inundated watercourses above a certain size appeared crucial, and mature caraiba trees were found only in the vicinity of watercourses with a width exceeding 8 m and, in ten 100 m sections of a creek, the maximum distance of trees from the nearest channel was only 17.6 m. Trees which did not grow in the watercourse or on the bank were found only in seasonally inundated areas. Also of importance was the presence of alluvial deposits, and all caraiba gallery woodlands were recorded in the middle and lower levels of creek systems, their absence in upper, rockier reaches perhaps due to unfavourable regeneration conditions.

That other habitats may have been visited, at least seasonally, is suggested by Sick (1993), who points out that in December 1974 sightings of these macaws were reported in the vicinity of the Rio Riachão, northwestern Bahia, where the dominant vegetation community was grassland savanna with scattered trees, including stands of *Mauritia* palms, but the reliability of these reports have been questioned.

Plate 39
Spix's Macaw *Cyanopsitta spixii*
LEFT adult RIGHT juvenile IN FLIGHT adult

F. KNIGHT 2010

MOVEMENTS Although many reports of this species from other areas or different habitats are most likely to be misidentifications of other small macaws, there are indications of local seasonal movements being undertaken. Sick (1993) notes that stands of *Mauritia* palms were visited during local movements. Collar (1997) points out that some wandering in response to rainfall seems to have taken place, with birds being absent from their known breeding area between December 1986 and March 1987, then reappearing when the rains commenced, and the last surviving wild bird sometimes disappeared for days or even weeks from its favoured area.

HABITS Spix's Macaws apparently were strongly traditional in their use of roosting and nesting trees, although local movements, possibly by young birds or unpaired adults without territories, could have been responsible for irregular occurrences in some districts. Roth tells me that in 1986 the three surviving birds occupied a restricted area along Melância Creek, where regularly they perched in the same few *Tabebuia* trees, invariably on the topmost branches, and local persons claimed that one of the two trees in which nesting was attempted in that year had been a breeding site for this species for more than 50 years (*in litt.* 1987). These macaws usually were seen in pairs, though in the past, when more numerous, they occurred also in parties of up to 15 or more. They were more timid than other macaws, and when approached, even from some distance, they emitted loud alarm calls and immediately flew off. Roth observed interactions with Blue-winged Macaws, which became very noisy when Spix's Macaws alighted in the same tree, and at times Blue-winged Macaws were displaced from the topmost branches. Nests of the two species have been found in the same tree, though hollows occupied by Spix's Macaws were noticeably larger.

The proportionately long tail produced a distinctive silhouette when these macaws were flying or perched on topmost, often leafless branches, and rather slow, flapping wingbeats characterised their flight.

CALLS Juniper (1991) recalls that when located in the company of Blue-winged Macaws, the last surviving Spix's Macaw immediately was identified by its resonant *kraa…kraa…kraa* call. Some screeching notes also were given and, although less powerful than the calls of other macaws, all notes were audible at some distance (Juniper and Parr 1998).

DIET AND FEEDING Food comprised fruits and seeds, with seeds of the very common favela *Cnidoscolus phyllacanthus* and pinhão *Jatropha pohliana* being particularly favoured. *Melanoxylon* seeds or fruits, and fruits of joazeiro *Zizyphus joazeiro* and *Maytenus rigida* also featured in the diet, while fruits of the large facheiro cacti *Cereus squamosus* may have been eaten. *Syagrus* palms were another possible food source, though it is doubtful whether the bill of this macaw was sufficiently strong to open the hard nuts (Juniper and Parr 1998). The pair of 'small, pale blue macaws' observed in the Angical district, southeastern Piauí, in early 1991, were feeding on *Hymenaea* fruits (in Olmos 1993).

BREEDING The breeding season was from November through to March, with some variation being brought about by climatic conditions (Roth *in litt.* 1987). Nestlings were removed by trappers in February. The nest was in a hollow limb or hole in a tree, normally in a caraiba tree *Tabebuia caraiba*, and the clutch comprised two or three eggs.

Virtually nothing is known of the nesting behaviour. Roth reports that in 1986 the surviving pair nested in a hollow in a *Tabebuia* tree traditionally used by the species, but attempts by trappers to capture these adults brought about abandonment of the site. A second attempt was made in another *Tabebuia* tree, approximately 4 km to 5 km away, but again repeated interference forced the pair to desert this second site and no further attempts were made. One of the eggs laid by this mixed pair contained a very early embryo, which subsequent DNA analysis revealed to be a hybrid (in Birdlife International 2016).

EGGS A single egg in the Natural History Museum at Tring, UK, measures 34.9 × 28.7 mm (Harrison and Holyoak 1970).

GENUS ***Rhynchopsitta*** Bonaparte

Rhynchopsitta Bonaparte, *Rev. et Mag. Zool.* (2), **6**, 1854, p. 149. Type, by monotypy, *Macrocercus pachyrhynchus* Swainson.

Two very closely allied, allopatric species belonging to this genus are large green, macaw-like parrots with long wings and a short, slightly graduated tail. The face is feathered, but with a conspicuous bare eyering, and the laterally compressed, strong bill is used to extract seeds from *Pinus* cones. There is no sexual dimorphism, and juveniles are duller than adults. A third species – *Rhynchopsitta phillipsi* – is extinct, and was described from Late Pleistocene cave deposits from Nuevo León, northeastern Mexico.

ENDANGERED

Thick-billed Parrot

Rhynchopsitta pachyrhyncha (Swainson)

Macrocercus pachyrhynchus Swainson, *Philos. Mag.* (n. s.), **1**, 1827, p. 439. (Tableland of Mexico).

DESCRIPTION Length 38 cm. Weight 300–350 g.
ADULTS General plumage bright green, yellowish on cheeks and ear-coverts; forehead, forecrown and broad superciliary stripe red; brownish marking in front of bare eyering; bend of wing, carpal edge and thighs red; greater underwing-coverts yellow; tail above green, below dusky grey; bill black; bare eyering brownish-yellow; iris orange-yellow; legs grey.
10 males: wing 254–276 (267.1) mm, tail 162–177 (170.4) mm, exp. cul. 36–41 (38.7) mm, tars. 21–24 (22.6) mm.
10 females: wing 254–265 (259.0) mm, tail 160–178 (169.4) mm, exp. cul. 37–39 (38.1) mm, tars. 22–24 (22.7) mm.
JUVENILES No red superciliary stripe; bend of wing and carpal edge green; less red on thighs; bill horn-coloured with grey at base of upper mandible; iris brown.

DISTRIBUTION Highlands of northwestern and central Mexico, in the Sierra Madre Occidental from western Chihuahua and eastern Sonora, or extralimitally Sinaloa, south over the central plateau to Michoacán or occasionally Jalisco; formerly farther south to Veracruz, southern Mexico, and north to central Arizona and southwestern New Mexico, USA. In 1986, as part of a failed reintroduction program, birds were released in southern Arizona.

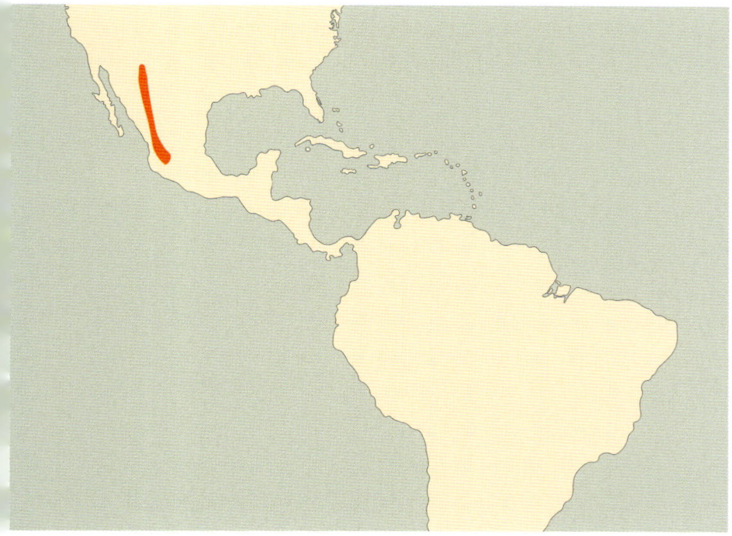

GENERAL NOTES I find the Thick-billed Parrot to be particularly appealing, and it always has been one of my favourite species. I recall how pleasing it was in 1988 to accompany Helen Snyder to pine forests at Mongollon Rim, in Arizona, where we observed a male from the reintroduced population perched watchfully near one of the two known nests. Regrettably, this nest was not successful.

STATUS Former occurrences of Thick-billed Parrots as far north as northern Arizona are indicated by references to parrots in an account of the Espejo Expedition of 1582–1583 and, to a lesser extent, by bones of four birds dating back to 1250 and recovered from Indian sites in the San Francisco Mountains (Wetmore 1931; Hargrave 1939). Because of an absence of breeding records and only irregular reports of large flocks, the parrots traditionally have been considered to be sporadic non-breeding, mainly winter visitors to Arizona and New Mexico, but a reassessment of records, particularly those claiming that some birds always could be found in the Chiricahua Mountains, southeastern Arizona, suggests that the species very likely did breed in the southwestern United States (Snyder *et al.* 1999). In this context, it should be noted that at present the northernmost nesting area in northern Mexico is only 50 km south of the border with New Mexico. Irruptions did occur, and historical sightings of large flocks at irregular intervals certainly represented incursions from northern Mexico, apparently appearing at times of drought and probably resulting from food shortages. In August 1904, a large flock, estimated to contain between 700 and 1000 birds, was observed by miners at Bonita Park, in the Chiricahua Mountains (Smith 1907). Wetmore (1935) documented another large-scale invasion that took place in 1917–18, commencing on 20 August 1917 with the appearance of a party of six or eight birds in Pinery Canyon, again in the Chiricahua Mountains, and numbers increased steadily until 1 September, when about 300 birds were present. Some of these birds disappeared in autumn as cold weather came on, but others remained throughout the entire winter, although at one time snow covered the ground for more than two weeks and the parrots were forced to seek food on the ground wherever the snow had been partly blown away. The last record from Pinery Creek was on 26 or 27 March 1918, when a flock of 10 or 12 was seen. To the south, in Rucker Canyon, the parrots were more abundant, being first recorded during the first week of July 1917, when a flock of 50 or 60 was seen, and numbers increased steadily until by early autumn there were 1000 or possibly 1500 birds present. By November numbers had decreased, but a few remained throughout the winter until March 1918, though they wandered much and occasionally were not observed for a week at a time, During the same period, that is July 1917 to March 1918, parrots were observed in neighbouring mountain ranges, including Animas Peak in southwestern New Mexico. In the Dragoon Mountains, a considerable flock arrived in Cochise Stronghold Canyon at the end of July and stayed for about six weeks. These birds were seen to fly directly out across the flats to the east each morning and return at night, so it is possible that they may have crossed each day to the Chiricahua Mountains to feed.

Extensive logging of pine forests in Sonora and Chihuahua, northern Mexico, and a resulting barrier preventing northward movements of the parrots have been identified as causes of the disappearance of the species from the southwestern United States, but persistent reports of widespread shooting suggest that excessive persecution may have brought about extirpation of the parrots from north of the Mexican border (Snyder *et al.* 1999). The last credible sightings in this northern sector of the range were of a flock seen in 1938 at the Chiricahua Monument, southeastern Arizona, and a small flock seen in 1964 in the Animas Mountains, southwestern New Mexico. Between September 1986 and September 1993, a determined effort was made to reintroduce Thick-billed Parrots to the southwestern United States, with 88 birds, in groups of six to 16, being released in the Chiricahua Mountains, southeastern Arizona (Snyder *et al.* 1994). Of the released birds, 65 had been obtained mainly as confiscations by the U.S. Fish and Wildlife Service and were presumed to have been trapped from the wild in Mexico. The remaining 23 birds were captive-bred, with 16 having been reared by the parents and seven having been reared by hand. Although parent-reared captive-bred birds fared better than hand-reared birds and young captive-bred birds fared better than older captive-bred birds, all captive-bred birds had poor survival rates due mainly to foraging deficiencies and an inability to withstand predation, especially by raptors. Conversely, birds obtained as wild-caught adults survived well and showed a readiness to breed, being involved in one confirmed successful nesting and several other attempts. In August 1989, in central Arizona, I observed a pair at their nest, but that breeding attempt failed when both eggs and the sitting female were taken by predators. Although these experimental releases did not result in establishment of a resident population, initial results were encouraging, with sightings of pale-billed juveniles confirming successful breeding by at least one pair, and it has been demonstrated that the only birds with good potential for re-establishment are translocated wild-caught adults. I suspect that a lone bird first observed at Armendaris Ranch, southwestern New Mexico, in May 2003, was a long-term survivor from the reintroduction program, though the possibility of its being a naturally occurring vagrant from neighbouring northern Mexico cannot be discounted.

There can be little doubt that a very significant decline in Mexican populations has resulted from extensive logging of high-elevation old-growth forests, and breeding pairs now are confined to undisturbed remote areas, disturbed areas with a few remaining suitable nesting sites, and isolated stands of forest on steep slopes or ridges that are inaccessible or cannot be logged economically (Snyder *et al.* 2000). It has been estimated that by 1995 nearly 99.5 per cent of *Pinus-Quercus* forest in the Sierra Madre Occidental had been at least partially logged, and only 22 km^2 of high-elevation old-growth forest, the preferred nesting habitat, remained intact (in Snyder *et al.* 1999). In some districts, the disappearance or near disappearance of these parrots and loss of the Imperial Woodpecker *Campephilus imperialis*, a sympatric species also closely associated with highland pine-oak forests, coincided with the almost complete logging of local pine forests. Surveys undertaken in the mid 1990s indicated that Thick-billed Parrots could survive in partially degraded forests, but in much lower densities than are found in undisturbed old-growth or near old-growth forests, and the loss of nesting sites in mature trees or snags poses a major threat to long-term survival of the species. Because they do not 'talk', Thick-billed Parrots rarely are kept as pets in Mexico, but increasing demands in other countries resulted in an upsurge in trapping for the export market, and this reached a peak in the mid 1980s, when there were large numbers of confiscations in the United States (in Snyder *et al.* 2000). The present population is estimated at 2000–2800 mature birds, though this may be an over-estimate, and there may be fewer than 100 active nests each year (in Birdlife International 2016).

A fundamental goal in any conservation program should be to increase the number of nesting areas (Monterrubio-Rico *et al.* 2006). This can be achieved by protecting stands of old-growth forest in all current and historical nesting areas, especially those at high elevations in Chihuahua and Durango. It may be some decades before trees large enough to support suitable nesting cavities become available, so to augment the availability of nesting sites it is recommended that trials with nestboxes be undertaken to ascertain whether these will be used by the parrots.

Thick-billed Parrots are legally protected in Mexico and the United States, and in Mexico efforts are being made to redress the inadequacy of habitat protection, with some key nesting areas now protected. Conservation efforts were advanced significantly in January 2000, when, after more than two years of delicate negotiations, agreement was reached with local communities to protect approximately 2000 ha of relatively undisturbed forest in the Bisaloachic-Cebadillas district, Chihuahua, which holds more than 100 nests possibly representing about 10 per cent of the total breeding population. In return for a 15 year moratorium on logging in the area, several national and international organisations will offset community income losses, initiate a sustainable forest management strategy, and develop ecotourism enterprises for the region.

The Thick-billed Parrot is listed on Appendix I of the Convention on International Trade in Endangered Species of Wild Fauna and Flora (CITES).

HABITATS Because of their strong dietary preference for *Pinus* seeds, Thick-billed Parrots are closely associated with highland conifer forests (Snyder *et al.* 1999). Mostly between 1200 m and 3500 m, they frequent mature *Pinus* and *Abies* forests or forests dominated by Douglas fir *Pseudotsuga menziesii* and mixed *Pinus-Quercus* forests or open woodlands on plateau-like tops of mountains or in valleys below high cliffs and rocky outcrops. At these high elevations they are well able to withstand subfreezing overnight temperatures in winter, often roosting in the snow zone and sometimes feeding during snowstorms or obtaining water by eating snow. Northern limits to the historical range may have been determined in large part by the presence of Chihuahua pines *Pinus leiophylla*, which are a reliable food source and tend to retain seeds throughout the year. Nesting occurs almost exclusively above 2000 m, and usually above 2400 m, possibly because of the elevated distribution of favoured food trees, including Mexican white pine *Pinus ayacahuite*, Arizona pine *P. arizonica* and Durango pine *P. durangensis*, or to avoid nest predation by arboreal snakes. There are records from lowland deciduous forests, and birds seen in these areas during the breeding season probably are non-breeders. Collar (1997) mentions a record of birds roosting in tropical lowland deciduous forest, but making daily flights to feed in highland pines.

MOVEMENTS Although the evidence is mostly indirect, regular seasonal migrations over long distances apparently are undertaken by some populations in Mexico (Snyder *et al.* 1999). Birds are said to be absent from Chihuahua and Sonora in winter, but are present throughout the year in some parts of Durango. There is a report of parrots at Cerro de Tancítaro, Michoacán, in mid July, which is well into the normal breeding season, but nesting has not been recorded from that state or from neighbouring Jalisco and Colima, so the region may be the wintering range for birds that breed in Chihuahua, Sonora and Durango. Although distances travelled are highly speculative, the usual absence of birds from Chihuahua and Sonora in winter suggests that some populations may migrate up to 1000 km between breeding and wintering ranges. In Chihuahua and Sonora, pairs return to their nesting sites and start inspecting hollows by late April or early May (Lanning and Shifflet 1983). Irregular nomadic movements in response to changes in food availability also are undertaken, and at times good numbers of parrots are observed well outside their normal range. Marshall (1957) reported that in the 1950s, during visits to the Sierra Madre Occidental in northwestern Chihuahua, each of a few weeks duration, he saw pairs, small parties and a couple of flocks of up to 50 or 60 birds, and it was apparent that there were local movements and fluctuations in numbers; several days could pass without any birds being seen, and only twice were they seen in the Sierra Madre Occidental in 1955, though flocks had been present a few years earlier, and in 1954 they were numerous in the Sierra Huanchinera, where none had been seen in the previous year.

A regular seasonal migration pattern adopted by the first released birds in southeastern Arizona followed a southeast to northwest orientation from the Chiricahua Mountains to the Mongollon Rim district, so paralleling the orientation of the Sierra Madre Occidental in Mexico and suggesting that seasonal migrations of at least some populations in Mexico normally may tend to follow mountain ranges (Snyder *et al.* 1994). These released birds generally remained in their wintering grounds in the Chiricahua Mountains throughout most or all of May and headed northwest during the first two weeks of June, with return to the Chiricahua Mountains taking place immediately after the end of breeding in September to mid October. Interruption of this pattern occurred in 1989, when birds left their breeding areas by early September after food shortages had caused the failure of nests, but little food was available in the Chiricahua Mountains so the parrots returned to the Mongollon Rim district and subsequently migrated back to the Chiricahua Mountains.

Birds were seen to depart on their spring migration flights in the early morning, and the 320 km distance between wintering and breeding areas apparently was covered in one day, with a direct southeast to northwest route being followed. Although not shown conclusively, migrating birds appeared to follow topographic features, especially mountain ranges, and juveniles learned migratory behaviour by following their parents (Snyder *et al.* 1999).

HABITS Thick-billed Parrots are highly social, and lone birds seldom are encountered (Snyder *et al.* 1999). They normally are seen in flocks comprising pairs or family groups of up to five birds, and paired birds stay close together, so making them readily discernible within the flocks. Quite large numbers often occupy traditional communal nighttime roosts in the tops of thickly foliaged trees, in tree cavities or in crevices in cliff-faces, usually at high elevations, though in the Cerro de Tancítaro area, Michoacán, western Mexico, birds were found roosting in tropical deciduous forest and flying each morning to feed in upland pine forests (in Blake and Hanson 1942). It is suspected that close clustering and physical contact within dense foliage enables roosting birds to withstand very cold overnight temperatures. If disturbed by predators during the night, roosting parrots rise into the air, calling loudly, and sometimes fly long distances in darkness before resettling. After such disturbances the parrots have not been observed to reuse the abandoned roosts. Early morning departure from the nighttime roost is by smaller flocks, which disperse widely, often travelling several kilometres to the feeding grounds. They normally remain in these tight flocks while feeding, methodically working their way down from the top of each pine tree. In the non-breeding season, a flock often feeds throughout the day in a patch of 10 or 12 trees, flying from one tree to the next after working only partway down from the top in each, and by staying in the uppermost branches the birds maximise their chances of escaping from raptors (in Snyder *et al.* 1999). In southeastern Arizona, the daily activities of a non-breeding flock were monitored for 13 days, and while at the feeding grounds this flock averaged 60 per cent of the time in foraging activities. With the onset of warmer weather in spring, the birds took to resting for longer periods at midday, and then resumed intensive feeding in the mid to late afternoon. Just prior to flying back to the roost, the parrots normally drank at a hillside spring, and occasionally they came to this same spring in the early morning. Observations revealed also that in the early morning the parrots returned to resume feeding in trees in which they fed the previous day, and usually to the tree in which they last fed before going to roost.

On warm, sunny days, at midday and again in the late afternoon, Thick-billed Parrots have been observed bathing at regular drinking sites atop waterfalls. While at rest, paired birds will preen each other, especially on the head and neck. While a flock is feeding, some birds commonly, though not always, act as sentinels, remaining alert while intently scanning surrounds from vantage perches, and regular changeovers occur as other individuals cease feeding to take over lookout duties. In response to danger, especially from approaching aerial or terrestrial predators, sharp alarm calls given by the sentinels cause the entire flock to immediately take flight.

Predation by raptors has been identified as the probable primary cause of mortality in free-flying birds, with the Red-tailed Hawk *Buteo jamaicensis* and the Northern Goshawk *Accipiter gentilis* being the principal predators (Snyder *et al.* 1999). These raptors rely on surprise or altitude advantage to succeed, for they normally fail to overtake parrots in full flight. Flying flocks attacked from above normally dive down towards the ground in rapidly accelerating flight, and individuals being chased by falcons sometimes take cover in the dense foliage of trees. Several instances of nocturnal predation by Ring-tailed Cats *Bassariscus astutus* on roosting parrots have been documented, and these cats also take incubating females in the nesting hollows. The strong flocking behaviour apparently serves as a defence against aerial predation and, in southeastern Arizona, a failure of released birds to form cohesive foraging flocks resulted in unsustainable levels of predation by Northern Goshawks (Snyder *et al.* 1994).

The fast, powerful and direct flight is characterised by regular, strong wingbeats. Flying flocks normally adopt a V-formation, like that commonly associated with waterfowl, or a stretched-out straight-line formation (Snyder *et al.* 1999). In southeastern Arizona, a radio-tracked group was followed by airplane on a non-stop flight of 100 km, and for seasonal migration from the Chiricahua Mountains an average flight speed of 50 km per hour would enable the parrots to reach their northern wintering grounds in a flight of approximately 6.5 hours.

CALLS Sounding like human laughter and audible at distance of up to 3 km, the flight call is a raucous, rollicking *kah-ha-ha-ha-ha-ha* repeated at intervals of five to 10 seconds. Also, a sharp *kuk-kuk-kuk-kuk* may be given by several parrots simultaneously to produce a noisy 'cackling' effect. Similar loud calls emitted more intermittently by feeding birds apparently are intended to persuade the flock to move to a new location (Snyder *et al.* 1999). Birds that have ceased feeding commence to call, are soon joined by others until the increased vocalisations reach a crescendo, and then the entire flock suddenly takes to the air. A 'gobbling' call is given by birds jostling for perching positions at a nighttime roost, defending a nest against other Thick-billed Parrots, or in other quarrelsome interactions. The alarm call is a sharp, shrill screech.

In July 2006, contact calls of flying and perched birds were recorded at the three main breeding sites in western Chihuahua, northern Mexico, to determine if geographical variation in vocalisation could constrain conservation strategies, such as translocation or reintroduction of populations (Guerra *et al.* 2008). Significant variation in vocalisation between the three sites was not detected, but there were significant differences among individuals within the sites. It is possible that geographic variation exists, or existed previously before range contraction, in the calls of these parrots, but on a larger scale than the one that was sampled, but this hypothesis is difficult to address, as the three sites sampled are the only known major breeding sites. The three sites are at an average distance of 120 km from each other, which is as far or farther apart than sites at which geographical variation in vocalisation has been detected in four other parrot species, and this suggests that any vocal variation among populations of Thick-billed Parrots occurs on a larger geographic scale than documented in any other parrot species to date.

DIET AND FEEDING Thick-billed Parrots feed primarily on *Pinus* seeds, and their plumage often becomes matted with resin. In western Mexico, the favoured food trees are Mexican white pine *Pinus ayacahuite*, Arizona pine *P. arizonica* and Durango pine *P. durangensis*, all of which are particularly important food sources during the breeding season, together with Apache pine *P. englemanni*, ocote *P. teocote* and weeping pine *P. lumholzi* (in Snyder *et al.* 1999). Stager (1954) reported that on the Barranca de Cobre, in the Sierra Madre Occidental, Chihuahua, parrots were seen eating terminal buds of Chihuahua pine *Pinus leiophylla* and weeping pine. In the Cerro de Tancítaro area, Michoacán, fruits of a local cherry tree *Prunus capuli* were eaten, and the crop and stomach of one parrot contained seeds of an unidentified leguminous plant (in Blake and Hanson 1942). In Chihuahua, parrots occasionally have been observed raiding acorn granaries of Acorn Woodpeckers *Melanerpes formicivorus*, and there are unconfirmed reports of their feeding at clay-licks (in Snyder *et al.* 1999).Wetmore (1935) noted that during the large-scale invasion into the Chiricahua Mountains, southeastern Arizona, in 1917–1918, the parrots immediately attacked the cones of Chihuahua pines and continued to eat the seeds until the entire crop had been consumed. There was some feeding on the seeds of yellow pine *Pinus brachyptera*, but this was rare. When the supply of pine seeds was exhausted, the parrots ate acorns, and these were the staple food during autumn and winter; at first they fed in the trees, but later descended to the ground to pick up fallen nuts. For birds released in southeastern Arizona between 1986 and 1993, the important food trees were Chihuahua pine, Arizona pine and ponderosa pine *Pinus ponderosa*, with Chihuahua pine being so dominant as to suggest that Thick-billed Parrots may experience difficulties in maintaining year-round populations north of its range (Snyder *et al.* 1994). The lone bird observed at Armendaris Ranch, southwestern New Mexico, in May 2003, was feeding on seeds of non-native mondell pines *Pinus eldarica* planted as a windbreak, and availability of these seeds at a time when native conifers were not producing cones may have been responsible for the bird remaining at this site.

With the shovel-like flange at the tip of the lower mandible, these parrots deftly extract seeds from *Pinus* cones. After being nipped free from the branch, a cone normally is held by the apex and, while either held up freely or steadied against a branch, it is rotated while in a long upward, spiralling motion the scales are prised off and the seeds extracted (in Snyder *et al.* 1999). Seeds usually are swallowed whole and the shredded cone dropped. Depending on the species of pine and size of the cone, processing of a single cone can take up to 20 minutes, with cones containing only a few large seeds, such as those of border piñon pine *Pinus discolor* and Mexican piñon pine, being processed quickly, but extracting seeds from large cones of Apache pine and Mexican white pine is especially difficult. These large cones usually are left attached to the branch while the parrots manoeuvre around them, often hanging upside down from a branch or the base of the cone to get at the seeds. If one of these large cones is removed from a branch, a parrot often is unable to hold on to it and so it falls to the ground intact. In southeastern Arizona, the average intake of food for a flock of released parrots was 19.5 g of seeds per bird per day, and this represented about 2080 seeds from an average of 109 cones for each bird per day (in Snyder *et al.* 1999).

In Chihuahua, northwestern Mexico, during 1996–1997, analyses were made of 102 crop samples from 64 nestlings in 35 nests, and seeds of Mexican white pine were found to be particularly important in the diet, being present in 63.7 per cent of samples and comprising 36.2 per cent of the diet by weight (in Snyder *et al.* 1999). This species is a consistent cone producer, with high energy levels from its large seeds possibly enhancing nesting success. Indeed, the presence of these high-altitude pines could be an important factor in determining nesting distribution. Separately or together, seeds of Arizona pine and Durango pine, also high-altitude species, were present in 70.6 per cent of crop samples and comprised 36.2 per cent of the diet by weight, while seeds of Mexican piñon pine were present in 21.6 per cent of samples and comprised 10.3 per cent of the diet by weight. *Quercus* acorns were present in 5.9 per cent of samples and comprised 4.2 per cent of the diet by weight. Although the dietary function of tree bark is unknown, its presence in 75.5 per cent of samples suggests that its intake was not accidental, and insect larvae, which were present in 18.6 per cent of samples, obviously were a regular, though minor, component of the diet.

BREEDING As the timing of nesting coincides with peak availability of favoured *Pinus* seeds there can be some variability from year to year, but usually it is from early June to early November. Some months prior to egg-laying, pairs can be observed inspecting potential nesting sites or enlarging hollows, and these activities can be at locations well away from where nesting eventually takes place (in Snyder *et al.* 1999). Lanning and Shiflett (1983) report the findings of a field study of the nesting ecology of Thick-billed Parrots carried out during April to October 1979, between 2300 m and 3070 m, in the Sierra Madre Occidental, from northwestern Chihuahua south to central-western Durango, northwestern Mexico, and all 55 monitored nests were in cavities in living or dead standing trees, with 58 per cent of hollows being in dead *Pinus* snags. Nesting densities were variable, and apparently were determined by the availability of suitable sites. On two occasions two active nests were found only 2 m apart in the same tree, and these were within 215 m of three other nests, which in turn were within 1 km of six more nests. Nest entrances were at heights of from 8 m to 28 m above the ground, while entrance holes were 8 cm to approximately 30 cm in diameter, with depths of hollows varying from 8 cm to 150 cm, and chambers at the bottom of hollows ranging from 23 cm to 43 cm in diameter. Egg-laying took place between mid June and late July, and clutches comprised two to four eggs, with 2.9 eggs being the mean clutch size for 21 nests. Incubating females were observed to spend between three and four hours on their eggs before leaving to take food from their mates, when they were out of the nest for periods of from five to nine minutes. Pairs usually perched within sight of the nesting tree when transferring food, and each evening both parents roosted in the nest. Eggs hatched between mid July and late August and, as the chicks grew, the female spent progressively less time at the nest. After the chicks were approximately 20 days old, both parents would leave the nest early in the morning, returning at intervals of two to five hours and each time remaining for 10 to 60 minutes. When young birds were nearly ready to fly, the adults spent longer periods perching in trees near to the nest, from where they would call to the chicks, which now were at the entrance. Young birds vacated the nests between early September and late October. Two young birds from a brood of three were seen to make their first flights three hours apart, and each flew strongly as it left the nest, though one paused momentarily on the ground before heading out over the valley; both parents joined these young birds, all the while exchanging calls with them, and escorting them out of sight.

Only five nesting areas now are known to remain in the

Plate 40
UPPER Thick-billed Parrot *Rhynchopsitta pachyrhyncha* (adult)
LOWER Maroon-fronted Parrot *Rhynchopsitta terrisi* (adult)

F. KNIGHT '15

breeding range, and field studies of nesting parrots undertaken between 1995 and 2001 at sites in the Sierra Madre Occidental in northwestern to central Chihuahua reveal that Thick-billed Parrots still exhibit a preference for snags, but the tendency to use pines has dropped enormously, relative to the use of other tree species (Monterrubio-Rico *et al.* 2002; Monterrubio and Enkerlin-Hoeflich 2004). In these studies, 187 nests were located, with 110 being in large snags and 77 being in living trees. Significant differences between results from these studies and the earlier study by Lanning and Shiflett in the frequency of tree species used for nesting can be attributed to widespread logging of pine forests; in the earlier study 42 of 55 nests were in living or dead pines, whereas in the recent studies pines and Douglas fir trees each accounted for 61 of the 157 identified nesting trees. Although for most tree species, nests were more frequent in snags than in living trees, the reverse was recorded for nests in Mexican white pines, white fir trees *Abies concolor* and oaks *Quercus sideroxyla*. In the best protected areas a slightly higher frequency of nesting sites occurred in pines, but in areas subjected to continuing logging pines were used much less for nesting than were *Pseudotsuga*, *Abies* or *Populus* trees, and in one study area all nests but one were in *Populus* trees. The mean diameter at breast height for all nesting trees was 75.2 cm, with diameters of 141 trees being within the range of 60 cm to 79 cm; 46 trees had diameters of less than 60 cm, and only two nests were in trees with diameters less than 40 cm. During these studies, nesting activities were well underway by 5 June of every year, with courtship and copulations being observed mostly between 5 and 25 June. Most egg-laying occurred between 5 and 14 July, and clutch sizes varied from one to five eggs, with 84 per cent of clutches comprising two or three eggs. Clutches of four and five eggs represented respectively 10 per cent and 2 per cent of total clutches, and most fledglings came from three-egg clutches. Incubation by the female lasted an average of 27 days, with a range of 25 to 32 days, and chicks hatched asynchronously at intervals of two to three days. Most hatchings occurred between 1 and 17 August. Of 187 clutches, 149 were successful, producing 405 nestlings from 510 eggs, and 307 chicks fledged. Overall, 80 per cent of nests fledged at least one chick, and 60 per cent of eggs laid produced fledglings. During the seven year period of these studies, egg-laying pairs fledged an average of 1.6 fledglings per nest, and nesting success ranged from a low of 60 per cent in 2000 to a high of 88 per cent in 1996. Of the 38 failed nests, 15 failed during incubation and 23 failed during the nestling stage, with desertion by the parents being the most frequent cause of all failures. Ectoparasites apparently caused the loss of 19 chicks from nine nests and predation was involved in the failure of nine nests; eight nests failed from flooding, tree fall or unknown causes, and one nest was lost to poaching.

At a nesting site near Madera, Chihuahua, at the eastern edge of the Sierra Madre Occidental, nearly all nests are in living or dead quaking aspens *Populus tremuloides*, and breeding activity was monitored from July to late October in 1998–2001 (Monterrubio-Rico *et al.* 2006). During this study, 147 nesting attempts were recorded in 72 different trees, and 10 different trees were recorded as roost sites. All but two of the 82 trees used for nesting or roosting were aspen, and the other two were Mexican white pines *Pinus ayacahuite*. Of 48 nesting trees monitored for at least two breeding seasons, 33 were reused in subsequent years. The majority of all 147 nesting attempts occurred in tree cavities that appeared to be old woodpecker holes, but 20 nesting attempts were in natural cavities formed by decay and detachment of large branches. Roosting was recorded in 30 cavities. Of the 147 nesting attempts, 45 involved more than one pair nesting in the same tree, with 18 attempts involving two pairs occupying different cavities in the same tree, and on three occasions three pairs were observed sharing different cavities in the same tree. Most nesting trees showed a clumped distribution pattern, forming aggregations or nest clusters in aspen stands. No agonistic behaviour was observed among neighbouring nesting pairs, which were in permanent contact. Synchronised foraging flocks formed every morning, and there was visual contact and loud calling among pairs. There were five instances of collective responses to raptors, in which the parrots rapidly formed a flock after sharp alarm calls had been given by birds that initially detected the intruders. It is suggested that this clumped nest distribution and multiple nests per tree may be a consequence of the existence of suitable tree cavities at high densities and the high level of sociality and tolerance of neighbouring nesting pairs. By selecting tree cavities near to those occupied by other pairs, nesting pairs also may experience lower rates of predation.

In central Arizona, at least one successful breeding by released birds occurred in the summer of 1988, yielding two fledglings, and in that same year another nesting attempt failed, apparently due to predation of the sitting female (Snyder *et al.* 1994). Subsequent attempts also failed, largely because of very poor food supplies during an unusually severe drought.

EGGS Ovate to rounded-ovate, glossy. Measurements of 20 eggs are 39.5 (37.6–42.0) × 30.6 (29.0–32.0) mm (Bent 1940).

Maroon-fronted Parrot

Rhynchopsitta terrisi Moore

ENDANGERED

Rhynchopsitta terrisi Moore, *Proc. Biol. Soc. Wash.*, **60**, 1947, p. 27. (Near Galeana, about 7500 feet, Sierra Potosí, Nuevo León, Mexico.)

DESCRIPTION Length 40 cm.
ADULTS General plumage dark green, paler on cheeks and ear-coverts; forehead, forecrown and broad superciliary stripe dark brownish-maroon; bend of wing, carpal edge and thighs dark red; greater underwing-coverts grey suffused olive; tail above dark green, below grey; bill black; bare eyering brownish-yellow; iris orange-yellow; legs grey.
4 males: wing 287–290 (287.8) mm, tail 185–204 (193.8) mm, exp. cul. 40–42 (41.3) mm, tars. 23–24 (23.8) mm.
1 female: wing 283 mm, tail 188 mm, exp. cul. 40 mm, tars. 24 mm.
JUVENILES No maroon superciliary stripe; bend of wing and carpal edge green; less red on thighs; bill horn-coloured with grey at base of upper mandible; iris brown.

DISTRIBUTION Northeastern Mexico, only along a 300 km section of the Sierra Madre Oriental in southeastern Coahuila, central-western Nuevo León and southwestern Tamaulipas, occasionally to Querétaro.

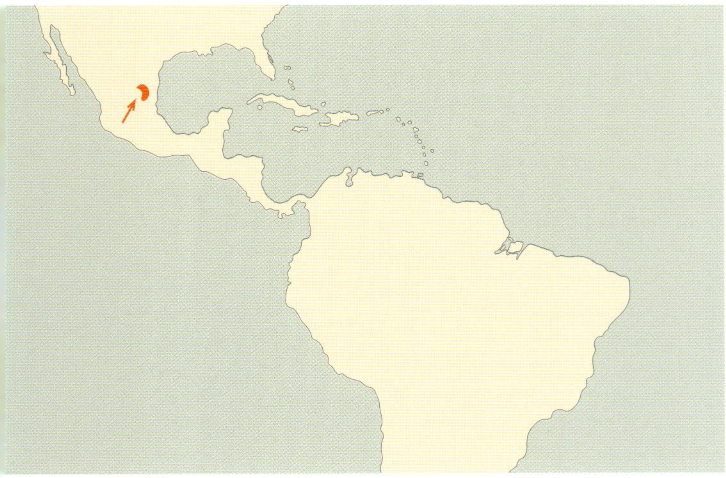

STATUS Being dependent on approximately 7000 km² of suitable habitat within a very restricted range makes the Maroon-fronted Parrot highly vulnerable. Habitat loss is the major threat, with significant areas of forest being destroyed or degraded by fire, logging operations, intensive grazing, and agriculture, the last being particularly prevalent in valleys and on lower slopes where vast apple orchards are being established. Increased frequency and intensity of forest fires is causing passive deforestation, for replanting with native species is occurring only on small parcels of privately owned lands, and naturally regenerating areas usually become covered with dense oak-chaparral vegetation, which is of no value to the parrots (in Snyder *et al.* 2000). Lanning and Lawson reported that in 1975, to the east of San Antonio de las Alazonas, Nuevo León, a major fire destroyed about 50 km² where parrots had been common, and logging subsequently occurred in the same area (*in litt.* 1977). In 1998, another 20 km² of habitat was destroyed by fire (in Birdlife International 2004). Forest fires are more prevalent during times of drought, when traditional watering places can dry up and, in 1994, at least 50 parrots drowned in a single incident while attempting to drink from a cement-walled water tank (in Snyder *et al.* 2000).

Two major nesting colonies are protected in El Taray Sanctuary and Cumbres de Monterrey National Park, though protection is not effective (in Birdlife International 2016). To safeguard the long-term survival of these parrots there is a need to provide effective protection for all major colonies and to implement effective fire-fighting measures.

From counts made in late March and early April 1977, along a 50 km north-facing ridge complex due east of Saltillo, southeastern Coahuila, Lanning and Lawson estimated the total population at 2000 to 3000 parrots (*in litt.* 1977). In 1996, the population again was estimated at 2500 to 3000 birds, so indicating that, although small, numbers had remained relatively stable. An estimate of approximately 3500 birds was proposed in 2008 but, as reproductive maturity is thought to be reached at about four years, there probably are less than 2500 mature individuals, and it is suspected that during the past 33 years, representing three generations, there has been a rapid decline (in Birdlife International 2016).

The Maroon-fronted Parrot is listed on Appendix I of the International Convention on Trade in Endangered Species of Wild Fauna and Flora (CITES).

HABITATS As expected, habitat preferences of the Maroon-fronted Parrot are very similar to those of its western counterpart, the Thick-billed Parrot *Rhynchopsitta pachyrhyncha*. Mostly between 2000 m and 3000 m, or occasionally down to 1300 m and up to 3700 m, this species favours highland *Pinus* forests and mixed *Pinus-Abies-Quercus* woodlands, especially near nesting and roosting sites in cliff-faces.

MOVEMENTS Outside the breeding season a few birds usually can be seen in the vicinity of major nesting sites in the northern breeding range in southeastern Coahuila and adjacent west-central Nuevo León, but a post-breeding exodus takes most of the population south to southern Nuevo León, where recorded mostly between May and July, and to southwestern Tamaulipas, where recorded in March and April (Howell and Webb 1995). During the winter of 1998, some birds were recorded in the Sierra Gorda, northern Querétaro, indicating that some wanderers occasionally come south of the Sierra Madre Oriental, and there is some debate as to whether old sight records from central Veracruz refer to this species or to the Thick-billed Parrot *Rhynchopsitta pachyrhyncha* (in Birdlife International 2004). Local, irregular or nomadic movements in response to changes in food availability also have been reported.

HABITS Much of what is known of the habits of Maroon-fronted Parrots comes from field studies undertaken in the 1970s by Dirk Lanning and Peter Lawson, and at that time they kindly forwarded information to me (*in litt.* 1977). Like the Thick-billed Parrot *Rhynchopsitta pachyrhyncha*, these parrots are daily commuters. Large flocks occupy traditional communal nighttime roosts in crevices in cliff-faces or in nearby trees, and in the early morning smaller flocks disperse over a wide area to feed; in the late afternoon they reassemble at their roosts for the night. Lanning and Lawson reported that on 4 April 1977, approximately 26 km east of Saltillo, southeastern Coahuila, 800 to 1000 parrots were observed coming to roost, while farther to the east, in scattered groups along the ridge, were an estimated 100 to 200 additional birds. During the day there were very few parrots in the vicinity of this roost, but about three hours before sunset small flocks commenced to come in, flying along the ridge. Arrivals were sporadic until about an hour before sunset when the rate increased significantly. As birds came in, they joined the main flock, circling noisily over the slopes, and then at sunset there was a final urgent rush when the entire flock settled down for the night. In the morning, at first light, all birds departed in the course of a few cacophonous moments. Small flocks repeatedly were observed to travel more than 40 km along a ridge between the roost and feeding sites, and within each flock birds usually flew in pairs. On two evenings, small flocks were seen to leave their feeding areas and fly out across 20 km of open grassland on their way back to the nighttime roost.

The flight is fast, powerful and direct, with speeds of more than 40 km per hour being recorded. The parrots use strong, flapping flight to cross valleys, or soaring flight to take advantage of up-draughts near cliffs and ridges, circling high above the surrounding land. The varied weather of the mountains affects their flight paths. In the early mornings, when there is no upward moving air, birds generally fly over the low slopes or valleys, but when the day warms up and air is moving up the canyons and slopes, they fly above the ridges and cliffs, often soaring to great heights. They are reluctant to travel in stormy weather when strong down-draughts are present or visibility is poor. They will route their flight around heavy clouds and thunderstorms. Lanning and Lawson reported that, on 23 May 1977, eight parrots were seen to come west around a ridge, only to find a massive thunderhead in their path and, after much indecision, six birds circled south to the next ridge, while two continued west, only to turn south a couple of minutes later (*in litt.* 1977).

CALLS A raucous *scronk* and a slightly softer *haw-haw-haw-haw-haw* usually are repeated every five to 10 seconds by one or two members of each flying flock, and less often while feeding or resting (Lanning and Lawson *in litt.* 1977); these calls can be heard at a distance of up to 3 km. All calls are similar to those of the Thick-billed Parrot *Rhynchopsitta pachyrhyncha*, and Ridgely tells me that he could not detect any differences (*in litt.* 1977).

DIET AND FEEDING The diet comprises mostly pine seeds, with *Pinus arizonica*, *P. gregii*, *P. teocote*, *P. montezumae*, and *P. cembroides* being among food trees exploited at different times of the year. *Abies* seeds, acorns, and flowers and seeds of *Agave macroculnis* also are eaten. Pairs and small flocks have been seen to alight among the outer branches of a pine tree and, using their bill and feet, clamber to a cone-bearing limb where they would shred and tear apart the cones to get at the seeds, with some partly shredded cones always falling to the ground (Lanning and Lawson *in litt.* 1977). From April through to August, flowering agaves were visited by the parrots, and as many as 30 birds have been counted on one plant; the parrots flew from one inflorescence to another, drinking the nectar without damaging the flowers. These parrots also have been seen drinking from a small seep high on the sheer face of a 100 m high cliff, picking up pieces of ice from around a pool at the base of a waterfall, and taking large bites of snow (Lanning and Lawson *in litt.* 1977).

In 12 visits of three days, between June and September 2005, observations were carried out at four clay-licks visited by Maroon-fronted Parrots (Valdés-Peña *et al.* 2008). These licks were well distributed in the breeding range, and it was believed that most licks in the region were located. Between 1600 m to 1800 m altitude, all four licks were near to valley bottoms in mud cliffs no higher than 40 m above seasonal streams. Parrots came to the licks from early morning until about 1500 hours, with peak attendance occurring in the late morning, between 0900 and 1100 hours, but at one lick the largest number of birds was recorded between 0700 and 0900 hours, and at another lick numbers were equal between 0700 and 0900 hours and between 0900 and 1100 hours. Emitting typical flight calls, the parrots usually arrived in small groups of one to five individuals. They perched in nearby trees before coming down to the clay cliffs, possibly to detect the presence of potential predators. The first bird to alight at the cliff generally showed nervous behaviour and did not commence feeding until joined by other birds, which usually arrived within the next couple of minutes. During feeding, parrots regularly flew to and from the lick to increase spacing with other parrots, and some birds remained watchful while others fed. Different individuals in turn adopted the watchful posture and occasionally the entire group took flight simultaneously in response to the emission of sharp alarm calls from one bird. Groups visiting the licks averaged nine individuals and spent an average of 18.3 minutes at the licks. It was estimated that the parrots flew up to 23 km from nesting colonies to the clay licks, and the known distribution of clay licks could serve all known breeding colonies.

BREEDING Maroon-fronted Parrots nest communally, and 23 nesting colonies are known (in Birdlife International 2016). They nest exclusively in crevices in limestone cliffs, and poor accessibility of most cliff-face nesting colonies gives the species a survival advantage over the tree-nesting Thick-billed Parrot *Rhynchopsitta pachyrhyncha*. Colonies may vary in size, with some comprising only a few nesting pairs while in others many pairs may nest in close proximity to each other. Nearly 100 breeding pairs, or almost 40 per cent of the total breeding population, nest along 28 km of north-facing cliffs in El Taray Sanctuary, near Monterrey, Nuevo León, making this the most important nesting site.

As with the Thick-billed Parrot, the timing of nesting is influenced by the local availability of pine seeds. Pairs normally arrive at the nesting sites between April and May, and egg-laying takes place in early July, with fledging of chicks occurring in October to November (in Birdlife International 2016). At El Taray Sanctuary, in 1978, Wiley observed that most young birds had fledged by 16 November, and a short while later all parrots had disappeared from the area (in Lawson and Lanning 1981). In 1981, again at this same site, most young birds fledged before 7 November (Nichols *in litt.* 1981).

Some entrances to nesting cavities are sufficiently large to allow simultaneous entry by both parents, but at smaller entrances one adult will enter directly while the partner briefly circles nearby and then enters. Little is known of the nesting behaviour, but pairs fledge one to three chicks, and the whole breeding population produces an estimated 95 to 150 fledglings per year (in Birdlife International 2016).

GENUS *Ognorhynchus* Bonaparte

Ognorhynchus Bonaparte, *Remarques Obs. Blanchard Caractères Psittacides, etc.*, 1857, p. 6. Type, by monotypy, *Conurus icterotis* Souancé.

The sole member of this genus is a large, macaw-like parrot with a long, strongly graduated tail and a large heavy bill with a notch in the upper mandible. The cere and face are feathered, but bare skin is exposed at the base of the lower mandible and there is a prominent bare eyering. The distinctive plumage colouration features elongated ear-coverts forming conspicuous broad 'tufts'. There is no sexual dimorphism, and juveniles resemble adults.

ENDANGERED

Yellow-eared Parrot

Ognorhynchus icterotis (Massena and Souancé)

Conurus icterotis Massena and Souancé, *Rev. et Mag. Zool.* (2), **6**, 1854, p. 71. (Ocaña, Colombia).

DESCRIPTION Length 42 cm. Weight 285 g.
ADULTS Upperparts, sides of neck, and lower cheeks dark green; forehead, lores, upper cheeks and elongated ear-coverts yellow; underparts yellowish-green, becoming darker green on thighs and under tail-coverts; underwing-coverts greenish-yellow; undersides of flight feathers dusky yellow; tail above green, below dusky orange-red; bill grey-black; bare eyering grey; bare skin at base of lower mandible pink-white (deep pink when alarmed or excited); iris orange; legs grey.
7 males: wing 218–233 (225.1) mm, tail 162–220 (192.3) mm, exp. cul. 33–37 (35.0) mm, tars. 20–22 (21.1) mm.
10 females: wing 216–231 (225.0) mm, tail 161–207 (186.2) mm, exp. cul. 33–37 (35.0) mm, tars. 20–23 (21.2) mm.
JUVENILES Like adults; shorter tail; iris brown.

DISTRIBUTION Confined to the Andean highlands in northwestern Ecuador, north from Pichincha and western Cotopaxi, and western Colombia, north to Antioquia and northwestern Norte de Santander; recent records are from only a few scattered localities in the Cordillera Central, Colombia, and western Cotopaxi, Ecuador.

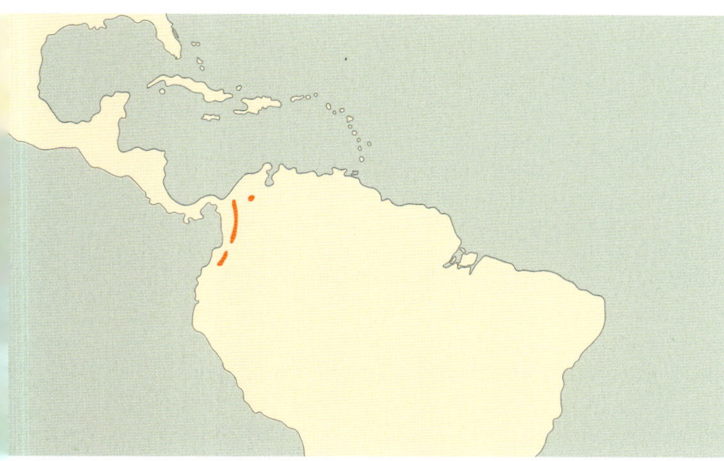

STATUS Formerly widespread in all three Andean cordilleras in Colombia, and common to locally abundant in much of their range, Yellow-eared Parrots suffered a catastrophic decline in numbers during the mid 1900s, and were brought to near extinction primarily by large-scale deforestation, with localised persecution probably being a contributing factor in some districts. In the early 1900s, Chapman and other fieldworkers recorded these parrots as being common and locally abundant in the Cordillera Central, in central Colombia (Chapman 1917). After seeing only a lone bird flying overhead early one morning during a visit of 10 days in July 1956 to the Moscopán region, Cauca, southwestern Colombia, where formerly the parrots had been relatively common, Lehmann (1957) warned of the deteriorating status of the species. This apparent decline could have been due to seasonal movements, because earlier visits by Lehmann had been made between December and April, but local people claimed that even during those months fewer birds were being seen. Subsequent searches by Robert Ridgely in the mid 1970s confirmed that the parrots no longer occurred at this locality, where the forest had been largely cleared, and his only sighting was in nearby Puracé National Park, on the eastern side of the cordillera, where a lone pair was seen on 18 May 1976. During the 1980s, fewer sightings of only single birds, pairs or small parties were made at a few localities in the Cordillera Central, and it became apparent that the species had disappeared from most of its range in Colombia. Searches for surviving birds then were undertaken in an historical stronghold in the Volcán Tolima-Ruiz massif, in the Cordillera Central, which encompassed possibly the largest remnant of suitable habitat with an extensive tract, albeit fragmented, of wax palms. Reliable reports from this area during 1995–1996 included sightings of lone pairs and one sighting of a flock of six to 12 birds, and then in October 1997 a flock of 24 birds was seen, followed in August 1998 by the sighting of a flock of approximately 20 birds.

These encounters in the Volcán Tolima-Ruiz massif prompted the setting up of 'Proyecto *Ognorhynchus*' with the aim of studying and protecting what probably was the last surviving flock in Colombia. Hopes for a recovery were boosted in 1999 when, on 18 April, two flocks of 31 and 30 parrots were encountered in a valley, and within a week an additional flock of 21 parrots was discovered roosting in an adjacent secluded valley. Successful nesting also was recorded at the sites and, in response to these encouraging developments, a multifaceted recovery program was initiated by Proyecto *Ognorhynchus* with the cooperation of local community organisations. Principal objectives of this program included the purchase of lands for protection of nesting and roosting sites, reforestation with wax palms and other important food trees, environmental education involving local communities and farmers, and field investigations focussing on movements and breeding behaviour. Initial successes included a virtual cessation of the destruction of wax palms for Palm Sunday processions and eventual use by the parrots of redesigned artificial nestboxes. Initial achievements have been spectacularly successful, with the population increasing to more than 600 birds in 2004, and the numbers were divided almost equally between two geographically separate sites – Jardín, in Antioquia, where the highest count was of 303 birds, and Roncesvalles, in Tolima, where 312 birds were present (see *Cyanopsitta*, no. 73: 17–18, 2004). Furthermore, in August 2003, at the end of the breeding season, the presence of many juveniles confirmed a much increased recruitment resulting from effective protection. More recent surveys have included monthly population censuses at these same two sites, and counts made in 2015 have confirmed that strong increases in numbers are continuing (Waugh 2016a). At the main roost site in Jardín, a maximum of 4251 birds was recorded in April and 1096 birds were present at another roost site in May, while in Roncesvalles a total of 949 birds was recorded. Waugh points out that even allowing for some possible movements of parrots between the sites, the population continues to increase rapidly. There are recent reports from new locations in western Colombia, including Apía in Risaralda, Tatamá National Park, and San Pedros de los Milagros in Antioquia, but many of these observations are likely to be of recent colonisers or fly-overs by dispersing birds (in Birdlife International 2016). Based on the observations of local villagers, a population estimated at about 84 birds is thought to have been resident for more than 30 years in the vicinity of San Luis de Cubbaral, westernmost Meta (in Birdlife International 2016).

It seems to me that throughout recent times Yellow-eared Parrots have not been plentiful at the southern extremity of the range in northwestern Ecuador, where during the past 50 years records from all but one of few localities have been of very small numbers. Krabbe and Sornoza Molina (1996) document occurrences of small groups in Imbabura and Carchi, and in neighbouring southwestern Colombia, in Nariño, during the late 1980s to early 1990s, pointing out that all records could refer to the same population, and they report also the finding in 1994, in the Caripero region of western Cotopaxi, of what may have been the last population in Ecuador. Local people claimed that this population had used the same communal roosting palm for decades and in previous years hundreds of birds had gathered there, but people from the entire area regularly had shot them for food. When discovered in 1994, this population was reduced to just 19 birds! There have been unconfirmed reports of flocks of about 20 birds in the Río Intag valley, Imbabura, since 2000, but searches undertaken in 2008 in the last confirmed strongholds in Imbabura and Carchi failed to locate any birds (in Birdlife International 2008).

Ridgely and Greenfield (2001a) note that the status of this species in northwestern Ecuador remains shrouded in mystery, but obviously it is extremely precarious, with any surviving populations being so small as perhaps not to be salvageable. Furthermore, areas where birds recently have been seen remain largely unprotected. While recognising that deforestation doubtlessly has played a major role in the decline of this species, Ridgely and Greenfield point out that in northwestern Ecuador the parrots now are so rare in many districts where formerly they occurred and where montane forest remains little altered as to suggest that additional factors may be involved. There is some evidence that hunting for food at traditional roosts and trapping may have impacted more severely on populations in northwestern Ecuador than in Colombia. At the roost in the Caripero region of western Cotopaxi, the parrots did not relocate despite persistent persecution, so it is likely that this population was decimated by hunters. Also, at this same site in 1994, a local farmer felled a nesting palm to obtain chicks for sale, but all four nestlings died when the tree came down (Krabbe and Sornoza Molina 1996). A bird discovered in the cagebird market in 1970 was reported to have been taken near Mindo, in Pichincha, and in the late 1980s a farmer was reported to have illegally trapped 40 parrots from a flock of 60 birds, but all died within two days of capture (in Kirwin *et al.* 1996; see *Cyanopsitta*, nos 52/53: 31–34, 1999). There seems to be no recent confirmed records from northwestern Ecuador, but Robert Ridgely tells me that there are continuing reports from local people of occasional individuals, pairs or very small flocks at a couple of localities, though none seem to be established at any single site, and it is difficult to determine whether these birds are wandering residents or possibly coming from neighbouring southwestern Colombia (*in litt.* 2016).

In summary, available evidence indicates that loss of its montane habitat was primarily responsible for near extinction of the Yellow-eared Parrot in Colombia, and probably also was the principal cause of its being possibly already extinct in northwestern Ecuador. In addition to the large-scale destruction of stands of wax palms, on which the parrots are so dependent, the loss of elfin forest deprived populations of crucial seasonal food sources, notably *Podocarpus* fruits (Salaman *et al.* 1999). These threats have been addressed in a concerted conservation program undertaken jointly by Fundación ProAves and the Loro Parque Fundación, and Waugh (2016a) points out that actions taken in Colombia to protect the two breeding populations, and more generally at regional and national levels, have achieved great success. Habitat restoration and reforestation has exceeded objectives, with 360 000 trees and wax palms planted across 920 ha, and 5000 ha of forest on 25 private farms have been designated as nature reserves. Some 16 km of fencing has been erected to protect forest fragments and young wax palms from cattle grazing, and 4585 ha of buffer zone to Las Hermosas National Park also is protected. In the newly established Yellow-eared Parrot Natural Bird Reserve of 188 ha, an additional 1045 seedlings of plants important for the parrots were planted in 2015. As an economic incentive for landowners to maintain forest on their lands, environmental service agreements have been set up with municipalities, whereby payments from users of water in the lower part of the watershed support the protection of forests in the upper regions. School students and local village communities have been targeted in an environmental awareness program, which includes local radio broadcasts and national radio and television broadcasts using the Yellow-eared Parrot and the wax palm as flagship species. Due to the positive effects of the program, there are no longer any reports of these parrots being killed or taken from the wild.

The Yellow-eared Parrot is listed on Appendix I of the Convention on International Trade in Endangered Species of Wild Fauna and Flora (CITES).

HABITATS In the upper subtropical and lower temperate zones, mostly between 1300 m and 3500 m, Yellow-eared Parrots frequent humid montane forest, elfin forest and partially cleared lands, where stands of *Ceroxylon* wax palms are prevalent. There is a close association between the parrots and these palms, which are utilised for feeding, roosting and nesting. On the western slope of the Andes in northwestern Ecuador, these parrots keep mainly to the canopy and borders of montane forest, but Ridgely and Greenfield (2001a) suggest that the association with wax palms may not always be as strong as has been claimed, for in recent years birds have been observed eating fruits from other trees or shrubs, and no birds have been recorded in some areas where wax palms are particularly plentiful. I would urge caution in interpreting any apparent underutilisation of seemingly suitable habitat in northwestern Ecuador, where the species is known to survive only in extremely low numbers, if at all, as evidence for downgrading the importance of dependence by the parrots on *Ceroxylon* wax palms, as confirmed during field studies undertaken at sites in the Volcán Tolima-Ruiz massif, in the Cordillera Central, central Colombia (see *Cyanopsitta*, nos 53/54: 30–34, 1999).

During the mid 1990s, pairs and small flocks were seen in the Volcán Tolima-Ruiz massif, in the Cordillera Central, central Colombia, an historical stronghold of the species, and all sightings were in areas of humid montane forest and elfin forest dominated by wax palms (Salaman *et al.* 1999). In this same region fieldwork and conservation efforts now are concentrated in a 50 km² area of montane forest and pasturelands intermixed with many dense groves of wax palms *Ceroxylon quindiuense*.

Plate 41
Yellow-eared Parrot *Ognorhynchus icterotis* (adults)

F KNIGHT '11

MOVEMENTS Yellow-eared Parrots are highly mobile, their well-documented seasonal and locally nomadic movements probably being in response to food availability, especially the fruiting of *Ceroxylon* wax palms. Historical records suggest to me that during the breeding season the parrots occupied traditional nesting areas, the key feature of which was a local abundance of wax palms, and there were post-breeding dispersals, possibly by smaller groups and along regular highland routes to non-breeding areas, where a wider variety of food sources was utilised. Between 1983 and 1989, at La Planada Nature Reserve, in Nariño, southwestern Colombia, a flock appeared almost every day in February, often staying until May and reaching a maximum of 35 birds in 1985, but no birds were seen in 1990 or thereafter, so presumably this flock now is extinct (in Snyder *et al.* 2000). Arndt reports that near Guallupe, in Imbabura province, northwestern Ecuador, a small group occurred annually between November and February, until at least the late 1980s or early 1990s (in Krabbe and Sornoza Molina 1996). Other records from northwestern Ecuador include a small group believed heard calling, though not seen, in forest above Maldonado, in Carchi province, on 3 January 1982, and a party of seven birds seen above Morán, in the same region, on 6 August 1991 (in Ridgely and Greenfield 2001a). At the communal roost discovered in 1994, in the Caripero region of western Cotopaxi, the parrots were present annually between July and November or early December, but it was not known where they spent the remainder of the year (Krabbe and Sornoza Molina 1996). Currently, birds are present throughout the year at study sites in the Tolima-Ruiz massif, in the Cordillera Central, central Colombia, and this sedentary behaviour probably reflects the almost total loss of preferred food sources, especially fruiting *Ceroxylon* wax palms, elsewhere in the range (see Birdlife International 2016).

HABITS Noisy and conspicuous when in flight, Yellow-eared Parrots may be heard well before they come into view, and they are highly social, especially at communal nighttime roosts. In the Volcán Tolima-Ruiz massif, in the Cordillera Central, central Colombia, pre-roosting socialising has been observed, with pairs and small groups gathering together to chatter, preen and frolic amidst the palm groves, sometimes flying to within a couple of metres of observers without showing any fear (see *Cyanopsitta*, nos 53/54: 30–34, 1999). Also at this roost, at dawn, the three flocks were seen to disperse far into surrounding highland forest on the main ridge of the Cordillera Central, before slowly returning, again feeding at lower levels and then arriving back at the roost by mid afternoon. Near Guallupe, in Imbabura province, northwestern Ecuador, flocks of 15 to 20 birds were seen to appear at about 0800 hours, remaining to feed during the day in low bushes in open country, and then departing at approximately 1600 hours to fly off towards higher forest, where presumably they roosted (Arndt 1992–1996). Attachment to traditional roosts is very strong, as evidenced by reports from the Caripero region of western Cotopaxi, northwestern Ecuador, where a flock roosted in the same palm for decades, and only when this palm fell in 1992 did the birds move to an adjacent palm (Krabbe and Sornoza Molina 1996).

The strong, direct flight is fast, and these parrots normally fly high across valleys or over mountain ridges. In the Volcán Tolima-Ruiz massif, in the Cordillera Central, central Colombia, two flocks were seen to descend from high flight in 'tightly synchronized barrel rolls' to alight in wax palms in pasture fields (see *Cyanopsitta*, nos 53/54: 30–34, 1999).

CALLS The call frequently given while in flight or when perched is a raucous, though somewhat nasal *raanh*, and often this is emitted more or less simultaneously by members of a flock to produce a loud, far-carrying cacophony (Ridgely and Greenfield 2001b). When heard from afar, the calls given by a flock resemble a musical, conversational babble (Hilty and Brown 1986).

DIET AND FEEDING Fruits of *Ceroxylon* wax palms are the principal food, and these are supplemented with seasonally available fruits, seeds, flower buds and bark from other trees or shrubs. In northwestern Ecuador and in the Cordillera Central, central Colombia, *Podocarpus* has been identified as a seasonal food source (Salaman *et al.* 1999). At the study site in the Volcán Tolima-Ruiz massif, in the Cordillera Central, central Colombia, the parrots feed primarily on fruits of wax palms, but 13 tree species, including four from the family Euphorbiaceae, also feature in the diet (see *Cyanopsitta*, no. 73: 17–18, 2004). In western Nariño, southwestern Colombia, birds were observed over a three month period to feed on *Sapium* fruits, while in the Caripero region of western Cotopaxi, northwestern Ecuador, feeding principally was on fruits of *Ceroxylon alpinum* at elevations between 1600 m and 1800 m, but fruits of *Saurauia tomentosa* also were eaten (Krabbe and Sornoza Molina 1996). Other recorded foods include seeds, flower buds and bark from *Ceroxylon*, *Cithorexylon*, *Podocarpus* and *Sapium* trees or shrubs, and the resident population at San Luis de Cubbaral, westernmost Meta, Colombia, is said to depend on the palm *Dictyocaryum lamakcianum* as well as wax palms (in Birdlife International 2016).

BREEDING A male with enlarged testes and a female that was laying were collected at Moscopán, Cauca, southwestern Colombia, in early March (YPM Collection). Chapman (1917) reported that along the Quindío Trail, above the Ríos Tochecito and Toché, in northern Tolima, central Colombia, nesting was recorded in May. In the Caripero region of western Cotopaxi, northwestern Colombia, nesting took place during those months when the parrots were present, that is July to November or early December (in Krabbe and Sornoza Molina 1996). Much of what is known of nesting behaviour comes from observations made at a study site in the Volcán Tolima-Ruiz massif, in the Cordillera Central, central Colombia, where, in 1999, a nest containing a chick being fed by adults was found on 26 April, this youngster fledging on 6 June, and confirmation of a second breeding cycle in that year came with the finding of six additional active nests in September (see *Cyanopsitta*, nos 53/54: 30–34, 1999). Pairs probably take advantage of a heavy fruiting of wax palms in some years and rear a second brood.

The nest is in a hollow excavated by the parrots high up in the trunk of a wax palm. Chapman noted that along the Quindío Trail, in northern Tolima, central Colombia, nest entrances were at heights of more than 25 m. 'Helpers' apparently assist with excavation of the nesting hollow and with feeding of the chicks. In the Caripero region of western Cotopaxi, northwestern Ecuador, local persons claimed that hollows were excavated by the entire flock, one bird excavating before being replaced by another and so on, with more than two birds carrying out excavation at each nest (in Krabbe and Sornoza Molina 1996). At the study site in the Volcán Tolima-Ruiz massif, in the Cordillera Central, central Colombia, some pairs successfully nested in artificial nestboxes, and at two nests under surveillance a 'helper' was seen to assist with feeding of the chicks (see *Cyanopsitta*, nos 53/54: 30–34, 1999). Also at this same study site, during the 2003 breeding season, half of the breeding population, in two breeding attempts, produced an average of two fledglings per nest, and in August, at the end of the breeding season, juvenile birds comprised 15 per cent of the total population (see *Cyanopsitta*, no. 73: 17–18, 2004).

GENUS ***Guaruba*** Lesson

Guaruba Lesson, *Traité d'Orn.*, livr. 3, 1830, p. 210. Type, by monotypy and tautonymy, *Guaruba lutea* Lesson = *Psittacus guarouba* Gmelin.

Previously included in *Aratinga*, the sole species in this genus now is known to be more closely allied to the small macaws, and particularly to *Diopsittaca* (see Tavares *et al.* 2004). It is a medium-sized parrot with a highly distinctive golden yellow plumage colouration. The proportionately massive bill, with a sharply pointed upper mandible, is similar to the diagnostic bill of the Southern Red-shouldered Macaw *Diopsittaca cumanensis*. There is no sexual dimorphism, but juveniles are much duller than adults and show extensive green markings on the head, back and tail.

VULNERABLE

Golden Parakeet

Guaruba guarouba (Gmelin)

Psittacus Guarouba Gmelin, *Syst. Nat.*, **1**, pt. 2, 1788, p. 320. (No locality = northeastern Brazil, *ex* Marcgrave).

OTHER NAMES Queen of Bavaria's Conure, Ararajuba (Brazil).

DESCRIPTION Length 34 cm. Weight 270 g.
ADULTS General plumage rich yellow; primaries, secondaries and outer upper wing-coverts dark green; undersides of flight feathers dusky yellow; tail yellow; bill horn-coloured; bare eyering pink-white; iris brown; legs pink.
10 males: wing 200–218 (208.8) mm, tail 141–165 (153.2) mm, exp. cul. 35–37 (36.2) mm, tars. 19–23 (21.2) mm.
10 females: wing 202–218 (211.8) mm, tail 146–162 (154.0) mm, exp. cul. 35–37 (35.9) mm, tars. 20–23 (21.1) mm.
JUVENILES (UKNHM 1950.25.1) Cheeks and ear-coverts dark green; crown; nape and upperparts yellow strongly marked dull green; underparts yellow, breast lightly marked dull green; uppersides of tail-feathers dark green margined yellow on inner webs; bill pink; bare eyering pale grey; legs pink-brown.

DISTRIBUTION Disjunctly distributed in northern Brazil, south from the Amazon River to about lat. 5°N between the Rio Tapajós, western Pará, and the Rio Turiacu, western Maranhão, and also the upper Rio Madeira and tributaries, northern Rondônia and northwestern Mato Grosso, where possibly isolated.

STATUS Laranjeiras and Cohn-Haft (2009) note that Golden Parakeets are known from approximately 70 localities, most of which are in Pará, but records are distributed across east-central Amazonia, from northern Rondônia east to western Maranhão and, despite the presence of apparently suitable habitat, there are no records from north of the Amazon River or west of the Rio Madeira. Until the mid 1980s these parrots were known only from 26 localities in Maranhão and Pará, but in the early 1990s records in northern Rondônia and northernmost Mato Grosso represented a significant westwards expansion of the range, and in 2007 the species was recorded in easternmost Amazonas. Laranjeiras and Cohn-Haft suspect that these new western records reflect increased research in areas where previously there had been little field investigation, and existence of the populations had been overlooked. However, there continues to be a significant distributional gap in west-central Pará between these western populations and those in the eastern part of the range. In the extreme east of the range there have been no records since the mid 1980s, and the parrots are presumed to have disappeared from this part of their historical range in response to increased deforestation that occurred in Maranhão and eastern Pará in the 1970s and 1980s. It is estimated that the current range is approximately 340 000 km², or 65 per cent of the total area known to have been occupied at one time or another.

Deforestation, hunting, persecution as a crop pest, and trapping for the live-bird market have been identified as serious threats, especially in the eastern sector of the range, where they have been responsible for extirpation of the parrots in some districts. Yamashita (2003) points out that habitat loss certainly is a serious threat, for the range coincides with regions in the Amazonian basin where the most extensive clearing and degradation of forests have been brought about by widespread legal or illegal logging, farming or ranching, industrial development, and increasing human settlement with supporting infrastructure development, including the construction of major dams for hydroelectricity generation and the opening of new roads or highways bisecting the region in both north to south and east to west directions. Separation of the main eastern and western populations of the parakeets is very recent, and a low genetic variability within these populations poses a

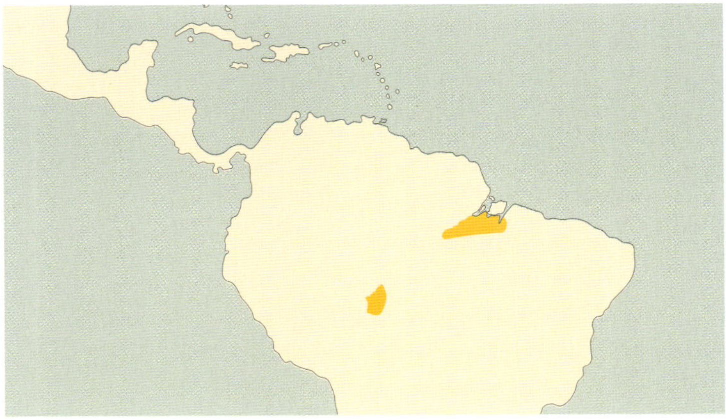

serious long-term threat as groups become further isolated and fragmented due to habitat loss and increasing pressures from advancing human settlement.

In the eastern sector of the range, extensive tracts of forest remain only in Gurupí Biological Reserve, in northwestern Maranhão, and in three Indigenous Reserves, but indigenous tribes in the Gurupí Biological Reserve are traditional trappers and have been primary suppliers of Golden Parakeets to the trade (Yamashita 2003). Illicit logging, particularly of mahogany *Swietenia macrophylla* remains a serious threat in these and other reserves, especially in the poorly protected Jamari National Forest, in northern Rondônia, which suffers constant pressure from loggers, squatters and poachers (Birdlife International 2016). Fabio Olmos tells me that these parakeets must be quite rare in Rondônia for he has never seen them during extensive fieldwork carried out there since 1998 (*in litt.* 2004).

Hunting and trapping take on increased significance as secondary pressures when impacting on populations already disadvantaged by habitat loss. The parakeets are killed for food, to obtain feathers for ceremonial head adornments, and to protect crops. To meet a very high demand in both domestic and international live-bird markets, the species is heavily targeted by trappers, and this is a complex problem with a long cultural history in the region, in both the indigenous and settler communities. For these communities the capture of birds is a long-standing and quite important source of income, with the Golden Parakeet being one of the most valuable species for trade. A traditional means of capture is to catch an entire family group at night in the roosting hollow, a risky operation requiring climbing skills to reach the high hollow, while another method is to use a decoy bird to lure wild birds to alight on branches coated with limestick. Felling nesting trees is a particularly damaging means of capture, for it destroys traditional nesting sites and often results in the death of both nestlings and adults. Most birds are sold on the domestic market, with local brokers taking advantage of the network of roads and highways to collect birds from trappers, but even with the benefit of faster transportation to markets in major urban centres there is very high post-capture mortality.

Laranjeiras (2008) reports that in his study area of approximately 4000 km² in and around Amazonia National Park, western Pará, where fieldwork was undertaken in 2007, the population was estimated at several hundred birds, but only a small proportion appeared to be breeding, and on average only two or three fledglings were seen in each flock. Laranjeiras and Cohn-Haft (2009) suggest that the current area of occurrence of Golden Parakeets probably does not exceed 340 000 km². In a recent survey along the Rio Tapajós, western Pará, an estimated 500 birds were recorded in a census strip of approximately 340 km, and these parakeets were as common as other, non-threatened parrots (in Birdlife International 2016). A highly conservative extrapolation of one bird per 16 km² across 174 000 km² of suitable habitat within the known extent of occurrence gives an estimate of about 10 875 individuals, and on this basis the population is placed in the band of 10 000–19 999 birds, assumed to include approximately 6600–13 400 mature individuals (Birdlife International 2016). Because of continuing deforestation and its susceptibility to hunting and capture for the live-bird market, the Golden Parakeet is expected to decline by a further 30 per cent over the next three generations (in Birdlife International 2016). It is for these reasons that I am uneasy with the downgrading from endangered status for this species.

To be effective, a conservation strategy for ensuring long-term survival of this species must embody a multifaceted approach addressing the multiplicity of pressures impacting adversely on populations. Habitat within reserves must be safeguarded against illicit logging, and in other areas current logging practices should be altered so that large, old trees are available for roosting and nesting. Also, there is a need to find alternative sources of income for indigenous tribes to replace cash returns currently obtained from the sale of trapped birds and from illicit logging. A public awareness campaign is underway, with particular attention being given to schoolchildren of the need to safeguard these parakeets as part of their cultural heritage.

The Golden Parakeet is listed on Appendix I of the Convention on International Trade in Endangered Species of Wild Fauna and Flora (CITES).

HABITATS Very much a bird of humid lowland rainforest up to about 500 m, the Golden Parakeet shows a marked preference for terra firme forest in hilly areas, only occasionally wandering into nearby varzéa forest, but regularly coming into forest clearings or more open country, including grasslands with scattered large trees. Oren and Novaes (1986) note that the only reported occurrence in seasonally inundated forest is at Arumatheua, near the west bank of the Rio Tocantins, eastern Pará, but here the habitat is limited and the parrots regularly wander in from surrounding extensive stands of upland forest. Larangeiras and Cohn-Haft (2009) point out that although terra firme forest appears to be preferred and the parrots do not occur regularly in seasonally flooded forests along rivers, records usually are within several kilometres of rivers, and not in regions farthest from them. They note also that barriers at northern and western extremities of the range are provided by the very wide Amazon and Madeira Rivers, both of which are flanked by broad expanses of a mosaic of várzea vegetation types, none of which are frequented by the parrots. Yamashita (2003) points out that proximity to feeding sites and a presence of large old trees or snags with hollows for roosting and nesting are key elements of the mosaic of habitats utilised by this species. Listing sight records made along the Trans-Amazonian highway between 1974 and 1980, Oren and Willis (1981) note that all observations were on hilly, upland sites, in forests, at forest edges, or in adjoining Brazil-nut plantations, and none were along rivers. During field studies undertaken in eastern Pará and northwestern Maranhão, Oren and Novaes recorded seasonal changes in habitat utilisation, with the parrots frequenting tall forest almost exclusively during the dry season of May to November, but during December to April they came to clearings adjacent to forest, including newly cleared fields, where isolated living or dead trees were selected for nesting and, following fledging of the young, all birds returned to the forest, though in areas where there was little disturbance to nesting, they could be seen sporadically near old nest sites during the dry season. Silveira and Belmonte (2005) report that during mid January to early February 2004, field studies were undertaken in the Grupo Agropalma Forest Reserve, in the Tailândia district, also in eastern Pará, where the habitat was selectively logged forest intermixed with natural clearings; this forest comprised stands of trees of varying ages, with an open understorey, and amongst the large, emergent trees were angelims *Dinizia excelsa*,

Plate 42
Golden Parakeet *Guaruba guarouba* (adults)

F. KNIGHT '11

which were particularly favoured by the parrots for nesting and roosting, and fruiting *Tetragastris* and *Protium* trees. Yamashita and França (1991) recall that in October–November 1989, in the Jamari National Forest, on the right bank of the Rio Madeira, northern Rondônia, birds were encountered in open tropical forest in hilly terrain. Martuscelli and Yamashita (1997) report that between July 1995 and March 1996, during three visits made to the Rio Tapajós region, northernmost Mato Grosso, all of the few sightings of this species were in upland evergreen forest with a 30 m high, multilayered canopy in which the dominant trees were Brazil-nut *Bertholletia excelsa*, *Dinzia excelsa*, *Pouteria* sp., *Euterpe precatoria*, *Astrocaryum vulgare* and *Bactris gasipaes*.

MOVEMENTS For the Golden Parakeet, I have difficulty reconciling behaviour typical of a sedentary, territorial species with claims of the occurrence of widespread, nomadic movements. Collar (1997) states that nomadic movements apparently occur, with birds seeming to wander widely, their presence at any one locality or in any particular season being unpredictable, though post-breeding dispersal regularly brings birds into vàrzea forest. Conversely, Yamashita (2003) points out that each family group is territorial within its home range, and is strongly attached to traditional nesting and roosting sites. Likewise, during monitoring studies conducted near Tailândia, eastern Pará, in the wet and dry seasons of 2003–2004, Silveira found these parrots to be sedentary, with the large territory occupied by each group containing multiple nesting and roosting sites (*in litt.* 2004). Similarly, Laranjeiras (2008) concludes that, although birds seemed to disappear for days from his study area in and around Amazonia National Park, westernmost Pará, they do not migrate seasonally or search nomadically for fruiting trees across large distances, and certain flocks at least occupy specific territories throughout the year. If nomadic movements occur at all, and there appears to be little or no evidence that they do, probably only bands of non-breeding, 'marginalised' birds are involved, but I suspect that reports of such movements can be attributed merely to localised wandering by groups within their large territories.

HABITS Despite their brilliant plumage colouration, Golden Parakeets can be overlooked while feeding quietly amidst dense foliage in the forest canopy, using the bill extensively when climbing among the branches, but in the tops of dead trees or in overhead flight they are noisy and highly conspicuous. Pairs or single birds seldom are encountered, for tightly cohesive groups are the basic social units, and it is uncertain whether these groups comprise multiple nesting pairs or a single breeding pair aided by offspring from successive breeding seasons. Yamashita (2003) points out that although some groups contain up to nine birds, it seems that seven is the optimal number, and additional individuals are forced to join up with other marginalised birds. Apparently some of these marginalised birds eventually join family groups, for DNA profiling has revealed the presence of unrelated birds in some groups. Groups probably are territorial, with the estimated home range for each group being approximately 10 km^2 of primary habitat, but groups will come together to form loose foraging flocks, and these flocks sometimes feed in the company of other parrot species. Groups are readily discernible within a foraging flock, and there can be noisy squabbling between groups. Listing sight records made along the Trans-Amazonian highway between 1974 and 1980, Oren and Willis (1981) point out that most were of small groups of from five to 12 birds seen feeding in the treetops or in flight over the canopy, but one notable exception was of an extraordinarily large flock of 27 birds seen flying over Tucuruí, 18 km east of the Rio Tocantins, eastern Pará. Similarly in December 1983, Paul Roth observed small groups of from nine to 14 birds in forests in the Gurupí district, northwestern Maranhão (*in litt.* 1987). Oren and Novaes (1986) report that during field studies carried out in eastern Pará and northwestern Maranhão, between 1981 and 1984, trios or flocks of up to more than 30 parakeets usually were observed between 0700 hours and 1000 hours, and again between 1530 hours and 1745 hours, as they flew just above the forest canopy, or occasionally up to 20 m above the treetops, with the discordant call-notes always denoting their passage; these flocks often passed along the same route in opposite directions in the morning and afternoon, so suggesting a regular use of the same feeding and roosting sites. Oren and Novaes also found groups of from four to nine birds roosting in tree hollows in forests and clearings during the non-breeding season, and these dormitory roosts often were changed on successive nights, though in the same general area and always within a few hundred metres of each other. Silveira and Belmonte (2005) recall that between mid January and early February 2005, a small group of six birds and a larger group of 10 birds were located at a study site in the Grupo Agropalma Forest Reserve, near Tailândia, eastern Pará. The larger group was tolerant of the presence of observers, and could be approached to within 5 m, especially while they were in the nesting tree. While either group foraged or rested, one or two birds acted as sentries, remaining watchful of approaching danger while perched in the topmost branches.

Laranjeiras reports that during field studies undertaken in and around Amazonia National Park, westernmost Pará, sightings were of single birds and pairs up to groupings of 50 individuals, but parties of five to seven birds were most common. Up to 20 birds were found roosting at night in the same tree cavity and, because the same number of individuals was observed occupying the same roost tree throughout the year, groups were identified as clans. Members of a clan remained loosely together during the entire day, separating in pairs to forage, mate, allopreen or for nest vigilance and defence. During the heat of the day, clans rested in shaded trees near to their roost sites. Some clans were hostile to the arrival of other Golden Parakeets at nesting or roosting sites, while in other areas clans intermingled peacefully. There were conspicuous social interactions amongst clan members, especially at nest sites in the early morning when individuals associated in pairs or trios on dead branches to allopreen. Playful actions included birds pecking at each other, at times hanging upside down until letting go of the branch and briefly falling before flying back to the same branch. Some birds that remained perched alone on higher branches may have been oldest flock members acting as sentinels. Aggressive behaviour by nesting birds towards other flocks of Golden Parakeets or other parrot species were observed on numerous occasions. Nest trees were occupied as dormitories in the non-breeding season.

The strong flight features deliberate, rhythmic wingbeats. Birds often fly rapidly from one tree to another (Arndt 1996).

CALLS In flight, these parrots give a high-pitched, discordant *kray*, either singly or in a series of three to four notes per second and then repeated after a pause (Oren and Novaes 1986). A prolonged, strident *kew-o* is reported to be emitted during courtship (Juniper and Parr 1998). Captive birds have been heard to give a shrill, metallic *keek-keek-keek*, repeated rapidly many times, and utter an occasional *keek* while feeding.

DIET AND FEEDING Foods comprise fruits, seeds, nuts, berries, buds and flowers, procured mostly in the treetops, and from an examination of museum specimens I noticed that the plumage of wild birds often becomes stained and matted with fruit juices, particularly on the underparts. Oren has observed a flock feeding in an isolated Brazil-nut tree *Bertholletia excelsa* in a three- to four-year-old plantation, and other flocks feeding on the fruits of an *Oenocarpus* palm and of an unidentified lauraceous tree (in Oren and Willis 1981). Oren and Novaes (1986) note that maize and some cultivated fruits are eaten, but foraging primarily is in tall forest, where recorded foods include *Symphonia* buds and flowers, together with fruits of *Anacardium spruceanum* and *A. occidentale*, *Protium* and *Tetragastris* spp., *Visnia gujanensis, Inga* and *Cecropia* spp., *Brysonima crassifolia, Carapa guianensis* and *Oenocarpus bacaba*. Commonly standing at forest borders or in gaps in the forest are two principal food trees – *Croton matoensis* reaches 8 m to 15 m in height and the fruits are targeted in 70 per cent of feeding bouts, while *Brysonima crispa* reaches 25 m in height and its fruits are taken in 20 per cent of feeding bouts (Yamashita 2003).

Kyle (2005b) reports that in April–May 2004, during field studies undertaken along the Rio Cupari, at the southern boundary of the Tapajós National Forest Reserve, and in Amazonia National Park, western Pará, mucurí trees *Byrsonima crassifolia* were identified as an important food source. The fruits of these trees are the shape and size of blueberries with three 5 mm-sized seeds encased in a hard black pod. In Amazonia National Park, two flocks of parrots fed daily in a 2 ha grove of murucí trees, and on one occasion the smaller flock of 10 birds chased the larger flock of 14 birds from a fruiting tree, suggesting that flocks may maintain and defend foraging territories. When feeding in these trees, the parakeets usually were confiding and would allow a close approach. Laranjeiras (2008) recalls that during field studies undertaken in and around Amazonia National Park, in 2007, Golden Parakeets were observed feeding in 11 different types of trees, and the birds moved from one type of fruiting tree to another species with the changing availability of fruits. In some trees they fed on fruit pulp or skin while in others they took seeds. The majority of observed feeding trees were in secondary growth, and it was not known which trees were utilised in primary forest. There were numerous fruiting trees visited by other parrot species but ignored by Golden Parakeets. Murucí trees again were identified as an important food source, especially during the post-fledging period, and large gatherings of up to 50 birds were seen in areas rich in these trees. On one occasion a flock of 19 birds fed in a grove of murucí trees for more than four hours. Also, in one grove of these trees a fledgling learning to feed by itself was seen to chew on flowers, holding each one in its bill for several seconds at a time.

Significant local damage can result when flocks attack ripening maize crops, and they also take ripening mangoes. These parakeets always drink from pools of water that build up in depressions or cavities in the forks of tall trees.

BREEDING Nesting takes place mostly during the wet months of November to February, but sometimes as early as October or as late as April. The nest is in a hollow in the main trunk or major branch of a tall tree, usually at 15 m to 20 m above the ground. Oren and Novaes (1986) recall finding four nests, one of which was in a living *Sclerobium* tree, while the others were in unidentified dead trees, and all of these nesting trees were within a few hundred metres of intact forest. In the Rio Gurupí area, northwestern Maranhão, on 13 December 1983, Paul Roth watched four birds visiting holes in a dead tree, and he was informed by local persons that chicks regularly are taken from nests at the end of January (*in litt.* 1987).

Although single pairs occasionally are reported to nest alone, nests with multiple attendants are the rule, and this indicates that the species regularly breeds communally, possibly with several females contributing to the clutch. Further evidence of this comes from hunters who claim that up to nine chicks, sometimes together with unhatched eggs, are found in nests with multiple attendants, but only two or three chicks are present in the nests of single pairs. Oren and Novaes note that in a captive breeding group of three males and three females, all flock members cared for the 14 chicks, which hatched over several days, the three females sharing incubation, but it was uncertain whether the males assisted. Another report of breeding in captivity refers to incubation being undertaken only by the female, though the male also roosted in the nestbox at night (Turner 1940). Another possibility is that only a single female lays, with helpers then participating in incubation and rearing of the chicks, and the laying female may adjust her clutch size in response to the level of assistance provided by the helpers. Arndt (1996) notes that incubation is of 28 to 30 days duration, with fledging occurring 55 to 60 days after hatching, and I presume that this information also is from breeding in captivity. Yamashita (2003) points out that during the first year fledglings remain with the family group, roosting at night in the nesting hollow, and young birds then take on the role of helpers within the family group.

Activities observed daily between 19 January and 5 February 2004, at a nest in the Grupo Agropalma Forest Reserve, near Tailândia, eastern Pará, are recorded by Silveira and Belmonte (2005). This nest was at a height of 30 m in a 41 m high *Dinizia* tree, which was one of four living trees in a group separated from an adjacent patch of forest by a highway. The nest was in a thick branch angled out at approximately 40° from the trunk, and the entrance hole was about 30 cm in diameter. At the time of observation, the nest contained two well-feathered chicks, which were fed by only four of 10 adults in the attending group, but it was not possible to confirm that feeding always was by the same four individuals. With much vocalisation, the entire group arrived in upper branches of the nesting tree between 0610 hours and 0630 hours each morning, and the nestlings responded by calling while appearing at the nest entrance. After some 10 minutes, four adults came down to feed the nestlings at the nest entrance. An average of eight feeding visits was made during the day, and while the nestlings were being fed by four adults other members of the group perched in the nesting tree, usually segregated in pairs or in groups of up to four to engage in mutual preening. Each visit by the group lasted approximately 35 minutes. At night, no adults roosted in the nest with the chicks, possibly because the cavity was too small for occupancy by adults as well as the full-sized nestlings. Near nightfall, between 1815 hours and 1835 hours, the group moved across to roost in another tree standing some 60 m from the nesting tree. On average, the two chicks appeared at the nest entrance 14 times during the day, staying for about eight minutes to observe the surroundings, often without any calling. On 29 January, one chick was seen to leave the nest for its first flight. At about 1700 hours, this chick came to the entrance and, in response to much calling from the adults, flew off towards the forest, followed by the adults. After some five minutes it returned to the canopy of the nesting tree, where it perched for about 20 minutes, being fed and preened by four attending adults. The young bird returned to the nest at 1725 hours. During subsequent days, flights from the nest by the fledgling averaged three times per day, each involving an absence of about 40 minutes, and always it was accompanied by the adults. The second, presumably younger nestling was not seen to leave the nest during the period of observation. All adults in the group participated in vigorous defence of the nest against other parrots and potential predators, especially toucans and aracaris. In response to appearance of a potential competitor or predator, the entire group flew directly at the intruder, separating in flight into two smaller groups which simultaneously attacked from the sides and rear. During the period of observation 22 defending attacks were recorded, and in all instances the intruders were expelled quickly, being pursued in the air for up to 20 m or more. Copulation and inspection of potential nesting sites by a pair from the group also was observed. At 0810 hours, a pair that was isolated from the group was observed to be preening each other, and this was followed by copulation that lasted about two minutes. The pair then moved to one of the adjacent *Dinizia* trees to inspect two hollows at a height of approximately 10 m and with entrances at an angle of about 45° from the trunk. During subsequent days, two birds, presumably the same pair, were seen to inspect these same hollows, spending between 15 and 20 minutes at each visit, one bird staying inside the cavity for most of the time while the other remained at the entrance or entered for brief periods. No information was obtained on any eventual nesting by this pair.

EGGS Rounded to oval, not glossy; 6 eggs, 33.8 (32.2–35.4) × 25.8 (25.1–26.5) mm (Walters 1974).

GENUS *Conuropsis* Salvadori

Conuropsis Salvadori, *Cat. Birds Brit. Mus.*, **20**, 1891, p. 146 (in key), p. 203. Type, by original designation and monotypy, *Psittacus carolinensis* Linnaeus.

The sole extinct species belonging to this genus is a midsized parrot with a long, graduated tail and long pointed wings. The bill is proportionately large and broad with a prominent notch in the upper mandible, and the cere is completely feathered. Bright yellow and orange markings and blue wing feathers are plumage features shared with *Aratinga* species, and a relationship with that genus is confirmed by molecular analyses (Kirchman *et al.* 2012). A fully feathered cere is present also in the Monk Parakeet *Myiopsitta monachus*, but is not indicative of a relationship with *Conuropsis*, the feature apparently having evolved independently as an adaptation for cold tolerance.

Wetmore (1926) described *Conuropsis fratercula* from a single complete humerus from the Middle Miocene lower-Sheep Creek beds of the Snake Creek Quarries, Sioux County, Nebraska. Wetmore noted that it is about three-quarters the size of the humerus of *C. carolinensis*, but otherwise is so similar 'that the two would be inseparable were it not for their disparity in size'. It remains the only reliable parrot fossil from North America, but its placement in *Conuropsis* was questioned by Olson (1985).

EXTINCT

Carolina Parakeet

Conuropsis carolinensis (Linnaeus)

Psittacus carolinensis Linnaeus, *Syst. Nat.*, ed. 10, **1**, 1758, p. 97. (Carolina and Virginia.)

DESCRIPTION Length 30 cm. Weight approximately 100 g (in Snyder and Russell 2002).
ADULTS General plumage green, paler and more yellowish on underparts; forehead and lores to upper cheeks orange; remainder of head to foreneck yellow; scapulars, greater wing-coverts and tertials tinged olive and margined greenish-yellow; primary-coverts deep green edged yellowish-green; outer webs of primaries yellow at bases; bend of wing, carpal edge and thighs yellow; undersides of flight feathers greyish; tail dark green above, greyish below; bill horn coloured; iris brown; legs flesh-brown.
21 males: wing 177–201 (187.0) mm, tail 131–158 (147.6) mm, exp. cul. 22–24 (23.5) mm, tars. 16–20 (17.8) mm.
13 females: wing 178–194 (185.5) mm, tail 125–157 (142.7) mm, exp. cul. 22–24 (23.3) mm, tars. 16–19 (17.9) mm.
JUVENILES Forehead, lores and around eyes tawny-orange; remainder of head to foreneck green; bend of wing and carpal edge yellowish-green; no yellow on thighs.

DISTRIBUTION Formerly occurred in the eastern United States of America; now extinct.

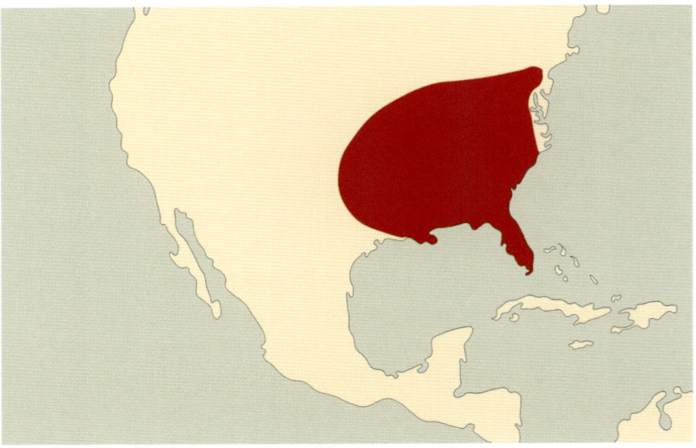

SUBSPECIES
1 *Conuropsis carolinensis carolinensis* (Linnaeus)
The nominate subspecies, as described above, formerly ranged throughout much of the southeastern United States, from Florida north to southern Virginia and occasionally as a visitor to Pennsylvania and possibly New York. Snyder (2004) points out

that the discovery of several bones at an archaeological site in Ontario dated at about 1100 AD may reflect a former natural occurrence of the species in extreme southern Canada or could be a consequence of trade in live birds and their skins between local native tribes in the region.

2 *Conuropsis carolinensis ludovicianus* (Gmelin)
Psittacus ludovicianus Gmelin, *Syst. Nat.*, **1**, pt. 1. 1788, p. 347. (Louisiana.)
ADULTS Like *carolinensis,* but green of rump, lower hindneck and sometimes wing-coverts decidedly more bluish; green of underparts less yellowish; greater wing-coverts, inner secondaries, and outer webs of primaries at bases more extensively brighter yellow.
6 males: wing 185–196 (190.2) mm, tail 121–163 (146.7) mm, exp. cul. 24–26 (24.7) mm, tars. 17–20 (18.0) mm.
5 females: wing 180–196 (187.6) mm, tail 124–157 (143.6) mm, exp. cul. 22–24 (23.2) mm, tars. 17–19 (18.0) mm.
1 unsexed: wing 190 mm, tail 145 mm, exp. cul. 24 mm, tars. 19 mm.
This poorly differentiated subspecies was formerly distributed throughout the Mississippi-Missouri Rivers drainage in the eastern interior of the United States from the Gulf of Mexico, between eastern Texas and Mississippi, or possibly western Alabama, north to the southern shores of the Great Lakes, from western New York to southern Wisconsin and to eastern Colorado, southern Nebraska, and possibly to South and North Dakota.

STATUS Extinction of the Carolina Parakeet, the only parrot native to the United States, was a tragedy. The last known specimen died at Cincinnati Zoo on 21 February 1918, and the last specimens taken in the wild apparently were collected in 1904, when one was taken opposite Platte County, at Potter, Atchison County, Kansas, in August, and four specimens were taken from a group of 12 or 13 birds encountered along Taylor Creek, Okeechobe County, Florida (Laycock 1969; in Snyder 2004). In his excellent book, *The Carolina Parakeet*: *Glimpses of a Vanished Bird*, Noel Snyder gives details of sightings made in Florida as late as the 1920s, including reports from Henry Redding, who encountered approximately 30 parakeets on Fort Drum Creek in February 1920 and saw six or seven birds near the village of Melbourne in the winter of 1928 (Snyder 2004). Much controversy has surrounded two particularly important late sightings, one in Florida and the other in South Carolina, and both accounts have been scrutinised meticulously by Snyder, who presents significant corroborative evidence in support of their validity.

In the spring of 1927, Charles Doe, a passionate egg-collector with questionable ethical standards, located three pairs of parakeets in Okeechobee County, Florida, and identified them as Carolina Parakeets; no specimens were collected, but he did take five eggs which now are in the University of Florida Museum. Doubts about the authenticity of these eggs were raised by Donald Nicholson, another egg-collector active at the same time in Florida, who suggested that the eggs could be those of a Mexican species, a number of which were reported to have escaped from captivity in Miami, and as no birds were collected the record could not be accepted (Nicholson 1948). Snyder identified the nesting locality as Gum Slough, some 27 km from Basinger, and was able to interview elderly residents who knew Charles Doe, were able to confirm the presence of Carolina Parakeets in the district during the 1920s, and recalled being told by Albert Loftin of the discovery of the nests. Loftin assisted Doe on local collecting trips, and it is likely that he discovered the nests. Loftin also claimed that the two nests were empty when first discovered, so he repeatedly checked them until eggs were present. I concur with Snyder in strongly doubting the likelihood of feral exotic parrots being present in the area in the 1920s, and in pointing out that the authenticity of these eggs could be verified by biochemical comparisons with known captive-laid eggs.

During the course of his research on Wild Turkeys *Meleagris gallopavo* along the lower Santee River in northern South Carolina, George Malamphy claimed to have seen Carolina Parakeets on multiple occasions, including a sighting on 19 or 20 January 1935 of seven birds at very close range from a hide as they came in to feed on sunflower seeds that he had put out to attract turkeys. Responding to these reports, the National Audubon Society initiated intermittent field observations in the area by two eminent ornithologists, Alexander Sprunt and Robert Porter Allen, between late 1935 and 1938, and two local woodsmen, Warren Shokes and his son Hollie, were hired to serve as wardens. A total of 27 reported sightings of parakeets were made over a four- to five-year period, but seven of these sightings, including all made by Allen, were of birds seen only in silhouette in fading light as they flew to or from an apparent nighttime roost on heavily forested Wadmacaun Island. In December 1937, after spending more than a week of fruitless searching in the area, a party of noted ornithologists, including Lulow Griscom and the executive director of the National Audubon Society, dismissed the reported sightings, and Griscom in particular was quite disparaging of observers making the reports. Although convinced of his identification at the time of the sightings, Allen later changed his mind and conceded that the birds probably were Mourning Doves *Zenaidura macroura*, but Sprunt always maintained that he had seen Carolina Parakeets. Warren and Hollie Shokes continued to make sightings of parakeets, and on 4 June 1938, at Bluff Landing, Warren Shokes encountered two adults and a fledgling. Subsequent assessments of these sightings by ornithologists and officials concluded that they were of feral exotic parrots. Again, I concur with Snyder in dismissing this explanation, for the Sun Conure *Aratinga solstitialis*, Jandaya Conure *A. jandaya* and Golden-capped Conure *A. auricapillus* are the only exotic species likely to be confused with the Carolina Parakeet, but *A. auricapillus* was little known in aviculture until the 1970s, and there are no records of feral populations of *A. solstitialis* or *A. jandaya* becoming established, even temporarily, in North America. I am convinced that Carolina Parakeets were seen along the lower Santee River and its tributaries up until 1938.

Snyder chronicles more recent reports of sightings, including a most intriguing claim made by Stanley Kolosky in a letter to Roger Tory Peterson that in the spring of 1944 he saw no less than 15 Carolina Parakeets along a five mile stretch of railroad track while a passenger on a troop train passing through forested swampland near Tabor City, in southern North Carolina, and not far from the Santee River in neighbouring northern South Carolina, where credible sightings were reported only six years earlier.

Finally, there is controversy surrounding a 16 mm colour film, of 50 seconds duration, and said to show living Carolina Parakeets in the wild. Presented by Roger Tory Peterson at the annual meeting of the American Ornithologists' Union in 1969, this film was a copy of footage of unknown origin and date supplied by

Dee Jay Nelson, an Audubon lecturer. It comprised three short clips of birds in a tree draped with what appears to be Spanish moss *Tillandsia usneoides*, and Nelson claimed that he had obtained the original film from a boat operator in Okefenokee Swamp of Georgia, but was unable to determine conclusively the identity of the photographer or location of the birds. Several birds, apparently all parrots, were visible in the footage and, although these birds showed some field marks typical of Carolina Parakeets, including yellow and orange head markings, there was no consensus among attendees at the 1969 meeting that the birds were Carolina Parakeets. The film was in colour, so indicating that the birds had been photographed no earlier than 1936 or 1937, when colour film became commecially available. A close examination of the film was made by Noel Snyder and colleagues, who concluded that, although it shows one living parrot in a setting that seems potentially appropriate for Carolina Parakeets, the birds in view probably are not living Carolina Parakeets (Snyder *et al.* 2010). The single bird that clearly seems to be alive exhibits relatively drab colouration, and appears to be some other parrot species artificially coloured to look somewhat like a Carolina Parakeet. Another two birds are more vividly coloured, and more closely resemble Carolina Parakeets, but these birds do not move naturally and very likely are not alive. Like the living bird, both could have been specimens of some other species artificially coloured to resemble Carolina Parakeets or may have been mounted specimens of Carolina Parakeets or models painted to resemble Carolina Parakeets. Multiple anomalous features of the film suggest that it is a deliberate, but clumsy hoax.

Although it is highly probable that Carolina Parakeets survived in the wild until the late 1930s or early 1940s, or possibly even until the early 1950s, I do not doubt that they now are extinct. The complete story behind their disappearance will never be known because documentation of their decline was sketchy, and no reliable estimates were made of total numbers at the time of European settlement. Early accounts describe very large flocks, sometimes comprising hundreds to thousands of birds, and it has been speculated that numbers may well have been in the millions. Snyder (2004) notes that although data are insufficient to confirm the accuracy of these estimates, it is apparent that there was little interaction with Native American culture, and it is reasonable to assume that the overall population was relatively stable prior to European settlement. Consequently, demise of the species can be attributed both directly and indirectly to the activities of European settlers. Greenway (1967) pointed out that during a period of about 90 years the range of *Conuropsis* gradually contracted from east to west, towards the Mississippi River, with dates of final records from various regions coinciding well with the spread of settlements and attendant deforestation. Persistent persecution and destruction of forested habitats are widely accepted as having been direct causes of the extirpation of these parakeets, but McKinley (1966) pointed out that it is too easy to claim that they were such pests of fruit and grain crops that they were relentlessly exterminated. In contrast to what we know of the widespread slaughter of Passenger Pigeons *Ectopistes migratorius*, which were driven to extinction at about the same time, there seems to have been little hunting of parakeets for food or sport, and I agree that more subtle primary causes were involved with their demise. However, the impact of persecution as a secondary pressure should not be underrated. The decline in numbers was quite perceptible by 1831, when Audubon wrote:

> *Our Parakeets are very rapidly diminishing in number; and in some districts, where twenty-five years ago they were plentiful, scarcely any are now to be seen..... I should think that along the Mississippi there is not now half the number that existed fifteen years ago.*

When their extreme rarity became evident, and Florida was proving to be the last stronghold, collectors and trappers eagerly sought out the remaining flocks. As is so often the case, rarity dramatically increased demand, and the final sprint to extinction was accelerated.

How significant was persecution as a cause of extinction? Was it so intense and widespread that mortality rates for the species exceeded recruitment levels throughout the entire range? Answers to these questions cannot be gleaned from the recorded evidence that we have, but it is known that the parakeets were shot and, because of their gregarious habits, it was possible to destroy large numbers with little effort. An account by Audubon probably is somewhat overdramatic, but it does give us a picture of what was happening in 1831, when he wrote:

> *Do not imagine, reader, that all these outrages are borne without severe retaliation on the part of the planters. So far from this, the Parakeets are destroyed in great numbers, for while busily engaged in plucking off the fruits or tearing the grain from the stacks, the husbandman approaches them with perfect ease, and commits great slaughter among them. All the survivors rise, shriek, fly round about for a few minutes, and again alight on the very place of most imminent danger. The gun is kept at work; eight or ten, or even twenty, are killed at every discharge. The living birds, as if conscious of the death of their companions, sweep over their bodies, screaming as loud as ever, but still return to the stack to be shot at, until so few remain alive, that the farmer does not consider it worth his while to spend more of his ammunition. I have seen several hundreds destroyed in this manner in the course of a few hours, and have procured a basketful of these birds at a few shots, in order to make choice of good specimens for drawing the figures by which this species is represented in the plate now under your consideration.*

Capture for the pet trade has been cited as another factor possibly contributing to the demise of these parakeets, and reports confirm that the birds were common and esteemed in European markets during the mid to late 19th century. Evidently, they were more popular in Europe than in North America, and presumably most birds came from Florida, which was a known stronghold of the species at that time. There are accounts of intensive trapping in Okeechobee County during the late 19th century, and Henry Redding is reported to have netted two flocks totalling 76 birds in February 1899 from roosting hollows in two cypress trees. Snyder notes that there is no evidence of trapping in central Florida continuing into the 1910s and 1920s, so he concludes that capture for the pet market probably was not a significant factor in the final disappearance of parakeets in this region. I suspect that a scarcity of birds or fragmentation of the population into scattered subpopulations brought about a cessation of trapping,

Plate 43
Carolina Parakeet *Conuropsis c. carolinensis*
UPPER adult CENTRE adult LOWER juvenile

F. KNIGHT '08

and it is possible that capture for the pet market did contribute significantly to the extirpation of Carolina Parakeets in central Florida.

It has been suggested also that in addition to the loss of nesting sites as a consequence of deforestation or logging operations, Carolina Parakeets may have faced competition for hollows from introduced honeybees. Snyder comments that bees probably favoured hollows larger than those selected by nesting parakeets, so conflict between bees and parakeets may have occurred at communal roosting sites rather than at nesting hollows.

Although he found little supporting evidence in early accounts, McKinley (1980) recognised the possibility of disease being a significant factor in demise of the Carolina Parakeet. Snyder points out that the susceptibility of parrots to a number of avian diseases is well documented, and disease does offer one of the more plausible explanations for disappearance of the last populations. A tendency to feed near to human structures and to roost in those structures was recorded by observers in central Florida, with three separate structures – a barn, an old log cabin and a blacksmith's shop – being identified specifically as places where parakeets habitually roosted. Such an association with human habitation in rural communities could have exposed the birds to diseases carried by poultry or other domestic animals.

In summary, it can be said that a lack of reliable evidence in early accounts prevents extinction of the Carolina Parakeet being attributed to any single cause, but these same accounts demonstrate forcefully that the species faced formidable adverse pressures associated with the spread of European settlement. A combination of intensive persecution and the loss of forested habitats required for successful nesting could have brought about a rapid decline in much of the range, and disease may have played an increasingly significant role as numbers reached low levels.

HABITATS Snyder and Russell (2002) point out that numerous early accounts indicate that Carolina Parakeets were associated closely with mature bottomland forests, especially cypress swamps in the Deep South and woodlands of sycamore *Platanus occidentalis* in more northerly regions. Although strongly attached to heavily forested river valleys, mature trees bordering streams, and extensive cypress swamps, they ranged far and wide in search of food, regularly coming into farmlands, weed-infested paddocks or roadside verges, and other disturbed habitats. Snyder (2004) notes that the predominance of heavily forested bottomlands in early records does not demonstrate a strong dependence of these parakeets on such habitats, and the importance of extensive clearing of these forests in the decline of Carolina Parakeets may have been overemphasised. I would caution against underestimating the impact of deforestation on populations, for there probably was a strong reliance on these forests for nesting sites, and availability of forested habitats in near proximity to feeding areas may have been a prerequisite for successful recruitment.

MOVEMENTS By mapping monthly records of Carolina Parakeets in regions north of Florida, McKinley (1977) attempted to determine if seasonal migratory movements were evident. Resulting patterns did indicate a widespread, though weak dispersal in winter, possibly constituting relatively small-scale migratory movements by some populations, but there were no indications of regular latitudinal migrations. McIlhenny claimed that the parakeets arrived in southern Louisiana in late April, when the black mulberries were ripening (in Bendire 1895).

Carolina Parakeets were known to be tolerant of cold weather, and in northern parts of the range often were encountered in snowstorms or subfreezing weather. Bendire (1895) noted that although rather restless at all times, they generally were resident, wandering locally in search of food, but usually returning at night to the same communal roost. This strong fidelity to nighttime roosts is well documented in early accounts. Snyder (2004) points out that reported data suggest that the parakeets were resident, though probably undertaking local or even long-distance wandering at times of food scarcity, and any sudden disappearance from an area, with no eventual return, more likely signalled extirpation of that population instead of a move to elsewhere.

HABITS In the early years of European settlement, Carolina Parakeets regularly were seen in large flocks, often comprising hundreds of birds, but as they became rare the sizes of flocks diminished and eventually only small parties, pairs or occasionally single birds were encountered. They were most active in the early morning and late afternoon, the remainder of the day being spent resting amongst the foliage of large, shady trees. In the early morning, just before sunrise, they would assemble in the roosting trees and, to the accompaniment of much chattering, would climb to the topmost branches. As soon as the sun rose, small flocks flew off in all directions to the feeding areas. After feeding for two to three hours they often went to a nearby stream to drink and bathe, and then retreated to a grove of trees where they sheltered during the hotter part of the day. In the late afternoon they returned to the feeding areas, remained until dark, and then made their way back to the roosting site, once again to the accompaniment of much screeching. They roosted communally, and a favoured roosting place was inside a large hollow stump or branch of a tree, where up to 20 or more parakeets would spend the night huddled together, each individual hanging attached to the wall by its bill and feet. This predilection for roosting as a massed group inside a hollow probably was an adaptation to withstand cold nights. If disturbed during the night, roosting birds were inactive, and one account states: 'The parrots make no attempt to fly off or to bite as they are gathered up like piles of wood'. Such accounts suggest that torpor may have been involved in overnight roosting, and this also could have been an adaptation for withstanding cold temperatures. Overnight roosting occurred also amidst the uppermost branches of tall trees and inside farm buildings. Snyder (2004) relates the experiences of Minor McGlaughlin, an elderly resident of Okeechobee County, central Florida, who remembered seeing Carolina Parakeets perching during the day on rafters inside his grandmother's barn, and again at night hanging by their bills lined up along the sides of the rafters; he remarked also on the inert state of these roosting birds at night. Another known roosting site was inside a blacksmith's shop near Fort Drum, central Florida, and in 1966, when this site was inspected by Alexander Sprunt and Glen Chandler, triangular marks made by parakeets' bills on the rafters were clearly visible (in Snyder 2004).

Recalling his encounter with Carolina Parakeets in eastern Florida, in March 1889, Chapman (1904) commented on an apparent regularity in their habits:

> *There was evident regularity in the habits of the birds we afterward observed – in all about fifty, in flocks of from six to*

twenty. At an early hour they left their roost in the "hummock" bordering the river and passed out into the pines to feed, always, so far as I observed, selecting thistle patches, and eating the seeds only when in the milky stage. At about ten o'clock they returned to the "hummock" and apparently to some favorite tree, here to pass the rest of the morning and early afternoon, when they again started out to feed, returning to the roost just before sunset. A flock of these birds feeding among the thistles is a most beautiful and animated sight; one is almost persuaded not to disturb them. There is constant movement as they fly from plant to plant, or, when securing thistles, they fly with them in their bills to a neighboring tree, there to dissect them at their leisure.

These parakeets appear to have been quite confiding, a trait sadly reflected in accounts of the ease with which large numbers could be shot. Minor McGlaughlin told Snyder that as a boy he was given the task of digging out green sandspur plants, a favoured food of the parakeets, from fields on his parents' farm, and as he worked in the fields large flocks of parakeets swirled overhead.

The flight was swift and direct, the flocks darting in and out through the trees with remarkable precision. When coming into feeding trees, they would spiral down from above until almost reaching the ground, and then rise up to alight on the branches.

Recalling his disturbance of a feeding flock of seven parakeets at a locality in eastern Florida, in March 1889, Chapman (1904) noted that '.... they darted like startled Doves through the pines, twisting and turning in every direction, and flying with such rapidity they were soon lost to view....'.

CALLS Carolina Parakeets were noisy, especially in flight when they emitted a loud, screeching *qui...qui...qui...qui-ii-ii-ii-ii*, each note terminating with an upward inflection and the last being drawn-out (in Bendire 1895). While feeding, the birds kept up a constant chatter, and subdued, twittering notes sometimes were uttered while at rest.

DIET AND FEEDING Snyder and Russell (2002) point out that it is evident from feeding records that the Carolina Parakeet was not a food specialist and fed on fruits, seeds and, to a lesser extent, buds and flowers of the most common plants. Because of observational biases, foods found in close association with human habitation are strongly over-represented in records, and fruits or seeds of forest trees probably are under-represented. Reflecting these biases is the predominance of seeds of *Xanthium* cockleburs and apples among food items compiled from literature records and interviews with elderly residents who were familiar with the parakeets in central Florida. Seeds of cockleburs, *Cenchrus* sandspurs, *Cirsium* thistles and other widespread weeds were the most commonly recorded foods, and it was suggested that the parakeets were more plentiful in the Mississippi River valley than in northeastern states mainly because cockleburs were more abundant in the valley (in Snyder 2004). Hard-shelled seed-balls of cypress *Taxodium distichum* evidently were an important winter food, and this probably accounts for frequent reports of an association between the parakeets and cypress swamps. *Pinus* cones, seed-balls of sycamores *Platanus occidentalis*, *Quercus* acorns, pecan nuts *Carya illinoensis*, seeds of *Acer* maples, berries of cabbage palm *Sabal palmetto*, and beech nuts *Fagus grandifolia* are among recorded wild foods, while there are many reports of the parakeets taking cultivated grain, mostly corn and wheat, and fruits, including apples, mulberries, oranges and cherries. When feeding on cockleburs, which are known to be poisonous to livestock, the parakeets may have been unpalatable, or even toxic to predators. This was noted by Audubon, and by Alexander Wilson who found that parakeets feeding on corn for some time apparently were harmless food for cats, but could be lethal when feeding on cockleburs (in Snyder 2004). Feeding occurred mostly in trees or bushes and low down among standing weed plants, but the parakeets occasionally came to the ground to pick up fallen seeds or to take sand and saline soil. Also they were reported to have come in large numbers, often in the company of myriads of Passenger Pigeons *Ectopistes migratorius* and herds of American bison, to drink saline water at well-established licks.

Chapman (1904) recounted observations of birds feeding on *Cirsium* thistles:

Late on the afternoon of our arrival we started a flock of seven Paroquets from a productive patch of the thistles (Cirsium lecontei) *which proved to be their favorite food........Two days passed before I again met Conurus, and this time to better advantage. It was a wet and drizzling morning when we found a flock of six birds feeding on thistles at the edge of a "prairie". Perched on the leafless branches of the tree before us, their brilliant green plumage showed to the best advantage. Several were skillfully dissecting the thistles they held in their feet, biting out the milky seed while the released fluffy down floated away beneath them.*

A similar technique, involving transfer of the food item to the foot, usually the left foot, is described by Seward when recalling observations of birds feeding on cockleburs (in Snyder 2004):

In eating, the bird picked up a burr with its beak, this was then delivered to one foot raised to receive it. Then one end of the burr was cut off with the sharp-ended under beak, the burr being held with the foot and the under side of the upper beak while two small kernels were extracted with the assistance of the tongue, and the husk was thrown away.

Although considered by some farmers to be beneficial because they fed so extensively on cockleburs and other serious weed pests, Carolina Parakeets were persecuted relentlessly because of their raids on cultivated fruits and grain crops. These depredations were described vividly by Audubon in 1831, when he wrote:

The Parrot does not satisfy himself with Cockle-burs, but eats or destroys almost every kind of fruit indiscriminately, and on this account is always an unwelcome visitor to the planter, the farmer or the gardener. The stacks of grain put up in the field are resorted to by flocks of these birds, which frequently cover them so entirely, that they present to the eye the same effect as if a brilliantly coloured carpet had been thrown over them. They cling around the whole stack, pull out the straws, and destroy twice as much of the grain as would suffice to satisfy their hunger. They assail the Pear and Apple-trees, when the fruit is yet very small and far from being ripe, and this merely for the sake of the seeds. As on the stalks of Corn, they alight on the Apple-trees of our orchards, or the Pear trees in the garden, in great numbers; and, as if through mere mischief, pluck off the fruits, open them up to the core, and, disappointed at the sight of the seeds, which are yet soft and of a milky consistence, drop the apple or pear, and pluck another, passing from branch to branch, until the trees which were before so promising, are left completely stripped, like the ship water-logged and abandoned by its crew, floating on the yet agitated waves, after the tempest has ceased. They visit the Mulberries, Pecan-nuts, Grapes, and even the seeds of the Dog-wood, before they are ripe, and on all commit similar depredations. The Maize alone never attracts their notice.

Cottam and Knappen (1939) report that stomach contents from a bird collected in Florida in 1885 comprised 32 seeds of loblolly pine *Pinus taeda*, two rabbit hairs, two fragments of the bird's own feathers, and two fragments of an ant; the feathers and rabbit hairs surely were ingested accidentally, possibly having been caught in cockleburs. Although the ant also could have been ingested accidentally, presence of these fragments prompted Snyder to note that in 1891 apparent feeding of insects was reported by David Thomas, who recorded his observations as follows:

A small cotton wood tree stands opposite the window where I am writing (with) dark excrescences on its branches like those which appear on this species in western parts of New York. It is well known that these blemishes are produced by the irritation of insects; - first by a puncture when the egg is deposited, and afterwards by the growth and motion of the worm. To procure this food, the parroquets have been busily employed, at times, through the day; but though they have become so familiar; and though they excel all the birds of this country in beauty of plumage, their scream is so discordant, and their fierceness of disposition so apparent, as to preclude every sensation of attachment.

BREEDING There are surprisingly few nesting records of Carolina Parakeets, and the authenticity of some reports is questionable. Snyder and Russell (2002) point out that, assuming the general validity of historical records, the seasonal timing of nesting apparently was quite uniform, with records of eggs for March and April and of nestlings or fledgling for June. In the United States National Museum, Washington, there is a clutch of two eggs taken in St Mary's Parish, Louisiana, in March 1878, and in the American Museum of Natural History, New York, there is a clutch of two eggs collected on 26 April 1855 at an unnamed locality in Georgia (in Bendire 1895). The eggs from Georgia were found on a layer of a few chips at the bottom of a hollow in a tree. With the notable exception of the two clutches taken by Charles Doe in central Florida in 1927, most of the other eggs in collections were laid in captivity.

As already mentioned, much controversy has surrounded the two clutches, one of three eggs and the other of two eggs, taken by Charles Doe on 30 April 1927 at Gum Slough, near Basinger, central Florida. Snyder (2004) details events leading to the discovery of these two nests, apparently by Alfred Loftin who assisted Doe in his collecting activities in the district. The account presented by Snyder, featuring interviews with elderly residents who knew both Doe and Loftin, is compelling and leaves me in no doubt about the authenticity of this extremely important nesting record.

Not so convincing are claims that in addition to nesting in hollows in trees, Carolina Parakeets also nested in open nests of twigs placed in the forks of trees. Snyder details some of these claims, noting that they were reported by ornithologists of considerable stature, and concludes that the possibilty of open twig nests should not be discounted. I would caution that these claims were not firsthand observations by reputed ornithologists, but were reports of secondhand records. Typical of such reports is the account by William Brewster (1889), who wrote:

While in Florida during February and March, 1889, I questioned everybody whom I met regarding the nesting of the Parrakeet. Only three persons professed any knowledge on this subject. The first two were both uneducated men – professional hunters of alligators and plume birds. Each of them claimed to have seen Parrakeets' nests, which they described as flimsy structures built of twigs and placed on the branches of cypress trees. One of them said he found a nest only the previous summer (1888), *while fishing. By means of his pole he tipped the nest over and secured two young birds which it contained.*

This account was so widely at variance with what has been previously recounted regarding the manner of nesting of this species that I considered it, at the time, as a mere fabrication, but afterwards it was unexpectedly and most strongly corroborated by Judge R. L. Long of Tallahassee. The latter gentleman, who, by the way, has a very good general knowledge of the birds of our Northern States, assured me that he had examined many nests of the Parrakeet built precisely as above described. Formerly, when the birds were abundant in the surrounding region, he used to find them breeding in large colonies in the cypress swamps. Several of these colonies contained at least a thousand birds each. They nested invariably in small cypress trees, the favorite position being on a fork near the end of a slender horizontal branch. Every such fork would be occupied, and he has seen as many as forty or fifty nests in one small tree. Their nests closely resembled those of the Carolina Dove (Mourning Dove), *being similarly composed of cypress twigs put together so loosely that the eggs were often visible from the ground beneath. The twigs of the cypress seemed to be preferred to those of any other kind of tree. The height at which the nests were placed varied from five or six feet to twenty or thirty feet. Mr Long described the eggs as being of a greenish white color, unspotted. He did not remember the maximum number which he had found in one nest, but thought it was at least four or five. He had often taken young birds from the nests to rear or to give to friends.*

It is conceivable that Carolina Parakeets may have excavated nesting cavities in large nests of twigs built by birds such as raptors or corvids, especially in districts where natural hollows were scarce, but I doubt very much that they built nests of twigs 'put together so loosely that the eggs were often visible from the ground beneath'.

Most credible records are of pairs nesting in natural hollows in trees or in holes made by woodpeckers. Included in notes made by Charles Doe on the collection of eggs at Gum Slough, central Florida, in April 1927, is the following brief description of the nesting hollows (in Snyder 2004):

Nesting in natural cavity in oaks, 30 and 40 ft high. #1 in cavity about a foot deep in a big limb, hole a rotted out one, hole went in at an angle, #2 in a hole 40 ft up formed by separation of two big limbs main part of tree, may have been dug by woodpeckers, both holes big enough to put my hand in.

In addition to nesting in colonies, as claimed by Judge Long in the account given by Brewster, there are references to communal nesting, most notably by Audubon (1831), who wrote:

> *Their nest, or the place in which they deposit their eggs, is simply the bottom of such cavities in trees as those to which they usually retire at night. Many females deposit their eggs together. I am of the opinion that the number of eggs which each individual lays is two, although I have not been able absolutely to assure myself of this. They are nearly round, and of a light greenish white. The young are at first covered with soft down, such as is seen on young Owls.*

Although uncommon in parrots, communal nesting involving two or more females laying in the same hollow and then sharing incubation, may be practised by the Golden Parakeet *Guaruba guarouba*, another highly social neotropical species, so the Carolina Parakeet may have nested communally. Another possibility is that helpers, probably offspring from previous seasons, helped with the rearing of nestlings, and this is more prevalent among parrots. No information on incubation or parental care was recorded from the wild, but a most interesting account of an encounter by Warren Shokes with two adults and an apparent fledgling on 4 June 1938, at Bluff Landing, on the lower Santee River, South Carolina, is recalled by Alexander Sprunt in a brief report of a trip he made to the same site on 17 November 1938 (in Snyder 2004):

> *Two adult birds made their appearance almost as soon as he arrived, and began circling overhead at a height of about fifteen to thirty feet, chattering loudly and much excited.... He sat down on a root, and the birds kept about so insistently that he began to think of young ones being about. He arose and walked toward the edge of the Bluff, when a young one left the bushes there and started to fly across the river toward Wadmacaun Island. Both adults ceased their circling over him, and flew out with the youngster. The latter seemed to be making heavy weather of it and kept getting nearer and nearer the water, but just did get into the growth on Wadmacaun.... Just after the young one disappeared, both adults came back across the river and circled and chattered again over Shokes' head for a few moments, then re-crossed the river and disappeared.*
>
> *The adults were described as "just like the birds in the picture" (the colored plate of the Paroquets which we sent them). Yellow heads, green bodies. The young was said to be a single color allover, very greenish. I asked him why he didn't report it at once, and he said that it might cause me, or someone else to come right up there, and he was afraid we couldn't find the birds again, and we would have that trouble for nothing!! He seemed to be chargrined at the failure last March to produce anything, that he said he didn't want to "stir up things again".*
>
> *I tell this exactly as he told it to me, and for what it is worth. He really seemed rather reluctant to tell it at all, but waxed more animated as he described it, particularly when the birds were almost over his head at close range. It must be recalled that he is rather deaf, and that the noise made by the birds must have been considerable for him to hear it. This seems much impressed on his mind. With his determined adherence to the colors of the birds, their actions etc. what they could have been other than Carolina Parakeets leaves one guessing. It is encouraging to know that what few are there are succeeding in raising a young bird now and then anyway.*

EGGS Ovate, slightly glossy; Bent (1940) listed 34.2 (32.1–37.0) × 27.8 (25.8–30.2) mm as the measurements of 24 eggs.

GENUS *Aratinga* Spix

Aratinga Spix, *Av. Bras.*, **1**, 1824, p. 29. Type, by subsequent designation, *Psittacus luteus* Boddaert = *Psittacus solstitialis* Linnaeus. (G. R. Gray, *Cat. Gen. Subgen. Bds*, 1855, p. 87).

Members of this genus are small to midsized parrots with a long, strongly graduated tail, a proportionately broad, heavy bill, unfeathered cere, and a prominent bare or partly feathered eyering, but the lores to upper cheeks are fully feathered. There is no sexual dimorphism, and juveniles are duller than adults.

Ribas and Miyaki (2004) point out that *Aratinga* formerly comprised 20 species distributed throughout Central America, the West Indies and South America, but molecular analyses fail to give support to the monophyly of the genus and, based on plumage and osteological characters, several groups of taxa can be differentiated. One of these groups includes the endangered Sun Parakeet *A. solstitialis*, which is the type species of *Aratinga*, so only that group, comprising four similar and very closely allied species in the '*solstitialis* complex' together with the more divergent *A. weddelli* and *A. nenday*, are retained in *Aratinga*. Plumage features identified by Ribas and Miyaki as uniting these species in *Aratinga* and differentiating them from other species formerly included in the genus are dark blue on the apex of the primaries, dark blue in the primary coverts, dark blue towards the tip of the upperside of the tail, and a dark underside of the tail.

Aratinga is disjunctly distributed in northern and central South America.

ENDANGERED

Sun Parakeet

Aratinga solstitialis (Linnaeus)

Psittacus solstitialis Linnaeus, *Syst. Nat.*, edn 10, **1**, 1758, p. 97. (Guinea, error = Cayenne substituted by Hellmayr, 1906.)

DESCRIPTION Length 30 cm. Weight 120 g.

ADULTS General plumage yellow, variably tinged orange on forehead, sides of head, lower abdomen, rump and lower back; under tail-coverts green tinged yellow; mantle, lesser and median upper wing-coverts and underwing-coverts yellow variably marked green; undersides of flight feathers dull grey; secondary-coverts green broadly tipped yellow; primary-coverts and outer webs of flight feathers dark blue; tail above olive, below olive-grey; bill grey-black; bare eyering greyish-white; iris dark brown; legs grey.

14 males: wing 146–160 (153.1) mm, tail 131–146 (137.6) mm, exp. cul. 21–25 (22.8) mm, tars. 16–18 (16.9) mm.
8 females: wing 150–162 (156.4) mm, tail 121–146 (133.1) mm, exp. cul. 19–24 (21.5) mm, tars. 16–18 (16.8) mm.
JUVENILES Crown marked green; upper back and scapulars green slightly edged yellow; rump and lower back tinged red; upper tail-coverts green tinged yellow; greenish throat; breast and abdomen orange; under tail-coverts green margined paler green; lesser and median upper wing-coverts green edged yellow; bill paler grey; cere and bare eyering white.

DISTRIBUTION Lower Amazon River basin, where confined to the upper Rio Branco region, northern Roraima, far northeastern Brazil, and central and southern Guyana east to the Kabalebo River in western Suriname; possibly extralimital in extreme southeastern Venezuela, where an unconfirmed sighting of a flying bird made south of La Gran Sabana in the late 1970s probably was of this species (see Hilty 2003). A sight record from Kourou, coastal French Guiana, is highly doubtful and probably refers to an escaped cagebird (see Tostain *et al.* 1992). Recent reports have come almost exclusively from the Karasabai district in the Rupununi River region, southern Guyana, and in and around Boa Vista, northern Roraima, Brazil.

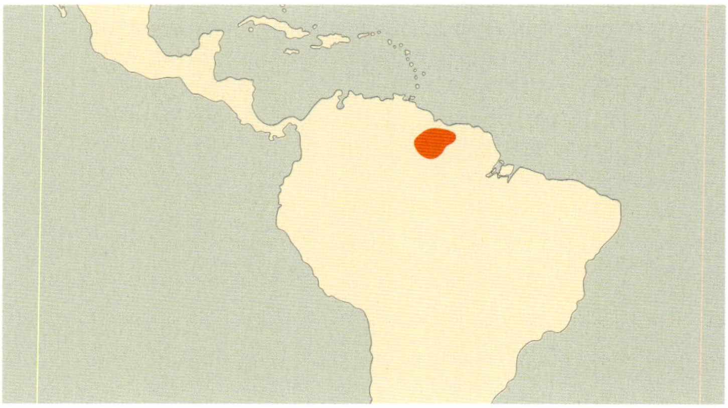

SUBSPECIES Currently, four similar allopatric taxa are included in the '*solstitialis* complex' as separate species, but previously three of these taxa – *A. solstitialis*, *A. jandaya* and *A. auricapillus* – have been treated variously as subspecies of *A. solstitialis* or as separate species in a single superspecies (see Dickinson and Remsen 2013). Ribas and Miyaki (2004) point out that treating the three taxa as subspecies has led to uncertainties about their vulnerability and to errors when referring to all birds, irrespective of where they occur, simply as *A. solstitialis*. They note also that molecular analyses reveal only a small amount of genetic divergence between them, so indicating a recent origin, but the results provide evidence that they are separate, well-defined species, and there is no evidence of interbreeding in the wild. Subsequently, *A. maculata* from north of the Amazon River in southern Amapá and northern Pará, northern Brazil, has been differentiated as a separate species and also included in the '*solstitialis* complex' (Silveira *et al.* 2005). I follow the arrangement of including all four taxa as separate species in a single superspecies, but am doubtful that specific differentiation between *A. solstitialis* and the very similar *A. maculata* is warranted.

STATUS Concern about the status of Sun Parakeets was raised by Joseph (1992), who found them to be less common than previously realised, and he suggested that they may indeed be adversely affected by trapping. Large-scale trapping for the live-bird market seems to have been the single most important threat responsible for catastrophic declines in populations of Sun Parakeets, especially in Guyana. In 1971, during a visit to Guyana, I was shown two captive birds at Georgetown Zoo, and was told that they probably were the first birds in captivity. Since the 1970s great numbers have been exported from Guyana, leading to near extirpation of the species in that country, and I am sure that now there are more birds in captivity than are left in the wild. An annual export quota of 600 birds was set by Guyana in the 1980s, and more than 2200 birds were imported into the United States between 1981 and 1985 (Jorgensen and Thomsen 1987). Gilardi points out that, because the parakeets are attracted from large distances to bait, it is easy to trap the entire population in an area (in Birdlife International 2016). It is reported that because of the scarcity of these parakeets in Guyana, trappers from that country and from French Guiana have travelled across the border to Brazil to buy birds for export.

Near Villa Pereira, northernmost Roraima, in June 1986, Leo Joseph found Sun Parakeets to be quite local, with sightings being recorded on four of the 12 occasions when searches were made (*in litt.* 1986). Since the early 1990s, records in northern Roraima have been limited to only nine localities, and the population is thought to comprise fewer than 1000 birds (in Birdlife International 2016). In savannas of the Rupununi River, southern Guyana, where these parakeets were fairly common in the 1970s, the surviving population near the village of Karasabai is estimated at approximately 200 birds (see Bergman 2009). Based on recent records, the total population is estimated to number 1000–2499 mature birds, which equates to 1500–4000 individuals (in Birdlife International 2016).

HABITATS Joseph (1992) notes that Sun Parakeets inhabit a broader range of habitats than previously realised, and are found in open savanna, savanna woodland, forested valleys and secondary growth. Silveira notes that although they use forest edges it seems that quite a large expanse of intact forest is required (in Birdlife International 2016). In June 1986, approximately 175 km and 200 km north of Boa Vista, northern Roraima, northeastern Brazil, Leo Joseph observed flocks in open savanna and in dry scrubland with rocky outcrops that supported thick vegetation (*in litt.* 1986). Birds were not encountered in extensive open plains where creeks were lined with *Mauritia* palms or in open savanna that lacked isolated, well-vegetated rocky outcrops. Bergman (2009) recalls that in an area of tall grasses and moist forests, near Karasabai, in southern Guyana where the savanna of the Rupununi River rises into the foothills of the Pakaraima Mountains, flocks were encountered in steep hillside forest surrounding a small farm, and at times birds came to feed in low bushes bordering a cassava field. An abundance of suitable habitat at this site provided ample evidence that trapping for the live-bird market, not habitat loss, was the cause of the rarity of these parakeets, and one of the 'toshaus' or captains of the village thought that, after the local population had been trapped out in the 1990s, repopulation may have originated with some birds

Plate 44
Sun Parakeet *Aratinga solstitialis* (adults)

that escaped from the trappers. In early September 1980, near Devis Vallen on the Kabalebo River, western Suriname, a bird was observed in rather open secondary forest (in Joseph 1992).

Diehl (2009) reports that in January 2008, a flock of 10 Sun Parakeets was encountered within the urban confines of Boa Vista, in northern Roraima, and it was suspected that this flock may have originated from birds that escaped from a holding cage after being confiscated as chicks.

MOVEMENTS Joseph (1992) notes that it is unclear whether movements by Sun Parakeets represent nomadism or a seasonal pattern based on food availability.

HABITS Sun Parakeets are spectacular birds, and Leo Joseph recalls a memorable sighting that caused him to consider them to be the most beautiful of neotropical parrots (*in litt.* 1986). When flushed from low bushes in which they were feeding, a party of seven parakeets flew up to alight on the topmost projecting leafless branches of a tree, where the early morning sun shone directly on to their plumage. At other times, these parakeets could be surprisingly inconspicuous as they fed quietly amidst shrubby vegetation, but in flight they were extremely noisy, and the screeching call-notes could be heard long before the birds came into view.

In the savanna of the Rupununi River, in southern Guyana, the village of Karasabai and its surrounds are frequented by some of the last flocks of Sun Parakeets, and Bergman (2009) recalls that, during a visit to the area, the striking beauty of these parakeets was most evident when they were seen in swift, direct flight before disappearing into dense forest, or when observed at close range. At a small farm, flocks of about 20 birds came each day from nearby hillside forest, arriving at about 0900 hours and remaining until mid-afternoon when they returned to the hills. They were confiding and could be watched at close quarters. Juveniles often were seen engaging in playful actions, including nipping at each other's feet, and begging from adults.

CALLS In flight these parakeets emit a loud, screeching *screek-screek*, repeated rapidly in short bursts. Also given is an unparrot-like, high-pitched repetitive wheezy note and, while feeding or when perched in clear sunlight, birds occasionally emit more typically parrot-like chuckling notes (Joseph 1992).

DIET AND FEEDING Seeds, fruits, nuts, berries and probably blossoms make up the diet. In early June 1986, near Villa Pereira, northernmost Roraima, Brazil, Leo Joseph watched birds that evidently were feeding on the red fruits of cacti growing out from between rocks, and on the ground underneath were found chewed pieces of the white pulp with numerous black seeds (*in litt.* 1986). Bergman (2009) recalls that in the savanna of the Rupununi River, southern Guyana, birds were seen feeding in low bushes beside a cassava field.

BREEDING Very little is known of the nesting of the Sun Parakeet in the wild. Diehl (2009) reports that in November 2008, in urban Boa Vista, northern Roraima, Brazil, a nest was found inside a concrete pole.

Nieremberg (1972) recorded details of a successful breeding in captivity. Four eggs were laid at intervals of two days, and incubation was undertaken solely by the female, during which time she left the nest only for brief periods of feeding. Three eggs hatched, the first four weeks after it was laid, and the chicks were fed by both parents. The chicks fledged approximately eight weeks after hatching. Seibels and McCullough (1978) recorded a captive breeding, when a single male paired repeatedly with two females, which reared successive broods in separate nestboxes.

EGGS The eggs are rounded, and measurements of three eggs are 29.5 (28.4–31.4) × 23.5 (22.8–24.0) mm (Schönwetter 1964). Arndt gives 26.7 (26.0–28.0) × 22.0 (22.0–22.0) mm as the measurements of seven eggs laid in captivity (*in litt.* 1986).

GENUS *Psittacara* Vigors

Psittacara Vigors, *Zool. Journ.*, **2**, 1825, p. 400. Type, by original designation, *Psittacus guianensis* 'Linnaeus' = *Psittacus guajanensis* Gmelin = *Psittacus leucophthalmus* S. Müller.

Previously included in *Aratinga*, parrots belonging to this genus are characterised by a predominantly green plumage colouration, often with red markings on the head, throat and underwing-coverts, a conspicuous white unfeathered eyering, and a proportionately large, robust horn-coloured bill (Silveira *et al.* 2005). The tail is uniformly green above, and olivaceous on the underside, and undersides of the flight feathers are yellowish. All are midsized parrots with long, graduated tails, and there is little or no sexual dimorphism.

Psittacara occurs in the West Indies, and is widespread throughout Central and South America. The writings of early voyagers suggest that additional species may have occurred in the West Indies, but their existence is not supported by fossil remains and it is not known how, or if, they differed from extant forms.

EXTINCT

Puerto Rican Parakeet

Psittacara maugei Souancé

Psittacara maugei Souancé, *Rev. et Mag. Zool.*(2), **8**, 1856, p. 59 (no locality = Puerto Rico).

DESCRIPTION Length 32 cm.

ADULTS General plumage green, slightly paler and more yellowish on underparts; carpal edge and lesser underwing-coverts dark scarlet; outer greater underwing-coverts pink-red; undersides of flight feathers and tail dull olive-yellow; bill horn coloured; legs brownish.

Plate 45
Puerto Rican Parakeet *Psittacara maugei* (adults)

F. KNIGHT '16

1 female: wing 163 mm, tail 153 mm, exp. cul. 27 mm, tars. 18 mm.
2 unsexed: wing 174 mm and 174 mm, tail 152 mm and 166 mm, exp. cul. 28 mm and 29 mm, tars. 19 mm and 21 mm.
JUVENILES Undescribed.

DISTRIBUTION Formerly occurred on Puerto Rico, nearby Mona Island and possibly nearby Vieques Island.

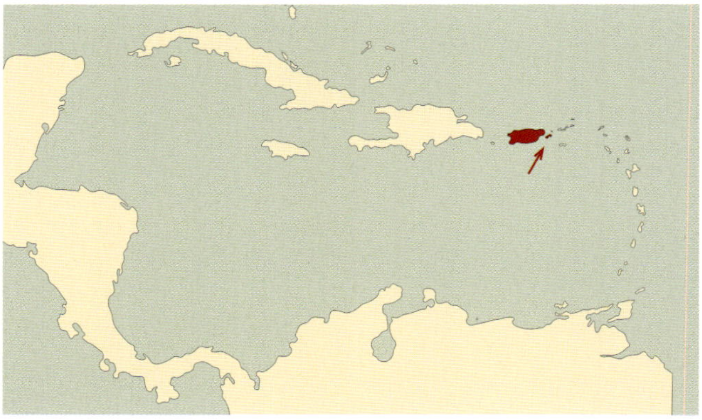

SUBSPECIES Hitherto the Puerto Rican Parakeet usually has been treated as a subspecies of the Hispaniolan Parakeet *Psittacara chloropterus* from Hispaniola. Olson (2015) points out that an examination of museum specimens and a comparison of skeletons with fossil and archaeological material from Puerto Rico show that it is a fully distinct species, differing in plumage and especially in bill morphology. Olson notes also that the fossil material establishes beyond any doubt that *P. maugei* was a distinct species and that it was resident on the main island of Puerto Rico. All of the fossil records come from the central region of Puerto Rico, but that is where most of the fossil localities exist, and there is no reason to assume that the parakeets were not distributed throughout the island. If they were capable of colonising remote Mona Island, then the report that they once occurred also on Vieques Island is plausible.

STATUS Olson (2015) notes that considering the probability that the first two known specimens of the Puerto Rican Parakeet were collected during the expedition of Nicolas Baudin in 1797 and almost certainly were obtained in mainland Puerto Rico, it may be assumed that the parakeets were present in reasonable numbers at that time. Wetmore (1927) noted that no parakeets were seen by C. Moritz, an entomologist and collector for the Zoological Museum in Berlin, who spent four months on Puerto Rico in 1835, and they were known to him only by reports. Perhaps paraphrasing comments made by Johannes Gundlach after his visits to Puerto Rico in 1873 and 1875, Stahl (1887) wrote that only the parents of old men had any recollection of parakeets (translation by Olson):

> [Psittacara maugei] *is long extinct. Existing old men remember hearing from their parents that this little parrot existed in ancient times and in places flocks occurred even in the plantations wreaking damage, but through untiring pursuit were soon eradicated.*

The parakeets evidently were present in reasonable numbers on Mona Island in about 1875, when wings were collected by Claudio Bloch, a Danish physician, and sent to Gundlach, but must have been almost extinct by 1892 when the last specimen was collected by Wilmot Brown. Olson points out that Brown was an assiduous and highly competent professional collector, and the parakeet would have been by far the most interesting and important bird that he encountered on the island, so his obtaining only a single specimen suggests that very few birds survived at that time.

Referring to the widespread environmental degradation on Puerto Rico, which has left only about 1 per cent of the original vegetation, Wiley (1991) notes that extinction of the Puerto Rican Parakeet and precipitous decline of the Puerto Rican Amazon *Amazona vittata* are not remarkable in view of the near-complete loss of habitat. Persecution of the parakeets because of their depredations on agricultural crops was mentioned by Gundlach, and it seems that there was a deliberate effort to eradicate the birds. Olson suggests that disease also may have contributed to the demise of these parakeets and, in addition to domestic poultry with their set of diseases, exotic parrots had been brought to Puerto Rico since the earliest times of Spanish colonisation.

HABITATS AND HABITS Apart from mention of their depredations on agricultural crops, nothing is recorded of the habitat preferences or habits of Puerto Rican Parakeets. Like the closely allied Hispaniolan Parakeet *P. chloropterus*, they probably favoured native forests or woodlands, and usually were seen in noisy, swiftly flying flocks travelling between roosting sites and feeding areas.

DIET AND FEEDING Olson (2015) notes that differences in bill morphology between Puerto Rican and Hispaniolan Parakeets are so profound as to suggest distinct differences in feeding habits, and in the bill of *P. maugei* the reduced notch in the tomium suggests the possibility of a preponderance of smaller or softer food items.

SPECIMENS AVAILABLE The holotype is held in the Museum National d'Histoire Naturelle (MNHN C.G. 2004–132), Paris, France, and other specimens are held in the Rijksmuseum van Natuurlijke Historie (RMNH 110.079), Leiden, Netherlands, Natural History Museum (UKNHM 1859.11.22.56 and 1859.11.22.3), Tring, UK, and the Field Museum of Natural History (FMNH 40314), Chicago, USA. The Field Museum specimen is a female collected on Mona Island on 25 February 1892, and is the only existing specimen with known provenance (Olson 2015).

GENUS *Pyrrhura* Bonaparte

Pyrrhura Bonaparte, *Naumannia*, **6**, 1856, *Consp. Psitt.* in Beilage no. 1, genus 14. Type, by subsequent designation, *Psittacus vittatus* Shaw (not of Boddaert) = *Psittacus frontalis* Vieillot.

Pyrrhura parrots are small to midsized birds with proportionately long, graduated tails and a rather broad bill with a notch in the upper mandible. The naked cere is prominent and in many species there is a conspicuous unfeathered eyering. Prominent scalloping or barring on the neck and breast is a plumage feature in many species. Similarities in plumage patterns have given rise to uncertainties about interspecific relationships, and in some species specific differentiation is manifested in quite subtle differences in patterns on the head and breast. Sexual dimorphism is absent and juveniles mostly resemble, though are duller than, adults. The little available data on nesting behaviour suggest that most if not all species breed cooperatively (see Gwynne *et al.* 2010).

Pyrrhura occurs in southern Central America, and is widespread in South America, but does not reach the West Indies. Molecular analyses have identified three main evolutionary lineages within *Pyrrhura* – one comprising only *P. cruentata* from the Atlantic Forest, and it is basal, another comprising the *picta-leucotis* complex, and the third comprising all other species that were sampled (Ribas *et al* 2006). All endangered species have restricted ranges, where they are threatened by habitat loss or degradation.

ENDANGERED

Maroon-faced Parakeet

Pyrrhura pfrimeri Miranda-Ribeiro

Pyrrhura pfrimeri Miranda-Ribeiro, *Rev. Mus. Paulista*, **12**, 1920, pt 2, p. 36 (Santa Maria de Taguatinga, Goiás).

OTHER NAMES Goiás Parakeet, Pfrimer's Parakeet.

DESCRIPTION Length 22 cm.
ADULTS General plumage green, slightly paler and more yellowish on underparts; forehead and sides of head, including ear-coverts, chestnut-red; crown and nape dull blue; throat and upper breast dull greenish-blue, feathers margined white to give prominently barred appearance; lower breast dull bluish-green, feathers margined white and subterminally banded dull brown to give barred appearance; centre of abdomen and patch on lower back to upper tail-coverts brownish-red; bend of wing red; under wing-coverts green; primary-coverts greenish-blue; outer webs of primaries and outer secondaries blue; tail above brownish-red marked green, below dull brownish-red; bill grey-brown; bare eyering palest grey; iris brown; legs grey.
3 males: wing 121–125 (123.3) mm, tail 105–115 (110.3) mm, exp. cul. 14–16 (15.0) mm, tars. 12–14 (13.0) mm.
3 females: wing 115–118 (117.0) mm, tail 107–119 (112.3) mm, exp. cul. 13–15 (14.0) mm, tars. 13–14 (13.3) mm.
JUVENILES Like adults.

DISTRIBUTION North-central Brazil, where known only from between the Serra Geral massif and Rio Paraná, northeastern Goiás and southeastern Tocantins.

SUBSPECIES The Maroon-faced Parakeet is one of three allopatric species from eastern Brazil which formerly were considered to be conspecific, but now are differentiated specifically in what is termed the '*leucotis-griseipectus-pfrimeri* species group'. Molecular analyses indicate that *leucotis* and *griseipectus* are very closely allied, but *pfrimeri* is well set apart (Ribas *et al.* 2006).

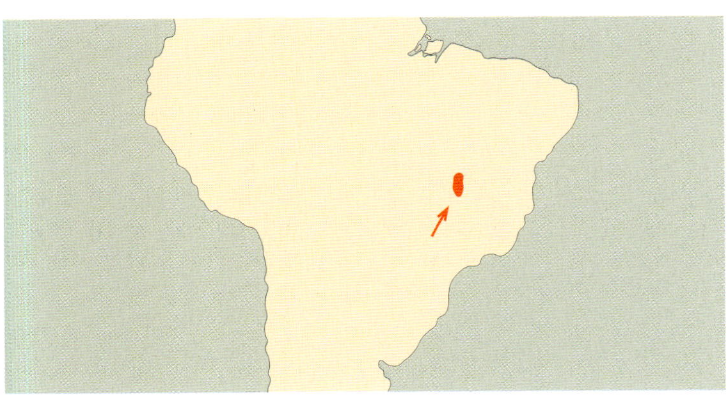

STATUS The Maroon-faced Parakeet is listed as endangered because it has an extremely small range which is severely fragmented and within which habitat loss and degradation are continuing (Birdlife International 2016). Deforestation is the principal threat, with patches of dry forest being cleared for pasture and selectively logged to meet a high demand for durable timbers used for fence poles (Olmos *et al.* 1997). In the 1990s, it was estimated that only 40 per cent of dry forest remained in the region, and logging was continuing. Widespread burning to improve poor pasture also destroys the dry forest habitat, and cement companies are beginning to target limestone outcrops in the area (in Birdlife International 2016). During surveys undertaken in June and October–November 1995, Maroon-faced Parakeets were found wherever dry forest persisted, even in disturbed and fragmented patches, suggesting that they are able to survive and even breed in small forest fragments, though in very reduced numbers, but it is possible that such isolated groups will be unable to persist in the long-term, especially as forest remnants are vulnerable to fire and continued selective logging (Olmos *et al.* 1997). There appears to be little persecution or capture for the live-bird market, though these parakeets have been offered by traders in Brasília, and the near proximity of the range to that city increases the potential for trafficking (Olmos *et al.* 1997).

Population density for these parakeets has been estimated at 11.7 birds per km^2, and the total population is estimated at 20 000–49 999 individuals but, because of strong reliance of the birds on forest habitats within a restricted range where there is a rapid rate of landclearance, the population is suspected to be declining rapidly (Birdlife International 2016). In northeastern Goiás, Maroon-faced Parakeets occur within the Terra Ronca State Park and surrounding environmental protection area and in the 2000 ha Mata Grande National Forest.

HABITATS Of the three species in the '*leucotis-pfrimeri-griseipectus* species group', the Maroon-faced Parakeet is the most disjunct and occupies the most distinctive habitat, being the only species restricted to dry forests and to occur outside the Atlantic Forest (Olmos *et al.* 2005). It is restricted to semi-deciduous or deciduous forests growing on limestone outcrops or limestone-derived soil patches in a stretch about 300 km long and, at some points, only 30 km wide along the foothills west of the Serra Geral massif (Olmos *et al.* 1997). Surrounded by expanses of drier cerrado vegetation, the more fertile limestone-derived soils support forests with an almost continuous 15–20 m high canopy dominated by pink trumpet tree *Tabebuia impetiginosa*, urunday *Myracrodruon urundeuva*, shaving-brush tree *Pseudobombax grandiflorum, Ceiba ventricosa*, barriguda *Cavallinesia arborea*, leopard tree *Caesalpinia ferrea*, earpod tree *Enterolobium contortisiliquum*, diesel tree *Copaifera langsdorffii* and *Piptadenia* sp., with occasional emergents, mostly *Cavallinesia arborea*, reaching 25 m or higher. In more disturbed patches there is a dense undergrowth of lianas or, in some places, of *Actinocladum* or *Guadua* bamboo. Several species of columnar cacti and terrestrial bromeliads occur on rocky outcrops. In simple terms, the habitat of Maroon-faced Parakeets represents a caatinga forest island amidst the predominating cerrado savanna.

During surveys undertaken in June and October–November 1995, Maroon-faced Parakeets were found in every patch of dry forest that was surveyed, even in some quite degraded patches, but they were not found in the surrounding cerrado vegetation, even where it abutted abruptly against the dry forest and where large numbers of Yellow-chevroned Parakeets *Brotogeris chiriri* came from the forest each day to feed in fruiting mango trees. They did come to farmlands to feed on fruits of cultivated guava *Psidium guajava*. All feeding bouts were observed in forests on less xeric, flatter ground, and none was observed during transects in the limestone outcrops, suggesting that the combination of flatter ground and foothill forest may be of critical importance for these parakeets. Peach-fronted Parakeets *Eupsittula aurea* were widespread in the cerrado, but apparently avoided areas of continuous dry forest, entering only fragmented and disturbed patches, so suggesting a degree of ecological segregation between this species and Maroon-faced Parakeets.

HABITS Maroon-faced Parakeets are fairly conspicuous, and usually can be observed quite easily. During surveys undertaken in June 1995 group sizes ranged from two to 32 birds, with a mean of 12.6 birds for 36 sightings, and the largest groups may have resulted from a coming together of smaller groups, with one aggregation of more than 50 birds recorded feeding on cultivated rice and more than 120 birds would congregate in the sole fruiting fig tree in an area (Olmos *et al.* 1997). They were noisy, and usually were heard before being seen. Fledglings were present in June, when the surveys were carried out, and their continuous begging was the best means of locating groups. Birds were seen to fly at or below canopy level, dashing among the trees, and perching both in the crowns and at lower levels. Groups of these parakeets were observed perched on very exposed small, dry trees standing in pasture, and birds flew at a comparatively low height of 6–9 m over such open areas.

Quite surprising for *Pyrrhura* parakeets was their regular feeding on the ground or at low levels, when a sentinel system was evident, with one or more birds perching atop a nearby tree, from where they gave alarm calls at the approach of any possible danger, and this caused the feeding group to take flight. They were particularly confiding while feeding on the flowers of *Tabebuia impetiginosa*, usually ignoring the approach of an observer right to the base of the food tree. On one occasion a group was seen to flee after alarm calls were given by a group of Jandaya Parakeets *Aratinga jandaya* distressed by a soaring Black-and-white Hawk-eagle *Spizaetus melanoleucus*, and another group with fledglings fled from approaching White-naped Jays *Cyanocorax cyanopogon*. At sunset, groups were seen to fly from feeding areas to roosting sites in the limestone hills and, although these nighttime roosts were not located, it was suspected that they were in tree hollows.

CALLS Calls are said to be similar to those of the White-eared Parakeet *Pyrrhura leucotis*, which are described as a sharp *teer-teer* repeated rapidly three or four times by birds in flight, and also a high-pitched, staccato *chee-cheet-chee* or *ki-ki*, and an occasional *teet* while perched.

DIET AND FEEDING During surveys undertaken in 1995, it was noted that Maroon-faced Parakeets used abundant food resources clumped both in time and space, with different groups congregating to share the resource, and showing a preference for soft items (Olmos *et al.* 1997). A total of 46 feeding bouts were recorded in June, and another 25 in October–November, with striking differences noted in food plants utilised in the two periods. In June, flowers of *Tabebuia impetiginosa* were the most important food items, being eaten in 43 per cent of all observed feeding bouts. The parakeets spent hours picking one flower at a time, chewing around its base before discarding it and then repeating the process. Eaten in 33 per cent of observed feeding bouts, the second most important food item was unripe seeds of *Hyptis* sp., an introduced ruderal herb very common along the dirt roads and tracks, and the parakeets came to the ground to take these seeds. They came to the ground also to feed on cultivated rice. Other food items eaten during the June period were *Cecropia* catkins, the fruits of *Ficus gomeleira* and cultivated guava *Psidium guajava*, and *Bauhinia* flowers, each featuring in 6 per cent of observed feeding bouts. Only the pulp of guava fruits was eaten, the seeds being discarded. All figs that were examined were heavily infested with larvae of Agaonidae wasps, and these probably added protein to the diet when parents were feeding fledglings. In October–November, the parakeets fed mainly on fruits of *Ficus gomeleira*, accounting for 72 per cent of observed feeding bouts, the pulp of *Pouteria* fruits, accounting for 12 per cent of observed feeding bouts, and inflorescences of old fustic *Maclura tinctoria*, accounting for 8 per cent of observed feeding bouts. Other food items were *Psittacanthus* shoots, featuring in 8 per cent of observed feeding bouts, and young leaves of *Myracrodruon urundeuva*, featuring in 4 per cent of observed feeding bouts. Figs certainly were the most important resource during this period, with all parakeets in the area feeding in a single fruiting tree.

BREEDING The observed presence of fledglings in June suggested that breeding occurs in April–May, which is the late rainy season to early dry season (Olmos *et al.* 1997). Mating was observed in June, when the male of a pair engaged in allopreening was seen to embrace the female with his left leg and wing before half mounting her. The female lowered her body and then raised and twisted her tail away from the vent of the male. The male did not grasp the female with his bill. The presence of several birds with worn tails in October–November suggested that they could have been nesting at that time, but no nest-sites were found. In July, the collection of a male with a worn tail that was encrusted with soil suggested that nests may be in rock cavities, which are very common in the area. Most groups that bred were seen to be accompanied by two fledglings, though there were some groups with three youngsters. Overall, fledglings comprised only 20.6 per cent of the six groups that could be identified and counted, suggesting that during the 1994–1995 breeding season most groups did not breed or were unsuccessful.

Plate 46
UPPER Grey-breasted Parakeet *Pyrrhura griseipectus* (adult)
LOWER LEFT Sinú Parakeet *Pyrrhura subandina* (adult)
LOWER RIGHT Maroon-faced Parakeet *Pyrrhura pfrimeri* (adult)

F KNIGHT '10

CRITICALLY ENDANGERED

Grey-breasted Parakeet

Pyrrhura griseipectus Salvadori

Pyrrhura griseipectus Salvadori, *Ibis*, 1900, p. 672 (locality unknown = Ceará, Brazil).

DESCRIPTION Length 22 cm.
ADULTS General plumage green, slightly paler and more yellowish on underparts; frontal band, cheeks, lores and around eyes brownish-maroon; ear-coverts white; forecrown to nape dark brown, variably spotted or streaked buff-brown; throat, sides of neck and breast dull grey, feathers broadly margined dull white to give prominently barred appearance; centre of abdomen and patch on lower back to upper tail-coverts brownish-red; bend of wing red; under wing-coverts green; primary-coverts greenish-blue; outer webs of primaries and outer secondaries blue; tail above brownish-red marked green, below dull brownish-red; bill grey-brown; cere and bare eyering palest grey; iris brown; legs grey.
4 males: wing 116–122 (119.3) mm, tail 106–110 (108.3) mm, exp. cul. 15–17 (15,5) mm, tars. 14–15 (14.3) mm.
4 females: wing 116–127 (120.3) mm, tail 107–125 (113.0) mm, exp. cul. 15–17 (15.5) mm, tars. 14–15 (14.3) mm.
JUVENILES Duller than adults with less pronounced barring on breast; scattered green feathers in brownish-red abdominal patch; cere and bare eyering white, less greyish.

DISTRIBUTION Northeastern Brazil, where currently known to occur at only three locations in Ceará – the Serra do Baturité, the Serra de Estevão, and another mountainside site where a small population was found in 2014; formerly occurred on the eastern slope of the Serra do Ibiapaba, in Ceará, and on the Serra Negra, in Pernambuco, but there have been no recent reports, and unconfirmed reports from Murici Ecological Station, in Alagoas, possibly refer to released birds (Birdlife International 2016).

SUBSPECIES The Grey-breasted Parakeet is another member of the '*leucotis-griseipectus-pfrimeri* species group', and formerly was treated as a subspecies of *P. leucotis*, with which it is closely allied.

STATUS Grey-breasted Parakeets are recorded historically from 15 locations in northeastern Brazil, but now are known to survive in only three localities in Ceará, and dramatic declines suffered by the extremely small population are continuing. In the Serra do Baturité, they seem to be very uncommon and appear to have been extirpated from several areas, but there are recent reports from the Baturité Mountains Environmental Protection Area, where surveys undertaken in 2007 in half of the remaining habitat revealed approximately 80 birds, so the population there now is estimated at about 250 birds (in Birdlife International 2016). Waugh (2010) reports that in March 2010, in the Serra do Estevão, some 70 km from the Serra do Baturité, in Quixadá municipality, a population estimated to comprise about 50 birds was discovered, so bringing the known population to an estimated 300 birds. In March 2014, a third population was discovered, when five birds were observed on another rocky hillside in Ceará (see *World Birdwatch* **36**(3): 9, 2014).

While the Grey-breasted Parakeet was considered to be a subspecies of the White-eared Parakeet *Pyrrhura leucotis*, little attention was focused on its status, but when separated as a distinct species field surveys were undertaken to map its distribution and to assess population levels. In 2007, a non-government conservation organisation (AQUASIS) commenced investigations in 16 sites that had been identified as locations where populations may occur, but it soon became apparent that widespread habitat loss or degradation, mainly through clearance for coffee plantations, and poaching for the national and international live-bird markets had brought about the extirpation of these parakeets from many parts of their former range. The largest area surveyed was the Serra do Ibiapaba, in western Ceará, where there was very good historical evidence of the occurrence of Grey-breasted Parakeets, and the area was divided into northern and southern sectors to allow field teams to investigate all significant remnant stands of moist forest. In addition to surveying these forest remnants, which were found to be very fragmented and degraded, a total of 147 interviews were conducted with local residents, but only two residents gave accurate descriptions of the parakeets, indicating that at this locality Grey-breasted Parakeets had been extinct for at least 20 years. Unfortunately, similar findings were recorded at other localities, and it became evident that the Serra do Baturité, in northeastern Ceará, is the last major refuge for these parakeets, so conservation efforts are being concentrated at this important site. In 1996, only 13 per cent of original forest remained in the Serra do Baturité, most having been cleared to make way for coffee plantations, and the remnant forest patches were degraded. Programs are being implemented to address this high level of habitat loss and to counteract continuing illegal poaching for the live-bird trade. At least 11 private reserves are being set up, and a wildlife reserve is being developed by AQUASIS, which has strengthened links with government agencies to influence policy decisions (in Birdlife International 2016). In a large-scale education and public awareness campaign the Grey-breasted Parakeet is promoted as a flagship species, and ecotourism enterprises are being fostered as a means of creating alternative sources of income for local residents. With the cooperation of landowners, nestboxes treated with insecticides to reduce bee and wasp infestations were provided to overcome a shortage of nesting sites in mature trees, and in the 2012 breeding season 16 of these nestboxes were occupied by the parakeets, resulting in 71 chicks hatching from 97 eggs laid. Studies are being undertaken to determine the survival rate of nestlings, and adults and juveniles are being colour-ringed. Captive populations are held in Brazil and elsewhere, and these could be used for future reintroductions.

HABITATS Grey-breasted Parakeets are restricted to enclaves of montane forest above 500 m in isolated massifs in the otherwise lowland dry caatingas of northeastern Brazil (Olmos *et al.* 2005). Peaking at 1114 m, the Serra do Baturité is a granite massif where the annual rainfall of about 1550–1700 mm is approximately three times that at lower altitudes. Because of their altitude and situation relative to the prevailing moisture-laden winds, some higher slopes of the Serra are clad in humid forest with a continuous canopy approximately 20 m in height and with some emergents. In less disturbed areas, common tree species are pedra-ume-caá *Myrica multiflora*, *Byrsonima sericea*, *Clusia nemorosa*, Guyanese wild coffee *Casearia guianensis* and palo yugo *Stryphnodendron purpureum*. These forests have been cleared for shade-coffee plantations, which is the most traditional activity in the Baturité area, and the coffee is shaded mainly by fast-growing *Inga* trees. At lower elevations, the humid forest is replaced by semi-deciduous forest. The parakeets frequent both humid and semi-deciduous forests, and occur also in shade-coffee plantations.

HABITS Grey-breasted Parakeets normally keep to the upper stages of forest trees, where they usually are seen in groups of four or five birds, or occasionally in flocks of up to 20 birds. In common with most *Pyrrhura* species, they are noisy when in swift flight through the canopy, but are fairly quiet while feeding amidst the foliage and can remain undetected.

CALLS Calls are said to be similar to those of the White-eared Parakeet *Pyrrhura leucotis*, which are described as a sharp *teer-teer* repeated rapidly three or four times by birds in flight, and also a high-pitched, staccato *chee-cheet-chee* or *ki-ki*, and an occasional *teet* while perched.

DIET AND FEEDING Reported foods are fruits of *Inga bahiensis*, *Croton* seeds and fruits of cultivated *Byrsonima* sp., jambul *Syzygium jambolanum* and loquat *Eryobotria japonica* (Olmos *et al.* 2005).

BREEDING In the Serra do Baturité, the breeding season is in February to May (Olmos *et al.* 2005). On 8 April, a nest was found at a height of 8 m in a hollowed *Inga bahiensis* tree with a trunk diameter of 30 cm, and the nesting hollow was 31 cm deep, 12 cm wide, with a round entrance hole 6 cm in diameter. A normal clutch comprises two to four eggs. Young birds attain adult plumage colouration after their second year.

Waugh (2010) reports that in March 2010, when a brood of four well-advanced nestlings in a nestbox was attacked by a swarm of wasps, two of the chicks managed to clumsily fly away, but two were trapped inside and one was stung severely while attempting to leave. Members of the field team who were monitoring the nest quickly removed the two remaining chicks and, with assistance from local people, repelled the wasps with smoke. An hour later the two chicks were returned to the nestbox and, after another hour, the parents returned to feed them. The next day, the two fledged chicks were flying around and both chicks in the nestbox were alive and active, but one had suffered such a severe attack that it subsequently died in the nestbox.

EGGS Measurements of a single egg are 25.1 × 19.9 mm (Olmos *et al.* 2005).

EXTINCT

Sinú Parakeet

Pyrrhura subandina Todd

Pyrrhura subandina Todd, *Proc. Biol. Soc. Wash.*, **30**, 1917, p. 6 (Jaraquiel, Bólivar, Colombia).

DESCRIPTION Length 21 cm.

ADULTS General plumage green, slightly paler and more yellowish on underparts; brownish-red patch on centre of abdomen and from lower back to upper tail-coverts; dull red frontal band, bordered above by faint dull blue; dull red in front of eyes; crown to nape and upper cheeks dark brown, suffused dull blue on lower cheeks; ear-coverts russet-brown; throat, breast and sides of hindneck dusky brown, feathers broadly margined greyish-buff to give prominently chevroned pattern; bend of wing green; primary-coverts greenish-blue; outer webs of primaries and outer secondaries blue; tail above brownish-red marked green, below dull brownish-red; bill grey-black; unfeathered eyering brownish-grey; iris yellow-orange; legs grey.
9 males: wing 111–117 (115.1) mm, tail 93–111 (104.1) mm, exp. cul. 14–16 (14.9) mm, tars. 12–14 (13.3) mm.
12 females: wing 112–124 (116.3) mm, tail 95–108 (103.2) mm, exp. cul. 14–16 (15.0) mm, tars. 12–14 (12.8) mm.
JUVENILES Undescribed.

DISTRIBUTION Formerly in northwestern Colombia, where known only from the lower Río Sinú valley, Córdoba.

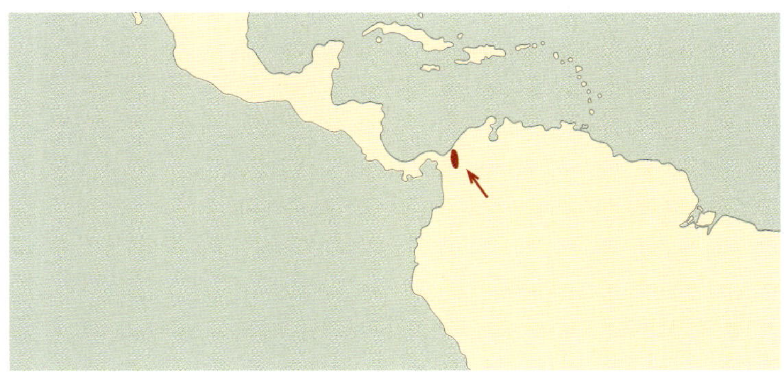

SUBSPECIES The Sinú Parakeet formerly was treated as a subspecies of the widespread Painted Parakeet *Pyrrhura picta* but, in a review of the *P. picta* and associated White-eared Parakeet *P. leucotis* complex, Joseph (2000) recommended that it be treated as a separate species, and that arrangement generally has been adopted.

STATUS Despite extensive searches, the Sinú Parakeet has not been recorded since 1949, and deforestation has occurred throughout a large proportion of its very restricted range, with all suitable habitat apparently having been lost at four sites with specimen records (Birdlife International 2016). Joseph and Stockwell (2002) note that climatic modelling did not show any regions beyond its recorded range with a high predicted likelihood of occurrence, and if it ever occurred in the lower Río Cauca and middle Río Magdalena valleys, as suggested by Hilty and Brown (1986), an absence of recent records may indicate that suitable habitat has been lost there and it is locally extinct. There seems to be very little likelihood of a small remnant population being discovered.

HABITATS As a subspecies of *P. picta*, the Sinú Parakeet is said to have frequented humid forest and forest margins, and there is an early unconfirmed record of birds being seen flying across cultivation from one forest patch to another.

HABITS Nothing is recorded of the habits of these parakeets, but presumably they are similar to the habits of other *Pyrrhura* species, and the birds probably associated in parties or small flocks.

BREEDING Birds in breeding condition were obtained during March to June (in Hilty and Brown 1986).

ENDANGERED

Perijá Parakeet

Pyrrhura caeruleiceps Todd

Pyrrhura caeruleiceps Todd, *Ann. Carnegie Mus.*, **30**, 1947, p. 336 (El Cauca, Colombia).

DESCRIPTION Length 22 cm.
ADULTS General plumage green, slightly paler and more yellowish on underparts; brownish-red patch at centre of abdomen and on lower back to upper tail-coverts; frontal band and lores maroon, becoming more brownish on upper cheeks and around eyes; forecrown blue; hindcrown brown suffused blue; blue nuchal collar; ear-coverts greyish-white; throat, breast and sides of hindneck dusky brown, feathers broadly margined greyish-buff to give prominently chevroned pattern; bend of wing red; primary-coverts greenish-blue; outer webs of primaries and outer secondaries blue; tail above brownish-red marked green, below dull brownish-red; bill grey-black; unfeathered eyering brownish-grey; iris palest yellow; legs dark grey.
3 males: wing 119–126 (123.0) mm, tail 102–120 (112.7) mm, exp. cul. 14–16 (15.0) mm, tars. 12–13 (12.7) mm.
3 females: wing 120–123 (121.0) mm, tail 113–116 (114.3) mm, exp. cul. 13–15 (14.3) mm, tars. 13–14 (13.3) mm.
JUVENILES Undescribed.

DISTRIBUTION Northeastern Colombia, where distributed along the western slope of the north Cordillera Oriental, from Norte de Santander and southern Cesar north through Los Motilones Mountains into the Sierra de Perija at the border with Venezuela.

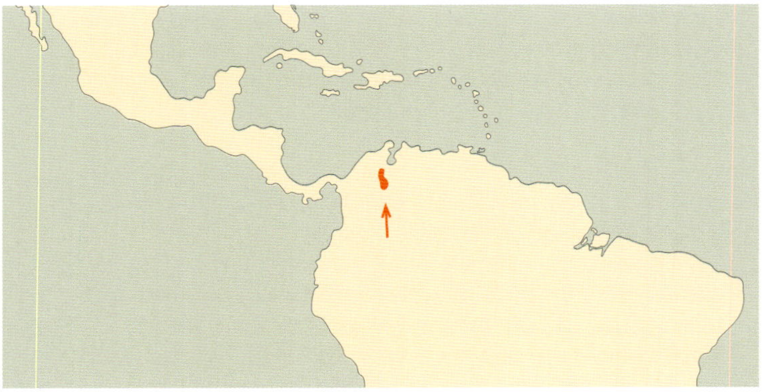

SUBSPECIES The Perijá Parakeet is another species that formerly was treated as a subspecies of the widespread Painted Parakeet *Pyrrhura picta*, but the recommendation by Joseph (2000) that it be differentiated as a separate species generally has been adopted.

STATUS Within its Colombian distribution, the Perijá Parakeet has lost some 70 per cent of its original habitat, and its area of occupancy is predicted to be less than 3700 km² (Botero-Delgadillo *et al.* 2012a). Only 20 per cent of this area of occupancy is within protected areas, so the loss of habitat is expected to continue. Tovar-Martinez (2010) identifies habitat destruction and poaching for the pet trade as the principal threats. Based on historical records and a high probability of a presence of parakeets, sites in the Chiriguaná and Becerril districts, both in the Los Motilones Mountains, and another site in the Perijá Mountains were surveyed in 2011 for Perijá Parakeets, and at the Chiriguaná and Becerril sites logging and forest clearance for cattle pastures appeared to be the most important pressures on habitat, with most forest fragments being separated by cattle pastures or transitional crops (Botero-Delgadillo *et al.* 2012b). It was evident also that pressures leading to deforestation decreased at higher elevations, where continuous remnants of primary forest persisted. Observed flight patterns suggested that the parakeets may be unwilling to overfly extensive cleared areas, so increasing their vulnerability to increased fragmentation of remnant forest (Botero-Delgadillo *et al.* 2013). Surveys at the same two sites in 2011 suggested that, aside from deforestation, proximity to human settlements or even farms is perhaps the main threat to local populations because human presence exerts additional pressures, mainly poaching for the pet market, as some local people value the parakeets as pets, and chicks are taken from nests to be raised before being sold. No parakeets were seen at the site in the Perijá Mountains, and three possible explanations for their absence were considered. It was possible that the birds were absent temporarily while undertaking local foraging movements, or the surviving population is so small that its presence could not be detected during a search of 10 days, or that severe habitat degradation associated with the dominant cattle pastures, semi-permanent crops and open-cut mining operations had brought about extirpation of the parrots in this region. Consequently, it was considered unlikely that populations inhabiting extensive remnants of primary forest on the Venezuelan side of the Perijá Mountains could be present also on the Colombian side of the range.

The overall status pattern is that the remaining population of Perijá Parakeets is thought to be very small and fragmented into extremely small subpopulations, which are undergoing continuing declines. In 2011, at two surveyed sites in the Los Motilones Mountains, northeastern Colombia, populations were estimated to comprise between 90 and 121 birds at Chiriguaná and between 31 and 65 birds at Becerril (Botero-Delgadillo *et al.* 2012b). In 2008, at a study site in the Ocaña district, Norte de Santander, a total of 286 parakeets were counted within an estimated area of 109 ha (see *Cyanopsitta*, no. 90, p. 16, 2008). A preliminary estimate of the total population is of 1000–2499 mature individuals, with fewer than 250 mature individuals in each subpopulation (Birdlife International 2016).

HABITATS Between 400 m and 2200 m, Perijá Parakeets frequent primary and secondary forest in both continuous stands or remnant patches interspersed with cultivation and pastures. Tovar-Martinez (2010) reports that during surveys undertaken in Norte de Santander, northeastern Colombia, these parakeets were found at two sites in the Ocaña and El Carmen districts, where they frequented premontane secondary forest and forest borders. In 2011, in the Los Motilones Mountains, northeastern Colombia, surveys were undertaken during 20 days in July at sites in the Chiriguaná district, where the dominant vegetation cover comprised fragments of secondary forests and shrubby vegetation separated mainly by cattle pastures, and during 20 days in August in the Becerril district, where most areas also were covered by secondary vegetation, but bordered by extensive pastures to the north and with primary forest fragments to the south

(Botero-Delgadillo *et al.* 2013). At both localities Perijá Parakeets frequented four habitat types:

(i) transient and permanent crops (including annual plants or shade-grown coffee),
(ii) forest–agricultural mosaics (mainly open areas with scattered trees or mosaics of crops and secondary vegetation),
(iii) native forests (fragmented secondary forest or riverine forest), and
(iv) herbaceous vegetation (early successional vegetation).

In the Chiriguaná district the parakeets used native forests more frequently than expected by chance, rarely were seen in crops or mosaics, and never were observed in herbaceous vegetation. In the Becerril district, habitat use also differed significantly between habitats, with most records obtained in forest-agricultural mosaics, a few from forests and successional areas, and none from croplands. Mean vegetation height for 13 plots established within frequented native forests was 15.6 m above ground, but height measurements within each plot showed considerable variation at some sites. Canopy height in most of the circular plots in the Chiriguaná district was above 10 m, but tended to be lower in the Becerril district. Vegetation cover across vertical strata also was irregular and variable.

Observations in the Chiriguaná and Becerril districts did not support the suggestion that Perijá Parakeets are restricted to native forests, for at Becerril a high proportion of records were in agricultural habitats or forest–agricultural mosaics (Botero-Delgadillo *et al.* 2013). Although this could be interpreted as an indicator of tolerance to forest clearance, it was acknowledged that further study is needed to determine whether non-forested lands are selected positively or negatively and whether agricultural habitats are utilised in proportion to their current availability. Also, it is noted that the observations indicate that, although these parakeets can overfly small expanses of open lands, they apparently depend on a continuous forest canopy for movements exceeding 20 km (Botero-Delgadillo *et al.* 2012b).

HABITS In 2011, at altitudes between 400 m and 2200 m, in the Los Motilones Mountains, northeastern Colombia, ecological data were compiled during field surveys of Perijá Parakeets undertaken during 20 days in July in the Chiriguaná district and during 20 days in August in the Becerril district (Botero-Delgadillo *et al.* 2013). Observations were carried out each day between 0630 and 1230 hours and again between 1430 and 1730 hours, covering an area of about 4.7 km² at Chiriguaná and approximately 8.1 km² at Becerril. A total of 393 minutes of cumulative observations was obtained at Chiriguaná and a total of 543 minutes was obtained at Becerril. At Chiriguaná there were 43 records of flying birds and 41 records of perched birds, and at Becerril there were 36 records of flying birds and 43 records of perched birds. At Chiriguaná, 81.4 per cent of records of flying birds were at less than 10 m above the forest canopy, and only 18.6 per cent were at heights of 10–20 m above the canopy. Similarly at Becerril, 83.3 per cent of observed flying flocks were below 10 m above the canopy and 16.7 per cent were at heights between 10 m and 15 m above the canopy. At Chiriguaná, 21.2 per cent of observations were of pairs, 63.3 per cent were of groups of 3–10 birds, 14.1 per cent were of flocks of 12–35 birds, and there was one flock of 90 birds. Likewise at Becerril, most observations were of parties of three to nine birds, and flocks of more than 20 birds accounted for only 3.5 per cent of observations. Birds resting with their eyes closed or simply staying still were observed mostly between 1100 and 1500 hours, when groups of five to 28 birds commonly were seen perched in scattered trees in cleared areas, emergent trees within the forest or dead trees at the forest edge. Preening and grooming birds were observed mostly between 1000 and 1200 hours, in groups varying from three to 28 birds, mainly in trees scattered across open areas or dead trees within native forest.

CALLS Tovar-Martinez (2010) reports that a *kiir-kiir* call is emitted by these parakeets while perched and a different *ki-ki-ki-ki* call is associated with vigilance behaviour.

DIET AND FEEDING During the surveys undertaken in 2011 in the Los Motilones Mountains, northeastern Colombia, 17 feeding records were obtained at Chiriguaná and seven records were obtained at Becerril, with five food plants being identified. Fruits of guava *Psidium guajava* were eaten in the majority of observed feeding bouts at both localities, followed by catkins of *Cecropia peltata*, *Inga* inflorescences, *Alchornea* fruits and fruits of *Ficus glabrata* (Botero-Delgadillo *et al.* 2013). At Chiriguaná, foraging time was spent mostly in fruiting guava trees, while at Becerril it was spent almost equally in *P. guajava* and *C. peltata* trees. *P. guajava* and *Cecropia peltata* also featured prominently in feeding records compiled in 2007 at sites in Norte de Santander, where *Trema*, *Croton* and *Ochroma* fruits also were included in the diet (Tovar-Martinez 2010).

BREEDING Behaviour associated with breeding was observed at Chiriguaná and Becerril, in the Los Motilones Mountains, northeastern Colombia (Botero-Delgadillo *et al.* 2013). At Chiriguaná, at 1015 hours on 29 June, two birds were seen apparently inspecting a cavity in a dead tree at the forest edge, and at 0730 hours the next day three birds were observed for 29 minutes while engaged in courtship-related activities, including mutual grooming and regurgitation feeding. At Becerril, on 6 August, at approximately 1030 hours, two parakeets were observed while engaged in mutual preening, then chasing each other and subsequently copulating. Nothing further is known of the nesting behaviour.

Azuero Parakeet

Pyrrhura eisenmanni Delgado

ENDANGERED

Pyrrhura eisenmanni Delgado, *Am. Orn. Un. Orn. Monogr.*, no. 36, 1985, p. 17 (Los Piraguales, El Cortezo de Tonosí , Los Santos Province, Azuero Peninsula, Panama).

DESCRIPTION Length 22 cm.

ADULTS General plumage green, slightly paler and more yellowish on underparts; brownish-red patch at centre of abdomen and from lower back to upper tail-coverts; narrow frontal band to lores and around eyes brownish-red, becoming more brownish on upper cheeks; forehead dull brown suffused dull blue; crown to nape sooty brown; ear-coverts buff-white; indistinct blue nuchal collar; sides of neck to throat and breast grey broadly edged white to give prominently scalloped appearance, and on centre of throat to centre of breast suffused buff; bend of wing narrowly edged red; primary-coverts greenish-blue; outer webs of primaries and outer secondaries blue; tail above brownish-red with green towards bases of outer feathers, below dull brownish-red; bill dull black; unfeathered eyering sooty brown; iris pale orange-brown; legs dark grey.

1 male (type): wing 117 mm, tail 105 mm, exp. cul. 15 mm, tars. 13 mm.
JUVENILES Undescribed.

DISTRIBUTION Restricted to southwestern Azuero Peninsula, south-central Panama.

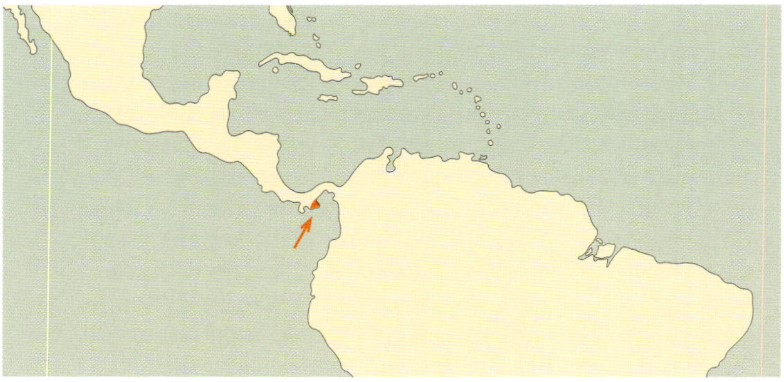

STATUS The Azuero Parakeet is known from only a single population occurring predominantly in Cerro Hoya National Park, and the total range is considered to be less than 700 km² (Montañez and Angehr 2007). Ridgely and Gwynne (1989) point out that within this very restricted range these parakeets are reported to be locally common, though much forest has been cleared and their survival is dependent on adequate protection of remaining stands.

The total population is estimated to comprise less than 2000 birds, so placing it in the 1000–2499 band, and it is suspected to be undergoing a continuing decline due to habitat loss and degradation (Birdlife International 2016).

HABITATS Azuero Parakeets frequent humid forest and forest margins in hilly country, and occasionally are encountered in adjacent partly cleared lands (Delgado 1985; Angehr and Dean 2010).

MOVEMENTS Delgado (1985) notes that these parakeets may undertake limited altitudinal movements. Local residents at the lower end of the altitudinal range reported that the parakeets are present only at certain seasons, when they visit bean plantations. That local movements in response to food availability also may occur is indicated by regular occurrences of these parakeets at a farm on the western border of Cerro Hoya National Park, during June and July, when they come to feed on nance fruits.

HABITS Azuero Parakeets are gregarious, and almost invariably are encountered in pairs, trios or flocks of up to 20 individuals (Delgado 1985). They keep to the mid and upper stages of the forest, and their flight is swift and direct. A favoured location for observing these parakeets is at a small farm on the western border of Cerro Hoya National Park, where noisy flocks regularly come each morning in June and July to feed in fruiting nance trees *Byrsonima crassifolia*.

CALLS Delgado (1985) reports that a short *eek* is uttered by birds in flight, and other noted vocalisations include a single, loud *peea* given by solitary birds attempting to relocate their flocks and a harsh, guttural *kleek-kleek* heard from perched birds. Perched birds also utter soft, barely perceptible calls, often while preening each other.

DIET AND FEEDING Local residents reported that these parakeets seasonally visit bean plantations (in Delgado 1985). At a small farm on the western border of Cerro Hoya National Park, flocks regularly come during June and July to feed on nance fruits *Byrsonima crassifolia*. There appear to be no additional reports of foraging behaviour.

BREEDING Delgado (1985) notes that specimens collected in January–February 1979 and January 1980 appeared to be in breeding condition, and also showed pre-breeding moult in the wings. Males had enlarged testes. Also in January 1980, at the southern end of the range, a group of three birds entered two holes on the underside of a large branch of a tall leafless wild cashew *Anacardium excelsum*: these birds peered out of each hole for a few minutes before flying off into the forest, and it seemed likely that they were prospecting for a nest site. An active nest was not found, but local residents claimed that eggs are laid in March, and fledglings can be seen in April or late March, at the beginning of the rainy season. There appear to be no other details of nesting behaviour.

Santarém Parakeet

Pyrrhura amazonum Hellmayr

Pyrrhura picta amazonum Hellmayr, *Bull. Brit. Orn. Club*, **19**, 1906, p. 8 (Obidos, Brazil).

DESCRIPTION Length 22 cm.
ADULTS General plumage green, slightly paler and more yellowish on underparts; brownish-red patch at centre of abdomen and from lower back to upper tail-coverts; crown to nape dark brown; blue band in front of eyes; narrow frons and face dark reddish-brown; ear-coverts brownish-buff; upper breast, throat and sides of neck greenish-brown, feathers broadly margined brownish-buff to give prominently chevroned pattern; lower breast greyish-green, feathers broadly margined yellowish-buff; bend of wing green; primary-coverts greenish-blue; outer webs of primaries and outer secondaries blue; tail above brownish-red with green towards bases of outer feathers, below dull brownish-red; bill grey-black; unfeathered eyering grey; iris orange-brown; legs grey.
15 males: wing 116–128 (121.3) mm, tail 90–112 (99.8) mm, exp. cul. 14–15 (14.7) mm, tars. 12–15 (13.5) mm.
11 females: wing 120–125 (122.6) mm, tail 90–111 (102.7) mm, exp. cul. 14–15 (14.4) mm, tars. 13–15 (13.5) mm.
JUVENILES Duller than adults; breast finely scalloped greyish-white; less extensive brownish-red patch on centre of abdomen; shorter tail; unfeathered eyering greyish-white.

DISTRIBUTION Eastern Amazonia in northern Brazil, where distributed along the north bank of the Amazon River from about Obidos east to Monte Alegre, Pará, and along the south

VULNERABLE

Plate 47
UPPER Perijá Parakeet *Pyrrhura caeruleiceps* (adult)
CENTRE Azuero Parakeet *Pyrrhura eisenmanni* (adult)
LOWER Santarém Parakeet *Pyrrhura amazonum* (adult)

F. KNIGHT 15

bank at Santarém and the Rio Tapajós drainage, Pará, east to Imperatriz, westernmost Maranhão, and south to northernmost Mato Grosso.

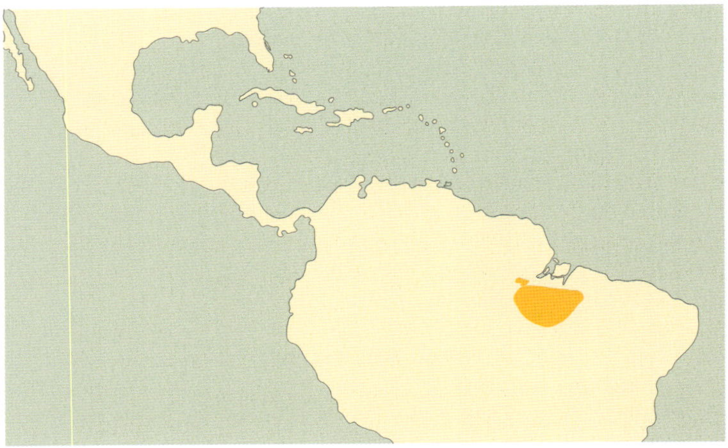

SUBSPECIES

1. *P. a. amazonum* Hellmayr

The nominate subspecies, as described above, occurs along the north bank of the Amazon River from about Obidos east to Monte Alegre, Pará.

2. *P. a. microtera* Todd

Pyrrhura picta microtera Todd, *Ann. Carnegie Mus.*, **30**, 1947, p. 335. (Santarém).

ADULTS Narrow frons and face darker maroon-brown; upper breast, throat and sides of neck dusky brown, feathers broadly margined greyish-buff to give overall darker chevroned pattern; smaller size.

15 males: wing 104–117 (110.8) mm, tail 81–102 (90.9) mm, exp. cul. 13–15 (13.9) mm, tars. 12–15 (13.1) mm.

10 females: wing 105–113 (109.9) mm, tail 83–97 (89.8) mm, exp. cul. 13–15 (13.9) mm, tars. 12–14 (12.9) mm.

This poorly differentiated subspecies is distributed along the south bank of the Amazon River at Santarém and the Rio Tapajós drainage, Pará, east to Imperatriz, westernmost Maranhão, and south to the Rio Aripuana and Rio Teles Pires, northernmost Mato Grosso.

STATUS Based on current levels of Amazonian deforestation, it is anticipated that the Santarém Parakeet will lose 43–52 per cent of suitable habitat over three generations, and it is susceptible to poaching for the live-bird market, so the population is likely to decline by more than 50 per cent over three generations (Birdlife International 2016). The primary threat is accelerating deforestation as land is cleared for cattle ranching and soy production, with consequent expansion of the roads network facilitating hunting and poaching activities.

HABITATS Up to 600 m, Santarém Parakeets frequent terra firme and dry deciduous forests, with a preference for less densely vegetated lower stages of the forest interior.

HABITS Between July 1977 and June 1979, at a study site on the upper Rio Aripuana, northwestern Mato Grosso, Paul Roth found that groups of five to 12 birds were the normal social units for Santarém Parakeets, and their preference for the less densely vegetated lower stages of the forest interior contrasted with that of Crimson-bellied Parakeets *Pyrrhura perlata* for thicker vegetation at the forest margins or in secondary growth (*in litt.* 1978). Roth observed also that if these two species met while foraging or when visiting favoured patches of bare ground to take mineral clay, Santarém Parakeets always gave way to the larger Crimson-bellied Parakeets. During the dry season, Santarém Parakeets came each day to watercourses to bathe, and their regularly used places were changed only when made unsuitable by receding water levels. On several occasions birds were seen at a rocky cliff bathing in spray from a nearby waterfall. Roosting at night was in tree hollows, with birds arriving mostly between 1730 and 1750 hours to enter hollows before 1800 hours; at the same hollow seven birds were seen by Roth to come to roost on 12 October 1977, six or seven on 16 October 1977, and at least five on 8 March 1979. The flight is swift and direct.

CALLS The calls are said to be similar to those of the closely allied Painted Parakeet *Pyrrhura picta*, which are described as a coarse, forceful *pik-pik* or *pik…pik…pik* given in flight.

DIET AND FEEDING The diet comprises seeds, fruits, nuts, berries, blossoms, vegetable matter, and probably insects and their larvae. Along the upper Rio Aripuana, northwestern Mato Grosso, Roth observed these parakeets taking small, berry-like fruits, including those of *Goupia glabra* and *Trema micrantha*, as well as larger soft fruits, such as those of *Bagassa guianensis*, and medium-sized hard *Euterpe* fruits. At times they ate blossoms, though not to the same extent as did Crimson-bellied Parakeets *Pyrrhura perlata*. In the company of large flocks of Golden-winged Parakeets *Brotogeris chrysoptera*, they were seen by Roth to come down on to overhanging branches to take algae that massed on the surface of deep pools.

Stomach contents from a specimen collected on the Rio Xingu, Pará, included 13 Diptera larvae, which probably were ingested in infested fruit (Schubart *et al.* 1965).

BREEDING Roth recalls that along the upper Rio Aripuana, northwestern Mato Grosso, on 28 June, at a height of 18 m in a fairly isolated tree, a pair of Santarém Parakeets was observed on a stout branch, approximately 30 cm in diameter, where they were inspecting a hole that was partly covered by lianas, and fledglings begging for food were seen between 19 September and 15 November, but in the intervening months pairs were so secretive that no occupied nests were located (*in litt.* 1978). Roth also observed copulating birds on 25 April and at other times outside the period of obvious breeding activity.

Santa Marta Parakeet

Pyrrhura viridicata Todd

ENDANGERED

Pyrrhura viridicata Todd, *Proc. Biol. Soc. Wash.*, **26**, 1913, p. 174 (San Lorenzo, Santa Marta Mountains, Colombia).

DESCRIPTION Length 25 cm.

ADULTS General plumage green, slightly paler and more yellowish on underparts; narrow red frontal band; ear-coverts purplish-

brown; on hindneck feathers conspicuously tipped purplish-brown; on throat, breast and sides of neck feathers very faintly tipped purplish-brown; abdomen variably and irregularly marked orange-red, sometimes forming broad abdominal band; bend of wing, carpal edge and underwing-coverts orange-red variably marked yellow; primary-coverts greenish-blue; outer webs of primaries and outer secondaries blue; tail above green, below dusky reddish-brown; bill horn-coloured; unfeathered eyering cream-white; iris brown; legs grey.
10 males: wing 140–146 (142.1) mm, tail 108–123 (114.2) mm, exp. cul. 16–17 (16.5) mm, tars. 14–16 (14.6) mm.
8 females: wing 137–143 (139.4) mm, tail 104–118 (113.0) mm, exp. cul. 15–17 (16.0) mm, tars. 14–16 (15.0) mm.
JUVENILES Undescribed.

DISTRIBUTION Restricted to the Sierra Nevada de Santa Marta, northeastern Colombia.

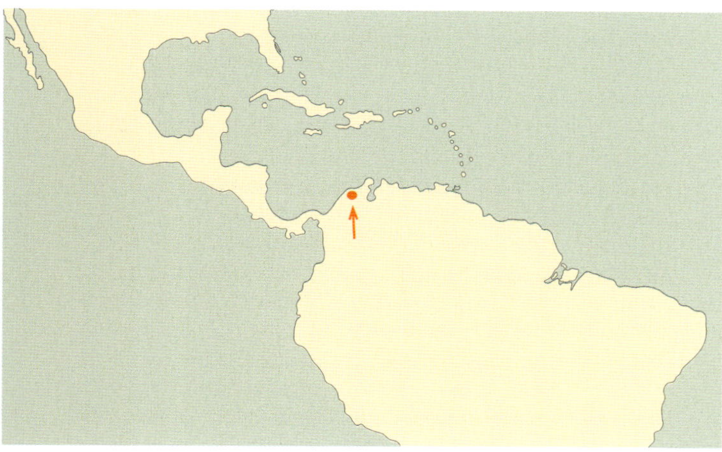

STATUS Although locally common in some districts, the Santa Marta Parakeet is seriously threatened by loss of habitat. Native forest has been cleared extensively to make way for *Pinus* and *Eucalyptus* plantations, and only 15 per cent of the original vegetation cover remains in the Sierra Nevada de Santa Marta, mostly on the north slope where these parakeets occur, and where less than 200 km² of primary forest may remain within their altitudinal range (in Birdlife International 2016). The Sierra Nevada de Santa Marta is protected by two national designations and is an International Biosphere Reserve, but this has not brought about effective habitat conservation. Strewe (2005) points out that all known sites for Santa Marta Parakeets are within Indian reservations, where indigenous people have management rights, and it is not possible to control management or hunting. Hunting of these parakeets occurs in the Río Frío valley, and in the San Pedro district birds have been shot in blackberry plantations. It is anticipated that habitat loss will continue and the population will undergo an ongoing moderate decline.

In 2006, national and international organisations were involved in establishing the El Dorado Nature Reserve, which encompasses 910 ha in the northwestern region of the Sierra Nevada de Santa Marta, and a conservation program is being implemented within this reserve to protect the parakeets (Birdlife International 2016). As part of this program a concerted effort is being made to eradicate invasive exotic plants, especially *Pinus* trees and saplings, and there is widespread planting of native trees to meet the feeding and nesting needs of the parakeets. Nestboxes are being provided, and these are being used by the parakeets, though there is some competition from Scarlet-fronted Parakeets *Psittacara wagleri*.

Prior to its being purchased and protected as the El Dorado Nature Reserve, the 910 ha area in the northwestern region of the Sierra Nevada de Santa Marta was identified as supporting the single largest surviving population of Santa Marta Parakeets, estimated at 120 individuals. Based on this estimate, and using forest cover estimates from satellite images to calculate remaining suitable habitat, the total population has been estimated at 4000–4500 birds (Strewe 2005). Other estimates are of a total population of 5000–10 000 individuals, roughly equating to 3300–6700 mature birds (Birdlife International 2016).

HABITATS Mostly between 1800 m and 2800 m, Santa Marta Parakeets frequent premontane and montane forests, including margins, and secondary growth, sometimes venturing into grassy slopes with scattered palms and shrubs. Field studies have indicated that habitats frequented for nesting are influenced by competition from more aggressive species, particularly Scarlet-fronted Parakeets *Psittacara wagleri*. Between December and April, when the Scarlet-fronted Parakeets, together with *Pionus* and *Amazona* species, were nesting, Santa Marta Parakeets occupied nestboxes only at sites with a complex vegetation structure, but during July to October, when the other parrots were not breeding, Santa Marta Parakeets occupied nestboxes and roosted in scattered palms in open areas (see *Cyanopsitta* no. 90, p. 16, 2008).

HABITS Like most *Pyrrhura* species, Santa Marta Parakeets usually are seen in noisy flocks twisting and turning in swift flight through the treetops or across open areas when travelling from one patch of forest to the next. Hilty and Brown (1986) note that a flock will alight suddenly in the crown of a tree, then pause momentarily before individuals clamber slowly along the branches while uttering soft call-notes. When disturbed, a feeding flock will dash from the tree, accompanied by much loud screeching, but sometimes the birds will wheel about and return to the same tree. At night they roost communally in a tree hollow or nestbox or among the fronds of a palm tree.

CALLS The screeching call-notes are said to be similar to those of the Painted Parakeet *Pyrrhra picta*, which are described as a coarse, forceful *pik-pik* or *pik...pik...pik* given in flight.

DIET AND FEEDING Santa Marta Parakeets routinely feed in the mid to upper stages of the forest, where presumably they take seeds, fruits, berries and possibly flowers or buds.

BREEDING Field studies undertaken in and around the El Dorado Nature Reserve have established that there is an almost all-year breeding season with peak periods in December to April and in July to October (see *Cyanopsitta* no. 90, p. 16, 2008). Hilty and Brown (1986) report that in September juveniles were seen and adults in breeding condition were collected, while in June a pair was seen at a hole 6 m up in a tree stub. The nest is in a tree hollow, but in and around El Dorado Nature Reserve increasing use is being made of nestboxes, and this has facilitated a monitoring of the successful fledging of chicks.

ENDANGERED

El Oro Parakeet

Pyrrhura orcesi Ridgely and Robbins

Pyrrhura orcesi Ridgely and Robbins, *Wilson Bull.*, **100**, 1988, pp. 173–182 (Buenaventura, El Oro, Ecuador).

DESCRIPTION Length 22 cm. Weight 65–75 g.
ADULTS General plumage green; male with lores and forehead bright red; female with lores green and less red on forehead; few scattered red feathers sometimes on forecrown; indistinct pale greyish-buff margins to feathers of breast, but more pronounced on sides of neck; on abdomen dull maroon bases to some feathers show through to produce slight indication of abdominal patch; bend of wing, forewing and primary-coverts bright red; outer webs of primaries and secondaries deep purplish-blue narrowly margined on outer webs and subterminally suffused bright green; tail above dark brownish-maroon, feathers edged green and becoming entirely green on inner webs towards bases; tail below dusky grey strongly suffused dark maroon on inner webs; bill horn-coloured; unfeathered eyering pinkish-white; iris dark brown; legs grey-black.
5 males: wing 120–132 (123.0) mm, tail 95–103 (100.5) mm, exp. cul. 16–17 (16.4) mm, tars. 14–14 (14.0) mm.
4 females: wing 117–124 (120.5) mm, tail 89–101 (94.0) mm, exp. cul. 16–16 (16.0) mm, tars. 13–15 (14.0) mm.
JUVENILES Like adults, but less red on forehead and bend of wing; no trace of maroon on abdomen; slightly less maroon on inner webs of tail feathers.

DISTRIBUTION On western slopes of the Andes in southwestern Ecuador, where disjunctly distributed in Cañar, Azuay, El Oro and Loja.

STATUS Although the El Oro Parakeet can be locally common in some patches of suitable habitat, overall numbers are quite low and the highland forest habitat is fragmented as a consequence of widespread deforestation in western Ecuador, where the rate of landclearance was 57 per cent per decade between 1958 and 1988, leaving a higher proportion of forest only at higher elevations with steeper terrain and harsher climate (Birdlife International 2016). In parts of the range of these parakeets around Manta Real, Azuay, and Piñas, El Oro, there were rapid rates of logging during the late 1980s and 1990s, and logged areas were burnt for cattle ranching. The favoured nesting tree – copal *Dacryodes peruviana* – frequently is targeted in logging operations.

In 1999, Fundación Jocotoco commenced purchase and protection of tracts of land at and around the type locality at Buenaventura, in El Oro Province, eventually leading to establishment of the 2260 ha Buenaventura Nature Reserve. Conservation efforts have focused on this reserve and surrounding areas, where monitoring in 2005 found 60 to 70 birds in an area of 4 km^2, with 15 to 20 of these birds being inside the reserve, and in the following year 20 to 30 parakeets again were counted inside the reserve and an additional 60 to 70 birds in adjacent forests. During a monitoring of nesting activities undertaken between January and mid-March 2008, the number of parakeets counted in the reserve and surroundings was 130 to 140, but in March it reached 227 individuals. In 2009, with funding from international sources, surveys were commenced at other sites within the range of the El Oro Parakeet, and up to August of that year 117 parakeets were counted at 10 sites, with the largest count of 85 birds being at a site near Buenaventura, so suggesting that El Oro Province is the stronghold for the species (Anon 2010). The total population is estimated at 350–1500 individuals but, due to the cooperative breeding system, it is possible that the number of breeding birds may be significantly lower (Birdlife International 2016).

Field studies and conservation efforts continue to be concentrated in the Buenaventura Nature Reserve, where an extensive reafforestation of cleared areas is being undertaken, artificial nestboxes are being set up to overcome the shortage of natural nest-sites, and a community awareness program is being conducted, especially among school children. DNA analyses of blood samples are being used to determine gene flow in the population as an indication of potential inbreeding and loss of genetic diversity within isolated subpopulations. It is suspected that both threats may be accentuated by the communal breeding system, with few individuals reproducing each year and the likelihood of closely related individuals pairing with each other. Results of these analyses have revealed a strong male-biased structure within the flocks, which always comprise only one female with two or more males, and a limited gene flow was found between flocks breeding in the northern part of the reserve and flocks breeding in the southern part, their separation being effected by a broad valley at lower altitude. The analyses also indicated a bottlenecked population in the reserve, which may be due to the extensive deforestation that has taken place in the region (Anon 2010). Successive years of providing nestboxes has raised the population within Buenaventura Nature Reserve to between 300 and 400 birds, but indications of inbreeding and loss of genetic diversity have highlighted the need to establish habitat corridors to connect cloud forest fragments to facilitate dispersal of the parakeets (Anon. 2015). Positive results have been achieved in discussions between Fundación Jocotoco and relevant government agencies in setting up a reafforestation program to create these corridors.

HABITATS Observed in the wild in August 1980, and identified as a new species after specimens were collected in 1985, the

Plate 48
UPPER Santa Marta Parakeet *Pyrrhura viridicata* (adult)
LOWER El Oro Parakeet *Pyrrhura orcesi* (adult)

F. KNIGHT 2010

El Oro Parakeet has been recorded at scattered localities in a narrow band of humid, upper tropical cloud forest between 600 m and 1200 m, occasionally or perhaps seasonally occurring up to 1500 m or 1550 m (Ridgely and Greenfield 2001a). Within its known range of approximately 750 km² remaining habitat is fragmented (Birdlife International 2016). Apparently it formerly occurred at lower altitudes, because an unrecognised specimen (UKNHM 53.68.105) was collected in early September 1939 at 300 m, at Piedras, also in El Oro, but all suitable forest habitat has long since been destroyed in this region, and the collection of only a single specimen could indicate that at the time these parakeets already were rare at this locality. Ridgely and Robbins (1988) report that at the type locality, about 900 m near Piñas, the forest canopy exceeds 20 m on more level ground, but is lower on steeper slopes, and from predawn until about midday clouds usually enshroud the forest, where trees and the relatively dense understorey are laden with epiphytes. Some habitat disturbance is tolerated, the parakeets being observed in secondary forest and found nesting in trees standing in cleared areas between forest patches. In 1991, a population at Buenaventura was found in orchards and gardens with only tiny remnant patches of wet forest (in Juniper and Parr 1998).

MOVEMENTS Occasional sightings of small flocks at both higher and lower elevations suggest that El Oro Parakeets may undertake seasonal altitudinal movements (in Ridgely and Greenfield 2001a; Birdlife International 2016).

HABITS El Oro Parakeets roost and nest communally, and normally are seen in small flocks of up to 12 birds, invariably with a strong preponderance of males, but larger flocks occasionally are encountered and there is a report of a flock of 60 birds being seen. Like other *Pyrrhura* species, they usually are seen feeding in the upper stages and canopy of forest trees or in swift flight through the treetops, their passage overhead accompanied by screeching call-notes. During a monitoring of nesting activities undertaken in Buenaventura Nature Reserve it was noted that at the end of the day almost all members of the group remain in the nest to sleep, and were still doing so almost one week after nesting had finished (Anon 2010).

CALLS The flight and contact calls comprise high-pitched raspy notes, not unlike those of other *Pyrrhura* species. A begging juvenile was heard to emit a shorter, slightly more high-pitched and more raspy call (Ridgely and Robbins 1988).

DIET AND FEEDING Recorded food items include seeds, fruits and blossoms, all of which are procured in the treetops. In late January 1998, in the Buenaventura district, El Oro, a male was observed eating *Cecropia* flowers (López-Lanús and Lowen 1999). Other records are of birds taking *Ficus* and *Hyeronima* fruits and fruits of balso blanco *Heliocarpus popayanensis* (Juniper and Parr 1998).

BREEDING Near Buenaventura, El Oro, presumed pre-nesting activity was observed on 14 December, when a bird was observed calling from within a hole in a tree at 1115 hours (López-Lanús and Lowen 1999). Standing on a 30° slope in a 10 ha grassy clearing between patches of forest, the 18 m high tree was approximately 75 m from the nearest forest border and 15 m from the nearest adjacent tree. With an entrance 50 mm wide by 40 mm high, the oval-shaped hole, apparently that of a woodpecker, faced east and was at a height of some 5 m above the ground. The hole was under observation from 1115 to 1305 hours, and the calling bird, which was identified as a male by its extensive red forehead, called continuously while regularly poking its head out before clambering fully out prior to flying off at 1225 hours. At 1211 hours an adult female momentarily poked its head out from the hole, but retreated inside and did not reappear for the remainder of the observation period. Although this possibly was an occupied nest, it was more likely that the male was warding off other birds while the female was preparing the nest, and the persistent, noisy, self-advertising by the male was thought to be typical of pre-breeding activity.

The main breeding season appears to be between November and March, and at Buenaventura Nature Reserve, field observations coupled with genetic analyses confirmed cooperative breeding and revealed that incubation is undertaken by both sexes. In the 2007 breeding season, 13 of 54 nestboxes were occupied by parakeets, and in the 2009 season 15 of 52 nestboxes were occupied. The breeding attempt by two flocks failed, but the remaining 10 flocks bred successfully, three of them only at the second attempt, and 26 chicks are presumed to have fledged (Anon. 2010). The loss of six broods was attributed to humidity in the nestboxes, though the effects of inbreeding may have contributed to mortality, and another two broods were lost to predation and nest poaching.

TRIBE **AMOROPSITTACINI** Brereton

The small, stocky green parrots belonging to this tribe have short rounded or acutely pointed tails, often with a prominent coloured band across the lateral feathers. The rather narrow bill lacks a notch in the upper mandible and the cere is unfeathered. In some species there is a prominent unfeathered eyering. Anatomical traits based on two genera (*Psilopsiagon* and *Bolborhynchus*) include a shallow, ill-defined and weakly muscled temporal fossa, a crescentic and posteriorly partly closed auditory meatus and a incompletely ossified orbital ring with a vestigial prefrontal extension (Schodde *et al.* 2013). A furcula is present and the uropygial gland is well developed. Some species are sexually dimorphic.

Representatives of this tribe are widely distributed in Central and South America, from southern Mexico to northern Argentina, and on Trinidad.

GENUS ***Touit*** G. R. Gray

Touit G. R. Gray, *Cat. Gen. Subgen. Birds,* 1855, p. 89. Type, by original designation, *Psittacus huetii* Temminck.

Parrots belonging to this genus collectively are termed parrotlets and, because of their small size, predominantly green plumage colouration and secretive habits in the rainforest canopy, they are poorly known, but the paucity of records may reflect difficulties in detecting them rather than any overall scarcity. Observing Lilac-tailed Parrotlets *Touit batavicus* in Guyana left me with an impression that the actions of these parrotlets are quite similar to the actions of *Cyclopsitta* fig parrots in Australia and New Guinea.

In most species the short, squarish tail is multicoloured, and sexual dimorphism is manifested in colours of the lateral feathers or bands across or at the tips of the lateral feathers. Known juveniles are duller than adults.

This genus is widespread from Costa Rica and Panama to northern South America, south to southeastern Brazil and northern Bolivia, and one species occurs on Trinidad.

ENDANGERED

Brown-backed Parrotlet

Touit melanonotus (Wied)

DESCRIPTION Length 15 cm.

ADULTS General plumage colouration green, paler and brighter on forehead, lores, cheeks and underparts; obscure brown marking on ear-coverts; mantle and tertials dark mauve-brown; scapulars and back brownish-black; rump and upper tail-coverts green; flanks and sides of lower breast pale bluish-grey, generally duller and less pronounced in females; primary-coverts brown tinged green; obscure dull reddish marking on carpal edge; underwing-coverts green; undersides of flight feathers dull bluish-green; central tail-feathers green, lateral tail-feathers red tipped black; bill yellowish-horn with grey towards base; prominent unfeathered eyering pale orange-brown; iris greyish; legs brownish-grey.
6 males: wing 107–116 (111.7) mm, tail 41–46 (43.0) mm, exp. cul. 14–15 (14.2) mm, tars. 11–13 (11.7) mm.
1 female: wing 111 mm, tail 43 mm, exp. cul. 14 mm, tars. 12 mm.

JUVENILES Like adults, but with paler unfeathered eyering and cere is flesh-coloured instead of grey.

DISTRIBUTION Confined to southeastern Brazil, from southern Bahia south to Paraná.

STATUS Although always considered to be rare, Brown-backed Parrotlets have been recorded more frequently in recent years, often in regions where it was presumed that they had been absent for very long periods or where earlier searches were unsuccessful, and locating them has been aided by recognition of their calls. Widespread deforestation for agriculture, plantations and urban development has resulted in fragmentation of the Atlantic forest, and these parrotlets now occur in small subpopulations mostly in nature reserves and national parks. In Bahia, at the northern extremity of the range, sightings from three sites in the late 1990s were the first records in that state since the 19th century, and at the southern extremity of the range the first records for Paraná were made in 1997, and now these parrotlets are known from seven localities in that state (in Birdlife International 2016).

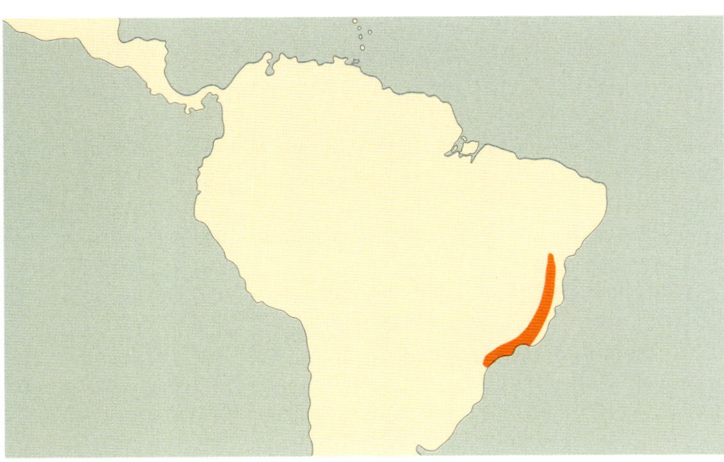

I refer to this species a record by Willis and Oniki (1981) of unidentified parrotlets observed in early June 1977 and late July 1979, during visits to Ubatuba Reserve, in coastal São Paulo state, a region where Brown-backed Parrotlets now are known to occur fairly regularly. A total of 32.2 hours of observations were undertaken during these visits, and a calculated density of 19 birds per 100 hours of observations was recorded.

The total population is estimated at less than 10 000 mature birds, with less than 1000 individuals present in each subpopulation, and a continuing moderate decline is likely because of habitat loss and degradation (in Birdlife International 2016).

HABITATS Lower montane evergreen forest between 500 m and 1200 m is favoured by Brown-backed Parrotlets, but they have been recorded up to 1400 m in Itatiaia National Park, at the Minas Gerais–Rio de Janeiro border, and down to near sea level in Bahia and São Paulo. Occasionally they venture into more open woodlands, and at times may be encountered in urban parks or gardens.

MOVEMENTS Seasonal migration or dispersal is suspected, though it may involve only short altitudinal movements.

HABITS Usually associating in small flocks of up to 20 birds, Brown-backed Parrotlets are more likely to be seen in flight, when their red tails aid identification, and they keep to the canopy or upper stages of forest trees, where they are easily overlooked while clambering stealthily among the foliage. A feeding group would be quiet and peaceful one minute, and then quite suddenly and without warning or provocation, the entire group would burst from the tree, and fly off to the accompaniment of loud calling.

Opportunities to observe and learn something of the habits of these secretive parrotlets arose at intervals between 2009 and 2011, when birds came into gardens at a condominium in urban Ubatuba, in the northern littoral of São Paulo state (Simpson and Simpson 2012). They had been reported sporadically at Ubatuba in previous years, and what is believed to be the first photograph of a wild bird was obtained there in 2006 when a bird flew into an office window at a city hotel and was photographed sitting on top of a computer screen. In February 2009, a flock of 40 birds was observed in a home garden, and in January 2011 a group was recorded feeding in condominium gardens, where they remained for several weeks, attracting the attention of observers from many parts of Brazil, and being seen regularly at three sites. The first site was in the uppermost branches of beach almond trees *Terminalia catappa*, at 10–15 m above the ground, where they were seen resting, preening and squabbling. They were observed also in a clump of 'mangue-bravo' trees *Clusia criuva* up to 5 m in height, and at a third site, where they fed and rested. At these sites the group size varied from six to 21 individuals, with a mean size of 12.75 birds over the entire period of observation, and in this time there was a gradual increase, reaching 10.8 birds in January and 16.33 birds in February. On one occasion 20 birds were perched when another 10 flew overhead without alighting, and there was one sighting of approximately 70 birds in flight. The population visiting the area was estimated at about 100–150 birds. The birds associated in smaller groups to feed or rest.

At Ubatuba and in the surrounds, morning and evening overhead flights were observed, and it seemed that the parrotlets roosted in forest at higher elevations on the Serra do Mar, descending each day to feed and rest at the urban sites, where they were most conspicuous in the December to March summer months. Local residents claimed that smaller numbers are present in October–November, and overflying birds can be seen and heard throughout the year. In the condominium gardens, while under observation, the flying parrotlets generally were noisy as they arrived, but at times they arrived silently or were already present when observations commenced. Much of the day was spent resting among the foliage on the topmost branches, where they were difficult to detect as they spent time preening and allopreening. They seemed to remain in pairs, but grouped together, and appeared to show less pair fidelity than is typical of other parrots, perhaps because the groups comprised several family parties and 'unattached' juveniles. Some individuals, possibly juveniles, were more active and occasionally joined pairs to preen, without attracting any aggression from the pairs. On multiple occasions one of these interfering birds was observed in an attempt to attract attention from an allopreening pair by hanging below the branch and then surreptitiously moving to the upper side of the branch, so forcing the pair apart, and it was thought that these interfering birds may have been offspring of the allopreening pair.

When preparing to come down lower in the canopy to feed, several birds were seen to become very active, moving quickly between branches, and occasionally making apparently playful biting actions at the legs of nearby birds. Movement through the foliage was slow and deliberate, the birds rarely flying between branches, but preferring to use their feet and bills to climb from one branch to the next, all without calling and so making detection difficult. On one occasion, a group of five vocalising Plain Parakeets *Brotogeris tirica* alighted in a *Terminalia* tree, causing a quiet perching group of Brown-backed Parrotlets to become agitated and commence calling, but when the parakeets departed the parrotlets quickly became silent again. A Great Kiskadee *Pitangus sulphuratus* was seen to chase an adult parrotlet from a *Clusia* tree to a nearby tree, where again it was attacked by the kiskadee, but the kiskadee did not follow when the parrotlet returned to the *Clusia* tree. It was learned that four parrotlets were killed by flying into windows at the condominium, and a dead bird seen to be carried by a cat was suspected to have been taken by the cat after it had been killed in this manner.

Following a brief rainshower, a bird was observed drinking water drops from leaves of a *Terminalia* tree, and another bird bathed in the droplets on a different leaf, passing the water into its feathers by rubbing against the leaves and then preening quickly for about 20 seconds.

Plate 49
Brown-backed Parrotlet *Touit melanonotus* (adults)

The flight is swift and direct, usually at or just below the forest canopy, and the birds invariably alight on a well-concealed branch. A security guard at the condominium in Ubatuba reported that the parrotlets usually left the area in one large flock at about 1700 hours, initially circling to gain height and then flying away to the south.

CALLS The most frequently heard call is a resonant, abrupt, ringing *tiriri*, given mostly in flight or sometimes while perched, and then apparently in agitation (Simpson and Simpson 2012). When darting out from a tree, these parrotlets call constantly. A low murmuring chatter, resembling that of the familiar Budgerigar *Melopsittacus undulatus*, is given between birds while resting, preening, or especially while allopreening, and a rasping squawk is emitted in aggression or stress. A persistent rasping note accompanies the up and down head-bobbing by juveniles when being fed by adults. It is noted that the call given in flight or while perched is more high-pitched than the call of the Golden-tailed Parrotlet *Touit surdus*, which occurs in part of the same range (Ridgely *et al.* 2016).

DIET AND FEEDING All feeding observations made in the urban gardens at Ubatuba were of birds eating unripened fruits of *Clusia criuva*, and the described feeding actions are very similar to feeding actions of Lilac-tailed Parrotlets *Touit batavicus* that I observed in Guyana, though I did not see any bird pluck a fruit by biting through the stalk and then carry that fruit to another branch to be eaten. This plucking technique was used by most birds at Ubatuba, and it may have been facilitated by the sticky surface of *Clusia* fruits, so aiding in their adhering to a branch (in Simpson and Simpson 2012). Typically, among the outermost branches, a bird would snip off a fruit with a small length of stalk, which was grasped to take the fruit to a more secure, heavier perch, where attempts were made to secure the fruit by balancing it and applying downward pressure, so taking advantage of its sticky surface. One bird was seen to drop two fruits while trying to grasp the stalks in its bill, but most birds were successful. The feet are not used during this process, and once the fruit was in place the bird seemed to inspect it from different angles, as if to make sure that it was securely positioned.

One bird was timed from when it entered a *Clusia* tree, and after 44 seconds it commenced eating the first fruit. During 35.5 minutes, it ate 11 fruits before departing with the flock. Some fruits were dropped before they were consumed completely, but fully consumed fruits were eaten in 2 minutes 35 seconds to 3 minutes 57 seconds, the mean time being 3 minutes 41 seconds. Intervals between the taking of fruits varied in relation to the proximity of the nearest fruits, but varied from two to 35 seconds, and feeding was continuous, the search for the next fruit starting immediately after the previous fruit was eaten or dropped. A fruit always was laid sideways and was prised open with the bill, the tongue being inserted into the fissure as if the bird was lapping up the seeds and pulp, an action that left considerable amounts of pulp on the bill, which was removed by wiping the bill against a branch. Some birds were observed with pulp remnants on the head, mantle, belly and wings.

Birds which ate fruits still attached to the stalk adopted various positions, even hanging upside-down to reach fruits, and from photographs it appeared that birds which used this technique were juveniles that may not have learned to feed like adults. Time taken to eat an entire fruit attached to the tree was considerably longer, averaging 4 minutes 17 seconds. On one occasion a

F. KNIGHT 15

second bird was seen to join the first in eating the same fruit still attached to the stem, and there was no observed interaction between these individuals.

Also at Ubatuba, in a *Terminalia* tree, two parrotlets were observed stripping flower buds along the stalk, but not biting off the stalk, and it was not obvious that they consumed the buds.

BREEDING Virtually nothing is known of the nesting behaviour of these parrotlets, though other *Touit* species are known to nest in holes in trees or in arboreal termitaria. At Ubatuba, adults were observed feeding begging juveniles in late January (in Simpson and Simpson 2012). Each adult ate several *Clusia* fruits, cleaned its bill against a branch, and then visited one of several small bromeliads, where it appeared to drink water, possibly to lubricate the fruit. The adult then perched beside a juvenile and regurgitated directly into its bill. During one recorded feeding bout, an adult regurgitated pulp five times.

TRIBE **ANDROGLOSSINI** Sundevall

This monophyletic lineage identified in molecular studies primarily is of short-tailed parrots and, not withstanding that tail length may not be fully informative as a taxonomic trait, it also is differentiated at tribal level (Joseph *et al.* 2012). Most species are sexually monomorphic, and head markings commonly are a prominent feature of plumage colouration. Anatomical features include a deeply defined, strongly muscled temporal fossa, an open, square-shaped auditory meatus, an orbital ring variably ossified by extension of the prefrontal, and a usually atrophied uropygial gland (Schodde *et al.* 2013). Prominent members of this tribe are the *Amazona* parrots and their allies, many of which are highly popular as cagebirds, and excessive capture for the live-bird trade had impacted adversely on local populations.

GENUS ***Brotogeris*** Vigors

Brotogeris Vigors, *Zool. Journ.*, **2**, 1825, p. 400. Type, by original designation, *Psittacus pyrrhopterus* Latham.

A narrow, protruding bill is the most conspicuous external feature of the small, arboreal green parrots belonging to this genus. There is a wide, rounded notch in the upper mandible, and the lower mandible is strongly curved. The short tail comprises narrow feathers with pointed tips, and in two species the central feathers are elongated. The cere and eyering are unfeathered. Osteological features of the skull include an incomplete orbital ring with a very small postfrontal process and large squamosal, and the presence of a large mandibular foramen. The sexes are alike and juveniles resemble adults.

Brotogeris is widely distributed in Central and South America, from southwestern Mexico to northern Argentina, and feral populations of two closely allied species are established in the southern United States.

ENDANGERED

Grey-cheeked Parakeet

Brotogeris pyrrhoptera (Latham)

Psittacus pyrrhopterus Latham, *Index Orn.*, Suppl., 1801. p. xxii (Brazil, error Guayaquil, Ecuador, substituted by Chapman, *Bull. Am. Mus. Nat. Hist.*, **55**, 1926, p. 261).

DESCRIPTION Length 20 cm. Weight 45–60 g.
ADULTS General plumage colouration green, becoming decidedly yellowish on sides of underbody; forehead, lores, chin and cheeks to ear-coverts pale grey; crown to nape pale blue with dull green bases to some feathers showing through; lesser and median wing-coverts suffused brownish; primary-coverts dark blue; primaries green, becoming mauve-blue on outer webs towards tips; underwing-coverts and axillaries orange; undersides of flight feathers bluish-green; tail green, feathers narrowly edged yellow on inner webs; bill horn-coloured slightly tinged orange; iris dark brown; legs pale pink.
10 males: wing 112–125 (117.8) mm, tail 65–75 (69.5) mm, exp. cul. 16–18 (17.2) mm, tars. 13–15 (13.9) mm.
8 females: wing 115–120 (117.3) mm, tail 63–70 (67.1) mm, exp. cul. 16–17 (16.3) mm, tars. 13–15 (13.8) mm.

JUVENILES Like adults, but crown to nape green with little or no pale blue suffusion.

DISTRIBUTION Western Ecuador, south from the Río Chone valley, in Manabí, and extreme northwestern Peru, in Tumbes and Piura.

STATUS It seems that the status of Grey-cheeked Parakeets has deteriorated significantly during the past 10 to 15 years, largely because of excessive exploitation for the live-bird trade. In response to uncertainties about the effects of habitat loss and trapping for the live-bird trade on these parakeets and on Red-masked Parakeets *Psittacara erythrogenys*, available data were reviewed to determine the status of both species, and it was found

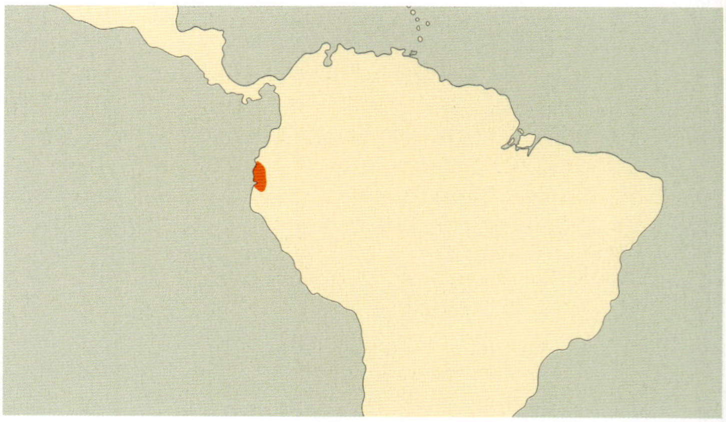

Plate 50
Grey-cheeked Parakeet *Brotogeris pyrrhoptera* (adults)

F KNIGHT 16

that the two main areas occupied by Grey-cheeked Parakeets were the Ecuadorian coastal provinces of Manabí and Guayas, and the southwestern corner of Ecuador and adjacent Tumbes department in northwestern Peru (Best *et al.* 1995). Evidence of their disjunct dispersal within the range being due probably to habitat loss came from records along the Río Babahoyo, in Guayas, Ecuador, where previously they were recorded from several sites, but no records were made during recent surveys, and their absence was attributed to large-scale conversion of the land to agriculture because only 50 km to the southwest they remained common in deciduous forest at Bosque Protector Cerro Blanco. The data indicated also that in far northwestern Peru the parakeets were scarcer than in Ecuador and, based on all available data and on field surveys, it was suspected that the total population was unlikely to exceed 15 000 birds, with the majority of these occurring within Ecuador. This estimate represented a decline of some 70 per cent in 10 years, although the parakeets remained locally common in remnants of suitable habitat (in Birdlife International 2016).

Concerns regarding the threatened status of Grey-cheeked Parakeets were questioned by Ridgely and Greenfield (2001a), who claimed that in western Ecuador numbers 'have held up relatively well', with populations persisting well in many mostly deforested agricultural areas, and pairs are resident in some city parks in Guayaquil. However, a necessity to monitor their status was acknowledged because of large numbers being trapped for the pet market (in Ridgely and Greenfield 2001a). Similarly in extreme northwestern Peru, they were considered to be locally fairly common, but declining (in Schulenberg *et al.* 2007). Transect counts in Amotape National Park and Tumbes National Reserve, northwestern Peru, revealed a decline of approximately 33 per cent between 1992 and 2008 (in Birdlife International 2016).

Continued illegal trapping for the live-bird trade, together with persecution and habitat destruction or fragmentation, have been identified as causes of a very rapid population decline that has occurred during the past 10 years. The rate of decline is expected to be slower during the next decade (Birdlife International 2016).

HABITATS In western Ecuador, Grey-cheeked Parakeets are recorded mainly below about 1000 m, though very small numbers have been observed as high as approximately 1500 m in the far south, and in far northwestern Peru they are found below 900 m (Ridgely and Greenfield 2001a; Schulenberg *et al.* 2007). They frequent deciduous and semihumid forests or woodlands, dry forests and arid *Acacia*-dominated scrublands, agricultural lands where a few large trees remain, and gardens or parklands in the lowlands (Best *et al.* 1995; in Birds International 2016). They range locally into more humid forest, but have not been found in the more humid forests of northwestern Ecuador, and overall appear to be most numerous in deciduous forest dominated by *Ceiba trichistandra*.

HABITS Grey-cheeked Parakeets usually are encountered in small groups of up to about 12 birds, which generally keep to the canopy or upper stages of trees, though sometimes descending lower at the forest margins or in clearings (Ridgely and Greenfield 2001b). Unusual records of larger flocks in western Ecuador include sightings of flocks of 50 birds seen at Bosque Protector Cerro Blanco, Guayas, in all months during 1992–1993, a flock of more than 50 birds seen between Arenillas and Puerto Pitthaya, Loja, in January 1991, and a flock of 60 birds seen at Tambo Negro, Loja, in August 1989, while in far northwestern Peru five flocks totalling 80 birds were recorded in Tumbes town, Tumbes, during February 1979, and flocks of 60 to 120 birds were seen at El Caucho, Tumbes, during February–March and in June in 1986 and again in 1988 (in Best *et al.* 1995). Flocks of up to 40 birds occasionally were seen during intensive fieldwork undertaken at Bosque Protector Cerro Blanco, between July and September 1996 (Pople *et al.* 1997). At times these parakeets associate in foraging groups with other parrots, including Red-masked Parakeets *Psittacara erythrogenys* and Bronze-winged Parrots *Pionus chalcopterus*.

In flight, the wings occasionally are withdrawn in near to the body, so producing an undulating pattern, and the birds often fly high overhead with a conspicuous flashing of the orange underwings.

CALLS These parakeets normally are quiet while resting during the middle of the day, but at other times are noisy, giving a variety of rather shrill, chattering calls, and they call more while perched than do many other parrots (Ridgely and Greenfield 2001b). Flight calls include a low, rolling *jurt* as well as a series of scratchy notes, and there are conversational vocalisations including calls given in rapid duets between paired individuals (Schulenberg *et al.* 2007).

DIET AND FEEDING Grey-cheeked Parakeets feed arboreally, and the diet comprises seeds, fruits, flowers and catkins, with a preference shown for *Ceiba* fruits, and small flocks have been recorded taking cultivated bananas and maize (in Birdlife International 2016). Recorded food items include *Erythrina* flowers and seeds, *Chorisa* flowers and seeds, flowers of pigio *Cavanillesia platanifolia*, *Ceiba* and *Ficus* fruits, and *Cecropia* catkins (in Best *et al.* 1995).

BREEDING In southwestern Ecuador, at Tambo Negro, in Loja, copulation was observed in late August 1989, and at the same location further observations of breeding behaviour, including copulation, allopreening, and inspections by pairs of potential nesting hollows, were recorded in February 1991, so nesting apparently takes place during and immediately after the January to March wet season (in Best *et al.* 1995). Most nests are in tree hollows, especially in *Ceiba* trees, but pairs nest also in tunnels excavated in arboreal termitaria.

Details of a successful breeding in captivity were given by Hampe (1939), who reported that five eggs were laid over a period of eight days, and four chicks hatched after incubation lasting approximately four weeks, with fledging occurring some five weeks after hatching. Arndt (1996) notes that in captivity a clutch of four to six eggs is laid, with incubation of 23 days duration commencing after laying of the last egg, to be followed by fledging some seven weeks after hatching, and other pairs or relatives may assist in brooding.

EGGS The eggs are rounded, and measurements of 22.8 (21.4–25.5) × 17.9 (16.7–18.6) mm are given for six eggs (Schönwetter 1964).

GENUS ***Amazona*** Lesson

Amazona Lesson, *Traité d'Orn.*, livr. 3, 1830, p. 189. Type, by subsequent designation, *Chrysotis farinosa = Psittacus farinosus Boddaert* (Salvadori, *Cat. Bds Brit. Mus.*, **20**, 1891, p. 268).

Collectively known as 'amazons', the arboreal parrots belonging to this genus are among the most familiar of neotropical species, and are easily recognisable because of strong similarities in body structure and plumage patterns, with a finely scalloped appearance in many species being derived from fine dark edging to the feathers. They are midsized to large, stocky parrots with strong, heavy bills and short, slightly rounded tails. The unfeathered cere and eyering are prominent, and there is a distinct notch in the upper mandible. The wings are broad and rounded, and in some species the flight feathers barely extend beyond the tertials when the wing is folded. Sexual dimorphism is absent or slight, and juveniles are similar to, or only slightly different from adults. The complete orbital ring and associated osteological features of the skull are similar to the arrangement present in closely allied *Pionus* (Thompson 1900). Molecular analyses also suggest an *Amazona-Pionus* sister group relationship (Rusello and Amato 2004).

Amazona parrots are widespread throughout Central and South America and in the Caribbean islands. Rusello and Amato note that molecular analyses indicate that species occurring in the Greater Antilles, including the endangered Puerto Rican Amazon *A. vittata*, constitute a monophyletic group with the White-fronted Amazon *A. albifrons* from Central America, but endemic species of the Lesser Antilles, including the endangered Imperial Amazon *A. imperialis*, are not each other's closest living relatives.

Two extinct *Amazona* species have been named from descriptions in accounts by early voyagers to the Lesser Antilles, but their existence is not supported by specimens or fossil remains. *A. violacea* (Gmelin) from Guadeloupe is known mainly from descriptions and drawings in a 17th century account by Jean-Baptiste Du Tertre and an 18th century account by Jean-Baptiste Labat. More poorly attested is *A. martinicana* Clark from Martinique, briefly described by Labat in 1742.

ENDANGERED

Lilac-crowned Amazon

Amazona finschi (Sclater)

Chrysotis finschi Sclater, *Proc. Zool. Soc. London,* 1864, p. 208 (Mexico).

OTHER NAME Lilac-crowned Parrot.

DESCRIPTION Length 33 cm. Weight 280–327 g.
ADULTS General plumage green, paler and more yellowish on underparts; feathers narrowly edged dusky black to give finely scalloped appearance; forehead, crown and lores deep maroon; hindcrown, sides of nape and hindneck bluish-mauve, feathers finely margined dusky black; cheeks and ear-coverts yellowish-green without dusky black margins to feathers; outer webs of primaries violet-blue, becoming green towards bases, inner webs greyish-black; broad red band on outer webs of S1 to S5, producing conspicuous red wing-speculum in flight; remaining secondaries green; underwing-coverts and undersides of flight feathers green; tail green tipped yellowish-green and outermost feathers edged blue on outer webs towards bases; bill horn-coloured; unfeathered eyering pale grey; iris orange; legs greenish-grey.
20 males: wing 191–215 (202.3) mm, tail 100–124 (114.6) mm, exp. cul. 29–32 (29.9) mm, tars. 21–24 (22.3) mm.
19 females: wing 185–208 (198.1) mm, tail 104–124 (114.2) mm, exp. cul. 27–31 (29.1) mm, tars. 21–23 (22.4) mm.
JUVENILES Like adults, but less red on forehead and forecrown; iris dark brown.

DISTRIBUTION Western Mexico, where distributed from extreme southeastern Sonora and southwestern Chihuahua south to Oaxaca; in USA, feral populations established in southern California and a small feral population in Florida.

SUBSPECIES Because of a less yellowish-green general plumage and a narrower, duller maroon frontal band, birds in the northern sector of the range, from extreme southeastern Sonora and southwestern Chihuahua south to central-eastern Sinaloa and Durango, at times are differentiated as *A. f. woodi* Moore, but these differences appear to be encompassed within individual variation and the likelihood of geographic variation is undermined by extensive movements undertaken by the parrots (see Dickinson and Remsen 2013).

STATUS Between 2001 and 2009, four research groups conducted 1475 extensive field survey sessions along the west coast of Mexico during both the January to May breeding season and the June to December non-breeding season, with results being compared to historical records for the period from 1900 to 1990. The current distribution of Lilac-crowned Amazons was estimated at 58 347 km², which represents a 72.6 per cent distribution reduction (Marín-Togo *et al.* 2012). The parrots appear to maintain a continuous distribution from southern Sonora along the foothills of the Sierra Madre Occidental to Michoacán and western Guerrero, but they have been extirpated from coastal areas below about 300 m in Nayarit and in all but one locality in Guerrero. They have been extirpated also from most coastal areas of Oaxaca as well as parts of Jalisco, Durango, Colima and Michoacán (in Birdlife International 2016). During the surveys it was found that Lilac-crowned Amazons had a low extent of only 850 km² of protected areas in their distribution,

and their presence was confirmed in only five protected areas. It was recognised that these parrots may not adapt to nesting in modified habitats, and the loss of preferred semi-deciduous forest habitat contributed to the distribution reduction, but the presence of up to 80 per cent of preferred habitat in areas where the birds no longer are present suggests that factors other than habitat loss are involved, and focus was directed at capture for the live-bird market. Lilac-crowned Amazons were the most captured *Amazona* species in the early 1980s, and continue to be one of the five most captured parrot species in Mexico, with an estimated 5400 birds captured illegally each year (Cantu *et al.* 2007). It is concluded that intense pressure for trade may have impacted on the current distribution, creating areas where the parrots no longer are present even though suitable habitat remains (Marín-Togo *et al.* 2012).

A population of 7000–10 000 individuals, roughly equating to 4700–6700 mature birds, was estimated from a recent survey conducted in 53 out of 77 sample sites of 50 × 50 km covering the entire range but, in view of the estimated 5400 birds captured illegally each year, it is suspected that the total population could be underestimated (Birdlife International 2016). Nevertheless, it is suggested that the high level of distribution decline indicates that the population is in very rapid decline.

In contrast to the decline of wild populations in western Mexico, feral populations in southern California are increasing, though they remain much less numerous than Green-cheeked Amazons *Amazona viridigenalis*. Froke (1981) recorded a minimum of 22 birds in the San Gabriel Valley between 1976 and 1978, with evidence of nesting, and in the late 1990s, I observed a few pairs associated with a flock of Green-cheeked Amazons at Sierra Madre, in the Pasadena district. Numbers have increased in the San Gabriel Valley, and in Orange County numbers have proved to be higher than in earlier reports (Mabb 2002). A small population recently has been reported from Santa Barbara County. Pranty and Garrett (2011) report that the total population in southern California perhaps comprises 500 or more birds. In Florida, Lilac-crowned Amazons are rare to locally uncommon, being recorded primarily in the Miami area, where 31 birds were reported in December 2006, but there was no known breeding record (in Pranty and Garrett 2011).

HABITATS Howell and Webb (1995) note that Lilac-crowned Amazons frequent semi-deciduous forest and margins, pine-oak forests and mangroves. They are most common in wooded foothills up to 1000 m, but are not present in coastal areas of Sinaloa and northern Nayarit, occurring mainly above 375 m in this part of their range. Renton (2001) points out that semi-deciduous forest produces greater food resource abundance in the early to mid dry season than in the rainy season, and is an important source of food during the dry season, when chicks are being reared. More particularly, semi-deciduous forest provides essential food resources during the late dry season months of May–June, so may be a key habitat at this critical time of the year. Nesting studies have shown that semi-deciduous forest is required as optimal breeding habitat, and local people reported that the parrots nest preferentially in low-lying areas of semi-deciduous forest (Renton and Salinas-Melgoza 1999). Conversely, deciduous forest may be important in providing abundant food at the end of the rainy season, prior to the onset of nesting. In addition to nesting and foraging habitats, Lilac-crowned Amazons require secure areas where they gather at communal nighttime roosts, and radiotelemetry studies have identified roost-sites at the top of ridge plateaus or near the summit of tall hills (Renton and Salinas-Melgoza 1999).

Renton and Salinas-Melgoza (1999) note that nesting studies were undertaken in the Chamela-Cuixmala Biosphere Reserve, Jalisco, where the hilly topography varies in elevation from 20 m to 520 m, and the dominant vegetation type on the slopes is tropical dry deciduous forest, with semi-deciduous forest in the larger drainages and more humid valleys, together with monospecific forests of guayabillo *Celaenodendron mexicanum* occurring as discontinuous patches within the tropical deciduous forest mosaic.

In southern California, Lilac-crowned Amazons frequent residential and suburban areas, sometimes occurring in stands of native oak trees, and nesting has been recorded in native coniferous forest in the San Gabriel Mountains (Garrett 1997).

MOVEMENTS Some altitudinal and latitudinal movements may be undertaken in response to seasonal changes in food availability, and records at altitudes above 1000 m could be attributed to these movements (in Birdlife International 2016).

HABITS Much of what is known of the habits of Lilac-crowned Amazons comes from observations made by Katherine Renton and colleagues in the late 1990s at study sites in the Chamela-Cuixmala Biosphere Reserve, Jalisco, western Mexico (see Renton and Salinas-Melgoza 1999). The primary social unit is the mated pair and pair-bonds are maintained throughout the year. Although pair-bonds are relatively permanent, it is likely that if one partner is lost the surviving individual will find a new mate. On one occasion, after the loss of an incubating female to predation, Renton observed the male returning to the nest-site each day and calling for the female; by about three weeks after her disappearance the male continued to return to the nest-site to call for the lost female, though now he was accompanied by another bird, which he appeared to feed, so formation of a new pair-bond seemed likely. Pairs and family groups may come together to form large foraging flocks, and peak times of foraging are during the first three hours after sunrise and the two to three hours before nightfall, when flights of flocks between communal nighttime roosts and foraging sites are conspicuous. The parrots are relatively inactive during the remainder of the day, spending much of the time among shaded foliage to seek refuge from the midday heat. Just before nightfall, the parrots again gather into flocks to return to the communal nighttime roosts, which at times can be occupied by very large numbers of birds. In November 1975, in wooded coastal hills above San Blas, Nayarit, near where I earlier had encountered small flocks in morning flights, Robert Ridgely recorded a roosting aggregation conservatively estimated at more than a thousand birds (*in litt.* 1977). Stager (1954) recalled making camp in a stand of oaks at 1470 m in the Barranca de Cobra, southwestern Chihuahua, and then finding that this particular grove of oaks must have been a communal roost of long standing because just before nightfall several hundred Lilac-crowned Amazons came streaming in from their feeding grounds and began settling in the trees for the night. Favoured sites for nighttime roosts are at the top of ridges or near the summits of tallest hills, and Renton has noted that a number of roosts may

Plate 51
UPPER Green-cheeked Amazon *Amazona viridigenalis* (adult)
LOWER Lilac-crowned Amazon *Amazona finschi* (adult)

F. KNIGHT '15

be used by parrots on any one night and individuals may change their use of a roost location after a few days. Departure from the roost is at first light.

Mabb (1997a) records observations made between May 1995 and August 1997 at nighttime roosts in Temple City and Arcadia, in the San Gabriel Valley of Los Angeles County, southern California, where Lilac-crowned and Green-cheeked Amazons were grouped together because of difficulties in distinguishing between them in shady trees in poor pre-sunset lighting, though Green-cheeked Amazons dominated at a ratio of up to 8:1. Numbers at the roosts were lowest in spring and early summer, and highest in autumn and winter. There was considerable variability in arrival times at the roosts, though in summer the parrots tended to frequent the roosting area approximately 30–60 minutes before sunset, and in winter they tended to arrive only five to 20 minutes before sunset. They were highly vocal until shortly before total darkness, when they became abruptly silent. Approximately five minutes before roosting, there was an increase in activity, with much calling, flying between branches or trees, and circling within and around the roosting site. From approximately 30 minutes before sunset until the time of roosting groups of 20 to 100 or more parrots regularly engaged in various social actions in nearby trees or on power lines, frequently preening or allopreening and performing various displays, including hanging upside-down, fanning their tails or raising feathers on the nape, spreading their wings, bowing their heads, dilating their pupils, indulging in feigned attacks with opened bills, and bill-wiping on branches. During summer and autumn, the parrots tended to prefer to roost in deciduous sweetgums *Liquidamber styracifolia* and western sycamores *Plantanus racemosa*, and when these trees lost their leaves in winter, spring preferences shifted to broadleaved evergreens, such as carrotwoods *Cupanopsis anacardiodes*, eucalypts or living oaks.

CALLS Howell and Webb (1995) note that the varied and raucous calls include a shrill *krih-krih* or *kreeih-kreeh*, a rolling *krreeeih*, a deeper *kyah'ha...kyah'ha* and an almost corvid-like *krra...krra*.

DIET AND FEEDING Renton (2001) reports that the diet and feeding preferences of Lilac-crowned Amazons were studied between January 1996 and July 1997 in the Chamela-Cuixmala Biosphere Reserve, Jalisco, where a total of 45 phenology transects of 200 m × 6 m were established, with 15 transects in each of three habitats – deciduous forest, semi-deciduous forest and watercourses. Transects were monitored during the second and third weeks of each month, excluding December 1996, and fruit- or seed-bearing trees exceeding 10 cm in diameter at breast height were recorded when detected within a distance of 3 m on either side of the transect to determine canopy variation in food availability. The diet was determined by observations of feeding recorded in each month when trails were walked in deciduous and semi-deciduous forest, and along watercourses, during the first four hours of the morning and the last three hours of the afternoon, the peak periods of foraging activity. The food plant was recorded, together with the number of parrots utilising that plant and the plant part eaten. In 137 feeding observations, Lilac-crowned Amazons were seen to consume 33 food items, with seeds comprising 81.8 per cent of the diet, fruits making up 8.8 per cent, insect larvae comprising 6.6 per cent and bromeliad stems comprising 2.9 per cent. Principal food items in the diet were unripe seeds of culebro *Astronium graveolens*, eaten by 207 parrots in 24 observed feeding bouts, unripe seeds of guayabillo *Celaenodendron mexicanum*, eaten by 127 parrots in 16 observed feeding bouts, seeds and fruits of ojite *Brosimum alicastrum*, eaten by 102 parrots in 14 observed feeding bouts, and unripe seeds of peacock flower *Caesalpinia pulcherrima*, eaten by 56 parrots in 12 observed feeding bouts. The diet varied considerably between the dry and rainy seasons, with little overlap or similarity in food items between the seasons. Main food items during the dry season were seeds of *Astronium graveolens*, *Brosimum alicastrum*, *Comocladia engleriana* and *Celaenodendron mexicanum*, and fruits of *Ficus insipida*. During the rainy season the diet comprised seeds of *Celaenodendron mexicanum* and *Eisenbeckia nesiotica* and fruits of *Sciadodendron excelsum*, together with seeds of physic tree *Jatropha standleyi*, *Jatropha malacophylla* and *Caesalpinia pulcherrima*. At no time were these parrots observed to eat flowers.

During field studies undertaken in southern California in the 1990s, Lilac-crowned Amazons were recorded utilising 21 food sources, but they did not come to take seed at garden bird-feeders (Garrett *et al.* 1997). Food plants and the items eaten are set out in Table 8.

BREEDING Field studies of the nesting behaviour were undertaken in the Chamela-Cuixmala Biosphere Reserve, Jalisco, during January to June in 1995-1998 (Renton and Salinas-Melgoza 1999). Nest searches were carried out in February during the nest prospecting and early incubation stages of the breeding cycle. A tree cavity was considered to be a potential nesting site if one of the adult parrots remained within the cavity for longer than 30 minutes, and reuse of a site was determined from the frequency of cavity occupation between years. Of 29 located nest sites, 15 were in living guayabillo trees *Celaenodendron mexicanum*, nine were in living culebro trees *Astronium graveolens* and two were in living *Tabebua* trees, with another site being in a dead tree and two were in unidentified trees. Reuse of sites was low, with only three of the 29 sites being occupied over more than one breeding season. One cavity was occupied in three of the four years, the only absence being in 1996, and another two cavities were reused once after a vacancy of one year. Mean measurements of 26 active nesting sites are listed in Table 9, the remaining three sites not being accessed because of safety reasons. Measurements with the least variation were height of the cavity entrance from the ground and width of the cavity entrance. Diameter of the nesting tree at breast height was fairly consistent and reflected a preference for large, mature trees characteristic of semi-deciduous forest. Greatest variation was in depth of the cavity and length of the cavity entrance.

The timing of egg-laying was highly synchronised between nests, with most pairs commencing incubation within 14 days of commencement of the first nest. The mean nest commencement date in 1996 was 6 February, with a range of 30 January to 13 February, and in 1997 it was 15 February, with a range of 10 to 23 February. Females sat very tightly during incubation, leaving the nest only once in the morning and once in the afternoon to be fed by the male, and at 20 nests total time spent out of the nest by the female in both the morning and afternoon averaged 39.4 minutes, with a range of 15 to 95 minutes. Conversely, males rarely were seen to enter the nesting cavity or to perch at the entrance. Daily activity periods were consistent between nests, with the male making two or three visits each day to feed the sitting female. For 45 observed morning visits by males the mean arrival time was 0824 hours, with a range of 0607–1008 hours, and for 40 observed afternoon visits the mean arrival time

Food plant	Items eaten	Number of observations
Platanus spp.		
sycamore/plane trees	fruits/seeds	more than 10
Ficus spp. figs	fruits/seeds	more than 10
Liquidamber styracifolia		
sweetgum	fruits/seeds	more than 10
Juglans regia walnut	fruits/seeds	five to 10
Quercus spp. oaks	fruits/seeds	five to 10
Carya illinoenensis	fruits/seeds	more than 10
Prunus armeniaca apricots	fruits/seeds	more than 10
Magnolia grandiflora		
magnolia	fruits/seeds	one to four
Melia azedarach chinaberry	fruits/seeds	one to four
Araceae palms	fruits/seeds	one to four
Olea europaea olive	fruits/seeds	one to four
Pinus spp. pine	fruits/seeds	one to four
Ceratonia siliqua carob	fruits/seeds	one to four
Bauhinia spp. orchid tree	fruits/seeds	one to four
Malus spp. apple	fruits/seeds	one to four
Persia americana avocado	fruits/seeds	one to four
Morus spp. mulberry	fruits/seeds	one to four
Citrus sinensis orange	fruits/seeds	one to four
Eucalyptus/Corymbia spp.		
eucalypts	flowers/nectar	five to 10
Jacaranda acutifolia jacaranda	flowers/nectar	five to 10
Acacia spp. acacia	flowers/nectar	one to four

Table 8. Food items observed being eaten by Lilac-crowned Amazon during studies undertaken in southern California, USA, in the 1990s; more than ten observations are indicative of an important food item (after Garrett *et al.* 1997).

Nesting site characteristics	Mean	Range
Tree diameter at breast height (cm)	43.1	27.7–66.3
Height of cavity above ground (m)	9.7	7.4–14.7
Width of cavity entrance (cm)	10.0	6.4–14.0
Length of cavity entrance (cm)	21.0	7.5–71.6
Internal diameter of cavity (cm)	19.9	10.5–35.0
Cavity depth (cm)	66.2	24.0–260.0
Circumference at cavity entrance (cm)	86.7	26.0–135.0
Nearest active nest (m)	948.9	25.0–2419.0

Table 9. Measurements of nesting sites located in the Chamela-Cuixmala Biosphere Reserve, Jalisco, during field studies undertaken in 1995-1998 (after Renton and Salinas-Melgoza 1999).

was 1814 hours, with a range of 1715–1903 hours. Each of 80 observed feeding visits lasted an average of 33.9 minutes, with a range of 5.0–113.0 minutes. The male usually called loudly on his approach to the nesting area, and perched in a tree adjacent to the nesting tree, where he emitted low contact calls until the female emerged. The pair gave a characteristic take-off squawk or bugle as the female flew from the nesting tree to join the male. Feeding by regurgitation from the male to the female occurred in a regularly used perching tree at an average 423 m from the nesting tree, and lasted an average of just 11.5 minutes, being the only time during the incubation period when both adults were away from the nesting area.

Duration of the incubation period was not determined, but the eggs hatched asynchronously, and females continued to brood nestlings during the day until the oldest nestling was about 20 days old. Females ceased roosting at night in the nest when the youngest chick was in its third week. During this early nestling period, the male continued to feed the female twice each day, and occasionally was seen to enter the nest for about two minutes on each feeding visit. When larger, the chicks were fed in the nest by both parents. The behaviour of nesting pairs changed when the female commenced to forage with the male. They became more secretive when in the vicinity of the nest, arriving and departing silently, and low, almost inaudible calls were emitted when in the nesting area. They approached the nest cautiously, and would not do so if any disturbance was detected. An average 2.6 visits per day were made by the parents to feed the nestlings, with each adult spending an average of four minutes inside the cavity on each visit. Prior to fledging, the chicks commenced to climb to the nest entrance, and were fed there by the adults. During this time, the parents spent more time perched near the nest entrance making low contact calls to the chicks. The mean age of chicks at fledging was 63.7 days, with a range of 56–68 days, and they fledged asynchronously. In 1996, all chicks fledged from observed nests in the 12 days between 6 and 18 May, while in 1997 all fledged in the 17 days between 11 and 28 May, and in 1998 all fledged in the 13 days between 2 and 15 May.

Froke (1981) reported successful nesting by a pair of Lilac-crowned Amazons in an old woodpeckers' cavity in a blue gum *Eucalyptus globulus* in Arcadia, Los Angeles County, southern California, where successful nests of Green-cheeked Amazons *Amazona viridigenalis* and Yellow-headed Amazons *A. oratrix* also were recorded. Mabb (1997b) reports that intermittent observations were made between 2 April and 2 June 1996 at a nest at a height of 9 m in a cavity in a utility pole at Temple City, also in Los Angeles County. This cavity was approximately a metre down from the top of the pole, on the north side, and directly underneath was a wooden cross-beam and transformer drum on which the nesting adults performed many courtship displays, such as tail-fanning, wing-spreading, raising of the nuchal feathers, and pupal dilation. They frequently engaged in allopreening, mate feeding by regurgitation, and sharing solid food held in the foot of one bird. The pair aggressively defended the nest, the utility pole and nearby power lines against other *Amazona* parrots, American Crows *Corvus brachyrhynchos* and Common Starlings *Sturnus vulgaris.* In early April, both adults were observed frequently entering and exiting the cavity, which presumably was an existing cavity that they had enlarged, and in the evening they usually disappeared into the cavity a minute or so before a nearby flock of some 350 *Amazona* parrots settled in their nighttime roost. Begging chicks were heard in the nest on 3 May, and an adult was seen to periodically peer out from the cavity entrance. The other adult made many short trips, each some five minutes in duration, to and from a foraging site, returning to put its head into the cavity entrance and pass food to the occupants. On 17 May and 2 June, 10–30 Lilac-crowned Amazons were observed feeding on fruits directly under and around the utility pole, but it could not be determined if any were interacting with the nest cavity.

Incubation periods of 28 and 29 days have been recorded in accounts of captive breeding (Lint 1952; Mann and Mann 1978).

EGGS Measurements of four eggs are listed as 37.0 (36.4–38.1) × 29.2 (27.9–30.2) mm (Schönwetter 1964).

ENDANGERED

Green-cheeked Amazon

Amazona viridigenalis (Cassin)

Chrysotis viridigenalis Cassin, *Proc. Acad. Nat. Sci. Philadelphia*, **6**, 1853, p. 371 (South America, error = northeastern Mexico).

OTHER NAME Red-crowned Parrot.

DESCRIPTION Length 33 cm. Weight 293–345 g.
ADULTS General plumage green, with fine dusky black margins to feathers giving scalloped appearance on upperparts, breast and flanks; cheeks to ear-coverts and sides of throat brighter green without dusky margins to feathers, and bordered by violet-blue band from behind eyes down sides of neck; lores, forehead and crown crimson, less extensive in female; broad red band on outer webs of S1 to S5, producing conspicuous red wing-speculum in flight; remaining secondaries green; outer webs of primaries deep blue becoming green towards bases and dusky black on inner webs; under wing-coverts and undersides of flight feathers green; tail green, broadly edged paler yellowish-green on outer webs and at tips of feathers; bill dull yellow; unfeathered eyering pale grey; iris yellow; legs greenish-grey.
8 males: wing 200–213 (207.5) mm, tail 103–116 (108.6) mm, exp. cul. 27–31 (28.9) mm, tars. 22–25 (23.8) mm.
8 females: wing 194–205 (200.4) mm, tail 97–109 (102.4) mm, exp. cul. 27–29 (27.6) mm, tars. 22–24 (22.9) mm.
JUVENILES Like adults, but crimson restricted to lores and forehead; iris pale grey.

DISTRIBUTION Northeastern Mexico, from eastern Nuevo León and southern Tamaulipas south through eastern San Luis Potosí to northern Veracruz and extreme northeastern Querétaro. Feral populations established elsewhere in Mexico, including the Monterrey district, in California and southeastern Florida, mainland USA, on Oahu, Hawaii, and on Puerto Rico; uncertain status of populations in Texas, USA.

STATUS Excessive exploitation for the cagebird market has contributed significantly to a rapid decline in wild populations of Green-cheeked Amazons in northeastern Mexico, so it is ironic that feral populations originating from escaped cagebirds are well established and increasing, especially in southern California. In northeastern Mexico, rapidly declining numbers in much of the range and total disappearance of the parrots from some districts are attributed to this excessive exploitation coupled with long-term habitat loss (Birdlife International 2016). Legal imports into the USA, between 1970 and 1982, totalled 16490 birds, and most were nestlings (in Birdlife International 2016). Enkerlin-Hoeflich and Hogan (1997) point out that illegal exports from Mexico and a pre-export mortality of approximately 50 per cent equates to 5000 birds per year. Trappers damage nests when taking nestlings, at times felling entire trees, so reducing nest-site availability and leading to permanent site abandonment (Snyder *et al.* 2000). Many gallery forests have been cleared or degraded, with more than 80 per cent of lowlands in Tamaulipas cleared for agriculture, and forest habitat now is patchily dispersed on cattle-ranches, where trapping pressure is intense (Enkerlin-Hoeflich and Hogan 1997).

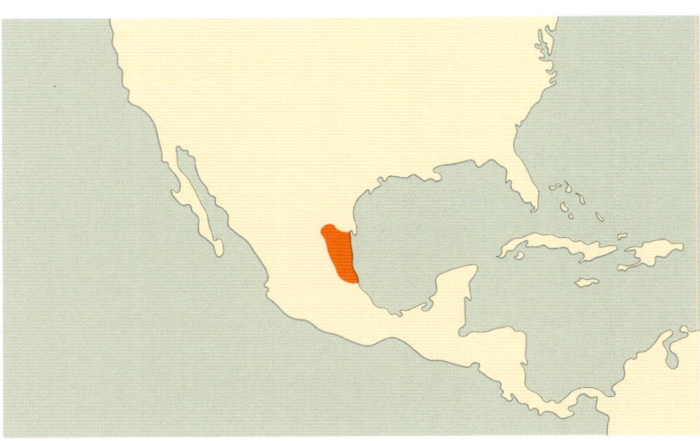

Green-cheeked Amazons remain locally fairly common to common in Tamaulipas and San Luis Potosí, with small colonies in extreme northeastern Querétaro, but there have been no recent records from Nuevo León and Veracruz. Recorded densities were 25.2 birds per km^2 at a locality in coastal Tamaulipas in the 1970s, falling to 5.7 birds per km^2 in one area in 1992–1994, indicating a decline of approximately 77 per cent in 20 years. Due to the ongoing threats of trapping and deforestation, the decline is suspected to be continuing at a rate exceeding 50 per cent over 10 years. The density of 5.7 birds per km^2 recorded in 1992–1994 indicates a total wild population of 3000–6500 birds, which equates roughly to 2000–4000 mature individuals (Birdlife International 2016).

First reports of feral Green-cheeked Amazons in the Los Angeles district, southern California, were made in the early 1960s, but there were no records of breeding. Froke (1981) recorded 50 birds and documented nesting in the San Gabriel Valley from 1973 to 1978, and an unspecified proportion of a mixed flock of *Amazona* parrots in West Los Angeles during that period also were of this species. In March 1986, I observed some 50 Green-cheeked Amazons with much smaller numbers of Lilac-crowned Amazons *Amazona finschi* coming to roost in street trees at Temple City, in the Los Angeles urban area, and a few weeks prior to my visit several hundred parrots had been seen at this roosting site, the fall in numbers being attributed to pairs having left the flock for nesting. Green-cheeked Amazons now are the most abundant parrots in the greater Los Angeles region, and together with increasing populations in the San Diego district, on Monterey Peninsula and possibly in the Sacramento area, it is estimated that 'thousands' of these parrots are present in urban parts of California (Garrett 1997; Pranty and Garrett 2011). Mabb (2002) suggests that more than 25 per cent of the world population now occurs in California.

Brush (2005) points out that Green-cheeked Amazons and Green Parakeets *Psittacara holochlorus* are the most numerous parrots in the lower Rio Grande valley, extreme southeastern Texas. It generally is assumed that populations of both species originated from escaped cagebirds, but Brush suggests that their arrival probably was part of a northward expansion of the ranges of several tropical species. Walker and Chapman (1992) also argue that populations of Green-cheeked Amazons in Texas represent a true or natural range expansion, noting that there are early records. In the 1920s and early 1930s, a flock roosted in an abandoned pump house in La Feria, and in 1956 at least four apparently wild birds were recorded on a home movie at an unknown locality in the lower Rio Grande valley. Large numbers suddenly appeared in the valley in early 1984, soon after severe frosts in December 1983 impacted on northeastern Mexico and southern Texas, and it was suspected that these freezing conditions forced a significant northward movement of the parrots. Although there is good evidence of irregular incursions of wild birds into southernmost Texas, I am hesitant to attribute the origin of present established populations solely to these incursions, because establishment seems to have occurred in the early 1980s, concurrent with the peak in trafficking, and there are claims that birds being transported illegally across the Mexico–USA border are released whenever traffickers perceive a risk of apprehension by customs officials. In the vicinity of Brownsville, breeding has occurred regularly since the 1980s, and the Green-cheeked Amazon has been named the city mascot.

The first record from Oahu, in the Hawaiian Islands, was a sighting of three birds in about 1969, near and on the grounds of the Honolulu Zoo, and at this same locality a small flock of adults tending a fledgling was reported in the early 1970s (Pyle and Pyle 2009). In 1982, three or four pairs escaped from an aviary during a hurricane and settled in a valley above Pearl City, where they commenced nesting, gradually increasing in numbers until more than 200 birds were present in December 2008. Introduced to Puerto Rico probably in the late 1990s, Green-cheeked Amazons occur in small numbers very locally around the coast, but up to 40 birds have been observed in a single flock near Salinas (Raeffaele *et al.* 1998).

The Green-cheeked Amazon is listed on Appendix I of the Convention on International Trade in Wild Fauna and Flora (CITES).

HABITATS In northeastern Mexico, Green-cheeked Amazons typically frequent diverse semi-deciduous tropical forest characterised by a relatively open canopy comprising frost-sensitive tropical tree species at about 15–20 m in height (Enkerlin-Hoeflich and Hogan 1997). Dominant canopy species include strangler fig *Ficus cotinifolia*, coma *Bumelia laetevirens*, ebony *Ebenopsis ebano*, gumbo-limbo *Bursera simaruba*, cerón *Phyllostylon brasilensis*, ojite *Brosimum alicastrum* and barreta *Hellieta parvifolia*. They occur also in Tamaulipan scrub dominated by thorn-bearing woody legumes of about 5–8 m in height, especially in ravines and riparian areas with larger trees and stands of dry oak-sweetgum *Quercus-Liquidamber styraciflua* forest. An early report from the Sierra de Tamaulipas, southern Tamaulipas, records their presence in the arid tropical lowlands, over the dry open pine-oak ridges and down into the tropical deciduous forests of the canyons (Martin *et al.* 1954).

In California, feral populations favour older urban and suburban areas, especially where there is a combination of exotic trees and shrubs, small orchards, and stands of native oaks *Quercus agrifolia* (Froke 1981; Garrett 1997). On Oahu, in the Hawaiian Islands, the Pearl City flocks roost primarily in *Eucalyptus* trees along ridges above the city and the foraging range includes mangroves at Pearl Harbour (Pyle and Pyle 2009).

MOVEMENTS Large flocks that form in the non-breeding season forage across wide areas, and these actions presumably are responsible for reports of regional movements of up to 'tens of kilometres' recorded for both wild populations in northeastern Mexico and feral populations in California and Puerto Rico (Enkerlin-Hoeflich and Hogan 1997).

HABITS Not unexpectedly, published accounts indicate that the habits of Green-cheeked Amazons are similar to those of the very closely allied Lilac-crowned Amazon *Amazona finschi*. In the non-breeding season they associate in large foraging flocks of 50 to 300 or more birds which wander widely in search of food, and during the breeding season pairs often nest in near proximity to form loose colonies in suitable habitat. Monitoring of photo-identified individuals indicates that lifelong pair-bonds are established, but pairs are not always discernible within the flocks. Peak levels of activity are in the morning and late afternoon, with the middle of the day spent resting in the shaded foliage of tall trees where their cryptic colouration and near total silence readily hides their presence. Also at this time there is a high degree of interaction among pair members, usually manifested in intense allopreening. A feeding flock suddenly may take flight to wheel about overhead in compact formation before alighting in the tops of other trees, all to the accompaniment of much screeching. Behaviour interpreted as interspecific socialising has been observed, particularly with Yellow-headed Amazons *Amazona oratrix*, and birds have been observed in the company of one or two other *Amazona* species feeding in the same tree. Nighttime roosts typically are in tall trees on ridges or slopes, and conspicuous high flights to and from feeding areas are accompanied by continual screeching. Also accompanied by loud calling are aerobatics and apparent friendly chases that often precede return flights to the nighttime roost. An early account from Sutton and Pettingill (1942) makes mention of roosting flocks being disturbed at night, but on one occasion a screeching flock that had not been disturbed flew past in the moonlight. Sutton and Pettingill also reported seeing an Ornate Hawk-Eagle *Spizaetus ornatus* sweeping down upon screaming flocks of these parrots, and parrot feathers were found under trees used regularly by the raptors.

Mabb (1997a) records observations made between May 1995 and August 1997 at nighttime roosts in Temple City and Arcadia, in the San Gabriel Valley of Los Angeles County, southern California, where Green-cheeked and Lilac-crowned Amazons were grouped together because of difficulties in distinguishing between them in shady trees in poor pre-sunset lighting, though Green-cheeked Amazons dominated at a ratio of up to 8:1. Numbers at the roosts were lowest in spring and early summer, and highest in autumn and winter. There was considerable variability in arrival times at the roosts, though in summer the parrots tended to frequent the roosting area approximately 30–60 minutes before sunset, and in winter they tended to arrive only five to 20 minutes before sunset. They were highly vocal until shortly before total darkness, when they became abruptly silent. Approximately five minutes before roosting, there was an increase in activity, with much calling, flying between branches or trees, and circling within and around the roosting site. From approximately 30 minutes before sunset until the time of roosting groups of 20 to 100 or more parrots regularly engaged in various social actions in nearby trees or on power lines, frequently preening or allopreening and performing various displays, including hanging upside-down, fanning their tails or raising feathers on the nape, spreading their wings, bowing their heads, dilating their pupils, indulging in feigned attacks with opened bills, and bill-wiping on branches. During summer and autumn, the parrots tended to prefer to roost in deciduous sweetgums *Liquidamber styracifolia* and western sycamores *Plantanus racemosa*, and when these trees lost their leaves in winter, spring preferences shifted to broadleaved evergreens, such as carrotwoods *Cupanopsis anacardiodes*, eucalypts or living oaks.

The flight is typical of that of *Amazona* species, featuring laborious, shallow wingbeats below body level, and the flight speed of birds returning to roost has been recorded at 43 km per hour (in Enkerlin-Hoeflich and Hogan 1997).

CALLS Howell and Webb (1995) describe the contact calls as a fairly mellow, rolling *rreeoo…rreeoo* or *keer-yoo…keer-you* and a slightly barking *rreh-rreh-rreh-rreh* or *rrak…rrak…rrak*, which often are combined in flight to resemble *cleeoo…cleeoo…ahk-ahk-ahk-ahk*. Also emitted are a quieter *rreah…rreah* and *clee-ik…clee-ik*.

Studies of vocalisation were undertaken at sites in northeastern Mexico, and it was observed that in the first few weeks after the fledging of chicks there were sessions of vocalisation in which young birds attempted to imitate the calls of adults (Enkerlin-Hoeflich and Hogan 1997). Calls of the young birds could be differentiated readily as coarse imitations of the calls of adults, and were more high-pitched in tone. The rackety begging calls of youngsters also were heard at this time. In a population there were individual and characteristic calls or versions of calls which allowed for individual identification. Several calls were identified as apparently being associated with special functions:

(i) **take-off squawk** – given in preparation to, or immediately at take-off; this call was relatively uniform among individuals,
(ii) **begging call** – a rackety *gaa-gaah-gaah* emitted by brooding females and young chicks to solicit feeding,
(iii) **individual flight call** – given during flight; there was considerable individual variation, and this may be an identification call,
(iv) **nest arrival call** – usually emitted by males as they approached the nest, before or when alighting, and usually followed by the nest exit prompt; this may be a shortened version of the individual flight call,
(v) **nest exit prompt** – given by the male in near proximity to the nest, could be repeated many times depending on the willingness of the female to leave the nest.

DIET AND FEEDING Recorded food items include seeds, fruits, flowers, leaves and buds, taken mostly in the upper stage to canopy of tropical trees, though at times the birds feed in lower trees or shrubs, and the actions of feeding birds are conspicuously slow and deliberate. Insects have been found in the crops of nestlings. Food items recorded opportunistically during breeding studies undertaken in Tamaulipas are listed in Table 10. Other recorded food items include chinaberries *Melia azedarach*, acorns and *Solanum* fruits, and cultivated pecans are a preferred food in the Monterrey district (in Enkerlin-Hoeflich and Hogan 1997). Pine seeds were present in the crop and stomach of a specimen collected in the Sierra de Tamaulipas, southern Tamaulipas (Martin *et al.* 1954).

During field studies undertaken in southern California in the 1990s, Green-cheeked Amazons were recorded utilising 24 food sources, but they did not come to take seed at garden bird-feeders (Garrett *et al.* 1997). Food plants and the items eaten are set out in Table 11.

BREEDING Much of what is known of the nesting of Green-cheeked Amazons comes from field studies undertaken by Ernesto Enkerlin-Hoeflich between 1992 and 1996 at a site in coastal Tamaulipas, northeastern Mexico. These studies indicated that pair-formation may occur among first-year birds during the non-breeding season, while the parrots are in large foraging flocks (in Enkerlin-Hoeflich and Hogan 1997). Flocks started to break up in early February, when pairs returned to their old nest-sites or searched for new sites. Some agonistic behaviour was observed early in the breeding season when pairs were searching for nest-sites. A 'wing-out' display, in which a parrot exhibited the red outer secondaries, sometimes was associated with the approach of intruders, but also was seen at times during intrapair interactions. Territorial defence appeared to be restricted to the immediate vicinity of the nest, and there were reports of two nests in the same tree. Copulation usually was limited to a period lasting from two to three weeks before egg-laying to no more than one month after the commencement of incubation. Males frequently used 'mock' regurgitation actions to stimulate females, and food-begging by the female with subsequent regurgitation feeding by the male usually occurred prior to copulation.

There is an early report of courtship from the Gomez Farias region, southwestern Tamaulipas, in mid March, when some birds obviously were paired, while others were observed squabbling and chasing each other, sometimes coming to grips in mid-air and fluttering upwards as if climbing an invisible wire (Sutton and Pettingill 1942). Males were seen to walk the entire length of a branch to present food items to their mates. In late March there was considerable bickering over nest sites and, after finally defeating other contenders, one pair was seen to occupy an old woodpeckers' hole at a height of about 20 m in a large cypress.

In March 1986, I observed obvious pre-nesting behaviour by Green-cheeked Amazons at and near a nighttime roost at Temple City, in the Los Angeles district. A colleague and resident of the district, Rae Anderson, told me that a few weeks prior to my arrival several hundred parrots were present at this roost, but only about 50 birds were seen by me, so confirming the suggestion that pairs had been leaving the flock for breeding. Pairs were most evident among the birds that I observed, and males were seen to display before their partners while perched on nearby telegraph wires, with copulation occasionally taking place.

The nest is in a natural cavity or an old woodpeckers' hole in a tree at varying heights from the ground. Both sexes participate in selection of the site, but the final decision may be with the female. At a study site in coastal Tamaulipas, these parrots appeared to prefer nest sites in open and wooded pastures rather than in heavily forested areas, though this may have been a consequence of the lower detectability of nests in forest (in Enkerlin-Hoeflich and Hogan 1997). Of 45 nests found between 1992 and 1996 at this study site, 16 were in coma *Bumelia laetevirens*, 13 were in ebony *Pithecellobium ebano*, nine were in strangler fig *Ficus cotinifolia*, four were in barreta *Hellieta parvifolia*, two were in dead snags and one was in gumbo-limbo *Bursera simaruba*. Finding more nests in coma *Bumelia laetevirens* than in ebony *Pithecellobium ebano* was unexpected, and the soft wood structure of coma could be responsible for this preference. However, any association with a particular tree is not strong and nesting has been recorded in many tree species and in artificial nestboxes. Characteristics of the 45 nests found between 1992 and 1996 at a study site in coastal Tamaulipas are set out in Table 12.

After selecting a site, both sexes are very vocal when in the vicinity of the nest and the approach of an adult, whether of the male coming to feed the female or of both birds arriving together, typically results in deliberate emission of a series of 'nest exit prompts' and 'nest arrival calls'. Birds often approach by first alighting in a nearby tree before flying to the nest tree. If the female is inside the nest when the male arrives, she usually exits promptly in response to calling from the male, and both birds give characteristic 'take-off squawks' as they fly to a nearby tree.

Eggs are laid at intervals of 48 hours, and mostly in the early morning. A clutch normally comprises two to five eggs, with the

Food plant	Parts eaten	Relative index of use	Relative presence in crop samples (%)	Relative importance by weight (%)
Cnidoscolus sp. mala mujer	fruits, flowers buds	rare	96	71
Acacia farnesiana mesquite	seeds	n/a	34	0
wood bark	various pieces	n/a	79	9
Ficus cotinifolia strangler fig	fruits	frequent	44	7
succulents	leaves	n/a	8	3
Passiflora sp. passionflower	seeds	n/a	3	<1
Ebenopsis ebano ebony	seeds in pod	n/a	10	<1
Bauhinia sp.	seeds in pod	n/a	3	<1
Brosimum alicastrum ojite	fruits	n/a	n/a	n/a
Bumelia laetevirens coma	drupes	frequent	n/a	n/a
Bursera simaruba gumbo limbo	drupes	rare	n/a	n/a
Myrcianthes fragans guayabillo	fruits	frequent	n/a	n/a
Solanum erithanum sosa	fruits	occasional	n/a	n/a
Wimmera concolor diente de tigre	alate seeds	occasional	n/a	n/a
Erethia elliptica	fruits	rare	n/a	n/a
Insects	whole	n/a	10	<1

Table 10. Food items recorded opportunistically during breeding studies undertaken at a site in Tamaulipas, northeastern Mexico, 1994–1995 (in Enkerlin-Hoeflich and Hogan 1997).

Food plant	Items eaten	Number of observations
Platanus spp. sycamore/plane trees	fruits/seeds	more than 10
Ficus spp. figs	fruits/seeds	more than 10
Liquidamber styracifolia sweetgum	fruits/seeds	more than 10
Juglans regia walnut	fruits/seeds	five to 10
Quercus spp. oaks	fruits/seeds	five to 10
Carya illinoenensis	fruits/seeds	more than 10
Prunus armeniaca apricots	fruits/seeds	more than 10
Magnolia grandiflora magnolia	fruits/seeds	one to four
Melia azedarach chinaberry	fruits/seeds	five to 10
Araceae palms	fruits/seeds	one to four
Olea europaea olive	fruits/seeds	one to four
Pinus spp. pine	fruits/seeds	one to four
Ceratonia siliqua carob	fruits/seeds	one to four
Bauhinia spp. orchid tree	fruits/seeds	one to four
Cupaniopsis anacardioides carrotwood	fruits/seeds	one to four
Alnus spp. alder	fruits/seeds	one to four
Prunus amygdalus almond	fruits/seeds	one to four
Betula spp. birch	fruits/seeds	one to four
Callistemon spp.	fruits/seeds	one to four
Rhamnus spp. coffeeberry	fruits/seeds	one to four
Thuja spp. golden arborvitae	fruits/seeds	one to four
Eucalyptus/Corymbia spp. eucalypts	flowers/nectar	more than 10
Eucalyptus citriodora lemon-scented gum	flowers/nectar	one to four
Jacaranda acutifolia jacaranda	flowers/nectar	one to four

Table 11. Food items observed being eaten by Green-cheeked Amazons during studies undertaken in southern California, USA, in the 1990s; more than 10 observations are indicative of an important food item (after Garrett *et al.* 1997).

Nest characteristics	Mean measurement	Measurement range
Tree diameter at breast height (cm)	26.65	38.2–229.5
Height above ground (cm)	677.83	330.0–1350.0
Number of entrances	1.0	1.0–2.0
Perimeter at nest entrance (cm)	131.79	66.5–330.0
Width of entrance (cm)	9.80	6.3–23.0
Length of entrance (cm)	16.16	7.9–49.0
Depth of cavity (cm)	102.32	25.5–479.0
Internal diameter of cavity	21.29	10.4–71.0

Table 12. Characteristics of 45 nests found between 1992 and 1996 at a study site in coastal Tamaulipas (in Enkerlin-Hoeflich and Hogan 1997).

mean size for 53 clutches at a study site in coastal Tamaulipas being 3.5 eggs (in Enkerlin-Hoeflich and Hogan 1997). Incubation by the female commences after laying of the first egg and lasts approximately 27 days. During incubation and while brooding newly hatched chicks, she sits very tightly, leaving the nest for only brief periods to be fed by the male. At the study site in coastal Tamaulipas, nesting pairs adopted distinctive daily routines, with the arrival of males to feed brooding females or pairs to feed chicks occurring approximately 70 minutes after sunrise and 130 minutes before sunset. Females usually left the nest promptly upon arrival of the male, and feeding by the male usually took place in a tree at 10 m to more than 100 m away from the nesting tree. These feeding bouts usually took place at the same location, though at times the location changed, and the food transfers varied from three to 16 per location and five to 16 per feeding bout. After being fed, brooding females returned promptly to the nest, usually within 15 minutes of their departure. Hatching occurs asynchronously, usually at intervals of 48 hours, and newly hatched chicks are sparsely covered with whitish down. Early and late phases can be differentiated in development of the chicks, the first lasts some 17 days after hatching and is characterised by rapid growth of the chicks and continual brooding by the females, while during the late phase there is no brooding by the female, weight gain by the nestlings is minimal and most feather growth occurs. Chicks are fed twice daily, the first feeding visit by the parents occurring about an hour after sunrise and the second occurring approximately 1.5 hours before sunset. Adults typically enter the nest only once to feed the chicks, normally entering one at a time, though occasionally both parents may be in the nest at the same time, and each feeding visit normally lasts a few minutes. Fledging occurs approximately 53 days after hatching, and young birds appear to be led by the parents to 'nursery sites', where they remain with low mobility and are fed by the parents for three to five weeks. Thereafter, the young birds follow the foraging adults, continuing to be fed by the parents in response to begging for at least 10 weeks after fledging.

At the study site in coastal Tamaulipas, one pair fledged seven youngsters in six breeding seasons, and overall the nesting success rate in the population was 1.45 fledglings per nest for 65 nests (in Enkerlin-Hoeflich and Hogan 1997). A flooding of nests in summer rains was thought to occasionally cause losses of chicks, and desertion by brooding females was suspected to be responsible for some complete nest failures.

Froke (1981) reported successful nesting by two pairs of Green-cheeked Amazons in a blue gum *Eucalyptus globulus* in Arcadia, Los Angeles County, southern California, where successful nests of Lilac-crowned Amazons *Amazona finschi* and Yellow-headed Amazons *A. oratrix* also were recorded. Mabb (1997b) reports that at Temple City, also in Los Angeles county, between 9 April and 23 June 1996, two mixed pairs of Green-cheeked Amazons × Lilac-crowned Amazons were observed at nests in cavities in silver maples *Acer saccharinum*, and on 5 May a recently fledged chick was seen to be accompanied by an adult Green-cheeked Amazon. This fledgling fluttered to the ground and when approached it was repeatedly pecked by the adult, also on the ground, until both took flight. The fledgling alighted several times and was pecked by the accompanying adult each time until both flew to the roof of a house where the begging fledgling immediately was fed by the adult, and then both departed the area.

EGGS A mean length of 37.41 mm and mean width of 29.29 mm were recorded for 16 eggs measured at no more than 10 days after laying during 1995 at a study site in coastal Tamaulipas (Enkerlin-Hoeflich and Hogan 1997). Slightly smaller measurements of 35.7 (35.0–36.4) × 27.5 (27.1–27.9) mm are recorded for two eggs, possibly laid in captivity (Schönwetter 1964).

ENDANGERED

Yellow-headed Amazon

Amazona oratrix Ridgway

Amazona oratrix Ridgway, *Man. North. Am. Bds*, 1887, p. 587. New name for *Chrysotis levaillantii* G. R. Gray 1859, not *Amazona levaillantii* Lesson 1831 (Petapa, Oaxaca, *fide* Nelson, *antea*).

OTHER NAME Yellow-headed Parrot.

DESCRIPTION Length 35 cm. Weight 517 g.
ADULTS General plumage green, slightly paler and more yellowish on underparts and upper tail-coverts; on mantle, back and breast feathers faintly margined dusky black to give lightly scalloped appearance; head and nape bright yellow, sometimes with scattered green feathers at sides of neck; bend of wing orange-red variably intermixed yellow, and carpal edge yellow variably marked pale orange-red; primaries and secondaries emerald green, becoming blue on outer webs towards tips, and broad red band on outer webs of S1 to S5, producing conspicuous red wing-speculum in flight; thighs yellow variably intermixed dull green; tail dark green with red at base and broadly tipped yellowish-green; bill pale yellowish-horn; unfeathered eyering white; iris orange; legs pink-grey.
15 males: wing 222–244 (232.7) mm, tail 104–135 (118.2) mm, exp. cul. 30–37 (33.9) mm, tars. 25–28 (26.9) mm.
13 females: wing 206–233 (222.4) mm, tail 105–126 (114.2) mm, exp. cul. 30–36 (31.7) mm, tars. 23–27 (25.5) mm.
JUVENILES Yellow only on crown and lores; less orange-red and yellow on bend of wing and carpal edge; thighs green; bill darker yellowish-horn colour with grey at base; iris brown.

DISTRIBUTION Central America, where occurring on the Islas Tres Marias, western Mexico, disjunctly distributed along the Pacific and Caribbean slopes of central Mexico, in Belize and El Petén in neighbouring northern Guatemala, and extreme northeastern Guatemala to northwestern Honduras. Feral populations are established in Florida and southern California, USA, and Puerto Rico, West Indies.

SUBSPECIES The Yellow-headed Amazon is part of what has been termed the '*Amazona ochrocephala* complex' comprising strongly polytypic and locally dimorphic forms distributed from central Mexico south through Central America to the Amazon River basin, and differentiated morphologically by patterns of yellow on the head, yellow and red on the wing, and bill colouration. Taxonomically, these forms usually have been treated variously as subspecies of *A. ochrocephala*, as three separate species, or as four species if *tresmariae* is given specific status. Previously, I treated all forms as subspecies of *A. ochrocephala*, though categorising them into three groups – the yellow-crowned or '*ochrocephala*' group, the yellow-naped or '*auropalliata*' group and the yellow-headed or '*oratrix*' group – but noting that separation between the *ochrocephala* and *auropalliata* groups is somewhat tenuous (see Forshaw 2006). There has been widespread acceptance of these three groups, sometimes termed the South American, Central American and Northern South American lineages, but molecular analyses have identified both the Blue-fronted Amazon *Amazona aestiva* and Yellow-shouldered Amazon *A. barbadensis* to be nestled within the *Amazona ochrocephala* complex (see Urantówka *et al.* 2014). It seems to me that although three lineages within the complex can

Plate 52
Yellow-headed Amazon *Amazona o. oratrix* (adult)

F. KNIGHT 16

be recognised, relationships between them are less evident, as is the relationship with *A. aestiva* and *A. barbadensis*, so further investigations are required. Pending those investigations and for the purpose of this book, I am treating the three lineages as separate species, as has been adopted by Dickinson and Remsen (2013) and by del Hoyo and Collar (2014), but am retaining subspecific differentiation for *tresmariae*.

1. *A. o. oratrix* Ridgway
The nominate subspecies, as described above, is disjunctly distributed in central Mexico, along the Pacific slopes from Jalisco to central Oaxaca, though recent reports are mainly from Michoacán and southern Jalisco, and along the Caribbean slopes from eastern Nuevo León and Tamaulipas to Tabasco and northern Chiapas. Feral populations are established in Florida and southern California, USA, and Puerto Rico, West Indies; irregularly recorded in extreme southeastern Texas, but no established population.

2. *A. o. tresmariae* Nelson
Amazona oratrix tresmariae Nelson, *Auk*, **17**, 1900, p. 256 (Maria Madre Island).
ADULTS Like *oratrix*, but yellow of head extending down on to neck and upper breast; upperparts slightly paler green; underparts slightly tinged blue; averages larger in size and with longer tail.
10 males: wing 226–245 (241.2) mm, tail 118–136 (129.2) mm, exp. cul. 32–35 (34.3) mm, tars. 26–28 (27.2) mm.
4 females: wing 229–232 (230.5) mm, tail 116–130 (124.3) mm, exp. cul. 31–33 (32.5) mm, tars. 25–26 (25.5) mm.
Confined to the Islas Tres Marías, off the coast of Nayarit, western Mexico.

3. *A. o. belizensis* Monroe and Howell
Amazona oratrix belizensis Monroe and Howell, *Occasional Papers of Museum of Zoology, Louisiana State University*, number 34, 1966, p.6 (Hill Bank, Orange Walk District, British Honduras).
ADULTS (yellow-faced and yellow-naped morphs) Yellow restricted to lores, forehead, crown and around eyes to ear-coverts and upper cheeks, together with (yellow-naped morph) or without (yellow-faced morph) yellow patch on nape to hindneck; little or no yellow on carpal edge of wing; unfeathered eyering greyish-white.
3 males: wing 216–222 (218.7) mm, tail 113–120 (117.7) mm, exp. cul. 34–36 (35.0) mm, tars. 25–27 (25.7) mm.
3 females: wing 205–217 (209.7) mm, tail 97–109 (103.7) mm, exp. cul. 32–33 (32.3) mm, tars. 25–27 (26.0) mm.
Occurs in central Belize to neighbouring El Petén in northwestern Guatemala (yellow-faced morph only) and disjunctly in northeastern Guatemala to extreme northwestern Honduras (both morphs).

4. *A. o. hondurensis* Lousada and Howell
Amazona oratrix hondurensis Lousada and Howell, *Bull. Brit. Orn. Club*, **117**, 1997, pp. 205–209 (along the Toloa canal at Urraco, about 12 miles northeast of La Lima, Departo. Cortes, Honduras).
ADULTS (variable plumage) Differ from *belizensis* by having yellow restricted to forehead and crown, thus approaching *A. ochrocephala*, but with or without yellow patch on nape to hindneck; on bend of wing variable red intermixed with yellow; carpal edge green.
No specimens examined.
1 female (type): wing 212 mm, tail 100 mm, exp. cul. 30 mm (Lousada and Howell 1997).
Known only from the lower Valle de Sula, northwestern Honduras.

STATUS Yellow-headed Amazons are very popular cagebirds, and the high levels of trapping or nest-robbing for the live-bird market has added to habitat loss in bringing about strong declines in many populations, especially in Mexico. There are conflicting assessments of the status of the population on the Islas Tres Marías, which has been estimated at fewer than 800 birds, but is said to be stable and is protected within the Islas Marías Biosphere Reserve (in Howell and Webb 1995; Birdlife International 2016). Hansen (1984) reported that on Maria Madre they were found to be quite plentiful in April 1983, when birds were concentrated in and around Balleto, the main village and seaport. The historical range of these parrots along the Pacific coast of Mexico was predicted by using ecological niche models, with presence records from museum collections and historical sightings overlain on vegetation maps, and this was compared with the current range, estimated with ecological niche models using presence–absence data from 798 field surveys undertaken in both breeding and non-breeding seasons between 2003 and 2008 at 437 localities, mostly up to 900 m, in all coastal regions from Jalisco to the Isthmus of Tehuantepec, Oaxaca (Monterrubio-Rico *et al.* 2010). From this comparison it was estimated that the range along the Pacific coast has contracted by 79 per cent, and now may cover only 18957 km² in three main areas – coastal Oaxaca, Guerrero-Michoacán and Jalisco. One of these areas is a small, isolated site on the coast of Jalisco, where the parrots may be vulnerable to extirpation or genetic endogamy, and there are only three small protected areas within the current range. Binford (1989) reports that in Oaxaca they are uncommon and local residents in two disjunct areas. Howell and Webb (1995) note that these parrots formerly were common along the Caribbean coast, but now are uncommon to rare. Ruth (2015) points out that several studies have estimated densities in Tamaulipas, but different methods have been used so making direct comparisons impossible. Based on the number of nests found and the area sampled, Enkerlin-Hoeflich estimated a density of 6.3 birds per 100 ha, but two earlier studies, using a variable circular plot method that included areas and habitats that the parrots did not use, estimated densities of 2.59 birds per 100 ha and 4.2 ± 2.1 birds per 100 ha. In 1994, the total population in Mexico was estimated at 7000 individuals, which is roughly equivalent to 4700 mature birds (Birdlife International 2016).

Jones (2003) notes that Yellow-headed Amazons are uncommon to common in Belize, but warns that, although the current population is stable, they are heavily persecuted for the pet trade and could easily be threatened with extinction if protection measures are not enforced. More recent data suggest that the decline in Belize has been quite dramatic, possibly as high as 90 per cent since the 1970s. They now are restricted primarily to central and northwestern areas, with much of the suitable coastal pine savanna habitat being outside the national protected system and under heavy development pressure (in Birdlife International 2016). Illegal fires pose a serious threat to the pine savanna, and poaching of nestlings for the local pet trade continues to be the major threat. These amazons are very rare vagrants in El Petén, northern Guatemala, where there is one early record and another from 1993 (Land 1970; in Birdlife International 2016). Eisermann (2003) reports that in 2001, in the Punta de Manabique Wildlife Refuge, on the Atlantic coast of northeastern Guatemala, a population estimate of 70 birds was made from counts conducted at the only known traditional roosting site, and a comparison with a roost census carried out there in 1994 suggested a population decline, largely because of nest-robbing for the pet trade. The rate of decline and the low number of surviving birds suggests that this population is facing extinction. Gallardo (2014) notes that the status of Yellow-headed Amazons in northwestern Honduras is unknown, but it is likely that they continue to decline because of continuing demands from the pet trade. In the Cuyamel area, Cortés Department, numerous pairs, including one with a nest, were observed during a survey undertaken in May 2014.

Robertson and Woolfenden (1992) report that small numbers of Yellow-headed Amazons are reported in south Florida, USA, but the population may not be self-sustaining in the long term. Similarly, in extreme southeastern Texas, there have been irregular records of small numbers in the lower Rio Grande valley, particularly prior to the 1950s, but a breeding population has not become established (Brush 2005). These parrots were introduced to Puerto Rico probably in the early 1970s, and breeding has been recorded, but they are rare, occurring very locally in only small numbers along the north coast (Raffaele *et al.* 1998). Garrett (1997) points out that early fieldworkers in southern California did not distinguish among the forms of the Yellow-headed Amazon, but reported 'yellow-headed' and 'yellow-crowned' parrots widely in the western San Gabriel Valley, west Los Angeles, and elsewhere. Flocks of 10 to 20 parrots were seen routinely in West Los Angeles in the 1970s, and their presence in urban Orange County dates from at least the early 1970s, but numbers in the greater Los Angeles region appear to have declined since that time. During studies undertaken between 1994 and June 1997, the largest flocks reported were of 12 in Mar Vista, West Los Angeles, and nine in Costa Mesa, and the overall population was estimated at probably less than 60 birds. Mabb (2002) reports that numbers have continued to decline, and in 2002 the overall population was estimated at only 27–31 birds. My sightings during visits to the Los Angeles district certainly indicate a marked decline, for I had far less sightings in the late 1990s than in the late 1980s and early 1990s.

HABITATS Along the Pacific coast of Mexico, up to about 900 m, Yellow-headed Amazons frequent dense deciduous, semi-deciduous and riparian forests, as well as semi-open areas with scattered trees (Monterrubio-Rico *et al.* 2010). During surveys undertaken in the Pacific lowlands between 2001 and 2009, the predominant vegetation type in the range of these parrots was conserved and modified tropical deciduous forest, recorded at 44.2 per cent, followed by transformed agricultural lands, recorded at 32 per cent, and conserved and modified tropical semi-deciduous forest, recorded at 19.4 per cent (Marin-Togo *et al.* 2012). During these surveys, the parrots were recorded up to 896 m, with 76 per cent of records below 400 m, mostly below 200 m, which accounted for 48 per cent of all records, and only 24 per cent of records were above 400 m. Most sightings were in tropical dry deciduous forests, accounting for 53 per cent of records, with 33 per cent of records in tropical semi-deciduous forest, and few records were in transformed agricultural lands. In westernmost Oaxaca, in late December 1995, a pair was observed in extensive xerophytic vegetation where the thorn-forest was fragmented by clearance for grazing (Roberson and Carratello 1997). Hansen (1984) notes that on Maria Madre, in the Islas Tres Marías, in April 1983, a concentration of these parrots was found in and around Balleto, the main village and seaport, presumably attributed to the availability of fruiting trees, and they were recorded also in a forested canyon where huge trees probably provided nesting sites. Semiarid regions are favoured in the northern Caribbean lowlands, but more humid savannas are preferred farther south (Birdlife International 2016).

In Belize, Yellow-headed Amazons occur in coastal pine savannas and occasionally in other open forested areas (Jones 2003). Eisermann (2003) recalls that in 2001, in the Punta de Manabique Wildlife Refuge, on the Atlantic coast of northeastern Guatemala, these parrots were found nesting in palm savanna and mangrove forest, and in April, during the breeding season, they were most numerous in open palm savanna dominated by royal palms *Roystonea oleracea* with an understorey of various grasses up to 2 m tall, and a traditional roosting site was in a stand of tall, mostly dead mangrove trees on the beach. Eisermann further points out that occurrence in *Roystonea* palm savanna seems to be a unique trait of the Manabique population, though the structure of this vegetation is similar to pine savanna. Gallardo (2005) notes that in northwestern Honduras, Yellow-headed Amazons frequent forested habitats, including borders, semi-open lands with scattered trees and coastal wetlands. Lousada and Howell (1996) recall that in the lower Valle de Sula, northwestern Honduras, they were found in 'coastal scrub habitat' and 'at a slightly elevated roosting site of patchy trees amid some cleared agricultural land'.

In the Los Angeles district, southern California, the feral population favours suburban parks and gardens. In 1986, during a visit to Los Angeles, the first birds that I observed were two Yellow-headed Amazons flying high above the freeway at the airport exit. In Puerto Rico, West Indies, the small feral population occurs in lowland second-growth forests (Raffaele *et al.* 1998).

HABITS Much of what is known of the habits of Yellow-headed Amazons comes from field studies undertaken by Ernesto Enkerlin-Hoeflich at sites in coastal Tamaulipas, northeastern Mexico, between 1992 and 1996, and his findings have been summarised by Ruth (2015). During the breeding season, when food was available uniformly, these and other *Amazona* parrots foraged independently in mated pairs, but in the non-breeding season pairs came together in larger groups to forage more widely, and at times were seen feeding in the company of other *Amazona* species. At least five pairs of Yellow-headed Amazons exhibited mate fidelity between two breeding seasons, and three of these pairs exhibited mate fidelity over three breeding seasons.

These parrots roost communally at night, and at these communal roosts, which may be in use for many years, numbers vary seasonally, being lowest during the breeding season when nesting pairs remain at or near their nests. Regrettably, the conspicuous daily flights of very large numbers between roosting and feeding sites described in early reports now are recorded very rarely. Dalquest recalled that in March 1947, near Jimba, Veracruz, eastern Mexico, large numbers were seen flying from their feeding grounds in the forest to roosting sites out on the coastal plain and, for an hour before dusk, from a few to a hundred or more parrots were in sight at all times, in pairs or small parties, all flying towards the east (in Lowery and Dalquest 1951). Stager (1957) reported that on Maria Cleofas, in the Islas Tres Marías, western Mexico, a flock of about 20 birds routinely came in the late afternoon to roost in a heavy stand of agaves and, after perching atop the tall flowering stems of the agaves for a short while, the birds would descend into the lower spiny leaves of the plants to within 2 m of the ground, where they remained until sunrise, when all individuals would reassemble as a flock and fly off towards the forested slopes of the island. Russell (1964) reported that at Hill Bank Lagoon, northern Belize, pairs and loose groups of several pairs of loudly screeching parrots flew across the lagoon every evening, making their way to the pine ridges, where they roosted in the highest branches of tall pines, one or two pairs in each tree. Eisermann (2003) notes that in 2001, in Punta de Manabique, northeastern Guatemala, at a traditional communal roost reported to have been in use for at least 10 years, pairs or trios arrived about 45–80 minutes after sunset, congregating in the upper canopy of seven to 10 mangrove trees on the beach some 10–50 m from the waterline, and during January to June a pair of Red-lored Amazons *Amazona autumnalis* and five Brown Pelicans *Pelecanus occidentalis* roosted in the same trees. There was a flying distance of approximately 13 km between this communal roost and nesting sites in palm savanna and mangrove forest. Lousada and Howell (1996) report that in the Valle de Sula, northwestern Honduras, Yellow-headed Amazons occupied what may have been a traditional roosting site in loose association with about 30 Red-lored Amazons.

All of my encounters with Yellow-headed Amazons have been with feral birds in the Los Angeles district, southern California, where usually they were seen singly, in pairs or trios typically in high flight overhead or feeding in the upper stages to canopy of fruiting and flowering trees. I did not observe flights to and from nighttime roosts, which are said to be accompanied by much calling, but during the day I noted that flying birds seldom called, an observation reported also by Howell and Webb (1995), and long flights invariably were very high. In the treetops they clambered among branches with the slow, deliberate actions that are typical of foraging amazons.

CALLS Howell and Webb (1995) describe vocalisation as comprising raucous though mellow, deep, rolled screams, often with a human quality, and including a rolled *kyaa-aa-aah* and *krra-aaaa-ow*, a deep, rolled *ahrrrr* or *ahrhrrrr* and *whoh-oh-ohr*, and a throaty, rolled *rrohrr*.

DIET AND FEEDING The diet comprises seeds, fruits, berries, nuts, blossoms and probably leaf buds. On Maria Madre, in the Islas Tres Marías, western Mexico, these amazons have been observed eating the seed pods of monkeypod *Pithecellobium dulce*, mango flowers and very small fruits (in Hansen 1984). Renton notes that they have been recorded eating the seeds of glassywood *Astronium graveolens*, garlic-pear tree *Crataeva tapia* and tempisque tree *Sideroxylon capiri* and the fruits of higueron *Ficus insipida*, and Enkerlin-Hoeflich reports that in Tamaulipas, northeastern Mexico, they were observed feeding on the seeds and fruits of the three most common tree species – coma *Bumelia laetevirens*, ebony *Ebenopsis ebano* and strangler fig *Ficus cotininfolia* – as well as seeds and fruits of guayabillo *Myrcianthes fragrans* and diente de tigre *Wimmeria concolor* (in Ruth 2015). Also in Tamaulipas, they frequently took seeds from *Solanum* and *Cnidoscolus* shrubs among disturbed vegetation in cattle pastures, and in riparian corridors they consumed seeds or fruits of copalchi *Croton niveus*. In the Los Angeles district, southern California, recorded food items taken by feral birds are seeds or fruits of walnut *Juglans regia*, olive *Olea europaea* and tulip tree *Liriodendron tulipifera* (Garrett *et al.* 1997).

BREEDING Much of what is known of the nesting behaviour comes from field studies undertaken by Ernesto Enkerlin-Hoeflich at sites in coastal Tamaulipas, northeastern Mexico, between 1992 and 1996, and his findings are summarised by Ruth (2015). There was very little annual variation in the commencement of egg-laying, with the mean commencement date in 1992, 1993 and 1994 being within a six-day period during the last week in March and first week in April. Nests were in cavities in trees in woodlots and in scattered trees in pastures, with those in trees in pastures being found more frequently, though this may have reflected an ease of detection. Of 18 nests found between 1992 and 1996 at a study site in coastal Tamaulipas, 11 were in coma *Bumelia laetevirens*, five were in ebony *Ebenopsis ebano* and two were in strangler fig *Ficus cotinifolia*, the three dominant tree species at the site. Characteristics of 17 nests found between 1992 and 1996 at this study site are set out in Table 13.

Pairs defended only the immediate vicinity of the nest or nesting tree, and most aggressive interactions were interspecific, typically occurring when another *Amazona* parrot landed or attempted to land in the nesting tree. Yellow-headed Amazons frequently approached near to the nests of other Yellow-headed Amazons and commenced prolonged vocalisation, involving a wide variety of calls, but did not exhibit open aggression, though occasional displacement behaviour may have indicated subtle aggression. Pairs tended to vocalise for long periods when near the nest cavity, but rarely when approaching the nesting tree, especially if they perceived the presence of observers. Clutches comprised two or three eggs, with a mean of 2.6 eggs for seven clutches, and incubation by the female commenced with laying of the first egg. The male normally visited the nest twice daily to feed the brooding female, there being a routine in his arrival and departure flight paths as well as times of the visits, with a mean arrival time of 0650 hours for 12 morning visits and of 1733 hours for 15 afternoon visits. In response to arrival of the male, the female sometimes left the nest promptly, but more frequently waited some time before coming out to accompany the male to another tree up to 100 m distant where feeding by regurgitation took place. The incubation period was not determined, but the chicks fledged approximately 57 days after hatching. A few days prior to fledging, the chicks would come to the cavity entrance in response to soft calling by the parents and a suspension of parental feeding.

In Belize, nests with eggs were found in pine trees in March and April, and a chick was present in a nest found in May (in Russell 1964). As part of the conservation program in Belize, artificial nestboxes are fitted into the trunks of pine trees to increase the availability of nesting sites. Eisermann (2003) reports that in Punta de Manabique, northeastern Guatemala, nests with chicks were found in April, mostly in holes in snags of *Roystonea* palms, but also in the trunks of *Salix* or other trees, and along the coastline three nests were in old red mangroves *Rhizophora mangle*, some up to 20 m in height. A majority of previously observed nests were destroyed by poachers, the entrances having been opened out,

Nest characteristics	Mean measurement	Measurement range
Tree diameter at breast height (cm)	80	42–234
Height above ground (cm)	559	280–1150
Width of entrance (cm)	11	6–19
Length of entrance (cm)	20	10–59
Depth of cavity (cm)	140	51–260
Internal diameter of cavity	27	14–52

Table 13. Characteristics of 17 nests found between 1992 and 1996 at a study site in coastal Tamaulipas (in Ruth 2015).

access holes having been chopped into the trunk at the bottom of the cavity, or the entire trunk having been felled.

At the study site in coastal Tamaulipas, northeastern Mexico, the dominant cause of recruitment failure was 'nest failure or abandonment', but the impact of predation may have been underestimated and, if suspected instances of predation had been taken into account, total losses to predation could have been as high as 30 per cent of all chicks and eggs (in Ruth 2015). Taking chicks for the pet trade is a major cause of low recruitment rates in many parts of the range, and it often involves a permanent loss of nesting sites.

EGGS Measurements of 38.8 × 30.5 mm are listed for a single egg of *A. o. oratrix* (Schönwetter 1964).

CRITICALLY ENDANGERED

Puerto Rican Amazon

Amazona vittata (Boddaert)

Psittacus vittatus Boddaert, *Table Pl. enlum.*, 1783, p. 49 (Santo Domingo, *ex* Daubenton, pl. 792 = Puerto Rico).

OTHER NAME Puerto Rican Parrot.

DESCRIPTION Length 29 cm. Weight 320 g.
ADULTS General plumage green, slightly paler and more yellowish on underparts; feathers finely edged dusky black, particularly on head and neck; lores and frontal band red; under tail-coverts yellowish-green; on abdomen feathers sometimes slightly tinged dull red; primary-coverts and primaries dark blue on outer webs, dusky black on inner webs; outer webs of outermost secondaries blue narrowly margined dull green; underwing-coverts green; undersides of flight feathers bluish-green; tail green narrowly tipped yellowish-green, lateral tail-feathers marked red at bases of inner webs and outermost feathers edged blue; bill yellowish-horn; unfeathered eyering white; iris brown; legs yellowish-brown.
8 males: wing 182–193 (188.5) mm, tail 90–103 (96.9) mm, exp. cul. 27–30 (28.5) mm, tars. 21–24 (22.1) mm.
5 females: wing 178–196 (185.6) mm, tail 93–104 (98.2) mm, exp. cul. 27–28 (27.2) mm, tars. 22–24 (23.0) mm.
1 unsexed: wing 203 mm, tail 105 mm, exp. cul. 32 mm, tars. 24 mm.
JUVENILES Like adults, but narrower orange-red frontal band; pale grey marking on sides of upper mandible.

DISTRIBUTION Now confined to Puerto Rico, but formerly also on nearby Culebra Island and possibly Vieques Island.

SUBSPECIES

1. *A. v. vittata* (Boddaert)
The nominate subspecies, as described above, occurs on Puerto Rico, and may have occurred on nearby Vieques Island.

EXTINCT

2. *A. v. gracilipes* Ridgway
Amazona vittata gracilipes Ridgway, *Proc. Biol. Soc. Wash.*, **28**, 1915, p. 106 (Culebra Island).
ADULTS Similar to *vittata*, but smaller and with relatively smaller, more slender feet.
2 males: wing 169 mm and 173 mm, tail 93 mm and 95 mm, exp. cul. 26 mm and 26 mm, tars. 20 mm and 21 mm.
1 female: wing 175 mm, tail 100 mm, exp. cul. 23 mm, tars. 20 mm.
1 unsexed: wing 173 mm, tail 90 mm, exp. cul. 27 mm, tars. 22 mm.
This poorly differentiated subspecies formerly occurred on Culebra Island, but is now extinct.

STATUS The last record of Puerto Rican Parrots on Culebra Island was in 1899, when A. B. Baker collected three specimens, all of which now are in the United States National Museum of Natural History, Washington. I do not know the origin of the specimen in the Naturhistorisches Museum, Vienna. Wetmore (1927) reported that on Vieques Island, in 1912, he was told that the parrots formerly were seen in heavy forests on the southern part of the island, and it was thought that they came from Puerto Rico. No specimens are known from Vieques Island and if the species did occur there it almost certainly has been extinct since the early 1900s.

On 17 February 1835, C. Moritz, an entomologist and collector for the Zoological Museum in Berlin, arrived on Puerto Rico, where he spent four months, and the reference to these parrots in his account of natural history observations is that in settled districts they 'destroy whole fields of corn in great flocks', but he found parrots only in the Cordillera Central, where large tracts of forest existed, and the parrot population there must have been little affected (in Wetmore 1927; in Snyder *et al.* 1987). In 1864, the British ornithologist, Edward Cavendish Taylor, noted that they were still common near San Juan, and it seems that until the end of the 19th century they remained plentiful, particularly in the interior. A warning of declining numbers came from Alexander Wetmore, who spent 10 months on Puerto Rico between December 1911 and September 1912, and he noted (in Wetmore 1927):

> *It appears that until thirty years ago the parrot was a common species in Porto Rico, but has disappeared except in a few localities, mainly in the northeastern portion of the island, on the mountain known as El Yunque de Luquillo and immediately around its base. From February 9 to 29, 1912, I found about twenty ranging through the dense swampy forests north of Mameyes (one taken), while perhaps fifty were seen between March 2 and 11 around the west fork of the Río Mameyes, or were heard calling in the dense forests covering the summit of El Yunque itself. (Eight, all adult, were taken here.) Parrots were said to be still fairly common around Preston's Ranch, above Naguabo. On the western end of the island they had practically disappeared, though birds were reported from*

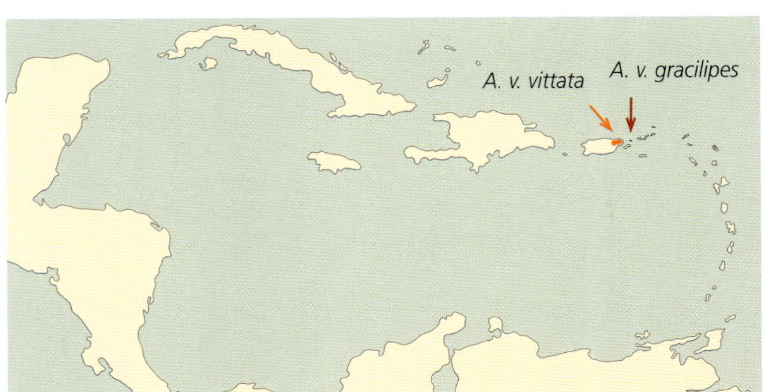

below San Sebastian, and Mr. Leopold B. Strude, of the Hacienda "Jobo", between Arecibo and Utuado, said that during the winter months fifty or sixty were found on his plantation, though none nested there. At Lares I was told that no parrots had been seen for ten years. ....The handful remaining is, however, too small to do any great damage, and the birds should be protected or they will be lost from the avifauna of the island.

In the course of a study undertaken between August 1953 and March 1956 in the Luquillo National Forest Reserve, eastern Puerto Rico, which by then was the last stronghold of the parrots, a flock of 200 parrots was the largest number recorded, and this same flock occurred at the same place at almost the same date in two consecutive years, so it seems likely that it contained most, if not all, of the parrots that were present in the forest (Rodríguez-Vidal 1959). There was a further drastic decline in numbers, and in December 1965 it was estimated that the surviving population probably did not exceed 50 birds (Recher and Recher 1966). During a survey undertaken by Victor Marquez over several months in 1966 the largest recorded flock contained 70 birds. At commencement of the intensive conservation program in 1968, a thorough census of the wild population was undertaken, and only 24 birds were located. By 1971, the count had dropped even further to 16 birds, but then the decline was arrested, with the population fluctuating for some time between 15 and 20 individuals.

Widespread landclearance and persecution were primary causes of the dramatic population decline, and as numbers reached low levels the impact of hurricanes, poaching for the live-bird market and competition or predation from introduced species became more significant. At commencement of the intensive conservation program, funded primarily by the United States Fish and Wildlife Service, the United States Forest Service and the Commonwealth of Puerto Rico, the surviving population had declined to only 13 birds, all in the eastern part of the island in virtually the only remaining old-growth forests in the Luquillo Mountains. The two approaches adopted in the program were to undertake intensive field studies of the wild population with a view to implementing a recovery strategy and to establish a captive-breeding program with an objective of releasing captive-bred birds into the wild population. Initially it seemed unlikely that a scarcity of nesting sites would be a major problem, for trees with natural cavities were reasonably common in the Luquillo forests, but systematic checking of these cavities revealed that very few were suitable for nesting and a number of pairs were unable to breed because of an inability to locate good sites. This scarcity of sites was addressed by providing artificial nest-sites, and modification of these artificial sites eventually overcame high levels of predation of eggs and chicks by Pearly-eyed Thrashers *Margarops fuscatus*. Predation of nest contents by these hole-nesting mimids was a major cause of nesting failure, and various measures were adopted to combat the thrashers, including direct guarding of parrot nests, artificial incubation of parrot eggs and manipulation of nest-sites. Parasitism by warble flies *Philornis pici* also was responsible for losses of some nestlings. A deeper artificial nest-site proved to be effective against entry by the thrashers, especially when their more preferred shallower nestboxes were provided in the immediate vicinity, and an additional benefit was that the parrots showed a preference for the deeper nest-sites. An exceedingly high population of Red-tailed Hawks *Buteo jamaicensis* occurs in the Luquillo forests, and predation by these raptors on adults and fledglings was another pressure restricting recovery of the population. Protection measures being implemented arrested the decline, but nesting success remained low and numbers increased very slowly.

As a safeguard and to provide birds for reintroduction to the wild, a captive breeding population was set up in the 1970s and, apart from a setback in April 2001 when some birds were stolen, this has been successful, enabling captive-produced chicks to be fostered into wild nests. This fostering technique, together with intensive guarding and management of wild nests brought about an increase in the population, which numbered almost 50 birds by the late 1980s. Releases of captive-bred birds to augment the wild population in the Luquillo forests occurred in June 2000, when 10 birds were released, and in May 2001, when another 16 birds were released. In response to strong indications that the Luquillo forests, with their high rainfall and high populations of both Pearly-eyed Thrashers and Red-tailed Hawks, may be suboptimal habitat for the parrots, preparations were made to establish wild populations at other sites by releasing captive-bred birds, and the first site chosen was the Río Abajo Commonwealth Forest, also in eastern Puerto Rico, which was known to have been frequented by the parrots before deforestation in the early 1900s. The first release of 20 birds took place at Río Abajo Forest in 2006, followed by the release of 26 birds in December 2007 and an additional 19 birds were released in December 2008. A nest with eggs was found in the forest in 2013 and, although these eggs did not hatch, it was evidence of attempted breeding by captive-bred birds. Success was achieved in 2014, when two chicks fledged from a nest in a natural tree cavity near to the Río Abajo Nature Reserve and again the parents were released captive-bred birds. The 2014 breeding season was very successful, with an additional 14 chicks fledging from artificial nest sites in the wild and 46 parrots being bred in captivity. Encouraged by the successful establishment of a wild population at the Río Abajo Reserve, a third population will be established in the Maricao State Forest, in western Puerto Rico.

At the time of writing more than 100 Puerto Rican Amazons are being monitored in the wild, and nearly 400 birds are held in captivity, so we can be optimistic that the sterling efforts by so many dedicated workers over so many years have saved these parrots from extinction. However, total numbers are not high, so warranting retention of endangered status for the species. It is unfortunate that so many other parrots, including other *Amazona* species, have been introduced to Puerto Rico because they are potential or real competitors for both food resources and nesting sites, and may impede colonisation of modified habitats by Puerto Rican Amazons. Also, the possibility of hybridisation is highlighted by a report of at least three captive-bred birds from Río Abajo Forest joining a large flock of feral Orange-winged Amazons *Amazona amazonica* (in Birdlife International 2016).

The Puerto Rican Amazon is listed on Appendix I of the Convention on International Trade in Endangered Species of Wild Fauna and Flora (CITES).

HABITATS Most forested areas of mountains and lowlands formerly were frequented by Puerto Rican Amazons (Raffaele *et al.*

Plate 53
Puerto Rican Amazon *Amazona v. vittata* (adults)

F. KNIGHT 11

1998). Wetmore (1927) recalled finding them in dense swampy forests north of Mameyes and in dense forests at the summit of El Yunque. The last surviving wild population was confined to forests in the Luquillo Mountains which rise steeply from the coastal plain to a height of little over 1000 m, so obstructing the northeasterly trade winds to produce a high annual rainfall often exceeding 500 cm at the highest peaks, but averaging well under 200 cm at the eastern base. The forests have been classified into four broad types that owe their distinctiveness to this high rainfall, which results in showers on most days of the year, mainly in brief downpours, together with wind conditions and soil variability (in Snyder *et al.* 1987). Characteristic of the lower mountain slopes, mostly below 600 m, is forest dominated by tabonuco trees *Dacryodes excelsa* which commonly grow to heights of 40 m or more and with trunk diameters occasionally exceeding 1.5 m. Before most large specimens were logged, these trees undoubtedly were of great importance to the parrots as a food source and in providing nesting sites. Also prominent in this forest type is cacao motillo *Sloanea berteriana*, but these trees normally do not grow sufficiently large to produce cavities suitable for nesting. Just above the tabonuco-dominated forest is a depauperate upland swamp of short-statured, shallow-rooted trees which rarely exceed 15 m in height, so allowing considerable light to penetrate to ground level, where cover comprises an interlaced tangle of roots and vines with a variety of herbaceous plants ranging from bromeliads and ferns to climbing bamboo *Arthrostylidum sarmentosum* and razor grass *Scleria canescens*. The most characteristic tree of this upland swamp forest is palo colorado *Cyrilla racemiflora*, which is susceptible to heart rot, so making it extremely valuable to the parrots as a source of nesting cavities. Occurring in dense patches within the altitudinal ranges of both tabonuco-dominated and palo colorado-dominated forests is forest dominated by sierra palms *Prestoea montana*, and it typically develops along streams and on steep unstable slopes. These palms provide an enormous abundance of fruits, which are heavily utilised by the parrots and their breeding season coincides with fruiting of the palms. Near ridge tops in the mountains trees become progressively shorter, blending into a forest of stunted vegetation, which in some places is less than 3 m in height, and in appearance this dwarf forest is a dim, foggy landscape where virtually all trunk and leaf surfaces are covered with epiphytes. The trees are not sufficiently large for the formation of nesting cavities, and the stunted trees only very rarely produce fruits, so the dwarf forests are not utilised by the parrots.

During the course of intensive field studies and conservation efforts over more than 40 years, it became increasingly evident that forests in the Luquillo Mountains may be occupied by the last surviving wild population of Puerto Rico Amazons not because they are optimal habitat, but because they are the only remaining old-growth forests. The high rainfall and prevalence of natural enemies obviously impacted adversely on nesting success, and the population appeared not to be self-sustaining in the long term. Further evidence came with success in establishing a population from the release of captive-bred birds in drier, second-growth forest at Río Abajo Commonwealth Forest, where there are significantly lower numbers of Pearly-eyed Thrashers and Red-tailed Hawks.

MOVEMENTS It seems that in early times, when they were widespread and common, Puerto Rican Amazons may have undertaken seasonal or local movements in response to weather conditions or food availability. Wetmore (1927) recalled being told that formerly the parrots were seen in heavy forests in the southern section of Vieques Island, and it was thought that they came from Puerto Rico as they were found in the rainy season. Referring to Puerto Rico, Wetmore also reported:

> *On the western end of the island they had practically disappeared, though birds were reported from below San Sebastian, and Mr. Leopold B. Strube, of the Hacienda "Jobo", between Arecibo and Utuado, said that during the winter months fifty or sixty were found on his plantation, though none nested there...........When it rained heavily on the mountain, many birds descended to the warmer valleys, returning when the weather cleared.*

HABITS Much of what is known of the habits of the Puerto Rican Amazon comes from field studies of the surviving wild population in the Luquillo forests, which have been ongoing since the late 1960s, but it is possible that data from these studies may not reflect fully the behaviour and activities of the parrots in early times, when they were widespread and plentiful. Wetmore (1927) recorded observations made between December 1911 and September 1912:

> *Near Mameyes they remained in the swamps during the day, coming out to the borders morning and evening to feed, but, when alarmed, flew back inland immediately. Sometimes a small flock was found feeding in company with white-crowned pigeons* [Patagioenas leucocephala] *in the tops of trees, when, due to their green coloration, the parrots were the less conspicuous of the two. Frequently the parrots hid in the trees and remained quiet for half an hour or more, and then flew out in sudden alarm.*
>
> *Around El Yunque the birds were somewhat tamer, as they were not hunted so much, and a few were secured about a large clearing above the Río Mameyes. The birds were in pairs or small bands during March and were then breeding. During the early morning and evening they fed in the forest and in passing often stopped in high, dead trees that were standing in the clearing. After heavy rains that came at intervals during the day, they alighted there also, to dry their feathers in the sun. The call-note, given on the wing, was a rapid strident* kar, kar, *that could easily be heard at the distance of a mile, and served to alarm the entire forest. When they were feeding, low chuckling notes were heard frequently. In the trees the birds went through the characteristic posturings of* Amazona *in captivity, swinging head down and then climbing to an upright position by the aid of their bills, or watched suspiciously to see what I was about.*

In the Luquillo forests, a pair is the basic unit of nearly all groupings of Puerto Rican Amazons and, of 413 observations of flying parrots during the non-breeding season of October 1968 to February 1969, 307 or 74 per cent were of pairs, with 73 or 18 per cent of single birds, 23 or 6 per cent of trios and only 10 or 2 per cent of larger flocks (in Snyder *et al.* 1987). Trios usually comprised a pair with an additional bird, and in many instances the additional bird was a juvenile of the pair. Many of the groups of four to six birds may also have been family groups, and flocks of more than six birds almost certainly included more than one family unit. Pairs were readily discernible within flocks and, except when females were in nests, partners in mated pairs remained close together throughout the year. Juveniles from the same nest tended to stay together during the first year, so some observed pairs were of siblings rather than mated adults. As is typical of *Amazona* parrots, and indeed of most parrots, members of pairs generally remained close together in flight, alternately drifting nearer together and farther apart, but rarely more than 3 m apart.

In the Luquillo forests these parrots were reported to be highly regular in their daily activities (Kepler and Kepler 1970). Throughout most of the year they called loudly for one or two hours at dawn, and then dispersed to feeding sites. Departure from the roosting sites was in a series of short, noisy flights, with intermittent perching for variable, but brief intervals and calling loudly from highly conspicuous locations. Some preening and occasional feeding took place during silent periods, but these were minor activities, and the birds appeared to be alert at all times. Much higher, direct flights to feeding areas took place at any time during the day, but normally followed the morning relocation movements. If food was available locally, the parrots flew to this same food source day after day, generally at the same time, and returned to their roosting places along the same route. A selected feeding site often was visited regularly over a period of weeks, despite the presence of fruiting trees of the same species at a nearer location. Regular flight paths were used when the parrots travelled from one valley to another, and it was observed that one path skirted the higher ridges extending up to El Yunque peak, the parrots obviously preferring to fly around, rather than over, these ridges, possibly because of the cloud cover that often envelops the higher slopes. Peak foraging times were between 0600 and 0900 hours and again between 1700 and 1900 hours, though some feeding occurred at other times, particularly during the breeding season when males were feeding incubating or brooding females and when pairs were feeding nestlings. After alighting among the upper branches of a fruiting tree, the parrots normally paused only momentarily before commencing to feed in silence. At other times of the day the parrots rested for a series of short periods rather than for a single extended period, and while at rest they were highly inconspicuous. Return flights in the evening to roosting sites followed the same pattern as morning departure flights, with intermittent conspicuous perching and calling before silently disappearing in the dense foliage. With the approach of darkness birds were observed making short flights from perch to perch, to the accompaniment of barely audible chuckles and mumbles, and then finally settling on a roosting perch well concealed in the foliage. No parrots were seen coming to the roosts after dark, and it was presumed that they would fly at night only if disturbed by predators.

CALLS Appearing to be critically important in coordinating daily movements are what has been termed the 'take-off squawks', which are emitted as the parrots take flight, and the 'bugle', a more musical flight call which normally follows the 'take-off squawks' and commonly continues intermittently during flight (in Snyder *et al.* 1987). 'Take-off squawks' are a raucous series of noisy squawks given only as a parrot leaves its perch and commences flight. The squawk syllable that forms the basis of this call appears to be relatively similar among all individuals, but there is considerable variation in the number of squawks uttered, their intensity and spacing. The most frequently heard call is the 'bugle', which is a relatively pure, loud call, descending about a minor third in pitch and carrying more than a kilometre under good conditions. Although a flying bird may give only a single 'bugle' note, 'bugles' normally are repeated in an extended series. When pairs are in flight, one bird usually emits 'bugles' while the other remains silent or responds in duet with chortles or soft squawks.

DIET AND FEEDING Wetmore (1927) reported that Puerto Rican Amazons were fond of *Ficus* fruits and fruits of jagua *Genipa americana*, and in swamps near Mameyes the ground underneath icaco trees *Chrysobalanus icaco* were strewn with discarded husks. Only vegetable matter was present in stomachs that he examined, with seeds of tabonuco *Dacryodes excelsa* and allied species in the same family predominating. A parrot collected near Mameyes had eaten seeds of icacos, and in 'birds obtained on El Yunque it was common to find the stomach filled predominantly with small seeds, while the crop was distended with larger fruits and drupes'.

In the Luquillo forests, Puerto Rican Amazons were observed taking fruits, seeds, flowers or leaves from at least 60 plant species, and these were procured mostly in the upper stages or canopy of the forest, though there were occasional records of birds feeding about 2 m above the ground (in Snyder *et al.* 1987). Listed in Table 14 are food plants featuring in 10 or more of 745 foraging observations in which the plant species was identified and, apart from the fruits of sierra palms *Prestoea montana*, which accounted for 22 per cent of observations, there was little evidence of a strong dependence on any single food source. The top 10 food plants featured in only 62 per cent of the foraging observations, and tabonuco fruits *Dacryodes excelsa*, the second most frequently observed food, featured in only 7 per cent of foraging observations. Recorded instances of selective feeding included feeding on the leaves of laurel sabino *Magnolia splendens*, with the parrots always consuming only the often still-furled fresh new foliage, each leaf being snipped off at the petiole and eaten from the base up, but never eating the tip. Although none of the leaves of caimitillo verde *Micropholis garciniaefolia* seen being eaten were fresh new leaves, feeding again was selective in that the parrots scraped off the leaf underside and left the upperside uneaten. Parrots were observed extracting nectar from fleshy bracts below flower clusters on bejuco de rana vines *Marcgravia sintenisii*, and one bird was seen to eat 18 marble-sized *Clusia* fruits in less than three minutes (Kepler and Kepler 1970). One parrot was seen to kill a giant green anole *Anolis cuvieri* that had entered its nesting cavity, but it was not known whether the lizard or any part of it was eaten because the bird flew off carrying the reptile in its bill (in Snyder *et al.* 1987).

While a group was feeding, some birds acted as sentinels, vigilantly scanning the surroundings from the highest vantage perches. Although feeding was undertaken in a slow deliberate manner, parrots in the Luquillo forests were able to consume quite large quantities in short periods. Feeding rates varied widely, but on average one sierra palm fruit was eaten every 23 seconds and one *Clusia* fruit every 21 seconds. Fruits normally were picked one at a time, but sometimes a cluster would be snipped off whole at the stem. The parrots appeared to favour taking items positioned at eye level, but at times they assumed contorted postures to reach fruits, even hanging upside down, and this gave them access to items within 30 cm in any direction. After being picked off, the fruit was rapidly mandibulated to strip away the pericarp, carried in the bill to a favoured perch, or held in the left foot while pieces were bitten out and swallowed. The normal foraging pattern was for fruits to be eaten on the spot and for all fruits within reach to be taken before the parrot shifted to a new feeding location.

BREEDING The breeding season extends from February to June, that is during the annual dry season and when palms are fruiting (Kepler and Kepler 1970). During field studies undertaken in the Luquillo forests between 1973 and 1978, recorded dates for

Food plant	Plant type	Food item	Part eaten (when known)	Number of observations (n = 745)
Prestoea montana sierra palm	tree	hard seed in thin pericarp	pericarp (32)	167
Dacryodes excelsa tabonuco	tree	hard seed in thick pericarp	pericarp (13)	49
Marcgravia sintenisii bejuco de rana	vine	fleshy fruit with tiny seeds	fruits (2), flower bracts (2), bark (1)	45
Clusia grisebachiana cupeillo	tree	fleshy pulp with tiny seeds	pulp (14), leaves (4)	45
Miconia sintenisii camasey	shrub	berry with many tiny seeds	fruit (1)	39
Casearia guianensis palo blanco	tree	capsule bearing single seed		34
Cupania triquetra guara blanca	tree	capsule bearing three seeds		25
Linociera domingensis hueso blanco	tree	drupe		24
Ilex nitida hueso prieto	tree	drupe with four nutlets		23
Byrsonima coriacea maricao	tree	juicy drupes with large stones		22
Ocotea spathulata nemocá	tree	fleshy berry with single large seed		21
Casearia decandra tostado	shrub	fleshy capsule with 2–4 seeds		17
Inga laurina guamá	tree	pods containing several seeds		16
Cordia sulcata moral	tree	drupe with sticky flesh and single stone		15
Clusia gundlachii cupey de altura	vine	fleshy pulp with tiny seeds	fruit (4)	10
Alchornea latifolia achiotillo	tree	capsule bearing two seeds	male flowers (1)	12
Cajan cajan gandules	shrub	pods with 4–7 seeds		10
Guarea ramiflora guaraguadillo	tree	hard capsule with up to 4 seeds		10
Sloanea berteriana motillo	tree	hard capsule with several seeds		10

Table 14. Foods of Puerto Rican Amazons recorded in 10 or more foraging observations in the Luquillo forests; for 'Part eaten' a number in parentheses is the number of observations in which the plant part was identified (after Snyder *et al.* 1987).

laying of the first egg ranged from 11 February to 19 March, with the first egg in a replacement clutch being laid in mid April (in Snyder *et al.* 1987). In early times, when the parrots were widespread and common, nests were in cavities in a wide variety of trees and in one region nesting commonly was in potholes in limestone cliffs. Of 25 nests monitored during field studies in the Luquillo forests, all but one were in palo colorados *Cyrilla racemiflora*, which were dominant at high and rainy elevations, and all were in five major nesting areas totalling only 125 ha. Preference for palo colorados was associated with the abundance of natural cavities resulting from a tendency for multiple cavities to form in individual trees, a relatively high frequency of cavities in smaller trees, and a higher number of trees in the largest size classes. Hollows selected for nesting generally were at heights exceeding 4.5 m above the ground, were at least 60 cm in depth, and were dry and flat at the bottom. Nesting was not recorded in cavities with an internal diameter less than 25 cm at the bottom, and the minimum entrance width was approximately 6 cm, though hollows with larger entrances often were used. Most natural hollows in the study area were wet, and this appeared to be the major factor limiting the availability of suitable nest sites.

In the Luquillo forests, most nests were reused many times, though pairs sometimes changed nest sites. One pair studied during 1968–1979 consistently shifted its nest site after each year in which chicks did not fledge, but reused the site after each year in which successful fledging occurred.

The pair-bond is strong, and is maintained throughout the year. In the Luquillo forests, pair formation was observed on two occasions, and it involved bowing displays by both sexes. Pairs were vigorously territorial, commonly engaging in vicious combat with other pairs, and territories were defended throughout the year. Territories defended by nesting pairs were centred on the nesting tree and varied in size, sometimes extending many metres out from the nesting tree but at other times being restricted to only part of the nesting tree. Some territorial pairs were without nest sites, so it appeared that territories sometimes were established before nesting hollows were selected. When searching for a nest site, females commonly were led by males, though it appeared that final selection of a site was made by the female. Both sexes uttered characteristic wailing calls as they approached potential nest sites, and they showed variable caution when inspecting a site, sometimes immediately entering the hollow or at other times inspecting the entrance for some days before entering. Prior to egg-laying, the selected nest site was inspected in the morning and evening, and sometimes at other times of the day, both sexes entering the hollow to spend time chewing the walls and bottom, but nearer to the time of laying the female spent more time inside the hollow, eventually spending most of the day there and being fed intermittently by the male. Copulation commonly, though not invariably, followed these feeding bouts.

Also in the Luquillo forests, incubation by the female commenced with, or very soon after laying of the first egg, and most clutches comprised three eggs, there being relatively few clutches of two or four eggs. During the incubation period of approximately 26 days, the female left the nest only to be fed by the male or to join with the male in territorial defence or in repelling intrusion by Pearly-eyed Thrashers *Margarops fuscatus*. At nests under observation, males arrived at the nest three to seven times during the day to feed sitting females, and these feeding bouts generally took place on perches within 30 m of the nesting hollow. Newly hatched chicks were fed by the female at frequent intervals throughout the day and night, the frequency declining as the chicks grew and the female spent more time foraging with the male, often at distance of 1.5 km or more from the nest. Asynchronous fledging took place at about nine weeks after hatching, and young birds apparently remained with their parents for several months.

Güivas (2015) reports that video monitoring with cameras inside nests in the wild and in the aviaries commenced in 2000, and the parrots have tolerated the presence of the cameras. This monitoring has revealed that when feeding nestlings both parents normally enter the nest together, and the chicks react to their entering by immediately moving their heads clumsily. Each parent feeds until all chicks are satisfied and, after pausing momentarily, both return to the forest. Judging from captive birds, the age of first egg-laying generally is four years, but there is some evidence to suggest that in the wild egg-laying sometimes may take place at three years.

EGGS Measurements of 48 eggs inspected during field studies undertaken in the Luquillo forests are listed as 32.8–41.8 (36.0) × 27.0–31.2 (28.1) mm (in Snyder *et al.* 1987).

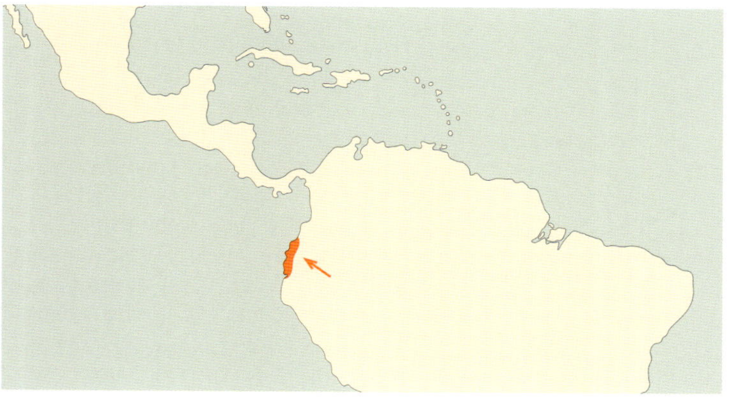

ENDANGERED

Lilacine Amazon

Amazona lilacina Lesson

Amazona (*Psittacus*) *Lilacina* Lesson, *Écho du Monde Sav.*, **11**, 1844, 2me semestre, col. 394, (Vicinity of Guayaquil, Ecuador).

OTHER NAME Lilacine Parrot.

DESCRIPTION Length 33 cm.

ADULTS General plumage green with dusky black margins to feathers of crown to mantle and breast giving a finely scalloped appearance; forehead and lores red, extending above eyes as a superciliary stripe; crown green, feathers broadly tipped lilac and finely margined reddish-violet; forecheeks greenish-yellow, becoming yellowish-green on hindcheeks and ear-coverts; carpal edge yellowish-green; primaries dusky black, on outer webs broadly edged violet-blue; secondaries green, on outer webs towards tips edged deep blue, with red at bases of S1 to S5 producing a conspicuous red wing-speculum in flight; tail green, lateral feathers broadly tipped yellowish-green and on inner webs red at bases, outermost feathers edged blue on outer webs; bill uniformly grey-black; unfeathered eyering white; iris orange; legs grey.

No specimens examined.

1 male (AMNH, Bull *in litt.* 1972): wing 201 mm, tail 133 mm, exp. cul. 22 mm.

1 unsexed (AMNH, Bull *in litt.* 1972): wing 197 mm, tail 121 mm, exp. cul. 25 mm.

1 unsexed (UKNHM, Goodwin *in litt.* 1972): wing 197 mm, tail 100 mm, exp. cul. 27 mm, tars. 20 mm.

JUVENILES Duller than adults; less extensive red and blue markings on head; bill grey, paler at base of upper mandible; iris brown.

DISTRIBUTION Confined to western Ecuador, south to El Oro; possible occurrence across the border in southernmost Nariño, southwestern Colombia, has not been confirmed.

SUBSPECIES The Lilacine Amazon formerly was considered conspecific with the closely allied Red-lored Amazon *Amazona autumnalis*, which is widely distributed from eastern Mexico south to northern and western Colombia, and subspecific differentiation may be appropriate. Plumage differences are very slight, and a low level of differentiation revealed in molecular analyses suggest that specific differentiation may not be warranted (see Russello and Amato 2004).

STATUS Ridgely and Greenfield (2001a) note that numbers of Lilacine Amazons have declined substantially as a result of deforestation and persecution. Between January and May 1994, a census was carried out at different vantage points along the coastline from Guayaquil, Guayas, and Puerto Lopez, southern Manabi, where only two populations comprising at least 327 birds were found at the end of the breeding season, and the total population was estimated at 400 to 600 birds (Kunz and Abs 1997). This same estimate was accepted by Juniper and Parr (1998). Berg and Angel (2006) report that at Puerto Hondo, Guayas, southwestern Ecuador, a large roost was censused between June 1999 and June 2000, and the mean number of birds leaving the roost each morning was 117, ranging from 31 on 10 May 2000 to 229 on 18 September 1999, with the largest count of 248 birds occurring on the evening of 18 September. Based on the mean of the five highest counts during the non-breeding season, the population at this roost was estimated at 214 birds. In surveys undertaken along more than 40 per cent of the coastline of Ecuador in November 2012 and January 2014 a minimum of 304 birds was recorded in the study area, and from this count the total population was estimated at 600 to 1700 birds, which appeared to exist in multiple subpopulations isolated by a lack of continuous habitat (in Birdlife International 2016). The largest subpopulation recorded in recent surveys comprised 139 birds, and it is assumed that no subpopulation comprised more than 250 mature birds. In the Puerto Hondo mangrove area, there has been an apparent reduction of between 18 and 35 per cent in numbers between 2006 and 2014, and the total population is assumed to be in continuing decline because of ongoing habitat destruction and unsustainable levels of exploitation (in Birdlife International 2016).

HABITATS A combination of coastal mangroves and contiguous dry tropical forest up to 700 m, or locally to 1300 m, is favoured by Lilacine Amazons, and this combination has been severely depleted by urbanisation and aquaculture development, so restricting the parrots to isolated subpopulations (Kunz and Abs 1997; in Birdlife International 2016). The parrots occupy communal nighttime roosts in the tidal forests of mangroves, where presumably they are better protected against predators, and they utilise forests and woodlands on western slopes of the Cordillera for both foraging and nesting. Berg and Angel (2006) note that at Puerto Hondo, in Guayas, where counts were made at a large nighttime roost, the area included a fishing village between a mangrove forest and the Cordillera de Chongón, a series of hills covered mainly in 40- to 80-year-old tropical dry forest and cattle pastures extending from near sea level to approximately 400 m. This is the only known roost site for several hundred kilometres and is in mangroves on an island on a tributary of the Estero Salado that converges with the Río Guayas, but for feeding and nesting the parrots travel several kilometres to the north to the forested hills of the Cordillera de Chongón, which is the only suitable habitat for many kilometres. In the Cerro Blanco Reserve, nesting occurred in dry tropical forest dominated by *Ceiba trichistandra*. Collar (1997) notes that along the western slopes of Ecuador these amazons persist also in a variety of forested habitats, dry scrublands and modified areas such as plantations. Low (1987) recalls that in October–November 1986, at localities in Guayas and Manabi, Lilacine Amazons were recorded up to 720 m, though more commonly in areas of lower rainfall below 300 m, where in some places the dry vegetation was characterised by small thorny shrubs and low deciduous trees.

HABITS Lilacine Amazons are poorly known, and the little available information on habits comes mostly from observations made during censuses undertaken in and around the Bosque Cerro Blanco Reserve in Guayas, southwestern Ecuador. Kunz and Abs (1997) report that at the nighttime roost these amazons were dispersed among several stands of mangrove, and shortly after sunrise they departed to fly in flocks along various routes to the Cerro Blanco Reserve. These flocks mostly comprised four to seven birds, and the highest count was of 90 birds seen flying together. Berg and Angel (2006) note that at a nighttime roost in the same district mean counts of 117 birds at dawn were significantly higher than mean counts of 98 at dusk, and the dawn counts probably were more accurate because the parrots flew from the roost at a height of about 10 m, but were flying at heights of 50 m to 500 m when coming back to the roost. The parrots flew in flocks typically of 10 to 30 individuals and, within these flocks, pairs usually were clearly discernible, comprising 91 per cent of all observations. Not unexpectedly, counts of pairs and singles differed significantly between the breeding season, when females remained at their nests, and the non-breeding season. Early in the morning, soon after arrival from the nighttime roost, and again in the late afternoon, shortly before returning to the roost, these parrots often perch conspicuously on high, exposed dead branches, but during the remainder of the day they are inconspicuous while resting or feeding in the leafy canopy, their presence betrayed only by the occasional falling of discarded food remnants or other debris (in Ridgely and Geenfield 2001b).

CALLS The typical flight call is described as a distinctive disyllabic or monosyllabic *cheekorák...cheekorák* or *cheekák...cheekák...cheekák* with a harsh quality, and perched birds emit a variety of other calls, some more musical and complex (Ridgely and Greenfield 2001b).

DIET AND FEEDING Fruits and seeds are procured in the upper stages and canopy of the forest, and it is likely that blossoms, buds and leaf shoots also are eaten. Low (1987) reports that an important item in the diet is fruits of muyuyo *Cordia lutea*, supplemented with *Minquartia* fruits, mangoes *Mangifera indica* and fruits of Spanish plum *Spondias purpurea*, while at lower, coastal elevations coffee beans and various cultivated fruits are eaten.

BREEDING Berg and Angel (2006) report that nesting takes place during the wet season in December to March, and fluctuations in counts at nighttime roosts in the Cerro Blanco region indicated that males continued to roost communally during the breeding season, but females remained at the nests. On 9 December 1999, a pair was first observed occupying a natural hollow in a living *Ceiba trichistandra* tree, and by 19 December the birds occupied the hollow throughout the day, so were assumed to have commenced incubation. A chick was first seen at the hollow entrance on 7 February 2000, and on 12 February both parents were seen regurgitating food to the young. On 10 March, two chicks were observed at the hollow entrance and likely had fledged by 25 March, the date after which they were not seen at the nest. On one occasion, the presumed male was seen flying in the direction of the communal roost just after sundown, and the presumed female typically was present at the nest in the early morning before commencement of the exodus from the communal roost, so confirming continued communal roosting by the male, but attendance at the nest at all times by the female.

From counts of single birds at the communal roost during two months immediately after the 1999–2000 breeding season, Berg and Angel estimated that about 17 birds may have been recruited into the roosting population, and this suggested a range of 0.36–2.36 fledglings per attempted nesting.

Diademed Amazon

ENDANGERED

Amazona diadema (Spix)

Psittacus diadema Spix, *Av. Bras.*, **1**, 1824, p. 43, pl. 32 (Rio Solimoës, Brazil).

OTHER NAME Diademed Parrot.

DESCRIPTION Length 33 cm.

ADULTS General plumage green with dusky black margins to feathers of crown to mantle and breast giving a finely scalloped appearance; forehead, lores and partly feathered cere red, not extending above eyes as a superciliary stripe; crown green, feathers broadly tipped lilac; occiput and nape green, feathers edged greenish-yellow; hindneck blue with feathers finely margined lilac; forecheeks yellowish-green; carpal edge yellowish-green; primaries dusky black, on outer webs broadly edged violet-blue; secondaries green, on outer webs towards tips edged deep blue, with red at bases of S1 to S5 producing a conspicuous red wing-speculum in flight; tail green, lateral feathers broadly tipped yellowish-green and on inner webs red at bases, outermost feathers edged blue on outer webs; bill dark grey and horn-coloured at base of upper mandible; unfeathered eyering white; iris orange; legs grey.

3 males: wing 223–237 (229.0) mm, tail 110–121 (114.0) mm, exp. cul. 31–32 (31.3) mm, tars. 24–26 (24.7) mm.

2 females: wing 235 mm and 239 mm, tail 130 mm and 135 mm, exp. cul. 32 mm and 35 mm, tars. 25 mm and 26 mm.

Plate 54
LEFT Lilacine Amazon *Amazona lilacina* (adult)
RIGHT Diademed Amazon *Amazona diadema* (adult)

F. KNIGHT 15

JUVENILES Like adults, but less extensive, paler red on lores and forehead; iris dark brown.

DISTRIBUTION Western Amazon River basin, northern Brazil, where confined to the lower Rio Negro and upper Amazon River region in Amazonas.

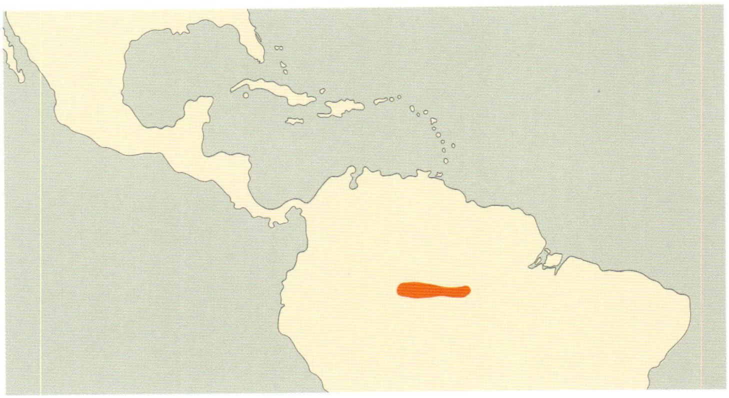

SUBSPECIES The Diademed Amazon is another species that formerly was treated as an isolated subspecies of the widespread Red-lored Amazon *Amazona autumnalis*, and again it is doubtful whether differentiation as a separate species is warranted. Plumage differences from *A. autumnalis salvini* are slight, and I expect that only a low level of differentiation would be indicated by molecular analyses.

STATUS There are no population estimates for the Diademed Amazon, and its endangered status is based on an anticipated continuation of accelerating deforestation in the Amazon River basin as land is cleared for cattle ranching and soy production, facilitated by expansion of the roads network (in Birdlife International 2016). A loss of 49–55 per cent of suitable habitat is projected to occur within its range during the next 45 years and, coupled with poaching for the domestic pet market, the population is likely to decline by 50 to 79 per cent during this period.

HABITATS Most records of these amazons come from the Ducke Forest Reserve, north of Manaus, where, in February to May 1986, Leo Joseph found them to be common in humid forest and at the forest margins (*in litt.* 1986).

HABITS There is virtually no recorded information on the habits of Diademed Amazons, but I suspect that, like the closely allied Red-lored Amazon *Amazona autumnalis*, they occupy communal nighttime roosts, where higher numbers congregate during the non-breeding season, and noisy conspicuous daily flights are undertaken to and from feeding areas. Peak levels of activity presumably are in the early morning and late afternoon, with shelter from the heat of the day being sought in the upper stages or canopy of forest trees, where their plumage would blend with the shaded foliage.

Plate 55
Red-browed Amazon *Amazona rhodocorytha* (adults)

CALLS Not known to differ significantly from the variety of harsh, discordant notes attributed to Red-lored Amazons *A. autumnalis*, which are described as a shrill *kiak...kiak...kiak* given in flight and a rapid repetition of abrupt *ack-ack* or *chek-chek* notes.

DIET AND FEEDING Fruits, seeds and nuts are procured in the upper stages of forest trees. In the Ducke Forest Reserve, Leo Joseph observed birds feeding on what almost certainly were *Tetragastris* fruits (*in litt.* 1986).

ENDANGERED

Red-browed Amazon

Amazona rhodocorytha (Salvadori)

Chrysotis rhodocorytha Salvadori, *Ibis*, 1890, p. 369 (Brazil). New name for *Psittacus dufresnianus* Kuhl 1820, not of Shaw 1812.

OTHER NAME Red-browed Parrot.

DESCRIPTION Length 35 cm. Weight 450 g.
ADULTS General plumage colouration green, with fine dusky black margins to feathers of neck to mantle and upper breast; forehead and crown red, some feathers of crown tipped dull greenish-blue; occiput dull reddish-purple, and feathers tipped dusky black and tinged blue; nape green, and feathers tinged dull reddish-purple and tipped dusky black; lores to forecheeks orange-yellow tinged red; throat and upper cheeks blue; carpal edge greenish-yellow; primaries dusky black, on outer webs broadly edged violet-blue; secondaries green, on outer webs towards tips edged deep blue, with red at bases of S1 to S5 producing a conspicuous red wing-speculum in flight; tail green, the lateral feathers broadly tipped yellowish-green and subterminally banded red, and the outermost feathers margined purple-blue; bill grey with pink-red at base of upper mandible; unfeathered eyering grey; iris orange; legs grey.
6 males: wing 215–229 (221.7) mm, tail 99–119 (111.5) mm, exp. cul. 30–36 (33.0) mm, tars. 24–26 (25.0) mm.
5 females: wing 209–226 (215.6) mm, tail 107–116 (110.4) mm, exp. cul. 29–32 (30.8) mm, tars. 25–26 (25.2) mm.
JUVENILES Forehead and forecrown red, on hindcrown becoming strongly suffused green; occiput to hindneck green, the feathers finely edged dusky black; lores yellow; ear-coverts to forecheeks and throat blue; less red at bases of only two outermost secondaries; much reduced red on lateral tail-feathers; bill pale pink tipped greyish-white; unfeathered eyering greyish-white; iris brown.

DISTRIBUTION Eastern Brazil, from eastern Alagoas, northeastern Bahia and eastern Minas Gerais to Espírito Santo and northern São Paulo.

STATUS Because of extensive deforestation in coastal Brazil and poaching for the cagebird market, Red-browed Amazons have disappeared entirely from parts of their range, and elsewhere are very locally dispersed in fragmented forest remnants. Early warnings of declining numbers came from Ridgely (1981), who pointed out that overall these amazons are quite rare and are considerably less numerous than either the Blue-throated Parakeet *Pyrrhura cruentata* or the White-eared Parakeet

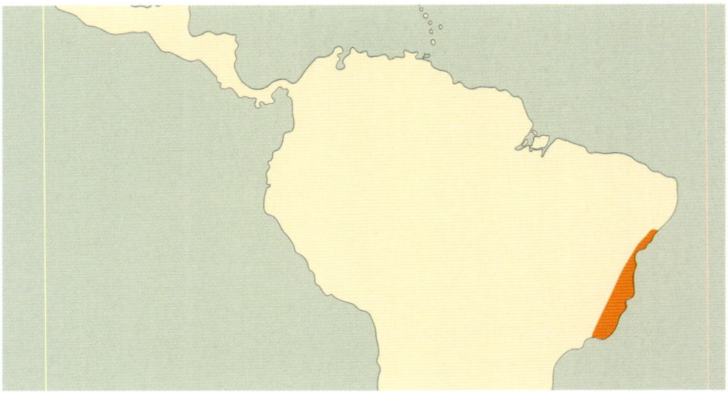

P. leucotis, two species from the same region that have received a good deal more attention. Only one isolated population is known to be present in Alagoas, at the northern extremity of the range, and where only 2 per cent of the original forest cover remains. Populations are known to occur at three sites in southeastern Bahia and at five sites in each of Minas Gerais and Rio de Janeiro (in Birdlife International 2016). The largest stands of forest remain in Espirito Santo, so the strongest populations now are found in that state, where 2295 birds were recorded during surveys undertaken between 2004 and 2006. Also in Espirito Santo, these amazons are still fairly common in Sooretama and adjacent Linhares Reserves. They are fairly common also on Ilha Grande, in Rio de Janeiro, but elsewhere in that state many important habitat fragments are being cleared (in Birdlife International 2016).

Illegal trafficking poses a serious threat to Red-browed Amazons in many parts of the range, and apparently is the overriding threat in Espirito Santo. In the 1998–1999 breeding season, 174 chicks were taken from nests, mostly in reserves, and 664 birds were recorded in captivity at a reintroduction centre in 2005–2006 (in Birdlife International 2016). Conflict with agriculture also is a problem, and in some districts these parrots are considered to be pests in papaya, coffee and cocoa plantations. Sweeney (1996) reports that souvenirs containing their feathers have been offered for sale outside Monte Pascoal National Park, in Bahia, so it seems that the birds are killed by hunters.

The total population is estimated at 1000–2500 birds, which equates to 600–1700 mature individuals, and an ongoing rapid decline is expected as a consequence of continuing habitat destruction and fragmentation coupled with poaching for the live-bird market.

The Red-browed Amazon is listed on Appendix I of the Convention on International Trade in Endangered Species of Wild Fauna and Flora (CITES).

HABITATS Red-browed Amazons occur primarily in humid lowland forests, but also range, at least seasonally, up to about 1000 m in adjacent montane areas, and there are reports of their associating with Orange-winged Amazons *Amazona amazonica* in winter movements into estuarine mangroves. Between March 1992 and January 1993, during studies of vertical stratification and diet of parrots in the Linhares Reserve, northern Espirito Santo, these amazons were recorded in both the dominant terra firme forest, comprising dense forest with trees up to 40 m in height, and in 'mussununga' forest, comprising dense forest with small trees and shrubs up to 15 m in height and growing on sandy soil (Simão *et al.* 1997).

HABITS Red-browed Amazons keep to the upper stages and canopy of the forest, when they can be difficult to detect during the day as pairs or small groups feed and rest amidst the foliage. Between March 1992 and January 1993, during studies of vertical stratification and diet of parrots in the Linhares Reserve, northern Espirito Santo, these amazons exhibited a strong preference for the superior strata of the forest, with 61 per cent of perching observations being at heights of 26–30 m, and only 4 per cent of observations were below a height of 16 m (Simão *et al.* 1997). At dusk they are more conspicuous and vocal when congregating in larger numbers at a communal nighttime roost, sometimes in the company of Orange-winged Amazons *Amazona amazonica*. In past years, when these amazons were locally plentiful, large numbers often congregated at nighttime roosts, and Pinto (1935) recalled that along the banks of the Rio Gongogy, not far from Bôa Vista, Bahia, large flocks were found roosting in the highest trees, and their raucous cries greeted the dawn of each new day.

CALLS In flight an almost incessant, raucous *caa-ua...caa-ua... caa-ua* is given, and other calls have been described as *cheee-ooo* and *nee-it* notes and a loud *koy-ok...koy-ok* or *kow-ow...kow-ow*.

DIET AND FEEDING The diet comprises seeds, fruits, berries and buds procured in the treetops, and birds have been recorded feeding in papaya, jackfruit, mango, cocoa, banana and coffee plantations (in Birdlife International 2016). Between March 1992 and January 1993, during dietary studies undertaken in the Linhares Reserve, northern Espirito Santo, these amazons were recorded eating the seeds of Brazil resintree *Protium heptaphyllum* and *Micropholis crassipedicelata* (Simão *et al.* 1997).

BREEDING There is little information on nesting, but pairs with fledglings have been observed in January, so the hatching of chicks probably takes place in about October (see Birdlife International 2016).

Mann (1982) gives details of breeding in captivity. The female sexually solicited by holding down her head and raising her vent, after which an excited quivering of wings by both sexes commenced prior to, and continued during copulation. The clutch comprised three eggs, which were laid at intervals of three days, and incubation of 24 days by the female commenced with laying of the second egg. The brooding female left the nest in the morning and early evening to be fed by the male. Newly hatched chicks were closely brooded by the female, and the male then spent much time in the nestbox. When approximately 30 days old the nestlings were removed for hand-rearing. Reinschmidt (2006) also records successful captive breeding, with three eggs being the average size, and one to five eggs the range of 13 clutches. At hatching chicks averaged 14.04 grams in weight, with a range of 12.2–16.8 grams.

EGGS Average measurements of 23 eggs laid in captivity are 39.4 × 30.6 mm (Reinschmidt 2006).

ENDANGERED

Vinaceous-breasted Amazon

Amazona vinacea (Kuhl)

Psittacus vinaceus "Prince Maximilian" Kuhl, *Nova. Acta. Acad. Caes. Leop. Carol.*, **10**, 1820, p. 77 (Brazil).

OTHER NAMES Vinaceous Amazon, Vinaceous-breasted Parrot.

DESCRIPTION Length 30 cm. Weight 370 g.
ADULTS General plumage green with dusky black margins to feathers giving scalloped appearance; lores and frontal band red; chin pink-red; elongated feathers of hindneck and sides of neck green broadly edged pale blue and strongly tipped dusky black; upper tail-coverts pale green; breast dull mauve-red variably suffused dull blue and feathers margined dusky black; upper abdomen green with feathers dull mauve-red at bases and tipped dusky black; lower abdomen yellowish-green, feathers finely tipped dusky black; carpal edge green variably marked yellow and red; primaries dusky black, on outer webs broadly margined deep blue; secondaries green with red at bases of S1 to S3 giving red wing-speculum in flight; underwing-coverts and undersides of flight feathers green; tail green narrowly tipped yellowish-green with red at bases of lateral feathers; bill dull pink-red with horn-coloured tip; unfeathered eyering grey; iris red; legs pale grey.
13 males: wing 204–220 (210.2) mm, tail 98–120 (109.9) mm, exp. cul. 24–29 (26.9) mm, tars. 21–24 (22.8) mm.
11 females: wing 203–217 (207.9) mm, tail 101–115 (108.2) mm, exp. cul. 24–29 (26.5) mm, tars. 22–24 (22.8) mm.
JUVENILES Less extensive duller red frontal band; breast dull mauve-red suffused green; carpal edge greenish-yellow; bill horn-coloured with dull pink-red at base of upper mandible; iris brown.

DISTRIBUTION Southeastern Brazil, from southern Bahía and western Espírito Santo south to northeastern Argentina, in Misiones and possibly northeastern Corrientes, and southeastern Paraguay.

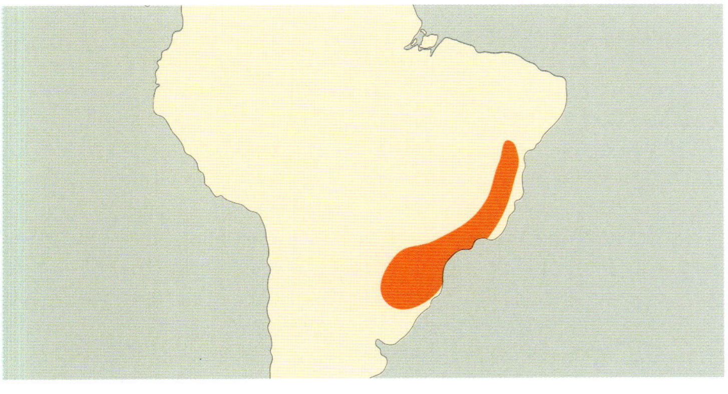

STATUS Widespread deforestation, the poaching of nestlings for the pet trade and local persecution as a pest in croplands have brought about a dramatic decline in numbers of Vinaceous-breasted Amazons, and they have been extirpated in parts of their range, especially in Argentina and Paraguay. Although common to abundant in the 19th century, they now are rare throughout their entire range, with highest numbers surviving in southeastern Brazil. Galetti (1997) notes that during studies of seasonal abundance and feeding ecology of parrots undertaken in 1994–1995 at Intervales State Park, São Paulo, southeastern Brazil, these amazons were observed so rarely that they were excluded from the findings. Farther north in eastern Brazil, in Rio de Janeiro, Espirito Santo and Bahia, they probably never were common, but now must be near to extinction (in Birdlife International 2016). Waugh (2016b) reports that the present predominantly southern occurrence was confirmed during a census undertaken throughout the range in 2015, when survey teams in Brazil recorded 1552 birds in Santa Catarina, 976 birds in Paraná, 282 in Rio Grande do Sul, 65 in São Paulo, and only two birds in Minas Gerais. During this same census, 143 birds were counted in northeastern Argentina and 133 birds were counted in eastern Paraguay, so giving an estimated total population of 3133 birds.

Evidence suggests that Vinaceous-breasted Amazons once were very common in western parts of the range, in northeastern Argentina and eastern Paraguay. White (1882) reported that at Concepcíon and San Javier, southern Misiones, northeastern Argentina, 'incredible numbers' were feeding in orange groves, and many were shot by local farmers. More recently, at Campo Viera and Montecarlo, also in Misiones, local persons recalled them as having been very abundant in the 1920s, when they attacked corn crops and were shot and poisoned by farmers (in Cockle *et al.* 2007). At Colonia Dorada, eastern Paraguay, in 1988, a local guide claimed to have seen 8000 or more birds feeding on mineral-rich soil in a swampy area in 1978, but at this site in 1988 most of the forest had been replaced by crops or pasture, the population of these amazons had been reduced to 'about 300', and local trappers reported catching 500 adult birds for export through Brazil (in Silva 1989). From 1997 to 2006, in this western part of the range, fieldworkers searched for Vinaceous-breasted Amazons during 1639 days of surveys, with 1150 days of searching at 28 localities in northeastern Argentina from March 1997 to May 2006, and 489 days of searching at 29 localities in eastern Paraguay from July 2000 to May 2006 (Cockle *et al.* 2007). The parrots were recorded on 568 of the 1150 days, or 49 per cent, in northeastern Argentina and on 94 of the 489 days, or 19 per cent, in eastern Paraguay. These records accorded with other recent records indicating that the parrots now occur only in central and eastern Misiones and at a cluster of sites in northeastern Paraguay, in departmento Canindeyú and adjacent Alto Paraná and Caaguazú, and they have disappeared from central-eastern Paraguay, southern Misiones in northeastern Argentina, and from the Río Paraná between Argentina and Paraguay. In northeastern Argentina, the largest simultaneous roost count was of 163 birds, with two other populations counted at other times totalling only about 40 birds, and in eastern Paraguay the largest count was of 167 birds at an evening roost, with few birds recorded at other sites at other times. Based on these counts the minimum total population in northeastern Argentina was estimated at 203 individuals and the minimum total population in eastern Paraguay was estimated at 220 individuals. These estimates accord quite well with numbers recorded during the census undertaken in 2015, and confirm that numbers have declined dramatically in western parts of the range.

At a forested property in São Paulo state, southeastern Brazil, the Lymington Foundation has established a facility for rehabilitating confiscated captive parrots for release into the wild, and Vinaceous-breasted Amazons are included in their program (Loewen 2015).

HABITATS In southeastern Brazil there is a close association between Vinaceous-breasted Amazons and *Araucaria* forests, and destruction of these forests has been identified as the principal reason for the dramatic decline in numbers. However, it appears that in other parts of the range the association with *Araucaria* forests is not so strong, and in some regions these forests are not present. In the Chapada Diamantina, at the northern extremity of the range in Bahia, eastern Brazil, these amazons frequent semihumid forests (Parrini *et al.* 1999). In both northeastern Argentina and eastern Paraguay, they historically were recorded in a variety of habitats, including highland *Araucaria* forest, lowland semideciduous forest and ecotones between forest and open campos or cerrado woodland, and they occurred in regions where *Araucaria* forest did not exist (Cockle *et al.* 2007). This was particularly so in eastern Paraguay, where the parrots formerly were widespread in lowland semideciduous forest and there was only one small patch of *Araucaria* forest. During surveys undertaken in eastern Paraguay, between July 2000 and March 2006, Vinaceous-breasted Amazons were recorded on 48 of 410 days, or 12 per cent, in large tracts of natural habitat, and on 49 of 75 days, or 65 per cent, in forest fragments, towns and farms, with all birds observed in the latter habitats being either in the canopy of a shade-grown crop or in trees in nearby pasture. In northeastern Argentina, there were large tracts of forest and mosaic landscapes of forest corridors on small subsistence farms with many clumps of trees and shrubby vegetation on abandoned agricultural lands, and these amazons were recorded on 34 of 404 days, or 8 per cent, in large tracts of natural habitat, and on 535 of 750 days, or 71 per cent, in forest fragments, towns and farms. They were seen perched, roosting and feeding in trees in towns, in isolated trees on subsistence farms, in forest fragments, and in large tracts of forest, and nesting in forest fragments on farms and at San Pedro township. They have disappeared from large protected areas in northeastern Argentina, including Iguazú National Park, where the most forest habitat remains, so it is possible that protecting forest habitat may not be enough to conserve remaining populations of these parrots (Cockle *et al.* 2007).

MOVEMENTS Evidence of possible seasonal movements comes from the Estancia Itabó Private Nature Reserve, departmento Canindeyú, eastern Paraguay, where Vinaceous-breasted Amazons were observed feeding on palmito fruits *Euterpe edulis* in August 1992, but in the summer of December 1994 the parrots were less common, and it was suggested that seasonal movements may occur, perhaps in response to the availability of palmito fruits (Loewen *et al.* 1995). Seasonal utilisation of different habitats may have occurred historically in much of the southern range, with lowland semideciduous forest tracts being sink habitats that relied on source populations in highland *Araucaria* forests (Cockle *et al.* 2007).

Collar (1997) notes that some post-breeding dispersal occurs in Rio Grande do Sul, southeastern Brazil, with many birds disappearing in January and all having departed by March, and then returning in April to remain during the rest of the year.

Plate 56
Vinaceous-breasted Amazon *Amazona vinacea* (adult)

HABITS Vinaceous-breasted Amazons are vocally and visually conspicuous, and during surveys undertaken in northeastern Argentina and eastern Paraguay, between 1997 and 2000, observed foraging groups averaged eight birds, but feeding by single birds and flocks of up to 80 birds were recorded (Cockle *et al.* 2007). While a group was feeding, one or more birds acted as sentinels, scanning the surrounds from vantage perches near the tops of nearby trees. Communal nighttime roosts were recorded at four sites in northeastern Argentina, all in *Araucaria* trees, many of which were isolated and in areas where the understorey and most trees had been cleared for farming. At one site, where there were 34 observations of up to 76 birds, flocks changed roosting places frequently, using specific roost trees for only one to five successive days, but often returning to them after one to two weeks. At the Estancia Itabó Private Nature Reserve, eastern Paraguay, where *Araucaria* trees were not present, nighttime roosting from October to December 2001 was in isolated pindó palms *Syagrus romanzoffiana* in a 500 ha pasture surrounded by secondary forest, and in March–April 2002 it was in unidentified trees, possibly also palms, in secondary forest approximately 2 km from the pasture. As expected, at all roosting sites numbers were consistently highest in March–April, during the non-breeding season.

These amazons are fairly tame while feeding, relying on camouflage to escape detection amidst treetop foliage. In southeastern Brazil, they have been seen feeding in the company of Red-spectacled Amazons *Amazona pretrei* and Scaly-headed Parrots *Pionus maximiliani*.

CALLS The call is described as a loud, guttural and rather harsh *how*, uttered singly or repeated irregularly (in Belton 1984).

DIET AND FEEDING Belton (1984) recalls that in late August 1975, near Vacaria, northeastern Rio Grande do Sul, southeastern Brazil, a pair of Vinaceous-breasted Amazons was observed eating *Araucaria* nuts and, while perched on spiny fronds of the tree, they chewed away a large section from the side of the cone until two or three nuts were exposed and then worked on the nuts while still fixed to the cone, eating leisurely with occasional rest intervals.

In northeastern Argentina and eastern Paraguay, during surveys undertaken between 1997 and 2006, these amazons were observed feeding mainly on fruits, seeds, leaves and young shoots of nine native plants and six exotic plants (Cockle *et al.* 2007). The native plants were parana pine *Araucaria angustifolia*, wood rosemary *Holocalyx balansae*, angico *Parapiptadenia rigida*, camboatá *Matayba elaegnoides*, *Ateleia glazioveana*, pindó palm *Syagrus romanzoffiana*, *Aechmea* sp., cambuí *Peltophorum dubium* and chal-chal *Allophylus edulis*. The exotic plants were white cedar *Melia azedarach*, Japanese raisin tree *Hovenia dulcis*, loquat *Eriobotrya japonica*, *Persea* sp., *Citrus* sp. and eucalypts. When feeding on *Eucalyptus* fruits the parrots scraped and mashed the outer skin, and they were observed chewing *Eucalyptus* bark. Also, they were seen to take unidentified food items from nests of tent caterpillars. Food plants recorded in the literature include palmito *Euterpe edulis*, *Guadua* sp., cedro batata *Cedrela fissilis*, pinheiro bravo *Podocarpus lambertii*, guava *Psidium longipetiolatum*, Brazilian cherry *Eugenia uniflora*, cherry of the Rio Grande *E. involucrata*, Brazilian coral tree *Erythrina falcata*, bracatinga *Mimosa scabrella* and *Laplacea fruticosa* (in Cockle *et al.* 2007).

F. KNIGHT 16

BREEDING Belton (1984) reports being told of a nest found in December 1978 in a hollow in an *Araucaria* tree at Aparados da Serra National Park, northeastern Rio Grande do Sul, southeastern Brazil. Von Ihering (1902) reported receiving an egg from a Sr. Ch. Enslen, of São Lourenco, Rio Grande do Sul, and noted that it was taken from a 2 m deep hollow high up in a huge myrtle tree.

During surveys undertaken in northeastern Argentina between 1997 and 2006, the earliest date on which signs of nesting were observed was 16 September, and local residents advised that chicks traditionally were taken from nests just prior to fledging 'at Christmas' (Cockle *et al.* 2007). Of six nests found along forest margins or in forest fragments ranging in size from 0.5 ha to 100 ha, four were in tree cavities formed by broken branches and two were in old nesting holes of woodpeckers at heights of 5–20 m, with an average of 12.8 m, in trees ranging from 10 m to 25 m in height, with an average of 18.4 m. Two nests were in parana pines *Araucaria angustifolia*, and one was in each of the following – angico *Parapiptadenia rigida*, *Prunus subcoriacea*, canjarana *Cabralea canjerana* and garapa *Apuleia leiocarpa*. Local residents reported that nests had been found in *Araucaria angustifolia*, *Parapiptadenia rigida*, moena *Ocotea puberula*, ibirápayé *Myrocarpus frondosus*, marmelero *Ruprechtia laxiflora* and cambuí *Peltophorum dubium*.

EGGS Oval-shaped, smooth and slightly glossy; measurements given for two eggs are 38.0 × 30.0 mm and 38.1 × 27.5 mm (Von Ihering 1902; Schönwetter 1964).

ENDANGERED

Imperial Amazon

Amazona imperialis Richmond

Amazona imperialis Richmond, *Auk*, **16**, 1899, p. 186. New name for *Psittacus augustus* Vigors 1837, not of Shaw 1792 (Dominica).

OTHER NAMES Imperial Parrot, Sisserou (local).

DESCRIPTION Length 50 cm.
ADULTS Forehead and lores deep brownish-purple, becoming more brownish on cheeks to ear-coverts; crown to neck dark maroon-purple variably suffused greenish-blue and the feathers finely edged dusky black; breast and abdomen deep mauve-purple and the feathers finely edged dusky black; lower underparts olive-green, feathers tipped greenish-blue; upperparts dark green, feathers finely edged dusky black; carpal edge red; secondaries green tipped violet-blue and with dark maroon at bases of S1 to S3; primaries dusky black, on outer webs dull violet-blue with dull green at bases and tipped brown; underwing-coverts green tipped blue; undersides of flight feathers green; tail dull reddish-brown tipped greenish-blue and suffused green on central feathers and at bases of lateral feathers; bill greyish-horn; eyering grey-brown; iris yellow to orange-red; legs grey.
8 males: wing 270–293 (285.8) mm, tail 158–176 (168.9) mm, exp. cul. 36–42 (39.4) mm, tars. 29–34 (31.3) mm.
10 females: wing 275–299 (284.0) mm, tail 153–179 (166.3) mm, exp. cul. 37–41 (39.2) mm, tars. 30–33 (31.9) mm.

JUVENILES Duller than adults; face dull rufous; occiput to hindneck green; hindcheeks tinged green; iris brown.

DISTRIBUTION Restricted to Dominica, in the Lesser Antilles.

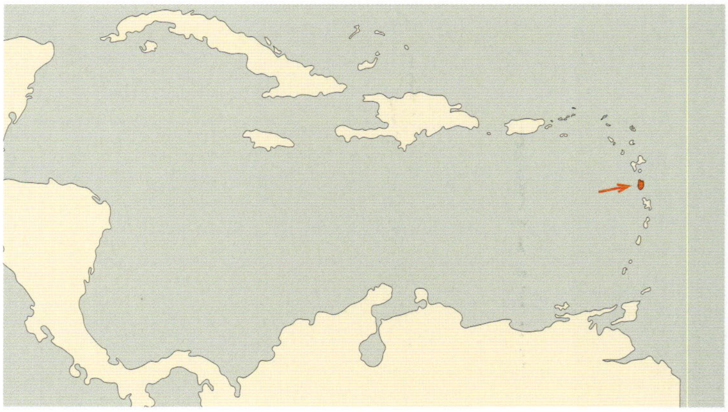

STATUS Of the two endemic *Amazona* parrots on Dominica, the Imperial Amazon always has been recorded in much lower numbers, and generally was considered to be scarce and locally dispersed in montane forests. In 1966, in response to my inquiry about numbers, David Wingate confirmed that it was scarce, but noted that numbers probably always were low because areas of high mountain forest, its preferred habitat, are of limited extent on the island (*in litt.* 1966). Estimating numbers is and always has been difficult because of the rugged terrain, coupled with the secretive habits of the parrots, but in the mid 1970s it was suggested that the total population comprised 150 birds. Loss and degradation of mountain forest, primarily because of conversion to plantations and damage by hurricanes, the latter being particularly severe in 1979 when Hurricane David brought about widespread destruction, together with hunting were identified as causes of a strong decline in numbers. In April 1980, during a brief visit to Dominica, I observed the devastating damage to mountain forests and came to the conclusion that there must have been a very serious post-hurricane depletion in numbers of Imperial Amazons. It was suspected that at this time as few as 50 birds survived in the Morne Diablotin area in the north of the island, and the small southern population in Morne Trois Pitons National Park had been extirpated. It seems that recovery was slow because in 1993 the population comprised only 80–100 birds (in Birdlife International 2016).

Reillo (2001) points out that since its near extinction by Hurricane David, recovery of the Imperial Amazon has been the subject of intense field research and a concerted conservation effort championed by Dominica's Forestry, Wildlife and Parks Division with support from international wildlife agencies. A very significant achievement occurred in January 2000 with establishment of Morne Diablotin National Park, encompassing 5300 ha of pristine mountain forest and a stronghold of the Imperial Amazon, and then in December 2000 a small population was located in the vicinity of Morne Trois Pitons National Park. In the past these parrots were shot for food or in the hope of obtaining injured birds for the illicit live-bird market, but this has been reduced considerably, if not eliminated, as a result of successful awareness and education programs emphasising the importance of the Sisserou as the national bird and the need to protect it. Some conflict with agriculture occurs, particularly in

Plate 57
Imperial Amazon *Amazona imperialis* (adult)

F. KNIGHT '15

citrus orchards near to the national parks and forestry reserves, where investigations are being undertaken to assess damage levels and to develop strains of fruit with less seeds, so making them less attractive to the parrots. Competition for nest-sites from the more numerous Red-throated Amazon *Amazona arausiaca* is likely to increase as lowland forest is lost and contact between the two species becomes more frequent and, as noted by Wingate so many years ago, habitat limitations may restrict increases in numbers. Despite these concerns, the long-term outlook is encouraging, and there is expectation of a continuation of the slow improvement in status of the Imperial Amazon already achieved with the conservation effort. In 2012, the total population was estimated at 250–350 birds, roughly equating to 160–240 mature individuals (in Birdlife International 2016).

The Imperial Amazon is listed in Appendix I of the Convention on International Trade in Endangered Species of Wild Fauna and Flora (CITES).

HABITATS Although there is some overlap, the two *Amazona* parrots on Dominica replace each other altitudinally, with the Imperial Amazon frequenting montane and elfin forest above 600 m, and its stronghold is in forests on the slopes of Morne Diablotin in the central-north of the island. Within these forests there is an association with mature gommier trees *Dacryodes excelsa*, which are favoured for nesting and feeding. Reillo (2001) reports that in early December 2000, in rocky terrain in the south of the island, a small population was located in protected forest at the base of Morne Watt and Morne John, and at the boundary of Morne Trois Pitons National Park birds were seen and heard in a valley beneath a ridgeline from Morne Prosper.

In April 1980, only months after the passage of Hurricane David, my only encounter with this species was along a tributary of the upper Picards River, on the slopes of Morne Diablotin, where a lone bird was seen in undamaged forest canopy in the deep river gorge, and showed a strong reluctance to move out into surrounding hurricane-damaged forest.

HABITS Imperial Amazons are the largest and arguably the most impressive of *Amazona* parrots. I had read that they take on a strangely raptor-like appearance in flight above the forest, and in April 1980, in a deep river gorge on the slopes of Morne Diablotin, a lone bird certainly took on an eagle-like appearance as I watched it glide above treetops in forest not damaged by Hurricane David. Reillo (2001) notes that densities of these parrots rarely exceed one bird per 180 ha, so they are encountered mostly singly, in pairs or in small groups. They are shy and secretive, so can be difficult to locate, especially when feeding silently among dense foliage in the canopy of tallest trees, and this has added to difficulties when undertaking field surveys. On the slopes on Morne Diablotin, during surveys undertaken in the 1970s, these amazons were found to be unusually quiet, calling very rarely during the day whether in flight or perched in the treetops, and only at dawn and in the half-hour before nightfall did they call extensively (Nichols *in litt.* 1977). After feeding, a pair was seen to sit, either inactively or preening, for several minutes before rising into the air and flying to the top of the forest canopy without emitting any flight call, and both birds perched a short distance away, again without any audible vocalisation.

CALLS The rarely emitted flight call is a distinctive, trumpeting *eeeee-er* cry downslurred at the end. A variety of shrieks, squawks or shrill whistling notes sometimes are given while perched, and a short, harsh screech is emitted when alarmed. All calls are more shrill and metallic than calls of the Red-throated Amazon *Amazona arausiaca*.

DIET AND FEEDING The diet comprises fruits and seeds, especially those of gommier trees *Dacryodes excelsa* and mountain palms, together with nuts, berries, blossoms and buds, including the leaf buds of mountain palms (in Wood 1924).

Nichols recorded details of a lengthy observation of a pair feeding on the berries of bois bande *Richeria grandis* (*in litt.* 1974). The birds browsed in typical *Amazona* fashion, frequently hanging upside-down to reach the berries, and walking or clambering from limb to limb, but making little or no vocalisations.

BREEDING Nichols reported that in 1974, in the Morne Anglais area, southern Dominica, an active nest was found in a 75 cm deep hollow at a height of 26 m in a *Slonea* tree (*in litt.* 1977). In 1975, Nichols found another active nest at a height of 22 m in a gommier tree on the ridge between upper forks of the Gillon River, also in the Morne Anglais area, and this nesting hollow was 90 cm deep, with a 45 cm wide entrance. On 12 April, the incubating female was observed in this hollow, and on 10 May there were two unfeathered chicks, but the nesting attempt failed.

Little is known of the nesting cycle, but Roland Seitre tells me that some information was obtained from nesting by a captive pair on Dominica (*in litt.* 2015). In 2005, nestboxes were installed in the aviary after the birds were observed displaying, and a clutch of two eggs was laid in April 2006, but proved to be infertile after three weeks incubation. Single-egg clutches were laid in 2007 and 2008, but again were infertile. In 2010, a single egg was laid on 7 April, and incubation by the female lasted 28 days, during which time she was fed by the male. The chick hatched on 5 May, and initially it was cared for by both parents, but on 21 May it was found to be in poor condition, apparently after being neglected by the parents, so was removed for hand-rearing.

EGGS An egg in the collection at the Natural History Museum, Tring, UK, measures 45.6 × 37.3 mm (Harrison and Holyoak 1970).

GENUS ***Hapalopsittaca*** Ridgway

Hapalopsittaca Ridgway, *Proc. Biol. Soc. Wash.*, **25**, 1912, p. 100. Type, by original designation and monotypy, *Psittacus amazoninus* Des Murs.

Midsized, stocky parrots belonging to this genus are poorly known inhabitants of highland forests in the northwest of South America, and it is ironic that a species which was lost for more than 90 years after its discovery probably now is the best known member of the genus. Morphological features include a short, slightly rounded tail and a proportionately small bill without a prominent notch in the upper mandible. The cere is bare, but the lores are fully feathered. The sexes are alike, and juveniles are duller than adults.

Hapalopsittaca is sporadically distributed from westernmost Venezuela and northern Colombia south to central Peru and central-western Bolivia. Within this range, only the southernmost, distinctively plumaged Black-winged Parrot *H. melanotis* is not considered to be threatened and, to the north, two of the three similarly plumaged, closely allied species are listed as vulnerable, with the third being critically endangered.

CRITICALLY ENDANGERED

Indigo-winged Parrot

Hapalopsittaca fuertesi (Chapman)

Pionopsitta fuertesi Chapman, *Bull. Am. Mus. Nat. Hist.*, **31**, 1912, p. 143 (Laguneta, 10 340 feet, Cauca, Colombia).

OTHER NAME Fuertes's Parrot.

DESCRIPTION Length 23 cm.
ADULTS General plumage colouration dull yellowish-green; sometimes, but not always, a very narrow, incomplete dull red frontal band comprising a few scattered feathers; forehead, lores and around eyes to ear-coverts greenish-yellow; chin and forecheeks green; crown blue; variable dull red suffusion in abdomen; bend of wing to lesser wing-coverts, carpal edge, and lesser underwing-coverts crimson; outer median wing-coverts violet-blue broadly edged pink; outer secondary-coverts green, on outer webs broadly edged violet-blue; primary-coverts, primaries and outer secondaries dusky black, on outer webs broadly edged violet-blue; inner secondaries green, on outer webs broadly edged blue; greater underwing-coverts and undersides of flight feathers greenish-blue; tail brownish-red tipped violet; bill horn-coloured with bluish-grey marking at base of upper mandible; iris greenish-yellow; legs dark grey.
4 males: wing 151–157 (153.5) mm, tail 77–94 (83.3) mm, exp. cul. 15–17 (15.8) mm, tars. 16–18 (17.3) mm.
1 female: wing 150 mm, tail 86 mm, exp. cul. 17 mm, tars. 17 mm.
JUVENILES Duller than adults; no red frontal band; crown green slightly suffused pale blue; bend of wing to lesser wing-coverts, carpal edge, and lesser underwing-coverts darker maroon; iris grey.

DISTRIBUTION Known only from below Nevado de Santa Isabel, on the western slope of the Cordillera Central, in the Quindio-Risaralda border region, western Colombia, with a possible sighting on the eastern slope, near Ibague, Tolima.

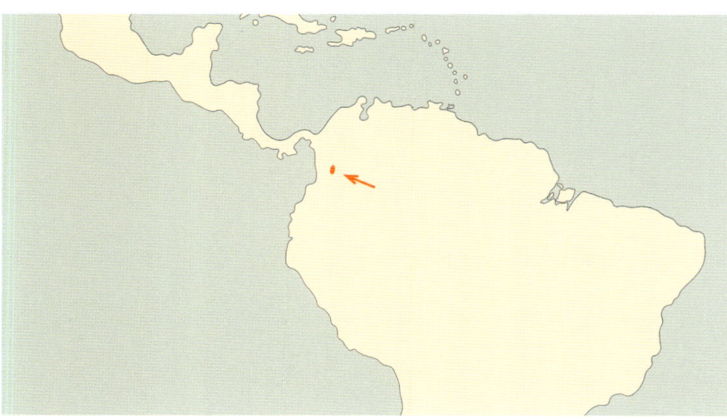

STATUS The Indigo-winged Parrot was described in 1912 from specimens collected in montane forests in the Cordillera Central, western Colombia, in August 1911 by Leo Miller and Arthur Allen from the American Museum of Natural History, and then it was 'lost' until 'rediscovered' in July 2002 by Jorge Velásquez and Alonso Quevedo, two Colombian ornithologists who found a flock of 14 birds in a small remnant stand of montane forest in the same region (Parr and Gilardi 2002). During subsequent searches at the time of the 'rediscovery', the largest observed group comprised 25 birds, and the total population was estimated at approximately 60 individuals (in Birdlife International 2016). A concerted conservation effort was initiated promptly, and this included protecting some 1500 ha of montane forest in the newly established El Mirador Nature Reserve and the setting up of more than 200 nestboxes to overcome a critical shortage of mature trees with cavities, a legacy of long-term selective logging of the forest. The shortage of suitable nesting hollows in mature trees became most evident when five of the nestboxes were occupied almost immediately. Additional reserves have been established, so that approximately 70 per cent of the known population now is protected in more than 7285 ha of key habitat (in Birdlife International 2016). These measures have stopped or slowed the decline in numbers, which in the past has been brought about by excessive deforestation, and the current population is estimated at 160 individuals, which equates to between 50 and 249 mature birds (Birdlife International 2016).

HABITATS From 2600 m to 3600 m, but mostly at 3300 m to 3500 m, Indigo-winged Parrots occur primarily in mature montane cloud forest, especially where there is a prevalence of mistletoe, which is an important food source, but have been seen also in moderately disturbed secondary forest and in fairly open, presumably disturbed forest with small palms and tree ferns. At 3200 m, rediscovery of these parrots occurred when birds were mistnetted in a small stand of cloud forest rich in epiphytes and tree ferns, but where tree-cutting had brought about a scarcity of mature trees.

HABITS The little available information on the habits of Indigo-winged Parrots suggest that their actions are similar to the actions of other *Hapalopsittaca* species. They are gregarious, usually associating in small flocks, which are conspicuous in flight above the forest canopy, but can be difficult to detect when feeding among the upper stages or canopy of the forest. In February, at the beginning of the breeding season, the flocks break up as pairs disperse for nesting and then reform after the chicks have fledged.

CALLS The flight call is described as a sharp cry.

DIET AND FEEDING Food is procured mostly among the upper branches of forest trees, and the parrots eat a variety of seeds, fruits and flower buds, primarily of olive and Romeron pine *Retrophyllum rospigliosii*. Fruits of mistletoe are said to be a favoured food, and it is reported that they also feed on thistles, so presumably must come down to near ground level to do so.

BREEDING Nesting takes place during January to May, and the flock of 14 birds found in July 2002 included three juveniles. The first recorded nest contained chicks, and was in a natural tree hollow at a height of 2.4 m above the ground. Martinez (2009) reports that 120 nestboxes were erected in December 2003 at the El Mirador Nature Reserve and nearby ranches, and during 2004 pairs nested in eight of these nestboxes. The use of these nestboxes was monitored during the 2005 breeding season, from January to July, when 11 nests were in nestboxes and a single nest was in a natural tree hollow, and the average duration of the breeding cycle was 86.6 days. In this season, 42 eggs were laid, of which 32 hatched and 25 chicks fledged, egg infertility being the main cause of losses. The average hatching success rate of 74 per cent was lower than that of other neotropical parrots, but the fledging success rate of 91 per cent, the average 2.1 chicks fledged per pair, and successful nesting attempts at 86 per cent were high, due mainly to a low incidence of nest predation. All three chicks in one nest were predated by a small mammal, two chicks in a brood of four chicks starved following the disappearance of one of the parents, and there were losses of some single chicks due to unknown causes.

During nesting studies undertaken in 2003, it was determined that the average clutch size is three eggs, and incubation is undertaken only by the female, but both parents care for the chicks (in Birdlife International 2016).

Plate 58
Indigo-winged Parrot *Hapalopsittaca fuertesi* (adults)

F. KNIGHT '16

REFERENCES CITED

Abbott I, Whitford K (2002) Conservation of vertebrate fauna using hollows in forests of south-western Australia: a strategic risk assessment in relation to ecology, policy planning, and operations management, *Pacific Conservation Biology* **7**, 240–255.

Acosta Hospitaleche CI, Tambussi CP (2006) *Cyanoliseus patagonopsis* nov. sp. (Aves, Psittaciformes) del Pleistoceno de Punta Hermengo, provincia de Buenos Aires, *Ameghiniana* **43**, 249–253.

Adams M, Baverstock PR, Saunders DA, Schodde R, Smith GT (1984) Biochemical systematics of the Australian cockatoos (Psittaciformes: Cacatuinae), *Australian Journal of Zoology* **32**, 363–377.

Althofer GW (1934) Birds of the Wellington district, New South Wales, *Emu* **34**, 105–112.

Amadon D (1942) Birds collected during the Whitney South Sea Expedition: notes on some non-passerine genera, *American Museum Novitates* no. 1176, 1–21.

Ameghino F (1891) Enumeracionde las aves fósiles de la Republica Argentina, *Revista Argentina de Historia Natural* **1**, 441–453.

Anderson A, Sand C, Perchey F, Worthy TH (2010) Faunal extinction and human habitation in New Caledonia: initial results and implications of new research at the Pindai Caves, *Journal of Pacific Archaeology* **1**, 89–109.

Andrews FW (1883) Notes on the Night Parrot, *Proceedings of the Royal Society of South Australia* **6**, 29–30.

Angehr GR, Dean R (2010) *The Birds of Panama: A Field Guide*. Cornell University Press, Ithaca.

Anon. (2010) Latest discoveries about the El Oro Parakeet, *Cyanopsitta* no. 95, 18–19.

Anon. (2015) Local government commits to the El Oro Parakeet ecological corridor, *Cyanopsitta* no. 106, 20–21.

Arndt T (1996) *Lexicon of Parrots*. Arndt-Verlag, Bretten.

Arredondo O (1984) Sinopsis de las aves halladas en depósitos fosilíferos pleisto-holocénicos de Cuba, *Instituto de Zoología, Reporte de Investigación* **17**, 1–35.

Atkinson IAE, Merton DV (2006) Habitat of kakapo (*Strigops habroptilus*) in the Esperance Valley, Fiordland, New Zealand, *Notornis* **53**, 37–54.

Audubon JJ (1831) *Ornithological Biography*, vol. I, Adam and Charles Black, Edinburgh.

Baird RF (1985) Avian fossils from Quaternary deposits in 'Green Waterhole Cave', south-eastern Australia, *Records of the Australian Museum* **37**, 353–370.

Baird RF (1986) Historical records of the Glossy Black Cockatoo *Calyptorhynchus lathami* and Red-tailed Black Cockatoo *C. magnificus* in south-eastern Australia, *South Australian Ornithologist* **30**, 38–45.

Baird RF (1991a) Avian fossils from the Quaternary of Australia. In *Vertebrate Palaeontology of Australasia* (Vickers-Rich P, Monaghan JM, Baird RF, Rich TH eds), pp. 809–870. Pioneer Design Studio, Melbourne.

Baird RF (1991b) Holocene avian assemblage from Skull Cave (AU-8), south-western Western Australia, *Records of the Western Australian Museum* **15**, 267–286.

Baird RF (1992) Fossil avian assemblage of pitfall origin from Holocene sediments in Amphitheatre Cave (G-2), south-western Victoria, Australia, *Records of the Australian Museum* **44**, 21–44.

Baird RF (1993) Pleistocene avian fossils from Pyramids Cave (M-89), eastern Victoria, Australia, *Alcheringa* **17**, 383–404.

Baker J (1997) The decline, response to fire, status and management of the Eastern Bristlebird, *Pacific Conservation Biology* **3**, 235–243.

Bangs O, Zappey WR (1905) Birds of the Isle of Pines, *American Naturalist* **39**, 179–215.

Barbour T (1923) The birds of Cuba, *Memoirs of the Nuttall Ornithological Club* **6**, 1–141.

Barnes C, Zillman E (2007) The lost world of the Ant-hill Parrot, *Wingspan* **17**(1), 45.

Barnwell PJ (1948) *Visits and Despatches 1598–1948*. Standard Printing Establishment, Mauritius.

Barré N, Theuerkauf J, Verfaille L, Primot P, Saoumoé (2010) Exponential population increase in the endangered Ouvéa Parakeet (*Eunymphicus uvaeensis*) after community-based protection from nest poaching, *Journal of Ornithology* **151**, 695–701.

Barrett GW, Silcocks AF, Cunningham R, Oliver DL, Weston MA, Baker J (2007) Comparison of atlas data to determine the conservation status of bird species in New South Wales, with an emphasis on woodland-dependent species, *Australian Zoologist* **34**, 37–77.

Basset Hull AF (1910) The birds of Lord Howe and Norfolk Islands, *Proceedings of the Linnaean Society of New South Wales* **34**, 636–693.

Beeton RJ (1977) The impact and management of birds on the Ord River, PhD Thesis, University of New England, Armidale.

Beggs JR, Wilson PR (1987) Energetics of South Island Kaka (*Nestor meridionalis meridionalis*) feeding on the larvae of kanuka longhorn beetles (*Oshrocydus huttoni*), *New Zealand Journal of Ecology* **10**, 143–147.

Beggs JR, Wilson PR (1991) The Kaka *Nestor meridionalis*, a New Zealand parrot endangered by introduced wasps and mammals, *Biological Conservation* **56**, 23–38.

Beissinger SR, Bucher EH (1992) Can parrots be conserved through sustainable harvesting? *Bioscience* **42**, 164–173.

Bell DT, Bell RC, Cousin JA (2010) Winter foraging patterns in the avifauna from south-western Australia with special reference to niche differentiation in the Acanthizidae and Meliphagidae, *Amytornis* **2**, 1–13.

Belton W (1984) Birds of Rio Grande do Sul, Brazil. Part I. Rheidae through Furnariidae, *Bulletin of the American Museum of Natural History* **178**(4), 369–636.

Bendire CE (1895) Life histories of North American birds, *Special Bulletin of the US National Museum* no. 5, 1–508.

Benstead P (2006) OBC in Action: conservation fund. Survey of distribution and status of nests of Yellow-crested Cockatoo *Cacatua sulphurea parvula* in Komodo National Park, *BirdingASIA*, 7–8.

Bent AC (1940) Life histories of North American cuckoos, goatsuckers, hummingbirds and their allies, *Bulletin of the U. S. National Museum* no. 176, 1–506.

Berg KS, Angel R (2006) Seasonal roosts of Red-lored Amazons in Ecuador provide information about population size and structure, *Journal of Field Ornithology* **77**, 95–103.

Berg KS, Horstman E (1996) The Great Green Macaw *Ara ambigua guayaquilensis* in Ecuador: first nest with young, *Cotinga* **4**, 53–54.

Berg KS, Socola J, Angel RR (2007) Great Green Macaws and the annual cycle of their food plants in Ecuador, *Journal of Field Ornithology* **78**, 1–10.

Bergman C (2009) The golden birds of the lost world, *PsittaScene* **21**(1), 5–7.

Berkunsky I, Daniele G, Kacoliris FP, Díaz-Luque JA, Silva Frias CP, Aramburu RM, Gilardi JD (2014) Reproductive parameters in the critically endangered Blue-throated Macaw: limits to the recovery of a parrot under intensive management, *PLoS One* **9**(6), e99941. doi:10.1371/journal.pone.oo99941.

Best BJ, Krabbe N, Clarke CT, Best AL (1995) Red-masked Parakeet *Aratinga erythrogenys* and Grey-cheeked Parakeet *Brotogeris pyrrhopterus*: two threatened parrots from Tumbesian Ecuador and Peru, *Bird Conservation International* **5**, 233–250.

Best HA (1984) The foods of Kakapo on Stewart Island as determined from their feeding sign, *New Zealand Journal of Ecology* **7**, 71–83.

Best HA, Powlesland RG (1985) *Kakapo*. John McIndoe and New Zealand Wildlife Service, Dunedin.

Binford LC (1989) A distributional survey of the birds of the Mexican state of Oaxaca, *Ornithological Monographs* no. 43, i–v, 1–418.

Birdlife International (2001) *Threatened Birds of Asia: The Birdlife International Red Data Book*, Birdlife International, Cambridge.

Birdlife International (2004) *IUCN Red List for Birds*, species factsheets downloaded from http://www.birdlife.org

Birdlife International (2016) *IUCN Red List for Birds*, species factsheets downloaded from http://www.birdlife.org.

Bishop KD (1992) Parrots in Indonesia: a brief review of their status and conservation. In *Issues in the Conservation of Parrots in Australasia and Oceania: Challenges to Conservation Biology, Proceedings of the RAOU/WWF Scientific Day and Workshop,* Sydney, 22–23 September 1990. *RAOU Report No. 83* (Joseph L, ed.), pp. 1–9. Royal Australasian Ornithologists Union, Melbourne.

Blake ER, Hanson HC (1942) Notes on a collection of birds from Michoacán, Mexico, *Publications of the Field Museum of Natural History, Zoology Series* **22**, 513–551.

Blakers M, Davies S JJF, Reilly PN (1984) *The Atlas of Australian Birds*. Melbourne University Press, Melbourne.

Blyth J (1996) Night Parrot (*Pezoporus occidentalis*) interim recovery plan for Western Australia, *CALM Interim Recovery Plan*, no. 4, 1–21.

Bohner F (1984) First breeding of the White-tailed Black Cockatoo, *Bird Keeping in Australia* **27**(2), 17–18.

Boles WE (1993) A new cockatoo (Psittaciformes: Cacatuidae) from the Tertiary of Riversleigh, northwestern Queensland, and an evaluation of rostral characters in the systematics of parrots, *Ibis* **135**, 8–18.

Boles WE (1998) A Budgerigar *Melopsittacus undulatus* from the Pliocene of Riversleigh, northwestern Queensland, *Emu* **98**, 32–35.

Boles WE, Longmore NW, Thompson MC (1994) A recent specimen of the Night Parrot *Geopsittacus occidentalis, Emu* **94**, 37–40.

Bond J (1956) *Check-list of the Birds of the West Indies*, 4th edition, Academy of Natural Sciences of Philadelphia, Philadelphia.

Bond J (1978) Derivations and continental affinities of Antillean birds. In *Zoogeography in the Caribbean* (Gill FB, ed.), Special Publications 13, Academy of Natural Sciences of Philadelphia, Philadelphia.

Bondin A, Comer S, Burbidge A (2011) Conserving Kyloring; saving the Western Ground Parrot from extinction, *Wingspan* **21**(2), 18–20.

Boon W-M, Kearvell JC, Daugherty CH, Chambers GK (2000) Molecular systematics of New Zealand *Cyanoramphus* parakeets: conservation of Orange-fronted and Forbes' Parakeets, *Bird Conservation International* **10**, 211–239.

Boon W-M, Kearvell JC, Daugherty CH, Chambers GK (2001) Molecular systematics and conservation of kakariki (*Cyanoramphus* spp.), *Science for Conservation* **176**, 1–46.

Boon W-M, Robinet O, Rawlence N, Bretagnolie V, Norman JA, Christidis L, Chambers GK (2008) Morphological, behavioural and genetic differentiation within the Horned Parakeet (*Eunymphicus cornutus*) and its affinities to *Cyanoramphus* and *Prosopeia*, *Emu* **108**, 251–260.

Botero-Delgadillo E, Páez CA, Bayly N (2012a) biogeography and conservation of Andean and trans-Andean populations of *Pyrrhura* parakeets in Colombia: modelling geographic distributions to identify independent conservation units, *Bird Conservation International* **22**, 445–461.

Botero-Delgadillo E, Páez CA, Sanabria-Mejía J (2012b) Discovery of two new localities for Todd's Parakeet *Pyrrhura picta caeruleiceps* using distribution models: enhancing knowledge of a little known neotropical bird, *Ardeola* **59**, 237–252.

Botero-Delgadillo E, Páez CA, Sanabria-Mejía J, Bayly NJ (2013) Insights into the natural history of Todd's Parakeet *Pyrrhura picta caeruleiceps* in north-eastern Colombia, *Ardeola* **60,** 377–383.

Boussekey M (1995) Conservation of the Red-vented Cockatoo, *PsittaScene* **7**(4), 7.

Boussekey M, Saint-Pie J, Morvan O (1991) Observations on a population of Red-fronted Macaws *Ara rubrogenys* in the Río Caine valley, central Bolivia, *Bird Conservation International* **1**, 335–350.

Bowler J, Taylor J (1989) An annotated checklist of the birds of Manusela National Park, Seram (Birds recorded on the Operation Raleigh Expedition), *Treubia* **4**, 3–29.

Boyer AG, James HF, Olson SL, Grant-Mackie JA (2010) Long-term ecological change in a conservation hot-spot: the fossil avifauna of Mé Auré Cave, New Caledonia, *Biodiversity and Conservation* **19**, 3207–3224.

Bregulla H (1993) Die papageien Neukaledoniens, *Gefiederte Welt* **117**, 310–313, 349–351, 381–385, 412–416.

Brewster W (1889) Nesting habits of the parrakeet (*Conurus carolinensis*), *Auk* **6**, 336–337.

Brodkorb P (1959) Pleistocene birds from New Providence Island, Bahamas, *Bulletin of the Florida State Museum, Biological Sciences* **4**, 349–371.

Brodkorb P (1971) Catalogue of fossil birds. Part 4: Columbiformes through Piciformes, *Bulletin of the Florida State Museum, Biological Sciences* **15**, 163–266.

Brodkorb P (1985) Preliminary report on the Plio/Pleistocene birds of east Africa. In *Proceedings of the 18th International Ornithological Congress*, Moscow, 16–24 August 1982 (Ilyichev VD, Gavrilov VM, eds), pp. 174–177. Nauka, Moscow.

Brooks T, Dutson G, King B, Magsalay PM (1995) An annotated check-list of the forest birds of Rajah Sikatuna National Park, Philippines, *Forktail* **11**, 121–134.

Brown DM, Toft CA (1999) Molecular systematics and biogeography of the cockatoos (Psittaciformes: Cacatuidae), *Auk* **116**, 141–157.

Brown PB (1979) The status of parrot species in western Tasmania, *Tasmanian Bird Report* no. 9, 4–11.

Brown PB (1989) *The Swift Parrot* Lathamus discolor: *A Report on Its Ecology, Distribution and Status, Including Management Considerations*. Tasmanian Department of Lands, Parks and Wildlife, Hobart.

Brown PB, Wilson RI (1980) A survey of the Orange-bellied Parrot *Neophema chrysogaster* in Tasmania, *Report to the World Wildlife Fund (Australia)*, Tasmanian National Parks and Wildlife Service, Hobart.

Brown PB, Wilson RI (1984) *The Orange-bellied Parrot Recovery Plan*. Tasmanian National Parks and Wildlife Service, Hobart.

Bruner PL (1972) *Field Guide to the Birds of French Polynesia*, Pacific Scientific Information Center, Bernice P. Bishop Museum, Honolulu.

Brush T (2005) *Nesting Birds of a Tropical Frontier: The Lower Rio Grande Valley of Texas,* Texas A & M University, College Station.

Bryant S (1994) Habitat and potential diet of the Ground Parrot in Tasmania, *Emu* **94**, 166–171.

Bucher E (1992) Neotropical parrots as agricultural pests. In *New World Parrots in Crisis: Solutions from Conservation Biology* (Beissinger SR, Snyder NFR, eds), pp. 201–219, Smithsonian Institution, Washington.

Buffon GLL, Comte de (1770–1783) *Histoire naturelle des oiseaux*. Imprimerie Royale, Paris [3 original editions with variable numbers of volumes, published over approximately the same dates – one in 10 volumes includes the *Planches Eluminées* by François-Nicolas Martinet].

Buller WL (1888) *A History of the Birds of New Zealand*, 2nd edition, The author, London.

Burbidge AH, Rolfe J, McNee S, Newbey B, Williams M (2007) Monitoring population change in the cryptic and threatened Western Ground Parrot in relation to fire, *Emu* **107**, 79–88.

Butchart SHM, Brooks TM, Davies CWN, Dharmaputra G, Dutson GCL, Lowen JC, Sahu A (1996) The conservation status of forest birds on Flores and Sumbawa, Indonesia, *Bird Conservation International* **6**, 335–370.

Butler D (1989) *Quest for the Kakapo*. Heinemann Reed, Auckland.

Cahill AJ, Walker JS, Marsden SJ (2006) Recovery within a population of the critically endangered Citron-crested Cockatoo *Cacatua sulphurea citrinocristata* in Indonesia after 10 years of international trade control, *Oryx* **40**, 161–167.

Cain AJ (1955) A revision of *Trichoglossus haematodus* and of the Australian platycercine parrots, *Ibis* **97**, 432–479.

Campbell AJ (1901) *Nests and Eggs of Australian Birds*, The author, Sheffield.

Campbell KE Jr (1976) The Late Pleistocene avifauna of La Carolina, southwestern Ecuador, *Smithsonian Contributions to Paleobiology* **27**, 155–168.

Campbell KE Jr (1979) The non-passerine Pleistocene avifauna of the Talara Tar Seeps, northwestern Peru, *Royal Ontario Museum Life Science Contribution* **118**, 1–203.

Cantu JC, Sanchez ME, Grosslet M, Silva J (2007), *The Illegal Parrot Trade in Mexico: A Comprehensive Assessment*, Defenders of Wildlife, Teyeliz.

Carlson LA (1999) Aftermath of a feast: human colonization of the southern Bahamian archipelago and its effects on the indigenous fauna. PhD dissertation, Department of Anthropology, University of Florida, Gainesville.

Carriker MA Jr (1910) An annotated list of the birds of Costa Rica including Cocos Island, *Annals of the Carnegie Museum* **6**, 314–970.

Carril J, Degrange FJ, Tambussi CP (2014) Jaw-muscle reconstruction of the Late Pliocene psittaciform *Nandayus vorohuensis* from Argentina, *Ameghiniana* **51**, 361–365.

Cattoi N (1957) Contribucion al estudio de la avifauna extinguida del Pleistoceno de la Republica Argentina, *Ameghiniana* **1**, 17–24.

Cayley NW (1938) *Australian Parrots: Their Habits in Field and Aviary*. Angus and Robertson, Sydney.

Chan K, Glover DR, Ramage CM, Harrison DK (2008) Low genetic diversity in the Ground Parrot (*Pezoporus wallicus*) revealed by randomly amplified DNA fingerprinting, *Annales Zoologici Fennici*, **45**, 211–216.

Chapman FM (1904) *Handbook of Birds of Eastern North America*, 6th edition, Appleton and Company, New York.

Chapman FM (1917) The distribution of bird-life in Colombia: a contribution to a biological survey of South America, *Bulletin of the American Museum of Natural History* **36**, 1–729.

Chapman P (1990) Realising a lifelong ambition: rare Golden-shouldered Parrots observed and filmed in their habitat, *Birds International* **2**, 23–33.

Charles KE (2012) Tree damage in Wellington as a result of foraging for sap and bark-dwelling invertebrates by the North Island Kaka (*Nestor meridionalis septentrionalis*), *Notornis* **59**, 171–175.

Charles KE, Linklater WL (2014) Selection of trees for sap-feeding by a native New Zealand parrot, the Kaka (*Nestor meridionalis*), in an urban landscape, *Emu* **114**, 317–325.

Chassot O, Arias GM (2002) Great Green Macaw: flagship species of Costa Rica, *PsittaScene* no. 53, 6–7.

Cheke A, Hume J (2008) *Lost Land of the Dodo*. Yale University Press, New Haven.

Cheneval J (2000) L'avifaune de Sansan, *Mémoires du Muséum national d'histoire naturelle*, **183**, 321–388.

Child P (1981) Birdlife of Aitutaki, Cook Islands, *Notornis* **28**, 29–34.

Chisholm AH (1922) The 'lost' Paradise Parrot, *Emu* **22**, 4–17.

Chisholm AH (1924) Seeking rare parrots, *Emu* **24**, 25–32.

Chisholm AH (1929) *Birds and Green Places: A Book of Australian Nature Gossip*. JM Dent and Sons, London.

Christiansen MB, Pitter E (1993a) Aspects of breeding behaviour of Red-fronted Macaws, *Ara rubrogenys*, in the wild, *Gerfaut* **82–83**, 51–61.
Christiansen MB, Pitter E (1993b) Aspects of behaviour of juvenile Red-fronted Macaws, *Ara rubrogenys*, in the wild, *Gerfaut* **82–83**, 63–69.
Christidis L, Schodde R, Shaw DD, Maynes SF (1991) Relationships among the Australo-Papuan parrots, lorikeets and cockatoos (Aves: Psittaciformes): protein evidence, *Condor* **93**, 302–317.
Clark AH (1905a) The Lesser Antillean macaws, *Auk* **22**, 266–273.
Clark AH (1905b) The genus *Conurus* in the West Indies, *Auk* **22**, 310–312.
Clark AH (1905c) The West Indian parrots, *Auk* **22**, 338–344.
Clark AH (1905d) The greater Antillean macaws, *Auk* **22**, 345–348.
Clout MN (2006) A celebration of kakapo: progress in the conservation of an enigmatic parrot, *Notornis* **53**, 1–2.
Clout MN, Craig JL (1995) The conservation of critically endangered flightless birds in New Zealand, *Ibis* **137**, S181–S190.
Coates BJ, Bishop KD (1997) *A Guide to the Birds of Wallacea: Sulawesi, the Moluccas and Lesser Sunda Islands, Indonesia*. Dove Publications, Alderley.
Cockle K, Capuzzi G, Bodrati A, Clay R, del Castillo H, Velázquez M, Areta JI, Fariña N, Fariña R (2007) Distribution, abundance, and conservation of Vinaceous Amazons (*Amazona vinacea*) in Argentina and Paraguay, *Journal of Field Ornithology* **78**, 21–39.
Cockrem JF (2006) The timing of breeding in the kakapo (*Strigops habroptilus*), *Notornis* **53**, 153–159.
Coetzer WG, Downs CT, Perrin MR, Willows-Munro S (2015) Molecular systematics of the Cape Parrot (*Poicephalus robustus*): implications for taxonomy and conservation, *PLos ONE* **10**(8), e0133376. doi:10.1371/journal.pone.0133376.
Collar NJ (1997) Family Psittacidae (Parrots). In *Handbook of the Birds of the World*, vol. 4: *Sandgrouse to Cuckoos* (del Hoyo J, Elliott A, Sargatal J, eds). Lynx Edicions, Barcelona.
Collar NJ, Marsden SJ (2014) The subspecies of Yellow-crested Cockatoo *Cacatua sulphurea*, *Forktail* **30**, 23–27.
Comer S, Berryman A, Burbidge A (2009) Turning down the heat: the challenges of managing the critically endangered Western Ground Parrot in a wildfire prone environment, *Wingspan* **19**(1), 10–13.
Condon HT (1942) The Ground Parrot in South Australia, *South Australian Ornithologist* **16**, 4–7.
Cooper JH, Armitage PL (2013) A parrot of the Caribbean? A remarkable find from a 17th-century Spanish shipwreck, *Bulletin of the British Ornithologists Club* **133**, 52–58.
Cooper R, McAllan I, Curtis B (2016) *An Atlas of the Birds of NSW and the ACT*, vol. 2, NSW Atlassers Inc., Woolgoolga.
Cottam C, Knappen P (1939) Food of some uncommon North American birds, *Auk* **56**, 138–169.
Cottam Y, Merton DV, Hendricks W (2006) Nutrient composition of the diet of parent-raised kakapo nestlings, *Notornis* **53**, 90–99.
Cottee-Jones HEW, Mittermeier JC, Purba EC, Ashuri NM, Hesdianti E (2014) An assessment of the parrot trade on Obi Island (Northern Moluccas, Indonesia) reveals heavy exploitation of the vulnerable Chattering Lory (*Lorius garrulus*), *Kukila* **18**, 1–9.
Courtney J (1996) The juvenile food-begging calls, food-swallowing vocalisation and begging postures in Australian cockatoos, *Australian Bird Watcher* **16**, 236–249.
Courtney J (1997) Age-related colour changes and behaviour in the Ground Parrot *Pezoporus wallicus*, *Australian Birdwatcher* **17**, 185–191.
Cracraft, J (1973) Systematics and evolution of the Gruiformes (Class Aves). 3 Phylogeny of the suborder Grues, *Bulletin of the American Museum of Natural History* **151**, 1–127.
Cracraft J (2001) Avian evolution, Gondwana biogeography and the Cretaceous-Tertiary mass extinction event, *Proceedings of the Royal Society of London B. Biological Sciences* **268**, 459–469.
Creel D, McKusick C (1994) Prehistoric macaws and parrots in the Mimbres area, New Mexico, *American Antiquity* **59**, 510–524.
Curtis LK, Dennis AJ, McDonald KR, Kyne PM, Debus SJS (eds) (2012) *Queensland's Threatened Animals*, CSIRO Publishing, Collingwood.
D'Aubenton EL (1765–1780) *Planches eluminées d'histoire naturelle*, 1008 plates, Paris.
Davis RA, Metcalf BM (2008) The Night Parrot (*Pezoporus occidentalis*) in northern Western Australia: a recent sighting from the Pilbara region, *Emu* **108**, 233–236.
Debus S (ed.) (2013) Historical article – Recollections of the last known sightings of the Paradise Parrot, by Eric Zillman, *Australian Field Ornithology* **30**, 166–168.
Delgado FS (1985) A new subspecies of the Painted Parakeet (*Pyrrhura picta*) from Panama, *AOU Ornithological Monographs* no. 36, 16–20.
del Hoyo J, Collar NJ (2014) *HBW and Birdlife International Illustrated Checklist of the Birds of the World*, vol. I: Non-passerines. Lynx Edicions, Barcelona.
Derrickson SR, Snyder NFR (1992) Potentials and limits of captive breeding in parrot conservation. In *New World Parrots in Crisis: Solutions from Conservation Biology* (Beissinger SR, Snyder NFR, eds), pp. 133–163, Smithsonian Institution, Washington.
Desenne P, Strahl SD (1991) Trade and the conservation status of the family Psittacidae in Venezuela, *Bird Conservation International* **1**, 153–169.
Dickinson EC, Remsen Jr VJ (eds) (2013) *The Howard and Moore Complete Checklist of the Birds of the World*, vol. 1: Non-passerines. Aves Press, Eastbourne.
Diehl LM (2009) Suns rise in Brazil, *PsittaScene* **21**(1), 8–9.
Dilger WC (1960) The comparative ethology of the African parrot genus *Agapornis*, *Zeitschrift für Tierpsychologie* **17**, 649–685.
Downs C, Hart L, Letsoalo D, Moya K, Symes C, Boyes RS, Coetzer R (2013) Cape Parrot: A South African endemic, *PsittaScene*, Winter 2013, 20–21.
Dubois S (1674) *Les voyages fait par le sieur D. B. aux isles Dauphine ou Madagascar et Bourbon ou Mascarenne es annés 1669, 70, 71 et 72*. Claude Barbin, Paris.
Duffield GE, Hesse AJ (1997) Ecology and conservation of the Blue-throated Macaw, *PsittaScene* **9**(2), 10–11.
duPont JE, Rabor DS (1973a) Birds of Dinagat and Siargao, Philippines: an expedition report, *Nemouria* no. 10, 1–11.
duPont JE, Rabor DS (1973b) South Sulu Archipelago birds, *Nemouria* no. 9, 1–63.
Dutson G (2011) *Birds of Melanesia: Bismarcks, Solomons, Vanuatu, and New Caledonia*, Princeton University Press, Princeton.
Dyck J (1971) Structure and spectral reflectance of green and blue feathers of the Rose-faced Lovebird (*Agapornis roseicollis*), *Det Kongelige Videnskabernes Selskab Biologiske Skrifter* **18**(2), 1–66.

Dyke GJ (2001) Fossil pseudasturid birds (Aves, Pseudasturidae) from the London Clay, *Bulletin of the Natural History Museum. Geology Series* **57**, 1–4.

Dyke GJ, Cooper JH (2000) A new psittaciform bird from the London Clay (Lower Eocene) of England, *Palaeontology* **43**, 271–285.

Dyke GJ, Mayr G (1999) Did parrots exist in the Cretaceous period? *Nature* **399**, 317–318.

Eason DK, Elliott GP, Merton DV, Jansen PW, Harper GA, Moorhouse RJ (2006) Breeding biology of kakapo (*Strigops habroptilus*) on offshore island sanctuaries, 1990–2002, *Notornis* **53**, 27–36.

Eaton JA, Shepherd CR, Rheindt FE, Harris JBC, van Balen S, Wilcove DS, Collar NJ (2015) Trade-driven extinctions of avian taxa in Sundaic Indonesia, *Forktail* **31**, 1–12.

Eckert J (1990) Orange-bellied Parrots feeding on a cultivated crop, *South Australian Ornithologist* **31**, 16–17.

Edwards HV (1921) A south-coastal Selborne, *Emu* **21**, 42–48.

Ehmke G, Tzaros C, Menkhorst P, Holdsworth M (2008) Trumped-up corella indeed! *Wingspan* **18**(3), 18–23.

Eisermann K (2003) Status and conservation of Yellow-headed Parrot *Amazona oratrix "guatemalensis"* on the Atlantic coast of Guatemala, *Bird Conservation International* **13**, 361–366.

Ekstrom JMM (2013) Greater Vasa Parrot *Coracopsis vasa* Perroquet vaza, Lesser Vasa Parrot *Coracopsis nigra* Perroquet noir. In *The Birds of Africa*, vol. VIII: *The Malagasy Region* (Safford RJ, Hawkins AFA, eds), pp. 525–528, 531–534, Christopher Helm, London.

Ekstrom JMM, Jones JPG, Willis J, Tobias J, Dutson G, Barré N (2002) New information on the distribution, status and conservation of terrestrial bird species in Grande Terre, New Caledonia, *Emu* **102**, 197–207.

Elliott GP (2006) A simulation of the future of kakapo, *Notornis* **53**, 164–172.

Elliott GP, Eason DK, Janson PW, Merton DV, Harper GA, Moorhouse RJ (2006) Productivity of kakapo (*Strigops habroptilus*) on offshore islands, 1990–2002, *Notornis* **53**, 138–142.

Emison WB, Beardsell CM, Norman FI, Loyn RH, Bennett SC (1987) *Atlas of Victorian Birds*, Department of Conservation, Forests and Lands, Melbourne.

Enkerlin-Hoeflich EC, Hogan KM (1997) Red-crowned Parrot (*Amazona viridigenalis*). In *The Birds of North America* (Rodewald PG, ed.), Cornell Laboratory of Ornithology, Ithaca. doi:10.2173/bna.292.

Española CP, Collar NJ, Marsden SJ (2013) Are populations of large-bodied avian frugivores on Luzon, Philippines, facing imminent collapse? *Animal Conservation* **16**, 467–479.

Evans TD, Magsalay P, Dutson GCL, Brooks TM (1993) The conservation status of the forest birds of Siquijor, Philippines, *Forktail* **8**, 89–96.

Falla RA (1937) *B.A.N.Z. Antarctic Research Expedition 1929–1931, Reports – Series B, Volume II, Birds*. B.A.N.Z.A.R. Expedition Committee, Adelaide.

Farnes RF (2007) *Birds of Portland District Victoria*. Portland Field Naturalists Club Inc., Portland.

Farrimond M, Elliott GP, Clout MN (2006a) Growth and fledging of Kakapo, *Notornis* **53**, 112–115.

Farrimond M, Clout MN, Elliott GP (2006b) Home range size of kakapo (*Strigops habroptilus*) on Codfish Island, *Notornis* **53**, 150–152.

Feduccia A (1975) Professor Hibbard's fossil birds, *University of Michigan Museum. Paleontology. Papers on Paleontology* **12**, 67–70.

Feduccia A, Martin LD (1976) The Eocene zygodactyl birds of North America (Aves: Piciformes). In *Collected Papers in Avian Paleontology Honoring the 90th Birthday of Alexander Wetmore* (Olson SL, ed.), pp. 101–110, *Smithsonian Contributions to Paleobiology* **27**.

Fjeldsa J, Krabbe N, Ridgely RS (1987) Great Green Macaw *Ara ambigua* collected in northwest Ecuador, with taxonomic comments on *Ara militaris*, *Bulletin of the British Ornithologists Union* **107**, 28–31.

Flannery T, Schouten P (2001) *A Gap in Nature: Discovering the World's Extinct Animals*. Text Publishing, Melbourne.

Fleming CA (1939) Birds of the Chatham Islands, *Emu* **38**, 380–413, 492–509.

Fletcher JA (1908) Bird notes from Cleveland, Tasmania, *Emu* **8**, 79–80.

Ford HA (1990) Research on Birds of crops in northern New South Wales, *Proceedings of the National Birds Pest Workshop*, 17–22.

Forshaw JM (1979) Eggs of the Double-eyed Fig Parrot, *Australian Bird Watcher* **8**, 98–99.

Forshaw JM (2006) *Parrots of the World: An Identification Guide*, Princeton University Press, Princeton.

Forshaw JM, Cooper WT (1989) *Parrots of the World*, 3rd (revised) edition, Lansdowne Editions, Willoughby.

Forshaw JM, Cooper WT (2016) *A Natural History of Australian Parrots: A Tribute to William T Cooper*, Nokomis Editions, Buxton.

Forshaw JM, Fullagar PJ, Harris JI (1976) Specimens of the Night Parrot in museums throughout the world, *Emu* **76**, 120–126.

Fortescue M, Ward R, Davidson RP (1999) *Draft Recovery plan for the Norfolk Island Green Parrot* Cyanoramphus novaezelandiae cookii *2000–2004*, Environment Australia, Canberra.

Froke JB (1981) Population, movements, foraging and nesting of feral *Amazona* parrots in southern California, M.S. Thesis, Humboldt State University, Arcata, CA.

Gala M, Lenoble A (2015) Evidence of the former existence of an endemic macaw in Guadeloupe, Lesser Antilles, *Journal of Ornithology* **156**, 1061–1066.

Gale RF (1983) From pets to parents, *AFA Watchbird* **19**(5), 19–22.

Galetti M (1997) Seasonal abundance and feeding ecology of parrots and parakeets in a lowland Atlantic forest of Brazil, *Ararajuba* **5**, 115–126.

Gallardo RJ (2014) *Guide to the Birds of Honduras*, Mountain-Gem Bird Tours, Tegucigalpa.

Garnett ST (1999) Cockatoos and peanuts at Lakeland, *PsittaScene* **11**(3), 14–15.

Garnett ST, Crowley G (1994) The plight of the parrot with gold on its wings, *Wingspan* no. 16, 21–23.

Garnett ST, Crowley G (1997) Golden-shouldered Parrot *Psephotus chrysopterygius*, *PsittaScene* **9**(2), 4–5.

Garnett ST, Crowley G (2000) *The Action Plan for Australian Birds 2000*, Environment Australia, Canberra.

Garnett ST, Crowley G, Duncan R, Baker N, Doherty P (1993) Notes on live Night Parrot sightings in north-western Queensland, *Emu* **93**, 292–296.

Garnett ST, Szabo JK, Dutson G (2011) *The Action Plan for Australian Birds 2010*, CSIRO Publishing, Collingwood.

Garrett KL (1997) Population, status and distribution of naturalized parrots in southern California, *Western Birds* **28**, 181–195.

Garrett KL, Mabb KT, Collins CT, Kares LM (1997) Food items of naturalized parrots in southern California, *Western Birds* **28**, 196–201.

Gartrell BD, Jones SM (2012) No evidence for synchrony of physiological or behavioural preparations for migration in a short-distance migratory parrot, *Emu* **112**, 1–8.

Gatter W (1997) *Birds of Liberia*, Pica Press, Mountfield.

Gilardi J (2001) Good news from the land of Lear's, *PsittaScene* **13**(3), 2–4.

Gosper DG (1995) Notes on the activities of Ground Parrots in northern NSW, *Australian Birds* **28**, 57–64.

Gosse PH (1847) *The Birds of Jamaica*, John Van Voorst, London.

Gorman ML (1975) Habitats of the land-birds of Viti Levu, Fiji Islands, *Ibis* **117**, 152–161.

Gould J (1865) *Handbook to the Birds of Australia*, vol. 2, The author, London.

Graves GR (1992) The endemic land birds of Henderson Island, southeastern Polynesia: notes on natural history and conservation, *Wilson Bulletin* **104**, 32–43.

Green RH (1969) The birds of Flinders Island, with references to other eastern Bass Strait islands and annotated lists of other vertebrate fauna, *Records of the Queen Victoria Museum* no. 34, 1–32.

Green RH (1984) The vegetation, fauna and archaeology of Ordnance Point, north-western Tasmania, *Records of the Queen Victoria Museum* no. 84, 1–67.

Greenway JCJr (1967) *Extinct and Vanishing Birds of the World*, 2nd revised edition, Dover Publications, New York.

Grimmett R, Inskipp C, Inskipp T (1999) *A Guide to the Birds of India, Pakistan, Nepal, Bangladesh, Bhutan, Sri Lanka and the Maldives*, Princeton University Press, Princeton.

Groom L (2012) *A Steady Hand: Governor Hunter & His First Fleet Sketchbook*, National Library of Australia, Canberra.

Groombridge JJ, Jones CG, Nichols RA, Carlton M, Bruford MW (2004) Molecular phylogeny and morphological change in the *Psittacula* parakeets, *Molecular Phylogenetics and Evolution* **31**, 96–108.

Grouard S (2002) Subsistence et mode de vie des premiers habitants de Guadeloupe (500 av – 1500 ap J.-C) *Préhistoires Méditerranéennes* **10–11**, 191–213.

Guerra JE, Cruz-Nieto J, Ortiz-Maciel SG, Wright TF (2008) Limited geographic variation in the vocalizations of the endangered Thick-billed Parrot: implications for conservation strategies, *Condor* **110**, 639–647.

Güivas BR (2015) Hidden lives: an intimate glimpse into the behavior of nesting Puerto Rican Amazons, *PsittaScene* winter 2015, 9–13.

Gwynne JA, Ridgely RS, Tudor G, Argel M (2010) *Wildlife Conservation Society Birds of Brazil: The Pantanal & Cerrado of Central Brazil*, Cornell University Press, Ithaca.

Gynther I (2013) Coxen's Fig-Parrot, *Australian Birdlife* **2**(4), 7.

Hachisuka M (1934) *The Birds of the Philippine Islands, With Notes on the Mammal Fauna*, Part 3, Witherby, London.

Hachisuka M (1953) *The Dodo and Kindred Birds, or the Extinct Birds of the Mascarene Islands*, Witherby, London.

Halse SA (1986) Parrot damage in apple orchards in south-western Australia – a review, *CALM Technical Report*, 1–75.

Hamilton N (1996) Notes on the captive breeding of the Long-billed White-tailed Black Cockatoo, *Australian Birdkeeper* **9**, 83–85.

Hampe H (1939) Brotogerys parrakeets, *Avicultural Magazine*, 5th series, **4**, 402–406.

Hansen W (1984) In search of the Tres Marias Double Yellow-headed Parrot, *AFA Watchbird* **11**(1), 14–23.

Hargrave LL (1939) Bird bones from abandoned Indian dwellings in Arizona and Utah, *Condor* **41**, 206–210.

Hargrave LL (1970) Mexican macaws: comparative osteology and survey of remains from the southwest, *Anthropological Papers of the University of Arizona* no. 20, 1–67.

Harper GA, Elliott GP, Eason DK, Moorhouse RJ (2006) What triggers nesting of kakapo (*Strigops habroptilus*)? *Notornis* **53**, 160–163.

Harrison CJO (1982) The earliest parrot: a new species from the British Eocene, *Ibis* **124**, 203–210.

Harrison CJO, Holyoak DT (1970) Apparently undescribed parrot eggs in the collection of the British Museum (Natural History), *Bulletin of the British Ornithologists Club* **90**, 42–46.

Harrison CJO, Walker CA (1972) The affinities of *Halcyornis* from the Lower Eocene, *Bulletin of the British Museum (Natural History), Geology* **21**, 153–170.

Harrison CJO, Walker CA (1977) Birds of the Lower British Eocene, *Tertiary Research, Special Papers* **3**, 1–52.

Hartley R (1998) Request for funding to protect newly discovered populations of Lear's Macaw and exploratory missions to find other unknown populations, *PsittaScene* **10**(4), 5.

Haverschmidt F (1968) *Birds of Surinam*, Oliver and Boyd, Edinburgh.

Heather B, Robertson H (2015) *The Field Guide to the Birds of New Zealand*, revised edition, Penguin Random House New Zealand, Wellington.

Heinsohn R, Webb M, Lacy R, Terauds A, Alderman R, Stojanovic D (2015) A severe predator-induced population decline predicted for endangered, migratory Swift Parrots (*Lathamus discolor*), *Biological Conservation* **186**, 75–82.

Heizmann EPJ, Hesse A (1995) Die mittelpleistozänen Vogel- und Säugetierefaunen des Nöordlinger Ries (MN 6) und des Steiheimer Beckens (MN 7) – ein Vergleich. In *Acta Palaeornithologica* (Peters S, ed.), *Courier Forschungsinstitut Senckenberg* **181**, 171–185.

Heptonstall R (2011) Invasions and reintroductions, *PsittaScene* **23**(1), 9.

Herrera M, Hennessey B (2007) Quantifying the illegal parrot trade in Santa Cruz de la Sierra, Bolivia, with emphasis on endangered species, *Bird Conservation International* **17**, 295–300.

Hesse AJ, Duffield GE (2000) The status and conservation of the Blue-throated Macaw *Ara glaucogularis*, *Bird Conservation International* **10**, 255–275.

Hicks J, Greenwood D (1989) Rescuing Norfolk Island's parrot, *Birds International* **1**(4), 35–47.

Higgins PJ (ed.) (1999) *Handbook of Australian, New Zealand and Antarctic Birds*, vol. 4, Oxford University Press, Melbourne.

Hilty SL (2003) *Birds of Venezuela*. Christopher Helm, London.

Hilty SL, Brown WL (1986) *A Guide to the Birds of Colombia*, Princeton University Press, Princeton.

Hindwood KA (1939) Nectar-feeding birds near Sydney, *Emu* **39**, 40–44.

Hindwood KA, Sharland M (1964) The Swift Parrot, *Emu* **63**, 310–326.

Hindwood KA, McGill AR (1958) *The Birds of Sydney (County of Cumberland)*, Royal Zoological Society of New South Wales, Sydney.

Hinsby KB (1947) The Orange-bellied Parakeet, *Emu* **47**, 67–68.

Holdaway RN, Worthy TH (1993) First North Island fossil record of Kea and morphological morphometric comparison of Kea and Kaka, *Notornis* **40**, 95–108.

Holdsworth M, Dettman B, Baker GB (2011) Survival in the Orange-bellied Parrot (*Neophema chrysogaster*), *Emu* **111**, 222–228.

Holmes G (1990) The biology and ecology of Coxen's Fig Parrot, *RAOU Report*, no. 65, p. 34.

Holyoak DT (1970) The relation of the parrot genus *Opopsitta* to *Psittaculirostris*, *Emu* **70**, 198.

Holyoak DT (1971) The supposed fossil of *Conurus* from the Pleistocene of Buenos Aires, *Ibis* **168**, 171.

Holyoak DT (1973a) Polynesian land birds face H-bombs and malaria, *New Scientist* **57**(832), 288–290.

Holyoak DT (1973b) An undescribed parrot from Mauritius, *Ibis* **115**, 417–418.

Holyoak DT (1974) *Cyanoramphus malherbi*, is it a colour morph of *C. auriceps*? *Bulletin of the British Ornithologists Club* **94**, 4–9.

Holyoak DT (1975) Les oiseaux des Iles Marquises (suite et fin), *Oiseau Revue francaise Ornithologie* **45**, 341–366.

Holyoak DT (1979) Notes on the birds of Viti Levu and Taveuni, Fiji, *Emu* **79**, 7–18.

Holyoak DT, Thibault J-C (1984) Contribution à l'étude des oiseaux de Polynésie orientale, *Memoires du Museum National d'Histoire Naturelle* Serie A: Zoologie **127**, 1–209.

Horne JFM (1987) Vocalisations of Mascarene Island birds. In *Studies of Mascarene Island Birds* (Diamond AW, ed.), pp. 101–150, Cambridge University Press, Cambridge.

Hoskin ES (1991) *The Birds of Sydney, County of Cumberland, New South Wales*, 2nd edition, Surrey Beatty and Sons, Chipping Norton.

Houde P, Olson SL (1988) Small arboreal nonpasserine birds from the Early Tertiary of western North America. In *Acta XIX Congressus Internationalis Ornithologici* (Ouellet H, ed.), pp. 2030–2036, University of Ottawa Press, Ottawa.

Howell SNG, Webb S (1995) *A Guide to the Birds of Mexico and Northern Central America*, Oxford University Press, Oxford.

Huber W (1933) Birds collected in north-eastern Nicaragua in 1922, *Proceedings of the Academy of Natural Sciences of Philadelphia* **84**, 205–249.

Hume JP (2003) The journal of the flagship *Gelderland* – dodo and other birds on Mauritius 1601, *Archives of Natural History* **30**, 13–27.

Hume JP (2007) Reappraisal of the parrots (Aves: Psittacidae) from the Mascarene Islands, with comments on their ecology, morphology, and affinities, *Zootaxa* **1513**, 1–76.

Hume JP (2013) A synopsis of the pre-human avifauna of the Mascarene Islands. In *Paleoornithological Research 2013* (Göhlich UB, Kroh A, eds), pp. 195–237, *Proceedings of the 8th International Meeting, Society of Avian Paleontology and Evolution*. Verlag Naturhistorisches Museum, Wien.

Hume JP, Walters M (2012) *Extinct Birds*, T & AD Poyser, London.

Hume JP, van Grouw H (2014) Colour aberrations in extinct and endangered birds, *Bulletin of the British Ornithologists Club* **134**, 168–193.

Hume JP, Winters R (2015) Captive birds on Dutch Mauritius: bad-tempered parrots, warty pigeons and notes on other native animals, *Historical Biology: An International Journal of Paleobiology*, doi:10.1080/08912963.2015.1036750.

Imansyah MJ, Purwadana D. Jessop T (2008) First records of Wallace's Hanging-parrot *Loriculus flosculus* from Rinca Island, Komodo National Park, Indonesia, *Forktail* 24, 131–132.

Ives NL (1971) Possible nest of Night Parrot in the Pilbara, WA, *Western Australian Naturalist* **12**, 11–12.

Jackson H, Jones CG, Agapow P-M, Tatayah V, Groombridge JJ (2015) Micro-evolutionary diversification among Indian Ocean parrots: temporal and spatial changes in phylogenetic diversity as a consequence of extinction and invasion, *Ibis* **157**, 496–510.

Jamieson BGM, Koehler L, Todd BJ (1995) Spermatazoal ultrastructure in three species of parrots (Aves, Psittaciformes) and its phylogenetic implications, *Anatomical Record* **241**, 461–468.

Janβen J (2010) Zeite Reise zu den fliegenden Edelsteinen der Südsee, Teil 2, *Papageien* **2/2010**, 59–64.

Jarman H (1965) The Orange-breasted Parrot, *Australian Bird Watcher* **2**, 155–167.

Jennings K (2011) Cyclone survival, *PsittaScene* **23**(1), 8.

Jeyarajasingam A, Pearson A (2012) *A Field Guide to the Birds of Peninsular Malaysia and Singapore*, 2nd edition, Oxford University Press, Oxford.

Johnstone RE, Storr GM (1998) *Handbook of Western Australian Birds*, vol. 1: Non-passerines (Emu to Dollarbird), Western Australian Museum, Perth.

Johnstone RE, Kirkby T (2008) Distribution, status, social organisation, movements and conservation of Baudin's Cockatoo (*Calyptorhynchus baudinii*) in south-west Western Australia, *Records of the Western Australian Museum* **25**, 107–118.

Jones CG (1980) Parrot on the way to extinction, *Oryx* **15**, 350–354.

Jones CG (1987) The larger land-birds of Mauritius. In *Studies of Mascarene Islands Birds* (Diamond AW, ed.), pp. 208–300, Cambridge University Press, Cambridge.

Jones CG (1999) Listening to echos and searching for ghosts: parrot conservation on Mauritius, *PsittaScene* **11**, 10–11.

Jones CG (2010) Back from the brink: the Echo Parakeet story, *PsittaScene* **22**, 3–5.

Jones CG, Malham J, Reuleaux A, Richards H, Raisin C, Tollington S, Zuel N, Chowrimootoo A, Tatayah V (2013) Echo Parakeet *Psittacula eques* Perruche de Maurice. In *The Birds of Africa*, vol. VIII: *The Malagasy Region* (Safford RJ, Hawkins AFA, eds), pp. 517–522, Christopher Helm, London.

Jones HL (2004) *Birds of Belize*, Christopher Helm, London.

Jones J, Marginson MA, Murray-Smith S (1970) South West Island, and other investigations in the Kent Group, *Victorian Naturalist* **87**, 344–371.

Jordan OC, Munn CA (1993) First observations of the Blue-throated Macaw in Bolivia, *Wilson Bulletin* **105**, 694–695.

Jorgenson A, Thompson JB (1987) Neotropical parrots imported by the United States, 1981–1985, *TRAFFIC (USA)* **7**, 3–8.

Joseph L (1988) A review of the conservation status of Australian parrots in 1987, *Biological Conservation* **46**, 261–280.

Joseph L (1992) Notes on the distribution and natural history of the Sun Parakeet *Aratinga solstitialis solstitialis*, *Ornitologia Neotropical* **3**, 17–26.

Joseph L (2000) Beginning an end to 63 years of uncertainty: the Neotropical parakeets known as *Pyrrhura picta* and *P. leucotis* comprise more than two species, *Proceedings of the Academy of Natural Sciences of Philadelphia* **150**, 279–292.

Joseph L, Stockwell D (2002) Climatic modeling of the distribution of some *Pyrrhura* parakeets of northwestern South America with notes on their systematics and special reference to *Pyrrhura caeruleiceps* Todd, 1947, *Ornitologia Neotropical* **13**, 1–8.

Joseph L, Toon A, Schirtzinger EE, Wright TF, Schodde R (2012) A revised nomenclature and classification for family-group taxa of parrots (Psittaciformes), *Zootaxa* **3205**, 26–40.
Juniper AT (1991) Last of a kind, *Birds International* **3**(1), 10–16.
Juniper AT (2002) *Spix's Macaw: The Race to Save the World's Rarest Bird*, Washington Square Press, New York.
Juniper AT, Yamashita C (1991) The habitat and status of Spix's Macaw *Cyanopsitta spixii*, *Bird Conservation International* **1**, 1–9.
Juniper AT, Parr M (1998) *Parrots: A Guide to the Parrots of the World*, Pica Press, Mountfield.
Kearvell JC (2002) Nest sites of sympatric Orange-fronted (*Cyanoramphus malherbi*) and Yellow-crowned Parakeets (*C. auriceps*), *Notornis* **49**, 261–263.
Kearvell JC (2013) Interspecific competition for nest sites between Common Starlings (*Sturnus vulgaris*) and Orange-fronted Parakeets (*Cyanoramphus malherbi*) in the Poulter valley, *Notornis* **60**, 188–190.
Kearvell JC, Briskie JV (2003) Similarity of chatter calls between Orange-fronted Parakeet *Cyanoramphus malherbi* and Yellow-crowned Parakeet *C. auriceps*, *Ibis* **145**, 333–336.
Kearvell JC, Steeves TE (2015) Evidence for assortative mating in sympatric populations of orange-fronted (*Cyanoramphus malherbi*) and yellow-crowned (*C. auriceps*) kākāriki, *Notornis* **62**, 71–75.
Kearvell JC, Farley MR (2016) Adult sex ratios in wild orange-fronted parakeet (*Cyanoramphus malherbi*): are there conservation implications? *Notornis* **63**, 26–31.
Kearvell JC, van Hal J (2016) Nest site selection by sympatric Orange-fronted (*Cyanoramphus malherbi*) and Yellow-crowned Parakeets (*C. auriceps*) in *Nothofagus* spp. forest: interspecific competition between two closely related species? *Notornis* **63**, 9–17.
Kearvell JC, Young JR, Grant AD (2002) Comparative ecology of sympatric Orange-fronted Parakeets (*Cyanoramphus malherbi*) and Yellow-crowned Parakeets (*C. auriceps*), South Island, New Zealand, *New Zealand Journal of Ecology* **26**, 139–148.
Kearvell JC, Grant AD, Boon WM (2003) The orange-fronted parakeet (*Cyanoramphus malherbi*) is a distinct species: a review of recent research into its taxonomy and systematic relationship within the genus *Cyanoramphus*, *Notornis* **50**, 27–35.
Kearvell JC, Connor C, Farley M (2014) Field identification of the Orange-fronted Parakeet (*Cyanoramphus malherbi*): pitfalls for the unwary, *Notornis* **61**, 200–204.
Kennedy RS, Gonzales PC, Dickinson EC, Miranda HC Jr, Fisher TH (2000) *A Guide to the Birds of the Philippines*, Oxford University Press, Oxford.
Kepler CB, Kepler AK (1970) Preliminary comparison of bird species diversity and density in Luquillo and Guánica forests. In *A Tropical Rain Forest* (Odim HT, ed.), pp. E183–E186, Division of Technical Information, U.S. Atomic Energy Commission, Washington.
Kershaw JA (1943) Concerning a rare parrot, *Victorian Naturalist* **59**, 196.
Kiernan C (1993) Paradise on earth? *Wingspan* no. 11, 24–25.
Kirchman JJ, Schirtzinger EE, Wright TF (2012) Phylogenetic relationships of the extinct Carolina Parakeet (*Conuropsis carolinensis*) inferred from DNA sequence data, *Auk* **129**, 197–204.
Kirk R, Powlesland R, Cork S (1993) Anatomy of the mandibles, tongue and alimentary tract of kakapo, with some comparative information from kea and kaka, *Notornis* **40**, 55–63.
Kirwan G, Marlow T, Coopmans P (1996) A review of avifaunal records from Mindo, Pichincha prov., north-western Ecuador, *Cotinga* **6**, 47–57.
Klau W, Langdon P (1994) Orange-bellied Parrot *Neophema chrysogaster* at Chinaman's Creek, *South Australian Ornithologist*, **32**, 22–23.
Klauke N, Schaefer HM, Bauer M, Segelbacher G (2016) Limited dispersal and significant fine-scale genetic structure in a tropical montane parrot species, *PLos ONE*, doi:10.1371/journal.pone.0169165.
Koenig CDE (1825) *Icones Fossilium Sectiles: Cenyuria Prima.* G B Sowerby, London.
Krabbe N, Sornoza Molina F (1996) The last Yellow-eared Parrots *Ognorhynchus icterotis* in Ecuador? *Cotinga* **6**, 25–26.
Ksepka DT, Clarke JA (2010) *Primobucco mcgrewi* (Aves: Coracii) from the Eocene Green River Formation: new anatomical data from the earliest constrained record of stem rollers, *Journal of Vertebrate Paleontology* **30**, 215–225.
Ksepka DT, Clarke JA (2012) A new stem parrot from the Green River Formation and the complex evolution of the grasping foot in Pan-Psittaciformes, *Journal of Vertebrate Paleontology* **32**, 395–406.
Ksepka DT, Clarke JA (2015) Phylogenetically vetted and stratigraphically constrained fossil calibrations within Aves, *Palaeontologia Electronica* 18.1.3FC, 1–25.
Ksepka DT, Clarke JA, Grande L (2011) Stem parrots (Aves, Hakcyornithidae) from the Green River Formation and a combined phylogeny of Pan-Psittaciformes, *Journal of Paleontology* **85**, 835–852.
Kuehler C, Lieberman A, Varney A, Unitt P, Sulpice RM, Azua J, Tehevini B (1997) Translocation of Ultramarine Lories *Vini ultramarina* in the Marquesas Islands: Ua Huka to Fatu Hiva, *Bird Conservation International* **7**, 69–79.
Kundu S, Jones CG, Prys-Jones RP, Groombridge JJ (2012) The evolution of the Indian Ocean parrots (Psittaciformes): extinction, adaptive radiation and eustacy, *Molecular Phylogenetics and Evolution* **62**, 296–305.
Kunz B, Abs M (1997) Status and ecology of the Lilacine Amazon in south-west Ecuador, *PsittaScene* **9**(4), 6–7.
Kyle T (2005a) Mountain macaws: an expedition to survey the breeding cliffs of Bolivia's Red-fronts, *PsittaScene* **17**(3), 4–8.
Kyle T (2005b) Gold rush: the search for Golden Conures in western Pará, Brazil, *PsittaScene* **17**(2), 2–4.
Kyle T (2006) Saving Bolivia's Blue-throated Macaw: highlights from the 2005 breeding season, *PsittaScene* **18**(1), 4–8.
Lack D (1976) *Island Biology Illustrated by the Land Birds of Jamaica*, Blackwell, Oxford.
Lambert F (1992) Present status of the Philippine Cockatoo in Palawan, *Species (IUCN, SSC Newsletter)* no. 18, 70.
Lambert F (1993a) Trade, status and management of three parrots in the North Moluccas, Indonesia: White Cockatoo *Cacatua alba*, Chattering Lory *Lorius garrulus* and Violet-eared Lory *Eos squamata*, *Bird Conservation International* **3**, 145–168.
Lambert F (1993b) Some key sites and significant records of birds in the Philippines and Sabah, *Bird Conservation International* **3**, 281–297.
Lambert F (1997) *Field Assessment of the Conservation Status of Red-and-Blue Lory* Eos histrio *in Indonesia,* IUCN Species Survival Commission, Bangkok.

Lambrecht K (1933) *Handbuch der Palaeornithologie*, Gebrüder Borntraeger, Berlin.

Land HC (1970) *Birds of Guatemala*, Livingston Publishing Company, Wynnewood.

Lanning DV (1991) Distribution and breeding biology of the Red-fronted Macaw, *Wilson Bulletin* **103**, 357–365.

Lanning DV, Shiflett JT (1983) Nesting ecology of Thick-billed Parrots, *Condor* **85**, 66–73.

Laranjeiras TO (2008) A golden gathering, *PsittaScene* **20**(2), 10–13.

Laranjeiras TO, Cohn-Haft M (2009) Where is the symbol of Brazilian ornithology? The geographic distribution of the Golden Parakeet (*Guarouba guarouba* – Psittacidae), *Revista Brasileira de Ornitologia* **17**, 1–19.

Lawson PW, Lanning DV (1981) Nesting and status of the Maroon-fronted Parrot. In *Conservation of New World Parrots* (Pasquier RF, ed.), *ICBP Technical Publications*, no. 1, 81–87.

Laycock G (1969) The last parakeet, *Audubon* **71**(2), 21–25.

Lea AM, Gray JT (1935) The food of Australian birds: an analysis of stomach contents, *Emu* **34**, 275–292; **35**, 63–98.

Leach JA (1926) *An Australian Bird Book, With Supplement: A Complete Guide to the Identification of Australian Birds*, 6th edition revised and enlarged. Whitcombe and Tombs, Melbourne.

Lee G (1935) Breeding Kuhl's Lory, *Vini kuhli* (*Psittacula kuhli*), *Aviculture* 3rd series **5**, 47–48.

Legault A, Chartendrault V, Theuerkauf J, Rouys S, Barré N (2011) Large-scale habitat selection by parrots in New Caledonia, *Journal of Ornithology* **152**, 409–419.

Lehmann FC (1957) Contribuciónes al estudio de la fauna de Colombia XII, *Novedades Colombianas* **3**, 101–156.

Lendon AH (1979) *Australian Parrots in Field and Aviary*, revised edition, Angus and Robertson, Sydney.

Lieberman A, McCormack G (2008) An island endemic: reintroduction of the Kuhl's Lory, *PsittaScene* **20**(4), 8–11.

Lin RS, Lee PF (2006) Status of feral populations of exotic cockatoos (Genus *Cacatua*) in Taiwan, *Taiwania* **51**, 188–194.

Lint KC (1952) Breeding of the Finsch's Amazon, *Avicultural Magazine* **58**, 23–24.

Littler FM (1910) *A Handbook of the Birds of Tasmania and Its Dependencies*, The author, Launceston.

Lloyd BD, Powlesland RG (1994) The decline of Kakapo *Strigops habroptilus* and attempts at conservation by translocation, *Biological Conservation* **69**, 75–85.

Loewen E (2015) Back to Brazil: stories of conservation in South America, Part 2, *PsittaScene*, autumn 2015, 10–15.

Long JL (1985) Damage to cultivated fruit by parrots in the south of Western Australia, *Australian Wildlife Research* **12**, 75–80.

López-Lanús B, Lowen JC (1999) Observations of breeding activity in El Oro Parakeet *Pyrrhura orcesi*, *Cotinga* **11**, 46–47.

López-Lanús B, Sócola J (2000) Estudio y conservación del *Ara ambigua* en la Reserva Cerro Blanco, sudeste de Ecuador, *Cotinga* **14**, 24–25.

Louchart A, Wesselman H, Blumenschine RJ, Hlusko LJ, Njau JK, Black MT, Asnake M, White TD (2009) Taphonomic, avian, and small-vertebrate indicator of *Ardipithecus ramidus* habitat, *Science* **326**, 66, 66e1–66e4.

Lousada SA, Howell SNG (1996) Distribution, variation, and conservation of Yellow-headed Parrots in northern Central America, *Cotinga* **5**, 46–53.

Low R (1980) *Parrots: Their Care and Breeding*, Blandford, Poole.

Low R (1985) Breeding the Tahiti Blue Lory, *Avicultural Magazine* **91**, 1–14.

Low R (1987) Status of the Lilacine Amazon Parrot *Amazona autumnalis lilacina*, *Avicultural Magazine* **93**, 223–225.

Low R (1993a) *Hancock House Encyclopedia of the Lories*, Hancock House Publishers, Surrey.

Low R (1993b) *Cockatoos in Aviculture*, Blandford, London.

Lowen JC, Clay RP, Brooks TM, Esquivel EZ, Bartrina L, Barnes R, Butchart SHM, Etcheverry NI (1995) Bird conservation in the Paraguayan Atlantic forest, *Cotinga* **4**, 58–64.

Lowery GHJ Jr, Dalquest WW (1951) Birds from the state of Veracruz, Mexico, *Publications of the University of Kansas Museum of Natural History* **3**, 531–649.

Loyn RH, Lane BA, Chandler C, Carr GW (1986) Ecology of Orange-bellied Parrots *Neophema chrysogaster* at their main remnant wintering site, *Emu* **86**, 195–206.

Loyn RH, Menkhorst P, Middleton D, Holdsworth M (2005) Mainland-bred Orange-bellied Parrots migrate to Tasmania, *Wingspan* **15**(1), 5–7.

Lydekker R (1891) *Catalogue of the Fossil Birds in the British Museum (Natural History)*, Museum Trustees, London.

Mabb KT (1997a) Roosting behavior of naturalized parrots in the San Gabriel Valley, California, *Western Birds* **28**, 202–208.

Mabb KT (1997b) Nesting behavior of *Amazona* parrots and Rose-ringed Parakeets in the San Gabriel Valley, California, *Western Birds* **28**, 209–217.

Mabb KT (2002) Naturalized (wild) parrots in California: a current assessment, 2002, *Home Contact Info Research CV Papers*.

Macgillivray W (1918) Ornithologists in north Queensland. Part III, *Emu* **17**, 180–212.

MacKinnon J (1988) *Field Guide to the Birds of Java and Bali*, Gadjah Mada University Press, Yogyakarta.

MacKinnon J, Phillipps K (2000) *A Field Guide to the Birds of China*, Oxford University Press, Oxford.

MacNally R, Horrocks G (2000) Landscape-scale conservation of an endangered migrant: the Swift Parrot (*Lathamus discolor*) in its winter range, *Biological Conservation* **92**, 21–24.

Manegold A (2013) Two new parrot species (Psittaciformes) from the early Pliocene of Langebaanweg, South Africa, and their palaeoecological implications, *Ibis* **155**, 127–139.

Manegold A, Louchart A, Carrier J, Elzanowski A (2013) The Early Pliocene avifauna of Langebaanweg (South Africa): a review and update. In *Palaeornithological Research 2013* (Göhlich UB, Kroh A eds), pp. 135–152, *Proceedings of the 8th International Meeting, Society of Avian Paleontology and Evolution*, Verlag Naturhistorisches Museum, Wien.

Mann REH (1982) Breeding the Red-topped Amazon Parrot, *Avicultural Magazine* **88**, 12–14.

Mann REH, Mann PD (1978) Breeding Finsch's Amazon Parrot, *Avicultural Magazine* **84**, 187–189.

Marien D, Koopman KF (1955) The relationships of the West Indian species of *Aratinga*, *American Museum Novitates* no. 1712, 14–15.

Marín-Togo MC, Monterrubio-Rico TC, Renton K, Rubio-Rocha Y, Macías-Caballero C, Ortega-Rodríguez JM, Cancíno-Murillo R (2012) Reduced current distribution of Psittacidae on the Mexican Pacific coast: potential impacts of habitat loss and capture for trade, *Biodiversity Conservation* **21**, 451–473.

Marsden SJ, Royle K (2015) Abundance and abundance change in the world's parrots, *Ibis* **157**, 219–229.

Marshall TJ Jr (1957) Birds of pine-oak woodland in southern Arizona and adjacent Mexico, *Pacific Coast Avifauna* no. 32, 1–125.

Martin LD (2010) Paleogene avifauna of the Holarctic. *Vertebrata Palasiatica* **48**, 367–374. [In Chinese.]

Martin PS, Robins CR, Heed WB (1954) Birds and biogeography of the Sierra de Tamaulipas, an isolated pine-oak habitat, *Wilson Bulletin* **66**, 38–57.

Martinez AET (2009) Parámetros reproductivos y anidación de la Cotorra Aliazul (*Hapalopsittaca fuertesi*) en cavidades artificiales, *Ornitologia Neotropical* **20**, 357–368.

Martuscelli P, Yamashita C (1997) Rediscovery of the White-cheeked Parrot *Amazona kawalli* (Grantsau and Camargo 1989), with notes on its ecology, distribution and taxonomy, *Ararajuba* **5**, 97–113.

Mathews GM (1917) *The Birds of Australia*, vol. 6, Witherby, London.

Mathews GM (1948) Eggs of the Spinifex or Night Parrot, *Emu* **48**, 84.

Mawson PR (1995) Observations on nectar feeding by Carnaby's Cockatoo *Calyptorhynchus latirostris*, *Western Australian Naturalist* **20**, 93–96.

Mawson PR, Johnstone RE (1997) Conservation status of parrots and cockatoos in Western Australia, *Eclectus* **2**, 4–9.

Mayr G (1998) A new family of Eocene zygodactyl birds, *Senckenbergiana Lethaea* **78**, 199–209.

Mayr G (2000a) New or previously unrecorded avian taxa from the Middle Eocene of Messel (Hessen, Germany), *Mitteilungen aus dem Museum fürNaturkunde in Berlin, Geowissenschafltiche Reihe* **3**, 207–219.

Mayr G (2000b) A new raptor-like bird from the Lower Eocene of North America and Europe, *Senckenbergiana Lethaea* **80**, 59–65.

Mayr G (2001) Comments on the systematic position of the putative Lower Eocene parrot *Pulchrapollia gracilis*, *Senckenbergiana Lethaea* **81**, 339–341.

Mayr G (2002a) A postcranial skeleton of *Palaeopsittacus* Harrison 1982 (Aves incertae sedis) from the Middle Eocene of Messel (Germany), *Oryctos* **4**, 75–82.

Mayr G (2002b) On the osteology and phylogenetic affinities of Pseudasturidae-Lower Eocene stem-group representatives of parrots (Aves, Psittaciformes), *Zoological Journal of the Linnean Society* **136**, 715–729.

Mayr G (2002c) Avian remains from the Middle Eocene of the Geiseltal (Sachse-Anhalt, Germany). In *Proceedings of the 5th Symposium of the Society of Avian Paleontology and Evolution, Beijing* (Zhou Z, Zhang F, eds), pp. 77–96, Science Press, Beijing.

Mayr G (2003) *Pseudasturides* n. gen., a replacement name for the stem group parrot *Pseudastur* Mayr 1998 (preoccupied by *Pseudastur* Blyth 1850), *Senckenbergiana Lethaea* **83**, 2.

Mayr G (2005a) The postcranial osteology and phylogenetic position of the Middle Eocene *Messelastur gratulator* Peters 1994 – a morphological link between owls (Strigiformes) and falconiform birds? *Journal of Vertebrate Paleontology* **25**, 635–645.

Mayr G (2005b) A *Fluvioviridavis*-like bird from the Middle Eocene of Messel, Germany, *Canadian Journal of Earth Sciences* **42**, 2021–2037.

Mayr G (2007) New specimens of Eocene stem-group psittaciform birds may shed light on the affinities of the first named fossil bird, *Halcyornis toliapicus* Koenig 1825, *Neues Jahrbuch für Geologie und Paläontologie-Abhandlungen* **244**, 207–213.

Mayr G (2009) *Paleogene Fossil Birds*, Springer Verlag, Berlin.

Mayr G (2010) Mousebirds (Coliiformes), parrots (Psittaciformes), and other small birds from the Late Oligocene/Early Miocene of the Mainz Basin, Germany, *Neues Jahrbuch für Geologie und Paläontologie-Abhandlungen* **258**, 129–144.

Mayr G (2011) Well-preserved new skeleton of the Middle Eocene *Messrlastur* substantiates sister group relationship between Messelasturidae and Halcyornithidae (Aves, ? PanPsittaciformes), *Journal of Systematic Palaeontology* **9**, 159–171.

Mayr G (2015a) A reassessment of Eocene parrotlike fossils indicates a previously undetected radiation of zygodactyl stem group representatives of passerines (Passeriformes), *Zoologica Scripta* **44**, 587–602.

Mayr G (2015b) Eocene fossils and the early evolution of frogmouths (Podargiformes): further specimens of *Masillapodargus* and a comparison with *Fluvioviridavis*, *Palaeobiodiversity and Palaeoenvironments* **95**, 587–596.

Mayr G (2016a) The world's smallest owl, the earliest unambiguous charadriiform bird, and other avian remains from the early Eocene Nanjemoy Formation of Virginia (USA), *Paläontologische Zeitschrift* **90**, 747–763.

Mayr G (2016b) *Avian Evolution. The Fossil Record of Birds and its Paleobiological Significance*, John Wiley and Sons Ltd, Chichester, UK.

Mayr G, Bertelli S (2011) A record of *Rhynchaeites* (Aves, Threskiornithidae) from the Early Eocene Fur Formation of Denmark, and the affinities of the alleged parrot *Mopsitta*, *Palaeobiodiversity and Palaeoenvironments* **91**, 229–236.

Mayr G, Daniels M (1998) Eocene parrots from Messel (Hessen, Germany) and the London Clay of Walton-on-the-Naze (Essex, England), *Senckenbergiana Lethaea* **78**, 157–177.

Mayr G, Göhlich UB (2004) A new parrot from the Miocene of Germany, with comments on the variation of hypotarsus morphology in some Psittaciformes, *Belgian Journal of Zoology* **134**, 47–54.

Mayr G, Mourer-Chauviré C, Weidig I (2004) Osteology and systematic position of the Eocene Primobucconidae (Aves, Coraciiformes *sensu stricto*), with first records from Europe, *Journal of Systematic Paleontology* **2**, 1–12.

Mayr G, Rana RS, Sahni A, Smith T (2007) Oldest fossil remains from the Indian subcontinental plate, *Current Science* **92**, 1266–1269.

Mayr G, Rana RS, Rose KD, Sahni A, Kumar K, Smith T (2010) *Quercypsitta*-like birds from the Early Eocene of India (Aves, ? Psittaciformes), *Journal of Vertebrate Paleontology* **30**, 467–478.

Mayr G, Rana RS, Rose KD, Sahni A, Kumar K, Smith T (2013) New specimens of the Early Eocene bird *Vastanavis* and the interrelationships of stem group Psittaciformes, *Paleontological Journal* **47**, 1308–1314.

McCormack G, Künzlé J (1996) The 'Ura or Rimatara Lorikeet *Vini kuhlii*: its former range, present status, and conservation priorities, *Bird Conservation International* **6**, 325–334.

McDougall A, Porter G, Mostert M, Cupitt R, Cupitt S, Joseph L, Murphy S, Janetzki H, Gallagher A, Burbidge A (2009) Another piece in an Australian ornithological puzzle – a second Night Parrot is found dead in Queensland, *Emu* **109**, 198–203.

McFarland DC (1988a) Fire and vegetation composition and structure of subtropical heathlands in south-eastern Queensland, *Australian Journal of Botany* **36**, 533–546.

McFarland DC (1988b) Geographical variation in the clutch size and breeding season of the Ground Parrot *Pezoporus wallicus*, *Australian Bird Watcher* **12**, 247–250.

McFarland DC (1991a) The biology of the Ground Parrot, *Pezoporus wallicus*, in Queensland, I. Microhabitat use, activity cycle and diet, *Australian Wildlife Research* **18**, 169–184.

McFarland DC (1991b) The biology of the Ground Parrot, *Pezoporus wallicus*, in Queensland, II. Spacing, calling and breeding behaviour, *Australian Wildlife Research* **18**, 185–197.

McFarland DC (1991c) The biology of the Ground Parrot, *Pezoporus wallicus*, in Queensland, III. Distribution and abundance, *Australian Wildlife Research* **18**, 199–213.

McFarland DC (1991d) Flush behaviour, catchability and morphometrics of the Ground Parrot *Pezoporus wallicus* in south-eastern Queensland, *Corella* **15**, 143–149.

McGill AR (1960) Parrots of the genus *Neophema* in New South Wales, *Emu* **60**, 39–46.

McGilp JN (1931) *Geopsittacus occidentalis*, Night-Parrot, *South Australian Ornithologist* **11**, 68–70.

McGregor RC (1909) *A Manual of Philippine Birds*, No. 1, Bureau of Science, Manila.

McKean JL (1985) Birds of the Keep River National Park (Northern Territory), including the Night Parrot *Geopsittacus occidentalis, Australian Bird Watcher* **11**, 114–130.

McKenzie RH (1979) Birds of the Hunua Ranges, *Notornis* **26**, 105–119.

McKinley D (1966) Conservation in grandfather's day: the gay life and sad end of America's parakeet, *Explorer* **8**(1), 20–23.

McKinley D (1977) Climatic relations, seasonal mobility, and hibernation in the Carolina Parakeet, *Jack-Pine Warbler* **55**, 107–124.

McKinley D (1980) The balance of decimating factors and recruitment in extinction of the Carolina Parakeet, Part I, *Indiana Audubon Q* **58**, 8–18, 50–61, 103–114.

Meijer HJ, Sutikna T, Saptomo EW, Awe RD, Jatmiko S, Wasisto S, James HF, Morwood MJ, Tocheri MW (2013) Late Pleistocene-Holocene non-passerine avifauna of Liang Bua (Flores, Indonesia), *Journal of Vertebrate Paleontology* **33**, 877–894.

Menkhorst PW, Isles AC (1981) The Night Parrot *Geopsittacus occidentalis* in north-western Victoria during the 1950s, *Emu* **81**, 239–240.

Menkhorst P, Ryan E (2015) C. H. McLennan ('Mallee Bird') and his Aboriginal informant Jowley: the source of early records of the Night Parrot *Pezoporus occidentalis* in Victoria? *Memoirs of Museum Victoria* **73**, 107–115.

Meredith CW (1984) The Ground Parrot *Pezoporus wallicus* Kerr, *RAOU Conservation Statement* no. 1, 1–12.

Meredith CW (1991) Vertebrate fossil faunas from islands in Australasia and the southwest Pacific. In *Vertebrate Palaeontology of Australasia* (Vickers-Rich P, Monaghan JM, Baird RF, Rich TH, eds), pp.1345–1381, Pioneer Design Studio, Melbourne.

Meredith CW, Isles AC (1980) A study of the Ground Parrot (*Pezoporus wallicus*) in Victoria, *Environmental Studies Division Publication*, no. 304.

Meredith CW, Jaremovic R (1990) Current status and management of the Ground Parrot in Victoria, *Arthur Rylah Institute Technical Report Series*, no. 58, 26 pp.

Meredith CW, Gilmore AM, Isles AC (1984) The Ground Parrot (*Pezoporus wallicus* Kerr) in south-eastern Australia: a fire-adapted species? *Australian Journal of Ecology* **9**, 367–380.

Merton DV (1985) Kakapo (*Strigops habroptilus* Gray). In *Readers Digest Complete Book of New Zealand Birds* (Robertson CJR, ed.), pp. 242–243, Readers Digest Services, Sydney.

Merton DV, Morris RD, Atkinson IAE (1984) Lek behaviour in a parrot: the Kakapo *Strigops habroptilus* of New Zealand, *Ibis* **126**, 277–283.

Meyer AB, Wiglesworth LW (1898) *The Birds of the Celebes and the Neighbouring Islands*, vol. 1, Friedländer, Berlin.

Millener PR (1991) The Quaternary avifauna of New Zealand. In *Vertebrate Palaeontology of Australasia* (Vickers-Rich P, Monaghan JM, Baird RF, Rich TH, eds), pp. 1317–1344, Pioneer Design Studio, Melbourne.

Milne-Edwards A (1867) Un psittacien fossile de l'Isle Rodrigues, *Annales des Sciences naturelles* (*Zoologie*) **8**, 145–156.

Milne-Edwards MA (1869–1871) *Recherches Anatomiques et Paléontologiques pour server a l'Histoire des Oiseaux Fossiles de la France*, vol. 2, G. Masson, Paris.

Milne-Edwards MA (1870) Observations sur la faune ornithologique du Bourbonnais pendant la period tertiare moyenne, during the Middle Tertiary Period, *Compte Rendus Hebdomadaires des Séances de l'Académie des Sciences* **70**, 557–559.

Minnis PE, Whalen ME, Kelley JH, Stewert JD (1993) Prehistoric macaw breeding in the North American southwest, *American Antiquity* **58**, 270–276.

Miskelly C, Empson R, Wright K (2005) Forest birds recolonising Wellington, *Notornis* **52**, 21–26.

Mittermeier JC, Cottee-Jones HEW, Purba EC, Ashuri NM, Hesdianti E, Supriatna J (2013) A survey of the avifauna of Obi Island, North Moluccas, Indonesia, *Forktail* **29**, 128–137.

Mlikovsky J (1998) A new parrot (Aves: Psittacidae) from the Early Miocene of the Czech Republic, *Acta Societatis Zoologicae Bohemicae* **62**, 335–341.

Montañez D, Angehr G (2007) Important bird areas of the neotropiocs: Panama, *Neotropical Birding 2007*, 12–19.

Monterrubio-Rico T, Enkerli-Hoeflich E, Hamilton RB (2002) Productivity and nesting success of Thick-billed Parrots, *Condor* **104**, 788–794.

Monterrubio-Rico T, Enkerlin-Hoeflich E (2004) Present use and characteristics of Thick-billed Parrot nest sites in northwestern Mexico, *Journal of Field Ornithology* **75**, 96–103.

Monterrubio-Rico TC, Cruz-Nieto J, Enkerlin-Hoeflich E, Venegas-Holguin D, Tellez-Garcia L, Marin-Togo C (2006) Gregarious nesting behavior of Thick-billed Parrots (*Rhynchopsitta pachyrhyncha*) in aspen stands, *Wilson Journal of Ornithology* **118**, 237–243.

Monterrubio-Rico TC, Renton K, Ortega-Rodriguez JM, Pérez-Arteaga A, Cancino-Murillo R (2010) The endangered Yellow-headed Parrot *Amazona oratrix* along the Pacific coast of Mexico, *Oryx* **44**, 602–609.

Montgomery SL, Gagné WC, Gagné BH (1980) Notes on birdlife and nature conservation in the Marquesas and Society Islands, *Elepaio* **40**, 152–155.

Moore JL (1985) Ensign Best's bird observations on Norfolk Island, *Notornis* **32**, 319–322.

Moorhouse RJ (1997) The diet of the North Island Kaka (*Nestor meridionalis septentrionalis*) on Kapiti Island, *New Zealand Journal of Ecology* **21**, 141–152.

Moorhouse RJ, Powlesland RG (1991) Aspects of the ecology of kakapo (*Strigops habroptilus*) liberated on Little Barrier Island (Hauturu), New Zealand, *Biological Conservation* **56**, 349–365.

Moreau RE (1948) Aspects of evolution in the parrot genus *Agapornis*, *Ibis* **90**, 206–239, 449–460.

Morris AK, McGill AR (1980) *Birds of New South Wales*, unpublished ms.

Mourer-Chauviré C (1992) Une nouvelle famille de perroquets (Aves: Psittaciformes) dans l'éocène Supérieur des phosphorites du Quercy, France, *Geobios, Mémoires Spécial* **14**, 169–171.
Mourer-Chauviré C, Bour R, Ribes S, Moutou F (1999) The avifauna of Réunion Island (Mascarene Islands) at the time of the arrival of the first Europeans. In *Avian Paleontology at the Close of the 20th Century: Proceedings of the 4th International Meeting of the Society of Avian Paleontology and Evolution,* Washington, DC, 4–7 June 1996 (Olson S, ed.), *Smithsonian Contributions to Paleobiology* **89**, 1–38.
Mourer-Chauviré C, Geraads D (2010) The Upper Pliocene avifauna of Ahl al Oughlam, Morocco. Systematics and biogeography. In *Proceedings of the VII International Meeting of the Society of Avian Paleontology and Evolution* (Boles WE, Worthy TH, eds), *Records of the Australian Museum* **62**, 157–184.
Mourer-Chauviré C, Pickford M, Senut B (2015) Stem group galliform and stem group psittaciform birds (Aves, Galliformes, Paraortygidae, and Psittaciformes, family incertae sedis) from the Middle Eocene of Namibia, *Journal of Ornithology* **156**, 275–286.
Murie J (1868) On the Nocturnal Ground Parrakeet (*Geopsittacus occidentalis*), *Proceedings of the Zoological Society of London,* 1868, 158–165.
Murphy S, Legge S (2007) The gradual loss and episodic creation of Palm Cockatoo (*Probosciger aterrimus*) nest-trees in a fire- and cyclone-prone habitat, *Emu* **107**, 1–6.
Murphy SA, Joseph L, Burbidge AH, Austin J (2010) A cryptic and critically endangered species revealed by mitochondrial DNA analyses: the Western Ground Parrot, *Conservation Genetics,* doi:10.1007/s10592-010-0161-1.
Murphy S, Night Parrot Recovery Team (2015) Shining a light: the research unlocking the secrets of the mysterious Night Parrot, *Australian Birdlife* **4**(3), 30–35.
Nandika D (2006) Recent observations of the critically endangered *sulphurea* subspecies of the Yellow-crested Cockatoo: Rawa Aopa Watumohai National Park, SE Sulawesi, Indonesia, *PsittaScene* **18**, 10–11.
Neilly B (2001) Barren Grounds: spring fever, *Wingspan* **11**(4), 29.
Newton A (1875) Note on *Palaeornis exsul, Ibis* 3rd series, **5**, 342–343.
Newton A, Newton E (1876) On the Psittaci of the Mascarene Islands, *Ibis* 3rd series **23**, 281–289.
Newton E (1867) On the land-birds of the Seychelles Archipelago, *Ibis* new series **3**, 335–360.
Newton E, Gadow H (1893) On additional bones of the Dodo and other extinct birds of Mauritius obtained by Mr. Théodore Sauzier, *Transactions of the Zoological Society of London* XIII, 281–302.
Nicholson DJ (1948) Escaped paroquets found nesting in Florida, *Auk* **65**, 139.
Nieremberg A (1972) The Kissi-Kissi, *Parrot Society Magazine* **6**, 136–139.
Nixon AJ (1981) The external morphology and taxonomic status of the Orange-fronted Parakeet, *Notornis* **28**, 292–300.
Nixon AJ (1994) Feeding ecology of hybridizing parakeets on Mangere Island, Chatham Islands, *Notornis* supplement **41**, 5–18.
North AJ (1896) Aves. In *Report on the Work of the Horn Scientific Expedition to Central Australia*, Part 2, Zoology, Dulau, London.
North AJ (1911) Nests and eggs of birds found breeding in Australia and Tasmania, vol. 3, pts 1 and 2, *Australian Museum Special Catalogue*, no. 1, 1–176.
O'Donnell CFJ, Dilks PJ (1989) Sap-feeding by the Kaka (*Nestor meridionalis*) in South Westland, New Zealand, *Notornis* **36**, 65–71.
Olah GS, Butchart HM, Symes A, Guzmán IM, Cunningham R, Brightsmith DJ, Heinsohn R (2016) Ecological and socio-economic factors affecting extinction rate in parrots, *Biodiversity and Conservation* **25**, 205–223.
Old P (2001) Observing macaws in Brazil, *PsittaScene* **13**(1), 8–9.
Oliver WRB (1955) *New Zealand Birds*, A. H. & A. W. Reed, Wellington.
Olmos F (1993) Birds of the Sierra da Capivara National Park in the 'caatinga' of north-eastern Brazil, *Bird Conservation International* **3**, 21–36.
Olmos F, Martuscelli P, de Silva RS (1997) Distribution and dry-season ecology of Pfrimeri's Conure *Pyrrhura pfrimeri*, with a reappraisal of Brazilian *Pyrrhura "leucotis"*, *Ornitologia Neotropical* **8**, 121–132.
Olmos F, Silva WAG, Albano C (2005) Grey-breasted Conure *Pyrrhura griseipectus*, an overlooked endangered species, *Cotinga* **24**, 77–83.
Olsen P (2007) *Glimpses of Paradise: The Quest for the Beautiful Parakeet*, National Library of Australia, Canberra.
Olsen SJ, Olsen JW (1974) The macaws of Grasshopper Ruin, *The Kiva* **40**, 67–70.
Olson SL (1985) The fossil record of birds. In *Avian Biology* (Farner DS, King JR, Parkes KC, eds), pp. 79–238, Academic Press, New York.
Olson SL (2015) History, morphology, and fossil record of the extinct Puerto Rican Parakeet *Psittacara maugei* Souancé, *Wilson Journal of Ornithology* **127**, 1–12.
Olson SL, Hilgartner HB (1982) Fossil and subfossil birds from the Bahamas. In *Fossil vertebrates from the Bahamas* (Olson SL, ed.), *Smithsonian Contributions to Paleobiology* **48**, 22–56.
Olson SL, Maíz López EJ (2008) New evidence of *Ara autochthones* from an archeological site in Puerto Rico: a valid species of West Indian macaw of unknown geographical origin (Aves: Psittacidae), *Caribbean Journal of Science* **44**, 215–222.
Olson SL, Suárez W (2008) A fossil cranium of the Cuban Macaw *Ara tricolor* (Aves: Psittacidae) from Villa Clara Province, Cuba, *Caribbean Journal of Science* **44**, 287–290.
Oren DC, Novaes FC (1986) Observations on the Golden Parakeet *Aratinga guarouba* in northern Brazil, *Biological Conservation* **36**, 329–337.
Oren DC, Willis EO (1981) New Brazilian records for the Golden Parakeet (*Aratinga guarouba*), *Auk* **98**, 394–396.
Orfila RN (1936) Los Psittaciformes Argentinos, *Hornero* **6**, 197–225.
Ortiz-Catedral L (2009) Some observations on the behaviour of the critically endangered orange-fronted parakeet (*Cyanoramphus malherbi*) on Maud Island, New Zealand, *Notornis* **56**, 165–166.
Ortiz-Catedral L (2012) Habitat use by the critically endangered orange-fronted parakeet (*Cyanoramphus malherbi*) on Maud Island: its relevance for future translocations, *Notornis* **59**, 148–152.
Ortiz-Catedral L, Brunton DH (2009) Notes on the diet of the critically endangered orange-fronted parakeet (*Cyanoramphus malherbi*) on Maud Island, *New Zealand Journal of Zoology* **36**, 385–388.

Ortiz-Catedral L, Kearvell JC, Hauber ME, Brunton DH (2009) Breeding biology of the critically endangered Malherbe's Parakeet on Maud Island, New Zealand, following the release of captive-bred individuals, *Australian Journal of Zoology* **57**, 433–439.

OSNZ (Ornithological Society of New Zealand Checklist Committee) (2010) *Checklist of the Birds of New Zealand, Norfolk and Macquarie Islands, and Ross Dependency, Antarctica*, 4th edition, Te Papa Press, Wellington.

Oustalet E (1897) Notice sur la faune ornithologique ancienne et moderne des Iles Mascareignes et en particulier de l'Ile Maurice, *Annales des Sciences naturelles* (*Zoologie*) series 8, **3**, 1–128.

Owen R (1846) *A History of British Fossil Mammals and Birds*, John van Voorst, London.

Owen R (1866) Evidence of a species perhaps extinct of large parrot (*Psittacus mauritianus*) contemporary with the Dodo in the island of Mauritius, *Ibis*, 168–171.

Pacifico de Assis E (2010) An inside job: reproductive biology of the Lear's Macaw, *PsittaScene* **22**(4), 3–6.

Paguntalan LM, Jakosalem PG (2008) Significant records of birds in forests on Cebu island, central Philippines, *Forktail* **24**, 48–56.

Parker SA (1980) Birds and conservation parks in the north-east of South Australia, *South Australian Parks and Conservation* **3**, 11–18.

Parmalee PW (1958) Remains of rare and extinct birds from Illinois Indian sites, *Auk* **75**, 169–176.

Parmalee PW (1967) Additional noteworthy records of bird from archaeological sites, *Wilson Bulletin* **79**, 155–162.

Parr M, Gilardi J (2002) Stop Press! "Missing" South American parrot rediscovered after 90 years, photographed for the first time, *PsittaScene* no. 52, 18.

Parrini R, Raposo MA, Pacheco JF, Carvalhães AMP, Melo TA Jr, Fonseca PSM, Minns JC (1999) Birds of the Chapada Diamantina, Bahia, Brazil, *Cotinga* **11**, 86–95.

Patten RA (1947) Observations on Kuhl's Ruffed Lory (*Vini kuhlii*) in captivity, *Avicultural Magazine* **53**, 40–43.

Pavia M (2014) The parrots (Aves: Psittaciformes) from the Middle Miocene of Sansan (Gers, southern France), *Paläontologische Zeitschrift* **88**, 353–359.

Pedersen B, Pedersen B (2002) Observations at an active nest of Blue-throated Macaws in the Beni Department of Bolivia, *PsittaScene* no. 50, 2–3.

Perrin MR (2011) A fossilised humerus of a lovebird tells little of the Pleistocene habitat of *Australopithecus robustus*, *South African Journal of Science* **107**, 15.

Perrin MR (2013) *Parrots of Africa, Madagascar and the Mascarene Islands*, Wits University Press, Johannesburg.

Peters DS (1994) *Messelastur gratulator* n. gen. n. spec., ein Greifvogel aus der Grube Messel (Aves: Accipitridae), *Courier Forschungsinstitut Senckenberg* **170**, 3–9.

Peters JL (1937) *Check-List of Birds of the World*, vol. 3, Harvard University Press, Cambridge.

Phillpotts A (1919) Notes on birds of southwest Otago, *Transactions and Proceedings of the New Zealand Institute* **51**, 222.

Pinner D, Bird L (1974) Birds of Hunter and its off-shore islands, *Tasmanian Naturalist* no. 39, 1–5.

Pinto OMdeO (1935) Aves da Bahia, *Revista Museu Paulista* **19**, 1–325.

Pitman H, Scott R, Stojanovic D, McLellan L (2007) Caring for Carnaby's: a recovery project seeks to secure the shrinking range of Carnaby's Black Cockatoo, *Wingspan* **17**(4), 24–25.

Pitter E, Christiansen MB (1995) Ecology, status and conservation of the Red-fronted Macaw *Ara rubrogenys*, *Bird Conservation International* **5**, 61–78.

Pocock TM (1969) Pleistocene fossil birds from Kromdraai and Sterkfontein, *Ostrich* **40** supplement 1, 1–6.

Podsiadlowski L, Gamauf A, Töfer T (2017) Revising the phylogenetic position of the extinct Mascarene Parrot *Mascarinus mascarin* (Linnaeus 1771) (Aves: Psittaciformes: Psittacidae), *Molecular Phylogenetics and Evolution* **107**, 499–502.

Pople RG, Burfield IJ, Clay RP, Cope DR, Kennedy CP, Läpez Lanãs B, Reyes J, Warren B, Yagual E (1997) Bird surveys and conservation status of three sites in western Ecuador, *Final Report of Project Ortalis '96*, CSB Publications, Cambridge.

Potter NS (1953) The birds of Calicòan, Philippine Islands, *Wilson Bulletin* **65**, 252–270.

Powlesland RG, Lloyd BD, Best HA, Merton DV (1992) Breeding biology of the kakapo (*Strigops habroptilus*) on Stewart Island, New Zealand, *Ibis* **134**, 361–373.

Powlesland RG, Roberts A, Lloyd BD, Merton DV (1995) Number, fate and distribution of Kakapo (*Strigops habroptilus*) found on Stewart Island, New Zealand, 1979–92, *New Zealand Journal of Zoology* **22**, 239–248.

Powlesland RG, Merton DV, Cockrem JF (2006) A parrot apart: the natural history of the Kakapo (*Strigops habroptilus*), and the context of its conservation management, *Notornis* **53**, 2–26.

Powlesland RG, Greene TC, Dilks PJ, Moorhouse RJ, Moran LR, Taylor G, Jones A, Wills DE, August CK, August ACL (2009) Breeding biology of the New Zealand kaka (*Nestor meridionalis*) (Psittacidae, Nestorinae), *Notornis* **56**, 11–33.

Pranty W, Garrett KL (2011) Under the radar, *Birding* **43**(5), 46–58.

Prassack KA (2010) Late Pliocene avifauna from the Hominid-bearing *Zinjanthropus* land surface at Olduvai Gorge, Tanzania. In *Proceedings of the VII International Meeting of the Society of Avian Paleontology and Evolution* (Boles WE, Worthy TH, eds), *Records of the Australian Museum* **62**, 185–192.

Pratt HD, Bruner PL, Berrett DG (1987) *A Field Guide to the Birds of Hawaii and the Tropical Pacific*, Princeton University Press, Princeton.

Pregill GK, Steadman DW, Olson SL, Grady FV (1988) Late Holocene fossil vertebrates from Burma Quarry, Antigua, Lesser Antilles, *Smithsonian Contributions to Paleobiology* **463**, 1–27.

Pregill GK, Steadman DW, Watters DR (1994) Late Quaternary vertebrate faunas of the Lesser Antilles: historical components of Caribbean biogeography, *Bulletin of the Carnegie Museum of Natural History* **30**, 1–51.

Pyle RL, Pyle P (2009) *The Birds of the Hawaiian Islands: Occurrence, History, Distribution and Status*, B. P. Bishop Museum, Honolulu.

Raffaele H, Wiley J, Garrido O, Keith A, Raffaele J (1998) *A Guide to the Birds of the West Indies*, Princeton University Press, Princeton.

Rand AL, Rabor DS (1960) Birds of the Philippine Islands: Siquijor, Mount Malindang, Bohol, and Samar, *Fieldiana: Zoology* **35**, 223–441.

Rasmussen PC, Anderton JC (2005) *Birds of South Asia. The Ripley Guide*, vol. 2, Smithsonian Institution and Lynx Edicions, Washington and Barcelona.

Raubenheimer D, Simpson SJ (2006) The challenge of supplementary feeding: can geometric analysis help save the kakapo? *Notornis* **53**, 100–111.

Rawlence NJ, Kennedy M, Scofield RP, Tennyson AJD, Boussès P, Chambers GK (2015) Taxonomic status of *Cyanoramphus* parakeets on the Auckland Islands and implications for the validity of the orange-fronted parakeet as a discrete species, *Journal of the Royal Society of New Zealand* **45**, 197–211.

Rea AM (1997) The indeterminate parrot of Nuevo Leon. In *The era of Alan R. Phillips: A festschrift* (Phillips AR, Dickerman RW, compilers), pp. 167–176, Horizon Communications, Robert Dickerman, Alberquerque.

Read A, McClelland P (1984) Orange-fronted Parakeets in the Hawdon valley, Arthur's Pass National Park, *Notornis* **31**, 266–267.

Recher HF (1985) Eucalypt forests, woodlands and birds: an introduction. In *Birds of Eucalypt Forests and Woodlands: Ecology, Conservation and Management* (Keast A, Recher HF, Ford HA, Saunders DA, eds), pp. 1–10, RAOU and Surrey Beatty and Sons, Sydney.

Recher HF, Recher JT (1966) A contribution to the knowledge of the avifauna of the Sierra de Luquillo, *Caribbean Journal of Science* **6**, 151–161.

Reichenow A (1880) Vogelreste aus dem Totenfelde von Ancona in West-Papua, *Ornithologisches Centralblatt* **5**, 187.

Reillo PR (2001) Imperial recovery: Dominica's flagship parrot on the comeback, *PsittaScene* **13**(4), 4–5.

Reinschmidt M (2006) Breeding the Red-browed Amazon, *Cyanopsitta* no. 50, 9–11.

Reis KR, Steadman DW (1999) Archaeology of Trants, Montserrat, Part 5, Prehistoric avifauna, *Annals of the Carnegie Museum* **68**, 275–288.

Renton K (2001) Lilac-crowned Parrot diet and food resource availability: resource tracking by a parrot seed predator, *Condor* **103**, 62–69.

Renton K, Salinas-Melgoza A (1999) Nesting behavior of the Lilac-crowned Parrot, *Wilson Bulletin* **111**, 488–493.

Rheindt FE, Christidis L, Kuhn H, de Kloet S, Norman JA, Fidler A (2014) The timing of diversification within the most divergent parrot clade, *Journal of Avian Biology* **45**, 140–148.

Ribas CC, Miyaki CY (2004) Molecular systematics in *Aratinga* parakeets: species limits and historical biogeography in the '*solstitialis*' group, and the systematic position of *Nandayus nenday*, *Molecular Phylogenetics and Evolution* **30**, 663–675.

Ribas CC, Joseph L, Miyaki CY (2006) Molecular systematics and patterns of diversification in *Pyrrhura* (Psittacidae), with special reference to the *picta-leucotis* complex, *Auk* **123**, 660–680.

Rich PV (1980) Preliminary report on the fossil avian remains from Late Tertiary sediments at Langebaanweg (Cape Province), South Africa, *South African Journal of Science* **76**, 166–170.

Richards H (2010) The 500 mark: a landmark season, *PsittaScene* **22**, 6–10.

Ridpath MG, Moreau RE (1966) The birds of Tasmania: ecology and evolution, *Ibis* **108**, 348–393.

Ridgely RS (1981) The current distribution and status of mainland neotropical parrots. In *Conservation of New World Parrots* (Pasquier RF, ed.), pp. 233–384, *ICBP Technical Publication*, no. 1.

Ridgely RS, Robbins M (1988) *Pyrrhura orcesi*, a new species from southwestern Ecuador, with systematic notes on the *P. melanura* complex, *Wilson Bulletin* **100**, 171–182.

Ridgely RS, Gwynn JA (1989) *A Guide to the Birds of Panama, with Costa Rica, Nicaragua and Honduras*, 2nd edition, Princeton University Press, Princeton.

Ridgely RS, Greenfield PJ (2001a) *The Birds of Ecuador: Status, Distribution and Taxonomy*, Cornell University Press, Ithaca.

Ridgely RS, Greenfield PJ (2001b) *The Birds of Ecuador: Field Guide*, Cornell University Press, Ithaca.

Ridgely RS, Gwynne JA, Tudor G, Argel M (2016) *Wildlife Conservation Society Birds of Brazil: The Atlantic Forest of Southeast Brazil Including São Paulo & Rio de Janeiro*, Cornell University Press, Ithaca.

Rijsdijk KF, Hume JP, Bunnik FD, Florens V, Baider C, Shapiro B, Plicht JvD, Janoo A, Griffiths O, Hoek Ostende LWvd, Cremer H, Vernimmen T (2009) Mid-Holocene vertebrate bone concentration-lagerstätte on oceanic island Mauritius provides a window into the ecosystem of the Dodo (*Raphus cucullatus*), *Quaternary Science Revues* **28**, 14–24.

Riley J (2002) Population sizes and the status of endemic and restricted-range bird species on Sangihe Island, Indonesia, *Bird Conservation International* **12**, 53–78.

Riley J (2003) Population sizes and conservation status of endemic and restricted-range bird species on Karakelang, Talaud Islands, Indonesia, *Bird Conservation International* **13**, 59–74.

Roberson D, Carratello R (1997) Updates to the avifauna of Oaxaca, Mexico, *Cotinga* **7**, 21–22.

Roberts RV (1953) Some birds of the Mokohinau Group, *Notornis* **5**, 197–198.

Robertson WB Jr, Woolfenden GE (1992) *Florida Bird Species: An Annotated List*, Florida Ornithological Society, Gainesville.

Robinet O, Salas M (1999) Reproductive biology of the endangered Ouvéa Parakeet *Eunymphicus cornutus uvaeensis*, *Ibis* **141**, 660–669.

Robinet O, Beugnet F, Dulieu D, Chardonnet P (1995) The Ouvéa Parakeet – state of knowledge and conservation status, *Oryx* **29**, 143–150.

Robinet O, Barré N, Salas M (1996) Population estimate for the Ouvéa Parakeet *Eunymphicus cornutus uvaeensis*: its present range and implications for conservation, *Emu* **96**, 151–157.

Robinet O, Craig JL, Chardonnet L (1998) Impact of rat species in Ouvéa and Lifou (Loyalty Islands) and their consequences for conserving the endangered Ouvéa Parakeet, *Biological Conservation* **86**, 223–232.

Robinet O, Bretagnolle V, Clout M (2003) Activity patterns, habitat use, foraging behaviour and food selection of the Ouvéa Parakeet (*Eunymphicus cornutus uvaeensis*), *Emu* **103**, 71–80.

Robson C (2000) *A Field Guide to the Birds of South-East Asia*, New Holland Publishers, London.

Rocamora G, Laboudallon V (2013) Seychelles Black Parrot *Coracopsis nigra* Perroquet noir des Seychelles. In *The Birds of Africa*, vol. VIII: *The Malagasy Region* (Safford RJ, Hawkins AFA, eds), pp. 529–531, Christopher Helm, London.

Rodríguez-Mahecha JV, Hernández-Camacho JI (2002) *Loros de Colombia*, Conservación Internacional Colombia, Bogotá.

Rodríguez-Vidal JA (1959) Puerto Rican Parrot study, *Monograph of the Department of Agriculture, Commonwealth of Puerto Rico*, no. 1, 1–15.

Rogers C (2010) Bird report 2007, *South Australian Ornithologist* **35**, 236–248.

Rojas A, Yucra E, Vera I, Requejo A, Tella J (2014) A new population of the globally endangered Red-fronted Macaw *Ara rubrogenys* unusually breeding in palms, *Bird Conservation International* **24**, 389–392.

Rothschild W (1905) Notes on extinct parrots from the West Indies, *Bulletin of the British Ornithologists Club* **16**, 13–15.

Rothschild W (1907) *Extinct Birds*, Hutchinson, London.

Rubenheimer D, Simpson SJ (2006) The challenge of supplementary feeding: can geometric analysis help save the Kakapo? *Notornis* **53**, 100–111.

Russell SM (1964) A distributional study of the birds of British Honduras, *Ornithological Monographs* no. 1, 1–95.

Russello MA, Amato G (2004) A molecular phylogeny of *Amazona*: implications for Neotropical parrot biogeography, taxonomy, and conservation, *Molecular Phylogenetics and Evolution* **30**, 421–437.

Ruth JM (2015) Yellow-headed Parrot (*Amazona oratrix*). In *Neotropical Birds Online* (Schulenberg TS, ed.), Cornell Laboratory of Ornithology, Ithaca, downloaded at http://neotropical.birds.cornell.edu/portal/species/overview?p_p_spp=23678.

Salaman PGW, López-Lanús B, Krabbe N (1999) Critically endangered: Yellow-eared Parrot *Ognorhynchus icterotis* in Colombia, *Cotinga* **11**, 39–41.

Salvadori T (1891) *Catalogue of Birds in the British Museum*, XX, Psittaci, British Museum Trustees, London.

Saunders DA (1974) The occurrence of the White-tailed Black Cockatoo *Calyptorhynchus baudinii*, in *Pinus* plantations in Western Australia, *Australian Wildlife Research* **1**, 45–54.

Saunders DA (1977) The effect of agricultural clearing on the breeding success of the White-tailed Black Cockatoo, *Emu* **77**, 180–184.

Saunders DA (1979a) Distribution and taxonomy of the White-tailed and Yellow-tailed Black Cockatoos, *Calyptorhynchus* spp., *Emu* **79**, 251–227.

Saunders DA (1979b) The availability of tree hollows for use as nest sites by White-tailed Black Cockatoos, *Australian Wildlife Research* **6**, 205–216.

Saunders DA (1979c) The biology of the short-billed form of the White-tailed Black Cockatoo *Calyptorhynchus funereus latirostris* Carnaby, PhD thesis, unpublished, University of Western Australia.

Saunders DA (1980) Food and movements of the short-billed form of the White-tailed Black Cockatoo, *Australian Wildlife Research* **7**, 257–269.

Saunders DA (1982) The breeding behaviour and biology of the short-billed form of the White-tailed Black Cockatoo *Calyptorhynchus funereus*, *Ibis* **124**, 422–425.

Saunders DA (1983) Vocal repertoire and individual recognition in the Short-billed White-tailed Black Cockatoo, *Australian Wildlife Research* **10,** 527–536**.**

Saunders DA (1986) Breeding season, nesting success and nestling growth in Carnaby's Cockatoo, *Calyptorhynchus funereus latirostris*, over 16 years at Coomallo Creek, and a method for assessing the viability of populations in other areas, *Australian Wildlife Research* **13**, 261–273.

Saunders DA (1991) The effects of land clearance on the ecology of Carnaby's Cockatoo and the inland Red-tailed Black Cockatoo in the wheatbelt of Western Australia, *Acta XX Congressus Internationalis Ornithologici*, 658–665.

Saunders DA, Rowley I, Smith GT (1985) The effects of clearing for agriculture on the distribution of cockatoos in the southwest of Western Australia. In *Birds of Eucalypt Forests and Woodlands: Ecology, Conservation and Management* (Keast A, Recher HF, Ford H, Saunders DA, eds), pp. 309–321, RAOU and Surrey Beatty and Sons, Sydney.

Saunders DA, Ingram JA (1995) *Birds of Southwestern Australia*, Surrey Beatty and Sons, Sydney.

Saunders DA, Mawson PR, Dawson R (2014a) Use of tree hollows by Carnaby's Cockatoo and the fate of large hollow-bearing trees at Coomallo Creek, Western Australia, *Biological Conservation* **177**, 185–193.

Saunders DA, Mawson PR, Dawson R (2014b) One fledgling or two in the endangered Carnaby's Cockatoo (*Calyptorhynchus latirostris*): a strategy for survival or legacy from a bygone era? *Conservation Physiology* **2**, doi:10.1093/conphys/cou001.

Saunders DL, Heinsohn R (2008) Winter habitat use by the endangered, migratory Swift Parrot (*Lathamus discolor*) in New South Wales, *Emu* **108**, 81–89.

Schauenberg P (1969) Note sur une perruche éteinte de Tahiti *Cyanoramphus zealandicus* (Latham 1781) conservée au Muséum de Genève, *Archives des Sciences* **22**, 645–649.

Schodde R (1997) Aves (Cacatuidae - Psittacidae). In *Zoological Catalogue of Australia* (Houston WWK, Wells A, eds) vol. 37.2, pp. 64–218, CSIRO Publishing, Melbourne.

Schodde R, Mason IJ (1980) *Nocturnal Birds of Australia*, Lansdowne Editions, Melbourne.

Schodde R, Mason IJ (1996) Birds. In *Murwillumbah Management Area Fauna Survey* (Schodde R and others, eds), pp. 1–161, CSIRO Division of Wildlife and Ecology, Canberra.

Schodde R, Remsen V Jr, Schirtzinger EE, Joseph L, Wright TF (2013) Higher classification of New World parrots (Psittaciformes: Arinae), with diagnoses of tribes, *Zootaxa* **3691**(5), 591–596.

Schönwetter M (1964) *Handbuch der oologie*, vol. 1, part 9, Akademie, Berlin.

Schubart O, Aguirre AC, Sick H (1965) Contribuição para o conhecimento da alimentação das aves Brasileiras, *Archivos de Zoologia do Estado de São Paulo* **12**, 95–249.

Schubel SE, Steadman DW (1989) More bird bones from archeological sites on Henderson Island, Pitcairn Group, South Pacific, *Atoll Research Bulletin* **325**, 1–13.

Schulenberg TS, Stotz DF, Lane DF, O'Neill JP, Parker TA III (2007) *Birds of Peru*, Princeton University Press, Princeton.

Schweizer M, Seehausen O, Gütert M, Hertwig ST (2010) The evolutionary diversification of parrots supports a taxon pulse model with multiple trans-oceanic dispersal events and local radiations, *Molecular Phylogenetics and Evolution* **54**, 984–994.

Schweizer M, Wright TF, Peñalba JV, Schirtzinger EE, Joseph L (2015) Molecular phylogenetics suggests a New Guinea origin and frequent episodes of founder-event speciation in the nectarivorous lories and lorikeets (Aves: Psittaciformes), *Molecular Phylogenetics and Evolution* **90**, 34–48.

Sclater PL (1867) Night Parrot, *Proceedings of the Zoological Society of London*, 1867, 891.

Scott JK, Black R (1981) Selective predation by White-tailed Black Cockatoos on fruit of *Banksia attenuata* containing the seed-eating weevil *Alphitopis nivea*, *Australian Wildlife Research* **8**, 421–430.

Seibels RE, McCullough WE (1978) Polygamy in the Sun Conure at Columbia Zoo, *International Zoo Yearbook* **18**, 105–106.

Seitre R (2015) Die Maidloris von Französisch-Polynesien, *Papageien* **11/2015** 386–393, **12/2015** 422–429.

Seitre R, Seitre J (1991) *Causes de disparition des oiseaux terrestres de Polynésie Française*, South Pacific Regional Environment Program (Occasional Paper Series 8), Nouméa.

Seitre R, Seitre J (1992) Causes of land-bird extinctions in French Polynesia, *Oryx* **26**, 215–222.

Sibson RB (1947) A visit to Little Barrier Island, *Notornis* (*N.Z. Bird Notes*) **2**, 134–144.

Sick H (1993) *Birds in Brazil: A Natural History*, Princeton University Press, Princeton.

Siebers HC (1930) Fauna Buruana: aves, *Treubia* **7** supplement, 165–303.

Silva T (1989) *A Monograph of Endangered Parrots*, Silvio Mattacchione and Company, Pickering.

Silveira LF, Belmonte FJ (2005) Comportamento reprodutivo e hábitos da ararajuba, *Guarouba guarouba*, no município de Tailândia, Pará, *Ararajuba* **13**, 89–93.

Silveira LF, de Lima FC, Höfling E (2005) A new species of *Aratinga* parakeet (Psittaciformes: Psittacidae) from Brazil, with taxonomic remarks on the *Aratinga solstitialis* complex, *Auk* **122**, 292–305.

Simão I, dos Santos FAM, Pizo MA (1997) Vertical stratification and diet of psittacids in a tropical lowland forest of Brazil, *Ararajuba* **5**, 169–174.

Simpson R, Simpson E (2012) Behavioural notes on Brown-backed Parrotlet *Touit melanonotus* in Ubatuba, São Paulo, Brazil, *Cotinga* **34**, 19–23.

Sinclair RG (1990) The economics of netting for bird control, *Proceedings of the National Birds Pest Workshop*, Armidale, 8–9 February, 134–141.

Sindel S, Lynn R (1989) *Australian Cockatoos*, Singil Press, Sydney.

Skead CJ (1964) The overland flights and the feeding habits of the Cape Parrot *Poicephalus robustus* (Gmelin) in the Eastern Cape Province, *Ostrich* **35**, 202–223.

Slud P (1964) The birds of Costa Rica: distribution and ecology, *Bulletin of the American Museum of Natural History* **128**, 1–430.

Smiet F (1985) Notes on the field status and trade of Moluccan parrots, *Biological Conservation* **34**, 181–194.

Smith AP (1907) The Thick-billed Parrot in Arizona, *Condor* **9**, 104.

Smith GA (1975) Systematics of parrots, *Ibis* **117**, 18–68.

Smith GT (1978) The status of Australian parrots. In *The Status of Australasian Wildlife* (Tyler MJ, ed.), pp. 101–108, Royal Zoological Society of South Australia, Adelaide.

Smithers CN, Disney HJ de S (1969) The distribution of terrestrial and freshwater birds on Norfolk Island, *Australian Zoologist* **15**, 127–140.

Snyder NFR (2004) *The Carolina Parakeet: Glimpses of a Vanished Bird*, Princeton University Press, Princeton.

Snyder NFR, Russell K (2002) Carolina Parakeet *Conuropsis carolinensis*. In *The Birds of North America* (Poole A, Gill F, eds), no. 667, Birds of North America Inc., Philadelphia.

Snyder NFR, Wiley JW, Kepler CB (1987) *The Parrots of Luquillo: Natural History and Conservation of the Puerto Rican Parrot*. Western Foundation of Vertebrate Zoology, Los Angeles.

Snyder NFR, Snyder HA, Johnson TB (1989) Parrots return to the Arizona skies, *Birds International* **1**(2), 41–52.

Snyder NFR, Koenig SE, Koschmann J, Snyder HA (1994) Thick-billed Parrot releases in Arizona, *Condor* **96**, 845–862.

Snyder NFR, Enkerlin-Hoeflich EC, Cruz-Nieto MA (1999) Thick-billed Parrot (*Rhynchopsitta pachyrhyncha*). In *The Birds of North America Online* (Poole A, Gill F, eds), no. 406, Cornell Laboratory of Ornithology, Ithaca.

Snyder N, McGowan P, Gilardi J, Grajal A (eds) (2000) *Parrots: Status Survey and Conservation Action Plan 2000–2004*. IUCN, Gland, Switzerland and Cambridge, UK.

Snyder NFR, Budney GF, Iñigo-Elias EE (2010) Evaluation of the Nelson Carolina Parakeet film, *Auk* **127**, 948–951.

Solano-Ugalde A (2011) Notes on the distribution and natural history of bird species in the Chocó bioregion of Ecuador, *Bulletin of the British Ornithologists Club* **131**, 249–255.

Somerville AD, Nelson BA, Knudson KJ (2010) Isotopic investigation of pre-Hispanic macaw breeding in northwest Mexico, *Journal of Anthropological Archeology* **29**, 125–135.

South Australian Ornithological Association (2008) *A Field List of the Birds of South Australia,* 4th edition, South Australian Ornithological Association Inc., Adelaide.

Spence T (1955) Breeding the Purple-naped Lory, *Avicultural Magazine* **61**, 14–17.

Spillman F (1942) Contribución al conocimiento de fósiles nuevos de la avifauna Ecuatoriana en el Pleistoceno de Santa Elena. In *Proceedings of the 8th American Scientific Congress,* Washington, 10–18 May 1940, vol. 4 (Oehser PH, ed.), pp. 375–389, American Scientific Congress, Washington DC.

Stager KE (1954) Birds of the Barranca de Cobre region of south-western Chihuahua, Mexico, *Condor* **56**, 21–32.

Stager KE (1957) The avifauna of the Tres Marías Islands, Mexico, *Auk* **74**, 413–432.

Stahl A (1887) Beitrag zur Vogelfauna von Portorico, *Ornis* **3**, 448–453.

Starks J (1992) Orange-bellied Parrot *Neophema chrysogaster* winter surveys in south-eastern Australia in 1988 and 1989, *RAOU Rep.*, no. 87.

Starks J, Brown P, Loyn R, Menkhorst P (1992) Twelve years of winter counts of the Orange-bellied Parrot *Neophema chrysogaster*, *Australian Bird Watcher* **14**, 305–312.

Steadman DW (1985) Fossil birds from Mangaia, southern Cook Islands, *Bulletin of the British Ornithologists Club* **105**, 58–66.

Steadman DW (1989) Extinction of birds in Eastern Polynesia: a review of the record, and comparisons with other Pacific Island Groups, *Journal of Archaeological Science* **16**, 177–205.

Steadman DW (1991) Extinct and extirpated birds from Aitutaki and Atiu, southern Cook Islands, *Pacific Science* **45**, 325–347.

Steadman DW (1992) Extinct and extirpated birds from Rota, Mariana Islands, *Micronesica* **25**, 71–84.

Steadman DW (1993) Biogeography of Tongan birds before and after human impact, *Proceedings of the National Academy of Sciences of the United States of America* **90**, 818–822.

Steadman DW (1995) Prehistoric extinctions of Pacific island birds: biodiversity meets zooarchaeology, *Science* **267**, 1123–1131.

Steadman DW (1999) The prehistory of vertebrates, especially birds, on Tinian, Aguiguan, and Rota, northern Mariana Islands, *Micronesica* **31**, 319–345.

Steadman DW (2006a) *Extinction and Biogeography of Tropical Pacific Birds*, University of Chicago Press, Chicago.

Steadman DW (2006b) A new species of extinct parrot (Psittacidae: *Eclectus*) from Tonga and Vanuatu, South Pacific, *Pacific Science* **60**, 137–145.

Steadman DW, Kirch PV (1990) Prehistoric extinction of birds on Mangaia, Cook Islands, Polynesia, *Proceedings of the National Academy of Sciences* **87**, 9605–9609.

Steadman DW, Kirch PV (1998) Biogeography and prehistoric exploitation of birds in the Mussau islands, Bismarck Archipelago, Papua New Guinea, *Emu* **98**, 13–22.

Steadman DW, Pahlavan DS (1992) Extinction and biogeography of birds in Huahine, Society Islands, French Polynesia, *Geoarchaeology* **7**, 449–483.

Steadman DW, Zarriello MC (1987) Two species of parrots (Aves: Psittacidae) from archeological sites in the Marquesas Islands, *Proceedings of the Biological Society of Washington* **100**, 518–528.

Steadman DW, Pregill GK, Olson SL (1984) Fossil vertebrates from Antigua, Lesser Antilles: evidence for Late Holocene human-caused extinctions in the West Indies, *Proceedings of the National Academy of Sciences of the United States of America* **81**, 4448–4451.

Steadman DW, Arroyo-Cabrales J, Johnson E, Guzman AF (1994) New information on the Late Pleistocene birds from San Joscito cave, Nuevo León, Mexico, *Condor* **96**, 577–589.

Steadman DW, White JP, Allen J (1999) Prehistoric birds from New Ireland, Papua New Guinea: extinctions on a large Melanesian island, *Proceedings of the National Academy of Sciences* **96**, 2563–2568.

Steadman DW, Franz R, Morgan GS, Albury NA, Kakuk B, Broad K, Franz SE, Tinker K, Pateman MP, Lott TA, Jarzen DM, Dilcher DL (2007) Exceptionally well preserved Late Quaternary plant and vertebrate fossils from a blue hole on Abaco, The Bahamas, *Proceedings of the National Academy of Sciences* **104**, 19897–19902.

Steinbacher G (1934) Zur kenntnis des magens blutenbesuchender der papageien, *Ornithologische Monatsberichte* **42**, 80–84.

Stephenson LH (1991) *The Orange-bellied Parrot Recovery Plan: Management Phase*, Department of Parks Wildlife and Heritage, Hobart.

Stidham TA (1998) A lower jaw from a Cretaceous parrot, *Nature* **396**(6706), 29–30.

Stidham TA (1999) Reply: did parrots exist in the Cretaceous period? *Nature* **399**, 318.

Stidham TA (2006) Parrots (Aves: Psittaciformes) from the Miocene Varswater Formation, Langebaanweg, South Africa, *African Natural History* **2**, 198–199.

Stidham TA (2009) A lovebird (Psittaciformes: *Agapornis*) from the Plio-Pleistocene Kromdraai B locality, South Africa, *South African Journal of Science* **105**, 155–157.

Stidham TA (2010) A small Pleistocene lovebird (Psittacidae: *Agapornis*) from Plovers Lake, South Africa, *Neues Jahrbuch für Geologie und Paläontologie-Abhandlungen* **256**, 123–128.

Stiles FG, Skutch AF (1989) *A Guide to the Birds of Costa Rica*, Cornell University Press, Ithaca.

Stojanovic D, Webb MH, Alderman R, Porfirio LL, Heinsohn R (2014) Discovery of a novel predator reveals extreme but highly variable mortality for an endangered migratory bird, *Diversity and Distributions*, 1–8, doi:10.1111/ddi.12214.

Stojanovic D, Terauds A, Westgate MJ, Webb MH, Roshier DA, Heinsohn R (2015) Exploiting the richest patch has a fitness pay-off for the migratory Swift Parrot, *Journal of Animal Ecology* **84**, 1194–1201.

Storr GM (1984) Revised list of Queensland birds, *Records of the Western Australian Museum* supplement **19**, 1–189.

Storr GM (1986) Birds of the south-eastern interior of Western Australia, *Records of the Western Australian Museum* supplement **26**, 1–60.

Storr GM (1991) Birds of the south-west division of Western Australia, *Records of the Western Australian Museum* supplement **35**, 1–150.

Strange M (2013) The Wallace Line according to Morten, *Nature Watch* October–November, 17–23.

Stresemann E (1914) Die vögel von Seran (Ceram), *Novitates Zoologicae* **21**, 25–153.

Stresemann E (1940) Die vögel von Celebes, Teil III. Systematik und biologie, *Journal für Ornithologie* **88**, 389–487.

Strewe R (2005) Aktuelle situation des Santa-Marta-Rotschwanzsittichs in Kolumbien, *Papageien* 3/2005, 94–97.

Sutton GM, Pettingill OS Jr (1942) Birds of the Gomez Farias region, south-western Tamaulipas, *Auk* **59**, 1–34.

Sweeney RG (1996) The Red-crowned Amazon Parrot EEP scheme, *Avicultural Magazine* **102**, 106–108.

Sweeney RG (1998) Breeding the Red and Blue Lory *Eos histrio* at Loro Parque, Puerto de la Cruz, *International Zoo Yearbook* **36**, 194–198.

Swinnerton K, Malijkovic A (2002) Red-throated Lorikeet *Charmosyna amabilis* in Fiji elusive or extinct? *Eclectus* **12**, 2–4.

Tambussi CP, Acosta Hospitaleche C, Rinderknecht A, Ubilla M (2009) Parrots (Aves, Psittaciformes) in the Pleistocene of Uruguay, *Ameghiniana* **46**, 431–435.

Tavares ES, Yamashita C, Miyaki C (2004) Phylogenetic relationships among some Neotropical parrot genera (Psittacidae) based on mitochondrial sequences, *Auk* **121**, 230–242.

Tavares ES, Baker AJ, Pereira SL, Miyaki CY (2006) Phylogenetic relationships and historical biogeography of Neotropical parrots (Psittaciformes: Psittacidae: Arini) inferred from mitochondrial and nuclear DNA sequences, *Systematic Biology* **55**, 454–470.

Tavistock Marquis of (1939) The breeding of the Ultramarine Lory, *Avicultural Magazine* 5th series **4**, 292–294.

Taylor A (2011) Perfect 10, *PsittaScene* **23**(3), 4–9.

Taylor FJ (1984) Birds on Aitutaki, Atiu and Mauke, Southern Cook Islands, *Notornis* **31**, 267–270.

Taylor RH (1975) Some ideas on speciation in New Zealand parakeets, *Notornis* **22**, 110–121.

Taylor RH (1979) How the Macquarie Island Parakeet became extinct, *New Zealand Journal of Ecology* **2**, 42–45.

Taylor RH (1985) Status, habits and conservation of *Cyanoramphus* parakeets in the New Zealand region, *ICBP Technical Publication* no. 3, 195–211.

Taylor RH (1998) A reappraisal of the Orange-fronted Parakeet (*Cyanoramphus* sp.) – species or colour morph? *Notornis* **45**, 49–63.

Taylor RH, Heatherbell EG, Heatherbell EM (1986) The Orange-fronted Parakeet (*Cyanoramphus malherbi*) is a colour morph of the Yellow-crowned Parakeet (*C. auriceps*), *Notornis* **33**, 17–22.

Temple SA (1978) The life histories of indigenous landbirds of Mauritius. Unpublished report.

Tennyson A, Martinson P (2006) *Extinct Birds of New Zealand: A Richly Illustrated View into the World of New Zealand's Lost Birds*, Te Papa Press, Wellington.

Theuerkauf J, Rouys S, Mériot JM, Gula R, Kuehn R (2009) Cooperative breeding, mate guarding, and nest sharing in two parrot species of New Caledonia, *Journal of Ornithology* **150**, 791–797.

Thibault J-C (1973) Notes ornithologiques Polynesiennes. II. Les Iles Marquises, *Alauda* **41**, 111–119.

Thomas DG (1970) Swift Parrots wintering in Tasmania, *Tasmanian Naturalist* no. 23, 4.

Thompson DW (1900) On characteristic points in the cranial osteology of the parrots, *Proceedings of the Zoological Society of London* [1899], 9–46.

Thomson DF (1935) *Birds of Cape York Peninsula*, Government Printer, Melbourne.

Thomson JSI (1912) *Voyages and Wanderings in Far-off Seas and Lands*, Headley Brothers, London.

Tipa R (2006) Kakapo in Maori lore, *Notornis* **53**, 193–194.

Todd WE, Worthington WW (1911) A contribution to the ornithology of the Bahama Islands, *Annals of the Carnegie Museum* **7**, 388–464.

Tonni EP (1972) *Cyanoliseus ensenadensis* (Cattoi) nov. comb. (Aves, Psittaciformes) del Pleistoceno medio de la provincia de Buenos Aires, *Ameghinana* **9**, 145–148.

Tonni EP, Noriega J (1996) Una nueva especies de *Nandayus* Bonaparte, 1854 (Aves: Psittaciformes) del Plioceno tardío de Argentina, *Revista Chilena de Historia Natural* **69**, 97–104.

Tostain O, Dujardin J-L, Erard C, Thiollay J-M (1992) *Oiseaux de Guyane: Biologie, Ecologie, Protection, Repartition*. Societé d'Etudes Ornitholgiques de France, Paris.

Tovar-Martinez AE (2010) Redescubrimiento y notas sobre la ecología y vocalizaciones del Periquito de Todd (*Pyrrhura picta caeruleiceps*) en el nororiente de Colombia, *Ornitología Colombiana* no. 9, 48–55.

Townsend CH, Wetmore A (1919) Reports on the scientific results of the expedition to the tropical Pacific in charge of Alexander Agassiz on the 'Albatross', 1899–1900. XXI. The birds, *Bulletin of the Museum of Comparative Zoology, Harvard* **63**, 151–225.

Trainor CR, Verbelen P, Johnstone RE (2012) The avifauna of Alor and Pantar, Lesser Sundas, Indonesia, *Forktail* **28**, 77–92.

Trevelyan R (1995) The feeding ecology of Stephen's Lory and nectar availability in its food plants, *Biological Journal of the Linnean Society* **56**, 185–197.

Triggs SJ, Daugherty CH (1996) Conservation and genetics of New Zealand parakeets, *Bird Conservation International* **6**, 89–101.

Turner GT (1940) The first European breeding of the Queen of Bavaria Conure, *Foreign Birds* **6**, 88–90.

Urantówka AD, Mackiewicz P, Strzala T (2014) Phylogeny of *Amazona barbadensis* and the Yellow-headed Amazon complex (Aves: Psittacidae): a new look at South American parrot evolution, *PLoS ONE* 9(5), e97228. doi:10.1371/journal.pone.0097228.

Valdés-Peña RA, Ortiz-Maciel SG, Valdez Juarez SO, Enkerlin-Hoeflich EC, Snyder NFR (2008) Use of clay licks by Maroon-fronted Parrots (*Rhynchopsitta terrisi*) in northern Mexico, *Wilson Journal of Ornithology* **120**, 176–180.

van Bemmel ACV, Voous KH (1948) A faunal list of the birds of the islands of Muna and Buton, S. E. Celebes, *Treubia* **21**, 27–104.

Vincent J (1967) Birds in danger of extinction. Threatened species of birds; general report, *Bulletin of the International Council of Bird Preservation* no. 10, 82–100.

Viney C, Phillipps K (1988) *Birds of Hong Kong*, Government Information Services, Hong Kong.

Voisin C, Voisin J-F, Didier M (1995) A fifth specimen of the Tahiti Parakeet, *Bulletin of the British Ornithologists Club* **115**, 262–263.

Von Ihering H (1902) Contribuições para conhecimento da ornithologia de São Paulo, II Descripção de novos ninhos e ovos, *Revista do Museu Paulista* **5**, 291–303.

Voous KH (1965) Specimens of Lear's Macaw in the Zoological Museum of Amsterdam, *Oiseau Revue français d'Ornithologie* **35** (no. spécial), 153–155.

Walker JS, Cahill AJ, Marsden SJ (2005) Factors influencing nest-site occupancy and low reproductive output in the critically endangered Yellow-crested Cockatoo *Cacatua sulphurea* on Sumba, Indonesia, *Bird Conservation International* **15**, 347–359.

Walker S, Chapman J (1992) Valley of the parrots, *Texas Parks and Wildlife* 1992(11), 37–41.

Walsh J, Wilson K-J, Elliott G (2006) Seasonal changes in home range size and habitat selection by kakapo (*Strigops habroptilus*) on Maud Island, *Notornis* **53**, 143–149.

Walsh S, Milner A (2011) *Halcyornis toliapicus* (Aves: Lower Eocene, England) indicates advanced neuromorphology in Mesozoic Neornithes, *Journal of Systematic Palaeontology* **9**, 173–181.

Walters M (1974) The eggs of the Golden Conure *Aratinga guarouba*, *Bulletin of the British Ornithologists Club* **94**, 71.

Walters M (1995) On the status of *Ara tricolor* Bechstein, *Bulletin of the British Ornithologists Club* **115**, 168–170.

Warner DW (1947) The ornithology of New Caledonia and the Loyalty Islands, PhD thesis (unpublished), Cornell University, New York.

Warren BH, Strasberg D, Bruggermann JH, Prŷs-Jones RP, Thébaud C (2010) Why does the biota of the Madagascar region have such a strong Asiatic flavour? *Cladistics* **26**, 526–538.

Waterhouse DM (2006) Parrots in a nutshell: the fossil record of Psittaciformes (Aves), *Historical Biology* **18**, 227–238.

Waterhouse DM, Lindow BE, Zelenkov NV, Dyke GJ (2008) Two new parrots (Psittaciformes) from the Lower Eocene fur formation of Denmark, *Palaeontology* **51**, 575–582.

Watling D (1983) Ornithological notes from Sulawesi, *Emu* **83**, 247–261.

Watling D (1995) Notes on the status of Kuhl's Lorikeet *Vini kuhlii* in the Northern Line Islands, Kiribati, *Bird Conservation International* **5**, 481–489.

Watling D (2001) *A Guide to the Birds of Fiji & Western Polynesia*, Environmental Consultants, Suva.

Watson AS, Plog S, Culleton BJ, Gilman PA, LeBlanc SA, Whitely PM, Claramunt S, Kennett DJ (2015) Early procurement of scarlet macaws and the emergence of social complexity in Chaco Canyon, NM, *Proceedings of the National Academy of Sciences* **112**, 8238–8243.

Waugh D (2010) A double boost for the Grey-breasted Parakeet, *Cyanopsitta* no. 97, 19–20.

Waugh D (2013) Schutz des Gelbwangenkakadus auf Sulawesi und Masakambing, *Papageien* **26**, 130–134.

Waugh D (2016a) Die rasante erholung der Gelbohrsittich-population, *Papageien* **29**, 204–206.

Waugh D (2016b) Loro Parque Fundación project confirms endangered status of the Vinaceous-breasted Parrot, *Aviarylife*, downloaded at www.aviarylife.com.au.

Weaver CM (1982) Breeding habitats and status of the Golden-shouldered Parrot *Psephotus chrysopterygius* in Queensland, *Emu* **82**, 2–6.

Weaver CM (1987) A comparison of temperatures recorded in nest chambers excavated in termite mounds by the Golden-shouldered Parrot, *Emu* **87**, 57–59.

Webb MH, Holdsworth C, Webb J (2012) Nesting requirements of the endangered Swift Parrot (*Lathamus discolor*), *Emu* **112**, 181–188.

Webb MH, Wotherspoon S, Stojanovic D, Heinsohn R, Cunningham R, Bell P, Terrauds A (2014) Location matters: using spatially explicit occupancy models to predict the distribution of the highly mobile, endangered Swift Parrot, *Biological Conservation* **176**, 99–108.

Webber LC (1948) The Ground Parrot in habitat and captivity, *Avicultural Magazine* **54**, 41–45.

Wetherbee DK (1985) The extinct Cuban and Hispaniolan macaws (*Ara*, Psittacidae), and description of a new species, *Caribbean Journal of Science* **21**, 169–175.

Wetmore A (1926) Descriptions of additional fossil birds from the Miocene of Nebraska, *American Museum Novitates* no. 211, 1–5.

Wetmore A (1927) The birds of Porto Rico and the Virgin Islands – Psittaciformes to Passeriformes, *Scientific Survey of Porto Rico and the Virgin Islands*, vol. 9, part 4, 409–598, New York Academy of Sciences, New York.

Wetmore A (1931) Early record of birds in Arizona and New Mexico, *Condor* **33**, 35.

Wetmore A (1935) The Thick-billed Parrot in southern Arizona, *Condor* **37**, 18–21.

Wetmore A (1937) Ancient records of birds from the island of St Croix with observations on extinct and living birds of Puerto Rico, *Journal of Agriculture of the University of Puerto Rico* **21**, 5–15.

Wetmore A (1938) Bird remains from the West Indies, *Auk* **55**, 51–55.

Wetmore A (1944) Remains of birds from the Rexroad fauna of the Upper Pliocene of Kansas, *University of Kansas Science Bulletin* **30**, 89–105.

White CMN, Bruce MD (1986) The birds of Wallacea (Sulawesi, the Moluccas & Lesser Sunda Islands, Indonesia), *BOU Check-list* no. 7, 8–524.

White EW (1882) Notes on the birds collected in the Argentine Republic, *Proceedings of the Zoological Society of London*, 591–629.

White NE, Phillips MJ, Gilbert MTP, Alfaro-Núñez A, Mawson PR, Spencer PBS, Bunce M (2011) The evolutionary history of cockatoos (Aves: Psittaciformes: Cacatuidae), *Molecular Phylogenetics and Evolution* **59**, 615–622.

Whitehead J (1899) Field-notes on birds collected in the Philippine Islands, Part III, *Ibis* 7th series **5**, 381–399.

Whitlock FL (1924) Journey to central Australia in search of the Night Parrot, *Emu* **24**, 248–281.

Widmann P, Lacerna ID, Diaz SH (2001) Aspects of biology and conservation of the Philippine Cockatoo *Cacatua haematuropygia* on Rasa Island, Palawan, Philippines, *Silliman Journal* **42**, 129–145.

Widmann P, Widmann IL (2008) The cockatoo and the community: ten years of Philippine Cockatoo Conservation Programme, *BirdingASIA* **10**, 23–29.

Wiley JW (1991) Status and conservation of parrots and parakeets in the Greater Antilles, Bahama Islands and Cayman Islands, *Bird Conservation International* **1**, 187–214.

Wiley JW, Kirwan GM (2013) The extinct macaws of the West Indies, with special reference to Cuban Macaw *Ara tricolor*, *Bulletin of the British Ornithologists Club* **133**, 125–156.

Wiley JW, Snyder NFR, Gnam RS (1992) Reintroduction as a conservation strategy for parrots. In *New World Parrots in Crisis: Solutions from Conservation Biology* (Beissinger SR, Snyder NFR, eds), pp. 165–200, Smithsonian Instutution, Washington.

Williams GR (1956) The Kakapo (*Strigops habroptilus*, Gray): a review and reappraisal of a near-extinct species, *Notornis* **2**, 29–56.

Williams ML, Steadman DW (2001) The historic and prehistoric distribution of parrots (Psittacidae) in the West Indies. In *Biogeography of the West Indies: Patterns and Perspectives*, 2nd edition (Wood CA, Sergile FE, eds), pp. 175–189, CRC Press, Boca Raton.

Williams S (2003) A successful season for the Lear's Macaws of Serra Branca, *PsittaScene* **15**(1), 6–7.

Willis EO, Oniki Y (1981) Levantamento preliminar de aves em treze areas do Estado de São Paulo, *Revista brasileira Biologica* **41**(1), 121–135.

Wilson DJ, Grant AD, Parker N (2006) Diet of kakapo in breeding and non-breeding years on Codfish Island (Whenua Hou) and Stewart Island, *Notornis* **53**, 80–89.

Wilson H (1937) Notes on the Night Parrot, with references to recent occurrences, *Emu* **37**, 79–87.

Wilson K-J (1993) Observations of the Kurämoó (*Vini peruviana*) on Aitutaki Island, Cook Islands, *Notornis* **40**, 71–75.

Wilson SB (1907) Notes on birds of Tahiti and the Society Group, *Ibis* 9th series **1**, 373–379.

Winge O (1888) *Fugle fra Knoglehuler i Brasilien*. (E Museo Lundii), Biancos Lunos, Kjøbenhavn.

Wood CA (1924) My quest of the Imperial Parrot, *Avicultural Magazine*, 4th series **2**, 57–59, 77–81.

Wood JR, Mitchell KJ, Scofield RP, Tennyson A, Fidler AE, Wilmshurst JM, Llamas B, Cooper A (2014) An extinct nestorid parrot (Aves, Psittaciformes, Nestoridae) from the Chatham Islands, New Zealand, *Zoological Journal of the Linnaean Society* **172**, 185–199.

Worthy TH, Holdaway RN (2002) *The Lost World of the Moa*, Canterbury University Press, Christchurch.

Worthy TH, Hawkins S, Belford S, Spriggs M (2015) Avifauna from the Teouma Lapita Site, Efate Island, Vanuatu, including a new genus of megapode, *Pacific Science* **69**, 205–254.

Worthy TH, Tennyson AJ, Scofield RP (2011) An Early Miocene diversity of parrots (Aves, Strigopidae, Nestorinae) from New Zealand, *Journal of Vertebrate Paleontology* **31**, 1102–1116.

Wragg GM, Weisler MI (1994) Extinctions and new records of birds from Henderson Island, Pitcairn Group, South Pacific Ocean, *Notornis* **41**, 61–70.

Wright T, Toft CA, Enkerlin-Hoeflich E, Gonzalez-Elizondo J, Albornoz M, Rodríguez-Ferraro A, Rojas-Suárez F, Sanz V, Trujillo A, Beissinger SR, Berovides-A V, Gálvez-A X, Brice AT, Joyner K, Eberhard J, Gilardi J, Koenig SE, Stoleson S, Martuscelli P, Meyers JM, Renton K, Rodríguez AM, Sosa-Asanza AC, Vilella FJ, Wiley JW (2001) Nest poaching in Neotropical parrots, *Conservation Biology* **15**, 710–720.

Wright T, Schirtzinger E, Eberhard J, Graves G, Sanchez J, Capelli S, Muller H, Scharpegge J. Chambers G, Fleischer R (2008) A multilocus molecular phylogeny of the parrots (Psittaciformes): support for a Gondwanan origin during the Cretaceous, *Molecular Biology and Evolution* **25**, 2141–2156.

Yamashita C (1987) Field observations and comments on the Indigo Macaw (*Anodorhynchus leari*), a highly endangered species from northeastern Brazil, *Wilson Bulletin* **99**, 280–282.

Yamashita C (2003) Field observations in Brazil on the biology and comments on the conservation of the Golden Conure (*Guaruba guarouba*), *AFA Watchbird* **30**, 38–40.

Yamashita C, França JT (1991) A range extension of the Golden Conure (*Aratinga guarouba*) to Rondônia state, western Amazonia, *Ararajuba* **2**, 91–92.

Yamashita C, Valle MP (1993) On the linkage between *Anodorhynchus* macaws and palm nuts, and the extinction of the Glaucous Macaw, *Bulletin of the British Ornithologists Club* **113**, 53–60.

Yamashita C, Machado de Barros Y (1997) The Blue-throated Macaw *Ara glucogularis*: characterization of its distinctive habitats in savannahs of the Beni, Bolivia, *Ararajuba* **52**, 141–150.

Yeap L, Shephard JM, Le Souef A, Holyoake C, Groom C, Dawson R, Kirkby T, Warren K (2015) Satellite tracking of rehabilitated wild Baudin's Cockatoos, *Calyptorhynchus baudinii*: a feasibility trial to track forest black cockatoos, *Pacific Conservation Biology* **21**, 163–167.

Yugovic JZ (1984) The grey glasswort (*Halosacrcia halocnemoides*) in coastal Victoria and some implications for the Orange-bellied Parrot, *Victorian Naturalist* **101**, 234–239.

Zelenkov NV (2016) The first fossil parrot (Aves, Psittaciformes) from Siberia and its implications for the historical biogeography of Psittaciformes, *Biology Letters* **12**, 20160717.

INDEX OF SCIENTIFIC NAMES

Page numbers in bold type denote illustrations.

INDEX OF ENGLISH NAMES

Page numbers in bold type denote illustrations.